Handbook of
Sodium-Ion Batteries

Handbook of
Sodium-Ion Batteries
Materials and Characterization

edited by
Rohit R. Gaddam
X. S. (George) Zhao

JENNY STANFORD
PUBLISHING

Published by

Jenny Stanford Publishing Pte. Ltd.
101 Thomson Road
#06-01, United Square
Singapore 307591

Email: editorial@jennystanford.com
Web: www.jennystanford.com

British Library Cataloguing-in-Publication Data
A catalogue record for this book is available from the British Library.

Handbook of Sodium-Ion Batteries: Materials and Characterization

ISBN 978-981-4968-15-7 (Hardcover)
ISBN 978-1-003-30874-4 (eBook)

Contents

Preface

The rapid increase in population and consumption of fossil fuels for energy has led to global eco-environmental problems such as climate change. The Intergovernmental Panel on Climate Change (IPCC) warns that carbon emissions be halved by 2030 to reach net-zero emissions by 2050 to have a 50% chance to limit the global temperature rise to 1.5°C above the pre-industrial levels. These global ambitions will only be possible by deploying renewable energy technologies and require advanced electrical energy storage technologies to solve the intermittency problem of renewables, such as solar and wind.

On the energy storage front, battery chemistries have evolved rapidly. The past decades have witnessed successful commercial uptakes of lithium-ion batteries (LIBs) for applications ranging from portable electronics and electric vehicles to grid-level energy storage. However, the economic considerations related to the rarity of the battery electrode materials and geopolitical concerns associated with the location of the resources restrict the application of LIBs for large-scale energy storage. In addition, it seems improvement of LIB cell performance has reached the limit. Therefore, alternative battery technologies that are cost-effective and based on sustainable resources have been extensively explored in recent years.

Sodium-ion batteries (NIBs) are a promising alternative that can overcome some of the setbacks related to LIBs. The configuration of NIB cells is similar to that of LIB cells. Thus, the infrastructure used for fabricating LIBs can be adopted by NIBs. However, there have been grand challenges in developing the NIB technology. The chemistry and electrochemistry of electrode materials for NIBs are sufficiently different from that of their lithium-ion counterparts although sodium and lithium are both Group 1A elements in the periodic table.

For instance, graphite, a commonly used anode material in LIBs, does not significantly allow sodium ions to intercalate in ester-based electrolytes (it should be noted that recent studies have shown that in ether-based electrolytes, graphite can store a relatively large amount of sodium ions via a co-intercalation mechanism). Instead, hard carbon with turbostratic graphite nanodomains and closed pores exhibits promising sodium-ion storage performance. However, the charge storage mechanism in hard carbon has not been completely understood. Hard carbon tends to have a low initial coulombic efficiency. Furthermore, the low cut-off potential of hard-carbon electrodes leads to safety concerns because sodium dendrites can easily form at such low potentials.

On the cathode side, layered transition metal oxides, metal phosphates and fluorophosphates, and Prussian blue and its analogs all hold great promise. However, a phase change from the O_3-phase to the P_3-phase occurs during cell discharge and charge in layered transition metal oxides, leading to multiple voltage plateaus and instability of the electrode materials. While metal phosphates and fluorophosphates display reasonable reversible capacity, the stability of redox couples during sodium-ion intercalation/deintercalation is a big issue. Prussian blue and its analogs have a low density, leading to low volumetric energy densities for NIBs compared to those using transition metal oxide cathodes.

The electrolyte also plays a major role in battery performance. A deeper understanding of the underlying electrified electrode–electrolyte interfaces is required for efficient sodium-ion storage. However, such interfaces are complex, involving the effects of the microstructure of the electrode, composition of the electrolyte, and other effects of electrode–electrolyte interactions. Therefore, a fundamental understanding of such underlying processes holds the key to developing efficient NIBs with tailored properties.

Other components in a NIB battery cell, including the separator and binder, also need to be optimized to make the NIBs an appearing energy storage technology. Such optimizations require understanding the charge–storage behavior.

This book contains a collection of the recent developments in electrode materials and characterization techniques for NIB research. With a brief account of the challenges and opportunities in NIBs in Chapter 1 by the editors, fundamental principles of battery electrochemistry are discussed in Chapter 2 by Prof. V. Ramadesigan. An overview of cathode materials for NIBs is presented by Prof. Guoxiu Wang and co-workers in Chapter 3, followed by an in-depth discussion on Prussian blue and its analogs as cathodes in Chapter 4 by Prof. Y. Jiang and co-workers. Contributed by Dr. Q. Zhao and Prof. M. Shao, Chapter 5 throws light on polymer electrode materials for NIBs. An important type of anode material, namely transition metal dichalcogenides is discussed in Chapter 6 by Prof. G. Singh. The effect of polymer binder materials on sodium-ion storage is discussed in Chapter 7 by Prof. S. Ramaprabhu and co-workers. In Chapter 8, Dr Q. Xia and Prof. X. S. Zhao provide an analysis of the different electrolytes for NIBs. In Chapter 9, Dr Y. Rangom et al. discuss cell failure mechanisms of NIBs in comparison with LIBs. Further, Prof. Q. Cai and co-workers look at electrolytes from computational perspectives in Chapter 10. Separators for fabricating NIB cells are reviewed by Prof. W. Chen in Chapter 11. Dr R. Knibbe and co-workers highlight electron microscope characterization techniques for NIB studies in Chapter 12. Finally, in the last chapter of this book, another advanced characterization technique, synchrotron radiation-based X-ray characterization methods, is presented for understanding the charge-storage mechanism by Prof. L. Song's group.

The book aims to provide a timely update on the research progress on advanced electrode materials and characterization techniques in developing NIBs. We acknowledge all contributors to this book for their support, collaboration, and fantastic content without which this project would have not been possible. We believe that the book would be useful for graduate students, scientists, and researchers who work on rechargeable batteries (in general) and sodium-ion batteries (in particular) to tackle fundamental problems across multidisciplinary fields spanning materials science, electrochemistry, physics, and energy.

Rohit R. Gaddam
X. S. (George) Zhao
July 2022

Chapter 1

Challenges and Opportunities in Sodium-Ion Batteries: An Introduction

Rohit R. Gaddam[a] and X. S. (George) Zhao[b,c]
[a]*Department of Chemical Engineering, Indian Institute of Science Education and Research – Bhopal, Madhya Pradesh 462066, India*
[b]*School of Chemical Engineering, The University of Queensland, St. Lucia, Brisbane 4072, Australia*
[c]*Institute of Materials for Energy and Environment, Qingdao University, Shandong 266071, China*
george.zhao@uq.edu.au

1.1 Importance of Batteries for Energy Storage

Energy consumption has increased manifold due to the rapidly growing technological innovations and a multifold increase in the use of energy-hungry devices due to exponential population growth [1, 2]. Several thousand terawatts of energy are consumed each day [3, 4], which has led to decades of research in generating energy via renewable sources (Fig. 1.1). Though some of these technologies have found success in recent years, still there are almost 2 billion people

Handbook of Sodium-Ion Batteries: Materials and Characterization
Edited by Rohit R. Gaddam and X. S. (George) Zhao

ISBN 978-981-4968-15-7 (Hardcover), 978-1-003-30874-4 (eBook)
www.jennystanford.com

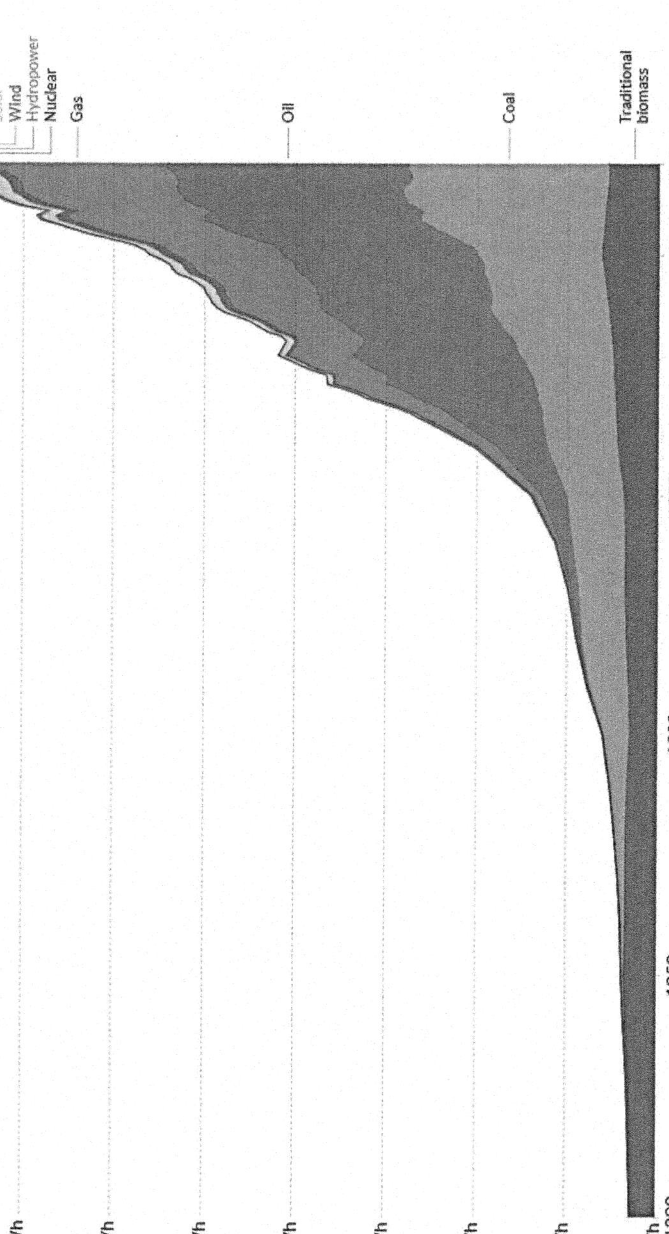

Figure 1.1 Global energy consumption from 1800 until 2019 [7]. Copyright https://ourworldindata.org/energy.

in the world who do not have access to electricity [5]. Such energy-poverty disproportionality affects the economy and quality of living. Also, achieving universal electrification via grid extension is not a very feasible solution for the near future. Therefore, non-continuous energy generation and requirement of access to electricity at remote places around the world (with no access to the grid) demand the necessity for better energy storage [6].

Energy storage could be electrical, chemical, mechanical, or electrochemical. However, a sustainable and fossil fuel independent storage device with a low carbon footprint is more feasible using electrochemical energy storage systems [8]. The underlying principles of such energy storage were known as early as the 1700s. These systems via redox reaction convert the chemical energy into electrical energy and vice versa [9]. The electrochemical devices comprise electrodes and an ionically conductive electrolyte. In this regard, supercapacitors and batteries represent one of the promising storage technologies for both grid-level storage and portable electronics [10]. Both these technologies have disparity in their energy and power densities [11] (Fig. 1.2). For instance, batteries involve redox reactions, which could be slower, while supercapacitors adsorb ions without any diffusion limitations, which could make fast charging feasible [12]. However, such surface-based adsorption of ions makes supercapacitors have lower energy densities.

With an upright balance between energy and power densities, lithium-ion batteries (LIBs) have revolutionized the energy storage industry for about three decades, leading to the electrification of vehicles and other portable electronics [2, 13–16]. LIBs have lowered the dependence on fossil fuel and are sought after as a replacement to Carnot-engine-based automobiles in the transportation sector. It is worth a mention that John Goodenough, Stanley Whittingham, and Akira Yoshino were awarded Nobel Prize in Chemistry for their pioneering work on LIBs [17]. The fundamental understanding of the physicochemical properties of materials has played a major role in the development in this field.

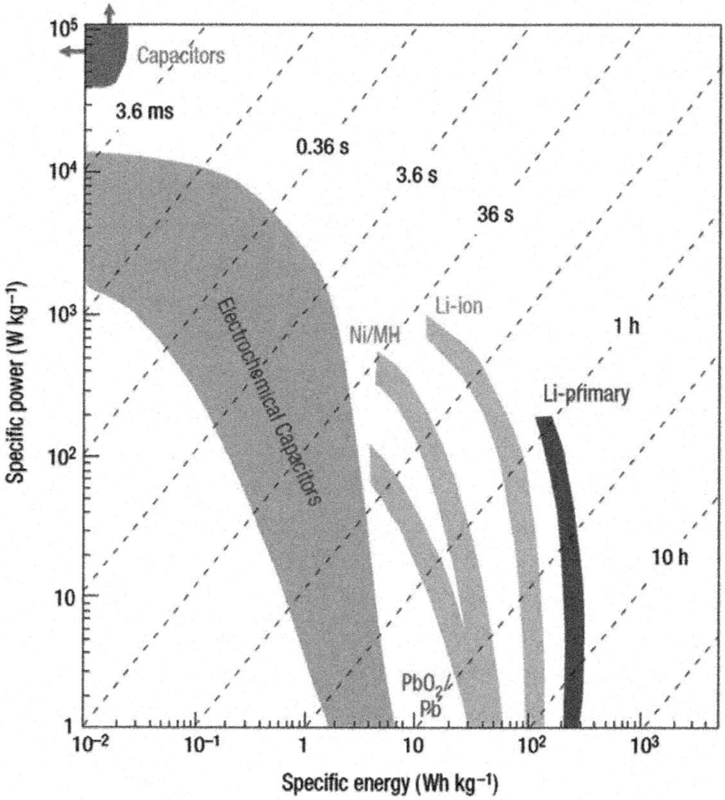

Figure 1.2 Ragone plot for energy storage devices [12]. Reprinted with permission from Ref. [12], Copyright 2008, Springer Nature.

Figure 1.3 represents a typical LIB operation. LIB comprises an anode (negative electrode), a cathode, a separator sandwiched between the electrodes, and an electrolyte (which is ionically conducting and electronically insulating). The current collectors help in delivering electrons emanating from the redox reactions to an external load. During the process of discharge, lithium-ions from the anode are inserted into the cathode and this process is reversed during charge. When the battery discharges, the cathode is reduced, as it accepts electrons, and the anode is oxidized. The reversible insertion and de-insertion of lithium ion indicate reversible charging/discharging of the battery, making the system rechargeable. The chemical reactions in a typical LIB with graphite as anode and $LiCoO_2$ as a cathode is as follows:

Anode

$$C + XLi + Xe^- \rightleftharpoons Li_XC_6$$

Cathode

$$LiCoO_2 \rightleftharpoons Li_{1-X}CoO_2 + XLi + Xe^-$$

The active materials selected for LIBs should possess good electrical conductivity, ionic conductivity, cycling stability, reversible capacity, along with being low cost and environmentally friendly. A non-aqueous electrolyte must be used in the battery, as lithium rigorously interacts with water molecules leading to safety concerns. Generally, the electrolyte consists of a lithium-salt solubilised in organic solvents like ethylene carbonate, dimethyl or diethyl carbonate. The most commonly used state-of-the-art cathodes in LIBs are $LiCoO_2$, $LiMn_2O_4$, $LiFePO_4$, etc., while graphite is the most commonly used anode material [18].

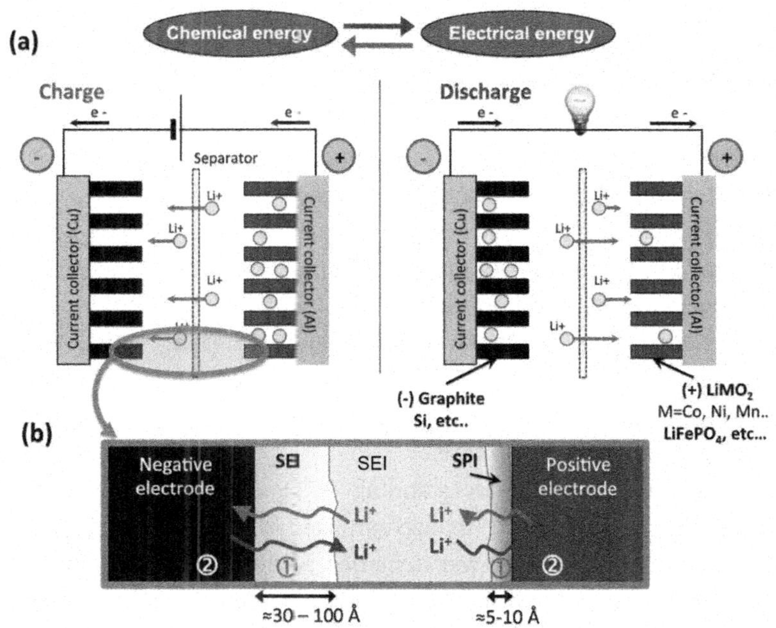

Figure 1.3 (a) Scheme for LIB operation. (b) Interphases formed on the electrode surface during LIB operation. Reprinted with permission from Ref. [19], Copyright 2016, IOP Publishing.

Until today LIBs have captured the interests of researchers and industries alike; however, there is a growing interest in alternative technology using sodium ion as a charge carrier due to the rarity of lithium precursors and its geographical presence at politically sensitive regions. Also considering the magnitude of use of LIBs, it can be conceived that the future cost of the battery could increase manifold. Though recycling of spent lithium is an option to be considered, the feasibility and the impact of using precursors of lithium continue. Therefore, sodium-ion batteries (NIBs) are a promising alternative to LIBs for grid-level storage that can be coupled with renewable energy sources.

1.2 Developing Sodium-Ion Battery

As compared to the understanding of LIBs, the current understanding of sodium-ion storage is at its nascent stage. Both Li and Na belong to group IA in the periodic table, thus sharing similar chemistries; therefore, the operation of the battery is similar with that of LIBs (Fig. 1.4, Table 1.1) [6, 20, 21]. Further, sodium-intercalation chemistry in case of cathode materials is very similar to that of lithium, making it feasible to use a similar structural configuration in case of NIBs. However, the mass and size of sodium being larger than that of lithium make sodium-ion based energy storage systems to fall short in terms of energy densities. However, in large-scale storage applications, sodium-ion systems could be competitive to the LIB market as it is more economical. It is to be noted that sodium-ion interaction with a host could be much different, which could influence the dynamics of NIBs. Therefore, this provides a variety of opportunities for the development of breakthrough technologies using NIBs for the near future.

The fundamental understanding of electrode and electrolyte properties is important to realize batteries with better performance in terms of energy and power densities. Some of the considerations for developing NIBs could be to shift toward using an aqueous electrolyte, which could circumvent the need for an inert and dry environment for battery fabrication. The use of aqueous electrolyte would make the battery cheaper and environmentally benign.

Aqueous electrolytes provide better ionic conductivity and act as a better heat sink than aromatic and aliphatic carbonates used in organic electrolytes. Hence, fluctuation-free battery performance and cost reduction could be expected. Current collectors, binders, and separators could indirectly influence the performance and cost of the battery.

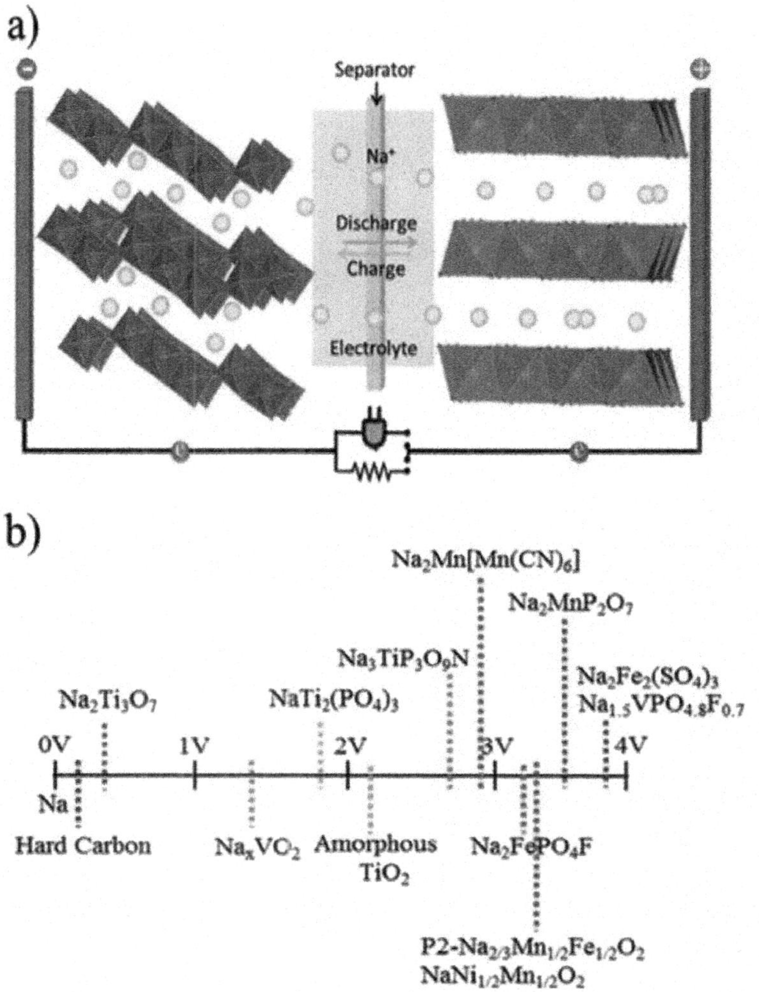

Figure 1.4 (a) Schematic representation of NIB operation. (b) Working potential of typical NIB electrodes. Reprinted with permission from Ref. [20], Copyright 2015, John Wiley & Sons.

Table 1.1 Characteristics of sodium versus lithium

Category	Lithium	Sodium
Relative atomic mass	6.94	22.99
Cation radius (Å)	0.76	1.02
$E°$ *vs.* SHE (V)	-3.04	-2.71
Melting point (°C)	180.5	97.7
First ionization energy (kJ mol^{-1})	520.2	495.8
Abundance in the earth crust (mg kg^{-1})	20	23.6×10^3
Distribution	70% in South America	Everywhere
Cost of carbonate ($ per ton)	5800	250–300
Theoretical capacity (mA h g^{-1})	3861	1166
Theoretical capacity (mA h cm^{-3})	2062	1131

Source: Reproduced with permission from Ref. [22], Copyright 2016, The Royal Society of Chemistry.

Researchers have reported several cathode materials; however, only some carbon-based anodes seem to have a comparable performance with that of commercial LIB anode (Fig. 1.5). Transition metal oxides, phosphates, and fluorides have been reported as viable cathode material for NIBs [21, 23–26]. Graphite is thermodynamically unfavorable to intercalate sodium; therefore, alternative hard-carbon-based anodes have been popularly researched for sodium-ion storage [1, 27, 28]. Metallic sodium, by itself as an anode, is not favorable as this could lead to dendrite formation. However, sodium dendrite formation in NIBs is not as remarkable as that of lithium as the melting point of sodium is much lesser than that of lithium, which could eventually melt below 100°C. In terms of electrolyte, aqueous, non-aqueous, ionic-liquid-based, and solid-state electrolytes have been actively investigated by researchers. In the following sections, electrodes, electrolytes, and other factors affecting NIB performance will be discussed.

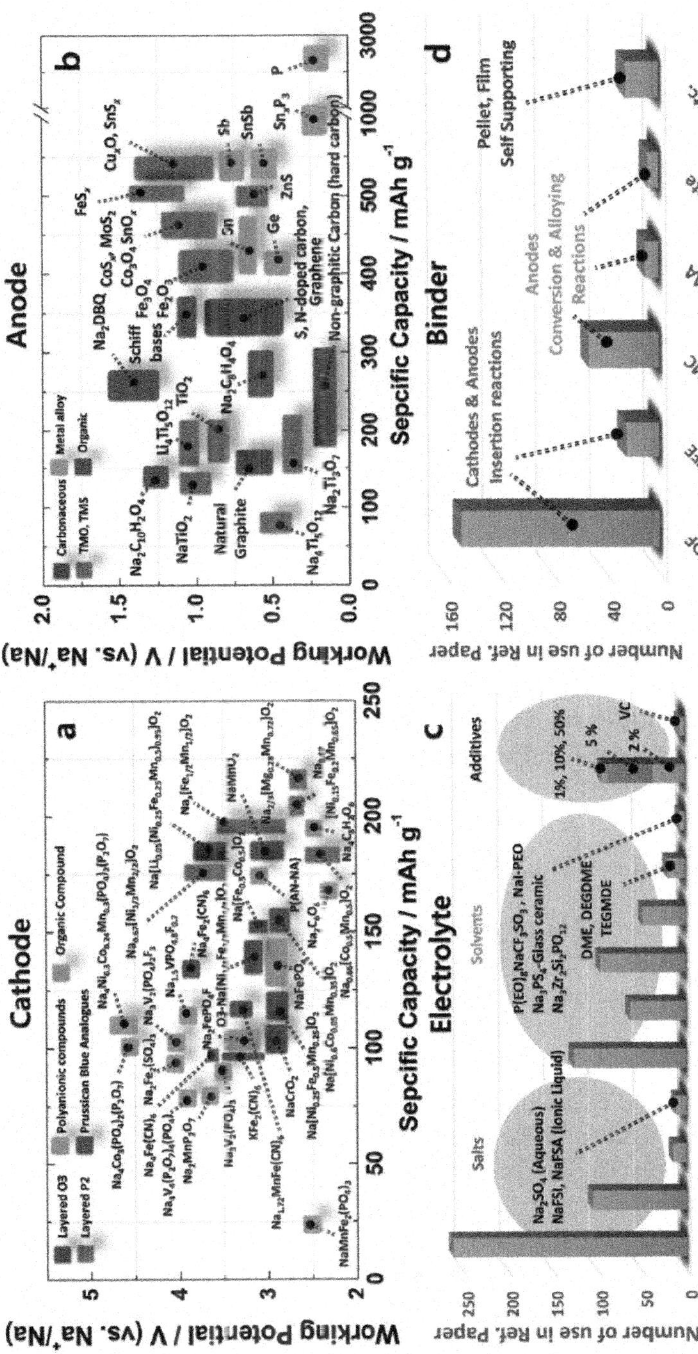

Figure 1.5 Consolidated view of various components of NIBs. Reproduced with permission from Ref. [21], Copyright 2017, The Royal Society of Chemistry.

1.3 Anodes

Finding a suitable anode material for NIB forms one of the major bottlenecks for sodium-ion commercialization for use in electric vehicles and grid-level storage. It is important to note that sodium-ion intercalation in graphite layers is thermodynamically not feasible as sodium plating occurs before it could intercalate [29]. This adds to the disadvantage of not being able to use the commercialized LIB anode, graphite. Further, similar to using metallic lithium anodes in LIBs, using pristine sodium metal is not feasible due to dendrite formation. However, as mentioned earlier, this is not as remarkable as that for lithium due to its lower melting point.

The anode part of the battery was investigated as early as the 1980s with TiS_2 as a promising candidate with reversible sodium-ion storage. However, the advent of hard-carbon materials as a promising anode, showcasing a capacity of ~300 mAh/g, led to renewed interest in anode material research. Many materials have been of interest as promising anodes, including carbon materials, transition metal oxides and sulfides based on conversion reactions, titanium-based composite materials, and organic compounds [30–34]. Many research groups have invested in understanding the mechanism of sodium-ion storage and challenges faced in commercial uptake of such anodes for NIBs.

1.3.1 Carbon-Based Anodes

Doeff et al. showed that sodium-storage capability in graphite was only around 31 mAh/g forming a configuration of NaC_{70} [31, 35]. However, commercial graphite when tested against lithium shows a capacity of 372 mAh/g [35]. Despite being disadvantaged in terms of capacity, carbon-based materials are important as anodes because of their economic production capabilities, widely available precursor materials, and easy tailoring of their microstructure [31]. Several carbon-based anodes have been reported so far for NIBs [27, 33, 36–44].

Though pristine graphite is unfavorable as anodes for NIBs, researchers have investigated the possibility of using reduced graphene oxide (RGO) as an electrochemically active material against sodium. It was investigated that such RGOs with a d-spacing of ~0.37

nm delivered a capacity of 174 mAh/g (current density of 40 mA/g) [45]. When the d-spacing was increased to 0.47 nm, a gravimetric capacity of 280 mAh/g (current density of 20 mA/g) was obtained [46]. Zhao et al. [47] further demonstrated a simple route to reduce graphene oxide by which the RGO anode material could deliver a specific capacity of 272 mAh/g at 50 mA/g.

To provide a better sodium uptake capability, the d-spacing between the graphene sheets was increased to 0.767 nm by placing porphyrin pillars (Fig. 1.6) [48]. The electrodes delivered a capacity of about 200 mAh/g at 100 mA/g with superior rate tolerance and cycling stability up to 700 cycles. Besides, the battery was placed at rest for 1 month, and negligible capacity loss was observed when recycled.

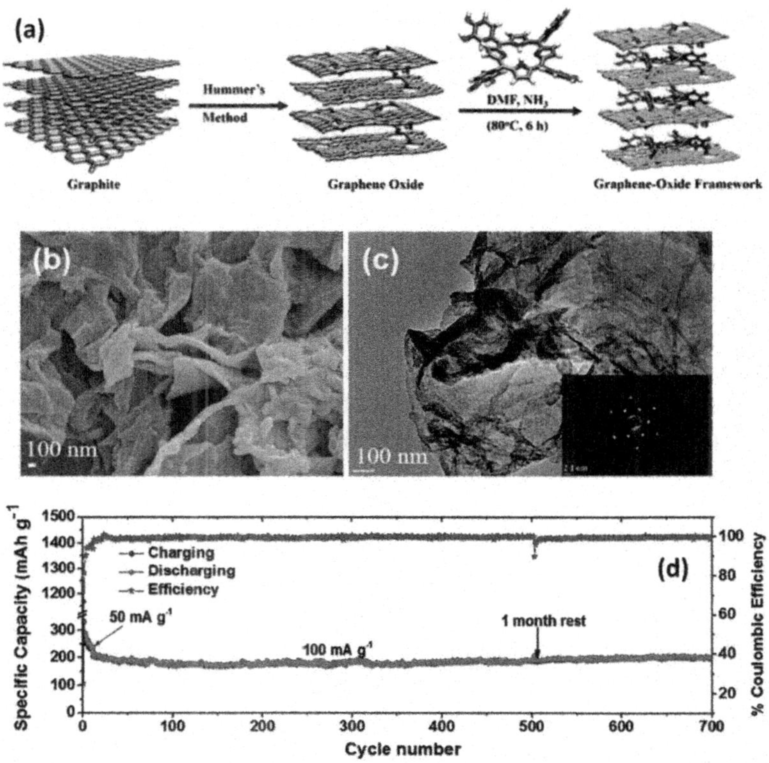

Figure 1.6 (a) Scheme for pillared RGO preparation. (b) FESEM and (c) TEM images of porphyrin-rich RCO preparation. (d) Cycling stability of the RGO. Reproduced with permission from Ref. [48], Copyright 2017, The Royal Society of Chemistry.

Recent studies show possible co-intercalation of solvated sodium ions into graphite. With electrolyte salt in ether-based solvents, an NaC_{20} configuration for graphite intercalation could be obtained [31, 49]. Also, the use of solvents like glyme produces a very thin layer of solid-electrolyte interphase leading to higher coulombic efficiencies in the initial cycles. This understanding of co-intercalation has made graphite a promising candidate for NIBs as well. This co-intercalation concept was not new and was also investigated for LIBs, but propylene carbonate solvation of lithium ions leads to exfoliation of graphite gallery, which was detrimental to the battery performance. However, in the case of NIBs, such exfoliations were not evident [31]. Despite research on graphite-based anode materials, their low specific energy has led to research on hard-carbon materials.

Hard carbons are disordered carbon materials with turbostratic nanodomains and larger interlayer spacing. Such carbons can be derived from biomass-based sources leading to easy upscalability for large-scale production [50, 51]. Further, the porosity and heteroatom content can boost sodium-ion storage capabilities. Dhan et al. reported a specific capacity of about 300 mAh/g for a hard carbon derived from glucose [52]. The same group also proposed a mechanism for sodium-ion storage via house of card model where sodium could be stored in the graphitic domain leading to a sloping profile in its galvanostatic charge–discharge curve [53]. The plateau regions corresponded to storage at nanovoids and defect sites. This theory was supported by *in situ* small-angle X-ray scattering measurements. However, the current understanding of sodium-ion storage mechanism in hard carbon is still an evolving subject for discussion. Various factors affect the performance of such hard carbon. Pyrolysis temperature of carbon precursors affects the carbon microstructure largely. The plateau capacity increases with an increase in carbonization temperature. One of the works reported hard carbon via carbonization of polymer precursor at high temperature [54]. Though longer plateau regions were observed, the overall surface area of carbon decreased with increase in temperature. Despite the observation of more plateau capacity, such sodium-ion storage at plateau region cannot be ascribed to pores, because with an increase in carbonization temperature, the nanopores decrease. Therefore, the plateau region could correspond to sodium-ion insertion into graphene layers, which is not in line

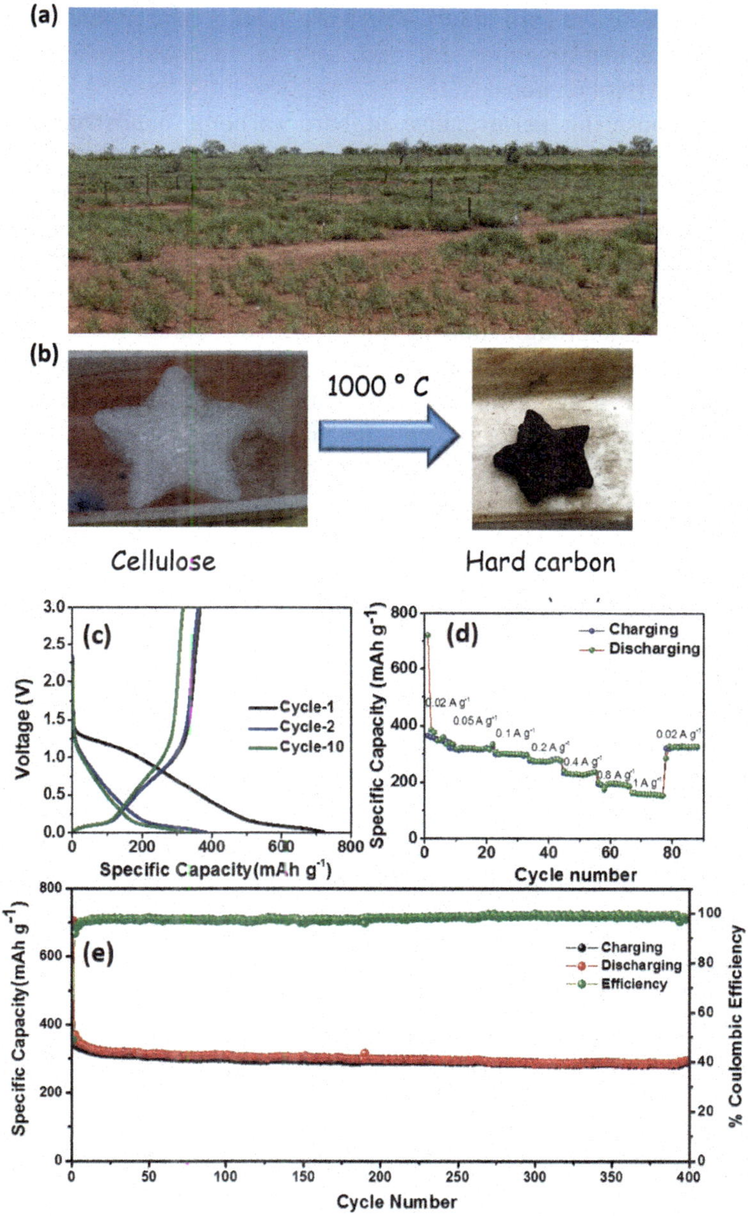

Figure 1.7 (a) Spinifex grassland in Queensland, Australia. (b) Nanocellulose obtained from spinifex carbonized at 1000°C. (c) Galvanostatic charge–discharge curve, (d) rate capability, and (e) cycling stability of spinifex-derived hard carbon. Reproduced with permission from Ref. [57], Copyright 2017, The Royal Society of Chemistry.

with the house of card model. Therefore, further experimental and theoretical investigations are necessary to elucidate sodium-ion storage in hard carbons.

To boost the performance of hard carbons, nanostructured carbon materials have been investigated. Carbon fibers, carbon spheres, nanotubes, etc. have been studied as NIB anode material. Zhao et al. investigated carbon nanomaterials synthesized via flame deposition method as potential anode with a specific capacity of 272 mAh/g against sodium [55]. The same group further investigated the effect of heteroatom dopant like nitrogen on sodium-ion storage. Such nitrogen-rich hard carbons displayed a high specific capacity of 520 mAh/g at 20 mA/g with stability over 1000 cycles [56]. Density functional theory calculations reveal that amide functional groups were responsible for better sodium-ion uptake. Therefore, the cost-effective heteroatom-rich hard carbon with long cycling life and rate tolerance is promising as an anode for NIB. In another interesting work, the same group produced a scalable hard carbon from spinifex-grass-derived nanocellulose (Fig. 1.7) [57]. During the freeze-drying processes, the cellulose nanofibers due to van der Waals force come together to form a sheet-like morphology. The carbons showed a specific capacity of 386 mAh/g at 20 mA/g. The performance was comparable with that of the graphite-based anode for lithium-ion batteries. The excellent electrochemical performance is attributed to the large interlayer spacing of the carbon (~0.39 nm), superior cycling stability, and high rate tolerance (326 mAh/g at 50 mA/g and 300 mAh/g at 100 mA/g). Hard carbons derived from such sustainable precursors are promising for next-generation rechargeable NIBs.

1.3.2 Carbon Alloy-Based Anodes

Several materials like Pb, P, Sn, Ge, etc. have exceptional sodium-ion storage capabilities, however, their use is hindered by large volume changes during charging and discharging [31, 33]. Therefore, the structural instability of these materials limits the battery performance, making such pristine materials impractical for real-time application in NIBs. Therefore, such materials are, in general, combined with carbon to buffer the volume changes that occur, as carbon offers better electrical contact.

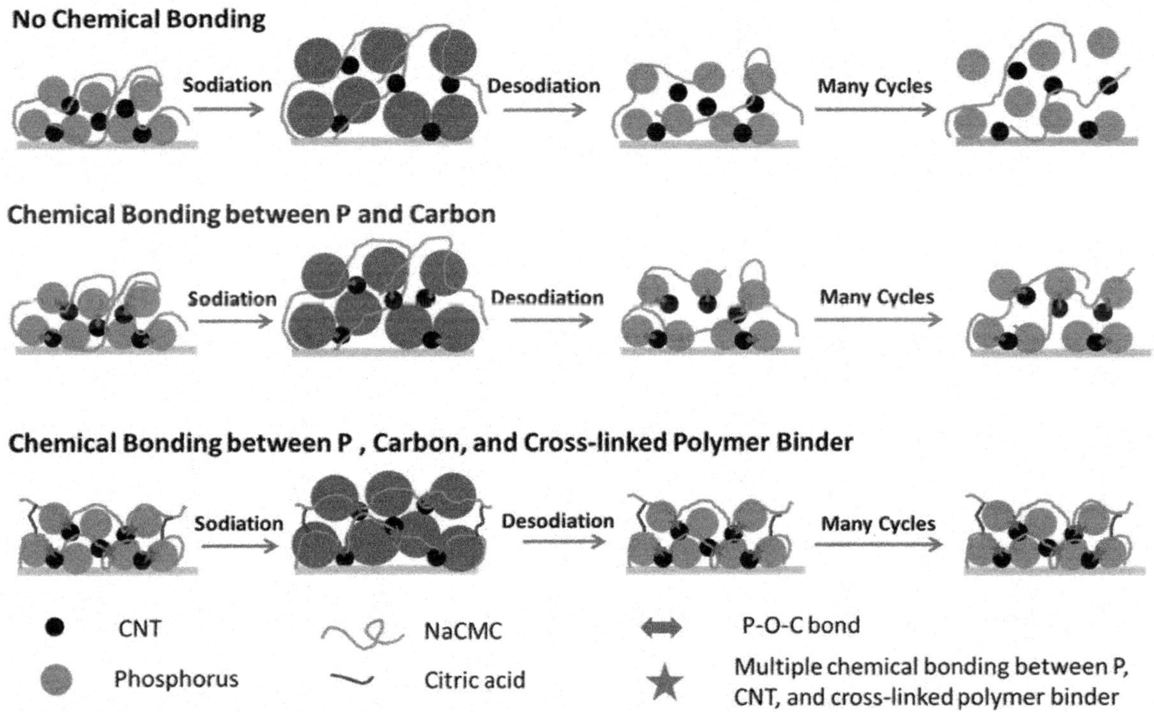

Figure 1.8 Representation of sodium interaction with phosphorus-based anodes during cycling. Reprinted with permission from Ref. [59], Copyright 2015, the American Chemical Society.

For instance, the theoretical capacity of phosphorous anode is 2596 mAh/g, but poor conductivity and large volume changes are detrimental to its electrochemical performance [32]. The sodiation in phosphorous occurs at a voltage as low as 0.4 V, hence making a larger voltage operable full cell. Some research groups have reported composite carbon materials with phosphorous to have capacities as high as 1700 mAh/g. Wang et al. [58] achieved a specific capacity of 550 mAh/g (current density of 0.5 A/g) in a carbon nanotube-phosphorus composite. The bonding between carbon-enabled phosphorous to buffer the volume changes during cycling, thus delivering a cycling stability of more than 200 cycles. Song et al. used a mechanochemical route to bind phosphorous to carbon nanotubes via ball milling, which delivered a discharge capacity of 1586.2 mAh/g [59]. The researchers highlighted the importance of binder in the electrode, which holds the composite material intact during cycling (Fig. 1.8).

Tin forms a stoichiometry of $Na_{15}Sn_4$, which gives a specific capacity of 847 mAh/g; however, it is accompanied by enormous volume change up to 420% [31]. Attempts have been made to realize Sn/C composites to buffer volume changes and provide stable performance. Ball milling with carbon materials including graphite has been a common method reported in the literature with a capacity ranging from ~400 to 500 mAh/g [31]. Similar to this, germanium- and antimony-based composites have also been investigated with carbon composites stabilizing the performance of anode.

1.3.3 Other Anode Materials

Materials, including metal oxides and sulfides, have been recently investigated as anodes for NIBs. Sodium interaction with metal oxides generally involves deoxygenation (to form sodium oxide) followed by sodium alloy formation with the metal [31]. For instance, the theoretical capacity of tin oxide is 1378 mAh/g [60]. However, the loss of electrical contact, volume expansion, and irreversible oxide formation with sodium lead to a lower capacity of ~667 mAh/g [61]. Hence, many researchers tried to stabilize SnO_2 by making carbon composites. Wang et al. [62] reported one such composite of SnO_2

with that of graphene, which resulted in a relatively larger specific capacity of 741 mAh/g at 20 mA/g, while the additional capacity was attributed to that of graphene. Apart from this, Fe-, Co-, and Ni-based oxides have been widely investigated by researchers. Some of the problems associated with such metal oxides are the higher sodiation potential and sluggish charge kinetics, which limit their application in NIBs. In one such attempt to enhance the performance of metal oxides like Nb_2O_5, fluorine-based heteroatom substitution was made [63]. Such fluorinated oxides improve electrochemical performance. Orthorhombic niobium oxyfluoride-carbon composite displayed a cycling stability of more than 10,000 cycles and a specific capacity of 292 mAh/g at 0.05 A/g. Two predominant storage mechanisms involving intercalation pseudo-capacitance and conversion reaction were responsible for the observed performance.

Metal sulfides, on the other hand, have advantages of greater energy density and reversibility of Na_2S stoichiometry than that of pristine metal or alloy. SnS_2, MoS_2, WS_2, CoS_2, etc. have been studied as anodes for NIBs. For instance, Wu et al. [64] reported cobalt sulfide nanoparticles encapsulated in a graphene cage via a simple one-step protocol. A specific capacity of 705 mAh/g at 0.01 A/g was obtained. *In situ* transmission electron microscopy was used to study the volume expansion that occurred in the cobalt sulfide particle, which revealed that the carbon layer buffers the volume changes that occur and provides stable performance (Fig. 1.9). Further, melamine was used for the growth of the graphene layer on the surface of cobalt sulfide, which doped nitrogen functional groups into graphene that enhanced the electrical conductivity of the material.

Apart from oxides and sulfides, organic electrodes have also been studied for NIBs. Sodium-terephthalate-based anodes provided a desodiation capacity of 300 mAh/g at 30 mA/g current density. The sodiation of this molecule leads to enolation of carboxyl groups. In another work, polyimides synthesized using a hydrothermal method, with a morphology of micro-flower, were reported by Zhao et al. [65]. A reversible discharge capacity of 125 mAh/g at a current density of 25 mA/g was obtained. The sodiation followed a two-step enolation reaction for the electrochemical reaction. The organic electrolyte, as such, has the flexibility to tailor the structure, and the sodium-ion storage is predominantly based on its interaction with functional groups. The sodiation potential depends on the redox reactions with such groups, which need to be taken into consideration before configuring a full cell.

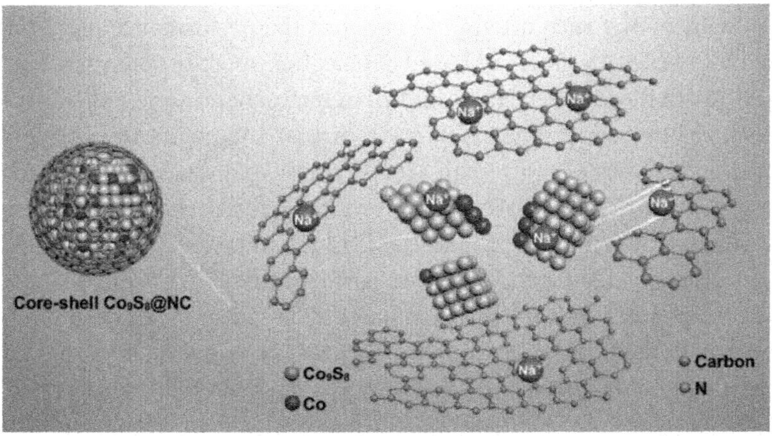

Figure 1.9 Schematic illustration of core–shell cobalt sulfide-graphene particles. Reprinted with permission from Ref. [64], Copyright 2020, Springer Nature.

1.4 Cathodes

From the preceding discussion, it is understood that the anode part provides higher specific capacity and lower voltage domains; therefore, research on cathode becomes important to boost the energy density. Transition metal oxides, Prussian blue analogies, phosphates, oxides, etc. have been investigated as potential cathodes (Fig. 1.10) [23–26, 66–69]. The design of such cathodes is predominantly based on lithium counterparts; however, lithium hosts need not necessarily host sodium ion. For instance, low reversible sodiation in spinel metal oxides and olivine-phosphates-based cathodes. Therefore, challenges still exist in developing high-performance cathodes with better sodium-ion transport and electrochemical performance.

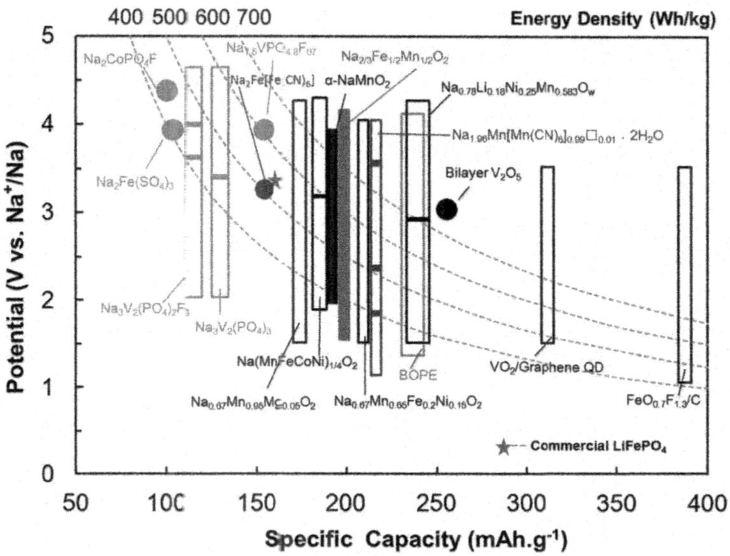

Figure 1.10 Cathode materials reported for NIBs, along with their specific capacity and energy density. Reprinted with permission from Ref. [69], Copyright 2016, John Wiley & Sons.

1.4.1 Fluoride-Based Cathodes

Fluorine-based compounds have been of keen interest to researchers given the high electronegativity of fluorine as well as the high theoretical capacity of metal fluorides. The low conductivity and large hysteresis in these materials have made researchers also to focus on nanoparticles to boost their performance [23]. Nishijima et al. carried out work on various metal fluorides of iron, titanium, manganese, vanadium, and cobalt [65]. In particular, FeF_3 was more interesting than others owing to its reversible charge storage and stability. FeF_3 was ball milled with acetylene black and tested against sodium half-cell. The sodium-ion storage occurred above 1.8 V via intercalation below 1.8 V by conversion, with a specific capacity of 145 mAh/g. Further, as electrical contact of active material is still an issue, carbon-based composites have also been investigated in metal fluorides.

In one report, FeF_3 composite with a porous carbon framework provided a specific capacity of 302 mAh/g at 15 mA/g [70]. Despite such interesting observations, the lack of sodium present in the cathode could hinder the actual applicability of such materials. Thus, sodium metal fluorides were synthesized via mechanochemical methods. The electrochemical performance was further improved by controlling the morphology of sodiated iron(III) fluoride and enhancing its crystallinity, thereby improving the specific capacity to 197 mAh/g [71]. Apart from this, Na_3FeF_6 was also produced by a mechanochemical method by ball milling fluorides of sodium and iron, which delivered a gravimetric capacity of ~111 mAh/g [72].

Another such fluoride material with theoretical capacities as high as 1400 mAh/g is sodium carbon fluoride, which could be synthesized in a cost-effective approach. The reversible formation of fluorides of sodium is key to such compounds. One of the studies shows the first discharge curve to possess along with a plateau at 2.4 V versus sodium with a high specific capacity of ~1000 mAh/g [73]. Further, some mechanistic insights were gained on charge storage behavior using scanning electron microscopy, X-ray photoelectron spectroscopy, and XRD. Other works further focused on reversible charging of such cathodes.

1.4.2 Prussian Blue Analogues

Among the various types of cathode materials available for NIBs, Prussian blue analogues (PBAs) are promising [74]. They offer high cycling stability, good rate tolerance, and low price for synthesis. Such PBAs have face-centered cubic structure with a representative formula of $A_xM^1[M^2(CN)^6]$ (where A is the cation and M^1 and M^2 are transition metals) (Fig. 1.11) [75]. PBAs studied till date containing a variety of intercalating ions (like Na, K) and transition metals (Fe, Co, Ni, Cu) have been utilized for aqueous and organic-electrolyte-based NIBs. However, they are unable to reach the standards for large-scale production.

An important consideration for large-scale utilization of NIBs is that the lifetime expectancy of at least a decade is required to reduce cost, which would amount a reversible ~4000 charge–discharge cycles at the lower end [76]. Despite the research being conducted in utilizing PBAs for NIBs, the actual mechanism underpinning the battery performance is yet to be fully comprehended. Hence, utilization of low-cost transition metals for PBA synthesis and understanding their sodium-storage mechanism would help to realize high-performance NIBs.

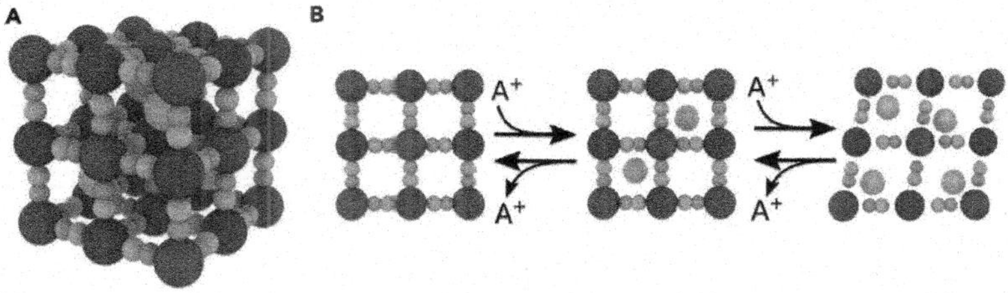

Figure 1.11 (a) Open framework lattice structure of PBAs. (b) Charge storage in PBAs. Reprinted with permission from Ref. [74], Copyright 2018, Elsevier.

There are some significant bottlenecks in knowledge about the use of PBAs as battery materials [74]. Tailoring of crystal structure and the electrochemical performance of PBAs would require some investigations on the mechanism of PBA formation in the solution. In general, PBAs are synthesized by a simple coprecipitation method, which makes the synthesis scalable for energy storage applications. Therefore, understanding ways to control their microstructure for tailored battery application is beneficial for scaling up the technology. The use of aqueous electrolyte demands the stability of water in PBA, which remains a mystery. Hence, understanding the performance of PBAs at various potentials will facilitate the development of PBAs with better sodium-ion storage. The interaction of sodium ions with PBA, their kinetics, and the role of aqueous media in ionic transport are poorly understood. Therefore, this research aims at the mechanistic study of sodium-ion interactions with PBA in aqueous media and understanding the role of water in ion kinetics.

1.4.3 Other Cathode Materials

Phosphates-based polyanions have been considered as promising cathodes for NIBs. These are available in NASICON or olivine configurations, which is promising for sodium-ion storage. Olivine $NaFePO_4$ shows a theoretical capacity of 154 mAh/g [6]. Zaghib et al. prepared $NaFePO_4$ by electrochemical synthetic routes and obtained closer to theoretical specific capacity (147 mAh/g); nevertheless, this compound has low reversibility [77]. Iron-, calcium-, and manganese-based olivine configurations were reported by Nazar et al. [78]. Further addition of fluorine has also been considered as a better choice. Sodium fluorophosphates, for instance, have shown growing interest. Sodium-based fluorophosphates of vanadium, iron, manganese are also considered promising candidates.

Another important cathode structure worth a mention is the NASICON type of materials. The electronic conductivity of such materials is poor; therefore, nanostructures of these are generally mixed with carbon or processed with additional conductive materials. From Fig. 1.12 one could visualize that larger NASICON-type materials are, in general, unfavorable for sodium-ion storage. Deliberate addition of carbon or having a conducting layer coating helps in better sodium uptake [79].

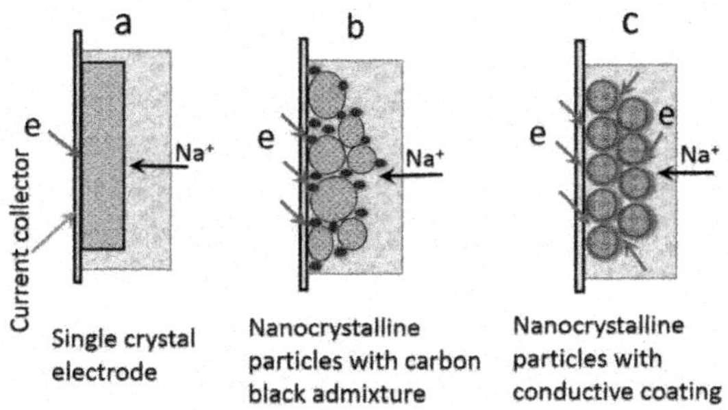

Figure 1.12 Sodium-ion interaction in NASICON-type material (a) in bulk, (b) nanocrystal mixed with carbon, and (c) nanostructure with conductive coating. Reprinted with permission from Ref. [79], Copyright 2017, John Wiley & Sons.

1.5 Electrolytes

Perchlorates, hexafluorophosphates, or bis(fluoromethahne) sulfoimides of sodium are used as an electrolyte salt in commonly used electrolytes. These are dissolved in carbonates like propylene carbonate (PC), ethylene carbonate (EC), or a mixture of other binary solvents with dimethyl carbonate (DMC) or diethycarbonates (DEC). Understanding the suitable electrolyte for NIBs is a tedious task. One of the important parameters to consider for an electrolyte is thermal stability and feasibility of usage at a larger potential range (Fig. 1.13) [80]. This is because the degradation of organic electrolytes within the battery could lead to the generation of

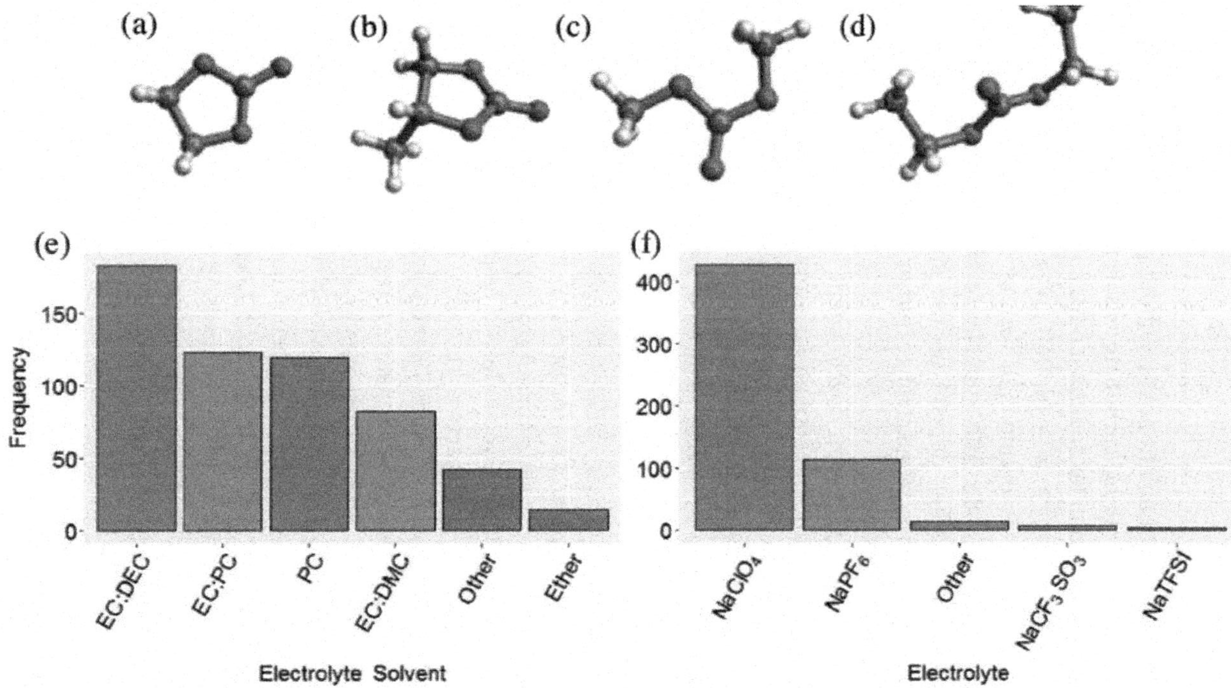

Figure 1.13 Structure of (a) EC, (b) PC, (c) DMC, and (d) DEC. (e) and (f) Solvents and electrolyte salts, respectively, with frequency of usage by researchers. Reprinted with permission from Ref. [80], Copyright 2018, John Wiley & Sons.

pressure inside the cell, eventually leading to battery explosion. One of the early compositions studied was EC: PC in the ratio of 1:1 using either $NaClO_4$ or $NaPF_6$ salt. Since EC by itself cannot be used as a solvent at room temperature, a combination with EC was essential. Molecular dynamics simulations carried out revealed that EC with PC and EC with ethyl methyl carbonate as solvent mixtures provided most favorable solvation for sodium ion [81]. However, EC with DEC was considered inferior to EC with DMC, which had better capacity retention. However, the opinions of various groups regarding the choice of electrolyte vary. The importance of solvent used in the electrolyte should rather depend on the stability of interphase formed on the surface of electrolyte rather than its sodium-ion solvation capabilities alone.

Another interesting recent development in the electrolyte part is the direct participation in sodium-ion storage. The so-called "co-solvation"-based electrolytes have recently gained interest due to their capability to co-intercalate into graphite [31, 49]. It is known that thermodynamical instabilities make sodium-ion intercalation in graphite unfavorable. However, solvated sodium ions show reversible co-intercalation in graphite-based anodes. The development of such co-intercalation-based anodes is quite significant in terms of utilization of graphite as a prospective anode for sodium-ion batteries. Also, higher cycling stability, better coulombic efficiency, and better rate capabilities are offered by such electrolytes.

Also, there has been active research in utilizing aqueous electrolytes. The most common electrolyte salts for aqueous rechargeable NIBs include sulphates, nitrates, and perchlorates of sodium [82]. It is known that organic electrolytes involve the formation of interphases on the surface of electrodes; however, there are seldom such observations for aqueous batteries. Another issue is that the operation window is just 1.23 V, which otherwise would lead to oxygen evolution at 0.81 V and hydrogen evolution at –0.42 V versus SHE [82]. In spite of this, the use of aqueous electrolyte would make the battery cheaper and environmentally benign. Aqueous electrolytes provide better ionic conductivity and act as a better heat sink than aromatic and aliphatic carbonates used in organic electrolytes. Hence, fluctuation-free battery performance and cost reduction could be expected. Super-concentrated electrolytes could help us overcome the voltage window issue, while removal of dissolved oxygen could help prevent side reactions that could occur with the electrode.

1.6 Summary and Future Perspective

NIBs are attractive as an alternative to LIBs for grid-level storage and even hybrid electric vehicles. Since the last decade, there have been exciting developments around various aspects of NIBs, including cathode, anode, electrolyte, binders, separators, and current collectors. To meet the demand for next-generation batteries, the present-day NIBs should outperform LIBs in terms of electrochemical performance. NIBs also face some similar problems associated with LIBs concerning safety and other technical barriers for high-performance electrodes. Though carbon-based anodes are currently being investigated as a promising one, still the understanding of sodium-ion storage is not well established. The arguments by many research groups on sodium-ion storage sites and mechanism are still not easy to pin down. Also, in the case of conversion-based electrodes, in general, they suffer from sluggish ion kinetics, poor ionic conductivity, and longer diffusion length, which need to be addressed. Further, such electrodes undergo large volume changes during sodiation and desodiation, which require them to be encapsulated to restrict such volume changes. Practical improvements in materials are only possible through understanding the underlying mechanism. This could be done by studying battery *in operando* using well-established techniques like XRD, XPS, Raman, NMR, TEM, etc. The distinct paths of redox reactions during charge and discharge, the phase transformation, and the voltage hysteresis should be well studied and understood for these materials. Understanding the physico-electrochemistry of materials is very much important, which can strongly impact battery performance.

Apart from complexity in terms of materials, selection of proper electrolyte is quite important, which determines the overall stability of the battery. The solvents need to be considered; it is found that ethylene carbonate with either propylene carbonate or diethylene carbonate shows no difference in terms of performance. Further, the importance of solid-electrolyte interphase (SEI) cannot be neglected, though its present understanding for NIBs is slim. The addition of fluoroethylene carbonate (FEC) provides stable interphase formation. Though FEC helps little with the coulombic efficiency in the first cycle, it does help in long-term cycling stability of the anodes. Apart from the selection of electrolytes, binders are quite important as well. Binder-free electrodes are impressive in terms of performance since binders themselves being insulating can lower the efficiency of the electrode. However, the synthesis strategies for binder-free electrodes are not very cost effective, therefore making it necessary for the binder to be present. PVDF, CMC, and PAA are most common binders for NIBs, though there are others that can be synthesized depending on the requirement. Further, understanding of the interaction of binders during cycling is quite essential. Though some of the important aspects are touched upon, new school of thoughts in various components of NIBs is essential to make it ready for commercial uptake.

References

1. Zhou Y, Yang Y, Jiao M, and Zhou Z. What is the promising anode material for Na ion batteries? *Science Bulletin*, 2018; **63**(3).
2. Zhang H, Hasa I, and Passerini S. Beyond insertion for Na-ion batteries: Nanostructured alloying and conversion anode materials. *Adv Energy Mater*, 2018; **8**(17).
3. Smil V. *Energy Transitions: Global and National Perspectives.* ABC-CLIO, 2016.
4. BBC. The biggest energy challenges facing humanity 2017 [Available from: https://www.bbc.com/future/article/20170313-the-biggest-energy-challenges-facing-humanity].
5. Hall P J and Bain E J. Energy-storage technologies and electricity generation. *Energy Policy*, 2008; **36**(12): 4352–4355.
6. Palomares V, Serras P, Villaluenga I, Hueso K B, Carretero-Gonzâlez J, and Rojo T. Na-ion batteries, recent advances and present challenges to become low-cost energy storage systems. *Energy Environ Sci*, 2012; **5**(3): 5884.
7. Data OWi. Energy 2020 [updated 2018. Web Page]. Available from: https://ourworldindata.org/energy.
8. Soloveichik, G L. Battery technologies for large-scale stationary energy storage. *Ann Rev Chem Biomol Eng*, 2011; **2**(1): 503–527.

9. Haas O J and Cairns E. Electrochemical energy storage. *Ann Rep Section C (Phys Chem)*, 1999; **95**(0): 163–198.
10. Gaddam R R, Kumar N A, Narayan R, Raju K V S N, and Zhao X S. Advanced carbon materials for electrochemical energy storage. In: Beeran Pottathara Y, Thomas S, Kalarikkal N, Grohens Y, and Kokol V (Eds), *Nanomaterials Synthesis*: Elsevier, 2019, pp. 385–418.
11. Simon P, Gogotsi Y, and Dunn B. Where do batteries end and supercapacitors begin? *Science*, 2014; **343**(6176): 1210–1211.
12. Simon P and Gogotsi Y. Materials for electrochemical capacitors. *Nat Mater*, 2008; **7**(11): 845–854.
13. Zhao C, Lu Y, Yue J, Pan D, Qi Y, Hu Y-S, et al. Advanced Na metal anodes. *J Energy Chem*, 2018; **27**(6): 1584–1596.
14. Slater M D, Kim D, Lee E, and Johnson C S. Sodium-ion batteries. *Adv Funct Mater*, 2013; **23**(8): 947–958.
15. Ellis B L and Nazar L F. Sodium and sodium-ion energy storage batteries. *Curr Opin Solid State Mater Sci*, 2012; **16**(4): 168–177.
16. Chevrier V L and Ceder G. Challenges for Na-ion negative electrodes. *J Electrochem Soc*, 2011; **158**(9): A1011.
17. Hu Y-S and Lu Y. 2019 Nobel Prize for the Li-ion batteries and new opportunities and challenges in Na-ion batteries. *ACS Energy Lett*, 2019; **4**(11): 2689–2690.
18. Goriparti S, Miele E, De Angelis F, Di Fabrizio E, Proietti Zaccaria R, and Capiglia C. Review on recent progress of nanostructured anode materials for Li-ion batteries. *J Power Sources*, 2014; **257**: 421–443.
19. Philippe B, Hahlin M, Edström K, Gustafsson T, Siegbahn H, and Rensmo H. Photoelectron spectroscopy for lithium battery interface studies. *J Electrochem Soc*, 2015; **163**(2): A178–A191.
20. Kundu D, Talaie E, Duffort V, and Nazar L F. The emerging chemistry of sodium ion batteries for electrochemical energy storage. *Angew Chem Int Ed*, 2015; **54**(11): 3431–3448.
21. Hwang J-Y, Myung S-T, and Sun Y-K. Sodium-ion batteries: Present and future. *Chem Soc Rev*, 2017; **46**(12): 3529–3614.
22. Kang H, Liu Y, Cao K, Zhao Y, Jiao L, Wang Y, et al. Correction: Update on anode materials for Na-ion batteries. *J Mater Chem A*, 2016; **4**(20): 7962.
23. Kim J, Kim H, and Kang K. Conversion-based cathode materials for rechargeable sodium batteries. *Adv Energy Mater*, 2018; **8**(17): 1702646.

24. You Y and Manthiram A. Progress in high-voltage cathode materials for rechargeable sodium-ion batteries. *Adv Energy Mater*, 2017; 8(2): 1701785.

25. Li W-J, Han C, Wang W, Gebert F, Chou S-L, Liu H-K, et al. Commercial prospects of existing cathode materials for sodium ion storage. *Adv Energy Mater*, 2017; **7**(24): 1700274-1-1700274-10.

26. Xiang X, Zhang K, and Chen J. Recent advances and prospects of cathode materials for sodium-ion batteries. *Adv Mater*, 2015; **27**(36): 5343–5364.

27. Saurel D, Orayech B, Xiao B, Carriazo D, Li X, and Rojo T. From charge storage mechanism to performance: A roadmap toward high specific energy sodium-ion batteries through carbon anode optimization. *Adv Energy Mater*, 2018; **8**(17): 1703268.

28. Lu Y, Lu Y, Niu Z, and Chen J. Graphene-based nanomaterials for sodium-ion batteries. *Adv Energy Mater*, 2018; **8**(17): 1702469.

29. Moriwake H, Kuwabara A, Fisher C A J, and Ikuhara Y. Why is sodium-intercalated graphite unstable? *RSC Adv*, 2017; **7**(58): 36550–36554.

30. Kim Y, Ha K H, Oh S M, and Lee K T. High-capacity anode materials for sodium-ion batteries. *Chem A Eur J*, 2014; **20**(38): 11980–11992.

31. Bommier C and Ji X. Recent development on anodes for Na-ion batteries. *Isr J Chem*, 2015; **55**(5): 486–507.

32. Kang H, Liu Y, Cao K, Zhao Y, Jiao L, Wang Y, et al. Update on anode materials for Na-ion batteries. *J Mater Chem A*, 2015; **3**(35): 17899–17913.

33. Balogun M S, Luo Y, Qiu W T, Liu P, and Tong Y X. A review of carbon materials and their composites with alloy metals for sodium ion battery anodes. *Carbon*, 2016; **98**: 162–178.

34. Luo W, Shen F, Bommier C, Zhu H, Ji X, and Hu L. Na-ion battery anodes: Materials and electrochemistry. *Acc Chem Res*, 2016; **49**(2): 231–240.

35. Doeff M M, Ma Y, Visco S J, and De Jonghe L C. Electrochemical insertion of sodium into carbon. *J Electrochem Soc*, 1993; **140**(12): L169–L170.

36. Li Y, Yuan Y, Bai Y, Liu Y, Wang Z, Li L, et al. Insights into the Na^+ storage mechanism of phosphorus-functionalized hard carbon as ultrahigh capacity anodes. *Adv Energy Mater*, 2018; **8**(18): 1702781.

37. Hou H, Banks C E, Jing M, Zhang Y, and Ji X. Carbon quantum dots and their derivative 3D porous carbon frameworks for sodium-ion batteries with ultralong cycle life. *Adv Mater*, 2015; **27**(47): 7861–7866.

38. Quan B, Jin A, Yu S-H, Kang S M, Jeong J, Abruña H D, et al. Solvothermal-derived S-doped graphene as an anode material for sodium-ion batteries. *Adv Sci*, 2018; **5**(5): 1700880.

39. Li J, Qi H, Wang Q, Xu Z, Liu Y, Li Q, et al. Constructing graphene-like nanosheets on porous carbon framework for promoted rate performance of Li-ion and Na-ion storage. *Electrochim Acta*, 2018; **271**: 92–102.

40. Zhang S-W, Lv W, Luo C, You C-H, Zhang J, Pan Z-Z, et al. Commercial carbon molecular sieves as a high performance anode for sodium-ion batteries. *Energy Storage Mater*, 2016; **3**: 18–23.

41. Cao B, Liu H, Xu B, Lei Y, Chen X, and Song H. Mesoporous soft carbon as an anode material for sodium ion batteries with superior rate and cycling performance. *J Mater Chem A*, 2016; **4**(17): 6472–6478.

42. Zhou D H, Peer M, Yang Z Z, Pol V G, Key F D, Jorne J, et al. Long cycle life microporous spherical carbon anodes for sodium-ion batteries derived from furfuryl alcohol. *J Mater Chem A*, 2016; **4**(17): 6271–6275.

43. Li W, Zeng L, Yang Z, Gu L, Wang J, Liu X, et al. Free-standing and binder-free sodium-ion electrodes with ultralong cycle life and high rate performance based on porous carbon nanofibers. *Nanoscale*, 2014; **6**(2): 693–698.

44. Wang Y, Wang C, Guo H, Wang Y, Huang Z. A nitrogen-doped three-dimensional carbon framework for high performance sodium ion batteries. *RSC Adv*, 2017; **7**(3): 1588–1592.

45. Wang Y X, Chou S L, Liu H K, and Dou S X. Reduced graphene oxide with superior cycling stability and rate capability for sodium storage. *Carbon*, 2013; **57**: 202–208.

46. Wen Y, He K, Zhu Y, Han F, Xu Y, Matsuda I, et al. Expanded graphite as superior anode for sodium-ion batteries. *Nat Commun*, 2014; **5**: 4033.

47. Kumar N A, Gaddam R R, Varanasi S R, Yang D F, Bhatia S K, and Zhao X S. Sodium ion storage in reduced graphene oxide. *Electrochim Acta*, 2016; **214**: 319–325.

48. Kumar N A, Gaddam R R, Suresh M, Varanasi S R, Yang D, Bhatia S K, et al. Porphyrin–graphene oxide frameworks for long life sodium ion batteries. *J Mater Chem A*, 2017; **5**(25): 13204–13211.

49. Jache B and Adelhelm P. Use of graphite as a highly reversible electrode with superior cycle life for sodium-ion batteries by making use of co-intercalation phenomena. *Angew Chem Int Ed*, 2014; **53**(38): 10169–10173.

50. Irisarri E, Ponrouch A, and Palacin M R. Review-hard carbon negative electrode materials for sodium-ion batteries. *J Electrochem Soc*, 2015; **162**(14): A2476–A2482.

51. Górka J, Vix-Guterl C, and Matei Ghimbeu C. Recent progress in design of biomass-derived hard carbons for sodium ion batteries. *C*, 2016; **2**(4): 24.

52. Stevens D A and Dahn J R. High capacity anode materials for rechargeable sodium-ion batteries. *J Electrochem Soc*, 2000; **147**(4): 1271–1273.

53. Stevens D A and Dahn J R. The mechanisms of lithium and sodium insertion in carbon materials. *J Electrochem Soc*, 2001; **148**(8): A803–A811.

54. Zhao J, Zhao L, Chihara K, Okada S, Yamaki J-I, Matsumoto S, et al. Electrochemical and thermal properties of hard carbon-type anodes for Na-ion batteries. *J Power Sources*, 2013; **244**: 752–757.

55. Gaddam R R, Yang D F, Narayan R, Raju K V S N, Kumar N A, and Zhao X S. Biomass derived carbon nanoparticle as anodes for high performance sodium and lithium-ion batteries. *Nano Energy*, 2016; **26**: 346–352.

56. Gaddam R R, Niaei A H F, Hankel M, Bernhardt D J, Nanjundan A K, and Zhao X S. Capacitance-enhanced sodium-ion storage in nitrogen-rich hard carbon. *J Mater Chem A*, 2017; **5**: 22186–22192.

57. Gaddam R R, Jiang E, Amiralian N, Annamalai P K, Martin D J, Kumar N A, et al. Spinifex nanocellulose derived hard carbon anodes for high-performance sodium-ion batteries. *Sustainable Energy Fuels*, 2017; **1**(5): 1090–1097.

58. Zhu Y, Wen Y, Fan X, Gao T, Han F, Luo C, et al. Red phosphorus–single-walled carbon nanotube composite as a superior anode for sodium ion batteries. *ACS Nano*, 2015; **9**(3): 3254–3264.

59. Song J, Yu Z, Gordin M L, Li X, Peng H, and Wang D. Advanced sodium ion battery anode constructed via chemical bonding between phosphorus, carbon nanotube, and cross-linked polymer binder. *ACS Nano*, 2015; **9**(12): 11933–11941.

60. Su D, Ahn H-J, and Wang G. SnO_2@graphene nanocomposites as anode materials for Na-ion batteries with superior electrochemical performance. *Chem Commun*, 2013; **49**(30): 3131–3133.

61. Wang Y, Su D, Wang C, and Wang G. SnO_2@MWCNT nanocomposite as a high capacity anode material for sodium-ion batteries. *Electrochem Commun*, 2013; **29**: 8–11.

62. Wang Y-X, Lim Y-G, Park M-S, Chou S-L, Kim JH, Liu H-K, et al. Ultrafine SnO_2 nanoparticle loading onto reduced graphene oxide as anodes for sodium-ion batteries with superior rate and cycling performances. *J Mater Chem A*, 2014; **2**(2): 529–534.

63. Wu Y, Fan X, Chen Y, Gaddam R R, Yu F, Xiao C, et al. Fluorine substitution enabling pseudocapacitive intercalation of sodium ions in niobium oxyfluoride. *J Mater Chem A*, 2019; **7**(36): 20813–20823.

64. Wu Y, Gaddam R R, Zhang C, Lu H, Wang C, Golberg D, et al. Stabilising cobalt sulphide nanocapsules with nitrogen-doped carbon for high-performance sodium-ion storage. *Nano-Micro Lett*, 2020; **12**(1): 48.

65. Zhao Q, Gaddam R R, Yang D, Strounina E, Whittaker A K, and Zhao X S. Pyromellitic dianhydride-based polyimide anodes for sodium-ion batteries. *Electrochimica Acta*, 2018; **265**: 702–708.

66. Islam M S and Fisher C A. Lithium and sodium battery cathode materials: computational insights into voltage, diffusion and nanostructural properties. *Chem Soc Rev*, 2014; **43**(1): 185–204.

67. Han M H, Gonzalo E, Singh G, and Rojo T. A comprehensive review of sodium layered oxides: Powerful cathodes for Na-ion batteries. *Energy Environ Sci*, 2015; **8**(1): 81–102.

68. Barpanda P. Pursuit of sustainable iron-based sodium battery cathodes: Two case studies. *Chem Mater*, 2016; **28**(4): 1006–1011.

69. Fang C, Huang Y H, Zhang W X, Han J T, Deng Z, Cao Y L, et al. Routes to high energy cathodes of sodium-ion batteries. *Adv Energy Mater*, 2016; **6**(5): 1501727.

70. Shen Y, Wang X, Hu H, Jiang M, Bai Y, Yang X, et al. Sheet-like structure FeF_3/graphene composite as novel cathode material for Na ion batteries. *RSC Adv*, 2015; **5**(48): 38277–38282.

71. Kitajou A, Komatsu H, Chihara K, Gocheva I D, Okada S, and Yamaki J-I. Novel synthesis and electrochemical properties of perovskite-type $NaFeF_3$ for a sodium-ion battery. *J Power Sources*, 2012; **198**: 389–392.

72. Shakoor R A, Lim S Y, Kim H, Nam K-W, Kang J K, Kang K, et al. Mechanochemical synthesis and electrochemical behavior of Na_3FeF_6 in sodium and lithium batteries. *Solid State Ionics*, 2012; **218**: 35–40.

73. Liu W, Li H, Xie J-Y, and Fu Z-W. Rechargeable room-temperature CFx-sodium battery. *ACS Appl Mater Interfaces*, 2014; **6**(4): 2209–2212.

74. Hurlbutt K, Wheeler S, Capone I, and Pasta M. Prussian blue analogs as battery materials. *Joule*, 2018; **2**(10): 1950–1960.

75. Bors R, Yun J, Marzak P, Fichtner J, Scieszka D, and Bandarenka A S. Chromium(II) hexacyanoferrate-based thin films as a material for

aqueous alkali metal cation batteries. *ACS Omega*, 2018; **3**(5): 5111–5115.

76. Marzak P, Yun J, Dorsel A, Kriele A, Gilles R, Schneider O, et al. Electrodeposited $Na_2Ni[Fe(CN)_6]$ thin-film cathodes exposed to simulated aqueous Na-ion battery conditions. *J Phys Chem C*, 2018; **122**(16): 8760–8768.
77. Zaghib K, Trottier J, Hovington P, Brochu F, Guerfi A, Mauger A, et al. Characterization of Na-based phosphate as electrode materials for electrochemical cells. *J Power Sources*, 2011; **196**(22): 9612–9617.
78. Lee K T, Ramesh T N, Nan F, Botton G, and Nazar L F. Topochemical synthesis of sodium metal phosphate olivines for sodium-ion batteries. *Chem Mater*, 2011; **23**(16): 3593–3600.
79. Chen S, Wu C, Shen L, Zhu C, Huang Y, Xi K, et al. Challenges and perspectives for NASICON-type electrode materials for advanced sodium-ion batteries. *Adv Mater*, 2017; **29**(48): 1700431.
80. Bommier C and Ji X. Electrolytes, SEI formation, and binders: A review of nonelectrode factors for sodium-ion battery anodes. *Small*, 2018; **14**(16): 1703576.
81. Kamath G, Cutler R W, Deshmukh S A, Shakourian-Fard M, Parrish R, Huether J, et al. In silico based rank-order determination and experiments on nonaqueous electrolytes for sodium ion battery applications. *J Phys Chem C*, 2014; **118**(25): 13406–13416.
82. Pahari D and Puravankara S. Greener, safer, and sustainable batteries: An insight into aqueous electrolytes for sodium-ion batteries. *ACS Sustainable Chem Eng*, 2020; **8**(29): 10613–10625.

Chapter 2

Principles of Electrochemistry

Nisha Garg, Venkatasailanathan Ramadesigan, and Sankara Sarma V. Tatiparti
Department of Energy Science and Engineering, Indian Institute of Technology Bombay, Mumbai 400076, India
venkatr@iitb.ac.in

Electrochemistry is a branch of physical chemistry that studies the chemical changes that occur due to the flow of electrical current. The flow of current, caused by a random chemical reaction or the chemical reaction caused by an external electricity supply, is an example of an electrochemical reaction. Electrochemical processes are the basis for large-scale chemical and metallurgical production of materials, metallic corrosion, power sources, and industrial metal finishing. Batteries and fuel cells are prime examples of electrochemical devices that convert chemical energy into electrical energy. Primary and secondary batteries are the backbone of today's engineering devices that require energy storage.

Energy storage systems are fast emerging as an essential part of the evolving clean energy systems of the 21st century. Energy storage

Handbook of Sodium-Ion Batteries: Materials and Characterization
Edited by Rohit R. Gaddam and X. S. (George) Zhao

ISBN 978-981-4968-15-7 (Hardcover), 978-1-003-30874-4 (eBook)
www.jennystanford.com

represents a huge economic opportunity and could help countries achieve their emission reduction targets by shifting to renewable-based generation. Most of the portable devices we see around us, such as laptops, mobile phones, watches, cars, and inverters, have one of these kinds of electrochemical storage systems. Therefore, it is the need of the hour to understand the basics of these electrochemical energy storage systems or specifically batteries. We discuss the underlying electrochemical principles of the working of batteries in this chapter. This understanding will eventually help us in designing better energy storage systems for future needs.

2.1 Electrochemical Cells

Electrochemical cells contain two electrodes, called the anode and the cathode. The electrode where oxidation occurs is called the anode, and the cathode is where reduction takes place. Electrodes are usually made from conductive materials, including metals, semiconductors, and graphite. The electrolyte is filled between these two electrodes, which contains ions that can freely move. A porous separator usually separates the two electrodes. An electrochemical cell can be of two types: galvanic or voltaic cell and electrolytic cell. In the case of galvanic/voltaic cells, the decrease in free energy during the spontaneous redox chemical reaction is converted into electrical energy. On the other hand, in an electrolytic cell, a nonspontaneous chemical reaction is driven by an external source of current, thus converting electrical energy into chemical energy.

Daniel cell (Fig. 2.1) is the oldest type of electrochemical cell consisting of two half-cells; the left half-cell contains zinc electrode dipped in $ZnSO_4$ solution, and the right half-cell consists of a copper metal electrode dipped in a solution of $CuSO_4$. These two half-cells are then connected by a salt bridge or a porous separator that prevents mixing of the two solutions and ensures electro-neutrality. The two electrodes are connected through an external circuit to allow the flow of electrons. The reactions occurring at electrode surfaces are called heterogeneous electron transfer reactions. The reactions occurring in the cell shown in Fig. 2.1 are written as half-cell reactions as shown:

$$Cu^{2+}_{(aq)} + 2e^{-} \rightarrow Cu_{(s)} \quad E^{\circ} = +0.34\text{ V} \tag{2.1}$$

$$Zn_{(s)} \rightarrow Zn^{2+}_{(aq)} + 2e^{-} \quad E^{\circ} = -0.76\text{ V} \tag{2.2}$$

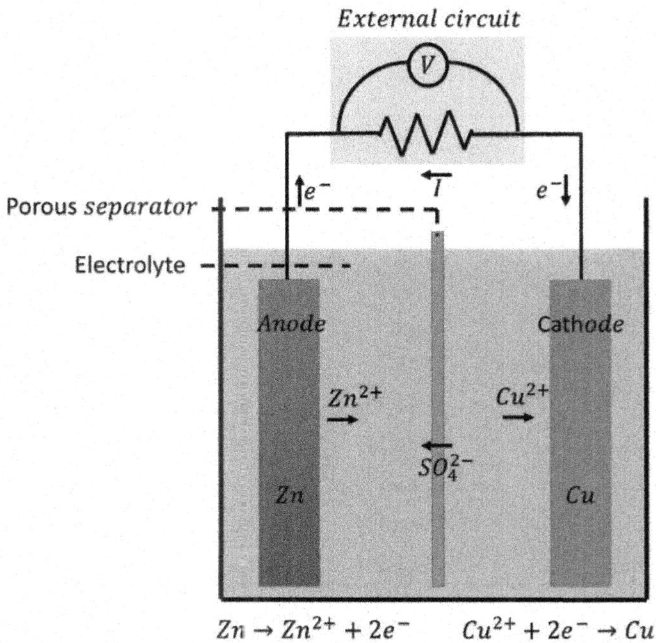

Figure 2.1 Schematic representation of a Daniel cell.

The zinc electrode undergoes oxidation to produce zinc ions and electrons, flowing from the zinc electrode to the copper electrode, where copper in the $CuSO_4$ solution is reduced to form metallic copper. The anode is where oxidation reaction (loss of electrons) occurs, and the cathode where reduction reaction (gain of electrons) occurs. The half-cell reactions of anode and cathode can be combined to give the full-cell reaction shown in Equation (2.3). The overall cell remains electrically neutral; i.e., a number of electrons produced in oxidation at the anode travel through the external circuit to the cathode and are consumed in the reduction reaction. The current flows in the opposite direction to the flow of electrons, as shown in Fig. 2.1.

$$Zn_{(s)} + Cu^{2+}_{(aq)} \rightarrow Zn^{2+}_{(aq)} + Cu_{(s)} \tag{2.3}$$

Let us understand some basic terms and units that are frequently used in electrochemistry. Apart from the base units within the International System of Units (SI units), as provided in Table 2.1, a few derived units are extensively used in the electrochemical systems.

Table 2.1 SI-derived units extensively used in electrochemistry

Quantity	Name of Unit	Symbol for Unit	Expression in SI Base Units
Electric charge	Coulomb	C	A·s
Potential	Volt	V	$m^2 \cdot kg \cdot s^{-3} \cdot A^{-1}$
Capacitance	Farad	F	$m^{-2} \cdot kg^{-1} \cdot s^4 \cdot A^2$
Resistance	Ohm	Ω	$m^2 \cdot kg \cdot s^{-3} \cdot A^{-2}$
Electrical conductance	Siemens	S	$m^{-2} \cdot kg^{-1} \cdot s^3 \cdot A^2$
Force	Newton	N	$m \cdot kg \cdot s^{-2}$
Pressure	Pascal	Pa	$kg \cdot m^{-1} \cdot s^{-2}$
Energy	Joule	J	$m^2 \cdot kg \cdot s^{-2}$
Power	Watt	W	$m^2 \cdot kg \cdot s^{-3}$
Frequency	Hertz	Hz	s^{-1}

When any metal is placed in a solution of its ions, a definite potential difference is developed between the metal and the solution due to de-electronation or electronation. This is referred to as the electrode potential (E). Each electrode has its potential due to the tendency to lose or gain electrons. If the tendency is to lose electrons and get oxidized, it is termed oxidation potential. On the contrary, if the tendency is to gain electrons, it is termed reduction potential. At standard temperature conditions and one molar solution concentration, this potential is called standard electrode potential ($E°$).

Let us now consider a species "A" undergoing reduction; n is the number of electrons involved in the reaction, as shown below.

$$A^{n-}_{(aq)} + ne^- \rightarrow A_{(s)} \quad (2.4)$$

The mass of species (m) either consumed or produced in the reaction can be expressed as

$$m = \frac{MQ}{nF} \quad (2.5)$$

where M is the molecular weight of A, Q is the total charge produced or consumed in the reaction, and F is Faraday's constant. Equation (2.5) is the mathematical representation of Faraday's law.

Changes in solution concentration and temperature can change the potential of an electrode. According to the law of mass action and Van't Hoff reaction isotherm, the Gibbs free energy (ΔG) can be expressed as

$$\Delta G = \Delta G^\circ + RT \ln Q \tag{2.6}$$

Substituting for charge Q from Equation (2.4) in terms of species concentration with the concentration of the solid:

$$\Delta G = \Delta G^\circ + RT \ln \frac{A_{(s)}}{A^{n+}_{(aq)}} \tag{2.7}$$

Concentration of a solid is taken as 1, i.e., $[A_{(s)}] = 1$. Equation (2.7) can be written as

$$\Delta G = \Delta G^\circ + RT \ln \frac{1}{A^{n+}_{(aq)}} \tag{2.8}$$

The decrease in free energy in a spontaneous redox reaction appears in the form of electrical energy. From the first law of thermodynamics, the relationship between a change in internal energy (dU) is given as

$$dU = \delta q + \delta w \tag{2.9}$$

where δq is the heat added to the system, and δw is the work done by the system. Also, both quantities may be expressed in terms of temperature (T), change in entropy (ds), and change in volume (dV), pressure (P), and electrical work done by the system ($\delta w_{\text{electrical}}$), respectively, as

$$\delta q = Tds \tag{2.10}$$

$$\delta w = -PdV - \delta w_{\text{electrical}} \tag{2.11}$$

Substituting Equations (2.10) and (2.11) in Equation (2.9) yields

$$dU = TdS - PdV - \delta w_{\text{electrical}} \tag{2.12}$$

At a given temperature and pressure, the change in Gibbs free energy can be written in terms of change in enthalpy and entropy of the system as

$$dG_{T,P} = dH_{T,P} - TdS \tag{2.13}$$

The change in enthalpy can be expressed in terms of change in internal energy and volume of the system at a given temperature and pressure. Thus, Equation (2.13) changes to

$$dG_{T,P} = dU_{T,P} + PdV - TdS \quad (2.14)$$

Now, substituting Equation (2.12) for change in internal energy in Equation (2.14) will give

$$dG_{T,P} = TdS - PdV - \delta w_{\text{electrical}} + PdV - TdS \quad (2.15)$$

Equation (2.15) can be further simplified and written as

$$dG_{T,P} = -\delta w_{\text{electrical}} \quad (2.16)$$

Hence the Gibbs function is at the heart of electrochemistry, and it identifies the amount of electrical work we can extract from a system. Now, we can write work in terms of net charge and potential of the system as

$$\delta w_{\text{electrical}} = VQ \quad (2.17)$$

Substituting $Q = nF$, where n is the number of electrons transferred in the system, in Equation (2.17) and further in Equation (2.16), we get

$$dG_{T,P} = -\delta w_{\text{electrical}} = -nFE \quad (2.18)$$

Therefore, the change in free energy is $\Delta G° = -nFE°$ at standard conditions and $\Delta G = -nFE$, otherwise, where E is the electrode potential. Thus, Equation (2.8) can be rewritten as

$$-nFE = -nFE° + RT \ln \frac{1}{[A^{n+}_{(\text{aq})}]} \quad (2.19)$$

Equation (2.19), upon simplification, gives us an important expression

$$E = E° - \frac{RT}{nF} \ln \frac{1}{[A^{n+}_{(\text{aq})}]} \quad (2.20)$$

Equation (2.20) is called the Nernst equation. From the Nernst equation, it is evident that the electrode potential is dependent on the temperature and molar concentration of ions in the solution. As described earlier, all galvanic cells are composed of two half-cells, one-half where oxidation occurs and the other half where reduction occurs. Therefore, one of the electrodes should have a higher potential than the other. The potential difference between the two electrodes is the driving force for the flow of electrons. The electrons flow from the electrode at a higher potential toward the electrode at a lower potential. This driving force due to the difference in potential is known as the electromotive force (EMF) of the cell or cell potential or voltage. The cell potential is expressed in terms of individual electrode potentials as

$$E^{\circ}_{\text{cell}} = E^{\circ}_{\text{cathode}} - E^{\circ}_{\text{anode}} \tag{2.21}$$

We know that the change in Gibbs free energy can be expanded in terms of other thermodynamic parameters, as shown in Equation (2.22).

$$\Delta G = \Delta H^{\circ} + \int_0^T \Delta C_{\text{P}} dT + \Delta\left(\sum L\right) - T\Delta S^{\circ} - \int_0^T \frac{\Delta C_{\text{P}}}{T} dT - T\Delta\left(\sum \frac{L}{T_{\text{L}}}\right) \tag{2.22}$$

where ΔH° and ΔS° are changes in the enthalpy and entropy, respectively, at STP and ΣL is the sum of molar latent heats corresponding to phase transitions occurring at T_{L}. $\Sigma(L/T_{\text{L}})$ defines the entropy due to phase transitions, and ΔC_{P} is the change in heat capacity due to the reaction, assuming that the reaction temperature change limits are T_1 and T_2 and no phase change occurs. Also, from Equation (2.13), we know $\Delta G = \Delta H - T\Delta S$. Hence, we can write changes in enthalpy and entropy as Equations (2.23) and (2.24), respectively:

$$\Delta H = \Delta H^{\circ} + \int_0^{T_1} \Delta C_{\text{P}} dT + \Delta(\sum L) \cong \Delta H^{\circ} + \int_0^{T_2} \Delta C_{\text{P}} dT + \Delta(\sum L) \tag{2.23}$$

$$\Delta S = \Delta S^{\circ} - T\int_0^{T_1} \frac{\Delta C_{\text{P}}}{T} dT - T\Delta(\sum \frac{L}{T_{\text{L}}}) \cong \Delta S^{\circ} - T\int_0^{T_2} \frac{\Delta C_{\text{P}}}{T} dT - T\Delta(\sum \frac{L}{T_{\text{L}}}) \tag{2.24}$$

Taking partial derivative of Gibbs free energy change with respect to temperature, we can further express the change in entropy and enthalpy in terms of cell potential as Equations (2.25) and (2.26), respectively, as:

$$\Delta S = -\frac{\partial \Delta G}{\partial T} = nF\frac{\partial E}{\partial T} \tag{2.25}$$

$$\Delta H = \Delta G + T\Delta S = -nFE + nF\frac{\partial E}{\partial T} \tag{2.26}$$

Thus, we can derive the thermodynamic parameters from cell potential measurements. However, cell measurements are themselves challenging due to the requirement of great precision in measuring instruments. Also, most of the reversible cell reactions have very low entropy changes associated with the reactions.

2.2 Alkali-Ion Batteries

A battery is a device that stores chemical energy for use as electricity at a later time. A battery consists of several cells assembled together to work as a source of electrical energy or electric power. A cell is a fundamental unit of a battery. Batteries can be categorized into two types, primary and secondary. A primary cell or battery is the one in which the chemical reaction is irreversible, and thus once the chemical energy is converted to electrical energy, it cannot be reversed back. These cells are discarded after the electrodes are fully consumed in the chemical reaction. A secondary cell or battery, on the other hand, is capable of reversible chemical reactions; i.e. the chemical energy can be converted into electrical energy, and electrical energy can be used to recharge the cell and store it as chemical energy. The conversion of chemical energy into electrical energy in secondary batteries is termed discharging, and the reverse process is called charging. Lithium-ion and sodium-ion batteries are examples of secondary batteries.

The energy level (or potential energy) of the outermost electron influences the electrochemical potential of metals. The reduction potential of a pure lithium metal electrode is –3.01 V, and that of a pure sodium metal electrode is –2.71 V. The corresponding reduction reactions are shown in Table 2.2. The electrode with the highest reducing power is suitable as an anode. Therefore, both these metals are suitable as anodes for an electrochemical cell. Li-ion batteries possess the highest energy densities and voltages but lack in terms of power density.

Table 2.2 Standard potentials of electrode reactions at 25°C

Electrode Reaction	$E°$(V)	Electrode Reaction	$E°$(V)
$Li^+ + e^- \leftrightarrow Li$	-3.01	$Tl^+ + e^- \leftrightarrow Tl$	-0.34
$Rb^+ + e^- \leftrightarrow Rb$	-2.98	$Co^{2+} + 2e^- \leftrightarrow Co$	-0.27
$Cs^+ + e^- \leftrightarrow Cs$	-2.92	$Ni^{2+} + 2e^- \leftrightarrow Ni$	-0.23
$K^+ + e^- \leftrightarrow K$	-2.92	$Sn^{2+} + 2e^- \leftrightarrow Sn$	-0.14
$Ba^{2+} + 2e^- \leftrightarrow Ba$	-2.92	$Pb^{2+} + 2e^- \leftrightarrow Pb$	-0.13
$Sr^{2+} + 2e^- \leftrightarrow Sr$	-2.89	$D^+ + e^- \leftrightarrow 1/2D_2$	-0.003
$Ca^{2+} + 2e^- \leftrightarrow Ca$	-2.84	$H^+ + e^- \leftrightarrow 1/2H_2$	0.000
$Na^+ + e^- \leftrightarrow Na$	-2.71	$Cu^{2+} + 2e^- \leftrightarrow Cu$	0.34
$Mg^{2+} + 2e^- \leftrightarrow Mg$	-2.38	$1/2O_2 + 2H_2O + 2e^- \leftrightarrow 2OH^-$	0.40
$Ti^+ + e^- \leftrightarrow Ti$	-1.75	$Cu^+ + e^- \leftrightarrow Cu$	0.52
$Be^{2+} + 2e^- \leftrightarrow Be$	-1.70	$Hg^{2+} + 2e^- \leftrightarrow Hg$	0.80
$Al^{3+} + 3e^- \leftrightarrow Al$	-1.66	$Ag^+ + e^- \leftrightarrow Ag$	0.80
$Mn^{2+} + 2e^- \leftrightarrow Mn$	-1.05	$Pd^{2+} + 2e^- \leftrightarrow Pd$	0.83
$Zn^{2+} + 2e^- \leftrightarrow Zn$	-0.76	$Ir^{3+} + 3e^- \leftrightarrow Ir$	1.00
$Ga^{3+} + 3e^- \leftrightarrow Ga$	-0.52	$Br_2^{2+} + 2e^- \leftrightarrow 2Br$	1.07
$Fe^{2+} + 2e^- \leftrightarrow Fe$	-0.44	$O_2 + 4H^+ + 4e^- \leftrightarrow 2H_2O$	1.23
$Cd^{2+} + 2e^- \leftrightarrow Cd$	-0.40	$Cl_2^{2+} + 2e^- \leftrightarrow 2Cl$	1.36
$In^{3+} + 3e^- \leftrightarrow In$	-0.34	$F_2^{2+} + 2e^- \leftrightarrow 2F$	2.87

Lithium as an anode has been widely used since the invention of Li-ion batteries. It has revolutionized the entire world due to its wide range of applications ranging from portable electronics to electric vehicles. The source of Li metal is limited and, hence, expensive. Therefore, other chemistries beyond lithium, such as sodium-, magnesium-, and calcium-ion batteries, are actively being explored by researchers. The working principle of a Na-ion battery is similar to that of Li-ion batteries. Besides, sodium is naturally abundant on earth (e.g. in seawater) and might be cheaper than its Li counterpart. In this section, we will understand the operations in Li-ion batteries that are well established and understood.

During charge and discharge, Li ions shuttle between the positive-layered oxide hosts (cathodes) and negative intercalation hosts (usually graphitic carbon-based anodes), as shown in Fig. 2.2. This is also termed "rocking chair" technology. Lithium ions in the positive electrode material pass through the separator and into the layers in the carbon electrode material, resulting in a charging current flow. Li-ions between the layers in the carbon electrode material pass through the separator and into the positive electrode material during discharging. The most common anode electrode materials used in Li-ion batteries are graphite (conventionally based on intercalation reaction), metals such as Si, Sn, Al (potential—based on alloying reactions), and $Li_4Ti_5O_{12}$. There are various cathode materials, primarily oxides and phosphates of lithium and transition metals such as $LiCoO_2$, $LiNi_{1/3}Mn_{1/3}Co_{1/3}O_2$, $LiMn_2O_4$, and $LiFePO_4$.

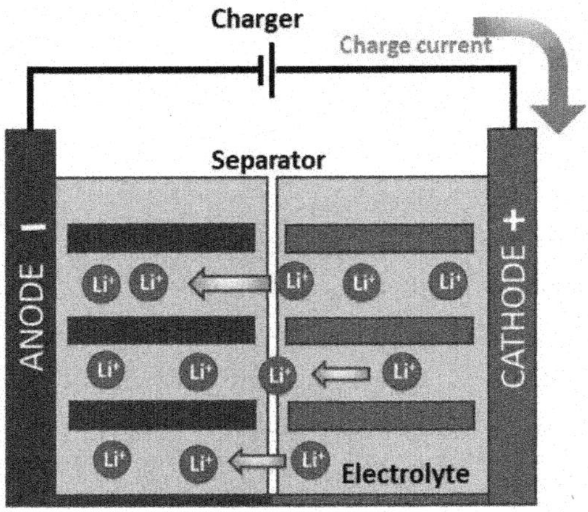

Figure 2.2 Schematic of working of a Li-ion battery.

An insertion-based electrode reaction may be represented as

$$\text{Host} + x\text{Li}^+ + xe^- \leftrightarrow \text{Li}_x\text{Host} \tag{2.27}$$

where the equilibrium potential E is given by

$$E = -\frac{\Delta G}{xF} = \frac{(E_{\text{Host}} + E_{\text{Li}} - E_{\text{Li}_x\text{Host}})}{xF} \tag{2.28}$$

The cell voltage in an alkali metal-ion battery is given by the energy gained in inserting the alkali atom A into the "de-alkaliated" host structure $A_{n-x}H$ to form A_nH, with the loss of the energy cost in extracting the alkali atom from the metal anode. The more stable the lithiated material, the greater the equilibrium voltage. The greater the energy released/needed upon insertion/removal of the electron and Li^+ ion, the greater the equilibrium voltage. This will depend on the energy level of the orbital to which the electron is going and the energy of the site where Li is being stored. In other words, the equilibrium voltage depends upon the bond formation between the host and Li. If the equilibrium potential is extremely negative, it signifies that the lithiated material is highly unstable.

In a spinal (AB_2O_4) structured $Li_xMn_2O_4$, Li is initially stored in tetrahedral sites up to $x = 1$. Beyond that, Li starts occupying the larger octahedral sites. The drop-in potential is very much visible in Fig. 2.3 despite the same $Mn^{4+/3+}$ redox reaction [1].

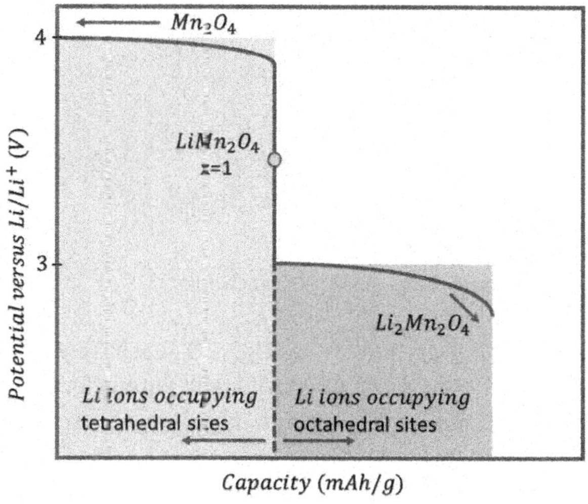

Figure 2.3 Potential drop in Li electrode [1].

In olivine structures of $LiFePO_4$, as shown in Fig. 2.4, O^- ions form a close packing array, with half of the octahedral sites occupied by Li ions/Na ions or Fe ions and one-eighth of the tetrahedral sites by P ions. The M–O bond is more ionic, which lowers the energy of M orbital corresponding to change in the oxidation state of M and hence raises the electrochemical potential [2].

In $LiMn_{1.5}Ni_{0.5}O_4$, the O^- ion framework is robust. There is minimal distortion during Li insertion/extraction; thus, the site energy remains the same. In $LiCo_{0.3}Ni_{0.3}Mn_{0.3}O_2$ and $Li_2MnO_3LiCo_{0.3}Ni_{0.3}Mn_{0.3}O_2$, the distortion is relatively greater, leading to a slight change in the site energy during Li insertion/extraction [3].

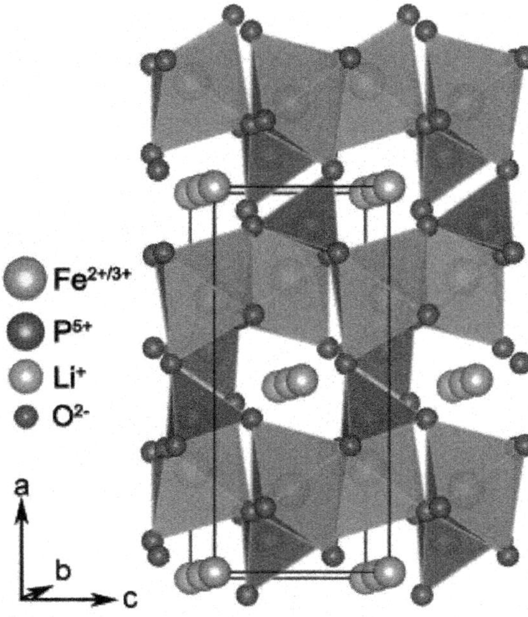

Figure 2.4 Olivine structures of $LiFePO_4$ electrodes.[1]

In $Na_2Fe_2(SO_4)_3$, as shown in Fig. 2.5, each FeO_6 octahedral shares edge with the crystallographically equivalent octahedral forming Fe_2O_{10} dimers. SO_4^{2-} anions interconnect these dimers to form a 3D structure with Na tunnels along the c-axis. The fairly flat profile ~3.80 V vs Na/Na^+ in the alluaudite structure shows good capacity retention and rate capability [4].

The specific capacity of the electrode is denoted in terms of the amount of charge stored per unit mass of the electrode (mAh/g). Capacity, $C = nF/3.6\ M$ mAh/g, where n is the number of moles of Li ions/number of moles of host atom, M is the molar mass of the host atom, and F is Faraday's constant. Another way of inferring the capacity is a capacity of 1000 mAh/g implies that if a current of 1000 mA is passed for 1 h, the electrode is expected to get filled with Li/Na to its theoretical capacity. The current is called C-rate; C/20 implies the current of 50 mA (1000/20) is supposed to fill the active electrode material with Li/Na in 20 h. However, the fraction

[1]Image generated by VESTA software, reproduced from https://crystallography365.wordpress.com/2014/04/29/lifepo4-the-unexpected-battery-success-story/

of electrodes used in the given time frame depends on its rate capability.

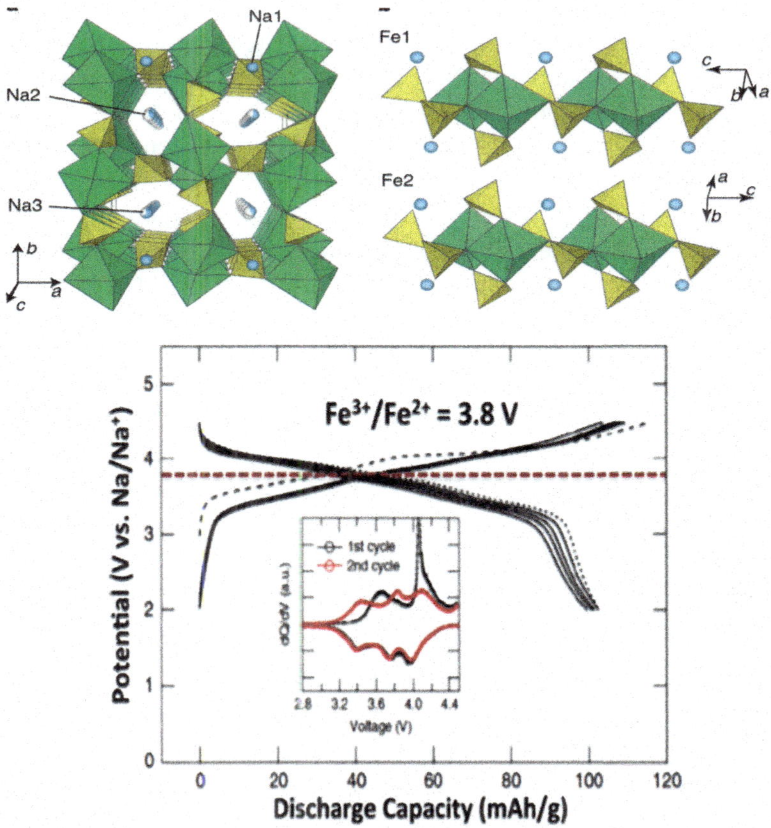

Figure 2.5 Structure of $Na_2Fe_2(SO_4)_3$ electrode.[2]

The mechanisms of operation and the design of sodium-ion batteries are analogous to the well-known lithium-ion batteries. However, despite all the similarities, the greatest challenge of sodium-ion batteries is finding suitable electrodes for the reversible ion diffusion process. The main reason is that the difference in the size of lithium and sodium ions (0.076 and 0.102 nm, respectively) poses a great hindrance in using similar electrode structures.

[2]Reproduced from Barpanda, P., Oyama, G., Nishimura, Si. et al. A 3.8-V earth-abundant sodium battery electrode. *Nat. Commun.* **5**, 4358 (2014). https://doi.org/10.1038/ncomms5358

2.3 Thermodynamics

A substance's unique propensity to contribute to a system's energy is its chemical potential or partial molar free energy. When the substance is a charged particle such as an electron or an ion, the chemical potential has an electrical component in the particle's response to the electrical field. This chemical potential is then known as electrochemical potential. The chemical energy of a cell is represented in the form of electrochemical potential. The electrochemical potential is a thermodynamic measure of one mole of atom/molecule to produce energy. The higher the electrochemical potential, the higher the amount of electrical energy (in Joules) converted from that chemical reaction. Thus, the cell potential measured in volts is defined as the electrochemical potential difference between two electrodes. The cell potential is the sum of cathodic and anodic potentials, as mentioned in Equation (2.21). This is also called the open-circuit voltage of an electrochemical cell.

The electrode and electrolyte phases of a battery are larger compared to the interfaces between them. Thus, the Gibbs free energy of a battery can be expressed in terms of the free energies of these individual phase components.

$$\begin{aligned} G(N_{\text{ion}}^{\text{C}}, N_{\text{ion}}^{\text{A}}, N_{\text{ion}}^{\text{E}}, N_{e^-}^{\text{C}}, N_{e^-}^{\text{A}}) &= G^{\text{C}}(N_{\text{ion}}^{\text{C}}, N_{e^-}^{\text{C}}) \\ &+ G^{\text{A}}(N_{\text{ion}}^{\text{A}}, N_{e^-}^{\text{A}}) + G^{\text{E}}(N_{\text{ion}}^{\text{E}}) \end{aligned} \tag{2.29}$$

where superscripts C, A, and E are for cathode, anode, and electrolyte phases, respectively. The preceding expression has implicit dependence on temperature, pressure, and the number of atoms. The electrochemical potential of ions and electrons in a phase α (anode, cathode, or electrolyte) can be expressed as

$$\eta_{\text{ion}}^{\alpha} = \frac{\partial G^{\alpha}}{\partial N_{\text{ion}}^{\alpha}} \tag{2.30}$$

$$\eta_{e^-}^{\alpha} = \frac{\partial G^{\alpha}}{\partial N_{e^-}^{\alpha}} \tag{2.31}$$

These electrochemical potentials will change with the addition/removal of each ion or electron, respectively. The chemical potential in any phase can be expressed in terms of the electrochemical potentials as:

$$\mu_{\text{metal}}^{\alpha} = \frac{\partial G^{\alpha}}{\partial N_{\text{ion}}^{\alpha}} = \frac{\partial G^{\alpha}}{\partial N_{\text{ion}}^{\alpha}} \frac{\partial N_{\text{metal}}^{\alpha}}{\partial N_{\text{ion}}^{\alpha}} + \frac{\partial G^{\alpha}}{\partial N_{e^-}^{\alpha}} \frac{\partial N_{e^-}^{\alpha}}{\partial N_{\text{metal}}^{\alpha}} = \eta_{\text{ion}}^{\alpha} + \eta_{e^-}^{\alpha} \quad (2.32)$$

When a battery is in its open state, no charge is drawn or added; the ions redistribute to minimize the total free energy of the system. The total number of ions in the battery is always constant:

$$N_{\text{metal}} = N_{\text{ion}}^{C} + N_{\text{ion}}^{A} + N_{\text{ion}}^{E} \quad (2.33)$$

Therefore, there are only two variables $N_{\text{ion}}^{\text{C}}$ and $N_{\text{ion}}^{\text{A}}$. Taking partial derivatives of G, equating them to zero yields

$$\left(\frac{\partial G}{\partial N_{\text{ion}}^{\text{C}}}\right)_{N_{\text{ion}}^{\text{E}}} = \frac{\partial G^{\text{C}}}{\partial N_{\text{ion}}^{\text{C}}} + \frac{\partial G^{\text{A}}}{\partial N_{\text{ion}}^{\text{A}}} \left(\frac{\partial N_{\text{ion}}^{\text{A}}}{\partial N_{\text{metal}}^{\text{C}}}\right) = \eta_{\text{ion}}^{\text{C}} + \eta_{\text{ion}}^{\text{A}} = 0 \quad (2.34)$$

$$\left(\frac{\partial G}{\partial N_{\text{ion}}^{\text{E}}}\right)_{N_{\text{ion}}^{\text{C}}} = \frac{\partial G^{\text{E}}}{\partial N_{\text{ion}}^{\text{E}}} + \frac{\partial G^{\text{A}}}{\partial N_{\text{ion}}^{\text{A}}} \left(\frac{\partial N_{\text{ion}}^{\text{A}}}{\partial N_{\text{metal}}^{\text{E}}}\right) = \eta_{\text{ion}}^{\text{E}} + \eta_{\text{ion}}^{\text{A}} = 0 \quad (2.35)$$

From this, we understand that electrochemical potential must be uniform throughout the battery at equilibrium, $\eta_{\text{ion}}^{\text{C}} = \eta_{\text{ion}}^{\text{E}} = \eta_{\text{ion}}^{\text{A}}$. This allows us to derive the Nernst equation:

$$E = -\frac{\eta_{e^-}^{\text{C}} - \eta_{e^-}^{\text{A}}}{e} = -\frac{\mu_{\text{ion}}^{\text{C}} - \mu_{\text{ion}}^{\text{A}}}{e} \quad (2.36)$$

where e is the charge of an electron. Substituting in Equation (2.36) and solving for E, we can get a general expression for the Nernst equation:

$$E = -\frac{\mu_{\text{A}}^{\text{C}} - \mu_{\text{A}}^{\text{A}}}{ne} \quad (2.37)$$

The above equation is a general expression for the Nernst equation, which shows that the potential of a battery is equal to the difference in chemical potential of the anode and cathode of the battery. The measurement of this potential can give insight into various thermodynamic entities of the electrode at different stages of charge.

The chemical potential of an ionic species in an electrode can be derived from its Gibbs free energy, and the dependence of species (ions) concentration in an electrode on Gibbs free energy gives us an expression between species concentration and electrode potential. The chemical potential of the species depends on the type of electrode, intercalation, alloy, or conversion. The species (Na or Li) concentration of an intercalation compound x is defined as the ratio of atoms to available interstitial sites, proportional to the number of host formula units. In the case of an alloy-type electrode, the concentration x refers to the relative fraction of species. In the conversion-type electrode, displacement reactions result in a redistribution of other species. Therefore, free energy is represented in terms of composition.

An electrostatic potential difference is formed at the electrode/ electrolyte interface due to the redistribution of charged ionic species (Na^+) between electrode and electrolyte phases to reach chemical equilibrium. This electrostatic potential difference results in the formation of a large electric field along with the interface. The change in electrostatic potential difference can be represented by Equation (2.40). The chemical potentials depend on temperature and pressure at the interface.

For anode,

$$-e\Delta\phi = \mu^{\mathrm{A}}_{\mathrm{metal}} - \mu^{\mathrm{E}}_{\mathrm{ion}} - \mu^{\mathrm{A}}_{e^-} \tag{2.38}$$

$$\Delta\phi = \phi^{\mathrm{A}} - \phi^{\mathrm{E}} \tag{2.39}$$

For cathode,

$$-e\Delta\phi = \mu^{\mathrm{C}}_{\mathrm{metal}} - \mu^{\mathrm{E}}_{\mathrm{ion}} - \mu^{\mathrm{C}}_{e^-} \tag{2.40}$$

$$\Delta\phi = \phi^{\mathrm{C}} - \phi^{\mathrm{E}} \tag{2.41}$$

The electrostatic potential difference ($\Delta\phi$) and the atomic structure of the compound across the interface influence the kinetics of electrochemical reactions at the interface. Therefore, to fully understand the reaction kinetics of alkali-ion batteries, it is very important to understand completely both these influencing factors at the interfaces. Many models have been proposed to understand the interfaces for Li-ion batteries. However, these areas are relatively new and unexplored for other chemistries (Na, K, Ca, Mg). The atomic structure at the electrode/ electrolyte interface is very complex due to the formation of layers of decomposition components. One such example is the solid electrolyte interface (SEI), commonly formed on the Li-ion anode/electrolyte interface. These complexities also pose challenges in developing computational models to understand the thermodynamics of the interfaces fully.

2.4 Electrode Reaction Kinetics

Electrochemical kinetics deals with the rates of chemical reactions. The reaction rate can be easily controlled by changing the potential. Usually, the electron transfer in an electrochemical reaction occurs on the metal surface, and it is easy to change the potential of metals. Another important aspect of electrochemistry is the reactions occurring at the interface between the metal electrode and electrolyte solution. In the previous sections, we have discussed the potential of an electrode at equilibrium. We will now see how the potential changes when the current is flowing in the system.

The overall chemical reaction taking place in an electrochemical cell is made up of two independent half-cell reactions. If one is interested in studying only one of these half-cell reactions, the electrode at which it occurs is referred to as the working electrode. The other half-cell is standardized (kept at constant potential) and acts as a reference electrode. For experimental work, it is desirable to have a reference electrode in the system to provide a known stable potential against which all other potentials can be measured. Therefore, a reference electrode should have stable and well-defined potential. The reaction occurring at the reference electrode should be reversible to maintain the electrode potential at its equilibrium. There are three usual standard reference electrodes:

Standard hydrogen electrode (SHE) or normal hydrogen electrode (NHE) consists of metal platinum on which hydrogen reacts, and hydrogen gas is bubbled around it, as shown in Fig. 2.6a. The hydrogen ion concentration in the solution is known. The pressure above the solution is a combination of hydrogen gas pressure and water vapor pressure. Usually, NHE is denoted as $Pt/H_2\ (a = 1)/H^+\ (a = 1, \text{aq})$.

The saturated calomel electrode (SCE) has a potential of 0.242 V versus NHE. The solution is kept saturated by the addition of KCl crystals to keep chloride ion concentration constant. The electrode consists of Hg(l), Hg_2Cl_2(s), and saturated KCl connected to electrolyte via a porous fit, as shown in Fig. 2.6b. The SCE is denoted as Hg/Hg_2Cl_2/KCl (saturated in water).

A silver–silver chloride electrode has a potential of ~0.19 V versus NHE. It consists of a simple silver wire upon which a silver chloride layer has been formed. Again, the concentration of chloride ions is kept constant by the addition of KCl crystals. A silver-silver chloride electrode can be denoted as Ag/AgCl/KCl (saturated with water).

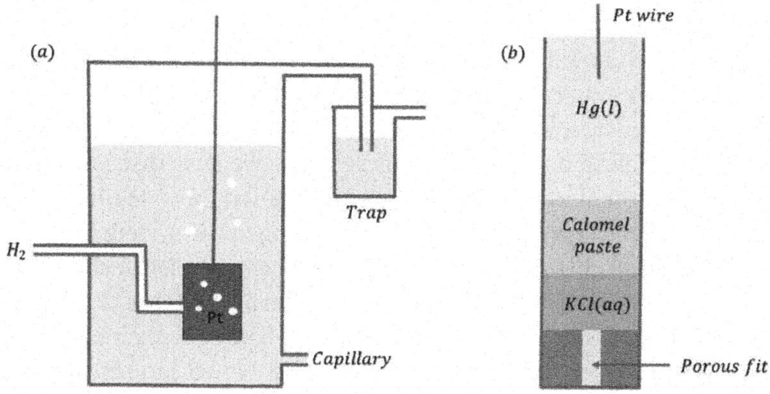

Figure 2.6 Schematic representation of (a) normal hydrogen electrode (NHE) and (b) saturated calomel electrode (SCE).

Usually, the potential energy of electrons in the working electrode is observed/controlled with respect to the reference electrode. If the potential of the working electrodes is made more negative, the energy of electrons is raised. They can reach a level high enough to move from the vacant electronic states of the species to the electrolyte. Thus, the current produced due to the flow of electrons from species to the solution is called reduction current. If the working electrode's potential is made more positive, the electrons in the solute will find favorable energy on the electrode and transfer there. The current thus produced is called oxidation current.

Electrode reaction kinetics involves the measurement of rates of chemical reactions. The rate of chemical reactions depends upon the concentration changes in the reactants and products. Chemical reactions can be exothermic or endothermic and occur in the bulk of the solution. Most electrode reactions involve multiple heterogeneous steps and depend upon charge transfer rates and mass transfer. Electrode reaction kinetics depends on the concentration of the reactants and products, temperature and electrode potential.

Electrode potential can give insight into determining various thermodynamic parameters. For a reaction to be electrochemical, at least one step should involve the transfer of charge. The rate of an electrochemical reaction is determined by the slowest step in the overall reaction. Due to the heterogeneous structure of electrodes, various factors influence the reaction rate, such as the double-layer structure, adsorption of reactants/products and electrolytes. Various techniques are implemented to study the kinetic parameters of electrode reactions. These techniques can be broadly classified into steady-state and transient techniques. Steady-state techniques are potentiostatic and galvanostatic, in which current and voltage, respectively, are kept constant. In transient techniques, the system is perturbed by a current or potential pulse, and change in current or potential with respect to time is measured. The transient methods can further be classified into steady-state and non-steady-state methods. In the steady-state method, any coupled chemical process, the interface structure, and the charge transfer process are kept steady throughout the measurement. The transient methods are generally used to study metal dissolution or deposition on the electrode surface.

Let us look at a simple electrochemical reaction occurring at an electrode surface, as shown in Fig. 2.7.

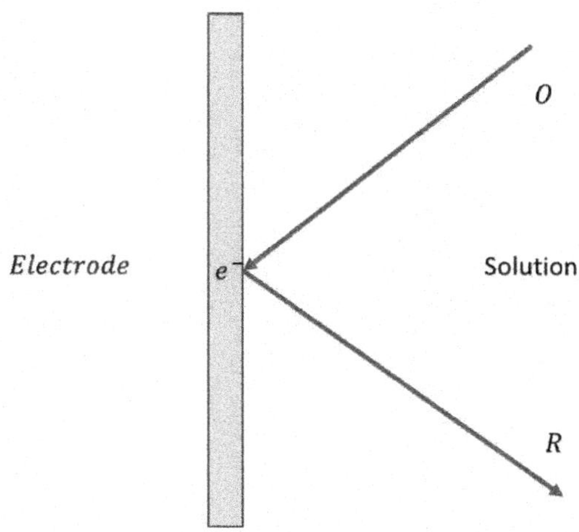

Figure 2.7 Electrode–electrolyte interface reduction reaction.

The reaction can be written as

$$O + ne \underset{k_b}{\overset{k_f}{\rightleftharpoons}} R \tag{2.42}$$

where O is the oxidized form, R is the reduced form, and n is the number of electrons involved. Here, k_f and k_b denote forward and backward reaction rates, respectively. Assuming the reaction to be first-order, the rates of forward and backward reactions are r_f and r_b, respectively:

$$r_f = k_f C_O \tag{2.43}$$

$$r_b = k_b C_R \tag{2.44}$$

C_O and C_R are the bulk concentrations of the oxidized and reduced species. Similarly, the current density for forward and backward reactions can be represented by i_f and i_b, respectively:

$$i_f = nFk_f C_O \tag{2.45}$$

$$i_b = nFk_b C_R \tag{2.46}$$

The rate constants are represented as follows:

$$k_f = k_f^0 e^{\frac{\alpha nFE}{RT}} \tag{2.47}$$

$$k_b = k_b^0 e^{\frac{(1-\alpha)nFE}{RT}} \tag{2.48}$$

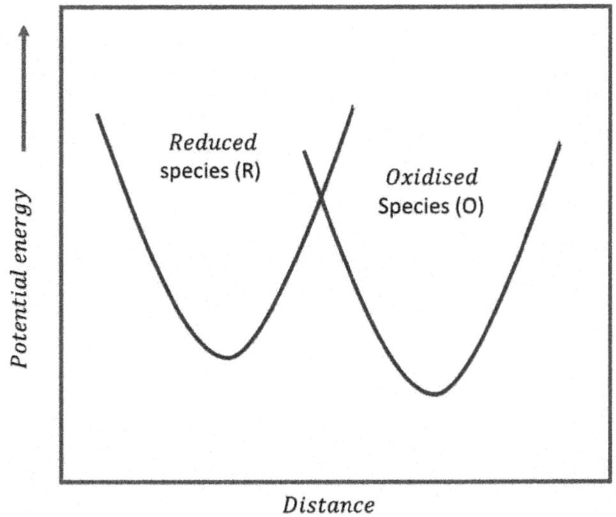

Figure 2.8 Potential energy diagram of oxidation–reduction reaction occurring at the electrode surface.

Figure 2.8 shows an approximate potential energy curve for the reaction shown in Equation (2.42). This curve is known as the Morse curve. E = 0 is the electrode potential, and k_f^0 and k_h^0 are the standard rate constants. At equilibrium, when $E = E^0$, the rate constants will be equal to $k_\mathrm{f} = k_\mathrm{b} = k^0$.

$$k_\mathrm{f} = k_\mathrm{f}^0 e^{\frac{\alpha nFE^0}{RT}} = k_\mathrm{b} = k_\mathrm{b}^0 e^{\frac{(1-\alpha)nFE^0}{RT}} = k^0 \tag{2.49}$$

At the equilibrium electrode potential, the net current must be zero.

$$i = nFv \tag{2.50}$$

$$i = nFk^0 \left[C_\mathrm{O} e^{\frac{-\alpha nF(E-E^0)}{RT}} - C_\mathrm{R} e^{\frac{(1-\alpha)nF(E-E^0)}{RT}} \right] \tag{2.51}$$

when $i = 0$,

$$C_\mathrm{O}^* e^{\frac{-\alpha nF(E-E^0)}{RT}} = C_\mathrm{R}^* e^{\frac{(1-\alpha)nF(E-E^0)}{RT}} \tag{2.52}$$

$$\frac{C_\mathrm{O}^*}{C_\mathrm{R}^*} = e^{\frac{\alpha nF(E_\mathrm{eq}-E^0)}{RT}} \tag{2.53}$$

Thus, even when the electrode is in equilibrium, a small amount of current flows in either direction (i^0).

$$i^0 = nFk^0 \left[C_\mathrm{O}^* e^{\frac{-\alpha nF(E_\mathrm{eq}-E^0)}{RT}} = C_\mathrm{R}^* e^{\frac{(1-\alpha)nF(E_\mathrm{eq}-E^0)}{RT}} \right] \tag{2.54}$$

$$i^0 = nFk^0 \left[C_\mathrm{O}^{*(1-\alpha)} C_\mathrm{R}^{*\alpha} \right] \tag{2.55}$$

when $C_\mathrm{O}^* = C_\mathrm{R}^* = C$, $i^0 = nFk^0C$. Dividing i/i^0, we get

$$\frac{i}{i_0} = \frac{C_\mathrm{O} e^{-\frac{\alpha nF(E-E^0)}{RT}}}{C_\mathrm{O}^{*(1-\alpha)} C_\mathrm{R}^{*\alpha}} - \frac{C_\mathrm{R} e^{-\frac{(1-\alpha)nF(E-E^0)}{RT}}}{C_\mathrm{O}^{*(1-\alpha)} C_\mathrm{R}^{*\alpha}} \tag{2.56}$$

At the electrode/electrolyte interface, electrochemical reactions occur during the charging and discharging of a battery. The driving force for these reactions is the difference in electrochemical potential between the electrode and electrolyte interface. The electrochemical reactions will occur at the interface when this electrochemical potential difference is disturbed. If the electrochemical potential difference becomes negative, the ions will flow from the electrolyte into the anode, and the reverse will occur if the electrochemical potential difference is positive. The negative/positive shift from the electrochemical equilibrium potential difference is called overpotential (η). The departure of electrode potential or cell potential from the equilibrium value upon passage of Faradaic current is termed polarization, and the extent of polarization is measured by overpotential. A particular current density is driven by certain overpotential, which is the sum of three overpotentials, mass transfer (η_{mt}), charge transfer (η_{ct}), and reaction overpotential (η_{rxn}).

$$i = \eta_{mt} + \eta_{ct} + \eta_{rxn} \tag{2.57}$$

On simplifying and substituting,

$$i = i^0 \left[\frac{C_O}{C_O^*} e^{-\frac{\alpha n F \eta}{RT}} - \frac{C_R}{C_R^*} e^{\frac{(1-\alpha) n F \eta}{RT}} \right] \tag{2.58}$$

$$\eta = E - E_{eq}^* \tag{2.59}$$

If there are no concentration gradients of electroactive species in the solution, the concentration is equal in the bulk and at all other points for the species O and R. Then the equation is further simplified, and the form is known as the Butler–Volmer equation (Equation (2.60)).

$$i = i^0 \left[e^{-\frac{\alpha n F \eta}{RT}} - e^{\frac{(1-\alpha) n F \eta}{RT}} \right] \tag{2.60}$$

Figure 2.9 shows the relationship between current and applied overpotential. The net current is the sum of the cathodic and the anodic currents.

Even though the net current is zero at equilibrium, there is balanced Faradaic activity that can be expressed in terms of exchange current density i^0, which is equal in magnitude to anodic as well as cathodic current ($i^0 = i_a = i_c = FAk^0C$). At small overpotentials, $\eta \ll$ small.

$$i = i^0 \left[\frac{nF\eta}{RT} \right] \tag{2.61}$$

Here η/i has the unit of resistance and is called the charge transfer resistance $R_{ct} = \frac{RT}{Fi^0}$. At large overpotentials, the linear relation is known as the Tafel equation (usually for the reversible process).

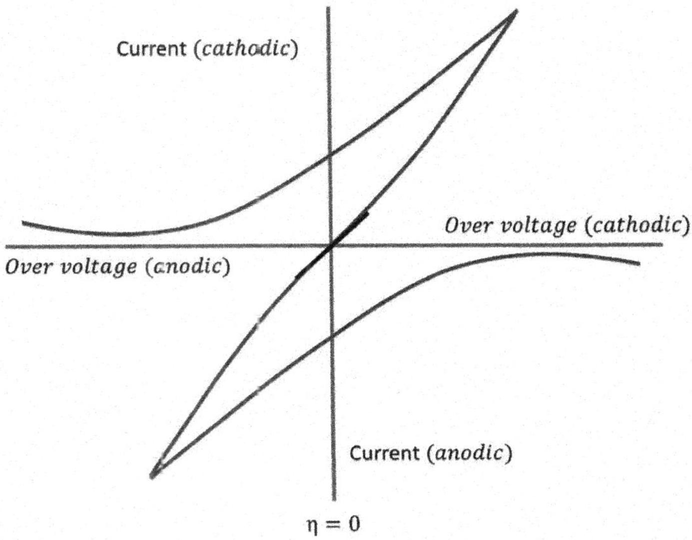

Figure 2.9 Relationship between overpotential and current.

$$i = i^0 \left[e^{-\frac{\alpha nF\eta}{RT}} \right] \tag{2.62}$$

$$\ln i = \ln i^0 - \frac{\alpha nF\eta}{RT} \tag{2.63}$$

Figure 2.10 is the plot of Equation (2.63), which can also be written in linear form, as shown in Equation (2.64).

$$\eta = \frac{RT}{\alpha nF} \ln i^0 - \frac{RT}{\alpha nF} \ln i \tag{2.64}$$

where intercept is $RT\ln(i^0)/\alpha nF$ and slope is $b = RT/\alpha nF$.

The exchange current density and the current depend on the concentration of the species at the surface. A species can be consumed or produced in an electrochemical reaction at the surface. This leads to a variation of species concentration at the surface. The potential difference depends on the species concentration at the interface and the thermodynamic variables such as temperature and pressure. The net flux of species concentration can be related to the potential difference by the following relation:

$$J = J^0 \left[e^{-\frac{\alpha\eta}{k_b T}} - e^{\frac{(1-\alpha)\eta}{k_b T}} \right] \quad (2.65)$$

where k_b is the Boltzmann constant, T is temperature, J^0 is exchange flux, and α is the symmetry factor. Both exchange current and symmetry factors are empirical parameters. At very small overpotentials,

$$J \approx J^0 \frac{\eta}{k_b T} \quad (2.66)$$

If the interface kinetics is very fast, compared to other reactions in bulk and electrolyte, η will be close to zero. Then the local equilibrium approximation becomes valid. The potential at the electrode will be approximately equal to that at the electrolyte.

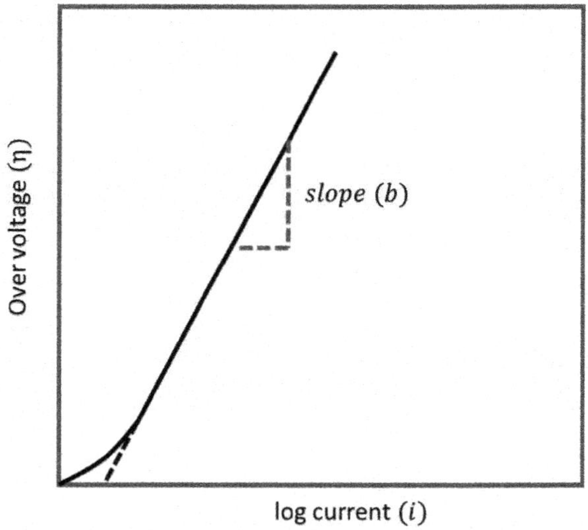

Figure 2.10 Schematic representation of Tafel equation.

Some other representations of the relation between net flux and overpotential, like those based on Marcus theory, exist. All these expressions have one thing in common: they have a few empirical parameters that are usually fitted with the help of experimental data. The transport of ionic species within electrodes depends on the mobility of ionic species within the compound structure, and the flux expression can be represented as shown in Equation (2.67).

$$J_{\text{species}} = -L\Delta\mu_{\text{species}} \tag{2.67}$$

The preceding expression can be converted to Fick's first law of diffusion using the chain rule of differentiation:

$$j_{\text{species}} = -D\Delta C \tag{2.68}$$

where C is species concentration in mol/m^3 and D is diffusion coefficient of the ionic species.

In electrodes that are not purely metallic, the electronic conductivity can be rate-limiting. Most of the electrodes used in alkali-ion batteries are either layered compounds or compounds having vacancies to host alkali ions. Such structures show sluggish electronic conductivities. Therefore, in such electrodes, net species flux and net electric flux are dependent on both ionic and electronic electrochemical potentials.

2.5 Electrode Reaction Mechanisms

Reactions at electrodes are both chemical and electrical and are, therefore, heterogeneous in type. Electrode reactions may be as simple as reducing metal or intercalation of an ion into the electrode structure. The overall mechanism of the simplest electrode reaction is complex. Electrode reaction mechanisms often involve multiple steps. These steps could be either diffusion of an electroactive species to the electrode surface or adsorption of electroactive material, or electron transfer step. These steps could also involve chemical reactions in the overall electrode reaction.

The usual experimental procedure is to measure the current density i at large overpotential values to study electrode reaction mechanisms. At large overpotentials, we can apply the Tafel relation and easily determine transfer coefficient and exchange current density. However, when the electrode reaction is fast (high i^0), it is difficult to accurately determine the preceding parameters from the Tafel region due to mass transfer interference.

The set of all intermediate steps involved in an electrochemical reaction is called the reaction pathway. A given reaction may occur by multiple pathways. In the case of reversible reactions, the pathway of forward reaction may not be the same for backward reaction. Let us consider an example to understand this. The reaction is

$$Mn^{2+} \rightarrow Mn^{4+} + 2e^{-} \tag{2.69}$$

which may occur by any one of the two pathways as given by Equations (2.70) and (2.71).

$$Mn^{2+} \xrightarrow{-e-} Mn^{3+} \xrightarrow{-e-} Mn^{4+} \tag{2.70}$$

$$2[Mn^{2+} \xrightarrow{-e-} Mn^{3+}]; \quad 2Mn^{3+} \rightarrow Mn^{2+} + Mn^{4+} \tag{2.71}$$

We can represent multistep electrode reactions involving more than a single pathway in the form, as shown in Table 2.3.

Table 2.3 Multistep electrode reaction pathway steps

k	Step	I_k	μ_k(I)	μ_k(II)
1	$Mn^{2+} \rightarrow Mn^{3+} + e^{-}$	1	1	2
2	$Mn^{3+} \rightarrow Mn^{4+} + e^{-}$	1	1	
3	$Mn^{3+} \rightarrow Mn^{2+} + Mn^{4+}$	0		1

Here I_k is the number of electrons in a step, and μ_k is the stoichiometric number of steps in pathways I and II, indicating how many times this step is repeated in a reaction. For an electrochemical reaction, $I_k = 1$, and for purely chemical reactions, $I_k = 0$. Also, $\Sigma\mu_k I_k = n$, where n is the total number of electrons. Purely chemical steps are denoted by C and electrochemical steps by letter E. Thus, the first pathway follows the *EE* scheme, and the second follows the *EC* scheme.

The number of particles produced in any step is equal to the number of particles reacting in the next step. Thus, the rates of all intermediate steps are dependent, as shown in Equation (2.72).

$$v = \frac{v_1}{\mu_1} = \frac{v_2}{\mu_2} = \ldots = \frac{v_k}{\mu_k} = \ldots = \frac{v_z}{\mu_z} \tag{2.72}$$

Each step has the same rate in a steady-state, which is equal to the overall reaction rate. Each intermediate step has its own kinetic rules and ways despite the same rate. To understand the overall electrochemical reaction, it is important to understand the kinetic parameters of intermediate steps. Consider a two-step reversible chemical reaction:

$$A \underset{(-1)}{\overset{(1)}{\rightleftharpoons}} B \underset{(-2)}{\overset{(2)}{\rightleftharpoons}} D \tag{2.73}$$

where B is the intermediate and $\mu_1 = \mu_2 = 1$. Assuming both steps are first order, that is

$$v_1 = k_1 c_A - k_{-1} c_B \tag{2.74}$$

$$v_2 = k_2 c_B - k_{-2} c_D \tag{2.75}$$

At steady state, $v_1 = v_2 = v$. Substituting Equations (2.74) and (2.75), we get

$$c_B = \frac{k_1 c_A + k_{-2} c_D}{k_{-1} + k_2} \tag{2.76}$$

For intermediate B (Equation (2.76)) and for the rate of overall reaction (Equation (2.77)),

$$v = \frac{k_1 k_2 c_A + k_{-1} k_{-2} c_D}{k_{-1} + k_2} \tag{2.77}$$

During discharging, the rate of overall reaction can be written as

$$v = k_0 c_A - k_{-0} c_D \tag{2.78}$$

Equations (2.77) and (2.78) give us the connection between kinetic parameters of the overall reaction and its steps:

$$k_0 = \frac{k_1 k_2}{k_{-1} + k_2}; \quad k_{-0} = \frac{k_{-1} k_{-2}}{k_{-1} + k_2} \tag{2.79}$$

The slowest step is the rate-determining step (RDS) of the reaction. The kinetic parameter of the RDS wholly determines the rate of the overall reaction. The ratio of k_{-1} and k_2 decides which of these is the RDS.

In the case of alkali metal-ion batteries, the exchange of ions occurs. Let us understand the mechanism of this exchange. There are three methods to categorize the exchange mechanism of alkali metal ions in the host structures: chemical, electrochemical, and combined. In the chemical mechanism of ion exchange, the reaction can be spontaneous or nonspontaneous. In case the chemical reaction is spontaneous, the displacement of stored alkali metal ions is thermodynamically favourable. This mechanism is usually observed in alkali metal halides. If the chemical reaction is nonspontaneous, the primary requirement is that the host material is partially oxidized/reduced with an oxidizing/reducing agent that might require multiple oxidation/reduction steps during deintercalation/intercalation of alkali metal ions.

In the electrochemical mechanism of ion exchange, the host material is oxidized by applying a positive current or voltage higher than the equilibrium potential of the cell. This leads to the extraction/deintercalation of metal ions from the host material. In the case of intercalation of metal ions in the host structure, a negative current or voltage lower than the equilibrium potential of the cell is applied. Li-ion and Na-ion battery electrodes are generally based on this electrochemical mechanism of intercalation/deintercalation of Li and Na ions, respectively, in the host electrode material structures.

In the combined ion-exchange mechanism, the host material is first chemically oxidized to remove the alkali ions, and then the host structure thus produced having "alkali-ion vacancies" is assembled into an electrochemical cell. This mechanism requires the host with "alkali-ion vacancies" to be stable during the electrochemical cell assembly. The driving force in the case of chemical mechanism could be one of these: low Gibbs free energy of the host compound for intercalation or an entropic effect caused by excess ions. In the electrochemical mechanism, the driving force is externally applied current or potential.

The rate of both chemical and electrochemical mechanisms can be limited by sluggish solid-state diffusion and large overpotential, respectively. Thus, the advantage of the electrochemical mechanism is that the rate can be directly controlled by changing the externally applied current or voltage. On the other hand, the disadvantage of the electrochemical mechanism is the poor stability of host structures. Out of these three mechanisms, the pure electrochemical alkali metal exchange in aqueous solutions has shown promising battery performance.

The intercalation/deintercalation of the Na-ion storage mechanism can be understood by dividing the host compounds into two categories: carbon-based and non-carbon based. In carbon-based compounds, hard carbon and graphitic carbon are the two most widely studied categories of anode materials for Na-ion batteries. Due to its disordered structure and large interlayer distance, hard carbon can store Na ion and has shown a reversible capacity of 300 mAh/g at low current density. The mechanism of Na-ion storage in hard carbon can be in one of these ways: intercalation of Na ions in the hard-carbon layers, adsorption of Na ions at defect sites, adsorption at the surface or filling in the nanopores. Graphitic carbons such as carbon nanotubes, graphene oxide, and carbon nanofillers have also been studied for Na-ion batteries. However, due to the large size of Na ions, these materials undergo large volume expansion and show poor cyclic stability. Many non-carbon compounds have been studied, for example, layered metal sulfides and metal oxides. Na ions intercalate into the layers on these compounds and are bound by weak van der Waals forces.

References

1. J. B. Goodenough and K.-S. Park, "The Li-ion rechargeable battery: A perspective," *J. Am. Chem. Soc.*, **135**, pp. 1167–1176, 2013.
2. F. Liu, S. Song, D. Xue, and H. Zhang, "Selective crystallization with preferred lithium-ion storage capability of inorganic materials," *Nanoscale Res. Lett.*, **7**, p. 149, 2012.
3. C. Liu, Z. G. Neale, and G. Cao, "Understanding electrochemical potentials of cathode materials in rechargeable batteries," *Mater. Today*, **19**, pp. 109–123, 2016.
4. P. Barpanda, G. Oyama, S.-I. Nishimura, S.-C. Chung, and A. Yamada, "A 3.8-V earth-abundant sodium battery electrode," *Nat. Commun.*, **5**, 4358, 2014, doi: 10.1038/ncomms5358

Chapter 3

Cathode Materials for Sodium-Ion Batteries

Xin Guo,[a] Shijian Wang,[a] Hong Gao,[a] Rui Zang,[b] Xiaogang Zhang,[b] Jian Yang,[c] Chengyin Wang,[c] and Guoxiu Wang[a]

[a]*Centre for Clean Energy Technology, School of Mathematical and Physical Sciences, Faculty of Science, University of Technology, Sydney, NSW 2007, Australia*
[b]*College of Material Science and Engineering, Nanjing University of Aeronautics and Astronautics, Nanjing 210016, China*
[c]*School of Chemistry and Chemical Engineering, Yangzhou University, Yangzhou, Jiangsu 225002, China*
guoxiu.wang@uts.edu.au

3.1 Introduction

Great success has been achieved in developing commercial lithium-ion batteries (LIBs) with satisfied safety, high energy density, and cycling stability in past decades [1, 2]. However, the limited lithium resources hardly meet the requirement for electric vehicles and large-scale energy storage [3]. Sodium-ion batteries (SIBs) have been

Handbook of Sodium-Ion Batteries: Materials and Characterization
Edited by Rohit R. Gaddam and X. S. (George) Zhao

ISBN 978-981-4968-15-7 (Hardcover), 978-1-003-30874-4 (eBook)
www.jennystanford.com

recently considered a promising energy storage system for replacing LIBs due to the natural abundance of sodium, low cost, and a similar reaction mechanism to LIBs [4, 5]. However, most of the advanced electrode materials for LIBs could not be directly applied to SIBs due to the larger radius (1.02 Å) of Na^+ than that of Li^+ (0.76 Å), which leads to colossal crystal lattice deformation and poor electrochemical performance [6]. Therefore, searching for suitable host materials for Na^+ storage is imperative for the development of practical SIBs.

Similar to LIBs, sodium cathodes play an essential role in determining the cost, power and energy density, cycle life, and safety of the SIBs. Cathodes deliver lower specific capacities of 100~180 mAh/g than the most important anodes (e.g., hard carbon), meaning that cathodes occupy a larger mass percentage in the practical SIBs. So far, a variety of cathode materials for SIBs have been reported. These include layer and tunnel-type transition metal oxides, polyanionic compounds, Prussian blue analogs, conversion-type materials, and organic polymers. This chapter will review the advances and limitations of each kind of cathode material, covering the synthetic strategies, mechanisms, electrochemical performance, and the corresponding modification strategies.

3.2 Sodium Layered Oxides

3.2.1 Structure and Properties of Layered Transition Metal Oxides

The chemical formula of layered transition metal oxides can be written as $Na_xM_yO_2$, where *M* represents single or multiple transition metal cations with different valence states. The conventional layer structure consists of a sheet of edge-sharing MO_6 octahedron, which forms a $(MO_2)_n$-layered structure, and the Na^+ can be coordinated between the $(MO_2)_n$ layers in an octahedral (*O*-phase), a tetrahedron (*T*-phase), or a prismatic (*P*-phase) site. These coordination environments depend on the stacking form of $(MO_2)_n$ layers. As shown in Fig. 3.1, in a close-packed structure of hexagonal sheets, oxygen atoms have three possible positions, e.g., A, B, and C [7, 8].

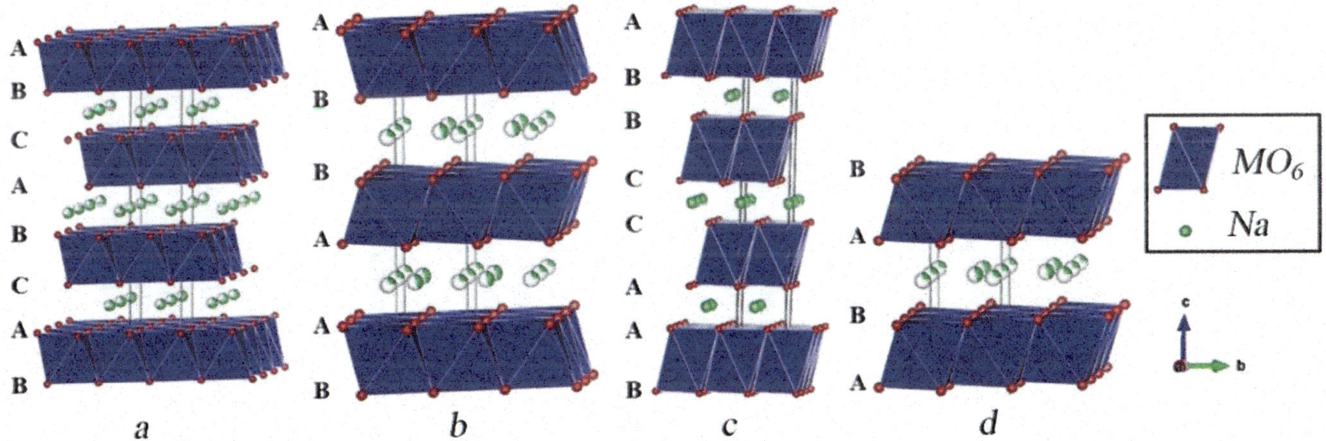

Figure 3.1 Typical structures of (a) *O*3-, (b) *P*2-, (c) *P*3-, and (d) *T*1-phase sodium layered oxides. Reprinted with permission from Ref. [8], Copyright 2015, Royal Chemical Society.

If the oxygen atoms adjacent to the B layer is in A or C positions, the Na ions are tetrahedrally or octahedrally coordinated. In this case, two oxygen packings are possible: ABAB (*T*1 phase) and ABCABC (*O*3 phase). In the former case, the Na ion is in a tetrahedron, and its chemical formula is estimated to be Na_2MO_2 because there are two tetrahedral positions in one octahedron [9]. However, this kind of structure cannot be realized because the tetrahedral vacancy is not big enough to accommodate the sodium ions. In the latter *O*3-phase structure, the Na ion is located at the octahedron center, and the chemical formula can be written as Na_xMO_2 ($x \leq 1$). The content of vacancies in this phase is relatively small, and a typical example is α-$NaFeO_2$ [10].

If the oxygen atom adjacent to the B layer is in a B position, the Na ions are prismatically coordinated. There are two possible types of oxygen stacking forms as well. One is the ABBA (*P*2-phase) structure, of which half of the triangular prisms share edges with the MO_6 octahedron, and the other half share faces. Na ions are more likely to occupy the voids of the co-edged prisms ($x \leq 1$) since the occupancy rate of the co-planar prisms is a bit lower ($x \approx 0.7$). The other form is the ABBCCA (*P*3-phase) type. In this structure, all prisms share one face, one MO_6 octahedron, and three edges with the MO_6 octahedron in the adjacent layer. Based on the experimental study results, the x value is 0.5. Unlike sodium ions, lithium ions cannot occupy the prism vacancy. In sodium-ion batteries, the layered transition metal oxide cathode materials mainly include three types: *O*3 phase, *P*2 phase, and *P*3 phase [11].

3.2.2 Na_xCoO_2 and Its Derivatives

Na_xCoO_2 has been widely investigated as the cathode material for SIBs since the successful application of its lithium counterpart [12]. Based on sodium content, the Na_xCoO_2 mainly contains three phases: *P*2, *O*3, and *P*3. The letter indicates the environment where Na is located (*O*: octahedral, *P*: prismatic), and the number indicates the number of unique interlayers surrounded by different oxide layers. The use of prime (′) indicates a distorted phase [13]. Through a simple solid-phase synthesis process and controlling the temperature at the range of 500~800°C, the Na_xCoO_2 species with $0.55 \leq x \leq 0.60$ (*P*′3), $0.64 \leq x \leq 0.74$ (*P*2), $x = 0.77$ (*O*′3) and

$x = 1$ ($O3$) are obtained [14]. The variation of sodium concentration gradient causes the slipping of the layer structure, which reflected by the phase change, such as $O3$–$O'3$–$P'3$ (starting from $O3$ phase) (Fig. 3.2a). The phase transition tends to occur in $O'3$ or $P'3$ phase [8], while the $P2$ phase is stable during the insertion/extraction of sodium ions (Fig. 3.2b) [15]. The dQ/dV curves of $P3$-$Na_{0.67}CoO_2$ and $O'3$-$Na_{0.83}CoO_2$ indicate that the $P3$ and $O'3$ phases have similar electrochemical behaviors during the charge/discharge process. When the voltage exceeds 2.7 V, the $P3$, $P2$, and $O'3$ phases have normal electrochemical behaviors. However, below 2.7 V, the electrochemical behavior of the $P2$ phase is very complicated due to the different arrangement order of sodium ions [16].

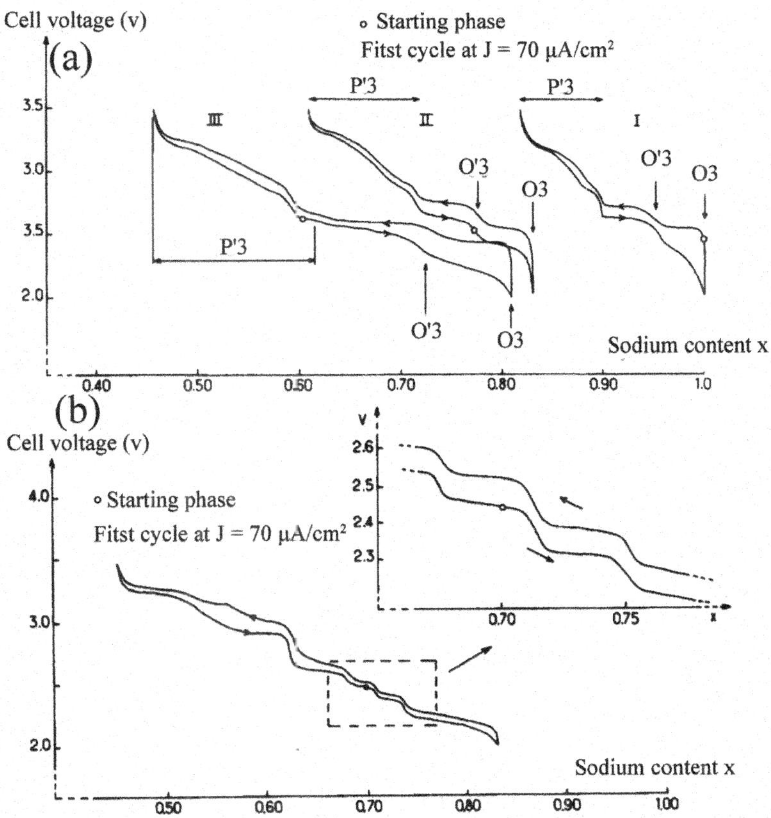

Figure 3.2 (a) The charge/discharge curves of $O3$-$NaCoO_2$ (I), $O'3$-$Na_{0.77}CoO_2$ (II), and $P'3$-$Na_{0.6}CoO_2$ (III). (b) The charge/discharge curves of $Na_{0.7}CoO_2$ ($P2$). Reprinted with permission from Ref. [14], Copyright 1981, Elsevier.

The redistribution of sodium ions reduces the electrostatic repulsion between Na^+ and Co^{3+} and the electron–electron interaction in the oxide layer [17]. Berthelot et al. studied the phase change of *P*2-Na_xCO_2 during the discharge process. As shown in Fig. 3.3a, a series

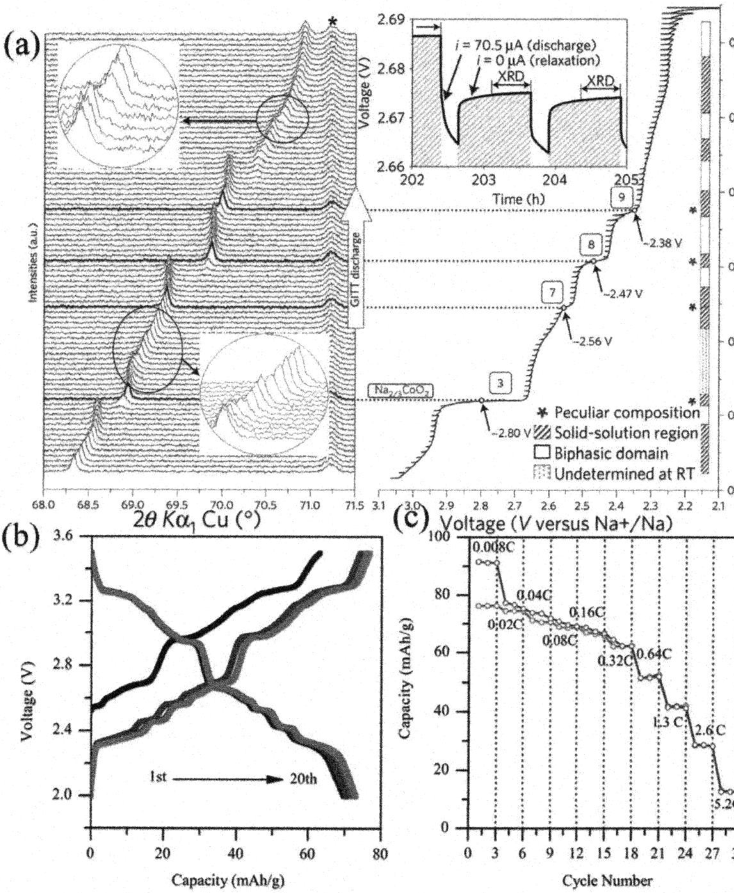

Figure 3.3 (a) *In situ* XRD patterns during sodium-ion intercalation in *P*2-Na_xCoO_2. The charge/discharge curves (b) and rate performance (c) of *P*2-$Na_{0.71}CoO_2$. Reprinted with permission from Ref. [20], Copyright 2014, Elsevier.

of ordered phases are observed: $Na_{1/2}CoO_2$ at 3.45 V, $Na_{4/7}CoO_2$ at 3.15 V, $Na_{2/3}CoO_2$ at 2.80 V, $Na_{0.72}CoO_2$ at 2.56 V, $Na_{0.76}CoO_2$ at 2.47 V, and $Na_{0.79}CoO_2$ at 2.38 V [17]. The distribution of sodium ions can be confirmed mainly through the repulsion between sodium ions and the interaction between Na^+ and Co^{3+}. When the shared faces (Na_f) and the shared edges (Na_e) are repelled minimally, the sodium ions are mainly distributed in these two positions. Besides, the content of sodium ions depends on the occupation states of Na_f and Na_e [18, 19]. The $P2$-$Na_{0.71}CoO_2$ electrode exhibits excellent reversibility, but the large particles lead to inferior rate performance, which can be further modified through nanoengineering (Fig. 3.3b,c) [20].

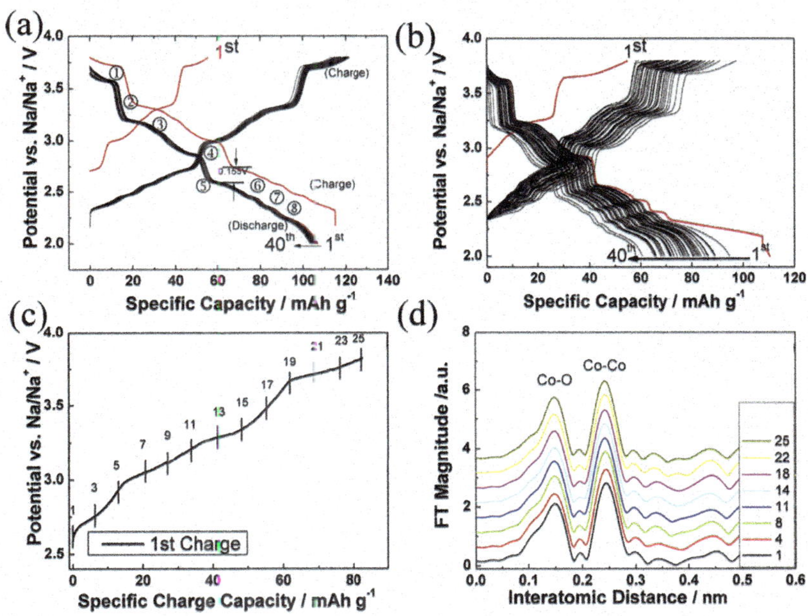

Figure 3.4 The charge/discharge curves of $P2$-$Na_{0.74}CoO_2$ electrode in (a) $NaPF_6$ electrolyte and (b) $NaClO_4$ electrolyte. (c) The charging curve of the $P2$-$Na_{0.74}CoO_2$ electrode. (d) The Fourier transform of the Co K-edge XANES spectra at different charge states. Reprinted with permission from Ref. [21], Copyright 2013, Elsevier.

Moreover, the used electrolytes significantly influence the electrochemical performances of the cathodes. Fu et al. investigated the electrochemical performances of $P2$-$Na_{0.74}CoO_2$ electrodes in different electrolytes [21]. Compared to the $NaClO_4$ electrolyte, a much better cycling performance was achieved using the $NaPF_6$ electrolyte, with a polarization voltage of about 150 mV (Fig. 3.4a,b). The reaction mechanism was studied by the *ex situ* X-ray absorption near edge structure (XANES) spectra, where during the charging process, the Co–O bond length decreases, indicating the increase in electron cloud density on the Co surface and the valence state of Co (oxidization from Co^{3+} to Co^{4+}). Meanwhile, similar to the $LiCoO_2$ charging process, the length of the Co–Co bond is basically unchanged (Fig. 3.4c,d) [22].

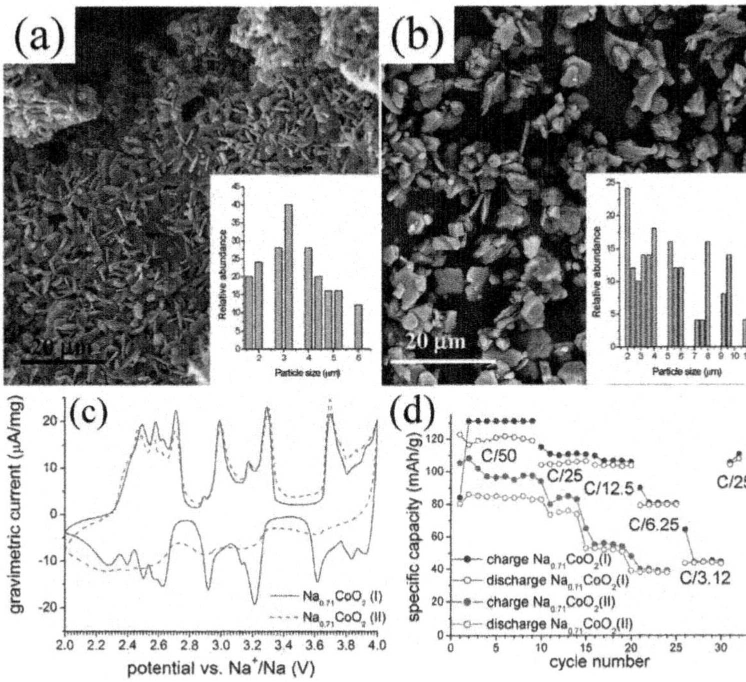

Figure 3.5 SEM images of *P*2-$Na_{0.71}CoO_2$ (a) using Na_2CO_3 (I) and (b) using NaOH (II). Insets are the corresponding particle-size distribution. (c) The CV curves and (d) rate performance of *P*2-$Na_{0.71}CoO_2$ using Na_2CO_3 (I) and NaOH (II). Reprinted with permission from Ref. [23], Copyright 2012, Royal Society of Chemistry.

The crystal structure, particle size, and architecture of Na_xCoO_2 have significant effects on their electrochemical properties [24–26]. For example, Wang et al. fabricated a partially amorphous $Na_{0.74}CoO_2$, which displayed an excellent Na^+ storage capacity (107.9 mAh/g at 0.1 C) compared to the crystalline one [27]. Polizzi et al. synthesized $Na_{0.71}CoO_2$ with different particle sizes by using different sodium sources [23]. The $Na_{0.71}CoO_2$ (I) obtained from the Na_2CO_3 precursor has smaller particles and more uniform particle distribution compared to the $Na_{0.71}CoO_2$ (II) produced from NaOH (Fig. 3.5a,b). The smaller particles indicate that the crystalline material has a shorter sodium-ion diffusion path, thereby accelerating the dynamic reaction rate. Hence, $Na_{0.71}CoO_2$ (I) shows a more obvious charge/discharge platform and better rate performance (Fig. 3.5c,d). Besides, Gao et al. found that the Na_xCoO_2 microspheres (s-NCO) exhibited much better electrochemical performance compared to the Na_xCoO_2 microparticles (i-NCO) in SIBs [28]. It can be attributed to the unique flower-like microspheres morphology, decreasing surface areas, mitigating side reactions, elevating the volumetric energy density, and achieving a more stable cycling performance.

In addition to the strategies mentioned earlier to improve the electrochemical performance of Na_xCoO_2, the doping techniques are also adopted, such as substituting cations [29], doping anions [31], and substituting transition metals [30, 32]. Matsui et al. improved the electrochemical performance of $P2$-$Na_{0.74}CoO_2$ electrodes, especially rate capability, by doping with Ca^{2+} ions (Fig. 3.6a) [29]. The Ca^{2+} ion occupies the Na^+ ion layer and eliminates the lattice mismatches of the two distinct phases in $Na_{2/3-x}Ca_xCoO_2$. After the cycling test, the $P2$-$Na_{0.74}CoO_2$ and $P2$-$Na_{5/8}Ca_{1/24}CoO_2$ cathodes show similar crystal structures (Fig. 3.6b). However, the intensity of (002) and (004) crystal planes in $Na_{0.74}CoO_2$ is much more robust, indicating that these two crystal planes of $P2$-$Na_{0.74}CoO_2$ show poor access to Na^+ ions. Besides, the (106) and (008) crystal planes of P2-$Na_{5/8}Ca_{1/24}CoO_2$ did not split, which indicates that the doped Ca^{2+} can inhibit the phase separation during the charging/discharging process and then improve the cycle performance.

Jugovic et al. improved the electrochemical performance of $P2$-$Na_{0.76}CoO_2$ electrodes by doping with F^-, leading to the decrease in particle size and compression of the CoO_2 layer in the c-axis [31]. The change in the CoO_2 layer usually means wider bands and enhanced electroconductivity of the compounds. Additionally, the partial replacement of Co ions with Mn ions could facilitate the Na diffusion and improve the electrochemical performance [33, 34]. Baster et al. confirmed that $Na_{0.7}Co_{0.7}Mn_{0.3}O_2$ behaved the best electrochemical performance among a series of P-$Na_{0.7}Co_{1-y}Mn_yO_2$ (y = 0, 0.2, 0.3, 0.4, 0.6, 0.8) [35]. Except for the single element substitution, multi-element substitution techniques are also investigated to improve

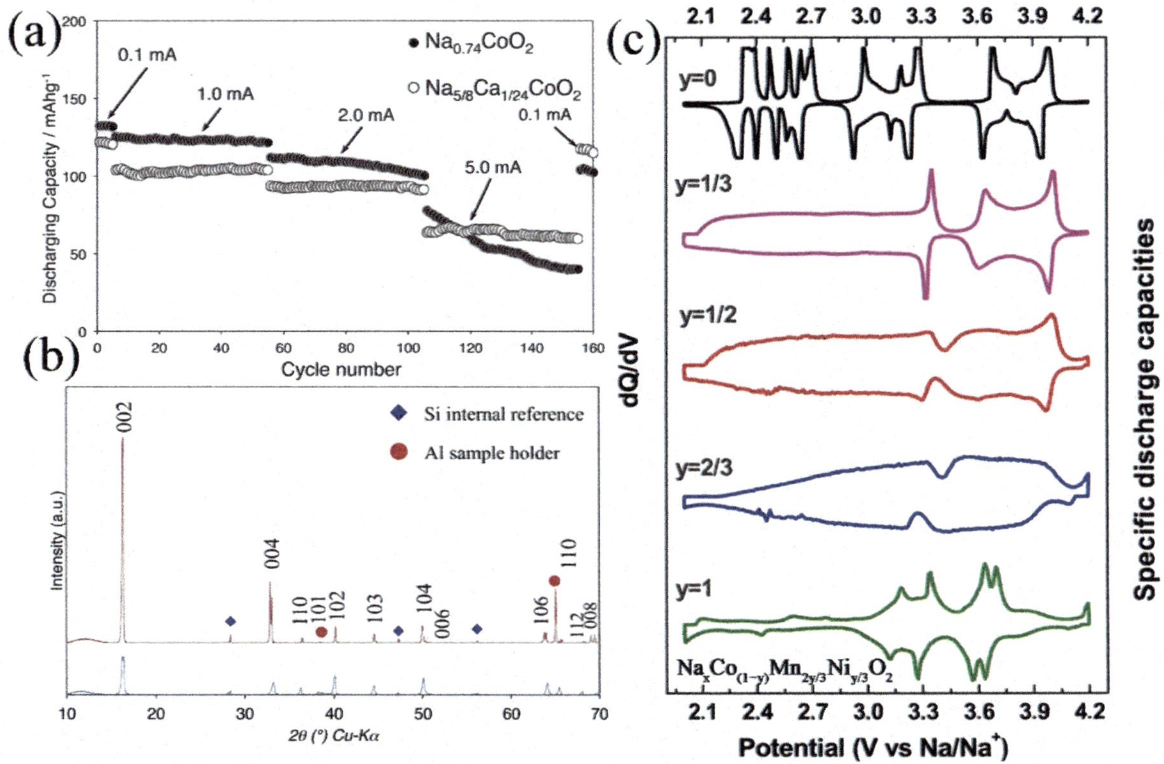

Figure 3.6 (a) Rate performance of *P*2-$Na_{0.74}CoO_2$ and *P*2-$Na_{5/8}Ca_{1/24}CoO_2$. (b) The XRD pattern of the *P*2-$Na_{0.74}CoO_2$ (I) and *P*2-$Na_{5/8}Ca_{1/24}CoO_2$ (II) after the cycles [29]. (c) The *dQ/dV* plots of *P*2-$Na_xCo_{(1-y)}Mn_{2y/3}Ni_{y/3}O_2$ [30]. Copyright 2018, American Chemical Society.

the electrochemical performance of $P2$-Na_xCoO_2 (Fig. 3.6c) [30]. During cycling, the ordered phase transition of $P2$-Na_xCoO_2 leads to an obvious voltage transition. For example, at about 2.8, 3.4, and 4.0 V, the voltage transition of $P2$-Na_xCoO_2 can be observed. With element substitution, the potential transition was significantly improved, indicating that element substitution could alleviate the phase transition and improve cycling performance.

3.2.3 Na_xMnO_2 and Its Derivatives

According to the crystal structure and stoichiometric ratios of elements, the Na_xMnO_2 can be mainly divided into three types: $Na_{0.7}MnO_{2+y}$ ($P2$), α-$NaMnO_2$ ($O'3$), and β-$NaMnO_2$ ($P2$). As early as 1985, Mendiboure et al. first studied their sodium storage properties (Fig. 3.7a–c) [36]. In 2011, Ma et al. re-evaluated the electrochemical performance of α-$NaMnO_2$ in the voltage range of 2.0~3.8 V. A reversible capacity of 185 mAh/g can be obtained, but the capacity drops rapidly during cycling [37]. Through TEM, synchrotron radiation XRD characterization, and theoretical calculation, Abakumov et al. demonstrated that numerous defects existed in α-$NaMnO_2$, which affects its magnetic and electrochemical properties. Moreover, it is predicted that high-concentration defects may distort α-$NaMnO_2$ to β-$NaMnO_2$ [38].

Compared with α- and β-$NaMnO_2$, $P2$-Na_xMnO_2 ($x = 0.7$) electrode shows better electrochemical performance. Caballero et al. disclosed that water molecules could enter the interlayer of the sodium layer in the $P2$-Na_xMnO_2 and increase the interlayer spacing by 0.25 nm. The electrode delivered a reversible specific capacity of 140 mAh/g, but the crystal structure gradually collapses into an amorphous state after several full cycles [39]. This distortion of $NaMnO_2$ structure is ascribed to the Jahn–Teller effect of high-spined Mn^{3+}. Doping or substitution of Mn with Al^{3+}, Li^+, Mg^{2+}, Ni^{2+}, Co^{3+}, Zn^{2+}, and Cu^{2+} can suppress this effect and obtain a more stable $P2$-phase material [40–49].

Yuan et al. synthesized $Na_{0.67}[Mn_{0.65}Ni_{0.15}Co_{0.15}Al_{0.05}]O_2$ by introducing Al^{3+} into the $P2$-type oxide [50]. The Al-doped electrode shows similar charge/discharge plateaus with the $Na_{0.67}[Mn_{0.65}Ni_{0.15}Co_{0.2}]O_2$ electrode (Fig. 3.7d,e). Although the Al-doped electrode displays a lower reversible capacity, its cycling

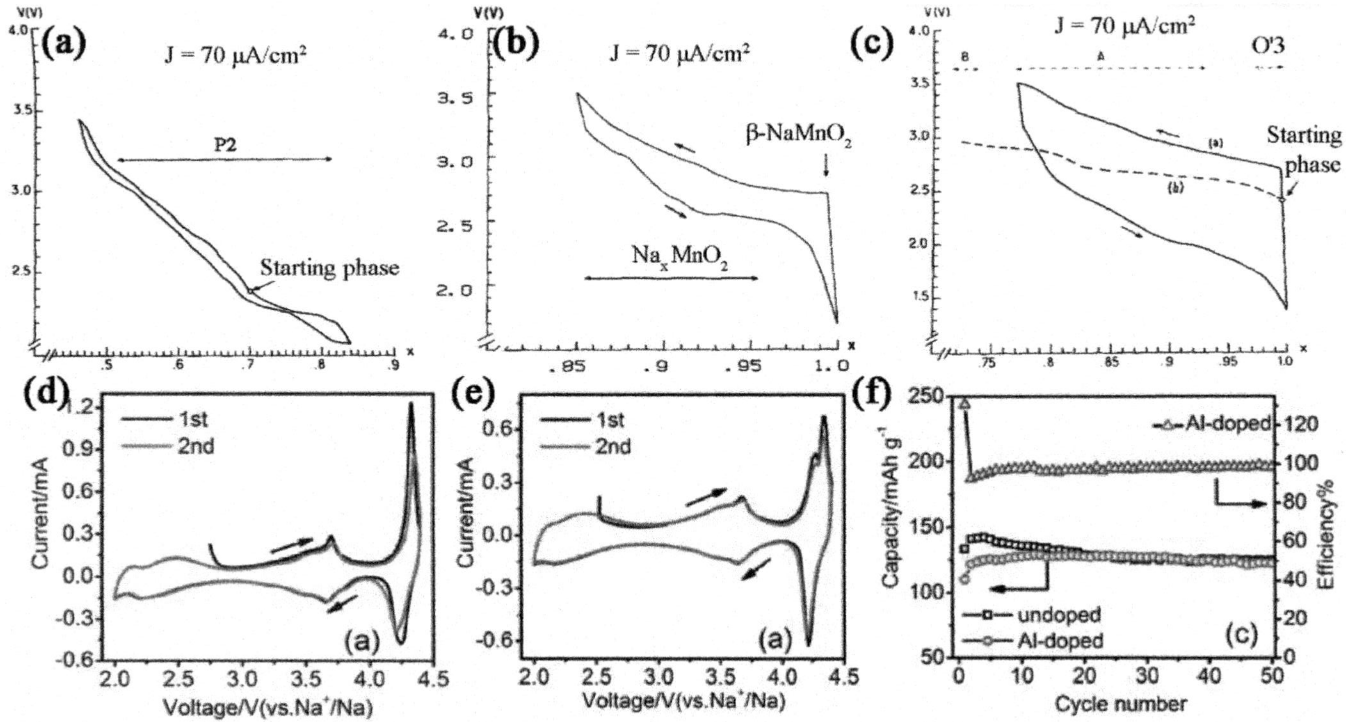

Figure 3.7 The charge/discharge curves of (a) *P*2-$NaMnO_2$, (b) β-$NaMnO_2$, (c) *α*-$NaMnO_2$. Reprinted with permission from Ref. [36], Copyright 1985, Elsevier. The CV curves of (d) $Na_{0.67}[Mn_{0.65}Ni_{0.15}Co_{0.2}]O_2$ and (e) $Na_{0.67}[Mn_{0.65}Ni_{0.15}Co_{0.15}Al_{0.05}]O_2$ electrodes. (f) Cycling performance of the $Na_{0.67}[Mn_{0.65}Ni_{0.15}Co_{0.2}]O_2$ and $Na_{0.67}[Mn_{0.65}Ni_{0.15}Co_{0.15}Al_{0.05}]O_2$ cathodes. Reprinted with permission from Ref. [50], Copyright 2013, Royal Society of Chemistry.

performance is much better than that of the $Na_{0.67}[Mn_{0.65}Ni_{0.15}Co_{0.2}]O_2$ electrode (Fig. 3.7f). Billaud et al. studied the effects of different Mg-doped contents on the electrochemical properties of $Na_{0.67}Mn_{1-x}Mg_xO_2$ ($0 \leq x \leq 0.2$) materials [51]. It was demonstrated that Mg dopant could improve the structural stability of the materials. However, the reversible capacity of the $Na_{0.67}Mn_{1-x}Mg_xO_2$ electrode would decrease when the content of Mg increases. The 5% Mg-doped electrode delivers a reversible specific capacity of 175 mAh/g.

In addition to element doping or substitution, the engineering of morphology, crystal structure, and interface is also an effective approach to improve the electrochemical performance of Mn-based oxide electrodes [48, 52–61]. For instance, Bucher et al. demonstrated that the spherical Na_xMnO_{2+z} with good electrolyte infiltration presented a better cycling performance than the flake-like Na_xMnO_{2+z} (Fig. 3.8a–c) [54]. Gao et al. compared the electrochemical properties of Mn-based oxides with different crystal structures, including a *P*2 and *T* hybrid phase of $Na_xCo_{0.1}Mn_{0.9}O_2$ (Fig. 3.8d) [58]. The *T*-phase species possess fast Na^+ ion diffusion and outstanding structural stability, while the *P*-phase counterparts exhibit large reversible capacity. The *P*2 and *T* hybrid phase $Na_xCo_{0.1}Mn_{0.9}O_2$ combines the advantages of the two phases and shows better electrochemical performance than pure *T*-phase or *P*2-phase $Na_xCo_{0.1}Mn_{0.9}O_2$ electrode (Fig. 3.8e). Alvarado et al. studied the effect of interface on the electrochemical properties of Mn-based oxides electrode [52]. The Al_2O_3-coated $Na_{2/3}Ni_{1/3}Mn_{2/3}O_2$ was synthesized by the atomic layer deposition (ALD) method (Fig. 3.8f). The charging and discharging plateaus remain, but the cycling performance is improved after coating (Fig. 3.8g). However, due to the incorporation of a less-conductive Al_2O_3 layer, the modified $Na_{2/3}Ni_{1/3}Mn_{2/3}O_2$ electrode shows an inferior rate performance (Fig. 3.8h).

In general, the electrochemical properties of the Na_xMnO_2 are negatively influenced by the structure change or phase transition. Doping or substitution with active or inert elements, optimizing the morphology, crystal structure, and interface can significantly inhibit the phase transition during the charge/discharge process and improve the electrochemical performance of the Na_xMnO_2 cathode materials.

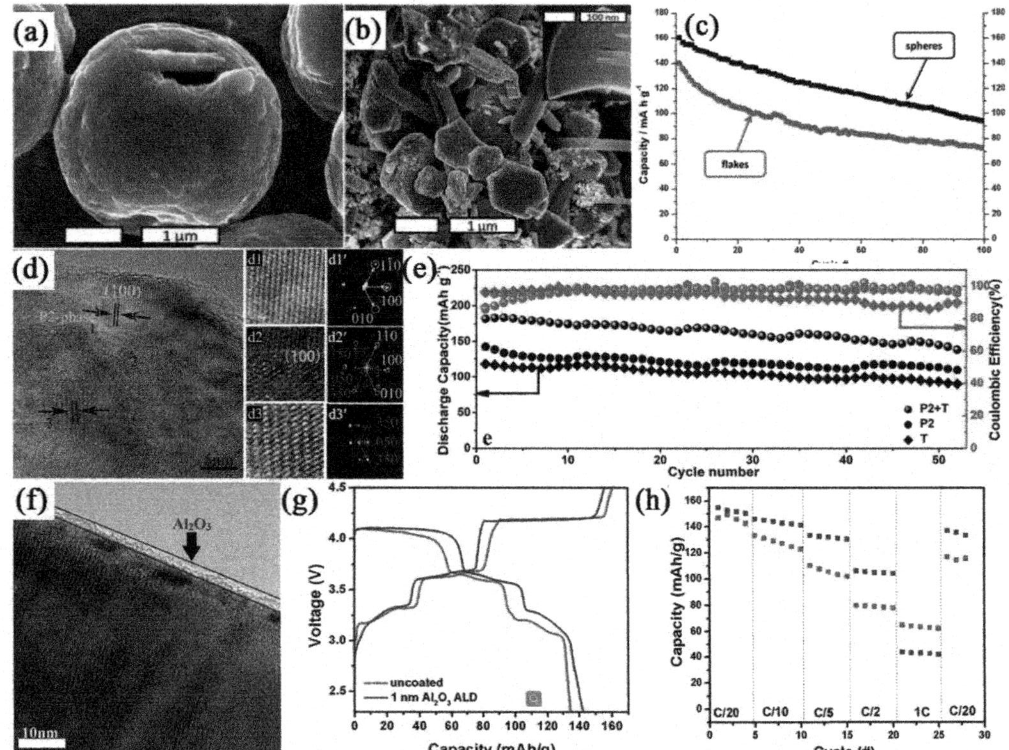

Figure 3.8 SEM images of (a) spherical and (b) flake-like Na_xMnO_{2+z}. (c) The cycling performance of the two electrodes. Reprinted with permission from Ref. [54], Copyright 2014, American Chemical Society. (d) HRTEM of the *P*2 and *T* hybrid phase $Na_xCo_{0.1}Mn_{0.9}O_2$. (e) The cycling performance of the *P*2, *T*, and hybrid phase $Na_xCo_{0.1}Mn_{0.9}O_2$ electrode. Reprinted with permission from Ref. [58], Copyright 2018, Royal Society of Chemistry. (f) TEM image of the Al_2O_3-coated $Na_{2/3}Ni_{1/3}Mn_{2/3}O_2$. (g) The charge/discharge curves and (h) rate performance of $Na_{2/3}Ni_{1/3}Mn_{2/3}O_2$ and Al_2O_3-coated $Na_{2/3}Ni_{1/3}Mn_{2/3}O_2$ electrodes. Reprinted with permission from Ref. [52], Copyright 2017, American Chemical Society.

3.2.4 Na_xFeO_2 and Its Derivatives

Both the two different phases (α and β phase) of $NaFeO_2$ have been investigated for sodium storage. The α-$NaFeO_2$ is a typical layered material with an *O*3 structure, and its electrochemical performance closely relates to the cut-off voltage. When the charging cut-off voltage exceeds 3.5 V, the excessive amount of sodium will cause the migration of the Fe^{4+} into the sodium layer, leading to an irreversible change in the structure and a decrease in the reversible capacity. When the cut-off voltage sets at 3.5 V, a reversible capacity of 100 mAh/g can be obtained (Fig. 3.9a) [10].

Lee et al. inspected the reason for electrochemical irreversibility of the α-$NaFeO_2$ electrode by *ex situ* Mössbauer spectroscopy and *in situ* XRD characterization [62]. It was found that more than 20% of Fe^{4+} would be spontaneously reduced to Fe^{3+} in the open-circuit state. This self-discharge behavior is accompanied by the electrolyte decomposition, resulting in increased impedance and decreased Coulombic efficiency of the SIBs. Moreover, a new phase was observed in the desodiation process through the *in situ* XRD measurement, and asymmetric structural changes were found during the cycling process. In the charging process, the Fe_3O_4 was formed, which prevents further insertion of Na^+ at the high voltage (Fig. 3.9b) [63].

Doping with other metal elements can also improve the stability of the $NaFeO_2$ electrode, for example the *O*3-Na_4FeRuO_6 with significantly enhanced cycling performance (Fig. 3.9c) [64]. Besides, the element Fe in $NaFeO_2$ can be replaced by Mn, Ni, Co, and other elements [66, 67]. For reducing the production cost, Mn replacement is an attractive option because of its abundance. Furthermore, Mn replacement behaves at a relatively higher capacity and cycling stability compared to other element doping.

Komaba et al. compared the electrochemical properties of *O*3-$NaFe_{1/2}Mn_{1/2}O_2$ and *P*2-$Na_{2/3}Fe_{1/2}Mn_{1/2}O_2$ electrodes. It was found that the *O*3 phase could release a capacity of 110 mAh/g at 4.3~1.5 V, while the *P*2 phase could achieve a reversible specific capacity of 190 mAh/g. Both Fe and Mn are active elements in SIBs, corresponding to the redox reaction of Mn^{3+}/Mn^{4+} and Fe^{3+}/Fe^{4+} redox couples during cycling [68]. However, problems still exist in these $Na_xFe_yMn_{1-y}O_2$-type materials: (1) the irreversible phase

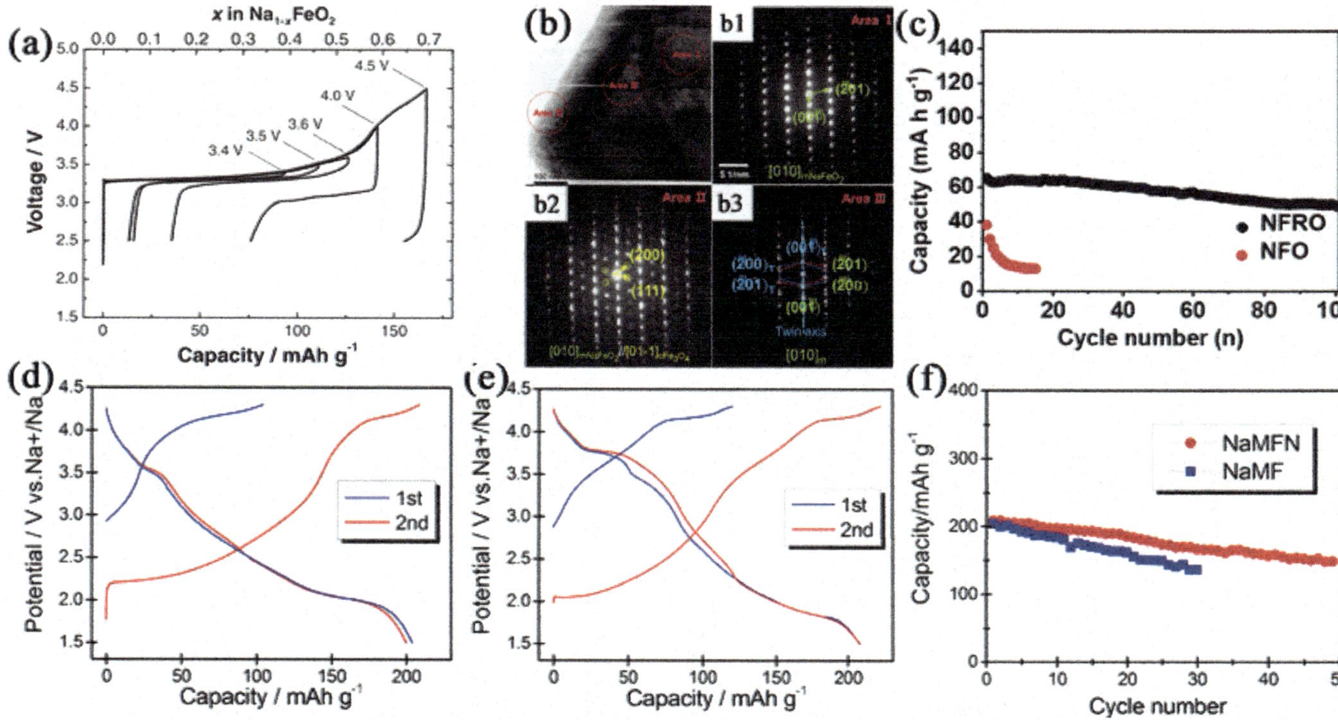

Figure 3.9 (a) The charge/discharge curves of $NaFeO_2$ with different cut-off voltage. Reprinted with permission from Ref. [10], Copyright 2012, Elsevier. (b) The HRTEM and corresponding selected-area diffraction patterns of $Na_{1-x}FeO_2$ after charging to 4.5 V. Reprinted with permission from Ref. [63], Copyright 2019, American Chemical Society. (c) The cycling performance of *O*3-Na_4FeRuO_6 (NFRO) and $NaFeO_2$ (NFO). Reprinted with permission from Ref. [64], Copyright 2020, Wiley-VCH. The charge/discharge curves of (d) the $Na_{0.67}Mn_{0.65}Fe_{0.35}O_2$ (NaMF) and (e) *P*2-$Na_{0.67}Mn_{0.65}Fe_{0.2}Ni_{0.15}O_2$ (NaMFN). (f) The cycling performance of NaMF and NaMFN. Reprinted with permission from Ref. [65], Copyright 2014, Elsevier.

transition of the α-$NaFeO_2$ electrode, especially at a charging voltage higher than 3.4 V, deteriorating the cycling stability; (2) the redox potential of $Mn^{2+/3+/4+}$ lowering the average output voltage; and (3) the dissolution and loss of Mn^{2+} in the electrolyte.

The cycle stability and average operating voltage can be improved by reducing the amount of poor-active Fe element or adding the element with high redox potential, such as Ni element. Yuan et al. [65] prepared and compared the electrochemical properties of *P*2-$Na_{0.67}Mn_{0.65}Fe_{0.35}O_2$ and $Na_{0.67}Mn_{0.65}Fe_{0.2}Ni_{0.15}O_2$ compounds. Both delivered the initial specific capacity of more than 200 mAh/g (Fig. 3.9d–f). The capacity retention rate of the $Na_{0.67}Mn_{0.65}Fe_{0.2}Ni_{0.15}O_2$ electrode is 71% after 50 cycles. Herein, the incorporation of Ni also improves the structural stability and mitigates the Jahn–Teller effect of Mn^{3+}.

Recently, Durai et al synthesized a novel β-$NaFeO_2$ nano-pebble, which showed excellent electrochemical performance. The β-$NaFeO_2$ electrode delivered high capacities of 335 mAh/g at 0.1C and 136 mAh/g after 150 cycles. The outstanding performance could attribute to the Fe^{3+}/Fe^{4+} redox couple during the charge/discharge process [69]. Besides, the CeO_2 coating is another strategy to improve the performance of $NaFeO_2$. The hetero-interfaces between $NaFeO_2$ and CeO_2 provide a fast ion-conducting path, leading to an improved ionic conductivity [70]. Moreover, the structural evolution, ion transport, and oxygen stability of $NaFeO_2$ could be studied by the theoretical calculation [71]. The lattice strains could modulate both ion transport and oxygen stability of $NaFeO_2$.

3.2.5 $NaNiO_2$ and Its Derivatives

Due to the Jahn–Teller effect of Ni^{3+}, the a/b value of *O*′3-$NaNiO_2$ changes from 1.73 to 1.86 [72]. In 1982, Hagenmuller et al. found that 0.2 Na can be released from $NaNiO_2$ at 3.5~2.0 V. The charge/discharge curve contains multiple plateaus, which correspond to different phase transitions [73]. Ceder et al. ascribed the excellent electrochemical performance of *O*′3-$NaNiO_2$ to the stable position of Ni^{3+}, which does not migrate to the Na position [74]. In the voltage range of 1.25~3.75 V, the first charge and discharge capacities of *O*′3-$NaNiO_2$ are 147 and 123 mAh/g, respectively. After 20 cycles, over 94% of reversible capacity can be maintained (Fig. 3.10a). A higher reversible capacity of 146 mAh/g was achieved in the expanded voltage range of 2.0~4.5 V, but the cycling performance dropped quickly due to the structural collapse of the *O*′3-$NaNiO_2$ electrode at the high voltage (Fig. 3.10b). Han et al. studied the electrochemical performance and structural evolution of $NaNiO_2$ by *in situ* XRD. The continuous structural changes from *O*′3, *P*′3, *P*″3, and *O*″3 to *O*‴3 correspond to the mutual transformation from $NaNiO_2$ to $Na_{0.91}NiO_2$, $Na_{0.84}NiO_2$, $Na_{0.81}NiO_2$, and $Na_{0.79}NiO_2$, respectively (Fig. 3.10c) [75].

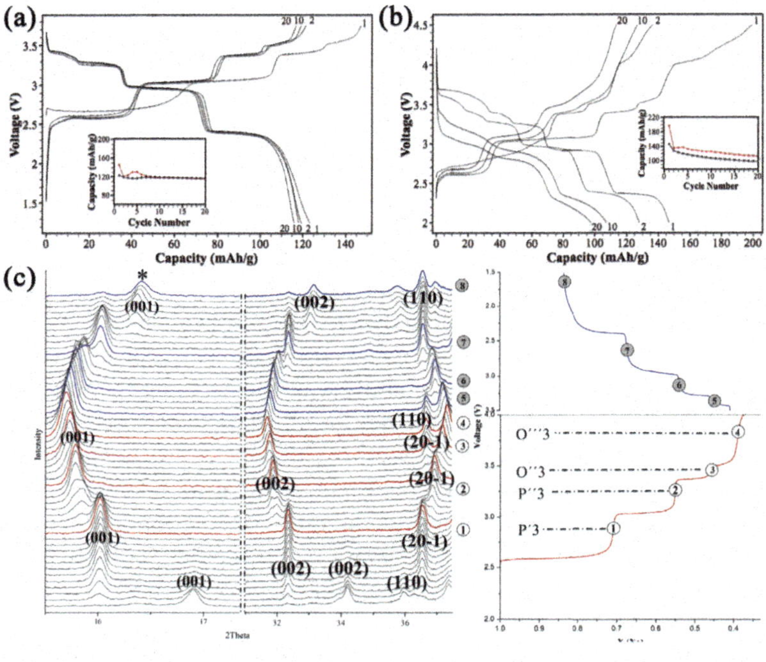

Figure 3.10 The charge/discharge curves of *O*′3-$NaNiO_2$ at (a) 1.25~3.75 V and (b) 2.0~4.5 V. Reprinted with permission from Ref. [74], Copyright 2012, the Electrochemical Society. (c) The *in situ* XRD of *O*′3-$NaNiO_2$. Reprinted with permission from Ref. [75], Copyright 2014, Elsevier.

Phase transition and Na rearrangement are also the main reasons to cause structural instability in the $NaNiO_2$ during the cycling process. The element substitution technique is an effective way to enhance the structural stability in $NaNiO_2$ [76–81]. The Ni valence state in $NaNiO_2$ will change from trivalent to bivalent after being replaced by Mn or Ti. Moreover, Ni can achieve a "bivalent–trivalent–quadrivalent" redox reaction, while Mn or Ti elements remain quadrivalent. The charge/discharge curves of the *O*3-$NaNi_{0.5}Mn_{0.5}O_2$

electrode exhibit distinguished plateaus, corresponding to the structure change of "*O*3–*O*′3–*P*3–*P*′3–*P*″3" in the charge/discharge curves. This indicates the existence of plane slippage and strong Jahn–Teller effect during the cycling process. By contrast, the *P*2-$NaNi_{1/3}Mn_{2/3}O_2$ electrode presents smoother charge/discharge curves, which can be ascribed to the inhibited phase transition in the *P*2-$NaNi_{1/3}Mn_{2/3}O_2$ electrode. Although the *O*3 phase contains more sodium than the *P*2 phase, the *P*2-phase electrode material shows better stability than the *O*3 phase [82, 83].

Yuan et al. synthesized a series of *O*3-$NaFe_x(Ni_{0.5}Mn_{0.5})_{1-x}O_2$ (x = 0, 0.1, 0.2, 0.3) materials by a sol-gel method. It was found that Fe substitution could inhibit the order of Na/vacancy, thereby improving the electrochemical performance. The $NaFe_{0.2}(Ni_{0.5}Mn_{0.5})_{0.8}O_2$ electrode delivered a reversible capacity of 131 mAh/g, with a capacity retention of 95% after 30 cycles and excellent rate performance at even 20 C (Fig. 3.11a). These behaviors are much higher than what the *O*3-$NaNi_{0.5}Mn_{0.5}O_2$ electrode shows, indicating that Fe substitution can inhibit phase transformation and increase cycle stability. Moreover, the structural evolution of $NaNi_{0.5}Mn_{0.5}O_2$ and $NaFe_{0.2}(Ni_{0.5}Mn_{0.5})_{0.8}O_2$ during the charge/discharge process was analyzed. In the charging profile, the $NaNi_{0.5}Mn_{0.5}O_2$ electrode shows the phase transition of *O*3–*P*3–*P*″3, while the $NaFe_{0.2}(Ni_{0.5}Mn_{0.5})_{0.8}O_2$ electrode experiences *O*3–*P*3–*OP*2 phase changes, which are beneficial to the structural stability in the high voltage [84].

Similar results can be obtained using the substitution or co-substitution with other elements, such as Ti, Co, Li, Te, Sr, and so on [85–87]. For example, Johnson et al. reported a *P*2-$Na_{0.85}Li_{0.17}Ni_{0.21}Mn_{0.64}O_2$ compound for SIBs, and a reversible capacity of 100 mAh/g could be realized with a discharge plateau at 3.4 V [88]. The replacement of Li can stabilize the transition metal layer and smooth the charge/discharge curve. At the same time, with the increase in Li content, the material gradually changes from *O*2-$Na_xLi_xNi_zMn_{1-y-z}O_2$ ($x > y$) to *P*2-$Na_xLi_xNi_zMn_{1-y-z}O_2$ ($x > y$), and then *O*2-$Li_yNa_xNi_zMn_{1-y-z}O$ ($x > y$). Through controlling the Li content, the cathode materials consisting of two or three phases can be obtained [89]. Guo et al. prepared an $Na_{0.7}Li_{0.3}Ni_{0.5}Mn_{0.5}O_2$ compound with the coexistence of *P*2 and *O*3 phase [90], which combines the advantages of *P*2 phase and *O*3 phase and realizes good cycling performance (*P*2 phase) and superior high capacity (*O*3 phase) (Fig. 3.11b).

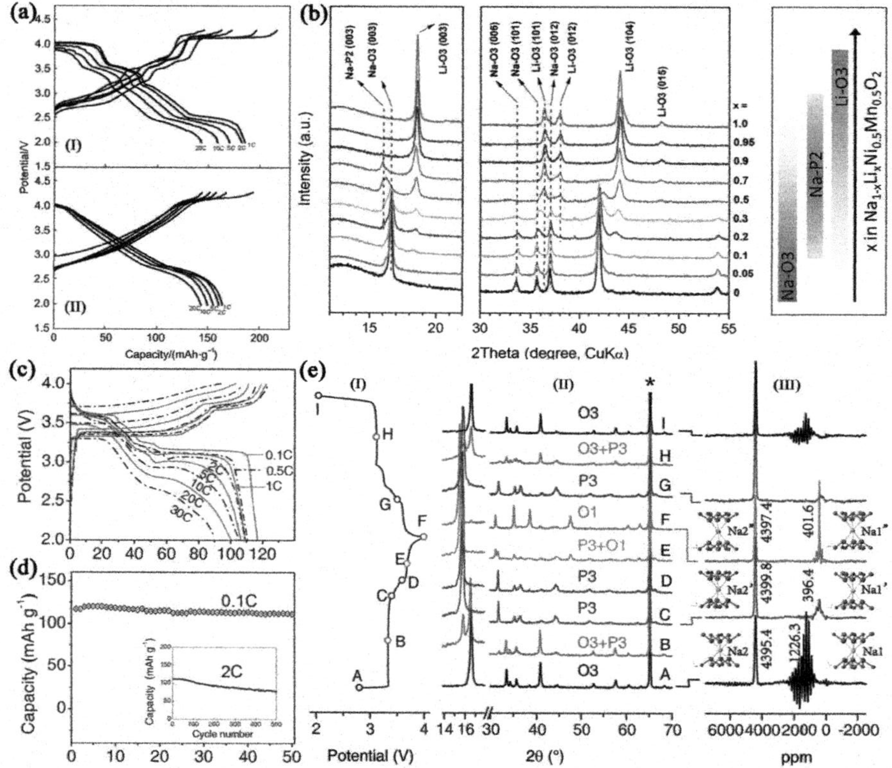

Figure 3.11 (a) The charge/discharge curves of the *O*3-$NaNi_{0.5}Mn_{0.5}O_2$ (I) and *O*3-$NaFe_{0.2}(Ni_{0.5}Mn_{0.5})_{0.8}O_2$ (II). Reprinted with permission from Ref. [84], Copyright 2015, American Chemical Society. (b) The XRD patterns of $Na_{1-x}Li_xNi_{0.5}Mn_{0.5}O_{2+\sigma}$. Reprinted with permission from Ref. [89], Copyright 2014, Wiley-VCH. (c) The charge/discharge curves of the *O*3-$Na_3Ni_2SbO_6$ at different current densities. (d) The cycling performance of the *O*3-$Na_3Ni_2SbO_6$ at 0.1C and 2C, respectively. (e) The XRD (II), NMR (III) of the *O*3-$Na_3Ni_2SbO_6$ at different charge/discharge process (I). Reprinted with permission from Ref. [91], Copyright 2014, Wiley-VCH.

Notably, a different type of layered transition metal oxide, honeycomb-ordered phase *O*3-$Na_3Ni_2SbO_6$, was reported by Yang et al. [91]. This layered structure, where each SbO_6 octahedron is surrounded by six NiO_6 edge-sharing octahedron forming the superstructure lattice, allows a fast Na^+ diffusion with easily changing their oxidation state. Thus, the *O*3-$Na_3Ni_2SbO_6$ electrode achieved a high initial reversible capacity of 117 mAh/g with the Coulombic efficiency of 95% (Fig. 3.11c). Moreover, the electrode exhibited capacity retention of 95% after 50 cycles (Fig. 3.11d). The electrode also delivered a capacity of 90 mAh/g at a high rate of 30 C (Fig. 3.11c). Meanwhile, the sodium storage mechanism and the structural stability of cathode materials were demonstrated by XPS, *ex situ* XRD, and *ex situ* NMR (Fig. 3.11e). Khalifah et al. also confirmed the advantages of this honeycomb layer oxide with high crystallinity [92].

3.2.6 $NiCrO_2$ and Its Derivatives

The investigation of the *O*3-Na_xCrO_2 electrode for SIBs was first reported in 1982, and 0.15 Na could reversibly intercalate/deintercalate during the cycling process, corresponding to the phase transformation of *O*3 to *P*3 [73]. The Na_xCrO_2 electrode showed a voltage of about 3 V and a reversible capacity of 120 mAh/g (0.48 Na) (Fig. 3.12a) [93]. Cr^{4+} migrates from the transition metal layer to the sodium layer in the high voltage, resulting in capacity attenuation [94, 95]. Carbon coating can improve the electrochemical performance of $NaCrO_2$, especially its rate performance [96, 97]. Besides, the $NaCrO_2$ electrode displayed a better performance in an ionic liquid electrolyte, where a reversible capacity of 113 mAh/g was achieved with a Coulombic efficiency of 99.6% [98]. After 100 cycles, 98% of capacity was retained.

The charging mechanism of $NaCrO_2$ has been extensively studied by Zhou et al. [99]. The *in situ* XRD results indicated that the structure of $NaCrO_2$ changed as follows: $O3_R \rightarrow O3_R + O3_M \rightarrow O3_M \rightarrow O3_M + P3_M \rightarrow P3_M$. The lattice constant *c* increased, but *a* and *b* decreased along with sodium insertion. *Ex situ* XAS analysis showed that Cr^{3+} was oxidized to $Cr^{3.5+}$ at 3.6 V, and the chromium ion was fixed at the octahedral position all the time. Sodium state changed from octahedral coordination to quasi-tetrahedral coordination and finally to prism coordination.

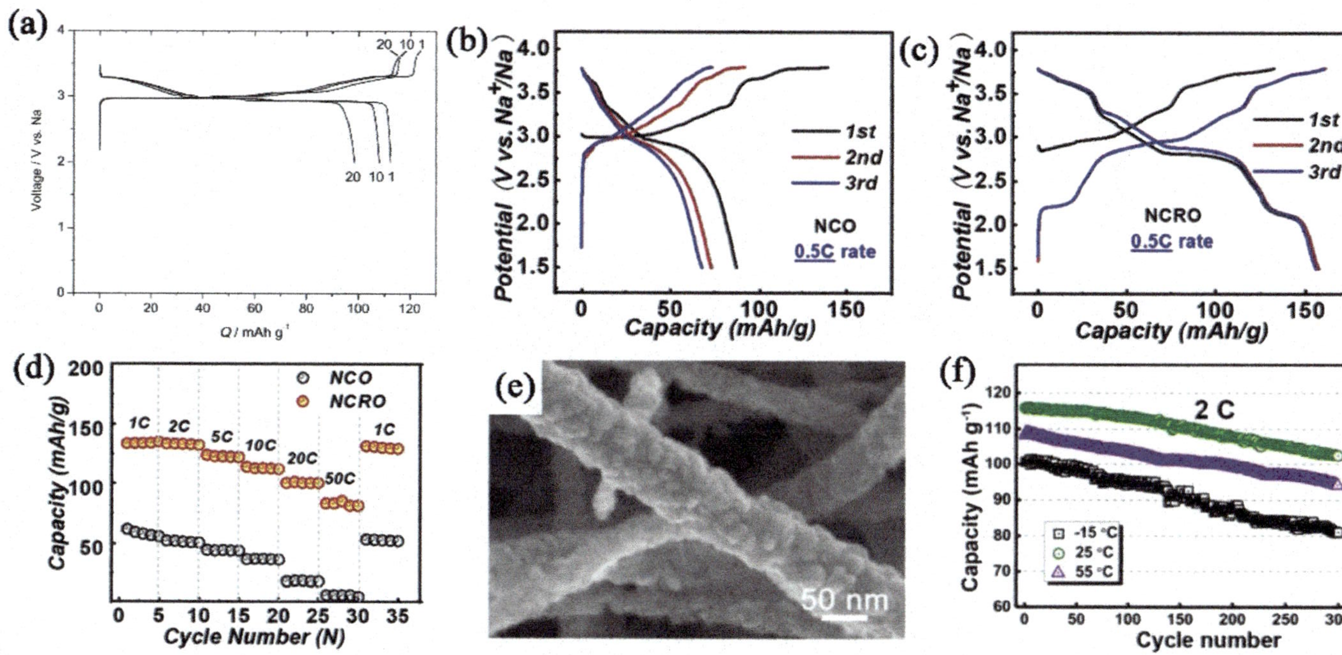

Figure 3.12 (a) The charge/discharge curves of $NaCrO_2$ electrode. Reprinted with permission from Ref. [93], Copyright 2010, Elsevier. The charge/discharge curves of (b) the $NaCrO_2$ (NCO) and (c) the $Na_{0.88}Cr_{0.88}Ru_{0.12}O_2$ (NCRO) at 0.5C. (d) The rate performance of NCO and NCRO. Reprinted with permission from Ref. [103], Copyright 2020, Elsevier. (e) SEM image of the as-prepared $NaCrO_2$ nanofiber. (f) The cycling performance of the $NaCrO_2$ nanofiber at 2 C under different temperatures. Reprinted with permission from Ref. [104], Copyright 2019, American Chemical Society.

Recently, Luo et al. found that the misplacement of Cr^{3+} would affect the capacity and cycling performance of *O*3-$NaCrO_2$ [100]. The effects of structural defects discovered in this study were interpreted based on the interference of Cr ions during Na^+ transport on the Na plane, in which the Cr ions are gradually migrating during cycling. It is shown that *O*3-$NaCrO_2$ crystals with a lower degree of misplacement of Cr^{3+} at Na sites and a longer Na–O bond length would deliver higher specific capacity, better capacity retention, and smaller cell impedance than the traditional *O*3-$NaCrO_2$.

Metal doping of $NaCrO_2$ is expected to significantly promote the Na^+ diffusion rate unraveled by first-principles calculations [101, 102]. For example, Zhou et al. synthesized an $Na_{0.88}Cr_{0.88}Ru_{0.12}O_2$ (NCRO) cathode, which shows an extended capacity of 156 mAh/g at the voltage range of 1.5~3.8 V [103]. The high-angle annular darkfield (HAADF) measurement revealed that Ru doping suppresses the irreversible chromium ions migration during sodium extraction, hence improving the electrochemical performance (Fig. 3.12b–d).

Nanoengineering is also employed to improve electrochemical performance. For instance, $NaCrO_2$ nanowires were prepared by electrospinning as cathodes for SIBs [104, 105]. The $NaCrO_2$ nanowire electrode processes directional and shortened electron/ion transport pathways and strong structural stability (Fig. 3.12e). Thus, the $NaCrO_2$ nanowire electrode shows an obvious charge/discharge plateau, high discharge/charge capacity (121.9/123.5 mAh/g at 1 C), and excellent cycling performance at various temperatures (Fig. 3.12f).

3.2.7 $NiVO_2$ and Its Derivatives

The physical and chemical properties of *O*3-$NaVO_2$ and *P*2-$Na_{0.7}VO_2$ have also attracted the interest of researchers [106, 107]. Delmas' group first revealed the structural evolution of P2-Na_xVO_2 system during sodium deintercalation (intercalation) through the *in situ* XRD experiments [108]. In the $0.5 \leq x \leq 0.9$ range, three single ordered phases (x = 1/2, 5/8, and 2/3) were evidenced. Hamani et al. also reported the sodium storage behavior of *O*3-$NaVO_2$ and *P*2-$Na_{0.7}VO_2$ electrodes [109]. *P*2-$Na_{0.7}VO_2$ shows a lower polarization than *O*3-$NaVO_2$, and the different *dQ*/*dV* curves demonstrate the different sodiation mechanisms of *O*3-$NaVO_2$ and *P*2-$Na_{0.7}VO_2$ electrodes (Fig. 3.13a,b).

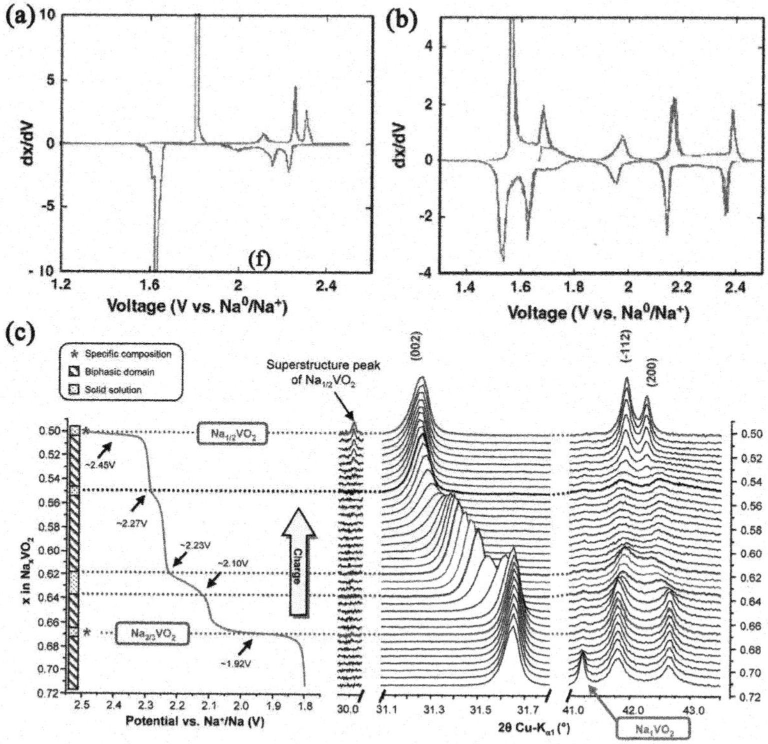

Figure 3.13 The *dQ/dV* curves of (a) *O*3-$NaVO_2$ and (b) *P*2-$Na_{0.7}VO_2$ electrodes. Reprinted with permission from Ref. [109], Copyright 2011, Elsevier. (c) The *in situ* XRD of *O*3-$NaVO_2$ electrode at different charge/discharge processes. Reprinted with permission from Ref. [110], Copyright 2012, Royal Society of Chemistry.

Moreover, the *in situ* XRD measurement shows that the *P*2-$Na_{0.7}VO_2$ electrode maintains the *P*2 phase during the whole electrochemical process, although the lattice constant changes slightly. Delmas et al. investigated the structural changes of *O*3-$NaVO_2$ electrodes during the charge/discharge process [110]. As shown in Fig. 3.13c, the single phases of $Na_{0.67}VO_2$ and $Na_{0.5}VO_2$ are detected at 1.92 and 2.45 V, respectively. While the voltage variation in the range of 2.1~2.23 V corresponds to a solid solution domain. The obvious broadening of the (112) and (200) peaks indicates a certain disorder of the structure in the meantime.

3.2.8 Other Na_xMO_2

Exploring new active-transition-metal-layer oxides is of great significance for the development of SIBs. Copper-based transition metal oxides are a promising family of cathodes because of their low cost and nontoxic properties, as well as the electrochemical activity of Cu^{2+}/Cu^{3+} redox couple [111, 112]. For example, the air-stable $P2$-$Na_{7/9}Cu_{2/9}Fe_{1/9}Mn_{2/3}O_2$ electrode shows the reversible capacity of 90 mAh/g and the discharge voltage of around 3.6 V [111]. Layered $NaTiO_2$ was also investigated as cathodes for SIBs. However, the relatively low average potential limits the energy density of the cells [113–118].

In summary, element substitution is an effective strategy to realize a stable crystal structure and improved charge/discharge mechanism of the layered transition metal oxides. So multielement layered oxides are promising to achieve a satisfactory electrochemical performance for practical SIBs. Currently, the research focus of element doping is on bivalent, trivalent, and quadrivalent elements. It might be possible to dope quinquevalent and hexavalent elements into layered oxides to improve battery performance.

3.3 Polyanionic Materials

Polyanionic-type cathode materials, $Na_xM_y(XO_4)_n$ (M = transition metal Fe, Co, Ni, Mn, V, etc.; X = S, P, Si, B, etc.), are composed of anionic $(XO_4)^{n-}$ tetrahedra or/and their derivatives $(X_mO_{3m+1})^{n-}$ with associated polyhedra MO_x. Among various cathode candidates for SIBs, polyanionic-type cathodes stand out for three unique advantages. (1) Due to the strong covalent bonds between oxygen atoms and transition metal atoms inside the MO_x polyhedra, polyanionic-type cathodes usually exhibit higher thermal stability than layered transition metal oxides, making them safer to be applied in large-scale sodium storage devices. (2) When introducing electronegative X, the unique inductive effect through M-O-X patterns can effectively improve the redox potential of M^{a+}/M^{b+} redox couples, ensuring the high energy density of polyanionic-type cathodes. (3) Some specific extra-stable frameworks (for example, NASICON) associated with large-sized MO_x polyhedra and XO_4 tetrahedra possess large cavities to store Na atoms inside and fast Na extraction

Table 3.1 Summary of representative polyanionic-type cathode materials

				Crystal Parameter				Electrochemical Performance						
								Capacity/ mAh/g		Cycling Stability			Rate Performance	
Type		Material	Synthesis Method	Space Group	Structure/ Symmetry	Redox Couples	Voltage/ V	Theor.	Expt.	Retention/ %	Cycles	Rate/ C	Capacity/ mAh/g	Rate/ C
Phosphates	**Olivine**	$NaFePO_4$ [124]	Ion exchange	*Pnma*	Triphylite	Fe^{3+}/Fe^{2+}	2.7	154	125	88	50	0.05	85	0.5
		$NaFePO_4$ [122]	Solid-state	*Pnma*	Maricite	Fe^{3+}/Fe^{2+}	2.6	154	142	92	200	0.05	90	1
	NASICON	$Na_3V_2(PO_4)_3$ [130]	Solid-state	*R3c*	Rhombohedral	V^{4+}/V^{3+}	3.4	117	107.7	93	80	0.1	–	–
		$Na_3V_2(PO_4)_3$ [129]	Soft template	*R3c*	Rhombohedral	V^{4+}/V^{3+}	3.37	117	116	89.7	3500	20	62	40
		$Na_3Fe_2(PO_4)_3$ [292]	Solid-state	*C*12*c*1	Rhombohedral	Fe^{4+}/Fe^{3+} Fe^{3+}/Fe^{2+}	3.0 2.5	115	109	96	200	1	23	5
		$Na_3Cr_2(PO_4)_3$ [293]	Solid-state	*R3c*	Rhombohedral	Cr^{4+}/Cr^{3+}	4.5	117	79	–	–	–	–	–
		$Na_3VCr(PO_4)_3$ [294]	Sol-gel	*R3c*	Rhombohedral	V^{5+}/V^{4+} V^{4+}/V^{3+}	4.1 3.4	117	90	95	200	0.5	–	–
		$Na_3V_{1.5}Al_{0.5}(PO_4)_3$ [295]	Solid-state	*R3c*	Rhombohedral	V^{5+}/V^{4+} V^{4+}/V^{3+}	3.95 3.37	121	120.8	–	–	–	–	–
		$Na_4MnV(PO_4)_3$ [131]	Sol-gel	*R3c*	Rhombohedral	V^{4+}/V^{3+} Mn^{3+}/Mn^{2+}	3.6 3.3	111	101	89	1000	1	90	5
		$Na_3FeV(PO_4)_3$ [131]	Sol-gel	*C*12*c*1	Monoclinic	V^{4+}/V^{3+} Fe^{3+}/Fe^{2+}	3.3 2.5	111	103	95	1000	1	90	10
		$Na_2VTi(PO_4)_3$ [132]	Sol-gel	*R3c*	Rhombohedral	V^{4+}/V^{3+} Ti^{4+}/Ti^{3+} V^{3+}/V^{2+}	3.4 2.1 1.6	178	147	77	500	10	49	20

			Crystal Parameter				Electrochemical Performance						
Type	Material	Synthesis Method	Space Group	Structure/ Symmetry	Redox Couples	Voltage/ V	Capacity/ mAh/g		Cycling Stability			Rate Performance	
							Theor.	Expt.	Retention/ %	Cycles	Rate/ C	Capacity/ mAh/g	Rate/ C
	$Na_3MnTi(PO_4)_3$ [296]	Spray-drying	*R*3*c*	Rhombohedral	Mn^{4+}/Mn^{3+} Mn^{3+}/Mn^{2+} Ti^{4+}/Ti^{3+}	4.0 3.5 2.1	177	160	92	500	2	129	2
	$Na_4MnCr(PO_4)_3$ [297]	Sol-gel	*R*3*c*	Rhombohedral	Mn^{4+}/Mn^{3+} Mn^{3+}/Mn^{2+}	4.15 3.52	111	108.4	71.5	250	2	70.8	10
Pyrophos-phate	α-$NaVP_2O_7$ [135]	Solid-state	$P2_1/c$	$NaMoP_2O_7$	V^{4+}/V^{3+}	3.4	108	38.4	–	–	–	20	0.2
	β-$NaVP_2O_7$ [136]	Soft chemistry	$P2_1/c$	$KAlP_2O_7$	V^{4+}/V^{3+}	4.0	108	104	95	100	1	77	50
	$Na_2FeP_2O_7$ [137]	Solid-state	*P*1	Triclinic	Fe^{3+}/Fe^{2+}	2.8	97	90	–	–	–	65	1
	$Na_2CoP_2O_7$ [138]	Solid-state	$Pna2_1$	Orthorhombic	Co^{3+}/Co^{2+}	4.3	96	95	86	30	0.1	65	4
	$Na_2MnP_2O_7$ [142]	Solid-state	*P*1	Triclinic	Mn^{3+}/Mn^{2+}	3.8	97	90	96	30	0.2	61	1
Fluorophos-phate	$NaVPO_4F$ [143]	Solvent-thermal	$I4/mnm$	Tetragonal	V^{4+}/V^{3+}	3.8	143	133	93.7	150	0.5	75	10
	$NaVPO_4F$ [145]	Molten-state-blending	$C2/c$	Monoclinic	V^{4+}/V^{3+}	3.4	143	135	90.4	1500	20	112.1	30
	$Na_3V_2(PO_4)_3F_3$ [147]	Sol-gel	$P4_2/mnm$	Tetragonal	V^{4+}/V^{3+}	3.8	128	130	50	3000	30	57	30
	Na_2FePO_4F [298]	Solid-state	*Pbcn*	Orthorhombic	Fe^{3+}/Fe^{2+}	3.0	124	117	85	1000	4	66.8	4
	Na_2CoPO_4F [151]	Solid-state	*Pbcn*	Orthorhombic	Co^{3+}/Co^{2+}	4.3	122	100	–	–	–	–	–

(*Continued*)

Table 3.1 *(Continued)*

				Crystal Parameter				Electrochemical Performance						
	Type	Material	Synthesis Method	Space Group	Structure/ Symmetry	Redox Couples	Voltage/ V	Capacity/ mAh/g		Cycling Stability			Rate Performance	
								Theor.	Expt.	Retention/ %	Cycles	Rate/ C	Capacity/ mAh/g	Rate/ C
		Na_2MnPO_4F [152]	Solid-state	$P2_1/n$	Monoclinic	Mn^{3+}/Mn^{2+}	3.66	125	120	–	–	–	–	–
	Mixed phosphate	$Na_4Fe_3(PO_4)_2P_2O_7$ [299]	Soft template	$Pn2_1a$	Orthorhombic	Fe^{3+}/Fe^{2+}	3.0	129	128.5	63.5	4000	10	79	100
		$Na_4Co_3(PO_4)_2P_2O_7$ [300]	Sol-gel	$Pn2_1a$	Orthorhombic	Co^{3+}/Co^{2+}	4.5	127	95	83	100	2	80	25
		$Na_4Mn_2Co(PO_4)_2P_2O_7$ [301]	Spray-drying	$Pn2_1a$	Orthorhombic	Mn^{3+}/Mn^{2+} Co^{3+}/Co^{2+}	3.86	129	96.1	76.4	150	1	41	10
	Metaphosphates	$NaFe(PO_3)_3$ [302]	Solid-state	$P2_12_12_1$	Orthorhombic	Fe^{3+}/Fe^{2+}	2.8	85	22	–	–	–	–	–
		$NaCo(PO_3)_3$ [303]	Solid-state	$Pa3$	Cubic	Co^{3+}/Co^{2+}	3.3	84	50	–	–	–	–	–
Sulfate	Fluorosulfate	$NaFeSO_4F$ [153]	Solid-state	$C2/c$	Maxwellite	Fe^{3+}/Fe^{2+}	3.5	138	6	–	–	–	–	–
		$KFeSO_4F$ [155]	Solid-state	$Pna2_1$	$KTiOPO_4$	Fe^{3+}/Fe^{2+}	3.6	128	120	–	–	–	–	–
	Alluaudites	$Na_2Fe(SO_4)_2$ [156]	Solid-state	$C2/c$	Monoclinic	Fe^{3+}/Fe^{2+}	3.6	91	82	84	100	0.5	50	2
		$Na_2Fe(SO_4)_2\cdot 2H_2O$ [157]	Soft chemistry	$P2_1/c$	Kröhnkite	Fe^{3+}/Fe^{2+}	3.25	82	70	86	20	0.05	–	–
		$Na_2Fe(SO_4)_2\cdot 4H_2O$ [158]	Soft chemistry	$P2_1/c$	Bloedite	Fe^{3+}/Fe^{2+}	3.3	73	50	–	–	–	–	–
		$Na_2Fe_2(SO_4)_3$ [159]	Solid-state	$C2/c$	Monoclinic	Fe^{3+}/Fe^{2+}	3.8	120	102	–	–	–	75	10
		$Na_{2.7}Fe_{1.65}(SO_4)_3$ [161]	Spray-drying	$C2/c$	Monoclinic	Fe^{3+}/Fe^{2+}	3.65	100	99	80.8	2000	30	72.2	80

				Crystal Parameter				Electrochemical Performance						
								Capacity/ mAh/g		Cycling Stability			Rate Performance	
	Type	Material	Synthesis Method	Space Group	Structure/ Symmetry	Redox Couples	Voltage/ V	Theor.	Expt.	Retention/ %	Cycles	Rate/ C	Capacity/ mAh/g	Rate/ C
Other Oxysalt	**Silicate**	Na_2FeSiO_4 [304]	Sol-gel	*F43m*	Cubic	Fe^{3+}/Fe^{2+}	1.9	276	106	96	20	0.02	40	0.7
		Na_2MnSiO_4 [305]	Soft chemistry	*Pn*	Monoclinic	Mn^{4+}/Mn^{3+} Mn^{3+}/Mn^{2+}	3.0	277	210	89.5	500	1	100	5
		Na_2CoSiO_4 [306]	Co-precipitation	*Pbca*	Orthorhombic	Co^{3+}/Co^{2+}	3.3	272	125	–	–	–	–	–
		Na_2CoSiO_4 [306]	Solid-state	*Pn*	Monoclinic	Co^{3+}/Co^{2+}	3.3	272	107	–	–	–	–	–
		$Na_2Fe_2Si_2O_7$ [307]	Solid-state	$P12_1/n1$	Monoclinic	Fe^{3+}/Fe^{2+}	2.6	82	20	–	–	–	–	–
	Carbono-phosphates	$Na_3MnPO_4CO_3$ [165]	Soft chemistry	$P2_1/m$	Sidorenkite	Mn^{4+}/Mn^{3+} Mn^{3+}/Mn^{2+}	4.0 3.4	191	125	80	10	0.01	60	1
		$Na_{2.24}FePO_4CO_3$ [166]	Hydrothermal	$P2_1/m$	Sidorenkite	Fe^{3+}/Fe^{2+}	2.6	191	120	79	50	0.05	60	1

channels, which dramatically promote the ionic diffusion kinetics and restrict volume variations during the sodiation/desodiation process. However, polyanionic-type cathodes suffer from intrinsic low electronic conductivity caused by the giant electron transfer barrier between XO_4 tetrahedra and the linked MO_x polyhedra via M–O–X–O–M patterns. Many reviews have summarized several general strategies to solve this problem, like building conductive carbon matrices, manufacturing optimal morphology, controlling particle size, and so on.

This section mainly introduces fundamental research on polyanionic-type cathodes in terms of the versatile crystal structure of some typical polyanionic-type cathode, such as phosphates and sulfates, with related ionic diffusion kinetics and corresponding electrochemical behaviors in SIBs. All details are briefly listed in Table 3.1 for comparison.

3.3.1 Phosphates

3.3.1.1 Olivine

In the history of developing LIBs, triphylite $LiFePO_4$ is the first polyanionic-type cathode being successfully commercialized because of its moderate potential (~3.6 V versus Li/Li^+), low cost, and long cycling life. Inspired by this great success, many researchers tried to explore the sodium storage ability of isostructural $NaFePO_4$. Unlike $LiFePO_4$, $NaFePO_4$ has two distinct orthorhombic (space group: *Pnma*) olivine polymorphs in terms of a metastable triphylite-type and a thermodynamically stable maricite-type phase (Fig. 3.14a). Triphylite-type $NaFePO_4$ shows high electrochemical reactivity since 1D Na^+ diffusion pathways parallel to the *b*-axis in frameworks, where layered units with slight distortion are built by corner-sharing FeO_6 octahedra, which link with PO_4 tetrahedra via edge-sharing mode. Due to the trend of an irreversible phase transition to maricite-type $NaFePO_4$ at high temperature (over 480°C) [119], triphylite-type $NaFePO_4$ can usually be synthesized through the chemical or electrochemical insertion of Na^+ into heterosite $FePO_4$ oxidized from isostructural $LiFePO_4$. In contrast, maricite-type $NaFePO_4$ can be directly obtained by a solid-state method but has no channels for Na^+ migration because of the infaust connecting mode between PO_4 tetrahedra and edge-sharing FeO_6 octahedra.

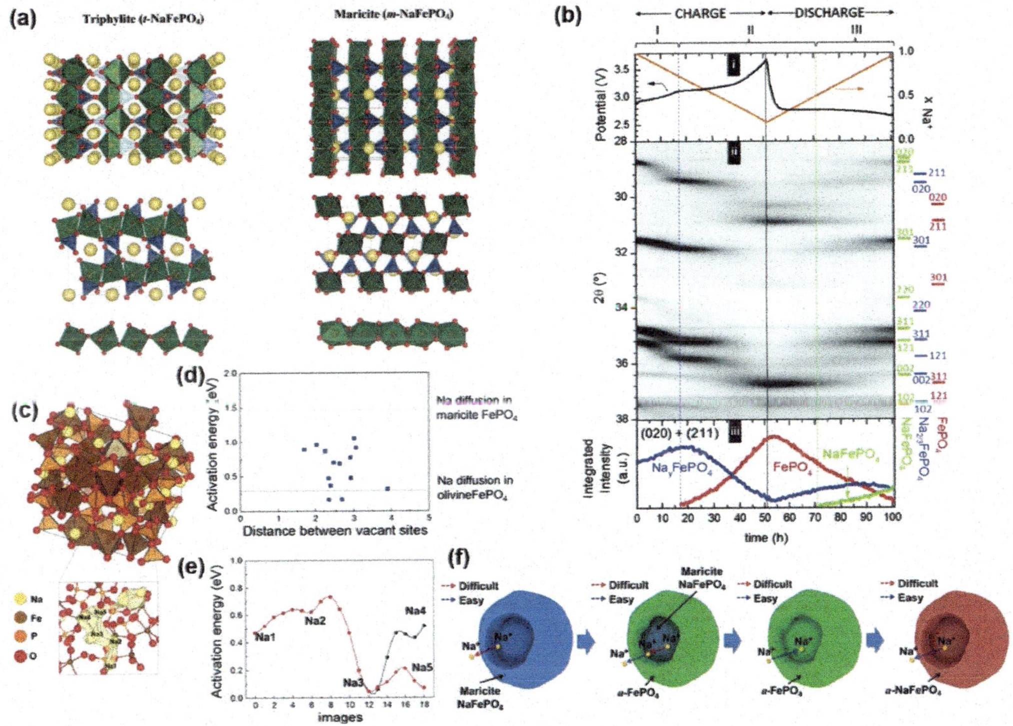

Figure 3.14 (a) Crystal structure of triphylite $NaFePO_4$ (left) and maricite $NaFePO_4$ (right). Reprinted with permission from Ref. [120], Copyright 2013, American Chemical Society. (b) *In situ* XRD experiment of olivine $NaFePO_4$. (i) Voltage versus time curve; (ii) 2θ versus time plot of the XRD patterns comprising a full cycle. The level of grey indicates the relative intensity (the darker, the more intense). Horizontal bars (right) indicate the position of the Bragg peaks. (iii) Sum of the (020) and (211) integrated intensity reflections for each phase involved versus time. Reprinted with permission from Ref. [121], Copyright 2014, Royal Society of Chemistry. (c) Plausible Na sites and diffusion pathways of *a*-$FePO_4$, (d) activation energies of Na hopping between Na sites as a function of the distance between Na sites, and (e) the activation energies for Na diffusion along Na1–Na2–Na3–Na4/Na5 diffusion pathways (over 10 Å) in *a*-$FePO_4$. The dashed lines indicate the activation energies of Na diffusion in olivine $FePO_4$ and maricite-type $FePO_4$. (f) Schematic representation of the electrochemical mechanism during charge/discharge cycling in maricite-type $NaFePO_4$. Reprinted with permission from Ref. [122], Copyright 2015, Royal Society of Chemistry.

Cabanas et al. have exhaustively investigated the sodiation/desodiation mechanism of triphylite-type $NaFePO_4$ [121]. During a full charge/discharge cycle (Fig. 3.14b), three regions are identified in detail by *in situ* X-ray diffraction (XRD) technology. In region I, the desodiation process involves an initial single-phase reaction where Na^+ ions are continuously extracted from Na_yFePO_4 $(1 < y < 0)$ until a stable intermediate phase $Na_{0.7}FePO_4$ appears at the voltage discontinuity. The following region II covers the symmetrically biphasic transformation of an Na-rich Na_yFePO_4 phase and an Na-deficient $FePO_4$ phase. Then a reversible sodiation reaction of triphylite-type $NaFePO_4$ occurs in region III. These are also proved by their density functional theory (DFT) calculations for the phase diagram, where the Na_xFePO_4 phase exhibits stabilization at $x = 2/3$ with a voltage drop of 0.16 V in charge profile [123]. Based on this sodium storage mechanism, triphylite-type $NaFePO_4$ can deliver a reversible capacity of over 120 mAh/g with an average redox potential at 2.8 V [124].

Without 1D Na^+ diffusion channels, maricite-type $NaFePO_4$ was initially regarded as an electrochemically inert material. However, Kim et al. found that nano-sized maricite-type $NaFePO_4$ shows unexpected sodium storage behaviors with a high reversible capacity of 142 mAh/g and 95% capacity retention over 200 cycles [122]. The crucial factor is the initial phase transformation from maricite-type $NaFePO_4$ to amorphous $FePO_4$ (a-$FePO_4$) (Fig. 3.14c). Quantum mechanics calculations prove that when Na hops along conventional pathways in a crystal maricite-type $NaFePO_4$, the activation barriers are too high to make Na^+ ions extracted compared with Na^+ diffusion in a-$FePO_4$ (Fig. 3.14d). However, if Na^+ ions transfer along the Na1–Na2–Na3–Na4/Na5 route in an amorphized maricite-type $NaFePO_4$, the activation energy can be reduced ~75% to 0.73 eV, giving a dramatic increase in Na^+ diffusion (Fig. 3.14e). Briefly, after initial activation, Na^+ ions are extracted/inserted with a reversible phase transformation between a-$FePO_4$ and amorphous $NaFePO_4$ (a-$NaFePO_4$) (Fig. 3.14f).

3.3.1.2 NASICON

Sodium superionic conductor (NASICON)-structured materials are defined as $Na_xM_2(XO_4)_3$ $(1 \le x \le 4)$, where M represents transition metals, like V, Fe, Ni, Mn, Ti, Cr, Zr, etc. and *X* consists of P, S, Si, Se, Mo, etc. NASICON-structured cathodes are famous for their excellent structural stability and superior ionic conductivity due to the unique crystal structure. For example, in a typical phosphate-type NASICON material, $Na_3V_2(PO_4)_3$, each VO_6 octahedra share an oxygen corner with PO_4 tetrahedra to form "lantern" units defined as $V_2(PO_4)_3$, which further build 3D frameworks with large interstices providing fast Na diffusion channels (Fig. 3.15a). In NASICON structures, both *M* cations and *X* anions are tunable and replaceable, which offers excellent versatility for cathode engineering, such as in phosphate-type cathodes, resulting in various compounds, like $Na_3V_2(PO_4)_3$, $Na_3Fe_2(PO_4)_3$, $Na_3Cr_2(PO_4)_3$, $Na_3VCr(PO_4)_3$, $Na_3V_{1.5}Al_{0.5}(PO_4)_3$, $Na_{3.5}V_{1.5}Fe_{0.5}(PO_4)_3$, $Na_4VMn(PO_4)_3$, $Na_3VFe(PO_4)_3$, $Na_3TiMn(PO_4)_3$, etc.

$Na_3V_2(PO_4)_3$, as a representative of the NASICON cathode, shows a flat plateau at 3.3 V, corresponding to the redox reaction of V^{3+}/V^{4+}, where two Na^+ ions can be extracted, contributing to a theoretical capacity of 117 mAh/g [129]. There are two distinct Na crystallographic sites in $Na_3V_2(PO_4)_3$, in terms of one six-fold coordinated Na1 site (6b) and two eight-fold coordinated Na2 sites (18e). From aberration-corrected annular-bright-field (ABF) scanning transmission electron microscopy (STEM), Chen et al. directly observed the atomic position changes of Na before and after extraction (Fig. 3.15b) [126]. After desodiation, all Na^+ ions at Na2 sites are extracted, leaving immobile Na^+ ions at Na1 sites. The formed $NaV_2(PO_4)_3$ can keep the skeleton NASICON structure stable with only 8.26% volume shrinkage, calculated from *in situ* XRD results [130]. As for Na-ion migration behaviors, two modes (Fig. 3.15c,d) have been put forward based on DFT calculations. Song et al. first provided a 3D ionic diffusion path along with *x* or *y* directions and a curved route from the voids/channels between adjacent PO_4 tetrahedron and VO_6 octahedron [127]. While Wang et al. suggested that Na1 sites are involved during the Na migration process in a stepwise ion-exchange route, one Na ion at the Na1 site migrates to a nearby Na2 site, and subsequently, the remaining vacancy is occupied by another adjacent Na ion at Na2 site [128].

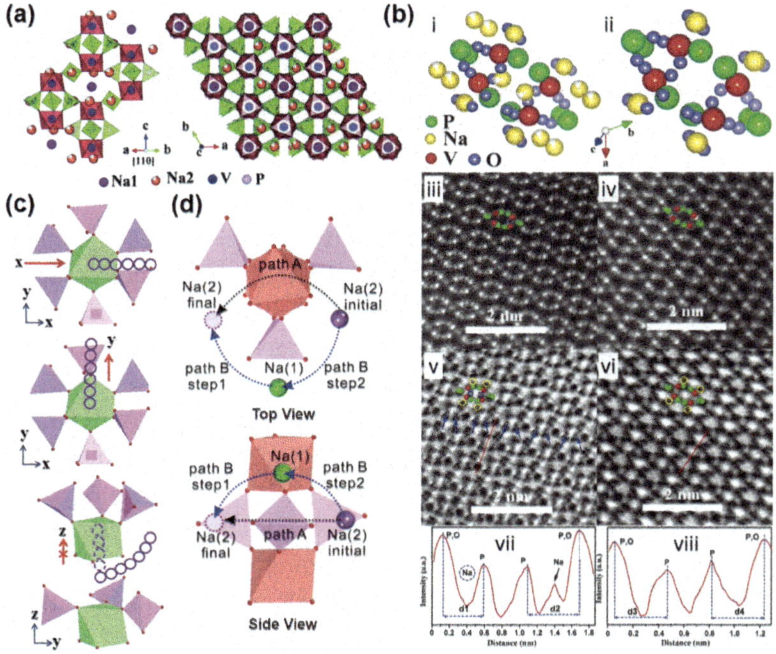

Figure 3.15 (a) Crystal structure of $Na_3V_2(PO_4)_3$. Reprinted with permission from Ref. [125], Copyright 2014, Royal Society of Chemistry. (b) Structural comparison of $Na_3V_2(PO_4)_3$ and $NaV_2(PO_4)_3$. The crystal structures for (i) $Na_3V_2(PO_4)_3$ and (ii) $NaV_2(PO_4)_3$; STEM HAADF images of (iii) $Na_3V_2(PO_4)_3$ and (iv) $NaV_2(PO_4)_3$ along with the [111] projection; STEM ABF images of (v) $Na_3V_2(PO_4)_3$ and (vi) $NaV_2(PO_4)_3$ along with the [111] projection (blue and yellow circles are Na atoms at Na1 and Na2 sites, respectively, the blue arrow indicates the Na atoms at Na2 site); line profiles along in the ABF images of (vii) $Na_3V_2(PO_4)_3$ and (viii) $NaV_2(PO_4)_3$. In the ABF line profiles, the contrast of the dark dot images is inverted and displayed as peaks. Reprinted with permission from Ref. [126], Copyright 2014, Wiley-VCH. (c) Possible Na ion migration paths in $Na_3V_2(PO_4)_3$ along *x*, *y*, and curved *z* directions. Reprinted with permission from Ref. [127], Copyright 2014, Royal Society of Chemistry. (d) Sketch map of the direct diffusion route (path A) and stepwise ion-exchange route (path B) for Na migration. The green and violet balls represent the Na atoms initially located at Na(1) and Na(2) sites. Reprinted with permission from Ref. [128], Copyright 2018, American Chemical Society.

Due to the flexible replaceability of transition metals, some researchers investigated alternatives of the toxic and expensive V element, such as the total replacement of V with Fe or Cr, resulting in $Na_3Fe_2(PO_4)_3$ and $Na_3Cr_2(PO_4)_3$. However, the low average working potential of $Na_3Fe_2(PO_4)_3$ and the poor cycling stability of $Na_3Cr_2(PO_4)_3$ limit their applications in SIBs. In contrast, the partial substitution of V takes effect. Goodenough et al. presented a deep understanding of a series of NASICON-structured $Na_xMV(PO_4)_3$ (M = Fe, Mn, Ni) cathodes [131]. In half-substituted NASICON structures, both M and V elements are distributed uniformly in trigonal $Na_4MnV(PO_4)_3$ ($R3c$) and monoclinic $Na_3FeV(PO_4)_3$. Aberration-corrected STEM identifies that both materials exhibit large Na-ion diffusion channels, resulting in their remarkable stability and rate capability. Notably, the combination between an Fe^{3+} ion and an Na3 site ion causes a monoclinic distortion in $Na_3FeV(PO_4)_3$ and thus getting a similar crystal structure as monoclinic $Na_3Fe_2(PO_4)_3$ ($C12c1$). Ti has also been introduced into $Na_3V_2(PO_4)_3$ to prevent crystal degradation and vanadium dissolution initially in aqueous SIBs. When half Ti-substituted $Na_3TiV(PO_4)_3$ is applied in organic systems, it can deliver a high specific capacity up to 147 mAh/g with three plateaus at 3.4, 2.1, and 1.6 V, corresponding to the redox of V^{4+}/V^{3+}, Ti^{4+}/Ti^{3+}, and V^{3+}/V^{2+}, respectively [132]. The extensive working potential extends the capability of $Na_3TiV(PO_4)_3$ as both cathode and anode. Except for the vanadium-based binary NASICON-structured cathode, manganese-based ($Na_3MnTi(PO_4)_3$, $Na_3MnZr(PO_4)_3$, and $Na_4MnCr(PO_4)_3$) and chromium-based ($Na_2CrTi(PO_4)_3$) analogs have also been reported.

3.3.1.3 Pyrophosphates

Pyrophosphates are prepared via exposing phosphates at high temperatures following the decomposition or oxygen evolution of PO_4 to P_2O_7. Typically, there are two kinds of pyrophosphate-type cathodes, Na-poor $NaMP_2O_7$ (M = Fe, V) and Na-rich $Na_2MP_2O_7$ (M = Fe, Co, Mn), reported in SIBs. Both are open-framework structures with sufficient and large Na^+ transport channels to enable the reversible sodiation/desodiation.

The first Na-poor pyrophosphate $NaFeP_2O_7$ was reported in 1982 [133]. However, there has been no further investigation in SIBs because of its poor electrochemical activity, low operating voltage, and sluggish sodium kinetics [134], whereas $NaVP_2O_7$ has the ability to store sodium ions reversibly. The first investigated $NaVP_2O_7$ polymorph is an $NaMoP_2O_7$-structured α-phase (monoclinic, space group $P2_1/c$, a = 7.3169(2) Å, b = 7.9350(2) Å, c = 9.567(2) Å,

β = 111.905(2)°), in which VO_6 octahedra share corners with five P_2O_7 bitetrahedral pyrophosphate groups, forming narrow Na^+ diffusion channels. Although it has a theoretical capacity of 108 mAh/g with a working potential of 3.4 V (versus Na/Na^+), only ~35% of reversible capacities are achieved in the reported publication [135]. Recently, Drozhzin et al. found a $KAlP_2O_7$-isostructural β-$NaVP_2O_7$ polymorph (monoclinic, space group $P2_1/c$, a = 7.1142(7) Å, b = 10.0709(7) Å, c = 8.0816(6) Å, β = 109.091(9)°) realizing not only higher reversible capacity of 104 mAh/g and average operating voltage of 3.9 V, but also superior rate capability at even 50 C. Based on nudged elastic band (NEB) calculations (Fig. 3.16a), these benefit from the larger Na^+ diffusion channels along the [001] and [110] directions compared with that for α-$NaVP_2O_7$, therefore resulting in a twice lower Na migration barrier in β-$NaVP_2O_7$ (0.25eV versus 0.5eV in α-$NaVP_2O_7$) [136].

Compared with Na-poor species, Na-rich pyrophosphates gained more research interest in energy storage applications. In 2012, Yamada et al. first proved the possibility of $Na_2FeP_2O_7$ as a cathode material in SIBs with a reversible capacity of 82 mAh/g and an average redox voltage of 3.0 V. This triclinic pyrophosphate is built by corner-sharing Fe_2O_{11} dimers interconnected with P_2O_7 diphosphate units, which constructs large tunnels along the [011] direction to reside in Na atoms [139]. Following closely, Kim et al. powerfully revealed the sodium storage mechanism of $Na_2FeP_2O_7$ [137]. They proved that over 0.86 sodium ions could be reversibly extracted and intercalated, associated with two distinct potential plateaus at 2.5 V and 3.0–3.25 V (Fig. 3.16b), which can be identified as a single-phase reaction and consecutive two-phase reactions, respectively, according to the calculated formation energy (Fig. 3.16c). Moreover, based on DFT calculations of Na^+ migration barriers (Figure 3.16d), they described that the single-phase reaction is associated with the first Na extraction from the most thermodynamically active Na1 site along the [001] channel direction, whereas the other plateaus indicate the further deintercalation of Na atoms from Na3–Na8 sites via 1D channels/2D pathways shown with red lines.

After discovering $Na_2FeP_2O_7$, $Na_2CoP_2O_7$, a Co-substituted pyrophosphate cathode is also naturally explored for SIBs by Yamada et al. [140]. $Na_2CoP_2O_7$ pyrophosphate was first disclosed in the early 1990s with three polymorphs associated with different

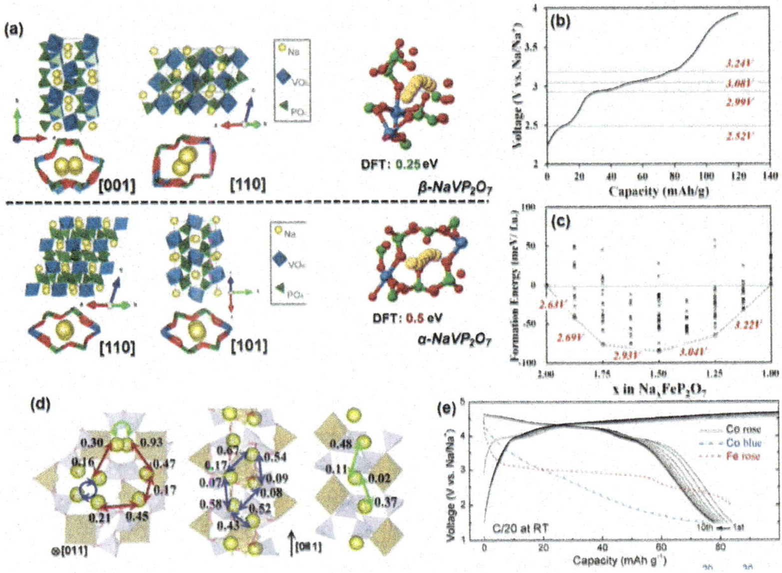

Figure 3.16 (a) Comparison of crystal structure and overlapped Na^+ positions from every image obtained from optimization with the NEB method for the shortest migration pathways for β-$NaVP_2O_7$ and α-$NaVP_2O_7$. Reprinted with permission from Ref. [136], Copyright 2019, American Chemical Society. (b) The quasi-equilibrium potentials obtained from the second cycle charge with the observed potentials denoted. (c) A calculated formation energy hull and a series of potentials for consecutive compositional intervals. (d) The calculated Na ion migration barriers (in eV) for the $Na_2FeP_2O_7$. The arrowhead and the number next to it indicate the direction of Na migration and the corresponding migration barrier (either forward or backward). For Na migrations between the two symmetrically equivalent sites, forward and backward migrations have the same barriers; thus, only one number is denoted in the middle of the arrow. Color code: red for 2D migration paths, and green (Na1 and Na2 sites) and blue (Na3–Na6 sites) for the outward channels. Migration barriers along the [011] channel direction (green and blue) are shown separately on the two rightmost panels for clarity. Reprinted with permission from Ref. [137], Copyright 2016, Wiley-VCH. (e) Galvanostatic cycles of the Co rose, Fe rose, and Co blue polymorphs. Reprinted with permission from Ref. [138], Copyright 2016, Wiley-VCH.

synthesis temperatures as well as thermal stabilities, in terms of orthorhombic (space group $Pna2_1$), triclinic (space group $P1$), and tetragonal (space group $P4_2/mnm$) phases [141]. Among these three phases, orthorhombic $Na_2CoP_2O_7$, known as the blue-form polymorph, is the most stable and the earliest applied in SIBs. The unique 2D Na^+ diffusion channels in its layered structure ensure the high reversibility of sodiation/desodiation with a reversible capacity of 80 mAh/g and an average working voltage of 3.0 V [140]. Triclinic $Na_2CoP_2O_7$ is thermodynamically less stable than the orthorhombic counterpart, which brings a significant challenge to the synthesis of the triclinic phase. Jung et al. found that the introduction of sodium defects could selectively promote the formation of the triclinic $Na_2CoP_2O_7$ phase [138]. Although the prepared nonstoichiometric compound exhibits a similar reversible capacity as the orthorhombic counterpart, the average operating voltage is dramatically enhanced to 4.3 V (Fig. 3.16e), showing the great potential of a cathode in SIBs.

Unlike $Na_2FeP_2O_7$ and orthorhombic $Na_2CoP_2O_7$ pyrophosphate cathodes, when Mn^{3+}/Mn^{2+} redox pairs are introduced into pyrophosphate polymorphs, the triclinic $Na_2MnP_2O_7$ (space group $P1$) can deliver an even higher operating potential up to 3.8 V versus Na^+/Na [142]. Besides, this type of pyrophosphate cannot maintain the intrinsic polymorphic structure during sodiation/desodiation processes as other transition metal pyrophosphates do, due to the existence of Mn^{3+}/Mn^{2+} redox pairs, which is well known to undergo Jahn–Teller distortion at the charge and discharge states. The DFT calculations prove that the resulted small-scaled atomic rearrangements help lower the barriers for electronic conduction and phase boundary migration and then enhance the kinetics of $Na_2MnP_2O_7$. Consequently, the $Na_2MnP_2O_7$ pyrophosphate cathode shows a remarkable reversible capacity of 90 mAh/g with good cycling stability (96% capacity retention for 30 cycles) and superior rate capability (61 mAh/g at 1 C).

3.3.1.4 Fluorophosphates

Typically, most polyanionic cathodes show a relatively lower redox potential than transition metal oxides, even with similar transition metal redox couples. Introducing electronegative anions is a practical method to improve the redox potential, such as the most electronegative F^- anions, which can lead to the inductive effect between electronegative anions and transition metal cations. The successfully worked fluorophosphate cathodes include $NaVPO_4F$, $Na_3V_2(PO_4)_3F_3$, Na_2MPO_4F (M = Fe, Co, Mn), and so on.

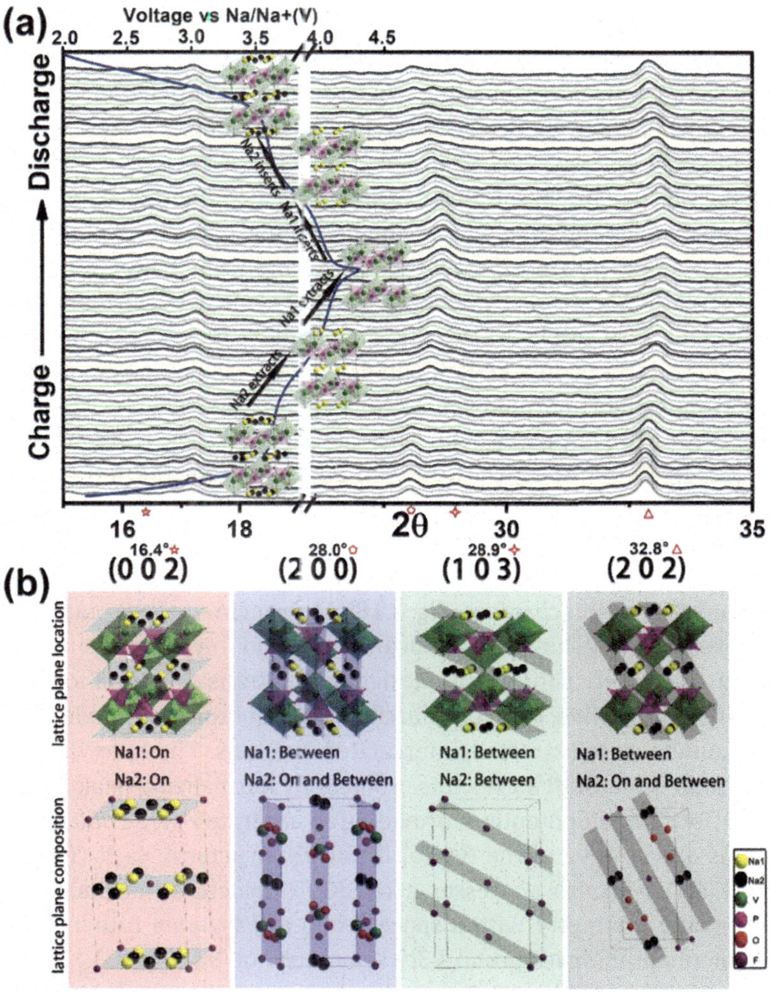

Figure 3.17 (a) *In situ* XRD pattern of tetragonal $NaVPO_4F$ and corresponding galvanostatic charge–discharge curve during a charge–discharge cycle at 0.5 C; (b) location of the selected crystal planes in the tetragonal $NaVPO_4F$ crystal structure. Reprinted with permission from Ref. [143], Copyright 2020, American Chemical Society.

The tetragonal (space group *I*4/*mnm*) structural $NaVPO_4F$, built by F vertex bridged VO_4F_2 octahedra with corner-shared PO_4 tetrahedra, was first applied to SIB in 2003 [144]. After coupling with a hard-carbon anode, this fluorophosphate cathode shows a high average working potential at 3.7 V with a reversible capacity of 82 mAh/g. Recently, Li et al. revisited this typical fluorophosphate cathode and pronounced a precise sodium storage mechanism through *in situ* XRD (Fig. 3.17a,b) [143]. There are two charge plateaus at 3.6, and 4.0 V corresponding to a two-step redox reaction of V^{4+}/V^{3+} redox pairs. By analyzing the changes in crystalline planes, it can be inferred that Na^+ ions are first extracted from Na1 sites, located on the (200) plane, at the lower voltage plateaus, and then from Na2 sites between (200) planes at the higher voltage plateaus. Except for the tetragonal polymorph, another phase, monoclinic (space group $C2/c$) $NaVPO_4F$, shows even better electrochemical performance in SIBs. Ling et al. applied a specific molten-state-blending technique to synthesize a nanoscale-laminated $NaVPO_4F$@C composite [145]. This monoclinic fluorophosphate can deliver an excellent reversible capacity of 135 mAh/g and superior rate capability (over 112 mAh/g at a high C-rate of 30 C). However, different from the tetragonal counterpart, monoclinic $NaVPO_4F$ has just one charge plateau at 3.4 V (Fig. 3.18a), corresponding to one-step oxidation of V^{3+} to V^{4+}. Both *in situ* XRD and *ex situ* solid-state NMR are adopted to investigate the sodium storage mechanism (Fig. 3.18b,c). From *in situ* XRD patterns, the reversible peak shifts indicate the good reversibility of Na extraction/intercalation without damage or transformation for the intrinsic monoclinic crystalline architecture, but this cannot identify the number x of extracted/intercalated Na^+ ions. The following *ex situ* solid-state NMR solves this problem by normalizing peak areas at various cut-off potentials, providing an accurate x value of 0.869.

As a derivative of the typical NASICON-structural $Na_3V_2(PO_4)_3$, $Na_3V_2(PO_4)_2F_3$ also possesses a NASICON structure but adopts a tetragonal symmetry with a space group of $P4_2/mnm$ rather than rhombohedral (space group $R3c$) symmetry for $Na_3V_2(PO_4)_3$. Here, PO_4 tetrahedra-bridged $V_2O_8F_3$ bi-octahedra build large Na^+ channels along the [110] and [110] directions, which benefit to excellent rate performance (survive under high C-rate of 15 C and 30 C) and ultralong cycling life (over 3000 cycles). Distinguished with one-plateau (3.3 V) $Na_3V_2(PO_4)_3$, $Na_3V_2(PO_4)_2F_3$ shows three working plateaus at 3.4, 3.7, and 4.2 V, corresponding to a high reversible capacity of 130 mAh/g [147]. There is a dispute on the sodium storage mechanism. Shakoor et al. claimed that sodium extraction and insertion is a one-phase reaction with negligible variation in crystal parameters, based on the *ex situ* XRD and DFT calculations [148]. While Bianchini et al. created a phase diagram illustrating the existence of four intermediate phases between $Na_3V_2(PO_4)_2F_3$ and the final product $NaV_2(PO_4)_2F_3$ via *operando* high-resolution synchrotron XRD [149].

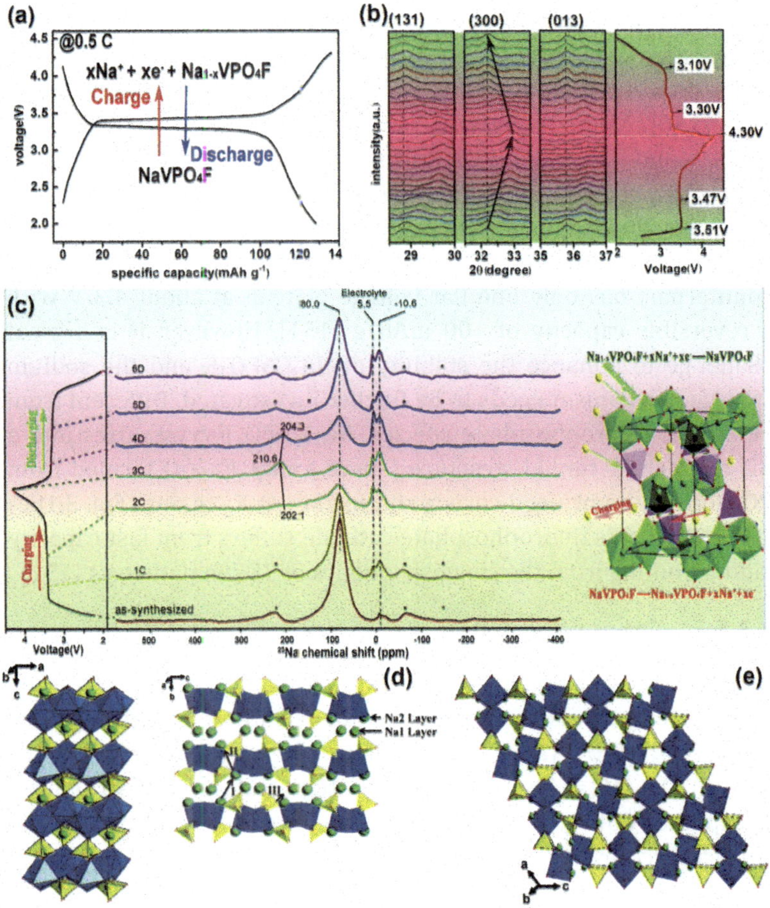

Figure 3.18 (a) Galvanostatic charge–discharge curve of monoclinic $NaVPO_4F$ at 0.5 C. (b) *In situ* XRD pattern of $NaVPO_4F$ during a charge–discharge cycle. (c) *Ex situ* ^{23}Na solid-state NMR spectra of $NaVPO_4F$@C cut-off at different charge–discharge states. Reprinted with permission from Ref. [145], Copyright 2018, Royal Society of Chemistry. Crystal structure of (d) Na_2FePO_4F view along [010] and along [100]. Ion transport pathways (I, II, and III) are marked. (e) Crystal structure of Na_2MnPO_4F. The transition metal octahedra are shown in blue, phosphate tetrahedra in yellow, and alkali ions in green. Reprinted with permission from Ref. [146], Copyright 2009, American Chemical Society.

Since the high toxicity and cost of vanadium, environmentally friendly transition metals like Fe, Co, Mn are introduced into fluorophosphate. Na_2FePO_4F and Na_2CoPO_4F have similar structures (Fig. 3.18d), adopting *Pbcn* orthorhombic space group, built from face-sharing $M_2O_4F_2$ (M = Fe and Co) bioctahedra chains linked by F atoms sharing corners with PO_4 tetrahedra. The formed 2D MPO_4F slabs can store Na atoms at Na1 and Na2, two distinct crystallographic sites [146]. As for Na_2FePO_4F, Yang et al. applied *ex situ* ^{23}Na NMR spectra and identified that during charge/discharge process, Na atom remains at the Na1 site, and only Na2 site involves in two two-phase reactions among Na_xFePO_4F (x = 1, 1.5, and 2), corresponding to the voltage plateaus at 2.91 and 3.06 V [150]. Whereas, Co-based counterpart has only one flat voltage plateau at about 4.3 V with a reversible capacity of 100 mAh/g [151]. However, it is a great challenge to enhance the stability of Na_2CoPO_4F, and the sodium storage mechanism needs to be further investigated. Different from Fe and Co fluorophosphate species, Na_2MnPO_4F crystallizes into a 3D monoclinic tunnel structure (space group $P2_1/n$) formed from PO_4 tetrahedra bridged corner-sharing $Mn_2O_8F_2$ chains (Fig. 3.18e) [146]. Also, this fluorophosphate cathode suffers from fast capacity fading, rooting from the sluggish kinetics of 3D ion pathways [152].

3.3.1.5 Other phosphates

Except for these widely investigated typical phosphates, some specific phosphates have also been reported, for example, mixed phosphates $Na_4M_3(PO_4)_2P_2O_7$ (M = Fe, Co, Mn, Ni, etc.), metaphosphates $NaM(PO_3)_3$ (M = Fe, Mn, Co, Cd, etc.), and layered-structured phosphate $VOPO_4$. Their crystal structure and electrochemical performance are listed in Table 3.1.

3.3.2 Sulfates

3.3.2.1 Fluorosulfates

Fluorosulfates ($NaMSO_4F$, M = Fe, Co, Mn, Ni, etc.) are crystallized by corner-sharing MO_4F_2 octahedra chains along the [001] direction bridged by isolated SO_4 tetrahedra. They possess monoclinic maxwellite structures with symmetry of $C2/c$, except for triplite-type $NaMnSO_4F$ [153]. However, among these fluorosulfate species, only

$NaFeSO_4F$ shows a slightly electrochemical activity with a potential plateau of 3.5 V in SIBs. The poor performance is ascribed to the high Na^+ diffusion barrier of the 1D zigzag-like ion pathway in $NaFeSO_4F$ and the considerable volume variation during sodiation/desodiation process [154]. However, potassium-substituted $KFeSO_4F$ can deliver a high reversible capacity over 120 mAh/g with an average operating potential at 3.5 V in SIB, due to the derivation of crystalline from maxwellite to $KTiOPO_4$ (KTP)-type orthorhombic structure (space group $Pna2_1$), which has sizeable Na^+ diffusion channels along the [011] and [011] directions [155].

3.3.2.2 Alluaudites

The most promising sulfate is alluaudite-type polymorphs with a general formula of $Na_2M(SO_4)_2 \cdot nH_2O$ (M = Fe, Co, Ni, Cu, Cr, Mn; n = 0, 2, 4). For example, in a typical alluaudite-type framework, $Na_2Fe(SO_4)_2$ (Fig. 3.19a), two FeO_6 octahedra share an edge to form an Fe_2O_{10} dimer unit, which connects with SO_4 tetrahedra along the c-axis. The formed framework provides three types of Na locations, although the SO_4 tetrahedra block the Na diffusion pathways from Na1 sites. According to the quantification of Fe^{2+} and Fe^{3+} peaks from *ex situ* XPS spectra during the charge/discharge process, Pan et al. proved that ~0.9 Na^+ ions were reversibly inserted and extracted with an operating voltage of 3.6 V and reversible capacity up to 82 mAh/g (theoretical capacity of 91 mAh/g) [156]. Moreover, *ex situ* XRD patterns show that only one phase is involved in the reaction with a tiny volume variation of 3.3%. Besides, alluaudite-type $Na_2Fe(SO_4)_2$ exhibits high thermal stability (580°C) and excellent moisture resistance (60 days in the air). These make $Na_2Fe(SO_4)_2$ a most striking candidate for SIBs. When increasing the amount n of H_2O molecules to 2, a Kröhnkite-type $Na_2Fe(SO_4)_2 \cdot 2H_2O$ (Fig. 3.19b) is derived, in which FeO_6 octahedra contribute four oxygen corners sharing with alternately linked SO_4 tetrahedra to form chains along the c-axis, and structural H_2O molecules substitute the other two oxygen atoms. The formed structure provides convoluted Na diffusion channels along the b-axis and a moderate working potential at 3.25 V [157]. However, Kröhnkite-type $Na_2Fe(SO_4)_2 \cdot 2H_2O$ is very sensitive to moisture, leading to poor air conditioning stability. After absorbing enough moisture (n = 4), a bloedite-type $Na_2Fe(SO_4)_2 \cdot 4H_2O$ (Fig. 3.19c) is formed and built by

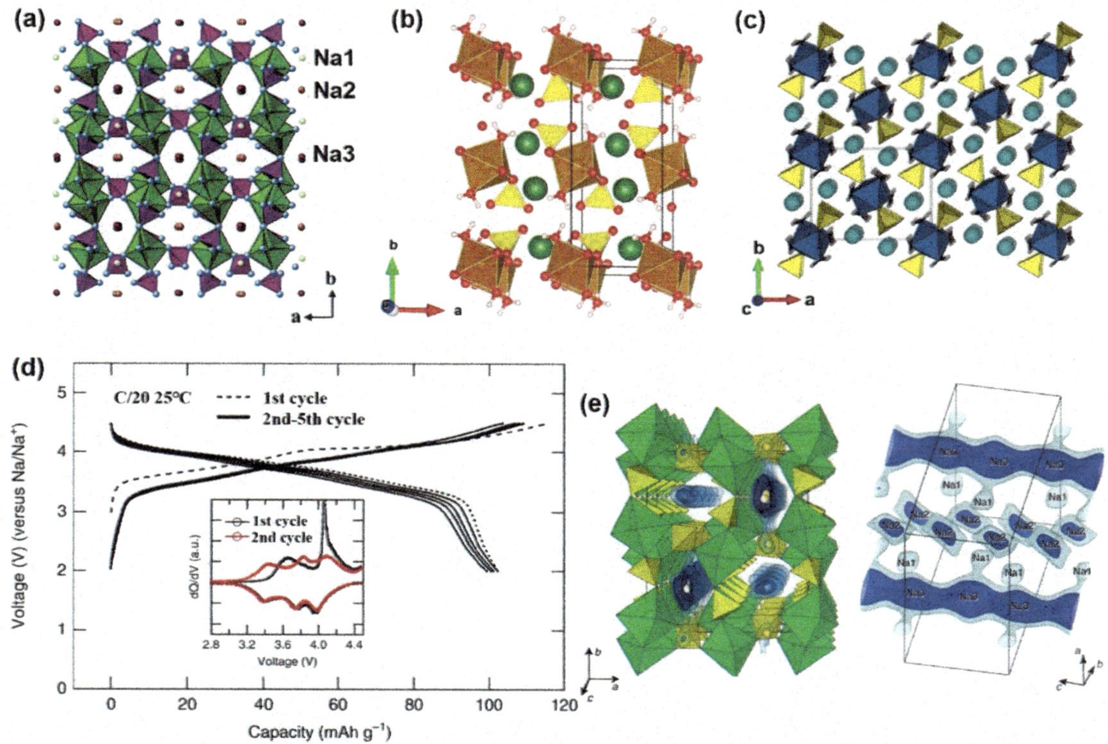

Figure 3.19 Crystal structure of (a) alluaudite-type $Na_2Fe(SO_4)_2$, reprinted with permission from Ref. [156], Copyright 2019, Royal Society of Chemistry; (b) Kröhnkite-type $Na_2Fe(SO_4)_2{\cdot}2H_2O$, reprinted with permission from Ref. [157], Copyright 2014, American Chemical Society; and (c) bloedite-type $Na_2Fe(SO_4)_2{\cdot}4H_2O$, reprinted with permission from Ref. [158], Copyright 2013, Electrochemical Society. (d) Galvanostatic charging and discharging profiles of $Na_{2-x}Fe_2(SO_4)_3$ cathode at the rate of C/20. (Inset) The differential galvanostatic profiles (dQ/dV) of $Na_{2-x}Fe_2(SO_4)_3$ cathode showing two distinct peaks during the first charge and broader three peaks upon subsequent discharging/charging processes. (e) Equi-value surface of the ΔBVS. The blue and light-blue surfaces are for $\Delta BVS = 0.2$ and 0.4, respectively. The inner side of the surface corresponds to accessible spaces for the Na ions. Green and yellow polyhedra are that of FeO_6 and SO_4, respectively. Reprinted with permission from Ref. [159], Copyright 2014, Nature Publishing Group.

isolated $Fe(SO_4)_2(H_2O)_4$ units with Na diffusion channels along the *c*-axis, where H_2O molecules replace four oxygen atoms in the FeO_6 octahedra [158]. Although nearly one Na^+ ion can be extracted at the first charge cycle, the poor cycling stability and inferior electronic conductivity limit the application in SIBs.

$Na_2Fe_2(SO_4)_3$ is another alluaudite-type sulfate cathode material first developed by Yamada et al. [159]. Different from typical NASICON-related $Fe_2(SO_4)_3$ with corner-sharing FeO_6 octahedral units, alluaudite-type $Na_2Fe_2(SO_4)_3$ adopting monoclinic framework (space group $C2/c$) is built from isolated edge-sharing FeO_6 octahedra forming Fe_2O_{10} dimers, similar to $Na_2Fe(SO_4)_2$. Benefited from the shortest bond length of Fe–Fe among iron-based polyanionic compounds, $Na_2Fe_2(SO_4)_3$ possesses a remarkably high working potential up to 3.8 V related with Fe^{2+}/Fe^{3+} redox (Fig. 3.19d). They applied the bond valence (BV) method to evaluate the validity of the crystal structure and demonstrate possible Na diffusion pathways through a map of the difference in the BV sum from the ideal value (ΔBVS) (Fig. 3.19e). It is clear to see that the Na3 site has a continuous diffusion channel along the [001] direction, corresponding to fast kinetics of the Na3 site, although all Na ions can be fully extracted in theoretical. However, in the practical application, only about 83% of theoretical capacities are achieved. They found the reason from *ex situ* Fe *K*-edge X-ray absorption near edge structure (XANES) spectra that the irreversible capacity was ascribed to the decomposition of electrolytes and the irreversible Fe migration at the first charge cycle [160]. In order to enhance the crystal stability and intercalation of Na^+ ions, researchers developed various off-stoichiometric $Na_{2+2x}Fe_{2-x}(SO_4)_3$. For example, Liu et al. synthesized an $Na_{2.7}Fe_{1.65}(SO_4)_3$@rGO composite with 90.8% and 85.9% capacity retention at the rates of 5 C and 10 C, respectively [161]. Except for off-stoichiometric $Na_{2+2x}Fe_{2-x}(SO_4)_3$, Mn, Co-substituted analogs, like $Na_{2.44}Mn_{1.79}(SO_4)_3$ [162] and $Na_{2.32}Co_{1.84}(SO_4)_3$ [163], have also been reported.

3.3.3 Other Oxysalts

3.3.3.1 Silicates

Among various polyanionic-type cathodes, sodium silicates Na_2MSiO_4 (M = Fe, Mn, Co, etc.) usually exhibit higher specific capacity beneficial from the lower molecular weight SiO_4 group. Besides, due to the strong covalent bonds of Si–O patterns, the thermal stability of silicates can reach 1000°C, and also, the extra-stable Si-O bonding ensures ultralow volume change after the full desodiation. Reversely, this causes a relatively weaker inductive effect between SiO_4 tetrahedra and MO_x polyhedra, resulting in lower redox potential in silicates than phosphates and sulfates, but this phenomenon ensures a two-electron transfer during the sodiation/desodiation process. Unlike other polyanionic-type cathodes, in silicate Na_2MSiO_4 structural framework, transition metal atoms are four-coordinated with oxygen forming small-sized MO_4 tetrahedra. Therefore, the 3D frameworks built from corner-sharing MO_4 tetrahedra and SiO_4 tetrahedra possess larger lattice holes to store and transfer Na, even without ionic diffusion channels as phosphates and sulfates. The electrochemical properties of some typical silicates, like Na_2FeSiO_4, Na_2MnSiO_4, Na_2CoSiO_4, and $Na_2Fe_2Si_2O_7$, are shown in Table 3.1.

3.3.3.2 Carbonophosphates

Carbonophoshpates are a novel family of polyanionic-type cathodes first discovered via a high-throughput *ab initio* computational calculation in 2012 [164]. To date, six carbonophosphates are synthesized with a general formula of $Na_3MPO_4CO_3$ (M = Mg, Mn, Fe, Co, Ni, Cu), but only Mn and Fe species have promising electrochemical activity in sodium storage. The most representative one is the sidorenkite-type $Na_3MnPO_4CO_3$, which adopts a monoclinic framework (space group $P2_1/m$), composed of isolated MnO_6 octahedra sharing four oxygen corners with PO_4 tetrahedra and one edge with planar CO_3 units [165]. Specifically, there are two distinct Na sites in $Na_3MnPO_4CO_3$ polymorph that can be extracted during a two-electron-transfer reaction corresponding to the oxidation of Mn^{2+} to Mn^{3+} and Mn^{3+} to Mn^{4+} with an average operating potential of 3.7 V. Therefore, $Na_3MnPO_4CO_3$ exhibits a high theoretical capacity of up to 190 mAh/g. Similarly, the isostructural $Na_3FePO_4CO_3$ also has two plateaus associated with Fe^{3+}/Fe^{2+} and Fe^{4+}/Fe^{3+} redox couples but performs a lower average operating potential 2.6 V in SIBs [166].

To sum up, we listed some classic polyanionic-type cathode materials such as phosphates and sulfates in this chapter. Their specific crystal structures and electrochemical performance in SIBs have been briefly introduced. Meanwhile, we demonstrated Na diffusion kinetics in these polymorphs through versatile advanced technologies, including *in situ* and (or) *ex situ* XRD, XPS, synchrotron techniques, various theoretical calculations, and simulations. We hope this chapter can guide the fundamental research on polyanionic-type cathodes for SIBs.

3.4 Prussian Blue Analogs

Prussian blue (PB) and its analogs (PBAs) are a large family of transition-metal hexacyanoferrates with an open framework structure [167, 168]. As we know, PB was first used as a pigment in the 18th and 19th centuries [169]. With further investigation in the past few decades, PB materials were extensively explored in more fields, such as hydrogen storage materials [170], biosensing [171], water treatment [172], and so on. Recently, PBAs got intensive attention on SIBs owing to their unique chemical properties and nanostructure. Their large interstitial sites in the lattice provide abundant channels for accommodating sodium cations and buffer extreme volume change during charging/discharging processes [173–175]. Furthermore, based on the theoretical two-electron redox reaction, the theoretical specific capacity of PBAs could achieve as high as 170 mA/g, corresponding to a high energy density of 496 Wh/kg, which is very close to that of olivine $LiFePO_4$ (≈530 Wh/kg) used for LIB [176, 177]. Therefore, benefiting from the high energy density and outstanding structure stability, PBAs are considered a candidate for SIB cathode.

3.4.1 Crystal Structure of Prussian Blue Analogs

PB and PBAs are a large family of transition-metal hexacyanoferrates with a perovskite-type, face-centered cubic structure (space group $Fm\bar{3}m$), which have been well studied by both X-ray and neutron diffraction techniques [178]. The generic formula of PB compounds can be roughly represented as $A_xM_a[M_b(CN)_6]_y{\cdot}nH_2O$ ($0 \leq x \leq 2, y \leq 1$), where M_a and M_b represent transition metal ions such as Mn, Fe, Co, Ni, Cu, Zn, etc., and A denotes an alkali such as Li, Na, K, etc. Based on its formula, a PBA can be called a metal hexacy-anometallic compound [179]. The $(C\equiv N)^-$ anions of the double-perovskite $M_a[M_b(CN)_6]$ framework are ordered to create high-spin (HS) M_aN_6 octahedra and low-spin (LS) M_bC_6, which form a 3D rigid framework containing open ionic channels and spacious interstitial spaces. Obviously, this large channel structure of PBAs allows reversible accommodation and facile transportation of large Na^+ ions for electrochemical insertion reaction. On the basis of precursors and preparation conditions, M_a and M_b presented a variety of combination of valence states. It is worth noting that the PBA samples were usually prepared by the simple chemical precipitation method, which leads to a hydrated PBA lattice with large amounts of $M_b(CN)_6$ vacancies. For example, the lattice of $Fe_4[Fe(CN)_6]_3 \cdot nH_2O$ Prussian blue structure is constructed in such a way that each Fe^{3+} ion is located in an octahedral nitrogen environment, while Fe^{2+} ions are surrounded by carbon (Fig. 3.20a,b) [180]. Due to the specific stoichiometry of this compound (3:4 ratio of Fe^{2+} and Fe^{3+} sites), the charge neutrality requirement results in a 25% vacancy of $[Fe^{2+}(CN)_6]^{4-}$ clusters. The resulting octahedral cavities are occupied by H_2O molecules, called the coordinated water. Six water molecules are attached to six high-spin Fe (Fe^{HS}) ions forming the corners of the octahedral cavity. Besides the coordinated water, the remaining water molecules called zeolitic water or interstitial water partially or fully occupy the eight possible 8*c* (1/4, 1/4, 1/4) sites of the unit cell.

The lattice vacancies and coordinated water in the crystal framework will severely impact the electrochemical performance of PBA cathodes. First, the presence of $Fe(CN)_6$ vacancies could introduce more coordinated water, which reduces Na content in PBA lattice, thus leading to lower Na-insertion capacity. Second, lattice water in the PB framework causes side reactions and nanostructure collapse, which results in poor cycling stability. Third, the presence of $Fe(CN)_6$ vacancies can decrease the electrical conductivity of PBAs, thereby causing poor rate capability. Therefore, it is of importance to eliminate lattice vacancies in preparing processes to promote sodium-ion storage performance of PBAs.

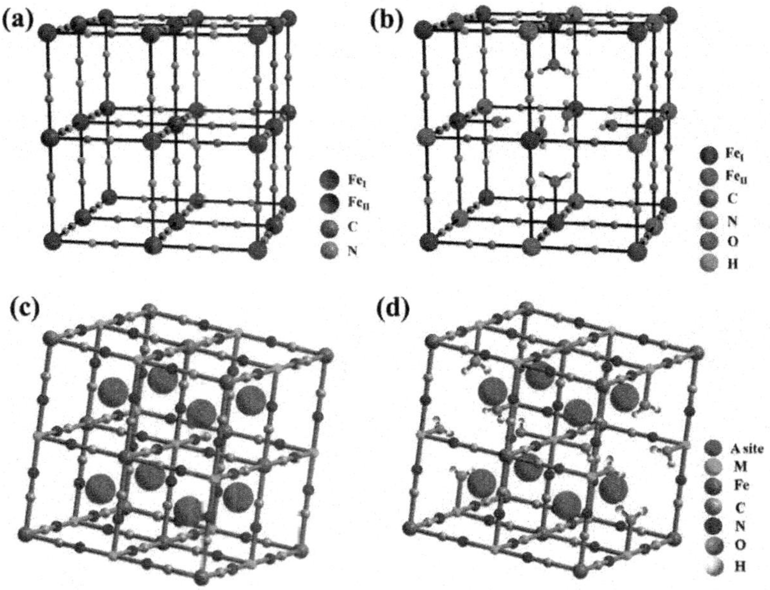

Figure 3.20 The PB crystal structure of (a) ideally vacancy-free $Fe_4[Fe(CN)_6]_3$ and (b) $Fe_4[Fe(CN)_6]_3$ with coordinated water. Reprinted with permission from Ref. [180], Copyright 2016, American Chemical Society. The PBA framework of (c) ideally $Na_2M_a^{II}[M_b^{II}(CN)_6]$ unit cell and (d) defect-rich $Na_2M_a^{II}[M_b^{II}(CN)_6]$. Reprinted with permission from Ref. [181], Copyright 2016, American Chemical Society.

PBA compounds exhibit various electrochemical properties via changing combinations of M_a and M_b transition metals and interstitial ions, which could adjust coordinated water content in the crystal structure. As shown in Fig. 3.20c and d, $Na_2M_a^{II}[M_b^{II}(CN)_6]$ exhibits a typical PBA structure, in which a combination of different valence states of M_a^{II} and M_b^{II} is possible, and the Na atoms are intercalated in 8*c* positions to avoid the formation of vacancies in the framework. It is obvious that PBA compounds contain two different redox-active centers: $M_a^{2+/3+}$ and $M_b^{2-/3+}$ couples, both of which can undergo a complete electrochemical redox reaction (when M = Fe, Co, Mn, etc.), totally contributing a two-electron transfer capacity through reversible 2 Na^+ insertion reaction/extraction process

$$Na_2M_a^{II}[M_b^{II}(CN)_6] \leftrightarrow Na_2M_a^{III}[M_b^{II}(CN)_6] + Na^+ + e^-$$

$$NaM_a^{III}[M_b^{II}(CN)_6] \leftrightarrow M_a^{III}[M_b^{III}(CN)_6] + Na^+ + e^-$$

Based on the formulas, $Na_2M_a^{II}[M_b^{II}(CN)_6]$ could achieve a specific capacity as high as 170 mAh/g, which is much higher than that of the transition metal oxides and phosphates, thus can fulfill the requirements for battery applications [176]. Although there are a few metals capable of occupying the M_b site, Fe is the most common element in PBAs used for Na-ion batteries. Therefore, hexacyanoferrate-based PB compounds could be abbreviated to M_aHCF. In addition, the color of PB compounds depends on the types of transition metals and the content of insertion cations, which leads to different names in turn. For instance, owing to the different sodium content, FeHCF shows colors from blue to white or green, and the corresponding names are Prussian blue, Prussian white, or Berlin green, respectively.

In earlier research, Goodenough et al. prepared a series of PBA compounds with different transition-metal ions and investigated their sodium storage performance [182]. However, open frameworks with large channels make PBAs suitable for sodium-ion diffusion, and the as-prepared PBAs show insufficient specific capacity due to low Na content and large numbers of lattice defects. Since then, many researchers have made efforts to acquire Na-rich PBAs with low crystal defect content. In recent years, various PBA materials have been successfully developed, and the obtained Na-rich PBA compounds show much higher reversible capacity with acceptable cycling stability.

3.4.2 Iron Hexacyanoferrate

FeHCF is the most widely investigated among PBA materials. Fe^{LS} and Fe^{HS} participate at different potentials in the redox reactions for sodium storage and enable a theoretical capacity of 170 mAh/g. However, the Fe^{LS} is less active, resulting in a practical lower capacity and discharge voltage. FeHCF also suffers from poor cycling stability due to the lattice vacancies in the crystal structure and side reactions between the material and electrolyte at a high voltage near 4 V. Sodium content in the as-prepared material is tightly related to the crystallinity of FeHCF.

Wu et al. first synthesized single-crystal $FeFe(CN)_6$ Prussian yellow nanoparticles for SIB cathode and studied the relationship between the electrochemical performance and crystal structure of the PBAs [183]. The $Fe^{III}Fe^{III}(CN)_6$ electrodes exhibit a high capacity of 120 mAh/g, an exceptional rate capability at 20 C, and superior cyclability with 87% capacity retention over 500 cycles (Fig. 3.21a,b). More importantly, the results reveal that increasing crystallinity of PBAs could decrease coordinated water in the crystal structure and thereby improve the structural stability and afford more active sites for sodium-ion storage.

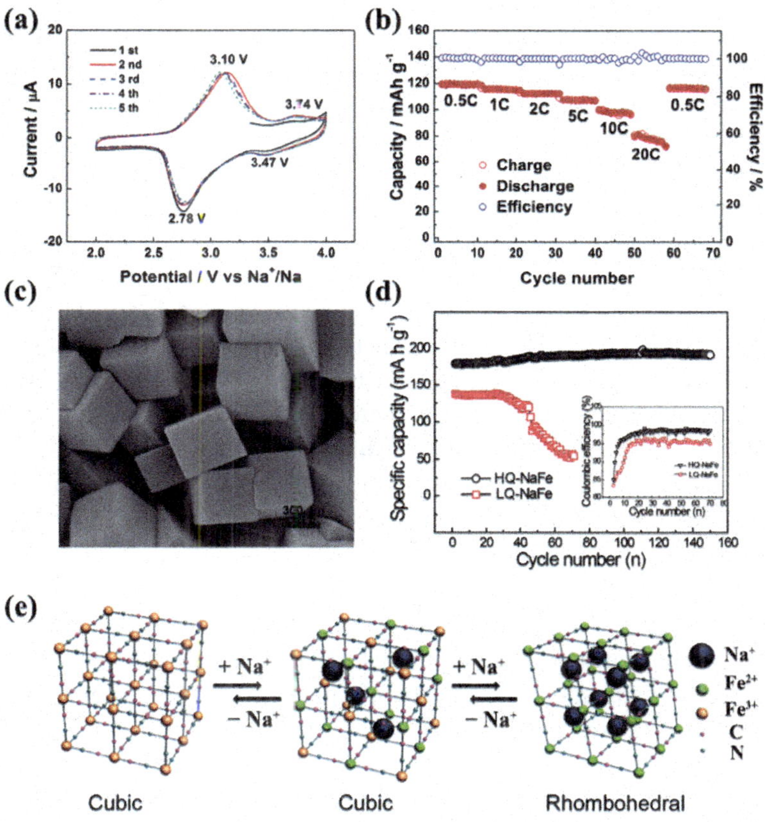

Figure 3.21 Electrochemical performance, morphology, and sodiation/desodiation mechanism of FeHCF. (a,b) CV curves and rate performance of $FeFe(CN)_6$. Reprinted with permission from Ref. [183], Copyright 2013, Royal Chemical Society. (c,d) Morphology and cycling performance of $Na_{0.61}Fe[Fe(CN)_6]_{0.94}$. (e) Schematic illustration of the redox mechanism of FeHCF. Reprinted with permission from Ref. [176], Copyright 2014, Royal Chemical Society.

In order to obtain FeHCF with high crystallinity, low lattice vacancies, and coordinated water, many synthesis strategies were employed so far. Guo et al. reported high-quality Prussian blue crystals with a small number of vacancies and a low water content using $Na_4Fe(CN)_6$ as the single iron-source precursor [176]. The obtained $Na_{0.61}Fe[Fe(CN)_6]_{0.94}$ with a smaller number of vacancies and a lower water content exhibits a high specific capacity of 170 mAh/g, excellent cycling stability without apparent capacity loss after 150 cycles (Fig. 3.21c–e). Also, compared with low-quality Prussian blue ($Na_{0.13}Fe[Fe(CN)_6]_{0.68}$), the superior rate capability of $Na_{0.61}Fe[Fe(CN)_6]_{0.94}$ results from its higher Na content and few defects, further proving that decreasing lattice vacancies is beneficial to increase conductivity and facilitate Na^+ diffusion. However, 0.6 Na per molecular formula is still too low to fulfill practical application in sodium-ion full batteries.

To increase Na content in the FeHCF lattice, Guo et al. prepared $Na_{1.63}FeFe_{0.89}(CN)_6$ by employing VC and N_2 protection simultaneously, which could protect $Fe^{2+}/[Fe^{II}(CN)_6]^{4-}$ from being oxidized to $Fe^{3+}/[Fe^{III}(CN)_6]^{3-}$, resulting in a decreased average valence of Fe and an increased content of Na^+ [184]. The obtained FeHCF with high Na content exhibits high discharge capacity (150 mAh/g), superior cycling performance (90% capacity retention over 200 cycles), and an impressive Coulombic efficiency (~100%) as a cathode material in room-temperature Na-ion batteries (Fig. 3.22a,b). Chou et al. prepared $Na_{1.56}Fe[Fe(CN)_6]$ sample through a facile and one-step method using $Na_4Fe(CN)_6$ as the precursor in highly concentrated NaCl solutions [185]. This Na-rich electrode shows a high specific capacity of more than 100 mAh/g and excellent capacity retention of 97% over 400 cycles (Fig. 3.22c,d). With increasing sodium ions entering the framework, the number of vacancies and coordinating water in the FeHCF decreased, resulting in enhanced structural stability. Goodenough et al. prepared Prussian white through hydrothermal processes at a reducing atmosphere [186]. The obtained Na-rich $Na_{1.92}Fe_2(CN)_6$ cathode with negligible water content (0.08 H_2O/f.u.) exhibits a high capacity of ~160 mAh/g in the first cycle and exhibits excellent cycling stability with 80% capacity retention even after 750 cycles (Fig. 3.22e,f). This research provides a facile synthetic strategy to obtain high-quality FeHCF crystals and could be extended to other PBAs.

Huang et al. systemically studied the effect of Na content in PBA framework on their electrochemical performance [187]. Moreover, the first-principle calculation was first given to investigate the sodium storage mechanism of Na-rich PBAs. The obtained Na-rich samples acquired better electrochemical performance as compared with the Na-poor samples, and theoretical calculation results indicate that Na^+ ions may intercalate into multiple sites, such as on 8*c* and 24*d*, in the large cavities of $Na_xFeFe(CN)_6$ framework. These findings offer new insights into the sodium storage mechanism for the attractive PB series cathode materials.

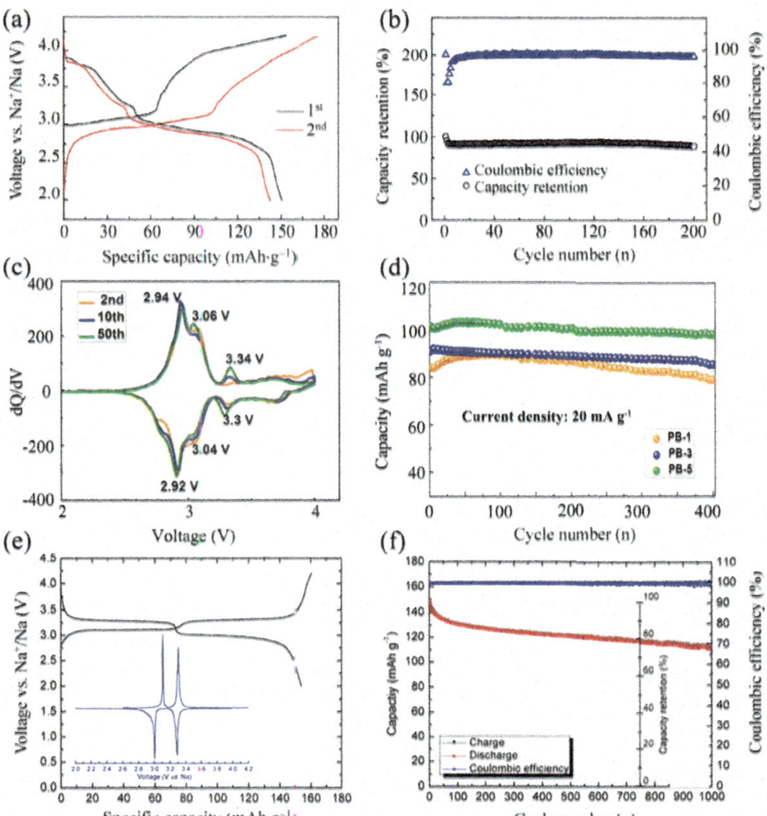

Figure 3.22 The charge/discharge curves, CV profiles, and cycling performance of Na-rich FeHCF. (a,b) $Na_{1.63}FeFe_{0.89}(CN)_6$. Reprinted with permission from Ref. [184], Copyright 2014, Springer. (c,d) $Na_{1.56}Fe[Fe(CN)_6]$. Reprinted with permission from Ref. [185], Copyright 2015, American Chemical Society. (e,f) $Na_{1.92}Fe_2(CN)_6$. Reprinted with permission from Ref. [186], Copyright 2015, American Chemical Society.

3.4.3 Manganese Hexacyanoferrate

MnHCF can be obtained by substituting one-half $Fe^{2+/3+}$ couples with $Mn^{2+/3+}$ couples and shows a similar lattice structure to the FeHCF with 2 Na^+ insertion mechanism. MnHCF is regarded as a promising cathode material for SIBs due to its low material cost, high energy density, and redox plateaus at high voltage. Moritomo et al. first electrochemically synthesized a thin film electrode of MnHCF ($Na_{1.32}Mn[Fe(CN)_6]_{0.83}\cdot 3.5H_2O$) for sodium-ion cathodes [188]. The as-prepared film electrode delivers a high discharge capacity of 109 mAh/g with a high average plateau of 3.4 V and excellent cyclability (90% capacity retention after 100 cycles).

As mentioned earlier, lattice vacancies and water aggravate the electrochemical performance of PBAs. Therefore, it is crucial to prepare MnHCF with low lattice water content and rational crystal structure. Goodenough et al. first reported a dehydrated $Na_{1.89}MnFe_{0.97}(CN)_6$ through drying under vacuum [189]. After removing interstitial H_2O, the monoclinic MnHCF changes to rhombohedral MnHCF (Fig. 3.23a). In contrast, water-free MnHCF shows only a single flat plateau at 3.53 V on charge and 3.44 V on discharge, corresponding to a higher cycle efficiency. Besides, the rhombohedral MnHCF cathode delivers an enhanced capacity of 150 mAh/g, superior rate capability as well as cycling performance (Fig. 3.23b,c).

The Jahn–Teller effect of Mn^{3+} is another problem to overcome for MnHCF cathode materials. It often causes the distortion of the lattice and the dissolution of Mn element into the electrolyte, leading to a poor cycling performance of MnHCF. By disassembling the cycled $Na_xMnFeCN_6$ half-cell, Chou et al. also found that the entire separator became brown, suggesting the dissolution of Mn in the electrolyte. Moreover, the volume changes of MnHCF during sodiation/desodiation processes may also cause capacity fading. Zhao et al. prepared $Na_xK_yMnFe(CN)_6$ ($x + y \leq 2$) grown on porous 3D carbon networks to stabilize the electrode structure [190]. Also, the introduction of K^+ and the unique 3D framework facilitate the reversible insertion/extraction of Na^+ ions, which jointly improved the electrochemical performance of the obtained PBA compounds (Fig. 3.23d,e). This is a good example that demonstrates that the structure and morphology design are two effective ways to overcome the limits of MnHCF.

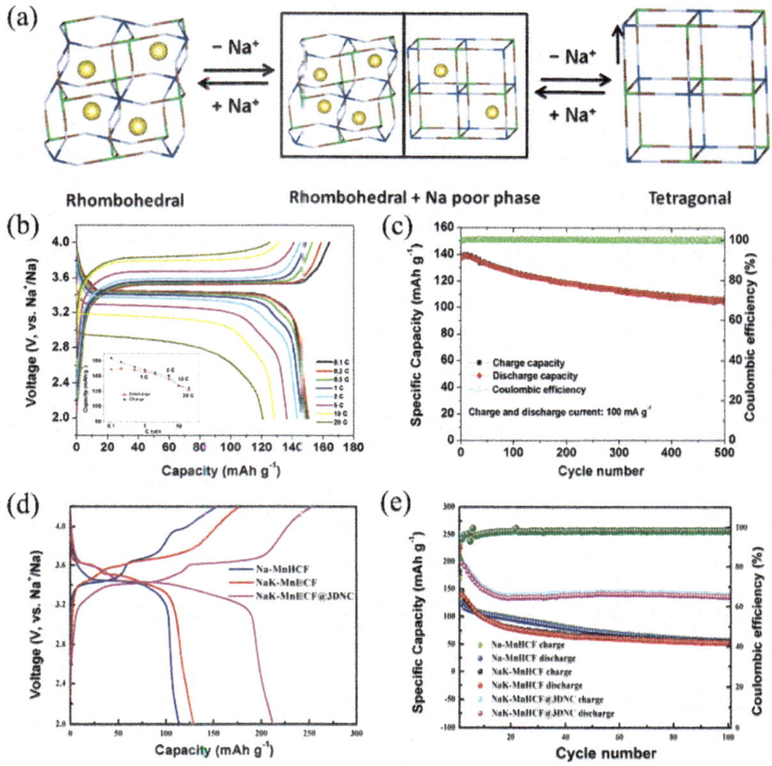

Figure 3.23 (a) Structural evolution of MnHCF without coordinated water. Charge/discharge curves and cycling performance of MnHCF. (b,c) $Na_{1.89}MnFe_{0.97}(CN)_6$. Reprinted with permission from Ref. [189], Copyright 2015, American Chemical Society. (d,c) $Na_xK_yMnFe(CN)_6$. Reprinted with permission from Ref. [190], Copyright 2019, Elsevier.

3.4.4 Cobalt Hexacyanoferrate

Cobalt element is often applied in commercial LIB cathode materials to promote structure stability during the charging/discharging processes. Recently, CoHCF has been studied as a cathode for SIBs, which shows good cycling performance. CoHCF also has two plateaus corresponding to Co^{HS} and Fe^{LS}. Moritomo et al. first synthesized a thin-film CoHCF ($Na_{1.60}Co[Fe(CN)_6]_{0.90}$·$2.9H_2O$) electrode, which exhibited a high capacity of 139 mAh/g with two plateaus at 3.4 and 3.8 V, and a slowly decreased capacity with 70% of the initial value after 100 cycles. Yang et al. prepared low-defect $Na_{1.85}Co[Fe(CN)_6]_{0.99}$ using facile citrate-assisted controlled crystallization method for SIBs [181]. Owing to its high crystallinity and suppressed $Fe(CN)_6$ defects, the as-prepared $Na_2CoFe(CN)_6$ material exhibits highly reversible 2-Na reactions with a high capacity of 150 mAh/g and superior long-term cyclability of ~90% capacity retention over 200 cycles (Fig. 3.24a). Although CoHCF exhibits superior cycling stability, the high cost of Co element limits the practical application of CoHCF cathode materials in commercial SIBs.

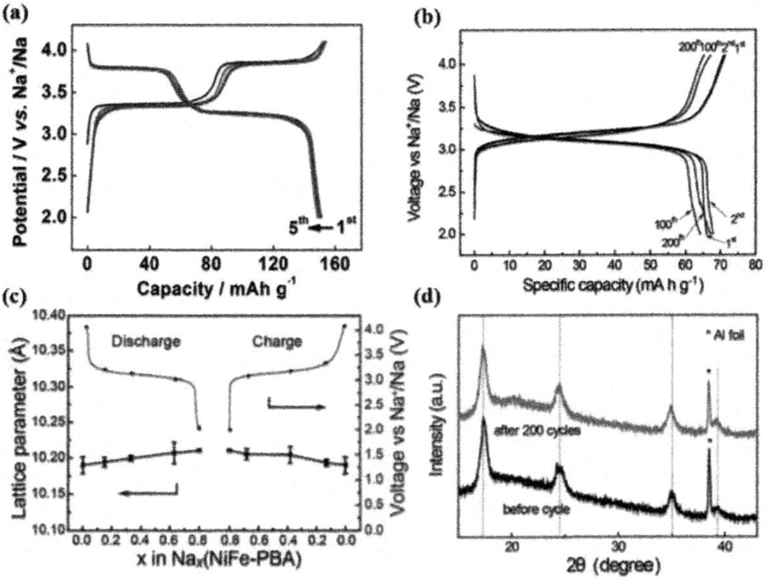

Figure 3.24 (a) Charge/discharge curves of CoHCF Reprinted with permission from Ref. [181], Copyright 2016, American Chemical Society. (b) Charge/discharge curves and (c) the corresponding lattice parameters of NiHCF derived from *ex situ* XRD patterns. (d) XRD patterns of the NiHCF electrode before and after 200 cycles. Reprinted with permission from Ref. [191], Copyright 2013, American Chemical Society.

3.4.5 Nickel Hexacyanoferrate

Compared with PBAs with 2 Na^+ insert mechanism, NiHCF only has one-electron reaction relying on the $[Fe(CN)_6]^{4-}/[Fe(CN)_6]^{3-}$ redox couple (3.07 V versus Na/Na^+) and shows much lower sodium storage capacity (Fig. 3.24b). Nevertheless, owing to the electrochemically inert Ni^{II} in the NiHCF framework, the lattice deformation of NiHCF can almost be neglected during charge/discharge processes, resulting in outstanding structure stability.

Guo et al. prepared NiHCF using a facile wet-chemical synthetic method. The as-prepared $K_{0.09}Ni[Fe(CN)_6]_{0.71}·6H_2O$ is demonstrated as a zero-strain insertion cathode material because the lattice parameters change negligibly (<1%) during Na^+ insertion/extraction as revealed by *ex situ* XRD measurements (Fig. 3.24c,d) [191]. Due to the excellent structural stability of NiHCF, it exhibits superior cycling stability with 99.7% capacity retention upon 200 cycles and high Coulombic efficiency (~100%) when used as a cathode material for room-temperature stationary SIBs. However, the relatively low capacity of NiHCF (~70 mAh/g) owing to the single-electron redox mechanism is hard to satisfy the requirements for practical application.

3.4.6 Other Hexacyanoferrate Compounds

Similar to NiHCF, CuHCF and ZnHCF are also zero-strain cathode materials because of the electrochemically inert Cu and Zn elements, which cause a low reversible capacity of 40~70 mAh/g. Jiao et al. prepared CuHCF cathode with the chemical formula $Cu_3[Fe(CN)_6]_2$ for SIBs [192]. The as-prepared sample shows a specific capacity of 44 mAh/g at a current density of 20 mA/g with 57.1% capacity retention after 50 cycles (Fig. 3.25a,b). The relatively low energy density and poor cycling performance are also attributed to the high content of lattice water. Choi et al. reported a ZnHCF PBA compound and investigated its Na^+ storage performance [193]. Through replacing the Fe^{3+} with Zn^{2+}, the obtained $Na_2Zn_3[Fe(CN)_6]_2$ shows a clear voltage plateau around 3.5 V corresponding Fe^{3+}/Fe^{2+} couple with a capacity of 56.4 mAh/g and good cycle life (Fig. 3.25c,d).

Wu et al. prepared $Na_{0.66}Ti[Fe(CN)_6]_{0.92}$ for SIB cathodes by replacing the transition metal with Ti through an aqueous reaction [194]. Thanks to $Ti^{2+/3+}$ and $[Fe(CN)_6]^{4-/3-}$ couples at 2.6/3.0 V and 3.2/3.4 V, respectively, the as-prepared TiHCF exhibits a relatively high reversible capacity of 92.3 mAh/g at 50 mA/g (Fig. 3.25e,f). Cui et al. first reported manganese hexacyanomanganate ($Na_2Mn^{II}[Mn^{II}(CN)_6]$) as cathode material for SIBs [195]. The as-synthesized MnHCMn has a monoclinic crystal structure composed of nonlinear Mn–N≡C–Mn bonds and contains eight large interstitial sites occupied by Na^+, resulting in a reversible capacity of 209 mAh/g at 0.2 C, which is unprecedented at that time. However, the relatively low charge voltage plateau (1.8 V versus Na^+/Na) significantly decreases the energy density of the final devices. Nevertheless, this novel PBA structure introduces a new sight for developing PBA compounds with improved sodium storage capacity.

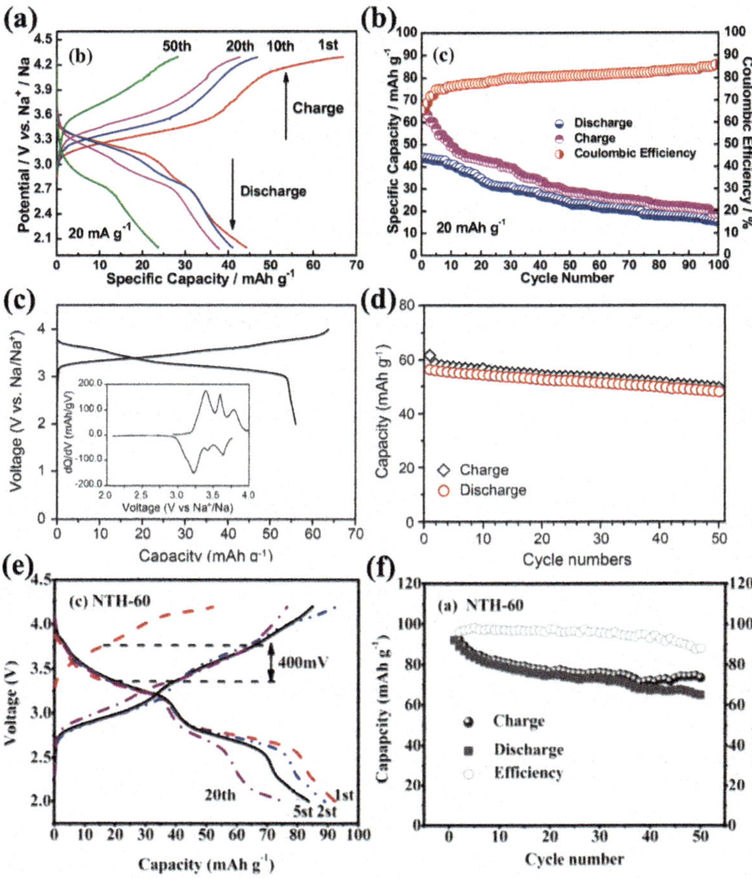

Figure 3.25 Charge/discharge profiles and cycling performance of other PBAs (a,b) CuHCF. Reprinted with permission from Ref. [192], Copyright 2017, Elsevier. (c,d) ZnHCF. Reprinted with permission from Ref. [193], Copyright 2012, Royal Chemical Society. (e,f) TiHCF. Reprinted with permission from Ref. [194], Copyright 2016, Elsevier.

3.4.7 Structural and Morphological Optimizations of PBAs

In summary, PBAs show promise as cathode materials for SIBs owing to their low cost, facile synthetic methods, environment-friendly properties, and high theoretical capacities. However, there are many obstacles impeding their practical applications, such as poor conductivity, unsatisfied structure stability, and low energy density caused by lattice water. Accordingly, various strategies are employed to optimize nanostructure and morphology to overcome these challenges and improve the electrochemical performance of PBAs.

Compositing PBAs with carbon-based materials is an effective way to improve the electrochemical performance of electrode materials due to remarkable electronic conductivity and low cost of carbon. Ma et al. synthesized reduced graphene oxide–Prussian blue composites (RGOPCs) without coordinated water for SIB cathodes [196]. After heat treatment of a graphene oxide–PB composite (GOPC), coordinated water was removed, and the obtained RGO significantly improved the electrical conductivity and structural stability of electrodes (Fig. 3.26a). Therefore, the as-prepared RGOPCs exhibit a high initial capacity of 149.7 mAh/g at 200 mA/g with 91.9% capacity retention after 500 cycles. Even at a high current density of 800 mA/g, the RGOPC electrode still achieves a high capacity of 112 mAh/g.

Dou et al. *in situ* prepared FeHCF@carbon (PB@C) nanocomposite using Ketjen black as a carbon source [197]. The nanosized PB was well dispersed on the surface of the carbon matrix and formed a 3D conductive network. The unique structure shortens the diffusion pathways of Na^+ and significantly improves the rate capability of PB materials. Moreover, PB crystals with low lattice defects can promote structure stability and Na^+ storage sites. Therefore, the PB@C composite shows an outstanding rate capability (77.5 mAh/g at 90 C) and excellent cycling stability (90% capacity retention at 20 C over 2000 cycles) (Fig. 3.26b).

Guo et al. prepared FeHCF/CNT composites with PB particles nucleated on CNT fibers, forming a continuous conductive network and contributing to fast electronic transport (Fig. 3.26c) [198]. The FeHCF/CNT cathodes have a specific capacity of 167 mAh/g at 25°C.

It is worth noting that the FeHCF/CNT cathode can still deliver a capacity of 142 mAh/g even at −25°C (0.1 C) and a negligible capacity loss and voltage polarization increase after 100 cycles.

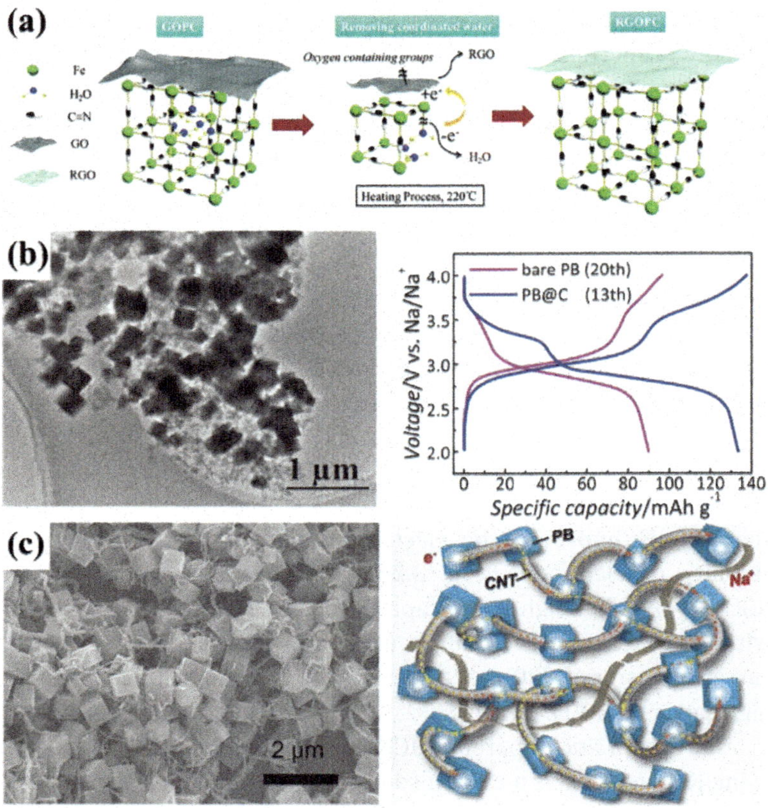

Figure 3.26 (a) Schematic illustration for preparing RGOPCs without coordinated water. Reprinted with permission from Ref. [196], Copyright 2015, Royal Chemical Society. (b) Morphology and charge/discharge curves of FeHCF@carbon. Reprinted with permission from Ref. [197], Copyright 2016, Wiley-VCH. (c) Morphology of PB/CNT and corresponding schematic illustration of the mixed conducting network. Reprinted with permission from Ref. [198], Copyright 2016, Wiley-VCH.

Conducting polymer coating is another effective way to enhance the surface conductivity of PBAs and buffer volume charge of electrode materials during Na^+ insertion/extraction. Chou et al. developed a soft chemistry approach to synthesize ClO_4^--doped polypyrrole-coated $Na_{1+x}MnFe(CN)_6$ composite as a cathode material (MnHCF@PPy) for SIBs (Fig. 3.27a) [199]. Beyond improving the electronic conductivity of MnHCF, the PPy also acts as a protective layer to reduce the dissolution of Mn in the electrolyte. Besides, the PPy doped with ClO_4^- can act as active materials to increase the capacity of the composite. As a result, the MnHCF@PPy electrode shows high energy density (428 Wh/kg), enhanced cycling performance (67% capacity retention after 200 cycles), and excellent rate capacity (46% capacity for 40 C rate) in SIBs. Another popular conductive polymer, polyaniline (PANI), was demonstrated to be effective for modifying the PB cathodes and enhancing battery performance as well (Fig. 3.27b) [200].

In addition to compositing with conductive materials, morphological engineering is also employed to improve the electrochemical performance of PBA electrodes. Chen et al. synthesized border-rich FeHCF through the inhibitor and temperature control [201]. This unique structure provides a large contact area between active materials and electrolyte, hence shortening Na^+ diffusion pathways (Fig. 3.27c). Similarly, Wang et al. prepared FeHCF hollow spheres as sodium cathode, which provides an increasing number of active sites and high diffusion dynamics for sodium ions (Fig. 3.27d) [202]. Both cathode materials delivered enhanced specific capacities, superior rate capability, and cycling stability.

Additionally, heteroatomic doping could optimize the nanostructure of PBAs and stabilize the PB structure, which leads to better long-term cycling performance. For example, Ni doping is considered an effective way to stabilize the framework of PBAs owing to the electrochemically inert Ni element for sodium-ion storage. Ma et al. reported an Ni-doped PBA compound with the formula $Na_{1.76}Ni_{0.12}Mn_{0.88}[Fe(CN)_6]_{0.98}$ (NiMnHCF) for sodium cathode [203]. During insertion/extraction of Na^+, the Ni sites were unreactive and could balance the tiny structural disturbances caused by the redox reactions of Mn sites. Thus, the NiMnHCF electrode delivers excellent cycling stability with 83.8% capacity retention after 800 cycles at 100 mA/g. Wu et al. prepared Ni-doped CoFeHCF and investigated the influence of Ni content on the electrochemical performance of PBA compounds [204]. The results show that

$Na_{1.67}Ni_{0.39}Co_{0.61}Fe(CN)_6$ has the best electrochemical performance with a high initial discharge capacity of 91 mAh/g and good cycling stability (85.9% capacity retention after 100 cycles).

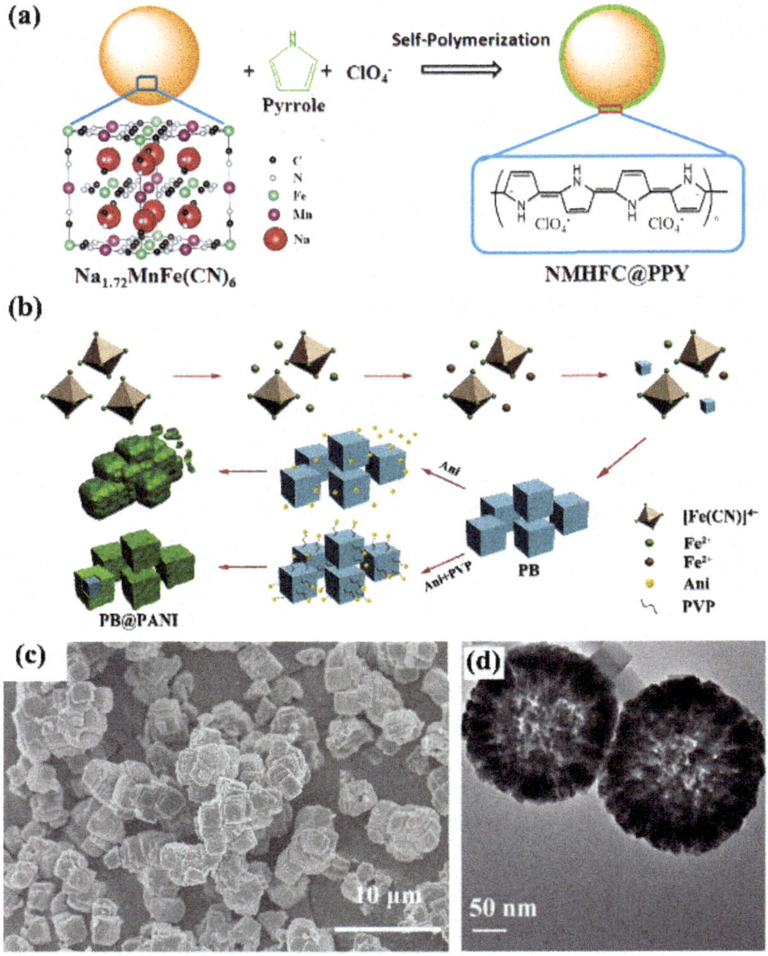

Figure 3.27 Schematic illustration for preparing PBAs/conducting polymer composites. (a) MnHCF@PPy. Reprinted with permission from Ref. [199], Copyright 2015, Elsevier. (b) PB@PANI. Reprinted with permission from Ref. [200], Copyright 2018, Elsevier. Morphologies of various nanostructured PBAs. (c) Border-rich FeHCF. Reprinted with permission from Ref. [201], Copyright 2017, Elsevier. (d) FeHCF hollow spheres. Reprinted with permission from Ref. [202], Copyright 2018, Springer.

In summary, we introduced the recent progress of PBAs as cathodes for SIBs. Although PBAs have revealed remarkable electrochemical performance as cathode for SIBs in some reports, unsatisfied gravimetric densities and limited electrochemical activities of PBAs hinder their further application in consumer electronics and electric vehicles.

3.5 Conversion-Based Cathode Materials

Conversion-based materials, such as metal halide and chalcogenide compounds, have attracted increasing attention owing to their higher theoretical energy densities for sodium storage than those of the aforementioned intercalation-based compounds. The high electronegativities of the F, S, and Se elements endow the corresponding cathodes with high redox potentials and enable more Na^+ to participate in the conversion reaction during the electrochemical reaction result in high capacities. Additionally, the relatively lower atomic mass of S or Se elements than those of the metal compound cathodes benefits the gravimetric capacity of SIBs [205]. However, the poor ionic and electronic conductivities, serious volume changes during sodiation and desodiation, as well as the low voltage plateaus (≈2 V versus Na^+/Na) of the conversion-based electrode materials hinder their further development in practical SIBs. Generally, nanoengineering of the active materials, combining them with the conductive matrix, and optimizing electrolyte configuration are three effective ways to alleviate the challenges mentioned earlier. In the following section, we introduce the advances of several kinds of conversion-type cathode materials, including fluorides, oxyfluorides, metal sulfides and selenides, and metal chlorides.

3.5.1 Metal Fluorides

Nishijima et al. initially reported perovskite-type transition metal fluorides (MF_3, M = Ti, V, Mn, and Co) for sodium-ion cathodes in 2009 [206]. The excellent electronegativity of fluorine endows the compounds with high voltages. Moreover, the fluoride compounds possess high theoretical gravimetric energy density relying on both intercalation reaction and conversion reaction. For example, the ball-milled FeF_3-C composite cathode obtained a discharge capacity of ~145 mAh/g at 0.2 mA/cm^2 for SIBs. The VF_3-C achieved a similar capacity at the same test conditions. However, the TiF_3-C, MnF_3-C, and CoF_3-C electrodes delivered much lower reversible capacities, which might be attributed to their high average voltage plateaus and the insufficient charge cutoff voltages in the measurement for preventing the electrolyte decomposition.

Besides the acetylene black that is normally used in the aforementioned work, other carbonaceous materials were also employed to ameliorate the electrical conductivities of metal fluoride electrodes, including graphene, CNTs, and amorphous carbon [207–213]. Although the incorporation of conductive materials can improve the electrochemical performance of metal fluoride electrodes to some extent, the slow Na-ion diffusion kinetics still hinders the full utilization of the potential of the cathodes.

He et al. initially studied the sodiation mechanism of FeF_2 nanoparticles via both *in situ* and *ex situ* TEM techniques [214]. The study demonstrated that the FeF_2 located on the particle surface took part in the conversion reaction and generated the NaF and Fe, while the rest of the FeF_2 located in the particle core transformed into Na_3FeF_6 and Fe via disproportionation reaction. The Fe crystals can be further fused into an interconnected network. As a result, a core-shell-structured Na_3FeF_6/NaF mixture phase was generated after sodiation (Fig. 3.28a,b).

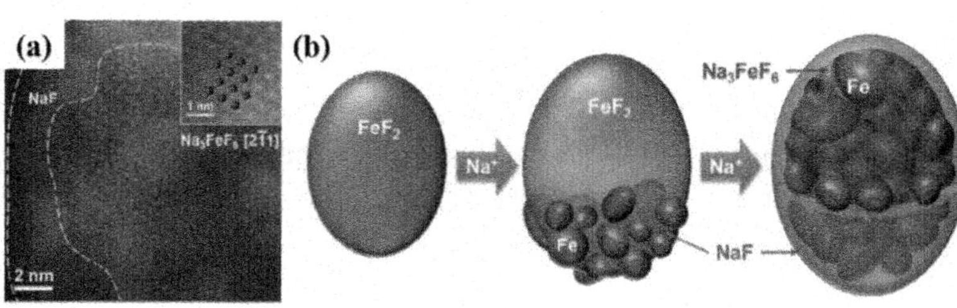

Figure 3.28 High-resolution STEM images showing (a) crystalline Na_3FeF_6 phase in the core of the sodiated particle, and the insets show close-up views matching with lattice models of FeF_2 (111) and Na_3FeF_6 (211). (b) Schematic illustration of sodiation by a heterogeneous reaction mechanism. Reprinted with permission from Ref. [214], Copyright 2014, American Chemical Society.

Li et al. constructed an open-framework $FeF_3{\cdot}0.33H_2O$ nanoparticle (~10 nm) with large 1D Na channels through a simple top-down approach [215]. During the synthesis process, a solid–solid phase transformation from $FeF_3 \cdot 3H_2O$ to $FeF_3 \cdot 0.33H_2O$ was realized by using 1-butyl-3-methylimidazolium tetrafluoroborate ($BmimBF_4$) ionic liquid (IL) as reaction medium and soft template (Fig. 3.29a). The single-wall carbon nanotubes (SWCNTs) wired $FeF_3 \cdot 0.33H_2O$ were also synthesized via adding SWCNTs in the IL in advance. The hybrid electrode with good conductivity and porous structure delivers an initial discharge capacity of 130 mAh/g at 0.1 C and retains a capacity of 74 mAh/g after 50 cycles (Fig. 3.29b,c).

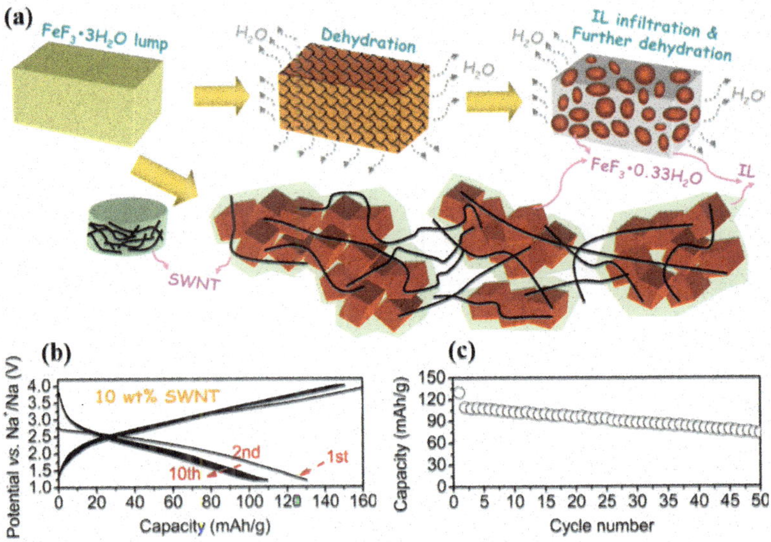

Figure 3.29 Scheme of the top-down fabrication process through IL-alleviating multistep dehydration, as well as typical SWNT-$FeF_3 \cdot 0.33H_2O$ mixed conductive network. (b) Voltage versus capacity profiles and (c) cycling performance of the $FeF_3 \cdot 0.33H_2O$ electrode wired by 10 wt% SWNTs at 0.1 C for Na storage. Reprinted with permission from Ref. [215], Copyright 2013, American Chemical Society.

Subsequently, this group reported another open-framework $FeF_3 \cdot 0.5H_2O$ with 3D Na channels via a similar IL-based precipitation strategy, where the different sizes of IL cations can remarkably affect the crystallinity, channel size, and surface defect concentration of the product [216]. A larger cavity in the pyrochlore-phase FeF_3 was achieved by choosing the 1-decyl-3-methylimidazolium tetrafluoroborate ($C_{10}mimBF_4$) as a structure-directing agent rather than the $BmimBF_4$ IL, which can be attributed to the larger $C_{10}mim^+$ moiety. In the charge/discharge profiles of the as-prepared $FeF_3{\cdot}0.5H_2O$ wired by 5 wt% SWNTs electrode, two voltage steps can be detected, corresponding to the intercalation reaction (1.2~4.0 V) and conversion reaction (0.8~1.2 V), respectively. Furthermore, an ultrahigh capacity of ~250 mAh/g was achieved in the voltage range of 0.8~4.0 V, but the capacity dropped rapidly along with cycling. Relatively stable cycling was observed when the cutoff voltage increased to 1.2 V (Fig. 3.30a,b).

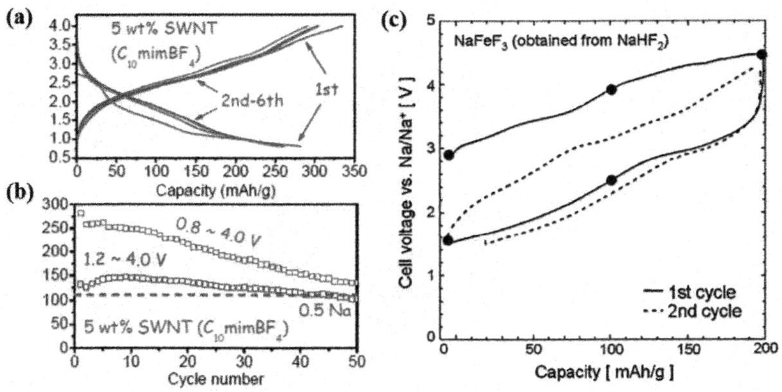

Figure 3.30 (a) Voltage versus capacity profiles and (b) cycling performance of the $FeF_3{\cdot}0.5H_2O$ cathodes (from $C_{10}mimBF_4$) wired by 5 wt% SWNTs at 0.1 C; A theoretically reversible capacity referring to 0.5 Na reaction is also labeled in dot line as a comparison. Reprinted with permission from Ref. [216], Copyright 2013, American Chemical Society. (c) First and second charge and discharge curves of $NaFeF_3$ (obtained from $NaHF_2$) at the rate of 0.076 mA/cm^2 between 1.5 and 4.5 V versus Na/Na^+. Reprinted with permission from Ref. [221], Copyright 2012, Elsevier.

Although some progress has been achieved to develop metal fluoride electrodes, the lack of Na in the fluoride compounds impedes them directly using as cathodes for full SIBs [217]. Thereby the $NaMF_3$ (M = Fe, Mn, and Ni) were fabricated by a mechanical ball-milling method with stoichiometric amounts of NaF and MF_2 as precursors [218, 219]. The as-prepared $NaFeF_3$ electrode behaved with an initial reversible capacity of 128 mAh/g at 0.2 mA/cm^2. By contrast, the $NaMnF_3$ and $NaNiF_3$ electrodes just delivered a capacity of ~40 mAh/g, which might be ascribed to the insulating characteristics and chemical instability of the corresponding charged phases. Nanoengineering and improving the crystallinity of $NaFeF_3$ are effective ways to alleviate these problems. The high-crystalline $NaFeF_3$ electrode delivered a much higher capacity of 197 mAh/g at 0.076 mA/cm^2 (Fig. 3.30c) [220, 221].

Beyond the $NaMF_3$ formula, a sodium iron fluoride compound Na_3FeF_6 fabricated by the ball-milling method was also reported for sodium cathodes, which delivered a reversible capacity of ~111 mAh/g in the voltage range of 0.5~4.5 V. NaF and Fe phases could be detected in the fully discharged electrode, suggesting the completion of the conversion reaction [222]. Besides, a series of manganese fluoride compounds, $NaMnF_3$, Na_2MnF_5, and $NaMnF_4$, have been investigated for sodium storage as well. The latter two compounds are less stable than the $NaMnF_3$, corresponding to lower capacities and poor cycling performance. Also, the insulating characteristics of manganese fluoride compounds lead to slow kinetics and large voltage polarization, thereby impeding their further development [223].

3.5.2 Carbon Fluorides

Sodium carbon fluoride (Na–CF_x) with high theoretical capacity (1410 mAh/g) and abundant natural resources of the raw materials has attracted great attention of researchers. Liu et al. first demonstrated the reversible formation and decomposition of NaF in the Na–CF_x batteries by *ex situ* XPS characterization [224]. Compared with the Li–CF_x cell, Na–CF_x shows a 10 times higher reversible capacity of 940 mAh/g (Fig. 3.31a). This can be attributed to the lower electromotive force needed to decompose NaF (~5.65 eV) than LiF (~6.1 eV), as well as the nanosize (~70 nm) of the crystalline NaF, formed on carbon during the discharge process. Shao et al. revealed a reversible reaction occurring between CF_x and NaF through soft X-ray absorption spectroscopy (XAS) and nuclear magnetic resonance (NMR) analyses [225].

Despite that the Na–CF_x chemistry is reversible in SIBs, a noticeable polarization between charge and discharge can be observed. The addition of conductive materials or structural engineering can improve the electrochemical activities of Na–CF_x batteries. For instance, a fluorinated MWCNT and a composite electrode consisting of CF_x, graphene, and FeF_3 were both demonstrated to have a lower voltage hysteresis and better cycle performance than that of the CF_x electrodes [226]. Besides, constructing highly disordered CF_x, compositing CF_x with graphene, mesoporous carbon, or polymers can also improve the trip-efficiency and cycling performance of batteries [227–229].

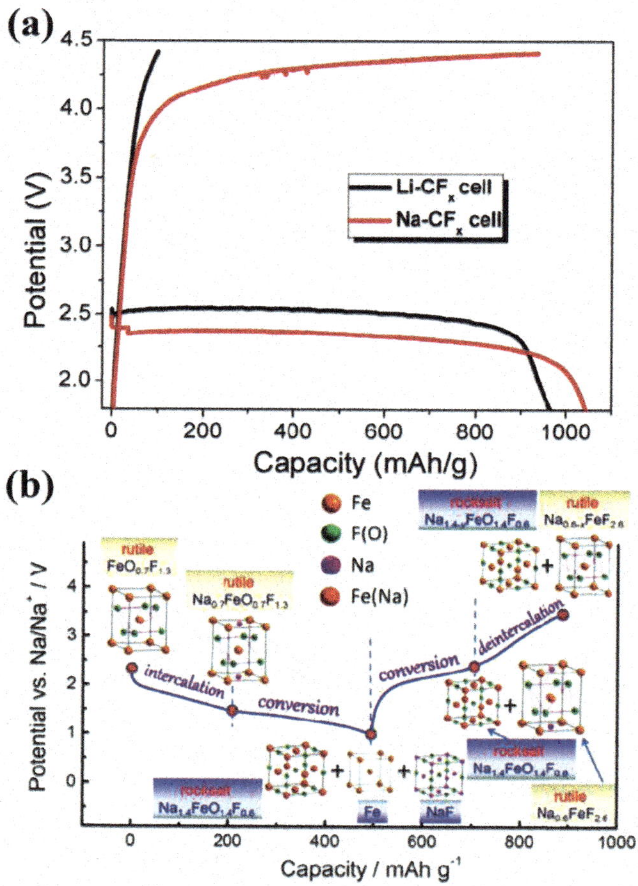

Figure 3.31 (a) First charge/discharge curves of Na–CF_x and Li–CF_x cells at 100 mAh/g. Reprinted with permission from Ref. [224], Copyright 2014, American Chemical Society. (b) Schematic of de/sodiation mechanism of $FeO_{0.7}F_{1.3}$/C during first cycle. Reprinted with permission from Ref. [230], Copyright 2015, Wiley-VCH.

3.5.3 Oxyfluorides

Incorporating a small amount of oxygen into the fluoride structure is expected to increase the conductivity and reversibility of the conversion reaction with sodium. Therefore, metal oxyfluoride compounds were synthesized and investigated as cathodes for SIBs [231, 232]. Nam et al. prepared an $FeO_{0.7}F_{1.3}/C$ composite via a solution-based process, and the hybrid cathode behaved an impressive capacity of ~360 mAh/g after 50 cycles in the voltage range of 1.0~3.5 V [230]. Figure 3.31b clearly demonstrates the first charge/discharge profiles and the corresponding reaction mechanism of the $FeO_{0.7}F_{1.3}/C$ electrode. The discharge products are identified as NaF, Fe, and rock salt phase $Na_{1.4}FeO_{1.4}F_{0.6}$. The recharging process looks reversible to a great extent, but with two different charge products. The competition between rock salt and rutile components during the charge might benefit the cyclability of the nanocomposite electrodes. Deng et al. fabricated FeOF nanorods by a solvothermal method using alcohol and $FeF_3 \cdot 3H_2O$ as starting reagents [231]. The nanorods assembly kept a specific capacity of 250 mAh/g for 20 cycles at 10 mA/g, indicating that nanoengineering is efficient for enhancing the electrochemical performance of conversion electrode materials.

3.5.4 Metal Sulfides

Sulfur (S) has been extensively investigated as electrodes for rechargeable batteries due to its high energy density and low cost. The high-temperature Na–S batteries have already been commercialized and applied in grid electricity storage [233]. However, the high chemical reactivity and corrosivity of molten sodium and sulfur cause severe safety concerns. Room-temperature Na–S batteries had become hot research objects in the past decades [234]. A more detailed description of the advances and challenges of Na–S batteries can be found in Ref. [235]. A massive amount of attention has been paid to this field, including many critical reviews, so we will not introduce them in detail. Herein, we mainly focus on the progress of metal sulfide cathode materials for SIBs.

Thanks to the weak transition metal–sulfur bonds, metal sulfides exhibit high theoretical capacities for SIBs owing to the intercalation-conversion reaction. Among the various sulfide candidates, iron sulfides (Fe_xS_y) stand out because of the natural abundance of Fe ore [236]. The reaction mechanism of the representative FeS_2 has been studied by first-principle calculations [237]. As displayed in Fig. 3.32, FeS_2 presents a theoretical capacity of 896 mAh/g, corresponding to a typical two-step electrochemical reaction listed as follow:

$$x\text{Na} + \text{FeS}_2 \rightarrow \text{Na}_x\text{FeS}_2 \quad (3.1)$$

$$(4-x)\ \text{Na} + \text{Na}_x\text{FeS}_2 \rightarrow 2\text{Na}_2\text{S} + \text{Fe} \quad (3.2)$$

It can be seen that 1.5 Na can be intercalated into FeS_2 at an average voltage of 1.37 V, and the conversion reaction occurs subsequently at an average voltage of 1.11 V.

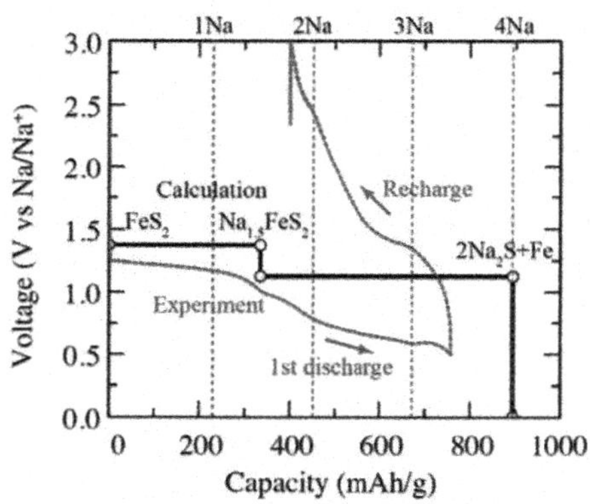

Figure 3.32 Theoretical voltage–capacity curve of Na/FeS_2 battery calculated using a two-step reaction formula (black solid line) and experimental curve of first discharge and recharge processes (red dotted line). Reprinted with permission from Ref. [237], Copyright 2015, The Physical Society of Japan.

As early as 2007, Ahn et al. initially employed a natural FeS_2 (pyrite) as cathode materials for SIBs [238]. Although a high capacity of 630 mAh/g with a voltage plateau of 1.3 V was detected in the first discharge process, the discharge plateau disappeared and converted into a slope line along with a severe capacity decay in the following cycles. Again, nanoengineering and carbon coating are two conventional ways to improve the electrochemical performance of the iron sulfide electrodes. For example, Pint et al. synthesized ultrafine FeS_2 nanoparticles (~4.5 nm), which benefit to sustain a reversible conversion reaction in SIBs [239]. This can be attributed to the nanoparticle size comparable to the Fe diffusion length, which forms thermodynamically reversible nanodomains of Fe and Na_xS (Fig. 3.33a). Thereby the ultrafine FeS_2 delivered a higher reversible capacity and better cycling performance than that of the bulk FeS_2.

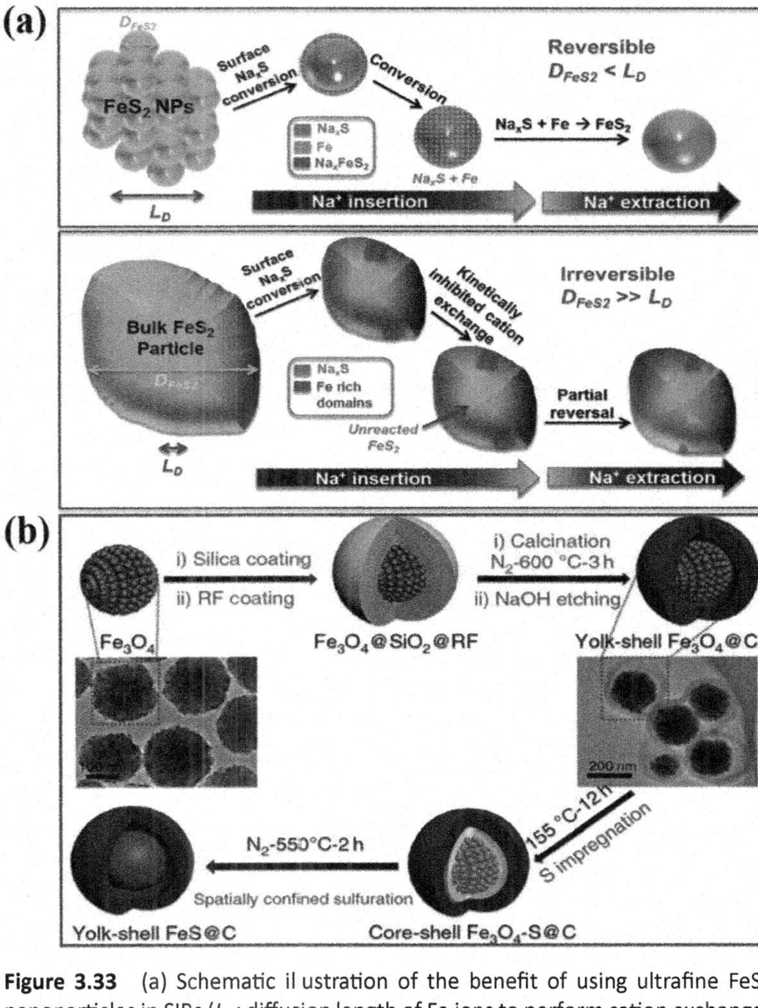

Figure 3.33 (a) Schematic il ustration of the benefit of using ultrafine FeS_2 nanoparticles in SIBs (L_D: diffusion length of Fe ions to perform cation exchange; D represents the diameter of the materials). Reprinted with permission from Ref. [239], Copyright 2015, American Chemical Society. (b) Illustration of the synthesis of yolk-shell FeS@C. Reprinted with permission from Ref. [240], Copyright 2015, Springer Nature.

Wang et al. prepared uniform yolk-shell FeS-carbon nanospheres, in which FeS yolk (170 nm) was encapsulated by carbon shell (30 nm thick) with a gap (~20 nm) between the yolk and shell (Fig. 3.33b) [240]. The as-prepared electrode delivered an average capacity of 621 mAh/g at 0.2 C and achieved 60% of the capacity at 5 C. Herein, the carbon coating not only enhances the electrical conductivity of the whole electrode and stabilizes the electrode structure but also suppresses the dissolution of sodium sulfides in the electrolyte and extends the cycle life of batteries.

Beyond structure engineering and carbon coating, optimizing electrolytes is another strategy to improve the electrochemical performance of iron sulfides for SIBs. It is demonstrated that the Na/FeS_2 cell can behave much more stable cyclability in ether-based electrolytes like $NaSO_3CF_3$/diglyme than in the carbonate-based electrolytes, such as $NaSO_3CF_3$/propylene carbonate and $NaSO_3CF_3$/ ethylene carbonate-diethylene carbonate [241, 242]. However, the relevant mechanism is yet to be disclosed clearly.

Except for iron sulfides, cobalt sulfides, copper sulfides, and nickel sulfides with high theoretical capacity have also been extensively investigated as electrode materials for SIBs [243]. They face similar issues of the conversion electrode materials, such as intrinsically low electronic and ionic conductivity, large volume expansion, and continuous side reactions between electrode materials and electrolytes. The mentioned strategies for modifying FeS_2 are all suitable for constructing high-performance metal sulfides materials [244, 245]. It should be noted that the excessive incorporation of carbonaceous material would lower the average voltage of the hybrid electrode, which is the reason why some of the sulfide materials are applied as anode materials in rechargeable batteries [246].

3.5.5 Metal Selenides

Compared to S, selenium (Se) has similar chemical properties and comparable volumetric capacities but higher electronic conductivity, which may lead to faster reaction kinetics. Metal selenides share similar chemistries with the sulfides introduced before; however, the relevant research is less active compared to the sulfide compounds. This might be ascribed to the higher molar weight and lower theoretical capacities of the metal selenides compared to the corresponding sulfide counterparts [247, 248]. Fu et al. synthesized Cu_2Se cathode materials for SIBs, which behaved a capacity of 250 mAh/g with a high voltage of 2.0 V [249]. *Ex situ* TEM and XAS analyses unraveled that the Cu_2Se transformed to Cu and Na_2Se after discharge. Besides, FeSe thin film synthesized via radio-frequency magnetron sputtering exhibited a reversible capacity of 300 mAh/g with a redox reaction occurring at 2.0 V [250]. Owing to the high reversibility of conversion reaction, the FeSe electrode also achieved a high capacity of 253 mAh/g after 200 cycles.

3.5.6 Other Conversion Cathode Materials

In addition to the prevailing metal fluorides and metal chalcogenides, metal chloride compounds such as CuCl, $CuCl_2$, and layered vanadium oxychloride (VOCl) were also evaluated as conversion-based electrodes for SIBs [251]. The CuCl electrode shows good reversibility during the conversion reaction, which transforms to NaCl and metallic Cu during the discharge reaction. As for the $CuCl_2$, which initially converts into CuCl and NaCl, and then transforms to NaCl and Cu. Both CuCl and $CuCl_2$ electrodes exhibit large discharge capacities, whereas serious capacity decay was observed during cycling due to the soluble Cl [252]. Polyanionic compounds $Cu_3(PO_4)_2$ were also investigated as a conversion cathode for SIBs. During the first discharge, the $Cu_3(PO_4)_2$ fully transforms into Cu and Na_3PO_4, but the Cu particles cannot be fully oxidized to Cu^{2+} upon charging [253].

In summary, conversion-based materials have higher theoretical capacities than the intercalation compounds, which potentially enable high-performance SIBs. However, they exhibit strong limitations, including large volume expansion that leads to poor cycling performance, low electronic and ionic conductivity, which causes the large overpotential during charging/discharging, and side reactions between active materials and electrolytes. Although several strategies have been under investigation to overcome these challenges, the development of conversion-based cathodes is still at the early stage of development.

3.6 Organic Cathode Materials

Organic materials have been considered promising alternative cathode candidates for SIBs. Compared to the inorganic materials mentioned earlier, organic materials are more competitive in the following aspects: (1) The production cost is low since most of the organic electrode materials consist of four abundant elements (C, N, H, O) and their synthetic processes are relatively simple. (2) Their structural versatility and high designability enable the possibility of the desired high-performance electrode materials. (3) The good mechanical flexibility can lead to bendable, lightweight, and portable SIBs for wearable electronic devices. (4) The organic electrode materials are recyclable and environmentally begin. So far, a wide catalog of organic materials is under research as cathode materials for SIBs, such as carbonyl compounds, organosulfur compounds, radical organic compounds, and conducting polymers. Generally, their redox mechanisms can be classified into anion-insertion type (p-type) and cation-insertion type (n-type). The p-type organic materials will be oxidized and then reacted with anions in the electrolyte during discharging, irrespective of the nature of cations. On the contrary, the n-type organic materials will be reduced and combined with Na^+. Most of the organic electrode materials belong to one of the two categories, but conductive polymers are an exception that can be either reduced or oxidized, depending on the applied voltage.

3.6.1 Carbonyl Compounds (C=O Reaction)

Currently, the carbonyl compounds are the most studied family of organic electrodes in SIBs due to their high reactivity, high capacity, and rich structural diversity. The reaction mechanism can be summarized in Eq. 3.3. According to the type of functional groups, carbonyl compounds can be categorized into three types: quinones and ketones, anhydrides, and imides. Thanks to the large structural variety of carbonyl materials, their redox behavior is able to be tuned by regulating the inductive and resonance effect. However, the main drawbacks of this type of materials are the dissolution of the cathode in the electrolyte, which leads to rapid capacity fade; their low electronic conductivity that causes poor rate performance; and low redox potentials and tap densities of some electrodes that would negatively influence the energy density of devices. It should be noted that a large number of carbonyl compounds can be employed as anode materials for SIBs as well, which will not be presented in the following discussion.

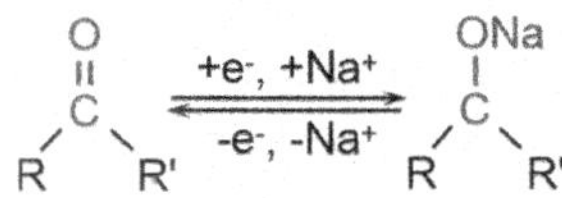

(3.3)

3.6.1.1 Quinones and ketones

Quinones are fully conjugated cyclic dicarbonyl structures that are formally derived from aromatic compounds by conversion of an even number of –CH= groups into –C(=O)– groups with any necessary rearrangement of double bonds. The archetypical member of the class is 1,4-benzoquinone, often simply called quinone ($C_6H_4O_2$). Some of the representative quinone structures for sodium storage are displayed in Fig. 3.34, and incorporating quinonyl into a polymer moiety is an effective approach to address its dissolution issues. Except for the high solubility, the main drawback of this class of materials is their relatively low redox potential as cathodes (<2 V), which is detrimental to high-energy-density SIBs (No. 1–4). Most of them were applied as anodes instead of cathodes for SIBs. For example, a disodium salt of 2,5-dihydroxyl-1,4-benzoquinone (Na_2DBQ) with two carbonyl groups can uptake two Na ions at an average voltage of 1.2 V during the discharge process [254].

The introduction of electronegative elements into the quinone structures can dramatically increase the sodium storage potential (No. 5–6). Kim et al. revealed that when the halogens substitute hydrogen atoms, the redox potential of benzoquinone follows this trend: $C_6F_4O_2 > C_6Cl_4O_2 > C_6Br_4O_2 > C_6H_4O_2$ [255]. And the rationally designed $C_6Cl_4O_2$ cathode presents an average voltage of 2.72 V, a capacity of 161 mAh/g, and an energy density of 420 Wh/kg ($C_6Cl_4O_2$/Na), which outperforms most of the organic and inorganic cathode materials.

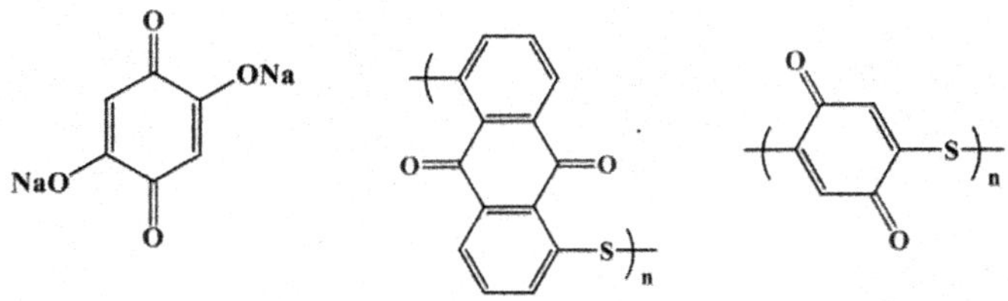

1. Na_2DBQ 2. PAQS 3. PBQS

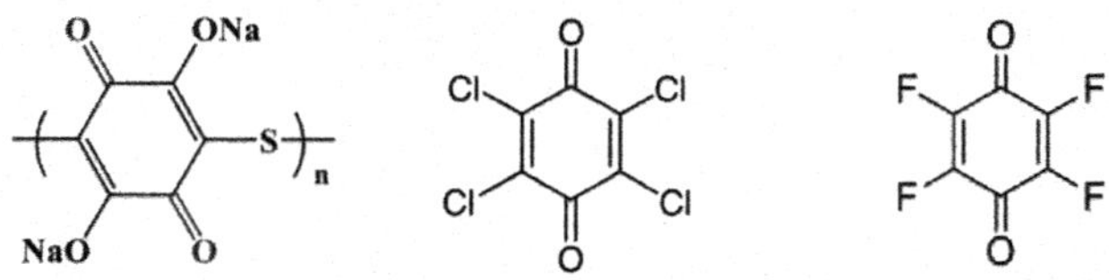

4. Na_2PDS 5. $C_6Cl_4O_2$ 6. $C_6F_4O_2$

Figure 3.34 Structures of quinones as electrode materials for SIBs. PAQS, poly(anthraquinonyl sulfide); PBQS, poly(benzoquinonyl sulfide); Na_2PDS, poly(2,5-dihydroxynp-benzoquinonyl sulfide).

Ketones share a structure that two carbon-containing substituents are directly attached to the carbonyl center. This catalog of materials can be reduced to the corresponding alcohols and reoxidized to the original state in the electrochemical reaction. Disodium rhodizonate (DSR, No. 7, $Na_2C_6O_6$, Fig. 3.35) is a representative cathode material of this family, which was first reported by Okada et al. in 2013 [256]. It shows an initial reversible capacity of 250 mAh/g at 18 mA/g in a 1 M $NaClO_4$/PC electrolyte, with an average voltage of 2.1 V. However, the cycling performance of DSR is limited to dozens of cycles due to the large volume change during discharging and charging.

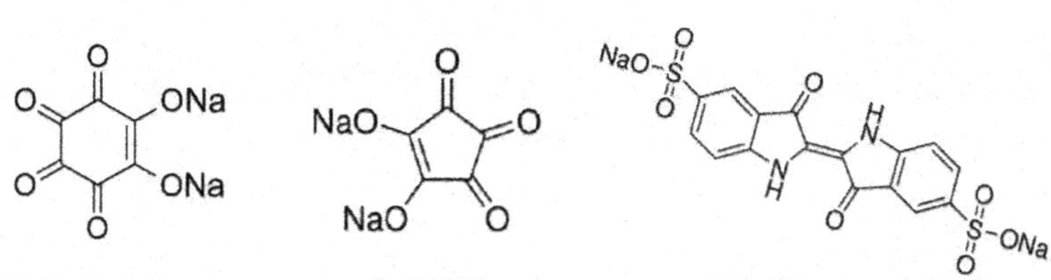

7. $Na_2C_6O_6$ 8. CADS 9. Indigo carmine

Figure 3.35 Structures of ketones as electrode materials for SIBs; CADS, croconic acid disodium salt.

Rational structure control and downsizing the particle size to the nanoscale can significantly improve its electrochemical performance. Yu et al. used a facile anti-solvent method to prepare microbulk, microrod, and nanorod DSR materials, which show strong size-dependent sodium storage properties (Fig. 3.36a–c) [257]. The DSR nanorods exhibit the highest capacity of ~190 mAh/g at 0.1 C and the best rate capability with 95 mAh/g at 10 C (Fig. 3.36d). DFT calculation results display that the DSR may undergo the sequential one-electron reduction process at about 2.4 V, 2.1 V, and 1.9 V, or the two-electron process at potential about 2.26 V and 2.05 V, which are in agreement with the stepwise plateaus. The positive impacts of morphological control on electrochemical performance were also demonstrated by Lei et al. [258]. They found that the flaky morphology of DSR facilitated the ionic diffusion and thereby benefited the fast-charge and fast-discharge properties of batteries. These results disclose that nanoengineering is a universal approach for designing high-performance organic electrode materials for SIBs.

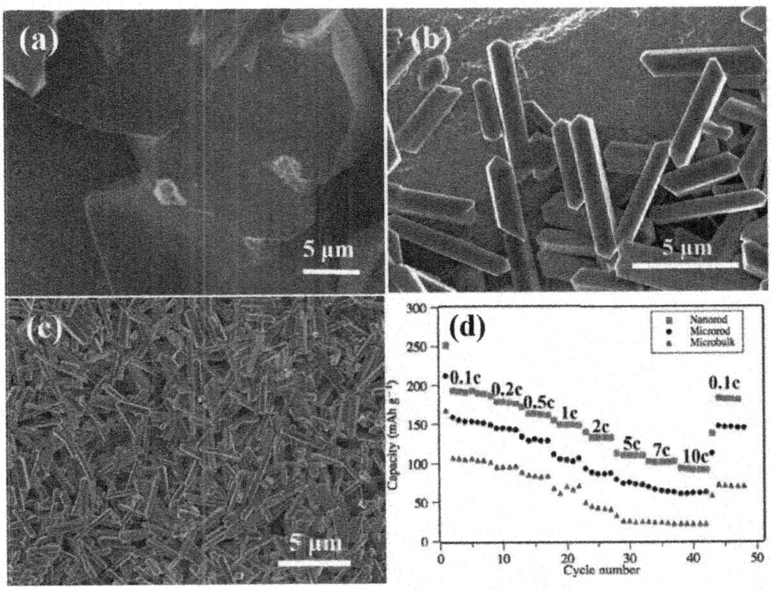

Figure 3.36 SEM images of SR microbulk (a), microrod (b), and nanorod (c) structured samples. (d) The rate performance of microbulk, microrod, and nanorod DSR samples at 25 mA/g. Reprinted with permission from Ref. [257], Copyright 2016, American Chemical Society.

Croconic acid disodium salt (CADS, No. 8) shared a similar structure with DSR and encountered the same volume expansion problem upon sodiation in the initial cycles. Wang et al. wrapped the nano-sized CADS with graphene oxide to enhance the integrity and conductivity of the whole electrode, achieving an improved capacity and cycling performance [259]. Additionally, the molecular design is also helpful to improve electrochemical performance. For instance, Yao et al. reported that indigo carmine (IC, 5,5′-indigodisulfonic acid sodium salt, No. 9) could act as a relatively stable positive electrode for SIBs owing to its very low solubility in organic solvents derived from the peripheral polar sulfonate groups [260].

3.6.1.2 Anhydrides and imides

The most studied anhydride-type organic cathode materials for SIBs are displayed in Fig. 3.37a. They are pyromellitic dianhydride (PMDA), 1,4,5,8-naphthalene-tetracarboxylic dianhydride (NTCDA), and perylene 3,4,9,10-tetracarboxylic dianhydride (PTCDA), respectively [261]. They share a similar structure in that two carbonyl groups bound with one oxygen to form O=C–O–C=O, and the carbon directly connects to an aromatic core, so their electrochemical behaviors are similar. Generally, they can deliver a capacity of around 130 mAh/g with an average output voltage of ~2.1 V in the potential window of 1~3 V. It should be noted that the anhydrides could accommodate more Na^+ ions and serve as anode materials when they were deeply

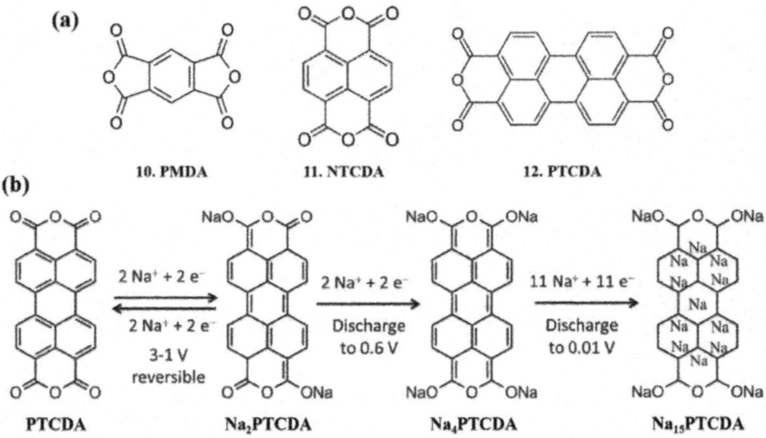

Figure 3.37 (a) Structures of anhydrides as cathode materials for SIBs. (b) The proposed electrochemical reactions during sodiation/desodiation of PTCDA. Reprinted with permission from Ref. [262], Copyright 2014, Wiley-VCH.

discharged. For example, the PTCDA can uptake 15 Na^+ through a three-step sodiation process when fully discharging to 0.01 V (Fig. 3.37b) [262]. But the excessive Na^+ insertion could destroy the crystal structure of PTCDA and result in poor reversibility.

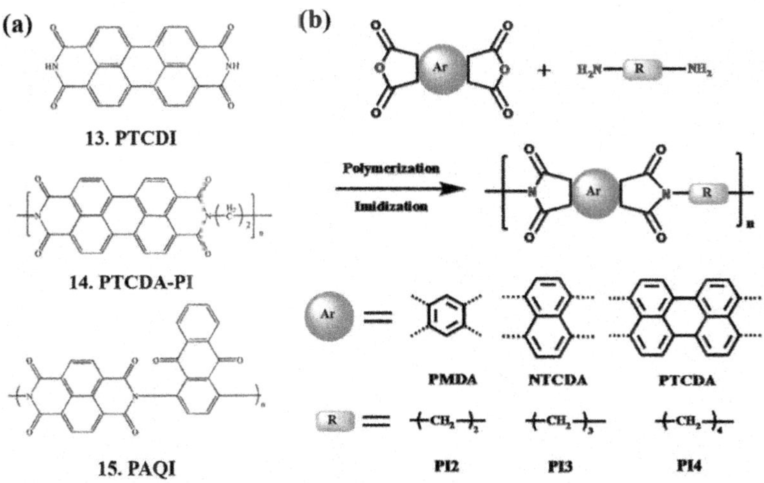

Figure 3.38 (a) Structures of representative imides as cathode materials for SIBs. (b) Synthesis route for various polyimides derived from anhydrides. Reprinted with permission from Ref. [265], Copyright 2014, Wiley-VCH.

The molecular structure of imides resembles the aforementioned anhydrides, but the N atom instead of the O atom connects the two carbonyl groups (Fig. 3.38a). Small molecule imides are barely used as electrode materials because of their high solubility in electrolytes [263]. One example is 3,4,9,10-perylene-bis(dicarboximide) (PTCDI), which is insoluble in organic electrolytes because of its stable π-conjugated structure [264]. PTCDI delivered a high capacity of 118 mAh/g at 100 mA/g via a two-electron redox reaction and retained 90% after 300 cycles. The majority are polyimides, consisting of an electroactive center and inactive linker to avoid harmful dissolution. Meanwhile, polyimides have advantages in terms of thermal stability, mechanical property, and cost compared to the corresponding monomers. A traditional approach to synthesize polyimides is through a stepwise polymerization and imidization of dianhydrides and diamine. Zhang et al. prepared a series of polyimides (PIs) by a one-pot solution polymerization route, where quinoline is employed to dissolve the starting materials and catalyze the polycondensation (Fig. 3.38b) [265]. The presence of the C=O and C–N bonds in the Fourier transform infrared (FTIR) spectra can confirm the completion of the polymerization and imidization processes.

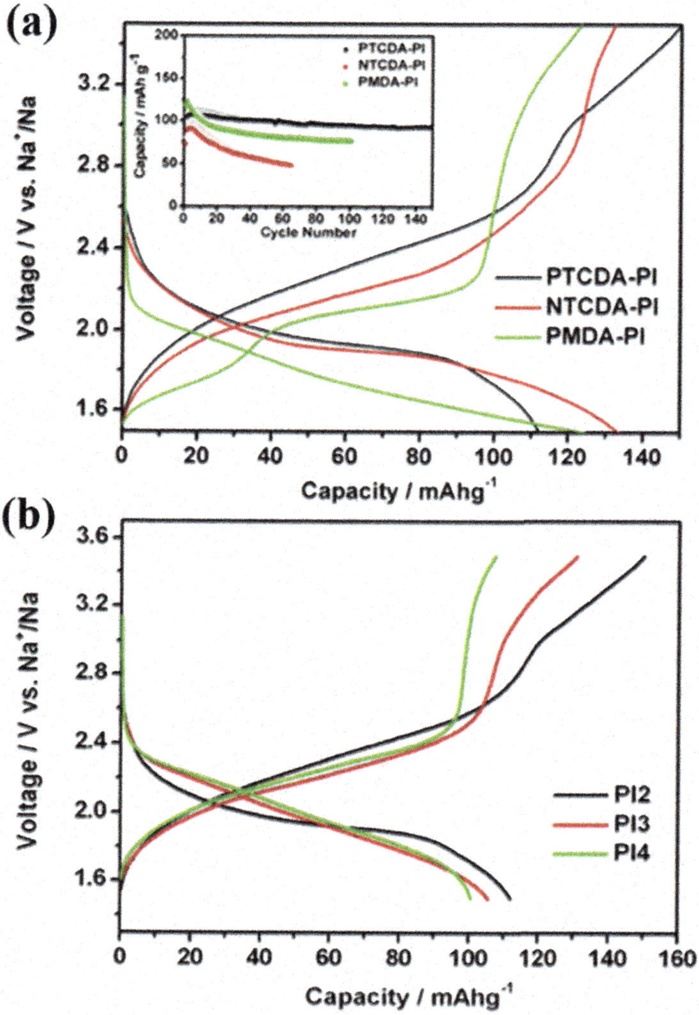

Figure 3.39 (a) Charge/discharge curves of different polyimide (PI) electrodes at a current density of 25 mA/g and cycle performance of the PI electrodes at 200 mA/g (inset). (b) Charge/discharge curves of different alkyl-chain-based PI electrodes at 25 mA/g. Reprinted with permission from Ref. [265], Copyright 2014, Wiley-VCH.

By changing the conjugated backbone from PMDA to NTCDA and PTCDA, the average discharge voltage increases from 1.73 to 1.89 and 1.94 eV, which could be ascribed to the difference in the lowest unoccupied molecular orbital energy stemming from the incorporation of the additional electron-withdrawing groups (Fig. 3.39a). The redox potential of PIs can be further improved by introducing stronger electron-withdrawing groups, such as cyano and bromine [266]. On the other hand, the specific capacity of the PIs depends on the molecular weight, so the decrease in the alkyl chain increases the gravimetric capacity (Fig. 3.39b). Another way to enhance the capacity is by replacing the inactive alkyl groups with redox-active fragments like anthraquinone (AQ) to link the anhydrides. Through this route, Yang et al. synthesized poly (anthraquinonyl imide) (PAQI, No. 15) from the PMDA and NTCDA, and the polymer shows a capacity of 190 mAh/g as Na-storage cathode [267].

Once again, it is worthy to mention that either diimides or polyimides can store more Na ions at a lower potential (<1.5 V), but it may also cause severe structural damage or even decomposition. Overall, carbonyl compounds can generate a decent capacity at an acceptable voltage range (>1 V). However, the moderate voltage (1.5~2.5 V) makes them less attractive compared to the high-voltage *p*-type organic materials [268].

3.6.2 Pteridine Derivatives (C=N Reaction)

Beyond the carbonyl compounds, pteridine derivatives with C=N redox centers are another promising *n*-type organic cathode material for sodium storage. Inspired by the redox cycling occurring at diazabutadiene of flavin cofactors in biological energy transduction, Kang et al. designed and synthesized three pteridine derivatives in the alloxazinic structures by removing the non-redox-active substitutes of the flavin molecules (Fig. 3.40a) [269]. On the basis of the reversible tautomerism of the alloxazinic and isoalloxazinic forms and the redox activity of the conjugated diazabutadiene region, lumichrome (LC, No. 16), alloxazine (ALX, No. 17), and lumazine (LMZ, No. 18) are capable of reversibly taking up and releasing two Na^+ in SIBs (Fig. 3.40b,c). The practical capacities of the pristine pteridine electrodes are lower than the theoretical

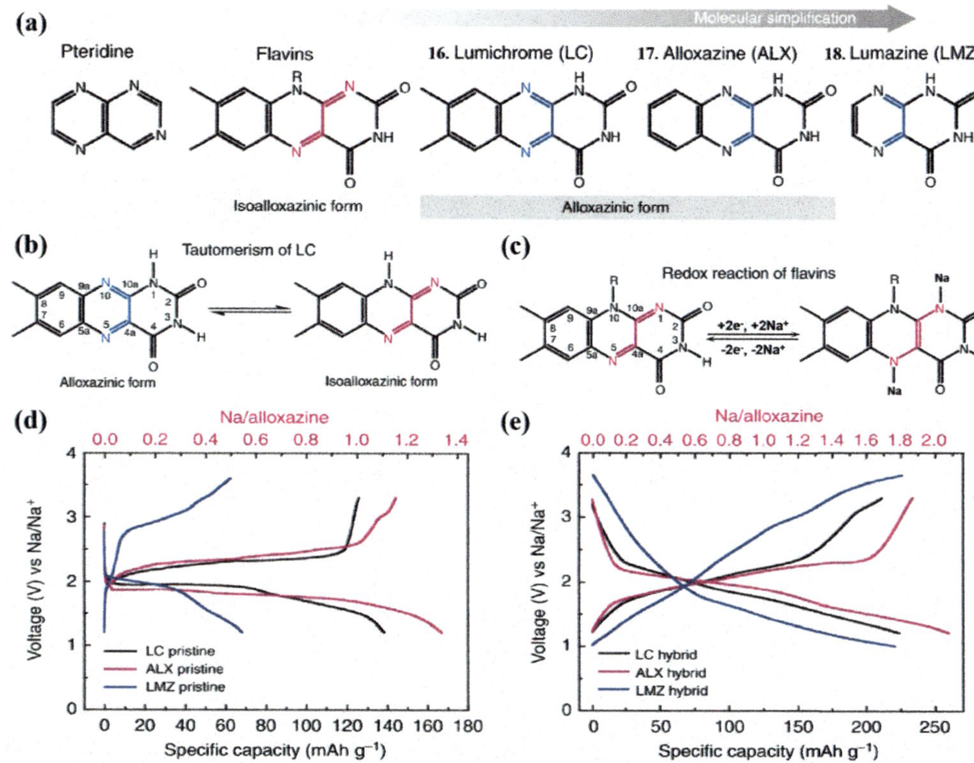

Figure 3.40 (a) Molecular structures of pteridine derivatives as bioinspired redox compounds for SIBs. (b) Reversible tautomerism of LC. (c) Redox mechanism of flavins for sodium storage inspired by biological energy transduction. (d) Pristine pteridine derivatives at 20 mA/g and (e) pteridine hybrid electrodes at 50 mA/g in SIBs. Reprinted with permission from Ref. [269], Copyright 2014, Springer Nature.

value, which can be significantly improved by incorporating CNTs during the electrode preparation. And the hybrids deliver a capacity of 222, 255, and 220 mAh/g for LC, ALX, and LMZ, respectively (Fig. 3.40d,e).

3.6.3 Polymers (Doping Reaction)

Unlike the high solubility of small organic molecules, polymers with extended backbones of repeating units are relatively insoluble in the conventional organic electrolytes, and thereby resulting in better cycling performance. Apart from the previously mentioned polyquinones and polyimides, other polymer cathode materials for SIBs include conductive polymers, nitroxyl radical polymers, microporous polymers, and organometallic polymers [270].

3.6.3.1 Conductive polymers

Conductive polymers, as the name implies, present high electronic conductivity, which is favorable for electrochemical energy storage. The research on conductive polymers for SIBs started in 1985 when Shackletter et al. first examined the possibility of using polyacetylene and polyparaphenylene as electrode materials [271]. Although both of them are capable of reversibly uptaking and releasing Na^+ in the appropriate electrolytes, they can only act as anode materials because of the low sodiation potential (below 1.0 V). Many conjugated conductive polymers can be employed as cathodes for SIBs; among them polyaniline (PAN, No. 19)- and polypyrrole (PPy, No. 20)-based polymers are two typical examples that have been widely studied in the past decade (Fig. 3.41). They are usually synthesized in the *p*-doped state and undergo anions insertion and extraction mechanism, which offer a relatively high output voltage (>3 V versus Na^+/Na). However, these *p*-doped polymers suffer from low doping levels and poor capacity utilization of the polymer chains, leading to a low specific capacity.

There are two strategies to cope with this challenge, and one is to construct porous nanoarchitecture to facilitate the accessibility of Na^+ and thereby enhance the utilization of the polymer chains [272]. For example, Feng et al. developed 2D mesoporous PPy nanosheets by a soft-template method for sodium cathodes [273]. The composite electrodes exhibit a much higher capacity of 123 mAh/g at 50 mA/g and better rate capability than that of the PPy-blank samples.

Another effective approach is to introduce redox-active moieties into the polymer chains to activate to contribute their redox capacity to the polymer materials. Yang et al. carried out extensive works in this field. They prepared the aniline/*o*-nitroaniline copolymer (PANNA, No. 21) by grafting electron-withdrawing *o*-nitroaniline groups onto polyaniline chains through a simple chemical oxidative polymerization [274]. The copolymer undergoes an electrochemical activation with an upgraded working potential of ~3.2 V and reversible capacity of 180 mAh/g compared to the pristine PPy.

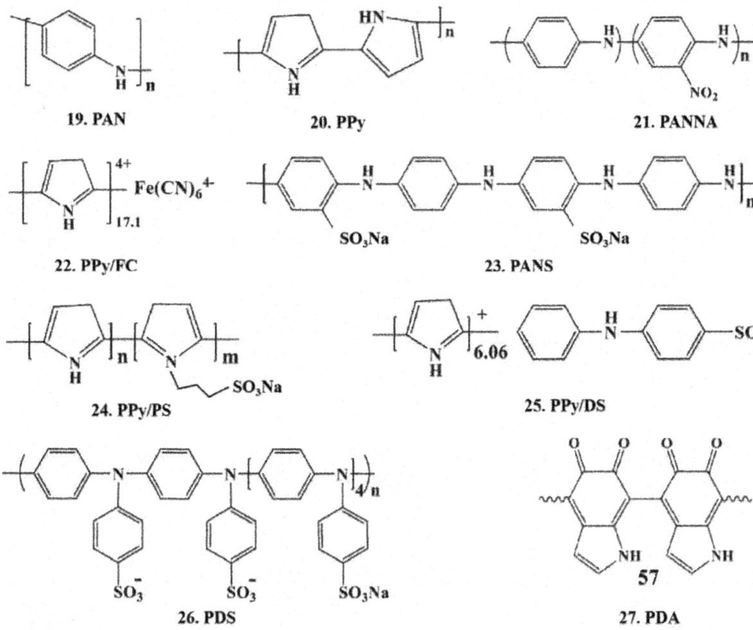

Figure 3.41 Structures of conductive polymers for SIB cathodes.

Additionally, doping (graft copolymerizing) polymers with anionic redox species can also change the reaction mechanism from the *p*-doping/dedoping of electrolyte anions to the combined redox processes with Na^+ insertion/extraction. The inorganic anions $Fe(CN)_6{}^{4-}$ doped polypyrrole (PPy/FC, No. 22) is a typical example for this combined redox mechanism, in which Na-insertion/extraction, $Fe(CN)_6{}^{3-/4-}$ redox and *p*-doping reaction together contribute to a high capacity of 135 mAh/g [275]. Besides, doping with organic redox species can enhance electrochemical performance as well. This type of organic materials include polyaniline-co-aminobenzenesulfonic sodium (PANS, No. 23) [276], poly(pyrrole-*co*-(sodium-3-(pyrrol-lyl) propanesulphonate)) (PPy-PS, No. 24) [277], diphenylamine-4-sulfonate-doped PPy (PPy/DS, No. 25) [278], poly(diphenylaminesulfonic acid sodium) (PDS, No. 26), and polydopamine (PDA, No. 27) [279].

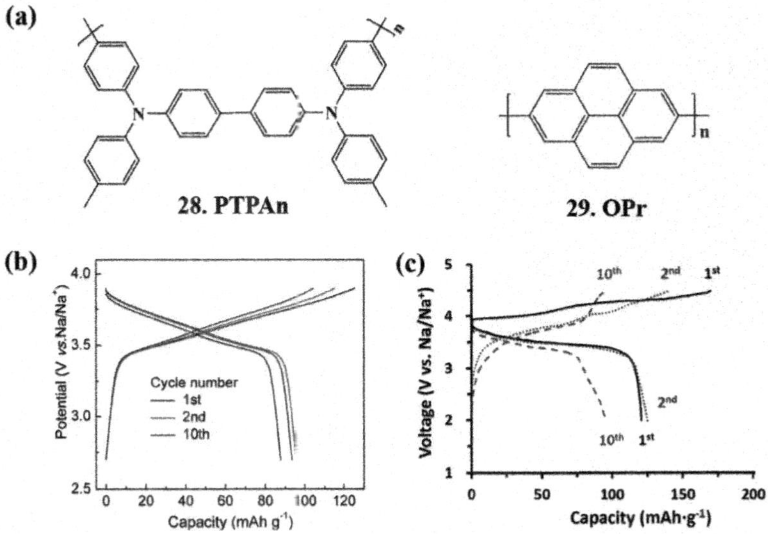

Figure 3.42 (a) Structures of high-voltage conductive polymer materials. Reprinted with permission from Ref. [280], Copyright 2014, Elsevier. Charge and discharge curves of (b) PTPAn and (c) OPr. Reprinted with permission from Ref. [281], Copyright 2013, Springer Nature.

Notably, polytriphenylamine (PTPAn, No. 28) and non-crystalline oligopyrene (OPr, No. 29) are two high-voltage conductive polymers for sodium storage relying on *p*-doping/dedoping reaction (Fig. 3.42a) [280]. PTPAn shows a structure of electroactive poly-aniline units embedded on highly conductive polyparaphenylene chains, and each triphenylamine unit can lose one electron and intercalate one PF_6^- upon oxidation at an average potential of 3.6 V (Fig. 3.42b) [281]. As cathode, it delivers a reversible capacity of 98 mAh/g at 0.5 C (1 C = 100 mA/g) and an excellent rate and cycling performance with 88 mAh/g at 5 C after 200 cycles. Amorphous OPr shows a plateau voltage at around 3.5 V, but the cycling and rate performances are much inferior to the PTPAn (Fig. 3.42c).

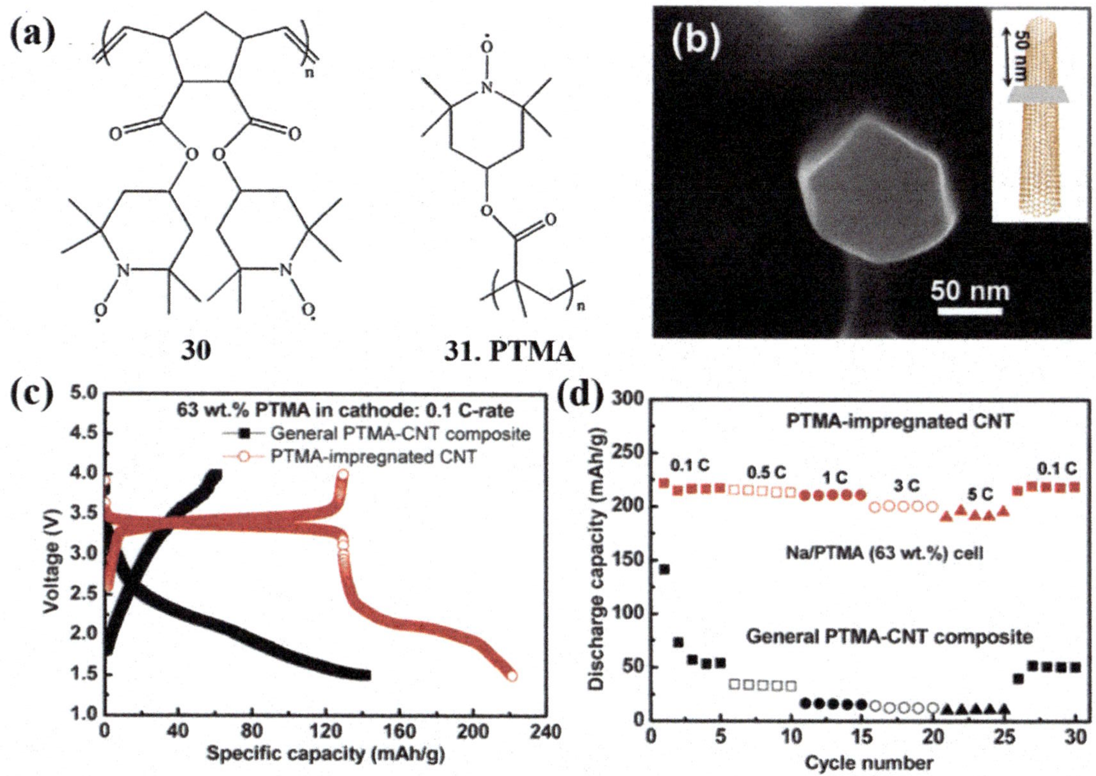

Figure 3.43 (a) Structures of nitroxyl radical polymers for sodium cathodes. (b) SEM image of the PTMA-impregnated CNT. (c) Initial charge–discharge curves and (d) rate performance of mixed PTMA-CNT composite and PTMA-impregnated CNT electrodes. Reprinted with permission from Ref. [283], Copyright 2016, Royal Chemical Society.

3.6.3.2 Nitroxyl radical polymer

So far, only two organic radical polymers, poly[norbornene-2,3-endo, exo-(COO-4-TEMPO)$_2$] (No. 30) and poly(2,2,6,6-tetramethylpiperidinyloxy-4-vinylmethacrylate) (PTMA, No. 31), are reported for sodium cathodes (Fig. 3.43a) [282]. Both of them contain the 2,2,6,6-tetramethyl-1-piperidinyloxy (TEMPO) radicals in the basic unit. During charging, the nitroxide radical moieties of the TEMPO are oxidized to oxoammonium cations at about 3.45 V; and the oxoammonium cations are reversibly reduced to nitroxide radicals during discharge. The PTMA can be further reduced to aminoxy anions when discharging to 1.5 V and deliver a high capacity of 222 mAh/g through the two-electron redox reaction. However, such radical polymer electrodes suffer from severe self-discharge and low practical capacity because they easily dissolve in organic electrolytes and form an insulating layer. To solve this problem, Kim et al. designed and encapsulated the PTMA into the CNTs, which plays a dual functionality for preventing the polymer dissolution and improving the electron transport (Fig. 3.43b) [283]. As a result, the PTMA composite exhibits a significantly improved capacity, rate performance, and cycle life (Fig. 3.43c,d).

3.6.3.3 Microporous polymers

Microporous polymers have attracted great attention because they combine the inherent merits of the π-conjugated skeletons with high surface area, which can facilitate electrolyte impregnation and provide more active sites for energy storage applications. The most studied microporous polymers are covalent organic frameworks (COF) and metal–organic frameworks (MOF), but they are mainly applied as anode materials for sodium storage because of the low average voltage [284]. Sakaushi et al. reported bipolar porous polymeric frameworks (BPPF) consisting of benzene rings and triazine rings as cathodes for SIBs (Fig. 3.44a, No. 32) [285]. The BPPF was synthesized through the trimerization of nitriles in a $ZnCl_2$ melt at 400°C [286]. It displays a sheet-like morphology with an interlayer distance of 0.34 nm. The porous structure of BPPF was investigated by nitrogen physisorption measurements, and the pore size distribution confirmed the existence of micropores of around 1.4 nm (Fig. 3.44b). Electron energy-loss spectroscopy (EELS) results

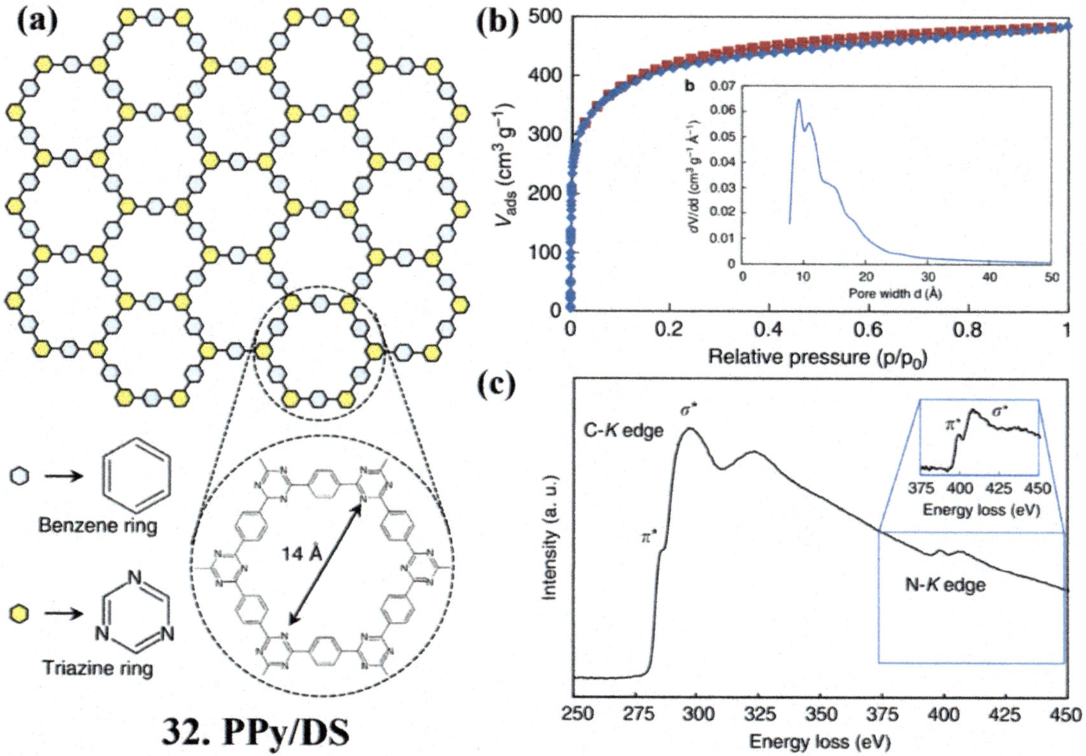

32. PPy/DS

Figure 3.44 (a) Schematic illustration of the BPPF structure having a pore diameter of 14 Å. (b) The nitrogen physisorption curves and pore-size distribution of the BPPF. (c) Electron energy loss spectrum of the carbon and nitrogen *K* edges recorded in the BPPF. The inset shows the N-*K* edge with the subtraction of the background. Reprinted with permission from Ref. [285], Copyright 2013, Springer Nature.

exhibit the sp^2-hybridization for carbon and nitrogen at 285 eV and 400 eV, respectively (Fig. 3.44c), which reveals the formation of a graphitic network constituted by triazine and benzene rings. Based on the mixed *p*-doping and *n*-doping reactions, the BPPF electrodes show a high capacity of 200 mAh/g and a quite good rate and cycling performance. Thanks to its broad working voltage range, an all-organic sodium-ion battery was assembled by utilizing the BPPF as both cathodes and anodes [287].

3.6.3.4 Organometallic polymers

Organometallic polymers are potential cathode candidates for rechargeable batteries due to their electroactive organometallic redox moieties. Poly(ferrocenyl-methylsilane) (PFS, No. 33, Fig. 3.45a) was synthesized by using a ring-opening polymerization method for cathode materials in both LIBs and SIBs [288]. Based on the Fe^{2+}/Fe^{3+} redox reaction accompanying the *p*-doping/dedoping behavior with ClO_4^- in the electrolyte, the PFS electrode presents a reversible capacity of 94 mAh/g with an average voltage of 3.2 V (Fig. 3.45b). The theoretical capacity of the ferrocene-based polymer is relatively low because of its high molecular weight.

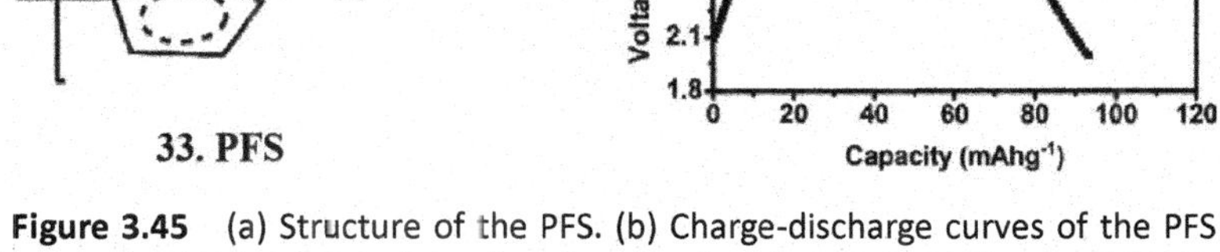

Figure 3.45 (a) Structure of the PFS. (b) Charge-discharge curves of the PFS in 1 M $NaClO_4$/EC+DMC electrolyte. Reprinted with permission from Ref. [288], Copyright 2014, Royal Chemical Society.

3.7 Conclusion

As discussed earlier, each type of cathode material has its unique features and limitations, summarized in Fig. 3.46. Among the variety of candidates, layered oxides, polyanionic compounds, and PBAs are more competitive considering the production cost, delivered energy density, cycle lifespan, and environmental impacts. Particularly, sodium layered oxides, which inherently manufacture the manufacture experience from the commercial lithium layer oxide materials, stand out for some prototype sodium-ion pouch cells [289].

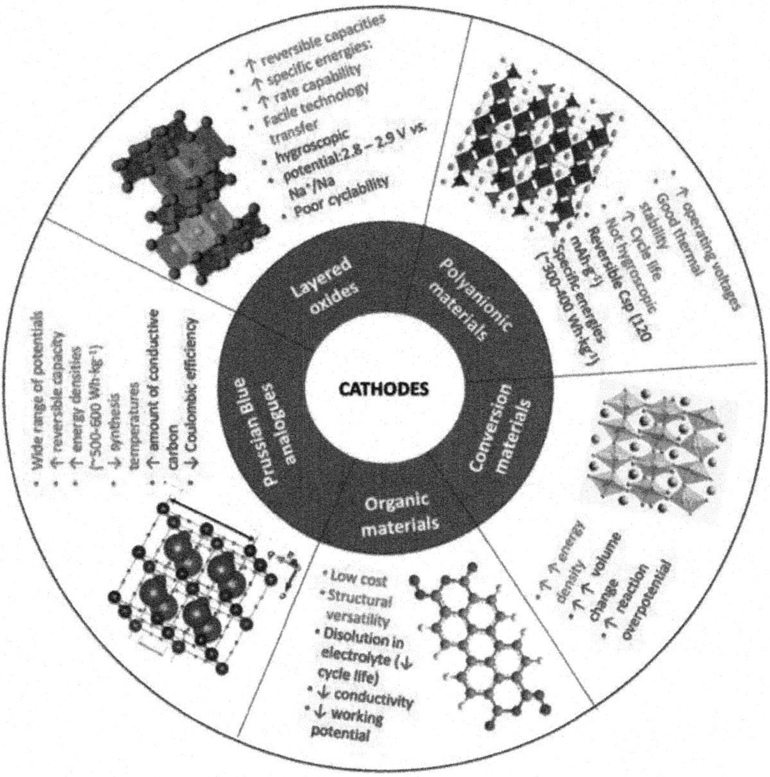

Figure 3.46 Advances (green) and challenges (red) of the five different types of cathode materials for SIBs. Reprinted with permission from Ref. [290], Copyright 2020, Wiley-VCH.

Mn- and Fe-based layered oxides with low toxicity, natural abundance, and high energy densities make them promise for SIBs aimed at large-scale energy storage. However, the issue of poor cycling stability should be addressed via either element doping/ substitution or other effective strategies before their practical application [291]. Polyanionic materials are advantageous in terms of energy and power density and cycle life. Nevertheless, most of them incorporate the vanadium element, which is toxic and more expensive compared to Fe-, Mn-, or Ni-based cathode materials. Besides, their specific capacities are still lower than those of the layered counterparts owing to the heavy weight of anions. The PBAs face the same problem; the relatively low gravimetric density and poor conductivity impede their widespread application, despite that they have good structural stability for sodium storage. The study of organic materials and conversion-type cathode materials is still at the laboratory stage, and many problems need to be solved before the real application. Therefore, more efforts are required in both theoretical and practical aspects to overcome these challenges and guide the rational design of sodium cathode materials in the future.

References

1. Goodenough JB and Park KS (2013). The Li-ion rechargeable battery: A perspective, *Journal of the American Chemical Society*, **135**, pp. 1167–1176.
2. Goodenough JB and Kim Y (2010). Challenges for rechargeable Li batteries, *Chemistry of Materials*, **22**, pp. 587–603.
3. Bruce PG, Scrosati B, and Tarascon JM (2008). Nanomaterials for rechargeable lithium batteries, *Angewandte Chemie-International Edition*, **47**, pp. 2930–2946.
4. Slater MD, Kim D, Lee E, and Johnson CS (2013). Sodium-ion batteries, *Advanced Functional Materials*, **23**, pp. 947–958.
5. Ellis BL and Nazar LF (2012). Sodium and sodium-ion energy storage batteries, *Current Opinion in Solid State & Materials Science*, **16**, pp. 168–177.
6. Jiang C, Shanshan Y, and Jang-Kyo K (2017). Recent progress in rational design of anode materials for high-performance Na-ion batteries, *Energy Storage Materials*, **7**, pp. 64–114.
7. Hwang JY, Myung ST, and Sun YK (2017). Sodium-ion batteries: Present and future, *Chemical Society Reviews*, **46**, pp. 3529–3614.
8. Han MH, Gonzalo E, Singh G, and Rojo T (2015). A comprehensive review of sodium layered oxides: Powerful cathodes for Na-ion batteries, *Energy & Environmental Science*, **8**, pp. 81–102.

9. Delmas C, Fouassier C, and Hagenmuller P (1980). Structural classification and properties of the layered oxides, *Physica B+C*, **99**, pp. 81–85.

10. Yabuuchi N, Yoshida H, and Komaba S (2012). Crystal structures and electrode performance of alpha-$NaFeO_2$ for rechargeable sodium batteries, *Electrochemistry*, **80**, pp. 716–719.

11. Liu ZB, Xu XJ, Ji SM, Zeng LY, Zhang DC, and Liu J (2020) Recent progress of P2-type layered transition-metal oxide cathodes for sodium-ion batteries, *Chemistry: A European Journal*, **26**(35), pp. 7747–7766.

12. Hilgenkamp H, Ariando, Smilde HJ, Blank DH, Rijnders G, Rogalla H, Kirtley JR, and Tsuei CC (2003). Ordering and manipulation of the magnetic moments in large-scale superconducting pi-loop arrays, *Nature*, **422**, pp. 50–53.

13. Miyazaki S, Kikkawa S, and Koizumi M (1983). Chemical and electrochemical deintercalations of the layered compounds LiMO2 (M = Cr, Co) and NaM′O2 (M′ Cr, Fe, Co, Ni), *Synthetic Metals*, **6**, pp. 211–217.

14. Delmas C, Braconnier J, Fouassier C, and Hagenmuller P (1981). Electrochemical intercalation of sodium in $NaxCoO_2$ bronzes, *Solid State Ionics*, **3–4**, pp. 165–169.

15. Niwa H, Higashiyama K, Amaha K, Kobayashi W, and Moritomo Y (2018). Electronic states in oxidized $NaxCoO_2$ as revealed by X-ray absorption spectroscopy coupled with ab initio calculation, *Journal of Power Sources*, **384**, pp. 156–159.

16. Shacklette LW (1988). Rechargeable electrodes from sodium cobalt bronzes, *Journal of The Electrochemical Society*, **135**, pp. 2669.

17. Berthelot R, Carlier D, and Delmas C (2011). Electrochemical investigation of the P2-$NaxCoO_2$ phase diagram, *Nature Materials*, **10**, pp. 74–80.

18. Platova TA, Mukhamedshin IR, Alloul H, Dooglav AV, and Collin G (2009). Nuclear quadrupole resonance and x-ray investigation of the structure of $Na_{2/3}CoO_2$, *Physical Review B*, **80**.

19. Hinuma Y, Meng YS, and Ceder G (2008). Temperature-concentration phase diagram of P2-$NaxCoO_2$ from first-principles calculations, *Physical Review B*, **77**.

20. Rai AK, Anh LT, Gim J, Mathew V, and Kim J (2014). Electrochemical properties of $NaxCoO_2$ (x~0.71) cathode for rechargeable sodium-ion batteries, *Ceramics International*, **40**, pp. 2411–2417.

21. Ding JJ, Zhou YN, Sun Q, Yu XQ, Yang XQ, and Fu ZW (2013). Electrochemical properties of P2-phase $Na_{0.74}CoO_2$ compounds as cathode material for rechargeable sodium-ion batteries, *Electrochimica Acta*, **87**, pp. 388–393.

22. Nakai I, Takahashi K, Shiraishi Y, Nakagome T, and Nishikawa F (1998). Study of the Jahn–Teller distortion in $LiNiO_2$, a cathode material in a rechargeable lithium battery, by in situ X-ray absorption fine structure analysis, *Journal of Solid State Chemistry*, **140**, pp. 145–148.

23. D'Arienzo M, Ruffo R, Scotti R, Morazzoni F, Mari CM, and Polizzi S (2012). Layered Na(0.71)CoO(2): A powerful candidate for viable and high performance Na-batteries, *Physical Chemistry Chemical Physics*, **14**, pp. 5945–5952.

24. Zhang W, Zhu K, Liu J, Wang J, Yan K, Liu P, and Wang Y (2018). Influence of the phase transformation in Na_xCoO_2 ceramics on thermoelectric properties, *Ceramics International*, **44**, pp. 17251–17257.

25. Olszewski W, Ávila Pérez M, Marini C, Paris E, Wang X, Iwao T, Okubo M, Yamada A, Mizokawa T, Saini NL, and Simonelli L (2016). Temperature dependent local structure of Na_xCoO_2 cathode material for rechargeable sodium-ion batteries, *The Journal of Physical Chemistry C*, **120**, pp. 4227–4232.

26. Hwang S, Lee Y, Jo E, Chung KY, Choi W, Kim SM, and Chang W (2017). Investigation of thermal stability of P2–Na_xCoO_2 cathode materials for sodium ion batteries using real-time electron microscopy, *ACS Applied Materials & Interfaces*, **9**, pp. 18883–18888.

27. Wang Z, Xiong F, Tao H, and Yue Y (2020). Revealing the role of the amorphous phase in $Na_{0.74}CoO_2$/C/N composite cathode, *Journal of Alloys and Compounds*, **815**, pp. 152616.

28. Gao L, Chen S, Hu H, Cheng H, Zhang L, and Yang X (2020). Hierarchical Na_xCoO_2 microspheres with low surface area toward high performance sodium ion batteries, *Materials Letters*, **260**, pp. 126965.

29. Matsui M, Mizukoshi F, and Imanishi N (2015). Improved cycling performance of P2-type layered sodium cobalt oxide by calcium substitution, *Journal of Power Sources*, **280**, pp. 205–209.

30. Doubaji S, Ma L, Asfaw HD, Izanzar I, Xu R, Alami J, Lu J, Wu TP, Amine K, Edstrom K, and Saadoune I (2018). On the P2-$Na_xCo_{1-y}(Mn_{2/3}Ni_{1/3})_yO(2)$ cathode materials for sodium-ion batteries: Synthesis, electrochemical performance, and redox processes occurring during the electrochemical cycling, *ACS Applied Materials & Interfaces*, **10**, pp. 488–501.

31. Jugovic D, Milovic M, Popovic M, Kusigerski V, Skapin S, Rakocevic Z, and Mitric M (2019). Effects of fluorination on the structure, magnetic and electrochemical properties of the P2-type Na_xCoO_2 powder, *Journal of Alloys and Compounds*, **774**, pp. 30–37.
32. Carlier D, Cheng JH, Berthelot R, Guignard M, Yoncheva M, Stoyanova R, Hwang BJ, and Delmas C (2011). The P2-$Na_{(2/3)}Co_{(2/3)}Mn_{(1/3)}O_2$ phase: Structure, physical properties and electrochemical behavior as positive electrode in sodium battery, *Dalton Transactions*, **40**, pp. 9306–9312.
33. Birgisson S, Shen YB, and Iversen BB (2017). In operando observation of sodium ion diffusion in a layered sodium transition metal oxide cathode material, P2 $Na_xCo_{0.7}Mn_{0.3}O_2$, *ChemComm*, **53**, pp. 1160–1163.
34. Cheng J-H, Pan C-J, Lee J-F, Chen J-M, Guignard M, Delmas C, Carlier D, and Hwang B-J (2014). Simultaneous reduction of Co^{3+} and Mn^{4+} in P2-$Na_{2/3}Co_{2/3}Mn_{1/3}O_2$ as evidenced by X-ray absorption spectroscopy during electrochemical sodium intercalation, *Chemistry of Materials*, **26**, pp. 1219–1225.
35. Baster D, Zając W, Kondracki Ł, Hartman F, and Molenda J (2016). Improvement of electrochemical performance of $Na_{0.7}Co_{1-y}MnO_2$—cathode material for rechargeable sodium-ion batteries, *Solid State Ionics*, **288**, pp. 213–218.
36. Mendiboure A, Delmas C, and Hagenmuller P (1985). Electrochemical intercalation and deintercalation of Na_xMnO_2 bronzes, *Journal of Solid State Chemistry*, **57**, pp. 323–331.
37. Ma X, Chen H, and Ceder G (2011). Electrochemical properties of monoclinic $NaMnO_2$, *Journal of The Electrochemical Society*, **158**, pp. A1307.
38. Abakumov AM, Tsirlin AA, Bakaimi I, Van Tendeloo G, and Lappas A (2014). Multiple twinning as a structure directing mechanism in layered rock-salt-type oxides: $NaMnO_2$ polymorphism, redox potentials, and magnetism, *Chemistry of Materials*, **26**, pp. 3306–3315.
39. Caballero A, Hernan L, Morales J, Sanchez L, Pena JS, and Aranda MAG (2002). Synthesis and characterization of high-temperature hexagonal P2-$Na_{0.6}MnO_2$ and its electrochemical behaviour as cathode in sodium cells, *Journal of Materials Chemistry*, **12**, pp. 1142–1147.
40. Konarov A, Kim HJ, Voronina N, Bakenov Z, and Myung ST (2019). P2-$Na_{2/3}MnO_2$ by Co incorporation: As a cathode material of high capacity and long cycle life for sodium-ion batteries, *ACS Applied Materials & Interfaces*, **11**, pp. 28928–28933.

41. Clement RJ, Billaud J, Armstrong AR, Singh G, Rojo T, Bruce PG, and Grey CP (2016). Structurally stable Mg-doped P2-$Na_{2/3}Mn_{1-y}Mg_yO_2$ sodium-ion battery cathodes with high rate performance: Insights from electrochemical, NMR and diffraction studies, *Energy & Environmental Science*, **9**, pp. 3240–3251.

42. Chen XL, Li N, Kedzie E, McCloskey BD, Tang HL, and Tong W (2019). High-capacity P2-type $Na_xLi_{0.25}Mn_{0.75}O_2$ cathode enabled by anionic oxygen redox, *Journal of the Electrochemical Society*, **166**, A4136.

43. Chen TR, Sheng T, Wu ZG, Li JT, Wang EH, Wu CJ, Li HT, Guo XD, Zhong BH, Huang L, and Sung S G (2018). Cu^{2+} dual-doped layer-tunnel hybrid $Na_{0.6}Mn_{1-x}Cu_xO_2$ as a cathode of sodium-ion battery with enhanced structure stability, electrochemical property, and air stability, *ACS Applied Materials & Interfaces*, **10**, pp. 10147–10156.

44. Birgisson S, Christiansen TL, and Iversen BB (2018). Exploration of phase compositions, crystal structures, and electrochemical properties of $Na_xFe_yMn_{1-y}O_2$ sodium ion battery materials, *Chemistry of Materials*, **30**, pp. 6636–6645.

45. Bao S, Luo SH, Wang ZY, Yan SX, and Wang Q (2018). Novel high-capacity hybrid layered oxides $Na_xLi_{1.5-x}Ni_{0.167}Co_{0.167}Mn_{0.67}O_2$ as promising cathode materials for rechargeable sodium ion batteries, *Ceramics International*, **44**, pp. 22512–22519.

46. Bai X, Sathiya M, Mendoza-Sanchez B, Iadecola A, Vergnet J, Dedryvere R, Saubanere M, Abakumov AM, Rozier P, and Tarascon JM (2018). Anionic redox activity in a newly Zn-doped sodium layered oxide P2-$Na_{2/3}Mn_{1-y}Zn_yO_2$ ($0 < y < 0.23$), *Advanced Energy Materials*, **8**.

47. Chen H, Wu ZG, Zhong YJ, Chen TR, Liu XH, Qu J, Xiang W, Li JT, Chen XC, Guo XD, and Zhong BH (2019). Boosting the reactivity of Ni^{2+}/Ni^{3+} redox couple via fluorine doping of high performance $Na_{0.6}Mn_{0.95}Ni_{0.05}O_{2-x}F_x$ cathode, *Electrochimica Acta*, **308**, pp. 64–73.

48. Chen T, Liu WF, Gao H, Zhuo Y, Hu H, Chen A, Zhang JW, Yan J, and Liu KY (2018). A P2-type $Na_{0.44}Mn_{0.6}Ni_{0.3}Cu_{0.1}O_2$ cathode material with high energy density for sodium-ion batteries, *Journal of Materials Chemistry A*, **6**, pp. 12582–12588.

49. Paulsen JM and Dahn JR (1999). Studies of the layered manganese bronzes, $Na_{2/3}[Mn_{1-x}M_x]O_2$ with M=Co, Ni, Li, and $Li_{2/3}[Mn_{1-x}M_x]O_2$ prepared by ion-exchange, *Solid State Ionics*, **126**, pp. 3–24.

50. Yuan DD, He W, Pei F, Wu FY, Wu Y, Qian JF, Cao YL, Ai XP, and Yang HX (2013). Synthesis and electrochemical behaviors of layered $Na_{0.67}Mn_{0.65}Co_{0.2}Ni_{0.15}O_2$ microflakes as a stable cathode material for sodium-ion batteries, *Journal of Materials Chemistry A*, **1**, pp. 3895–3899.

51. Billaud J, Singh G, Armstrong AR, Gonzalo E, Roddatis V, Armand M, Rojo T, and Bruce PG (2014). $Na_{0.67}Mn_{1-x}Mg_xO_2$ ($0 \leq x \leq 0.2$): A high capacity cathode for sodium-ion batteries, *Energy & Environmental Science*, **7**, pp. 1387–1391.

52. Alvarado J, Ma CZ, Wang S, Nguyen K, Kodur M, and Meng YS (2017). Improvement of the cathode electrolyte interphase on P-2-$Na_{2/3}Ni_{1/3}Mn_{2/3}O_2$ by atomic layer deposition, *ACS Applied Materials & Interfaces*, **9**, pp. 26518–26530.

53. Aragon MJ, Ortiz GF, Lavela P, Tirado JL, Oliveira JDC, Motta FV, and Bomio MRD (2019). On the use of guanidine hydrochloride soft template in the synthesis of $Na_{2/3}Ni_{1/3}Mn_{2/3}O_2$ cathodes for sodium-ion batteries, *Journal of Alloys and Compounds*, **789**, pp. 1035–1045.

54. Bucher N, Hartung S, Nagasubramanian A, Cheah YL, Hoster HE, and Madhavi S (2014). Layered Na_xMnO_{2+z} in sodium ion batteries-influence of morphology on cycle performance, *ACS Applied Materials & Interfaces*, **6**, pp. 8059–8065.

55. Buchholz D, Chagas LG, Vaalma C, Wu LM, and Passerini S (2014). Water sensitivity of layered P2/P3-$Na_xNi_{0.22}Co_{0.11}Mn_{0.66}O_2$ cathode material, *Journal of Materials Chemistry A*, **2**, pp. 13415–13421.

56. Chen H, Wu ZG, Zheng Z, Chen TR, Guo XD, Li JT, and Zhong BH (2018). Tuning the component ratio and corresponding sodium storage properties of layer-tunnel hybrid $Na_{0.6}Mn_{1-x}Ni_xO_2$ cathode by a simple cationic Ni^{2+} doping strategy, *Electrochimica Acta*, **273**, pp. 63–70.

57. Dang RB, Li Q, Chen MM, Hu ZB, and Xiao XL (2019). CuO-coated and Cu^{2+}-doped Co-modified P2-type $Na_{-2/3}$ $Ni_{1/3}Mn_{2/3}O_{-2}$ for sodium-ion batteries, *Physical Chemistry Chemical Physics*, **21**, pp. 314–321.

58. Gao GF, Tie D, Ma H, Yu HJ, Shi SS, Wang B, Xu SM, Wang LL, and Zhao YF (2018). Interface-rich mixed P2+T phase $Na_xCo_{0.1}Mn_{0.9}O_2$ ($0.44 <= x <= 0.7$) toward fast and high capacity sodium storage, *Journal of Materials Chemistry A*, **6**, pp. 6675–6684.

59. Chen J, Li LJ, Wu L, Yao Q, Yang HP, Liu ZS, Xia LF, Chen ZY, Duan JF, and Zhong SK (2018). Enhanced cycle stability of $Na_{0.9}Ni_{0.45}Mn_{0.55}O_2$ through tailoring O3/P2 hybrid structures for sodium-ion batteries, *Journal of Power Sources*, **406**, pp. 110–117.

60. Chen TR, Wu ZG, Xiang W, Wang EH, Wu CJ, Chen MZ, Guo XD, and Zhong BH (2017). The influences of sodium sources on the structure evolution and electrochemical performances of layered-tunnel hybrid $Na_{0.6}MnO_2$ cathode, *Ceramics International*, **43**, pp. 6303–6311.

61. Aragon MJ, Lavela P, Ortiz G, Alcantara R, and Tirado JL (2017). Nanometric P2-$Na_{2/3}Fe_{1/3}Mn_{2/3}O_2$ with controlled morphology as

cathode for sodium-ion batteries, *Journal of Alloys and Compounds*, **724**, pp. 465–473.

62. Lee E, Brown DE, Alp EE, Ren Y, Lu J, Woo JJ, and Johnson CS (2015). New insights into the performance degradation of Fe-based layered oxides in sodium-ion batteries: Instability of Fe^{3+}/Fe^{4+} redox in alpha-$NaFeO_2$, *Chemistry of Materials*, **27**, pp. 6755–6764.

63. Susanto D, Cho MK, Ali G, Kim J-Y, Chang HJ, Kim H-S, Nam K-W, and Chung KY (2019). Anionic redox activity as a key factor in the performance degradation of $NaFeO_2$ cathodes for sodium ion batteries, *Chemistry of Materials*, **31**, pp. 3644–3651.

64. Xu JL, Han Z, Jiang KZ, Bai PL, Liang Y, Zhang XY, Wang P, Guo SH, and Zhou HS (2020). Suppressing cation migration and reducing particle cracks in a layered Fe-based cathode for advanced sodium-ion batteries, *Small*, **16**, e1904388.

65. Yuan DD, Hu XH, Qian JF, Pei F, Wu FY, Mao RJ, Ai XP, Yang HX, and Cao YL (2014). P2-type $Na_{0.67}Mn_{0.65}Fe_{0.2}Ni_{0.15}O_2$ cathode material with high-capacity for sodium-ion battery, *Electrochimica Acta*, **116**, pp. 300–305.

66. Wang X, Liu G, Iwao T, Okubo M, and Yamada A (2014). Role of ligand-to-metal charge transfer in O3-type $NaFeO_2$–$NaNiO_2$ solid solution for enhanced electrochemical properties, *The Journal of Physical Chemistry C*, **118**, pp. 2970–2976.

67. Singh G, Aguesse F, Otaegui L, Goikolea E, Gonzalo E, Segalini J, and Rojo T (2015). Electrochemical performance of NaFe $(Ni_{0.5}Ti_{0.5})_1$–O2 (x = 0.2 and x = 0.4) cathode for sodium-ion battery, *Journal of Power Sources*, **273**, pp. 333–339.

68. Yabuuchi N, Kajiyama M, Iwatate J, Nishikawa H, Hitomi S, Okuyama R, Usui R, Yamada Y, and Komaba S (2012). P2-type $Na_{(x)}[Fe_{(1/2)}Mn_{(1/2)}]O_2$ made from earth-abundant elements for rechargeable Na batteries, *Nature Materials*, **11**, pp. 512–517.

69. Durai L, Gopalakrishnan A, and Badhulika S (2020). One-step solid-state reaction synthesis of β-$NaFeO_2$ nanopebble as high capacity cathode material for sodium ion batteries, *Materials Letters*, **270**, pp. 127739.

70. Xing Y, Akbar M, Yousaf M, Shah MAKY, Xia C, Gao J, and Wang X (2020). CeO_2 coated $NaFeO_2$ proton-conducting electrolyte for solid oxide fuel cell, *International Journal of Hydrogen Energy*. DOI: 10.1016/j.ijhydene.2020.05.219.

71. Gao Y, Wang Z, and Lu G (2019). Atomistic understanding of structural evolution, ion transport and oxygen stability in layered $NaFeO_2$, *Journal of Materials Chemistry A*, **7**, pp. 2619–2625.

72. Molenda J (1990). Electronic and electrochemical properties of nickel bronze, $NaxNiO_2$, *Solid State Ionics*, **38**, pp. 1–4.

73. Braconnier JJ, Delmas C, and Hagenmuller P (1982). Etude par desintercalation electrochimique des systemes $NaxCrO_2$ et $NaxNiO_2$, *Materials Research Bulletin*, **17**, pp. 993–1000.

74. Vassilaras P, Ma X, Li X, and Ceder G (2012). Electrochemical properties of monoclinic $NaNiO_2$, *Journal of The Electrochemical Society*, **160**, pp. A207–A211.

75. Han MH, Gonzalo E, Casas-Cabanas M, and Rojo T (2014). Structural evolution and electrochemistry of monoclinic $NaNiO_2$ upon the first cycling process, *Journal of Power Sources*, **258**, pp. 266–271.

76. Yuan DD, Wang YX, Cao YL, Ai XP, and Yang HX (2015). Improved electrochemical performance of Fe-substituted $NaNi_{0.5}Mn_{0.5}O_2$ cathode materials for sodium-ion batteries, *ACS Applied Materials & Interfaces*, **7**, pp. 8585–8591.

77. Zheng L, Fielden R, Bennett JC, and Obrovac MN (2019). Hexagonal and monoclinic $NaNi_{0.8}Co_{0.15}Al_{0.05}O_2$ (Na-NCA) for sodium ion batteries, *Journal of Power Sources*, **433**, pp. 226698.

78. Zhou P, Liu X, Weng J, Wang L, Wu X, Miao Z, Zhao J, Zhou J, and Zhuo S (2019). Synthesis, structure, and electrochemical properties of O′3-type monoclinic $NaNi_{0.8}Co_{0.15}Al_{0.05}O_2$ cathode materials for sodium-ion batteries, *Journal of Materials Chemistry A*, **7**, pp. 657–663.

79. Choi J, Kim KH, Jung CH, and Hong SH (2019). A P2-type $Na_{0.7}(Ni_{0.6}Co_{0.2}Mn_{0.2})O_{2}$ cathode with excellent cyclability and rate capability for sodium ion batteries, *ChemComm*, **55**, pp. 11575–11578.

80. Fielden R and Obrovac MN (2015). Investigation of the $NaNi_xMn_{(1-x)}O_{(2)}$ (0 <= x <= 1) system for Na-ion battery cathode materials, *Journal of the Electrochemical Society*, **162**, pp. A453–A459.

81. Maletti S, Sarapulova A, Schokel A, and Mikhailova D (2019). Operando studies on the $NaNi_{0.5}Ti_{0.5}O_2$ cathode for Na-ion batteries: Elucidating titanium as a structure stabilizer, *ACS Applied Materials & Interfaces*, **11**, pp. 33923–33930.

82. Komaba S, Yabuuchi N, Nakayama T, Ogata A, Ishikawa T, and Nakai I (2012). Study on the reversible electrode reaction of $Na_{1-x}Ni_{0.5}Mn_{0.5}O_2$ for a rechargeable sodium-ion battery, *Inorganic Chemistry*, **51**, pp. 6211–6220.

83. Wang H, Yang B, Liao X-Z, Xu J, Yang D, He Y-S, and Ma Z-F (2013). Electrochemical properties of P2-$Na_{2/3}[Ni_{1/3}Mn_{2/3}]O_2$ cathode material for sodium ion batteries when cycled in different voltage ranges, *Electrochimica Acta*, **113**, pp. 200–204.

84. Yuan DD, Wang YX, Cao YL, Ai XP, and Yang HX (2015). Improved electrochemical performance of Fe-substituted $NaNi_{0.5}Mn_{0.5}O_2$ cathode materials for sodium-ion batteries, *ACS Applied Materials & Interfaces* **7**, pp. 8585–8591.

85. Shanmugam R and Lai W (2014). $Na_{2/3}Ni_{1/3}Ti_{2/3}O_2$: "Bi-functional" electrode materials for Na-ion batteries, *ECS Electrochemistry Letters*, **3**, pp. A23–A25.

86. Gupta A, Mullins CB, and Goodenough JB (2013). $Na_2Ni_2TeO_6$: Evaluation as a cathode for sodium battery, *Journal of Power Sources*, **243**, pp. 817–821.

87. Saadoune I, Maazaz A, Ménétrier M, and Delmas C (1996). On the $Na_xNi_{0.6}Co_{0.4}O_2$ system: Physical and electrochemical studies, *Journal of Solid State Chemistry* **122**, pp. 111–117.

88. Kim D, Kang S-H, Slater M, Rood S, Vaughey JT, Karan N, Balasubramanian M, and Johnson CS (2011). Enabling sodium batteries using lithium-substituted sodium layered transition metal oxide cathodes, *Advanced Energy Materials*, **1**, pp. 333–336.

89. Lee E, Lu J, Ren Y, Luo X, Zhang X, Wen J, Miller D, DeWahl A, Hackney S, Key B, Kim D, Slater MD, and Johnson CS (2014). Layered P2/O3 intergrowth cathode: Toward high power Na-ion batteries, *Advanced Energy Materials*, **4**, pp. 1400458.

90. Guo SH, Liu P, Yu HJ, Zhu YB, Chen MW, Ishida M, and Zhou HS (2015). A layered P2-and O3-type composite as a high-energy cathode for rechargeable sodium-ion batteries, *Angewandte Chemie-International Edition*, **54**, pp. 5894–5899.

91. Yuan DD, Liang XM, Wu L, Cao YL, Ai XP, Feng JW, and Yang HX (2014). A honeycomb-layered $Na_3Ni_2SbO_6$: A high-rate and cycle-stable cathode for sodium-ion batteries, *Advanced Materials*, **26**, pp. 6301–6306.

92. Ma J, Bo SH, Wu LJ, Zhu YM, Grey CP, and Khalifah PG (2015). Ordered and disordered polymorphs of $Na(Ni_{2/3}Sb_{1/3})O_{-2}$: Honeycomb-ordered cathodes for Na-ion batteries, *Chemistry of Materials*, **27**, pp. 2387–2399.

93. Komaba S, Takei C, Nakayama T, Ogata A, and Yabuuchi N (2010). Electrochemical intercalation activity of layered $NaCrO_2$ vs. $LiCrO_2$, *Electrochemistry Communications*, **12**, pp. 355–358.

94. Kubota K, Ikeuchi I, Nakayama T, Takei C, Yabuuchi N, Shiiba H, Nakayama M, and Komaba S (2014). New insight into structural evolution in layered $NaCrO_2$ during electrochemical sodium extraction, *The Journal of Physical Chemistry C*, **119**, pp. 166–175.

95. Bo SH, Li X, Toumar AJ, and Ceder G (2016). Layered-to-rock-salt transformation in desodiated Na_xCrO_2 (x 0.4), *Chemistry of Materials*, **28**, pp. 1419–1429.

96. Ding J-J, Zhou Y-N, Sun Q, and Fu Z-W (2012). Cycle performance improvement of $NaCrO_2$ cathode by carbon coating for sodium ion batteries, *Electrochemistry Communications*, **22**, pp. 85–88.

97. Yu CY, Park JS, Jung HG, Chung KY, Aurbach D, Sun YK, and Myung ST (2015). $NaCrO_2$ cathode for high-rate sodium-ion batteries, *Energy & Environmental Science*, **8**, pp. 2019–2026.

98. Chen C-Y, Matsumoto K, Nohira T, Hagiwara R, Fukunaga A, Sakai S, Nitta K, and Inazawa S (2013). Electrochemical and structural investigation of $NaCrO_2$ as a positive electrode for sodium secondary battery using inorganic ionic liquid NaFSA–KFSA, *Journal of Power Sources*, **237**, pp. 52–57.

99. Zhou Y-N, Ding J-J, Nam K-W, Yu X, Bak S-M, Hu E, Liu J, Bai J, Li H, Fu Z-W, and Yang X-Q (2013). Phase transition behavior of $NaCrO_2$ during sodium extraction studied by synchrotron-based X-ray diffraction and absorption spectroscopy, *Journal of Materials Chemistry A*, **1**, pp. 11130.

100. Luo M, Ortiz AL, and Shaw LL (2020). Enhancing the electrochemical performance of $NaCrO_2$ through structural defect control, *ACS Applied Energy Materials*, **3**, pp. 7216–7227.

101. Wei J, Shaw L, and Chen W (2020). First-principles prediction of Na diffusivity in doped $NaCrO_2$ layered cathode materials with van der Waals interactions, *The Journal of Physical Chemistry C*, **124**, pp. 12239–12248.

102. Wang YS, Cui PX, Zhu W, Feng ZM, Vigeant MJ, Demers H, Guerfi A, and Zaghib K (2019). Enhancing the electrochemical performance of an O-3-$NaCrO_2$ cathode in sodium-ion batteries by cation substitution, *Journal of Power Sources*, **435**, 226760.

103. Xi KY, Chu SF, Zhang XY, Zhang XP, Zhang HY, Xu H, Bian JJ, Fang TC, Guo SH, Liu P, Chen MW, and Zhou HS (2020). A high-performance layered Cr-based cathode for sodium-ion batteries, *Nano Energy*, **67**, 104215.

104. Liang L, Sun X, Denis DK, Zhang J, Hou L, Liu Y, and Yuan C (2019). Ultralong layered $NaCrO_2$ nanowires: A competitive wide-temperature-

operating cathode for extraordinary high-rate sodium-ion batteries, *ACS Applied Materials & Interfaces*, **11**, pp. 4037–4046.

105. Liang L, Zhang W, Denis DK, Zhang J, Hou L, Liu Y, and Yuan C (2019). Comparative investigations of high-rate $NaCrO_2$ cathodes towards wide-temperature-tolerant pouch-type Na-ion batteries from −15 to 55°C: Nanowires vs. bulk, *Journal of Materials Chemistry A*, **7**, pp. 11915–11927.

106. McQueen TM, Stephens PW, Huang Q, Klimczuk T, Ronning F, and Cava RJ (2008). Successive orbital ordering transitions in $NaVO_2$, *Phys Rev Lett*, **101**, pp. 166402.

107. Onoda M (2008). Geometrically frustrated triangular lattice system Na_xVO_2: Superparamagnetism in x = 1 and trimerization in x ≈ 0.7, *Journal of Physics: Condensed Matter*, **20**, pp. 145205.

108. Guignard M, Didier C, Darriet J, Bordet P, Elkaim E, and Delmas C (2013). P2-$Na_{(x)}VO_2$ system as electrodes for batteries and electron-correlated materials, *Nature Materials*, **12**, pp. 74–80.

109. Hamani D, Ati M, Tarascon J-M, and Rozier P (2011). Na_xVO_2 as possible electrode for Na-ion batteries, *Electrochemistry Communications*, **13**, pp. 938–941.

110. Didier C, Guignard M, Darriet J, and Delmas C (2012). O'3-$Na_{(x)}VO_2$ system: A superstructure for $Na_{(1/2)}VO_2$, *Inorganic Chemistry*, **51**, pp. 11007–11016.

111. Li Y, Yang Z, Xu S, Mu L, Gu L, Hu YS, Li H, and Chen L (2015). Air-stable copper-based P2-$Na_{7/9}Cu_{2/9}Fe_{1/9}Mn_{2/3}O_2$ as a new positive electrode material for sodium-ion batteries, *Advanced Science (Weinheim)*, **2**, pp. 1500031.

112. Mu LQ, Xu SY, Li YM, Hu YS, Li H, Chen LQ, and Huang XJ (2015). Prototype sodium-ion batteries using an air-stable and Co/Ni-free O-3-layered metal oxide cathode, *Advanced Materials*, **27**, pp. 6928.

113. Dhariwal M, Pisani L, and Maitra T (2014). Competing electronic states in high temperature phase of $NaTiO_{(2)}$, *Journal of Physics Condensed Matter*, **26**, pp. 205501.

114. Biswas S, Mukherjee D, De S, and Kathiravan A (2016). Probing the highly efficient electron transfer dynamics between zinc protoporphyrin IX and sodium titanate nanosheets, *Journal of Physical Chemistry A*, **120**, pp. 7121–7129.

115. Clarke SJ, Fowkes AJ, Harrison A, Ibberson RM, and Rosseinsky MJ (1998). Synthesis, structure, and magnetic properties of $NaTiO_2$, *Chemistry of Materials*, **10**, pp. 372–384.

116. Vasileiadis A and Wagemaker M (2017). Thermodynamics and kinetics of Na-ion insertion into hollandite-TiO_2 and O3-layered $NaTiO_2$: An unexpected link between two promising anode materials for Na-ion batteries, *Chemistry of Materials*, **29**, pp. 1076–1088.

117. Wu D, Li X, Xu B, Twu N, Liu L, and Ceder G (2015). $NaTiO_2$: A layered anode material for sodium-ion batteries, *Energy & Environmental Science*, **8**, pp. 195–202.

118. Yoshida H, Yabuuchi N, Kubota K, Ikeuchi I, Garsuch A, Schulz-Dobrick M, and Komaba S (2014). P2-type $Na_{2/3}Ni_{1/3}Mn_{2/3-x}Ti_xO_2$ as a new positive electrode for higher energy Na-ion batteries, *ChemComm*, **50**, pp. 3677–3680.

119. Moreau P, Guyomard D, Gaubicher J, and Boucher F (2010). Structure and stability of sodium intercalated phases in olivine $FePO_4$, *Chemistry of Materials*, **22**, pp. 4126–4128.

120. Avdeev M, Mohamed Z, Ling CD, Lu J, Tamaru M, Yamada A, and Barpanda P (2013). Magnetic structures of $NaFePO_4$ maricite and triphylite polymorphs for sodium-ion batteries, *Inorganic Chemistry*, **52**, pp. 8685–8693.

121. Galceran M, Saurel D, Acebedo B, Roddatis VV, Martin E, Rojo T, and Casas-Cabanas M (2014). The mechanism of $NaFePO_{(4)}$ (de)sodiation determined by in situ X-ray diffraction, *Physical Chemistry Chemical Physics*, **16**, pp. 8837–8842.

122. Kim J, Seo D-H, Kim H, Park I, Yoo J-K, Jung S-K, Park Y-U, Goddard III WA, and Kang K (2015). Unexpected discovery of low-cost maricite $NaFePO_4$ as a high-performance electrode for Na-ion batteries, *Energy & Environmental Science*, **8**, pp. 540–545.

123. Saracibar A, Carrasco J, Saurel D, Galceran M, Acebedo B, Anne H, Lepoitevin M, Rojo T, and Casas Cabanas M (2016). Investigation of sodium insertion-extraction in olivine Na_xFePO_4 (0 </=x</= 1) using first-principles calculations, *Physical Chemistry Chemical Physics*, **18**, pp. 13045–13051.

124. Oh S-M, Myung S-T, Hassoun J, Scrosati B, and Sun Y-K (2012). Reversible $NaFePO_4$ electrode for sodium secondary batteries, *Electrochemistry Communications*, **22**, pp. 149–152.

125. Jian Z, Han W, Liang Y, Lan Y, Fang Z, Hu Y-S, and Yao Y (2014). Carbon-coated rhombohedral $Li_3V_2(PO_4)_3$ as both cathode and anode materials for lithium-ion batteries: Electrochemical performance and lithium storage mechanism, *Journal of Materials Chemistry A*, **2**, pp. 20231–20236.

126. Jian Z, Yuan C, Han W, Lu X, Gu L, Xi X, Hu Y-S, Li H, Chen W, Chen D, Ikuhara Y, and Chen L (2014). Atomic structure and kinetics of NASICON $Na_xV_2(PO_4)_3$ cathode for sodium-ion batteries, *Advanced Functional Materials*, **24**, pp. 4265–4272.

127. Song W, Ji X, Wu Z, Zhu Y, Yang Y, Chen J, Jing M, Li F, and Banks CE (2014). First exploration of Na-ion migration pathways in the NASICON structure $Na_3V_2(PO_4)_3$, *Journal of Materials Chemistry A*, **2**.

128. Wang Q, Zhang M, Zhou C, and Chen Y (2018). Concerted ion-exchange mechanism for sodium diffusion and its promotion in $Na_3V_2(PO_4)_3$ Framework, *The Journal of Physical Chemistry C*, **122**, pp. 16649–16654.

129. Saravanan K, Mason CW, Rudola A, Wong KH, and Balaya P (2013). The first report on excellent cycling stability and superior rate capability of $Na_3V_2(PO_4)_3$ for sodium ion batteries, *Advanced Energy Materials*, **3**, pp. 444–450.

130. Jian Z, Han W, Lu X, Yang H, Hu Y-S, Zhou J, Zhou Z, Li J, Chen W, Chen D, and Chen L (2013). Superior electrochemical performance and storage mechanism of $Na_3V_2(PO_4)_3$ cathode for room-temperature sodium-ion batteries, *Advanced Energy Materials*, **3**, pp. 156–160.

131. Zhou W, Xue L, Lu X, Gao H, Li Y, Xin S, Fu G, Cui Z, Zhu Y, and Goodenough JB (2016). $Na_xMV(PO_4)_3$ (M = Mn, Fe, Ni) structure and properties for sodium extraction, *Nano Letters*, **16**, pp. 7836–7841.

132. Wang D, Bie X, Fu Q, Dixon D, Bramnik N, Hu YS, Fauth F, Wei Y, Ehrenberg H, Chen G, and Du F (2017). Sodium vanadium titanium phosphate electrode for symmetric sodium-ion batteries with high power and long lifespan, *Nature Communication*, **8**, pp. 15888.

133. Gabelica-Robert M, Goreaud M, Labbe Ph, and Raveau B (1982). The pyrophosphate $NaFeP_2O_7$: A cage structure, *Journal of Solid State Chemistry*, **45**, pp. 389–395.

134. Barpanda P, Nishimura S-I, and Yamada A (2012). High-voltage pyrophosphate cathodes, *Advanced Energy Materials*, **2**, pp. 841–859.

135. Kee Y, Dimov N, Staikov A, Barpanda P, Lu Y-C, Minami K, and Okada S (2015). Insight into the limited electrochemical activity of $NaVP_2O_7$, *RSC Advances*, **5**, pp. 64991–64996.

136. Drozhzhin OA, Tertov IV, Alekseeva AM, Aksyonov DA, Stevenson KJ, Abakumov AM, and Antipov EV (2019). β-$NaVP_2O_7$ as a superior electrode material for Na-ion batteries, *Chemistry of Materials*, **31**, pp. 7463–7469.

137. Kim H, Shakoor RA, Park C, Lim SY, Kim J-S, Jo YN, Cho W, Miyasaka K, Kahraman R, Jung Y, and Choi JW (2013). $Na_2FeP_2O_7$ as a promising iron-based pyrophosphate cathode for sodium rechargeable batteries: A combined experimental and theoretical study, *Advanced Functional Materials*, **23**, pp. 1147–1155.

138. Kim H, Park CS, Choi JW, and Jung Y (2016). Defect-controlled formation of triclinic $Na_2CoP_2O_7$ for 4 V sodium-ion batteries, *Angewandte Chemie International Edition*, **55**, pp. 6662–6666.

139. Barpanda P, Ye T, Nishimura S-i, Chung S-C, Yamada Y, Okubo M, Zhou H, and Yamada A (2012). Sodium iron pyrophosphate: A novel 3.0 V iron-based cathode for sodium-ion batteries, *Electrochemistry Communications*, **24**, pp. 116–119.

140. Barpanda P, Lu J, Ye T, Kajiyama M, Chung S-C, Yabuuchi N, Komaba S, and Yamada A (2013). A layer-structured $Na_2CoP_2O_7$ pyrophosphate cathode for sodium-ion batteries, *RSC Advances*, **3**.

141. Avdeev M, Wang CW, Barpanda P, Fujii K, and Yashima M (2019). Polymorphism and temperature-induced phase transitions of $Na_2CoP_2O_7$, *Inorganic Chemistry*, **58**, pp. 16823–16830.

142. Park CS, Kim H, Shakoor RA, Yang E, Lim SY, Kahraman R, Jung Y, and Choi JW (2013). Anomalous manganese activation of a pyrophosphate cathode in sodium ion batteries: A combined experimental and theoretical study, *Journal of the American Chemical Society*, **135**, pp. 2787–2792.

143. Ling M, Lv Z, Li F, Zhao J, Zhang H, Hou G, Zheng Q, and Li X (2020). Revisiting of tetragonal $NaVPO_4F$: A high energy density cathode for sodium-ion batteries, *ACS Applied Materials & Interfaces*, **12**, pp. 30510–30519.

144. Barker J, Saidi MY, and Swoyer JL (2003). A sodium-ion cell based on the fluorophosphate compound NaVPO[sub 4]F, *Electrochemical and Solid-State Letters*, **6**.

145. Ling M, Li F, Yi H, Li X, Hou G, Zheng Q, and Zhang H (2018). Superior Na-storage performance of molten-state-blending-synthesized monoclinic $NaVPO_4F$ nanoplates for Na-ion batteries, *Journal of Materials Chemistry A*, **6**, pp. 24201–24209.

146. Ellis BL, Makahnouk WRM, Rowan-Weetaluktuk WN, Ryan DH, and Nazar LF (2010). Crystal structure and electrochemical properties of A_2MPO_4F fluorophosphates (A = Na, Li; M = Fe, Mn, Co, Ni)†, *Chemistry of Materials*, **22**, pp. 1059–1070.

147. Liu Q, Wang D, Yang X, Chen N, Wang C, Bie X, Wei Y, Chen G, and Du F (2015). Carbon-coated $Na_3V_2(PO_4)_2F_3$ nanoparticles embedded in a

mesoporous carbon matrix as a potential cathode material for sodium-ion batteries with superior rate capability and long-term cycle life, *Journal of Materials Chemistry A*, **3**, pp. 21478–21485.

148. Shakoor RA, Seo D-H, Kim H, Park Y-U, Kim J, Kim S-W, Gwon H, Lee S, and Kang K (2012). A combined first principles and experimental study on $Na_3V_2(PO_4)_2F_3$ for rechargeable Na batteries, *Journal of Materials Chemistry*, **22**.

149. Bianchini M, Fauth F, Brisset N, Weill F, Suard E, Masquelier C, and Croguennec L (2015). Comprehensive investigation of the $Na_3V_2(PO_4)_2F_3$–$NaV_2(PO_4)_2F_3$ system by operando high resolution synchrotron X-ray diffraction, *Chemistry of Materials*, **27**, pp. 3009–3020.

150. Li Q, Liu Z, Zheng F, Liu R, Lee J, Xu GL, Zhong G, Hou X, Fu R, Chen Z, Amine K, Mi J, Wu S, Grey CP, and Yang Y (2018). Identifying the structural evolution of the sodium ion battery Na_2FePO_4F cathode, *Angewandte Chemie International Edition*, **57**, pp. 11918–11923.

151. Kubota K, Yokoh K, Yabuuchi N, and Komaba S (2014). Na_2CoPO_4F as a high-voltage electrode material for Na-ion batteries, *Electrochemistry*, **82**, pp. 909–911.

152. Kim SW, Seo DH, Kim H, Park KY, and Kang K (2012). A comparative study on Na_2MnPO_4F and Li_2MnPO_4F for rechargeable battery cathodes, *Physical Chemistry Chemical Physics*, **14**, pp. 3299–3303.

153. Barpanda P, Chotard JN, Recham N, Delacourt C, Ati M, Dupont L, Armand M, and Tarascon JM (2010). Structural, transport, and electrochemical investigation of novel $AMSO_4F$ (A = Na, Li; M = Fe, Co, Ni, Mn) metal fluorosulphates prepared using low temperature synthesis routes, *Inorganic Chemistry*, **49**, pp. 7401–7413.

154. Tripathi R, Gardiner GR, Islam MS, and Nazar LF (2011). Alkali-ion conduction paths in $LiFeSO_4F$ and $NaFeSO_4F$ tavorite-type cathode materials, *Chemistry of Materials*, **23**, pp. 2278–2284.

155. Recham N, Rousse G, Sougrati MT, Chotard J-N, Frayret C, Mariyappan S, Melot BC, Jumas J-C, and Tarascon J-M (2012). Preparation and characterization of a stable $FeSO_4F$-based framework for alkali ion insertion electrodes, *Chemistry of Materials*, **24**, pp. 4363–4370.

156. Pan W, Guan W, Liu S, Xu BB, Liang C, Pan H, Yan M, and Jiang Y (2019). $Na_2Fe(SO_4)_2$: An anhydrous 3.6 V, low-cost and good-safety cathode for a rechargeable sodium-ion battery, *Journal of Materials Chemistry A*, **7**, pp. 13197–13204.

157. Barpanda P, Oyama G, Ling CD, and Yamada A (2014). Kröhnkite-type $Na_2Fe(SO_4)_2{\cdot}2H_2O$ as a novel 3.25 V insertion compound for Na-ion batteries, *Chemistry of Materials*, **26**, pp. 1297–1299.

158. Reynaud M, Boulineau S, Sougrati M T, Melot B C, Rousse G, Chotard J-N, and Tarascon J-M (2013). Bimetallic sulfates $A_2M(SO=)_2{\cdot}nH_2O$ (A = Li, Na and M = Transition Metal): As new attractive electrode materials for Li- and Na-ion batteries, *ECS Transactions*, **50**, pp. 11–19.

159. Barpanda P, Oyama G, Nishimura S, Chung SC, and Yamada A (2014). A 3.8-V earth-abundant sodium battery electrode, *Nature Communications*, **5**, pp. 4358.

160. Oyama G, Pecher O, Griffith KJ, Nishimura S-I, Pigliapochi R, Grey CP, and Yamada A (2016). Sodium intercalation mechanism of 3.8 V class alluaudite sodium iron sulfate, *Chemistry of Materials*, **28**, pp. 5321–5328.

161. Liu X, Tang L, Xu Q, Liu H, and Wang Y (2019). Ultrafast and ultrastable high voltage cathode of $Na_{2+2x}Fe_{2-x}(SO_4)_3$ microsphere scaffolded by graphene for sodium ion batteries, *Electrochimica Acta*, **296**, pp. 345–354.

162. Dwibedi D, Araujo RB, Chakraborty S, Shanbogh PP, Sundaram NG, Ahuja R, and Barpanda P (2015). $Na_{2.44}Mn_{1.79}(SO_4)_3$: A new member of the alluaudite family of insertion compounds for sodium ion batteries, *Journal of Materials Chemistry A*, **3**, pp. 18564–18571.

163. Dwibedi D, Gond R, Dayamani A, Araujo RB, Chakraborty S, Ahuja R, and Barpanda P (2016). $Na_{2.32}Co_{1.84}(SO_4)_3$ as a new member of the alluaudite family of high-voltage sodium battery cathodes, *Dalton Transactions*, **46**, pp. 55–63.

164. Chen H, Hautier G, and Ceder G (2012). Synthesis, computed stability, and crystal structure of a new family of inorganic compounds: Carbonophosphates, *Journal of the American Chemical Society*, **134**, pp. 19619–19627.

165. Chen H, Hao Q, Zivkovic O, Hautier G, Du L-S, Tang Y, Hu Y-Y, Ma X, Grey CP, and Ceder G (2013). Sidorenkite ($Na_3MnPO_4CO_3$): A new intercalation cathode material for Na-ion batteries, *Chemistry of Materials*, **25**, pp. 2777–2786.

166. Huang W, Zhou J, Li B, Ma J, Tao S, Xia D, Chu W, and Wu Z (2014). Detailed investigation of $Na_{2.24}FePO_4CO_3$ as a cathode material for Na-ion batteries, *Scientific Reports*, **4**, pp. 4188.

167. Paolella A, Faure C, Timoshevskii V, Marras S, Bertoni G, Guerfi A, Vijh A, Armand M, and Zaghib K (2017). A review on hexacyanoferrate-based

materials for energy storage and smart windows: Challenges and perspectives, *Journal of Materials Chemistry A*, **5**, pp. 18919–18932.

168. Zakaria MB and Chikyow TJCCR (2017). Recent advances in Prussian blue and Prussian blue analogues: Synthesis and thermal treatments, **352**, pp. 328–345.

169. Kraft AJBHC (2008). On the discovery and history of Prussian blue, *Bulletin for the history of chemistry / Division of the History of Chemistry of the American Chemical Society*, **33**, pp. 61–67.

170. Kaye SS and Long JR (2005). Hydrogen storage in the dehydrated Prussian blue analogues M_3 $[Co(CN)_6]_2$ (M = Mn, Fe, Co, Ni, Cu, Zn), *Journal of the American Chemical Society*, **127**, pp. 6506–6507.

171. Fiorito PA, Gonçales VR, Ponzio EA, and de Torresi SIC (2005). Synthesis, characterization and immobilization of Prussian blue nanoparticles. A potential tool for biosensing devices, *ChemComm*, **3**, pp. 366–368.

172. Thammawong C, Opaprakasit P, Tangboriboonrat P, and Sreearunothai P (2013). Prussian blue-coated magnetic nanoparticles for removal of cesium from contaminated environment, *Journal of Nanoparticle Research*, **15**, pp. 1689.

173. Yu DYW, Prikhodchenko PV, Mason CW, Batabyal SK, Gun J, Sladkevich S, Medvedev AG, and Lev O (2013). High-capacity antimony sulphide nanoparticle-decorated graphene composite as anode for sodium-ion batteries, *Nature Communications*, **4**, pp. 7.

174. Wang J, Luo C, Gao T, Langrock A, Mignerey AC, and Wang C (2015). An advanced MoS_2/carbon anode for high-performance sodium-ion batteries, *Small*, **11**, pp. 473–481.

175. Liang Y, Lai WH, Miao Z, and Chou SL (2018). Nanocomposite materials for the sodium-ion battery: A review, *Small*, **14**, 1702514.

176. You Y, Wu X-L, Yin Y-X, and Guo Y-G (2014). High-quality Prussian blue crystals as superior cathode materials for room-temperature sodium-ion batteries, *Energy & Environmental Science*, **7**, pp. 1643–1647.

177. Scrosati B and Garche J (2010). Lithium batteries: Status, prospects and future, *Journal of Power Sources*, **195**, pp. 2419–2430.

178. Maspoch D, Ruiz-Molina D, and Veciana J (2007). Old materials with new tricks: Multifunctional open-framework materials, *Chemical Society Reviews*, **36**, pp. 770–818.

179. Qian J, Yang H, and Wu C (2017). Recent progress and challenges in the development of Prussian blue analogues as new intercalation cathode materials, *Scientia Sinica Chimica*, **47**, pp. 603–613.

180. Wu X, Shao M, Wu C, Qian J, Cao Y, Ai X, and Yang H (2016). Low defect $FeFe(CN)_6$ framework as stable host material for high performance Li-ion batteries, *ACS Applied Materials & Interfaces*, **8**, pp. 23706–23712.

181. Wu X, Wu C, Wei C, Hu L, Qian J, Cao Y, Ai X, Wang J, and Yang H (2016). Highly crystallized $Na_{(2)}CoFe(CN)_{(6)}$ with suppressed lattice defects as superior cathode material for sodium-ion batteries, *ACS Applied Materials & Interfaces*, **8**, pp. 5393–5399.

182. Lu Y, Wang L, Cheng J, and Goodenough JB (2012). Prussian blue: A new framework of electrode materials for sodium batteries, *ChemComm*, **48**, pp. 6544–6546.

183. Wu X, Deng W, Qian J, Cao Y, Ai X, and Yang H (2013). Single-crystal $FeFe(CN)_6$ nanoparticles: A high capacity and high rate cathode for Na-ion batteries, *Journal of Materials Chemistry A*, **1**.

184. You Y, Yu X, Yin Y, Nam K-W, and Guo Y-G (2014). Sodium iron hexacyanoferrate with high Na content as a Na-rich cathode material for Na-ion batteries, *Nano Research*, **8**, pp. 117–128.

185. Li W-J, Chou S-L, Wang J-Z, Kang Y-M, Wang J-L, Liu Y, Gu Q-F, Liu H-K, and Dou S-X (2015). Facile method to synthesize Na-enriched $Na_{1+x}FeFe(CN)_6$ frameworks as cathode with superior electrochemical performance for sodium-ion batteries, *Chemistry of Materials*, **27**, pp. 1997–2003.

186. Wang L, Song J, Qiao R, Wray LA, Hossain MA, Chuang YD, Yang W, Lu Y, Evans D, Lee JJ, Vail S, Zhao X, Nishijima M, Kakimoto S, and Goodenough JB (2015). Rhombohedral prussian white as cathode for rechargeable sodium-ion batteries, *Journal of the American Chemical Society*, **137**, pp. 2548–2554.

187. Liu Y, Qiao Y, Zhang W, Li Z, Ji X, Miao L, Yuan L, Hu X, and Huang Y (2015). Sodium storage in Na-rich $Na_xFeFe(CN)_6$ nanocubes, *Nano Energy*, **12**, pp. 386–393.

188. Matsuda T, Takachi M, and Moritomo Y (2013). A sodium manganese ferrocyanide thin film for Na-ion batteries, *ChemComm*, **49**, pp. 2750–2752.

189. Song J, Wang L, Lu Y, Liu J, Guo B, Xiao P, Lee JJ, Yang XQ, Henkelman G, and Goodenough JB (2015). Removal of interstitial H_2O in hexacyanometallates for a superior cathode of a sodium-ion battery, *Journal of the American Chemical Society*, **137**, pp. 2658–2664.

190. Mao Y, Chen Y, Qin J, Shi C, Liu E, and Zhao N (2019). Capacitance controlled, hierarchical porous 3D ultra-thin carbon networks

reinforced prussian blue for high performance Na-ion battery cathode, *Nano Energy*, **58**, pp. 192–201.

191. You Y, Wu X-L, Yin Y-X, and Guo Y-G (2013). A zero-strain insertion cathode material of nickel ferricyanide for sodium-ion batteries, *Journal of Materials Chemistry A*, **1**.

192. Jiao S, Tuo J, Xie H, Cai Z, Wang S, and Zhu J (2017). The electrochemical performance of $Cu_3[Fe(CN)_6]_2$ as a cathode material for sodium-ion batteries, *Materials Research Bulletin*, **86**, pp. 194–200.

193. Lee H, Kim YI, Park JK, and Choi JW (2012). Sodium zinc hexacyanoferrate with a well-defined open framework as a positive electrode for sodium ion batteries, *ChemComm (Camb)*, **48**, pp. 8416–8418.

194. Xie M, Huang Y, Xu M, Chen R, Zhang X, Li L, and Wu F (2016). Sodium titanium hexacyanoferrate as an environmentally friendly and low-cost cathode material for sodium-ion batteries, *Journal of Power Sources*, **302**, pp. 7–12.

195. Lee H-W, Wang RY, Pasta M, Woo Lee S, Liu N, and Cui Y (2014). Manganese hexacyanomanganate open framework as a high-capacity positive electrode material for sodium-ion batteries, *Nature Communications*, **5**.

196. Yang D, Xu J, Liao XZ, Wang H, He YS, and Ma ZF (2015). Prussian blue without coordinated water as a superior cathode for sodium-ion batteries, *ChemComm (Camb)*, **51**, pp. 8181–8184.

197. Jiang Y, Yu S, Wang B, Li Y, Sun W, Lu Y, Yan M, Song B, and Dou S (2016). Prussian blue @C composite as an ultrahigh-rate and long-life sodium-ion battery cathode, *Advanced Functional Materials*, **26**, pp. 5315–5321.

198. You Y, Yao HR, Xin S, Yin YX, Zuo TT, Yang CP, Guo YG, Cui Y, Wan LJ, and Goodenough JB (2016). Subzero-temperature cathode for a sodium-ion battery, *Advanced Materials*, **28**, pp. 7243–7248.

199. Li W-J, Chou S-L, Wang J-Z, Wang J-L, Gu Q-F, Liu H-K, and Dou S-X (2015). Multifunctional conducing polymer coated $Na_{1+}MnFe(CN)_6$ cathode for sodium-ion batteries with superior performance via a facile and one-step chemistry approach, *Nano Energy*, **13**, pp. 200–207.

200. Zhang Q, Fu L, Luan J, Huang X, Tang Y, Xie H, and Wang H (2018). Surface engineering induced core-shell Prussian blue@polyaniline nanocubes as a high-rate and long-life sodium-ion battery cathode, *Journal of Power Sources*, **395**, pp. 305–313.

201. Huang Y, Xie M, Zhang J, Wang Z, Jiang Y, Xiao G, Li S, Li L, Wu F, and Chen R (2017). A novel border-rich Prussian blue synthetized by inhibitor control as cathode for sodium ion batteries, *Nano Energy*, **39**, pp. 273–283.

202. Tang X, Liu H, Su D, Notten PHL, and Wang G (2018). Hierarchical sodium-rich Prussian blue hollow nanospheres as high-performance cathode for sodium-ion batteries, *Nano Research*, **11**, pp. 3979–3990.

203. Yang D, Xu J, Liao X-Z, He Y-S, Liu H, and Ma Z-F (2014). Structure optimization of Prussian blue analogue cathode materials for advanced sodium ion batteries, *Chemical Communications*, **50**, pp. 13377–13380.

204. Xie M, Xu M, Huang Y, Chen R, Zhang X, Li L, and Wu F (2015). $Na_2Ni_xCo_{1-x}Fe(CN)_6$: A class of Prussian blue analogs with transition metal elements as cathode materials for sodium ion batteries, *Electrochemistry Communications*, **59**, pp. 91–94.

205. Wu F and Yushin G (2017). Conversion cathodes for rechargeable lithium and lithium-ion batteries, *Energy & Environmental Science*, **10**, pp. 435–459.

206. Nishijima M, Gocheva ID, Okada S, Doi T, Yamaki J-I, and Nishida T (2009). Cathode properties of metal trifluorides in Li and Na secondary batteries, *Journal of Power Sources*, **190**, pp. 558–562.

207. Ma D-l, Wang H-G, Li Y, Xu D, Yuan S, Huang X-l, Zhang X-B, and Zhang Y (2014). In situ generated FeF_3 in homogeneous iron matrix toward high-performance cathode material for sodium-ion batteries, *Nano Energy*, **10**, pp. 295–304.

208. Shen Y, Wang X, Hu H, Jiang M, Bai Y, Yang X, and Shu H (2015). Sheet-like structure FeF_3/graphene composite as novel cathode material for Na ion batteries, *RSC Advances*, **5**, pp. 38277–38282.

209. Bao T, Zhong H, Zheng H, Zhan H, and Zhou Y (2015). One-pot synthesis of FeF_3/graphene composite for sodium secondary batteries, *Materials Letters*, **158**, pp. 21–24.

210. Zhang L, Ji S, Yu L, Xu X, and Liu J (2017). Amorphous FeF_3/C nanocomposite cathode derived from metal–organic frameworks for sodium ion batteries, *RSC Advances*, **7**, pp. 24004–24010.

211. Shen Y, Wang X, Hu H, Jiang M, Yang X, and Shu H (2015). A graphene loading heterogeneous hydrated forms iron based fluoride nanocomposite as novel and high-capacity cathode material for lithium/sodium ion batteries, *Journal of Power Sources*, **283**, pp. 204–210.

212. Ali G, Oh SH, Kim SY, Kim JY, Cho BW, and Chung KY (2015). An open-framework iron fluoride and reduced graphene oxide nanocomposite

as a high-capacity cathode material for Na-ion batteries, *Journal of Materials Chemistry A*, 3, pp. 10258–10266.

213. Ali G, Lee J-H, Cho BW, Nam K-W, Ahn D, Chang W, Oh SH, and Chung KY (2016). Probing the sodiation-desodiation reactions in nano-sized iron fluoride cathode, *Electrochimica Acta*, **191**, pp. 307–316.
214. He K, Zhou Y, Gao P, Wang L, Pereira N, Amatucci GG, Nam K-W, Yang X-Q, Zhu Y, Wang F, and Su D (2014). Sodiation via heterogeneous disproportionation in FeF_2 electrodes for sodium-ion batteries, *ACS Nano*, **8**, pp. 7251–7259.
215. Li C, Yin C, Mu X, and Maier J (2013). Top-down synthesis of open framework fluoride for lithium and sodium batteries, *Chemistry of Materials*, **25**, pp. 962–969.
216. Li CL, Yin CL, Gu L, Dinnebier RE, Mu XK, van Aken PA, and Maier J (2013). An $FeF_3 \cdot 0.5H_{(2)}O$ polytype: A microporous framework compound with intersecting tunnels for Li and Na batteries, *Journal of the American Chemical Society*, **135**, pp. 11425–11428.
217. Martin A, Doublet M-L, Kemnitz E, and Pinna N (2018). Reversible sodium and lithium insertion in iron fluoride perovskites, *Advanced Functional Materials*, **28**, pp. 1802057.
218. Gocheva ID, Nishijima M, Doi T, Okada S, Yamaki J-I, and Nishida T (2009). Mechanochemical synthesis of $NaMF_3$ (M = Fe, Mn, Ni) and their electrochemical properties as positive electrode materials for sodium batteries, *Journal of Power Sources*, **187**, pp. 247–252.
219. Dimov N, Nishimura A, Chihara K, Kitajou A, Gocheva ID, and Okada S (2013). Transition metal $NaMF_3$ compounds as model systems for studying the feasibility of ternary Li-MF and Na-MF single phases as cathodes for lithium–ion and sodium–ion batteries, *Electrochimica Acta*, **110**, pp. 214–220.
220. Yamada Y, Doi T, Tanaka I, Okada S, and Yamaki J-I (2011). Liquid-phase synthesis of highly dispersed $NaFeF_3$ particles and their electrochemical properties for sodium-ion batteries, *Journal of Power Sources*, **196**, pp. 4837–4841.
221. Kitajou A, Komatsu H, Chihara K, Gocheva ID, Okada S, and Yamaki J-I (2012). Novel synthesis and electrochemical properties of perovskite-type $NaFeF_3$ for a sodium-ion battery, *Journal of Power Sources*, **198**, pp. 389–392.
222. Shakoor R, Lim SY, Kim H, Nam K-W, Kang JK, Kang K, and Choi JW (2012). Mechanochemical synthesis and electrochemical behavior of Na_3FeF_6 in sodium and lithium batteries, *Solid State Ionics*, **218**, pp. 35–40.

223. Nava-Avendaño J, Arroyo-de Dompablo ME, Frontera C, Ayllón JA, and Palacín MR (2015). Study of sodium manganese fluorides as positive electrodes for Na-ion batteries, *Solid State Ionics*, **278**, pp. 106–113.

224. Liu W, Li H, Xie J-Y, and Fu Z-W (2014). Rechargeable room-temperature CF_x-sodium battery, *ACS Applied Materials & Interfaces*, **6**, pp. 2209–2212.

225. Shao Y, Yue H, Qiao R, Hu J, Zhong G, Wu S, McDonald MJ, Gong Z, Zhu Z, and Yang W (2016). Synthesis and reaction mechanism of novel fluorinated carbon fiber as a high-voltage cathode material for rechargeable Na batteries, *Chemistry of Materials*, **28**, pp. 1026–1033.

226. Liu W, Shadike Z, Liu Z-C, Liu W-Y, Xie J-Y, and Fu Z-W (2015). Enhanced electrochemical activity of rechargeable carbon fluorides–sodium battery with catalysts, *Carbon*, **93**, pp. 523–532.

227. Liu W, Zhan BX, Shi B, Li Y, Guo R, Pei HJ, and Xie JY (2017). Sandwich-like structure of carbon fluoride/graphene oxide/polyacrylonitrile cathode for lithium and sodium batteries, *ChemElectroChem*, **4**, pp. 436–440.

228. Liu W, Li Y, Zhan B-X, Shi B, Guo R, Pei H-J, Xie J-Y, and Fu Z-W (2016). Amorphous, highly disordered carbon fluorides as a novel cathode for sodium secondary batteries, *The Journal of Physical Chemistry C*, **120**, pp. 25203–25209.

229. Liu W, Wang Y, Li Y, Shi B, Huang P, Guo R, Pei H, Zheng Y, Lu J, and Xie J (2018). Long-life sodium/carbon fluoride batteries with flexible, binder-free fluorinated mesocarbon microbead film electrodes, *ChemComm*, **54**, pp. 2341–2344.

230. Zhou YN, Sina M, Pereira N, Yu X, Amatucci GG, Yang XQ, Cosandey F, and Nam KW (2015). $FeO_{0.7}F_{1.3}$/C nanocomposite as a high-capacity cathode material for sodium-ion batteries, *Advanced Functional Materials*, **25**, pp. 696–703.

231. Zhu J and Deng D (2015). Wet-chemical synthesis of phase-pure FeOF nanorods as high-capacity cathodes for sodium-ion batteries, *Angewandte Chemie International Edition*, **54**, pp. 3079–3083.

232. Deng D (2017). Transition metal oxyfluorides for next-generation rechargeable batteries, *ChemNanoMat*, **3**, pp. 146–159.

233. Dunn B, Kamath H, and Tarascon J-M (2011). Electrical energy storage for the grid: A battery of choices, *Science*, **334**, pp. 928–935.

234. Wang Y, Zhou D, Palomares V, Shanmukaraj D, Sun B, Tang X, Wang C, Armand M, Rojo T, and Wang G (2020). Revitalising sodium–sulfur batteries for non-high-temperature operation: A crucial review, *Energy & Environmental Science*, **13**, pp. 3848–3879.

235. Wang Y-X, Zhang B, Lai W, Xu Y, Chou S-L, Liu H-K, and Dou S-X (2017). Room-temperature sodium–sulfur batteries: A comprehensive review on research progress and cell chemistry, *Advanced Energy Materials*, **7**, pp. 1602829.

236. Zang R, Li P, Guo X, Man Z, Zhang S, Wang C, and Wang G (2019). Yolk-shell N-doped carbon coated FeS_2 nanocages as a high-performance anode for sodium-ion batteries, *Journal of Materials Chemistry A*, **7**, pp. 14051–14059.

237. Momida H, Kitajou A, Okada S, Yamashita T, and Oguchi T (2015). Discharge reaction mechanisms in Na/FeS_2 batteries: First-principles calculations, *Journal of the Physical Society of Japan*, **84**, pp. 124709.

238. Kim T, Choi J, Ryu H, Cho G, Kim K, Ahn J, Cho K, and Ahn H (2007). Electrochemical properties of sodium/pyrite battery at room temperature, *Journal of Power Sources*, **174**, pp. 1275–1278.

239. Douglas A, Carter R, Oakes L, Share K, Cohn AP, and Pint CL (2015). Ultrafine iron pyrite (FeS_2) nanocrystals improve sodium–sulfur and lithium–sulfur conversion reactions for efficient batteries, *ACS Nano*, **9**, pp. 11156–11165.

240. Wang Y-X, Yang J, Chou S-L, Liu HK, Zhang W-x, Zhao D, and Dou SX (2015). Uniform yolk-shell iron sulfide–carbon nanospheres for superior sodium–iron sulfide batteries, *Nature Communications*, **6**, pp. 1–9.

241. Hu Z, Zhu Z, Cheng F, Zhang K, Wang J, Chen C, and Chen J (2015). Pyrite FeS_2 for high-rate and long-life rechargeable sodium batteries, *Energy & Environmental Science*, **8**, pp. 1309–1316.

242. Zhu Y, Suo L, Gao T, Fan X, Han F, and Wang C (2015). Ether-based electrolyte enabled Na/FeS_2 rechargeable batteries, *Electrochemistry Communications*, **54**, pp. 18–22.

243. Liu Y, Li Y, Kang H, Jin T, and Jiao L (2016). Design, synthesis, and energy-related applications of metal sulfides, *Materials Horizons*, **3**, pp. 402–421.

244. Xiao Y, Hwang J-Y, Belharouak I, and Sun Y-K (2017). Superior Li/Na-storage capability of a carbon-free hierarchical CoS_x hollow nanostructure, *Nano Energy*, **32**, pp. 320–328.

245. Shi B, Liu W, Zhu K, and Xie J (2017). Synthesis of flower-like copper sulfides microspheres as electrode materials for sodium secondary batteries, *Chemical Physics Letters*, **677**, pp. 70–74.

246. He Q, Rui K, Yang J, and Wen Z (2018). Fe_7S_8 nanoparticles anchored on nitrogen-doped graphene nanosheets as anode materials for high-

performance sodium-ion batteries, *ACS Applied Materials & Interfaces*, **10**, pp. 29476–29485.

247. Luo MH, Yu HX, Hu FY, Liu TT, Cheng X, Zheng RT, Bai Y, Shui M, and Shu J (2020). Metal selenides for high performance sodium ion batteries, *Chemical Engineering Journal*, **380**, pp. 32.

248. Hu Z, Liu Q, Chou S-L, and Dou S-X (2017). Advances and challenges in metal sulfides/selenides for next-generation rechargeable sodium-ion batteries, *Advanced Materials*, **29**, pp. 1700606.

249. Yue J-L, Sun Q, and Fu Z-W (2013). Cu_2Se with facile synthesis as a cathode material for rechargeable sodium batteries, *ChemComm*, **49**, pp. 5868–5870.

250. Chen G-Y, Sun Q, Yue J-L, Shadike Z, Yang Y, Ding F, Sang L, and Fu Z-W (2015). Conversion and displacement reaction types of transition metal compounds for sodium ion battery, *Journal of Power Sources*, **284**, pp. 115–121.

251. Gao P, Wall C, Zhang L, Reddy MA, and Fichtner M (2015). Vanadium oxychloride as electrode material for sodium ion batteries, *Electrochemistry Communications*, **60**, pp. 180–184.

252. Klein F, Jache B, Bhide A, and Adelhelm P (2013). Conversion reactions for sodium-ion batteries, *Physical Chemistry Chemical Physics*, **15**, pp. 15876–15887.

253. Zhao W, Zhong G, McDonald MJ, Gong Z, Liu R, Bai J, Yang C, Li S, Zhao W, and Wang H (2016). $Cu_3(PO_4)_2$/C composite as a high-capacity cathode material for rechargeable Na-ion batteries, *Nano Energy*, **27**, pp. 420–429.

254. Wu X, Jin S, Zhang Z, Jiang L, Mu L, Hu Y-S, Li H, Chen X, Armand M, Chen L, and Huang X (2015). Unraveling the storage mechanism in organic carbonyl electrodes for sodium-ion batteries, *Science Advances*, **1**, pp. e1500330.

255. Kim H, Kwon JE, Lee B, Hong J, Lee M, Park SY, and Kang K (2015). High energy organic cathode for sodium rechargeable batteries, *Chemistry of Materials*, **27**, pp. 7258–7264.

256. Chihara K, Chujo N, Kitajou A, and Okada S (2013). Cathode properties of $Na_2C_6O_6$ for sodium-ion batteries, *Electrochimica Acta*, **110**, pp. 240–246.

257. Wang Y, Ding Y, Pan L, Shi Y, Yue Z, Shi Y, and Yu G (2016). Understanding the size-dependent sodium storage properties of Na2C6O6-based organic electrodes for sodium-ion batteries, *Nano Letters*, **16**, pp. 3329–3334.

258. Wang C, Fang Y, Xu Y, Liang L, Zhou M, Zhao H, and Lei Y (2016). Manipulation of disodium rhodizonate: Factors for fast-charge and fast-discharge sodium-ion batteries with long-term cyclability, *Advanced Functional Materials*, **26**, pp. 1777–1786.

259. Luo C, Zhu Y, Xu Y, Liu Y, Gao T, Wang J, and Wang C (2014). Graphene oxide wrapped croconic acid disodium salt for sodium ion battery electrodes, *Journal of Power Sources*, **250**, pp. 372–378.

260. Yao M, Kuratani K, Kojima T, Takeichi N, Senoh H, and Kiyobayashi T (2014). Indigo carmine: An organic crystal as a positive-electrode material for rechargeable sodium batteries, *Scientific Reports*, **4**, pp. 3650.

261. Yin X, Sarkar S, Shi S, Huang Q-A, Zhao H, Yan L, Zhao Y, and Zhang J (2020). Recent progress in advanced organic electrode materials for sodium-ion batteries: Synthesis, mechanisms, challenges and perspectives, *Advanced Functional Materials*, **30**, pp. 1908445.

262. Luo W, Allen M, Raju V, and Ji X (2014). An organic pigment as a high-performance cathode for sodium-ion batteries, *Advanced Energy Materials*, **4**, pp. 1400554.

263. Renault S, Mihali VA, Edström K, and Brandell D (2014). Stability of organic Na-ion battery electrode materials: The case of disodium pyromellitic diimidate, *Electrochemistry Communications*, **45**, pp. 52–55.

264. Deng W, Shen Y, Qian J, Cao Y, and Yang H (2015). A perylene diimide crystal with high capacity and stable cyclability for Na-ion batteries, *ACS Applied Materials & Interfaces*, **7**, pp. 21095–21099.

265. Wang H-g, Yuan S, Ma D-l, Huang X-l, Meng F-l, and Zhang X-b (2014). Tailored aromatic carbonyl derivative polyimides for high-power and long-cycle sodium-organic batteries, *Advanced Energy Materials*, **4**, pp. 1301651.

266. Banda H, Damien D, Nagarajan K, Raj A, Hariharan M, Shaijumon MM (2017). Twisted perylene diimides with tunable redox properties for organic sodium-ion batteries., *Advanced Energy Materials*, **7**, pp. 1701316.

267. Xu F, Wang H, Lin J, Luo X, Cao S-A, and Yang H (2016). Poly(anthraquinonyl imide) as a high capacity organic cathode material for Na-ion batteries, *Journal of Materials Chemistry A*, **4**, pp. 11491–11497.

268. Poizot P, Gaubicher J, Renault S, Dubois L, Liang Y, and Yao Y (2020). Opportunities and challenges for organic electrodes in electrochemical energy storage, *Chemical Reviews*, **120**, pp. 6490–6557.

269. Hong J, Lee M, Lee B, Seo D-H, Park CB, and Kang K (2014). Biologically inspired pteridine redox centres for rechargeable batteries, *Nature Communications*, **5**, pp. 5335.

270. Zhao Q, Whittaker AK, and Zhao X (2018). Polymer electrode materials for sodium-ion batteries, *Materials*, **11**, pp. 2567.

271. Shacklette LW (1985). Polyacetylene and polyphenylene as anode materials for nonaqueous secondary batteries, *Journal of The Electrochemical Society*, **132**, pp. 1529.

272. Su D, Zhang J, Dou SX, and Wang G (2015). Polypyrrole hollow nanospheres: Stable cathode materials for sodium-ion batteries, *ChemComm*, **51**, pp. 16092–16095.

273. Liu S, Wang F, Dong R, Zhang T, Zhang J, Zhuang X, Mai Y, and Feng X (2016). Dual-template synthesis of 2D mesoporous polypyrrole nanosheets with controlled pore size, *Advanced Materials*, **28**, pp. 8365–8370.

274. Zhao R, Zhu L, Cao Y, Ai X, and Yang HX (2012). An aniline-nitroaniline copolymer as a high capacity cathode for Na-ion batteries, *Electrochemistry Communications*, **21**, pp. 36–38.

275. Zhou M, Zhu L, Cao Y, Zhao R, Qian J, Ai X, and Yang H (2012). $Fe(CN)_6^{-4}$-doped polypyrrole: A high-capacity and high-rate cathode material for sodium-ion batteries, *RSC Advances*, **2**, pp. 5495–5498.

276. Zhou M, Li W, Gu T, Wang K, Cheng S, and Jiang K (2015). A sulfonated polyaniline with high density and high rate Na-storage performances as a flexible organic cathode for sodium ion batteries, *ChemComm*, **51**, pp. 14354–14356.

277. Zhu L, Shen Y, Sun M, Qian J, Cao Y, Ai X, and Yang H (2013). Self-doped polypyrrole with ionizable sodium sulfonate as a renewable cathode material for sodium ion batteries, *ChemComm*, **49**, pp. 11370–11372.

278. Zhou M, Xiong Y, Cao Y, Ai X, and Yang H (2013). Electroactive organic anion-doped polypyrrole as a low cost and renewable cathode for sodium-ion batteries, *Journal of Polymer Science Part B: Polymer Physics*, **51**, pp. 114–118.

279. Shen YF, Yuan DD, Ai XP, Yang HX, and Zhou M (2014). Poly(diphenylaminesulfonic acid sodium) as a cation-exchanging organic cathode for sodium batteries, *Electrochemistry Communications*, **49**, pp. 5–8.

280. Han SC, Bae EG, Lim H, and Pyo M (2014). Non-crystalline oligopyrene as a cathode material with a high-voltage plateau for sodium ion batteries, *Journal of Power Sources*, **254**, pp. 73–79.

281. Deng W, Liang X, Wu X, Qian J, Cao Y, Ai X, Feng J, and Yang H (2013). A low cost, all-organic Na-ion battery based on polymeric cathode and anode, *Scientific Reports*, **3**, pp. 2671.

282. Dai Y, Zhang Y, Gao L, Xu G, and Xie J (2009). A sodium ion based organic radical battery. *Electrochemical and Solid State Letters*, **13**, pp. A22.

283. Kim J-K, Kim Y, Park S, Ko H, and Kim Y (2016). Encapsulation of organic active materials in carbon nanotubes for application to high-electrochemical-performance sodium batteries, *Energy & Environmental Science*, **9**, pp. 1264–1269.

284. Zhang S, Huang W, Hu P, Huang C, Shang C, Zhang C, Yang R, and Cui G (2015). Conjugated microporous polymers with excellent electrochemical performance for lithium and sodium storage, *Journal of Materials Chemistry A*, **3**, pp. 1896–1901.

285. Sakaushi K, Hosono E, Nickerl G, Gemming T, Zhou H, Kaskel S, and Eckert J (2013). Aromatic porous-honeycomb electrodes for a sodium-organic energy storage device, *Nature Communications*, **4**, pp. 1485.

286. Kuhn P, Antonietti M, and Thomas A (2008). Porous, covalent triazine-based frameworks prepared by ionothermal synthesis, *Angewandte Chemie International Edition*, **47**, pp. 3450–3453.

287. Sakaushi K, Hosono E, Nickerl G, Zhou H, Kaskel S, and Eckert J (2014). Bipolar porous polymeric frameworks for low-cost, high-power, long-life all-organic energy storage devices, *Journal of Power Sources*, **245**, pp. 553–556.

288. Zhong H, Wang G, Song Z, Li X, Tang H, Zhou Y, and Zhan H (2014). Organometallic polymer material for energy storage, *ChemComm*, **50**, pp. 6768–6770.

289. Choi JW and Aurbach D (2016). Promise and reality of post-lithium-ion batteries with high energy densities, *Nature Review Materials*, **1**, pp. 16013.

290. Goikolea E, Palomares V, Wang S, de Larramendi IR, Guo X, Wang G, and Rojo T (2020). Na-ion batteries: Approaching old and new challenges, *Advanced Energy Materials*, **10**, 2002055.

291. Zhao C, Wang Q, Yao Z, Wang J, Sánchez-Lengeling B, Ding F, Qi X, Lu Y, Bai X, Li B, Li H, Aspuru-Guzik A, Huang X, Delmas C, Wagemaker M, Chen L, and Hu Y-S (2020). Rational design of layered oxide materials for sodium-ion batteries, *Science*, **370**, pp. 708–711.

292. Rajagopalan R, Chen B, Zhang Z, Wu XL, Du Y, Huang Y, Li B, Zong Y, Wang J, Nam GH, Sindoro M, Dou SX, Liu HK, and Zhang H (2017). Improved reversibility of Fe^{3+}/Fe^{4+} redox couple in sodium super

ion conductor type $Na_3Fe_2(PO_4)_3$ for sodium-ion batteries, *Advanced Materials*, **29**.

293. Kawai K, Zhao W, Nishimura S-I, and Yamada A (2018). High-voltage Cr^{4+}/Cr^{3+} redox couple in polyanion compounds, *ACS Applied Energy Materials*, **1**, pp. 928–931.

294. Liu R, Xu G, Li Q, Zheng S, Zheng G, Gong Z, Li Y, Kruskop E, Fu R, Chen Z, Amine K, and Yang Y (2017). Exploring highly reversible 1.5-electron reactions ($V^{3+}/V^{4+}/V^{5+}$) in $Na_3VCr(PO_4)_3$ cathode for sodium-ion batteries, *ACS Applied Materials & Interfaces*, **9**, pp. 43632–43639.

295. Lalère F, Seznec V, Courty M, David R, Chotard JN, and Masquelier C (2015). Improving the energy density of $Na_3V_2(PO_4)_3$-based positive electrodes through V/Al substitution, *Journal of Materials Chemistry A*, **3**, pp. 16198–16205.

296. Zhu T, Hu P, Wang X, Liu Z, Luo W, Owusu KA, Cao W, Shi C, Li J, Zhou L, and Mai L (2019). Realizing three-electron redox reactions in NASICON-structured $Na_3MnTi(PO_4)_3$ for sodium-ion batteries, *Advanced Energy Materials*, **9**.

297. Zhang W, Li H, Zhang Z, Xu M, Lai Y, and Chou SL (2020). Full activation of Mn^{4+}/Mn^{3+} redox in $Na_4MnCr(PO_4)_3$ as a high-voltage and high-rate cathode material for sodium-ion batteries, *Small*, **16**, pp. e2001524.

298. Deng X, Shi W, Sunarso J, Liu M, and Shao Z (2017). A green route to a Na_2FePO_4F-based cathode for sodium ion batteries of high rate and long cycling life, *ACS Applied Materials & Interfaces*, **9**, pp. 16280–16287.

299. Pu X, Wang H, Yuan T, Cao S, Liu S, Xu L, Yang H, Ai X, Chen Z, and Cao Y (2019). $Na_4Fe_3(PO_4)_2P_2O_7$/C nanospheres as low-cost, high-performance cathode material for sodium-ion batteries, *Energy Storage Materials*, **22**, pp. 330–336.

300. Nose M, Nakayama H, Nobuhara K, Yamaguchi H, Nakanishi S, and Iba H (2013). $Na_4Co_3(PO_4)_2P_2O_7$: A novel storage material for sodium-ion batteries, *Journal of Power Sources*, **234**, pp. 175–179.

301. Tang L, Liu X, Li Z, Pu X, Zhang J, Xu Q, Liu H, Wang YG, and Xia Y (2019). CNT-decorated $Na_4Mn_2Co(PO_4)_2P_2O_7$ microspheres as a novel high-voltage cathode material for sodium-ion batteries, *ACS Applied Materials & Interfaces*, **11**, pp. 27813–27822.

302. Gond R, Meena SS, Yusuf SM, Shukla V, Jena NK, Ahuja R, Okada S, and Barpanda P (2017). Enabling the electrochemical activity in sodium iron metaphosphate [$NaFe(PO_3)_3$] sodium battery insertion material:

Structural and electrochemical insights, *Inorganic Chemistry*, **56**, pp. 5918–5929.

303. Gond R, Rao RP, Pralong V, Lebedev OI, Adams S, and Barpanda P (2018). Cubic sodium cobalt metaphosphate [$NaCo(PO_3)_3$] as a cathode material for sodium ion batteries, *Inorganic Chemistry*, **57**, pp. 6324–6332.

304. Li S, Guo J, Ye Z, Zhao X, Wu S, Mi JX, Wang CZ, Gong Z, McDonald MJ, Zhu Z, Ho KM, and Yang Y (2016). Zero-strain Na_2FeSiO_4 as novel cathode material for sodium-ion batteries, *ACS Applied Materials & Interfaces*, **8**, pp. 17233–17238.

305. Law M, Ramar V, and Balaya P (2017). Na_2MnSiO_4 as an attractive high capacity cathode material for sodium-ion battery, *Journal of Power Sources*, **359**, pp. 277–284.

306. Treacher JC, Wood SM, Islam MS, and Kendrick E (2016). Na_2CoSiO_4 as a cathode material for sodium-ion batteries: Structure, electrochemistry and diffusion pathways, *Physical Chemistry Chemical Physics*, **18**, pp. 32744–32752.

307. Panigrahi A, Nishimura S-i, Shimada T, Watanabe E, Zhao W, Oyama G, and Yamada A (2017). Sodium iron(II) pyrosilicate $Na_2Fe_2Si_2O_7$: A potential cathode material in the $Na_2O–FeO–SiO_2$ System, *Chemistry of Materials*, **29**, pp. 4361–4366.

Chapter 4

Prussian Blue Analogues as Cathode Materials for Sodium-Ion Batteries

Yinzhu Jiang, Yao Huang, and Yuting Gao
School of Materials Science and Engineering, State Key Laboratory of Clean Energy Utilization, ZJU-Hangzhou Global Scientific and Technological Innovation Centre, Zhejiang University, Hangzhou, Zhejiang 310027, P. R. China
yzjiang@zju.edu.cn

4.1 Introduction

Sodium-ion batteries (SIBs) are one of the most promising candidates for lithium-ion batteries (LIBs) due to their similar chemical nature to that of LIBs, low cost, and abundant resources [1, 2]. The advantages make them superior especially in large-scale energy storage systems. In general, the cathode materials provide reversible Na^+ insertion/extraction during charge/discharge of SIBs, which directly affects energy/power density, lifespan, and safety issues. Generally speaking, the cathode materials of SIBs need to meet the following requirements:

- High redox potential, which is less affected by the sodium content to ensure high and stable output voltage;

Handbook of Sodium-Ion Batteries: Materials and Characterization
Edited by Rohit R. Gaddam and X. S. (George) Zhao

ISBN 978-981-4968-15-7 (Hardcover), 978-1-003-30874-4 (eBook)
www.jennystanford.com

- A large number of reversible insertion/extraction sites for sodium ions to ensure high specific capacity;
- Enough sodium-ion diffusion channels and good electronic conductivity to reduce polarization and enhance rate capability;
- Small variation in the structure during cycling to ensure good structural stability and long cycle life;
- Easy to prepare, abundant resources, low cost, environmentally benign, and capable of long-term large-scale development.

Based on the preceding requirements, the primary cathode materials used in SIBs can be divided into three categories: transition metal oxides, polyanion compounds, and Prussian blue analogs (PBAs). PBAs show rich physicochemical properties and have been applied in kinds of research fields, such as hydrogen storage [3], biosensing [4], cancer therapy [5], water purification [6], and information recording [7]. In 2011, PBAs, including $K_{0.6}Ni_{1.2}Fe(CN)_6 \cdot 3.6H_2O$ (NiHCF) [8] and $K_{0.71}Cu[Fe(CN)_6]_{0.72} \cdot 3.7H_2O$ (CuHCF) [9] were first studied as cathodes for aqueous SIBs. NiHCF possesses excellent rate performance (39 mAh/g at 41.7 C) and cycling stability (no obvious decay after 5000 cycles), while CuHCF maintains high capacity retention of 83% after 40,000 cycles and high rate capability with the specific capacity of 40 mAh/g at 83 C [9]. Followed by that, Goodenough et al. synthesized a series of PBAs by coprecipitation and studied the sodium storage performance in nonaqueous SIBs, which raises the curtain on the application of PBAs in aprotic SIBs. PBAs allow multiple tuning of their chemical composition and stoichiometry remaining the original framework structure. The rigid open framework and its small interaction with interstitial cations make PBAs remarkable emerging cathode materials for SIBs [10].

4.2 Structure and Working Principle of PBAs

4.2.1 Typical Structures and Phases of PBAs

Prussian blue (PB) is a trade name derived from its birthplace, which is the same material as Turnbull's blue in fact. According to the different chemical compositions, the chemical names of the material having the same crystal structure as Prussian blue are named PBAs, which can also be called metal hexacyanometallic compounds (HCFs). For instance, PB can be referred to as iron hexacyanoferrate (FeHCF). The generic chemical formula of PBAs can be represented as $A_xM_{Ay}[M_B(CN)_6]_{1-z}\gamma_z \cdot nH_2O$, where M_A and M_B represent the transition metals, including Mn, Fe, Co, Ni, Cu, Zn, etc., and A is usually the alkaline earth elements (Li, Na, K). M_A and M_B can be distinguished according to their spin states, which are determined by the strength of ligands around the central ion. As Fig. 4.1 illustrates, M_A is connected with N to form M_AN_6 octahedra,

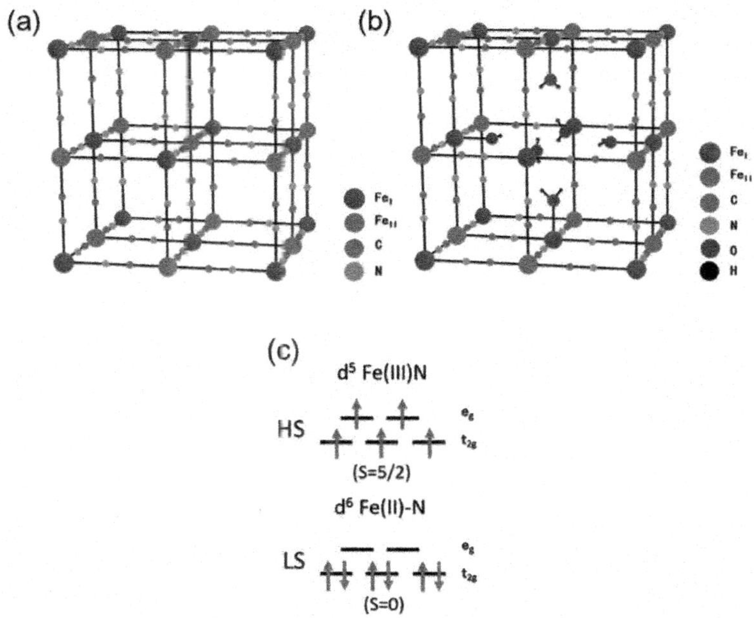

Figure 4.1 (a, b) Schematic diagram of the crystal structure. (c) Schematic diagram of the electronic configuration of high-spin Fe^{III} and low-spin Fe^{II} in PBAs.

and M_B is connected with C to form M_BC_6 octahedra. Since N-M_A coordinates weakly and C-M_B coordinates strongly in crystal fields, M_A is in the state of high spin (HS) while M_B is low spin (LS). Through C≡N bridging, these two different octahedra are interconnected alternately to form a three-dimensional (3D) open framework [11]. Generally, there are $M_B(CN)_6^{n-}$ vacancies in the framework distributed randomly. For example, $Fe_4[Fe(CN)_6]_3 \cdot nH_2O$ (n = 6~14) is one of the typical PBs, in which the valence states of Fe^{LS} and Fe^{HS} ions are +2 and +3, respectively. The specific stoichiometry of Fe^{2+} and Fe^{3+} sites is a ratio of 3:4, that is, 25% $[Fe(CN)_6]^{4-}$ vacancy exists due to the charge neutrality [12]. The content of $M_B(CN)_6$ vacancies mainly depends on the stoichiometry of the compound, which is related to the oxidation states of M_A and M_B. Once the vacancies are generated, oxygen atoms from H_2O will coordinate to the M_B ions, and the octahedral cavities generated by $M_B(CN)_6$ vacancy will be occupied by coordinated H_2O molecules [12–14]. Besides the coordinated water, the open framework structure with zeolitic sites is able to host species with an ionic radius smaller than 1.6 Å. The ideal species in zeolitic sites is Na for SIBs. However, zeolitic/interstitial H_2O will partly or fully occupy these eight possible sites. A-free PBAs, whose Na^+ sites are fully occupied by H_2O, generally have a high water content (more than 20 wt%) and possess the cubic phase, such as $Fe_4[Fe(CN)_6]_3$ [15] and $FeFe(CN)_6$ [16–18].

The presence of defects, including vacancies and water molecules, will generally cause adverse effects on electrochemical performance. The randomly distributed vacancies break the M_B-C≡N-M_A bridges, making the framework structurally unstable and more likely to suffer structure collapse during cycling [19]. In addition, the vacancies will cut off electronic conduction along the M_B-C≡N-M_A bridge [20], leading to a blockage of Na^+ insertion reaction in the microdomains of the lattice and the increased ohmic polarization [1]. The structural water molecules reduce the sites to host sodium ions resulting in the loss of capacity [13], and the high interstitial water ratio might impede the Na^+ diffusion along the 3D tunnel pathways in the PBA framework structure, leading to large overpotential and thus low energy efficiency [21]. The introduction of H_2O also aggravates the change in lattice volume during cycling [22] and may react with organic electrolytes, resulting in poor stability and safety hazards.

Table 4.1 Reported crystal phase structures and space groups of typical PBAs

	MnHCF	FeHCF	CoHCF	NiHCF	CuHCF	ZnHCF
Cubic (Fm $\overline{3}$m)	✓	✓	✓	✓	✓	✓
Rhombohedral (R $\overline{3}$, R $\overline{3}$c)	✓	✓				✓
Monoclinic (P21/n)	✓	✓	✓	✓	✓	

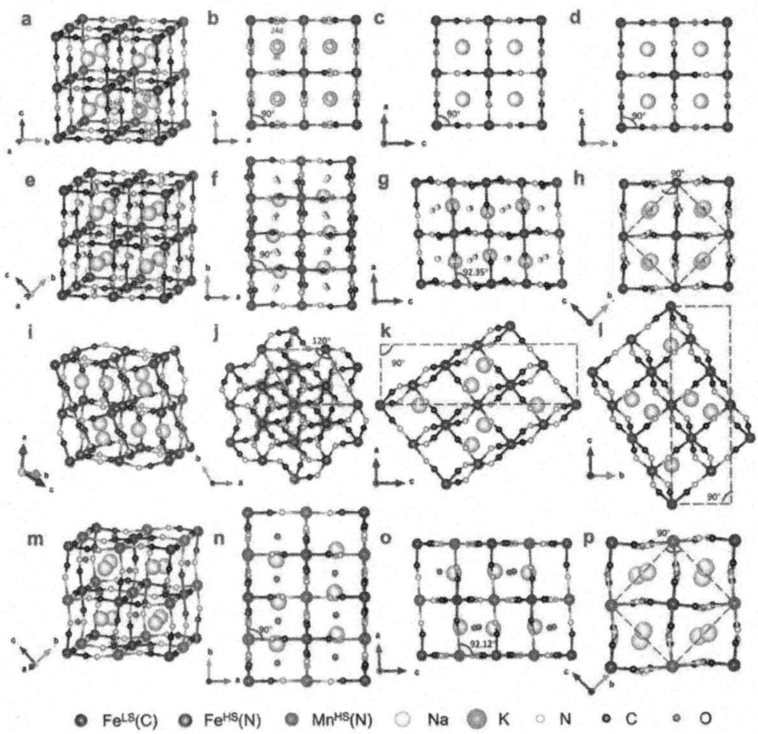

Figure 4.2 Schematic illustration of (a–d) cubic NaFeHCF [17], (e–h) monoclinic NaFeHCF [23], (i–l) rhombohedral NaFeHCF [23], and (m–p) monoclinic NaMnHCF. Reproduced with permission from Ref. [24]. Copyright ©2020, WILEY–VCH.

Depending on the utilization of various precursors and preparation conditions, there will be variation in valence states and a variety of M_A and M_B lead to multiple crystal structures, as shown in Table 4.1. The symmetry of FeHCF usually corresponds to *Fm-3m*, which is used as a reference for other crystal structures. Normally, increasing the sodium content will enlarge the lattice parameter continuously and decrease the crystal symmetry. For example, when the sodium content (x) reaches around 1.6 to 1.8 with the existence of water, the phase will be distorted from cubic (*Fm-3m*) to monoclinic (*P21/n*) [25–27]. Compared to the cubic phase, a (10.43 Å) increases slightly in the monoclinic phase, while b (7.52 Å) and c (7.27 Å) reduce obviously and β is distorted from 90° to 92.35° [11, 23, 26], resulting in 6.5% volume expansion. If the Na content decreases continuously and interstitial water is removed by thermal annealing, the phase will change from monoclinic to rhombohedral. As shown in Table 4.1 and Fig. 4.2, *c* (18.88 Å) is elongated heavily, and *a* and *b* (6.54 Å) are both shortened with $\gamma = 120°$, with a shrunk volume by 12.5% compared to the cubic phase [26]. The large distortion causes the large relative rotation of ∠Fe^{HS}–N≡C from 173° to 145°.

The structure and phase transitions of MnHCF are pretty similar to FeHCF but a little different in the lattice parameters and unit cell volumes. The reason causing the difference is the larger radius of Mn^{2+}(0.83 Å) than that of Fe^{2+} (0.78 Å). The Mn^{HS}–N bonds are more "ionical" and longer than those of Fe^{HS}–N, leading to more obvious local distortion. Furthermore, according to crystal field theory, $Mn^{HS}(3d^5 4s^2)$–N octahedrons will suffer from the Jahn–Teller effect more seriously than $Fe^{HS}(3d^6 4s^2)$–N octahedrons. Both in monoclinic and rhombohedral phases of MnHCF, the lattice parameter in the *c*-axis will be more elongated than that of FeHCF.

Sodium-ion content (*x*) and water content (*n*) play important roles in the structural evolution of PBAs. Based on the density functional theory (DFT) calculation, the diverse phases are observed due to three competing interactions: Na–N Coulomb attraction, d–π orbital overlap, and Pauli repulsion [32]. Among them, the Na–N Coulomb attraction tends to reduce the volume of the lattice, while the d–π orbital overlap and Pauli repulsion tend to restrain the unit volume. The introduction of sodium ion increases the Na–N Coulomb attraction, while the introduction of water molecule increases the Pauli repulsion and prevents the formation of the rhombohedral structure.

Table 4.2 Crystal structures and lattice parameters of some typical HCFs for SIBs

PBAs	Water [wt.%]	Structure	Lattice Parameters	Volume [Å^3]	Refs.
$Fe[Fe(CN)_6]_y$	20	C	$a = b = c =$ 10.18 Å, $\beta = 90°$	527.45	[16]
$Na_{0.69}Fe[Fe(CN)_6]_{0.93}$	18	C	$a = b = c =$ 10.22 Å, $\beta = 90°$	533.85	[11]
$Na_{1.81}FeFe(CN)_6$	12	M	a = 10.43 Å, b = 7.51 Å, c = 7.27 Å, $\beta = 92.35°$	568.53	[23]
$Na_{1.88}FeFe(CN)_6$	1	R	$a = b$ = 6.54 Å, c = 18.88 Å, $\gamma = 120°$	466.37	[23]
$Na_{1.38}Mn[Fe(CN)_6]_{0.91}$	15	C	$a = b = c =$ 10.53 Å, $\beta = 90°$	583.75	[22]
$Na_{1.90}Mn[Fe(CN)_6]_{0.99}$	10	M	a = 10.59 Å, b = 7.54 Å, c = 7.34 Å, $\beta = 92.12°$	584.95	[22, 28]
$Na_{1.86}Mn[Fe(CN)_6]_{1.02}$	0.2	R	$a = b$ = 6.58 Å, c = 18.93 Å, $\gamma = 120°$	473.16	[22, 28]
$Na_{0.67}Co[Fe(CN)_6]_{0.67}$	15	C	$a = b = c =$ 10.3 Å, $\beta = 90°$	546.35	[29]
$Na_{1.44}Co[Fe(CN)_6]_{0.76}$	7	M	a = 10.33 Å, b = 7.46 Å, c = 7.19 Å, $\beta = 91.27°$	553.45	[30]
$Na_{0.68}Ni[Fe(CN)_6]_{0.72}$	23	C	$a = b = c =$ 10.23 Å, $\beta = 90°$	534.65	[31]
$Na_{1.48}Ni[Fe(CN)_6]_{0.89}$	16	M	a = 10.28 Å, b = 7.39 Å, c = 7.15 Å, $\beta = 92.13°$	542.18	[31]

Source: Reproduced with permission from Ref. [24]. Copyright ©2020, WILEY–VCH. (C = cubic, M = monoclinic, R = rhombohedral).

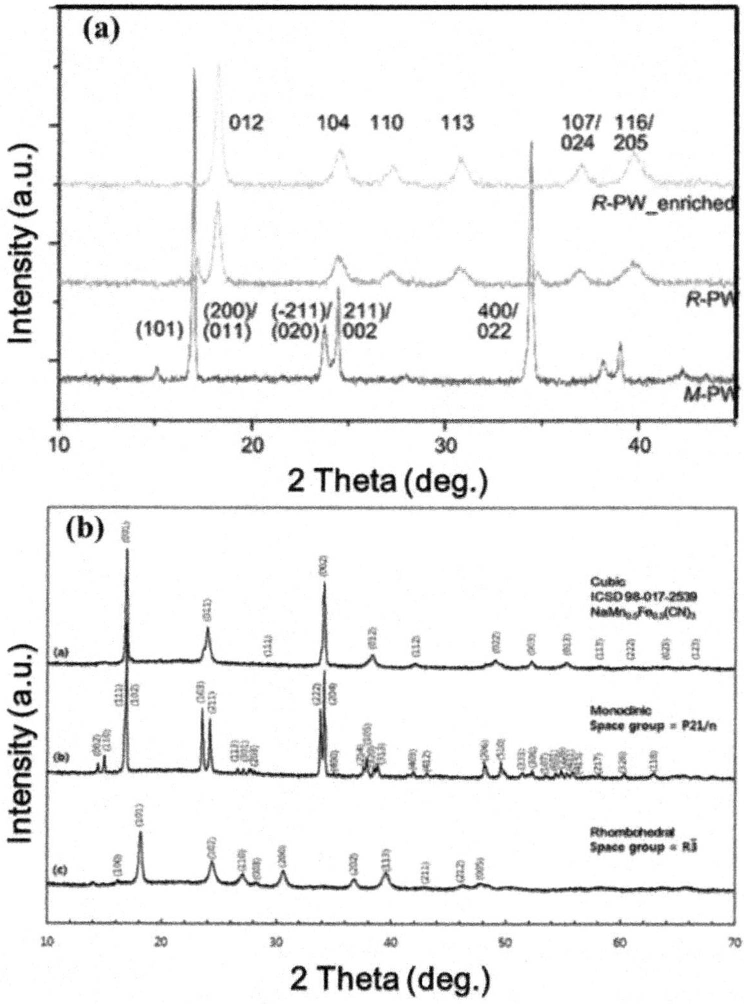

Figure 4.3 XRD patterns of different phases of (a) FeHCF [23]. Reproduced with permission from Ref. [23]. Copyright ©2019, American Chemical Society. (b) MnHCF [28]. Reproduced with permission from Ref. [28]. Copyright ©2017, Elsevier.

Upon increasing the sodium content and removing water, the Coulomb attraction is enhanced, and the Pauli repulsion is reduced, which shortens the Na–N distance and bends the ∠Mn–N≡C angle, leading to the structure changing to a smaller rhombohedral phase.

The phase evolution between monoclinic and rhombohedral is generally reversible: when $x = 2$, removing interstitial H_2O induces the monoclinic to rhombohedral structural variation, but when exposed to wet air for several hours at 25°C, the rhombohedral phase will change back to the monoclinic one [22].

X-ray diffraction (XRD) analysis can be conducted to distinguish the variation in the phase of PBAs. No matter FeHCF or other PBAs, the peak around 25° is split in the monoclinic structure. Controversy still exists about the XRD pattern of the rhombohedral phase. A new view suggests that due to the higher strain in the structure induced by lattice distortion and volume reduction, the peaks of the rhombohedral phase shift toward higher angles remarkably, and the diffraction peaks are broadened and weakened compared with the monoclinic and cubic ones [23, 24]. Some PBAs reported with the rhombohedral structure might actually be of monoclinic phase from this point of view. As shown in Table 4.1, only cubic and monoclinic phases of NiHCF [31, 33, 34] and CoHCF [35, 36] have been reported. The sizes of Co^{2+} (0.75 Å) and Ni^{2+} (0.69 Å) make the structure of CoHCF and NiHCF similar to FeHCF and MnHCF of the same phase but different in lattice parameters. Almost all reports of CuHCF are corresponding to the cubic phase, and only one study on the monoclinic structure of CuHCF has been reported recently [37].

4.2.2 Redox Reaction and Electric Energy Storage Mechanism

Due to the weak bonding between Na ions and C≡N groups, PBAs generally exhibit fast ion diffusivity, which is beneficial to sodium storage kinetics. If M_A and M_B both feature redox activity, PBAs could enable two Na^+ insertions of per unit formula theoretically. As shown in Table 4.3, the theoretic specific capacity can reach about 170 mAh/g, which is advantageous compared with other kinds of cathodes.

Table 4.3 Theoretical specific capacity of different PBAs (mAh/g)

	MnHCF	FeHCF	CoHCF	NiHCF	CuHCF
Na storage	171.3	170.7	169.6	84.9	83.4

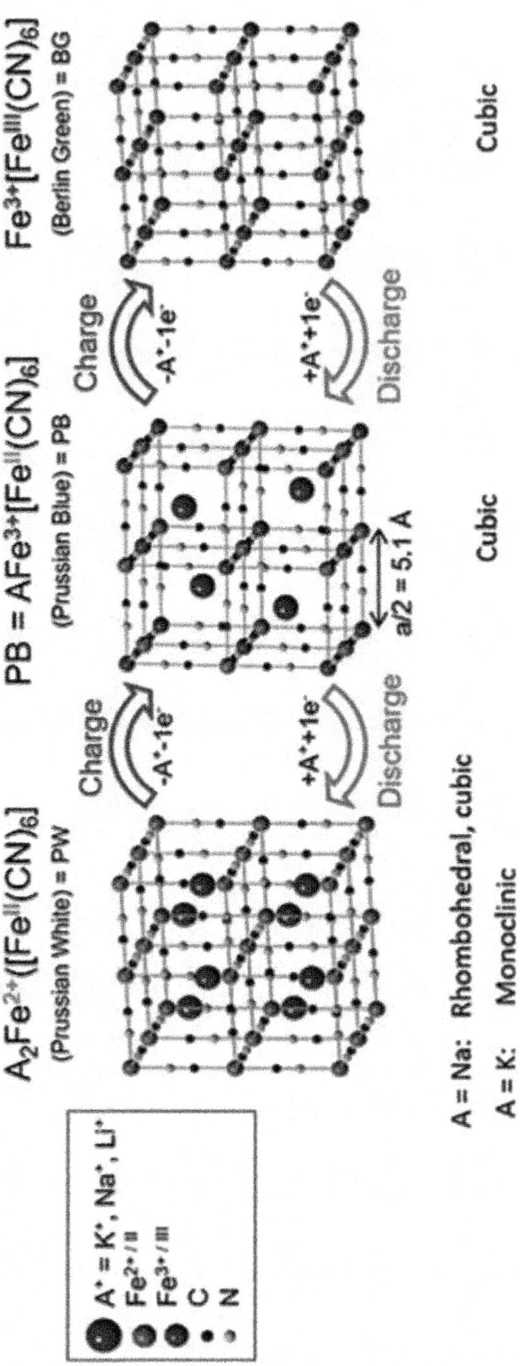

Figure 4.4 The phase transition and redox mechanism of FeHCF [38]. Reproduced with permission from Ref. [38]. Copyright ©2018, María José Piernas Muñoz, Elizabeth Castillo Martínez.

The electrochemical energy storage mechanism of PBAs is based on the ion intercalation along with the corresponding redox reaction of the transition metals in PBAs. The color variation and phase evolution of PBAs are always observed during ions intercalation, and the schematic illustration of the corresponding phase transition and redox process of NaFeHCF during Na intercalation/deintercalation is displayed in Fig. 4.4 [19]. Assuming an ideal stoichiometry of FeHCF ($x = 2, y = 1, z = 0, n = 0$), the ion intercalation and redox reaction can be described as follows [16]:

$$A_2M_A^{II}M_B^{II}(CN)_6 \rightarrow A^+ + AM_A^{III}M_B^{II}(CN)_6 \text{ (charge 1, redox O1)} \quad (4.1)$$

$$AM_A^{III}M_B^{II}(CN)_6 \rightarrow A^+ + M_A^{III}M_B^{III}(CN)_6 \text{ (charge 2, redox O2)} \quad (4.2)$$

$$M_A^{III}M_B^{III}(CN)_6 + A^+ \rightarrow AM_A^{III}M_B^{II}(CN)_6 \text{ (discharge 2, redox R2)} \quad (4.3)$$

$$AM_A^{III}M_B^{II}(CN)_6 + A^+ \rightarrow A_2M_A^{II}M_B^{II}(CN)_6 \text{ (discharge 2, redox R1)} \quad (4.4)$$

Through theoretical calculation and experimental analysis, the sequence of the redox reactions of FeHCF has been verified. During the extraction of sodium ions, M_A^{HS} is oxidized first and the phase is changed from rhombohedral to cubic. Upon extracting the second sodium ion, M_B^{LS} is oxidized without a phase transition. pDOS analysis and Fe L-edge soft X-ray absorption (sXAS) both prove the preceding sequence postulate [39, 40].

It is worth mentioning that the variation in the oxidation states of transition metals benefits the energy storage in PBAs, which, however, requires vacant Na storage sites for the intercalation of sodium ions. As discussed earlier, it is hard to gain an ideal crystal without defects such as vacancies and water molecules. The presence of $Fe(CN)_6$ vacancies reduces the M_B redox-active sites, and the interstitial water would occupy the Na storage sites, both of which will degrade the specific capacity of PBAs. Besides, H_2O and Na ions can react during cycles forming Na_2O on the separator surface, thus resulting in low initial CE [28].

As shown in Tables 4.3 and 4.4, Cu and Ni are electrochemically inert within the operating window of traditional organic electrolytes. As a result, PBAs like CuHCF and NiHCF just have one redox-active site per unit formula. In contrast, FeHCF, MnHCF, and CoHCF possess two redox-active sites because of the electrochemical activity of both M_A and M_B. Normally, the redox potentials of M_A and M_B are different even when composed of the same elements. Therefore, two potential plateaus are usually observed in the charge–discharge profiles. Exceptions are also reported for MnHCF, which exhibits one voltage plateau at around 3.5 V when the interstitial H_2O is completely removed from the rhombohedral framework [22].

Table 4.4 Discharge voltage plateaus of typical PBAs in organic electrolyte SIBs

	FeHCF	MnHCF	CoHCF	CuHCF	NiHCF
Voltage (V versus Na/Na^+)	Fe^{LS}: 3.23~3.80 Fe^{HS}: 2.60~2.89	Mn^{HS}: 3.48~3.60 Fe^{LS}: 3.26~3.40	Fe^{LS}: ~3.73 Co^{HS}: 3.04~3.12	Fe^{LS}: ~3.32	Fe^{LS}: 2.95~3.33

4.2.3 Na^+ Diffusion

There are four possible positions for the location of sodium ions in the framework of FeHCF: **8c** (body center), **24d** (face center), **32f** (displaced from body center along an [111] axis toward the N-coordinated corner), and **32f′** (displaced from body center along an [111] axis toward C-coordinated corner) [39]. It is suggested that Na^+ will tend to occupy the 8c sites when the material is Na poor, while when the sample is Na rich, extra Na^+ can continue to occupy the empty 24d sites [16, 26]. DFT calculation gives the possible Na^+ diffusion pathways in the cubic FeHCF [41], which reveals that energetically **24d** site with open <100> channels of 3.59 Å in diameter is preferred for Na^+ diffusion. In the calculation, four possible Na^+ diffusion paths can be divided into two types: single diffusion and cooperative diffusion (as shown in Fig. 4.5). In the single diffusion, Fig. 4.5b,f shows that one Na^+ diffuses along the [100] axis through the body center of the structure, and the migration energy barrier reaches 0.64 eV at the 8c position. Figure 4.5c,g shows that one Na^+ migrates through a W-shaped path and bypasses the body center with the highest energy barrier of 0.58 eV. In the cooperative diffusion, Fig. 4.5d,h shows that one Na^+ migrates from a **24d** site to another nearby **24d** site and another Na^+ migrates to the next empty **24d** site. As Fig. 4.5l illustrates, the migration energy barrier has been decreased to 0.49 eV. The route with the lowest migration energy barrier (0.28 eV) is called collective migration; all Na^+ passes through a 1–2–3–4 hop sequence one by one, shown in Fig. 4.5m.

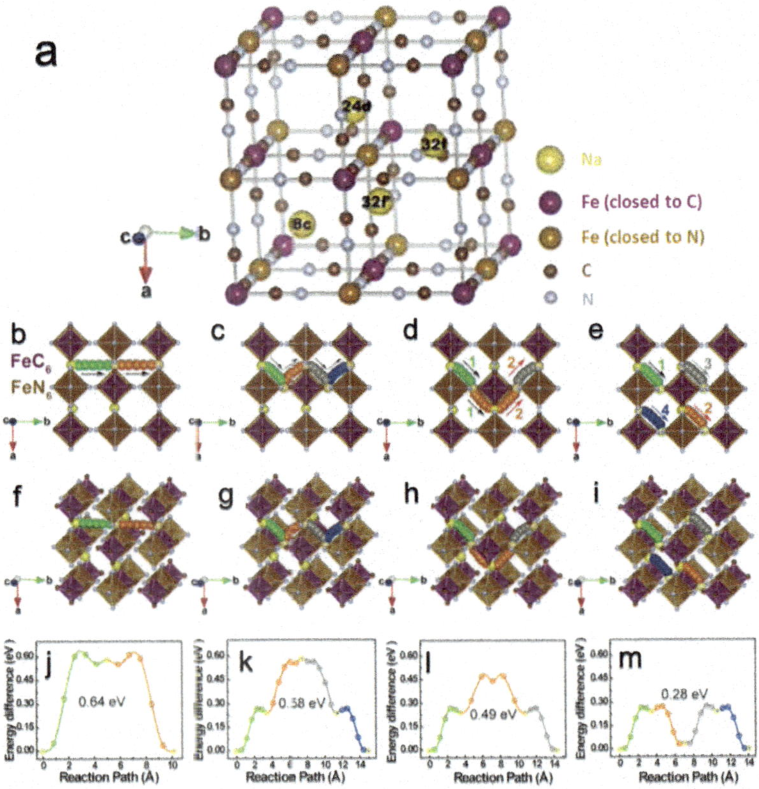

Figure 4.5 (a) Possible positions of Na^+ in the cubic framework structure. (b–e) Possible Na^+ diffusion paths between 24d sites from front views. (f–i) From side views. (j–m) Calculated migration energy of the above four paths [41]. Reproduced with permission from Ref. [41]. Copyright ©2016, Wiley-VCH.

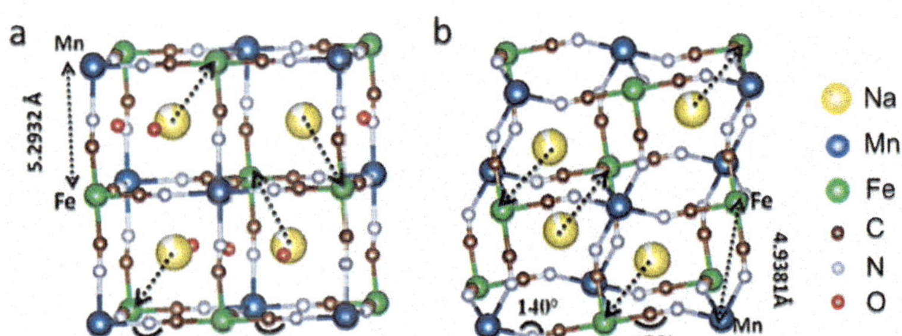

Figure 4.6 The Na^+ storage site and migration direction of (a) M-Na2−δMnHCF and (b) R-Na2−δMnHCF [22]. Reproduced with permission from Ref. [22]. Copyright ©2015, American Chemical Society.

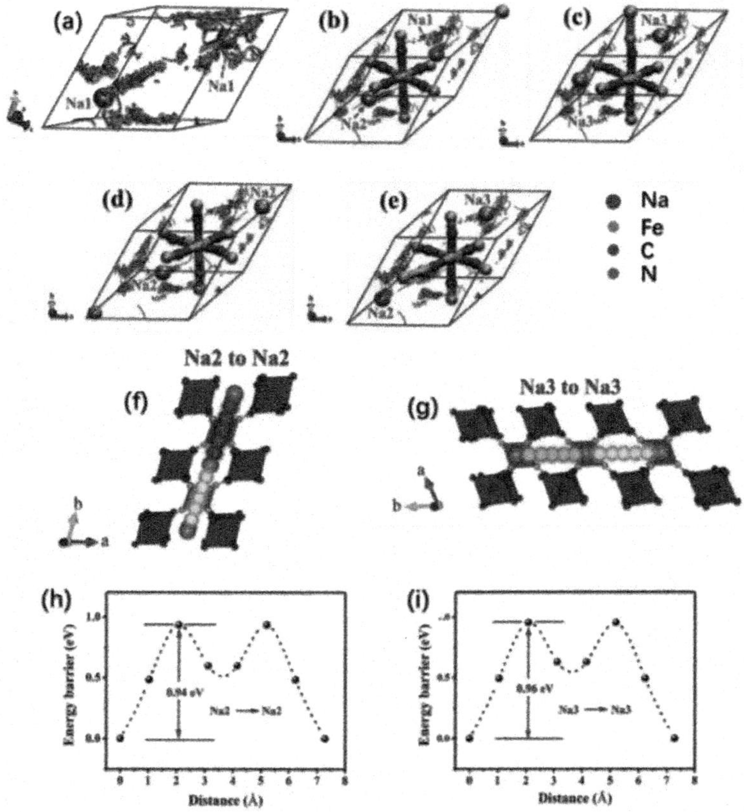

Figure 4.7 (a) Possible Na^+ migration paths in rhombohedral FeHCF (gray tube trace are migration paths). (b–e) Three kinds of Na^+ location during the migration process in rhombohedral FeHCF. (f) Na^+ migration path of Na2 to Na2 positions in multi-unit cells, and (g) corresponding energy barrier for Na^+ migration. (h) Na^+ migration path of Na3 to Na3 positions in multi-unit cells and (g) corresponding energy barrier for Na^+ migration [42]. Reproduced with permission from Ref. [42]. Copyright ©2020, Elsevier

Along the original cubic [111] axis, the Na^+ cooperatively displaces toward the Fe vertex, and a synergic rotation of the linear $(C{\equiv}N)^-$ anions brings the N atoms toward the displacement axis to make contact with the Na^+ ions in its new position, which results in the rhombohedral structure [22]. Further study on the Na^+ diffusion in the rhombohedral phase was carried out by the methods of *ab initio* molecular dynamics (AIMD) and climbing image nudged elastic band (CINEB) calculations [42]. As can be seen from Fig. 4.7,

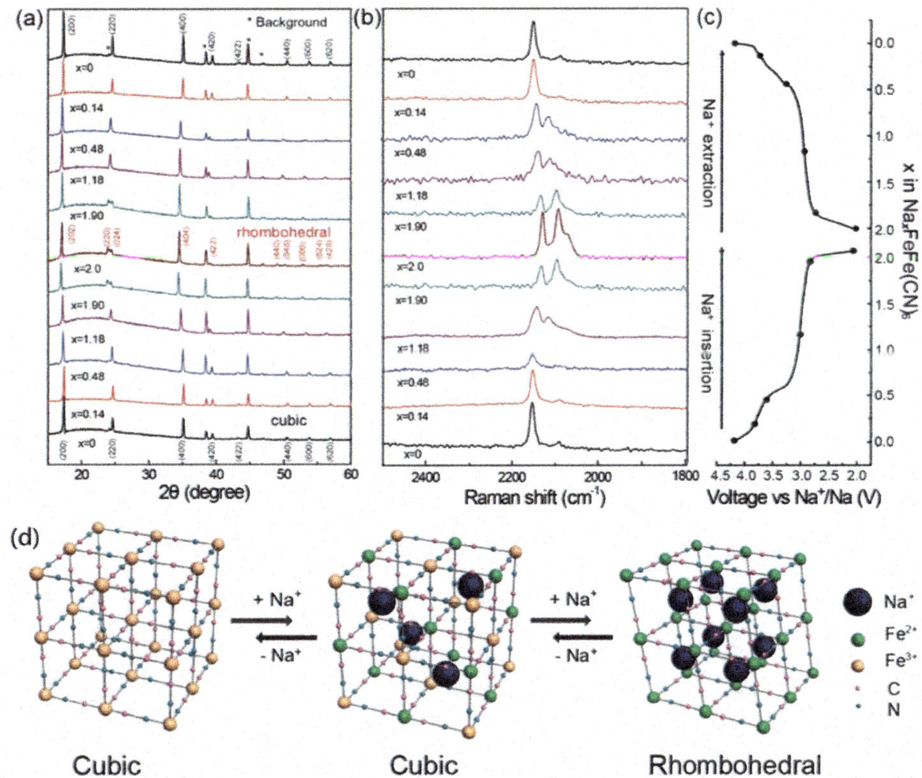

Figure 4.8 (a) *Ex situ* XRD patterns of HQ-NaFeHCF. (b) *Ex situ* Raman spectra of HQ-NaFeHCF at various discharge and charge states. (c) Typical charge/discharge profiles of HQ-NaFeHCF electrodes. (d) Schematic illustration of the redox mechanism of HQ-NaFeHCF [19]. Reproduced with permission from Ref. [19]. Copyright ©2014, The Royal Society of Chemistry.

Na^+ positions are divided into Na1 (interstitial sites), Na2 (edge), Na3 (face), and all the possible Na^+ diffusions are indicated by gray noodle traces. Na^+ in Na1 prefers to move to Na2 or Na3, while the Na^+ in Na2 and N3 would randomly migrate to each other's position. Two types of diffusion paths (Na2 to Na2 and Na3 to Na3) are shown in Fig. 4.7f,g, and the diffusion energy barriers are 0.94 and 0.96 eV, respectively.

4.2.4 Phase Transition During Charge/Discharge

As for the pristine cubic FeHCF, Guo et al. [19] reported that the peaks in the XRD pattern shift toward a smaller angle with structure changing from cubic to rhombohedral when Na content (x) is increased to 2.0, as shown in Fig. 4.8. When x returns to 0, the structure changes back to be cubic reversibly. The result demonstrates that high Na content will reduce the symmetry of FeHCF, and the structure change is highly reversible, during the Na^+ insertion/extraction process.

The structure evolution is different when the pristine phase is rhombohedral [42]. In the charging (Na^+ extraction) process, the peaks corresponding to (012) and (024) reflections stay still in the beginning and then shift gently toward higher angles. There is a turning point at the voltage of ~3.2 V with the doublet of the (110)/(104) reflection planes merging into a single peak, which confirms the formation of the cubic phase. As more Na^+ ions are extracted, the structure further changes from cubic to tetragonal, indicated by the peak shift toward a lower angle. At the end of the initial cycle, the positions of the (012) and (024) peaks change slightly higher as a result of Pauli repulsion caused by the extraction of lattice water. The revolution of the structure is also highly reversible, as shown in Fig. 4.9a. A similar structural evolution has also been verified in other PBAs, such as MnHCF [21, 22, 43], CoHCF [44], and NiHCF [33].

4.3 Synthesis Methods

The common synthesis methods of PBAs are coprecipitation, single iron source, and other methods. The coprecipitation method has

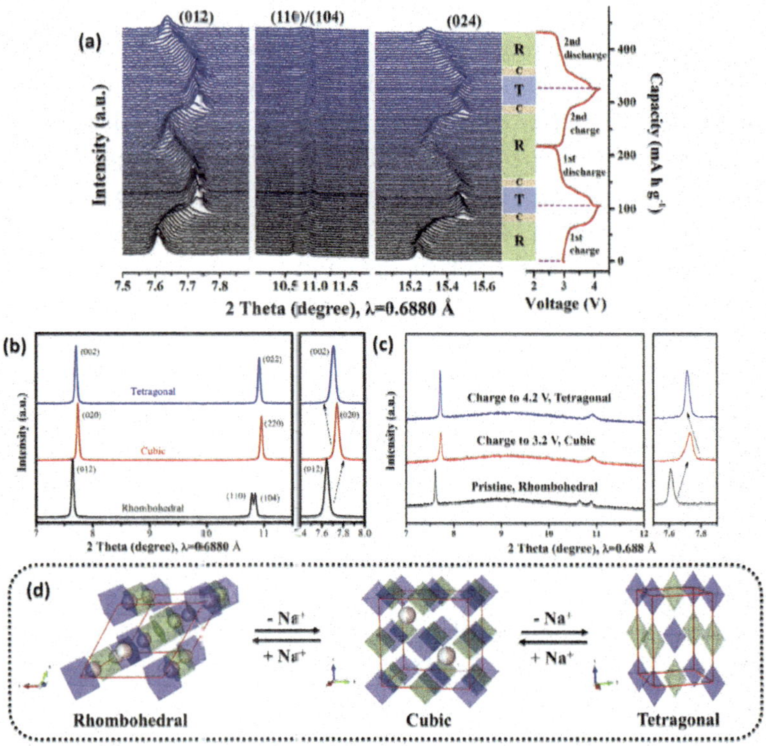

Figure 4.9 (a) *In situ* synchrotron PXRD patterns of NaFeHCF at different discharge and charge states, (b) simulated PXRD for the three different phases, (c) synchrotron *in situ* PXRD patterns at different voltage states during the first charge process, and (d) schematic illustration of phase evolutions of NaFeHCF [42]. Reproduced with permission from Ref. [42]. Copyright ©2020, Elsevier.

the advantages of a simple synthesis process, facile temperature treatment, and easy access to pure phase products [19]. The key to the preparation of PBAs by coprecipitation is the control of reaction conditions [45]. The factors that affect the morphology and crystal shape include the type of precipitant, feeding mode, reaction temperature, aging time, pH value, and additives. In addition, other relevant studies have also reported the introduction of chelating agents, reducing agents, and atmosphere control. In the early period, $K_3Fe(CN)_6$ was usually used during coprecipitation. Even if the as-prepared cathode material shows good electrochemical performance, it cannot be used in the practical full cell due to its absence of sodium ions [16]. In the years after 2013, most researchers replaced $K_3Fe(CN)_6$ with $Na_4Fe(CN)_6$ as the precursor during coprecipitation. Using $FeCl_2$ instead of $FeCl_3$ will also greatly improve the sodium content of the material and further improve the electrochemical performance of the material. For example, in 2015, Huang et al. [26] first used $FeCl_2$ and $Na_4Fe(CN)_6$ to synthesize sodium-rich $Na_{1.70}FeFe(CN)_6$ by the coprecipitation method, with a specific capacity of 120.7 mAh/g (at 200 mA/g) and 73.6 mAh/g (at 1200 mA/g).

In the single iron source synthesis, $[Fe(CN)_6]^{4-}$ is decomposed into Fe^{3+}/Fe^{2+}, which reacts with undecomposed $[Fe(CN)_6]^{4-}$ to prepare FeHCF. The decomposition of $[Fe(CN)_6]^{4-}$ must be triggered by one or multiple acids. The protons provided by the acid can attack the CN^- anion, weakening its ability with Fe^{2+} and thus resulting in the release of free Fe^{2+}. If there is an oxidant in the solution, the free Fe^{2+} will be partially oxidized to Fe^{3+}. When the Fe^{2+}/Fe^{3+} reaches a certain concentration, it will nucleate with the original $[Fe(CN)_6]^{4-}$. With the continuous release of Fe^{2+} and the precipitation reaction, the crystal nucleus will gradually grow and finally, a solid precipitate with good crystallinity can be obtained. The structure, composition, and morphology of FeHCF prepared by the single iron source method strongly depend on the synthesis parameters, such as solution acidity, reaction atmosphere, and temperature.

In addition, there are several unconventional preparation methods for PBAs. In the ball grinding method, the reactants are mixed directly, and the chemical bonds of reactants are broken and regenerated during high-energy ball grinding, leading to the preparation of nanometer-sized PBAs. However, the PBAs prepared by this method generally have irregular morphology and unsatisfied cycle stability. The photon-assisted precipitation method has also been applied for the synthesis of PBAs. Metal ions are generated indirectly under the assistance of photons, and a precipitation reaction occurs to produce PBAs. For example, Zhu et al. [46] synthesized FeHCF flexible electrodes with photosensitizer solution as the precursor, which is composed of ammonium ferric citrate aqueous solution and $K_3Fe(CN)_6$ solution. Under ultraviolet (UV) light, the iron ions of ammonium ferric citrate were reduced to Fe^{2+} ions, and the resulting Fe^{2+} ions can immediately coordinate with the existing excess $Fe(CN)_6^{4-}$ ions to form FeHCF precipitate. It is worth noting that due to the insufficient supply of Fe^{2+} ions, the nucleation and growth rate of FeHCF crystals were inhibited, thus resulting in an improved crystallinity and good dispersion of FeHCF particles.

4.4 Typical Hexacyanoferrate

4.4.1 Nickel Hexacyanoferrate (NiHCF)

NiHCF is the most studied single-electron redox PBA material, which exhibits an ultra-long lifespan and outstanding rate performance. The structures of NiHCF are commonly cubic or monoclinic depending on the sodium and water content. NiHCF shows only one pair of voltage platforms at 3.3 V corresponding to the Fe^{LS} redox. During sodium-ion insertion/extraction, there are negligible changes in lattice parameters (<1%), which ensures long-term cycling stability. Some reported NiHCFs delivered a reversible capacity of more than 80 mAh/g, which is close to the theoretical value, displayed good rate capability (>50 C), and achieved an ultra-long lifetime (>2000 cycles). Guo et al. [47] first investigated NiHCF as a cathode for nonaqueous SIBs. The as-prepared NiHCF delivers a capacity of 66 mAh/g with a capacity retention of 99.7% after 200 cycles at 20 mA/g, which is verified with the zero-strain property as shown in Fig. 4.10. Dai et al. [48] synthesized mesoporous NiHCF, which provided large channels for facile Na^+ diffusion and thus showed good rate performance (52 mAh/g at 500 mA/g). Different from the previous reports Han et al. [49] reported highly crystalline NiHCF, which exhibited a high capacity of 63 mAh/g at 850 mA/g, with a high capacity retention of 99% after 1200 cycles. In addition, Hu et al. [50] obtained nanoframe-like NiHCF, which exhibited a capacity of 53 mAh/g at a high rate of 11.68 A/g in an aqueous electrolyte. More detailed information is given in Table 4.5.

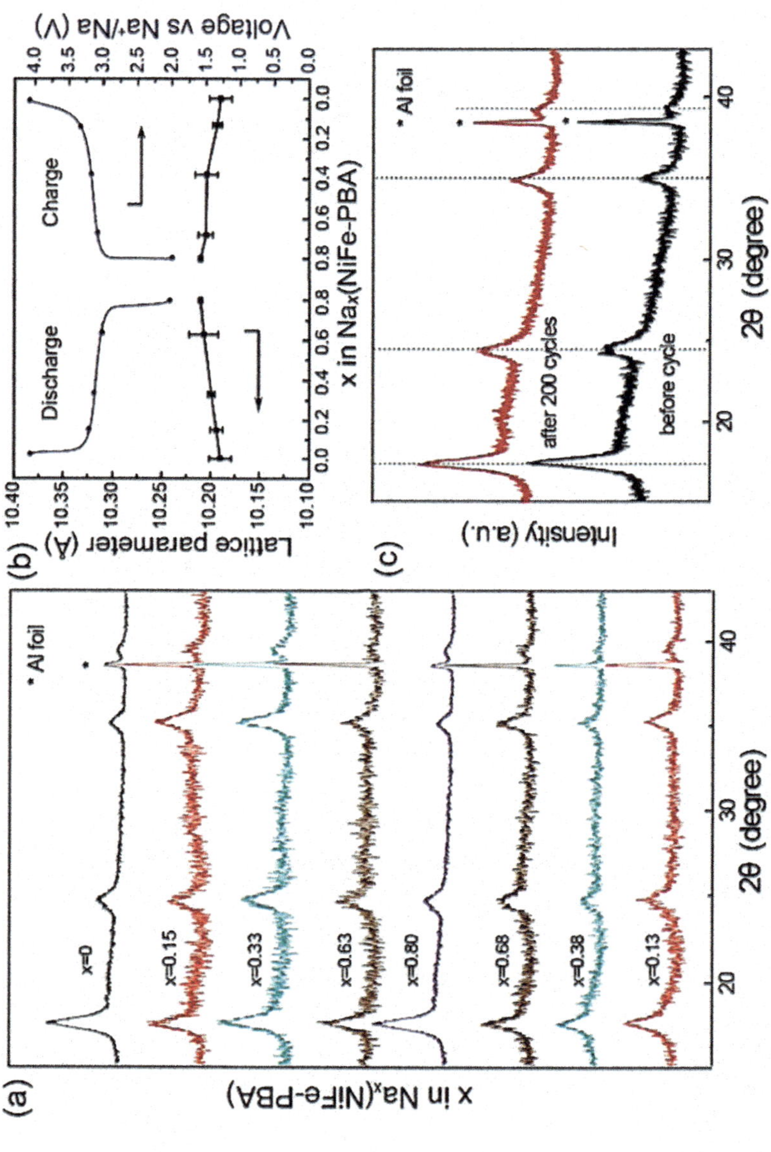

Figure 4.10 (a) *Ex situ* XRD patterns of NiFeHCF at various charge/discharge states. (b) Lattice parameter variation as a function of *x* in Na_xNiFeHCF. (c) XRD patterns of NiFeHCF before and after 200 cycles [47]. Reproduced with permission from Ref. [47]. Copyright ©2013, The Royal Society of Chemistry.

Table 4.5 Representative performance of reported NiHCF in SIBs

Active Material	Water [wt%]	Struc-ture	Size and Morphology	V_{disch}	Rate Capability [mAh/g@ mA/g]	Voltage Window	Cycle Life [Retention, Cycles, Rate]	Electrolyte	Ref.
$Na_{0.84}Ni[Fe(CN)_6]_{0.71}$	6	C	20–50 nm particle	3.07	66@20	2–4.1	99.7%, 200@20	1 M NP in EC:PC = 1:1	[47]
$Na_{1.46}Ni[Fe(CN)_6]_{0.83}$	12	M	300–500 nm cube	3.25	67/59@160/480	2–3.9	96%, 200@80	1 M NC in EC:DEC = 1:1	[33]
$Na_{1.11}Ni[Fe(CN)_6]_{0.80}$ (alkali etching)	4.2	M	100–200 nm nanoflower	3.2/3.3	90/71@99/3996	2–4.0	83.2%, 5000@495	1 M NC in EC:DMC = 1:1 with 5% FEC	[51]
$Na_xNi[Fe(CN)_6]_{0.80}$ (acid etching)	N/A	M	200 nm nanoframe	0.36 (Ag/AgCl)	61/53@80/11680	0–1 (Ag/AgCl)	94.3%, 500@5000	1 M $NaNO_3$	[50]
$(K/Na)_xNi[Fe(CN)_6]_{0.97}$/ CNTs+BCP	4	C	70–100 nm cube	3.2	128/98@168/2520	2.2–4	62.7%, 500@84	1 M NC in EC:DEC = 1:1	[52]

(Continued)

Table 4.5 (*Continued*)

Active Material	Water [wt%]	Structure	Size and Morphology	V_{disch}	Rate Capability [mAh/g @ mA/g]	Voltage Window	Cycle Life [Retention, Cycles, Rate]	Electrolyte	Ref.
$Na_{1.48}Ni[Fe(CN)_6]_{0.89}$	18	M	50 nm cube	3.25	85.7/66.2@8.5/4250	2–4.2	78%, 1200@300	1 M NC in EC:DEC = 1:1 with 2% FEC	[31]
$Na_{1.67}Ni[Fe(CN)_6]_{0.87}$	9	C	500 nm cube	3.14	80/63@15/850	2–4.0	99%, 1200@170	1 M NC in EC:DEC = 1:1 with 2% FEC	[49]
$Na_{2.01}Ni[Fe(CN)_6]_{0.85}$	9.2	M	300 nm particle	3.17	86.3/66.4@17/4250	2–4.2	90%, 800@170	1 M NC in EC:DEC = 1:1 with 2% FEC	[53]

Note: M = monoclinic, C = cubic, R = rhombohedral; NP = $NaPF_6$, NC = $NaClO_4$

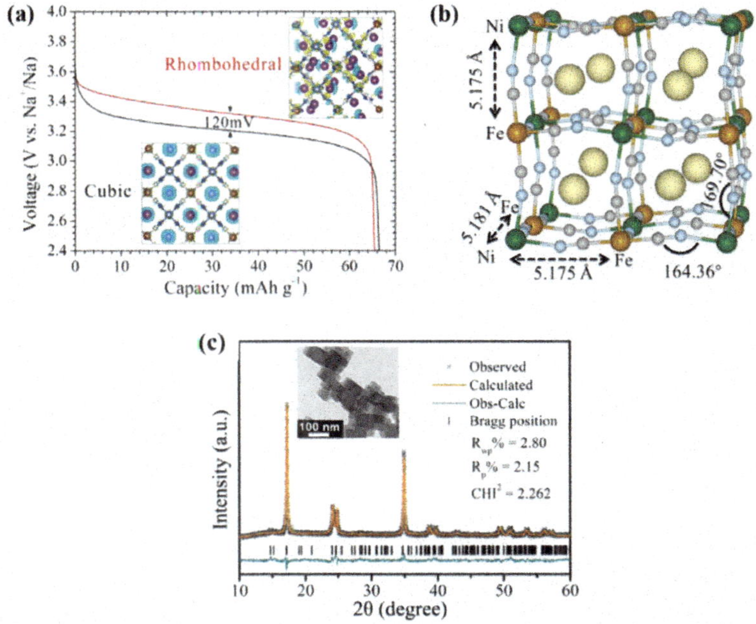

Figure 4.11 (a) Discharge curve of m-NiHCF and c-NiHCF [33]. Reproduced with permission from Ref. [33]. Copyright ©2016, American Chemical Society. (b) Local structures of m-NiHCF. (c) The Rietveld refinements of m-NiHCF [31]. Reproduced with permission from Ref. [31]. Copyright ©2018, WILEY-VCH.

NiHCF crystallized in the monoclinic phase was reported to show superior electrochemical performance. Han et al. [33] adopted the citrate-assisted coprecipitation method to obtain monoclinic NiHCF, which exhibits a 120 mV higher voltage plateau compared to that of the cubic one. And the mechanism was unraveled by DFT simulations. A similar phenomenon was also reported by Fang et al. [31]. Increasing the specific capacity is another challenge for NiHCF. Mai et al. [51] reported surface-activated NiHCF using the acid etching approach to increase the specific surface area and activate more Na^+ storage sites. The as-prepared NiHCF with high specific surface area, which boosts Na^+ diffusion kinetics, delivered a high specific capacity of 90 mAh/g, an outstanding rate performance of 71.0 mAh/g at 44.4 C, and an ultra-long lifetime over 5000 cycles. Wang et al. [52] obtained hybrid electrode CNT@NiHCF-BCP, which showed a surprisingly high initial discharge specific capacity of 194 mAh/g and long cyclability, and excellent rate capability, making it a promising candidate for the cathode of SIBs. The impressive cyclability and rate capability have triggered a wide research interest in NiHCF, which has become the most widely studied PBA material for SIBs.

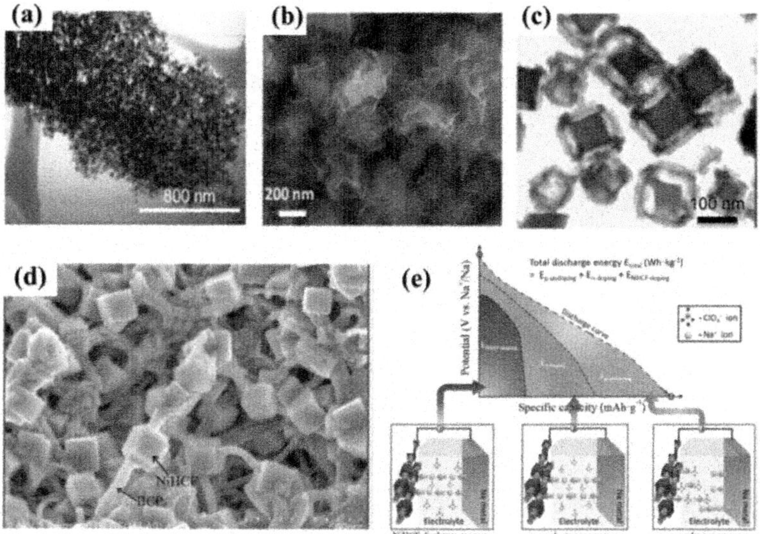

Figure 4.12 Morphologies of (a) mesoporous Na_2NiFe-PB particles. Reproduced with permission from Ref. [48]. Copyright ©2014, WILEY-VCH. (b) NiHCF-etch particles. Reproduced with permission from Ref. [51]. Copyright ©2017, American Chemical Society. (c) Na_2NiFe-PB nanoframes. Reproduced with permission from Ref. [50]. Copyright ©2018, MDPI. (d) CNT@NiHCF-BCP [52]. (e) Energy storage mechanism of the CNT@NiHCF-BCP cathode in the discharge process. Reproduced with permission from Ref. [52]. Copyright ©2017, The Royal Society of Chemistry.

4.4.2 Iron Hexacyanoferrate (FeHCF)

FeHCF is the most studied two-electron redox PBA material, which features two discharge voltage plateaus at 2.60~2.89 V and 3.23~3.80 V corresponding to Fe^{HS} and Fe^{LS}, and a high theoretical specific capacity of 170 mAh/g. However, the Fe^{LS} redox is generally not activated completely, resulting in capacity loss compared with the theoretical one. Besides, different from NiHCF, FeHCF suffers from poor cyclability because of the existence of defects and side reactions between the cathode and the electrolyte, especially at high voltages. Therefore, increasing sodium content and reducing defects, which are heavily associated with the crystallinity of FeHCF, are the keys to promote the capacity and lifetime of FeHCF.

FeHCF can be synthesized via two main different routes: (1) single iron source method and (2) coprecipitation method. Generally speaking, the nucleation and grain growth rate of FeHCF can be slowed down via the single iron source method. Guo et al. [19] first used the single iron source method to obtain high-quality $Na_{0.61}Fe[Fe(CN)_6]_{0.94}$. The synthesis process is facile with hydrochloric acid (37%) solution added to release the Fe^{2+} ions. The as-prepared FeHCF shows cube-like morphology (300–600 nm), which delivers a high specific capacity of 170 mAh/g at 25 mA/g and good cycling stability without obvious decay after 150 cycles. However, the sodium content of $Na_{0.61}Fe[Fe(CN)_6]_{0.94}$ is still too low to be used as a cathode in full cells. Many studies have been conducted to modify the single iron source method to improve the sodium content of FeHCF. Employing a reducing agent (VC) and an inert atmosphere (N_2), Guo et al. [27] improved the sodium content from 0.83 (air; without VC) to 1.63 (N_2; with VC) per formula unit, while the Fe^{III} content reduces from 76.8% to 41.4%. On account of the addition of N_2 and VC, Na-rich FeHCF, which crystallized in monoclinic with a size of 2–3 μm, delivered a capacity of 150 mAh/g and a longer lifetime (90% retention after 200 cycles). Based on the aforementioned approach, Brant et al. prepared monoclinic $Na_2FeFe(CN)_6{\cdot}2H_2O$ by adjusting the synthesis temperature to 107°C, which contains almost no $Fe(CN)_6$ vacancies and shows an excellent cyclability/rate capability within the voltage window of 3.9–2.0 V [54]. Brant et al. [23] also reported that monoclinic $Na_{1.81}FeFe(CN)_6{\cdot}2.3H_2O$ can be fabricated under 80°C, and the monoclinic FeHCF will transform into rhombohedral phase with further vacuum drying under 120°C. When the reaction temperature reached 140°C and HCl was replaced by ascorbic acid, Lu et al. [40] obtained $Na_{1.92}FeFe(CN)_6{\cdot}0.08H_2O$, the stoichiometry of which was close to ideal. The obtained FeHCF retains 80% capacity over 750 cycles. It is usually reported that a large size of grain (>1 μm) will influence the capacity and rate capability of FeHCF, but $Na_{1.92}FeFe(CN)_6{\cdot}0.08H_2O$ still delivers high capacity (157 mAh/g at 15 mA/g) and outstanding rate performance (145 mAh/g at 1500 mA/g) when the size of the grain is around 2–8 μm. More detailed information on typical FeHCF can be seen in Table 4.6.

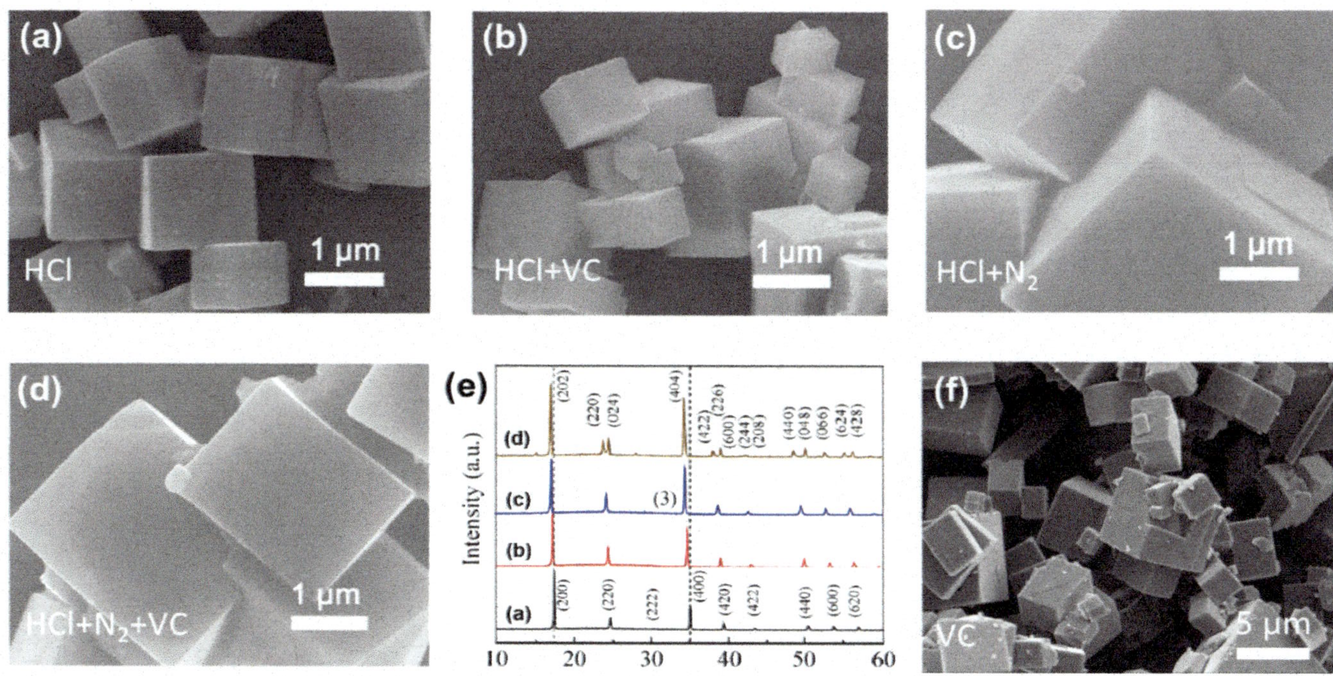

Figure 4.13 SEM images of FeHCF under different self-decomposition synthesis conditions (a) HCl; (b) HCl + VC; (c) HCl + N_2; (d) HCl + N_2 + VC. (e) XRD patterns of different FeHCF. Reproduced with permission from Ref. [27]. Copyright ©2014, Springer Nature. (f) SEM images of FeHCF obtained with only VC. Reproduced with permission from Ref. [40]. Copyright ©2015, American Chemical Society.

Table 4.6 Representative performance of reported FeHCF in SIBs

Active Material	Water [wt.%]	Struc-ture	Size and Morphology	V_{disch}	Rate Capability [mAh/g @ mA/g]	Voltage Window	Cycle Life [Retention, Cycles, Rate]	Electrolyte	Ref.
				Single iron source method					
$Na_{0.61}Fe[Fe(CN)_6]_{0.94}$	15.7	C	300–600 nm cube	3	170/70@25/600	2–4.2	100%, 200@25	1 M NP in EC:DEC = 1:1	[19]
$Na_{1.63}Fe[Fe(CN)_6]_{0.89}$	N/A	R	2.5–4 μm cube	2.9	150@25	2–4.2	90%, 200@25	1 M NP in EC:DEC = 1:1	[27]
$Na_{1.56}Fe[Fe(CN)_6]$	15.5	C	0.5–1 μm cube	2.92/3.42	103.6@20	2–4.0	97%, 400@25	1 M NC in EC:DEC = 1:1	[76]
$Na_{1.92}FeFe(CN)_6$	0.5	R	2–8 μm cube	3.0/3.29	157/145@15/1500	2–4.2	80%, 750@75	1 M NP in EC:DEC = 1:1 with 5% FEC	[40]
FeHCF-H2	13.1	C	900 nm cube with concave	2.9/3.4	93.1/82@200/5000	2–4.2	74%, 500@200	1 M NC in PC:EC = 1:1 with 5% FEC	[62]
$Na_{3.1}Fe_4[Fe(CN)_6]_3$	8.8	C	Hierarchical nanotubes	2.9	115/83@240/6000	2–4.0	65%, 10,000@1200	1 M NC in EC:DMC = 1:1 with 5% FEC	[64]

(*Continued*)

Table 4.6 (*Continued*)

$Na_{1.02}Fe[Fe(CN)_6]_{0.95}$	5.6	C	200 nm yolk shell	2.73/3.29/3.77	135/50@14/1400	2–4.0	95.8%, 500@140	N/A	[77]
$Na_{0.39}Fe[Fe(CN)_6]_{0.82}$	8.8	C	0.5–2 μm cube with nanopores	2.78/3.64	115.2/77.2 @50/400	2–4.2	99.6%, 200@50	1 M NP in EC:DEC = 1:1	[63]
FeHCF@PANI	11	C	500 nm cube	2.88/3.34	108.3/63.3 @100/5000	2–3.6	93.4%, 500@100	1 M NP in EC:DEC:DMC = 1:1:1	[78]
$Na_{0.51}Fe[Fe(CN)_6]_{0.82}$ @PEDOT:PSS	9	C	1 μm Cube	2.9/3.4	131/113@12/1220	2–4.35	96%, 100@50	0.7 M NC in EC:DEC = 1:1	[79]
				Coprecipitation method					
$Na_{1.92}FeFe(CN)_6$	13.2	C	50 nm cube	2.8/3.4	120/71@200/1200	2–4.2	70%, 100@200	1 M NC in PC:EC = 1:1	[26]
$Na_{1.95}Fe[Fe(CN)_6]_{0.93}$	10.8	M	3–5 μm cube	2.9/3.3	115/79@50/800	2–4.2	78%, 280@50	1 M NP in EC:DEC = 1:1 with 5% FEC	[25]
$Na_{1.73}Fe[Fe(CN)_6]$	18	R	1.5 μm cube	2.8/3.32	108/89/70 @10/500/2000	2–4.0	75%,500@200	1M NC in PC:EC=1:1 with 3% FEC	[59]

$K_{0.33}FeFe(CN)_6$@RGO	2.2	C	300 nm cube	2.83/3.42	158/126@60/2400	2–4.0	92.2%, 1000@60	1 M NC in EC:DMC = 1:1	[66]
$Na_xFeFe(CN)_6$@ZnO	4.6	C	1 μm cube	2.9/3.4	100/86@200/10000	2–4.2	74.2%, 500@200	1 M NC in PC:EC = 1:1 with 5% FEC	[80]
$Na_{1.54}Fe[Fe(CN)_6]_{0.96}$ @NiHCF	N/A	C	2 μm cube	2.71/3.12/3.32/3.67	93/60@25/2000	2–4.2	78%, 800@200	1 M NC in PC:EC = 1:1	[81]
$Na_{1.54}Fe[Fe(CN)_6]_{0.96}$ @PPy	12.4	C	2 μm cube	2.82/3.66	113/75@25/3000	2–4.2	79%, 500@200	1 M NC in PC:EC = 1:1	[82]
$Na_{1.36}FeFe(CN)_6$ @PDA	9.9	C	20–200 nm cube	2.72/2.91/3.28/3.44	116.7/81.7 @100/5000	2–4.2	77.4%, 500@200	1 M NC in EC:DMC = 1:1	[83]
Ball-milling method									
$Na_{0.90}Fe[Fe(CN)_6]_{0.96}$	13	C	<40 nm irregular	2.8/3.3	129/106 @34/5100	2–4.2	80%, 500@1700	1 M NC in PC:EC = 1:1 with 5% FEC	[84]

Note: M = monoclinic, C = cubic, R = rhombohedral; NP = $NaPF_6$, NC = $NaClO_4$

Although coprecipitation is more facile and efficient, the FeHCF obtained via this route will suffer from more defects, including vacancies and crystal water, as a result of the fast nucleation and grain growth. In order to control the precipitation rate, many factors need to be considered, such as the reaction temperature, the concentration of precursors, and chelating agents. Ma et al. [55] systematically studied the effect of reaction temperature at 0, 40, and 80°C. The $Na_{1.73}Fe[Fe(CN)_6]_{0.98}$ prepared at 0°C shows higher Na content and better crystallinity, which delivers better electrochemical performance with a capacity of 123 mAh/g at 100 mA/g and 73% retention after 200 cycles. In earlier studies, NaCl [56] and NaI [57] were used as extra sodium sources to enrich the Na content of the PBA precipitate. Through a chemical reduction method, Qian et al. [57] synthesized $Na_{1.33}Fe[Fe(CN)_6]_{0.82}$ nanoparticles by mixing and stirring the $Fe[Fe(CN)_6]_{0.87}$ precursor and NaI solution for 6 h. However, the addition of NaCl or NaI cannot slow down the growth process to obtain a larger size of the grain but only increased the sodium content to a certain extent [58].

In order to further improve crystallinity and enrich the Na content, chelating agents, which can play the role of sodium source and have a high complexation ability with transition metal ions, have become the focus of studies. As Fig. 4.14 illustrated, on account of the strong complex ability of the ligand, the TM ligand can release TM ions slowly, which slows down the nucleation and growth process. Take sodium citrate as an example; the citrate ions with a negative charge may even absorb on the nucleus surfaces to suppress the growth and inhibit crystal aggregation. The release of TM ions, which is gradual and sustainable, subsequently enables these nuclei to grow into monodispersed FeHCF nanocubes with good morphology and fewer defects [36]. Sodium citrates (Na_3CA) are the most reported chelating agents in PBAs [25, 26, 59]. Zhang et al. [26] first reported the Na-rich FeHCF using citrates-assisted precipitation. On account of the introduction of Na_3CA, the morphology of FeHCF changes from irregular nanoparticles to nanocubes. With the increase in Na_3CA, the capacity of the as-prepared FeHCF increases from 99.8 to 120 mAh/g at 200 mA/g and the rate performance improves remarkably (27.6 versus 71.2 mAh/g at 1200 mA/g). Chou et al. [59] investigated the controllable citrates-assisted precipitation method systematically. The results show that high-crystalline FeHCF ($Na_{1.73}Fe[Fe(CN)_6]\cdot 3.8H_2O$) delivers 108/89/70 mAh/g at

10/500/2000 mA/g and 75% retention over 500 cycles at 200 mA/g. Ethylenediaminetetraacetic acid (EDTA) is another kind of chelate that has been confirmed to be efficient [60, 61]. The complexation ability of EDTA is much stronger than Na_3CA; therefore, $[Fe(CN)_6]^{2-}$ cannot get TM ions from TM-EDTA directly. Because of that, acid is introduced to reduce the complexation ability to release TM ions. Cao et al. [60] studied the effect of solution pH and EDTA concentration on the preparation of FeHCF. The result shows that higher pH results in slower nucleation kinetics leading to a larger size of the grain, less vacancies, and crystal water, which exhibits more stable cycling as cathodes for SIBs.

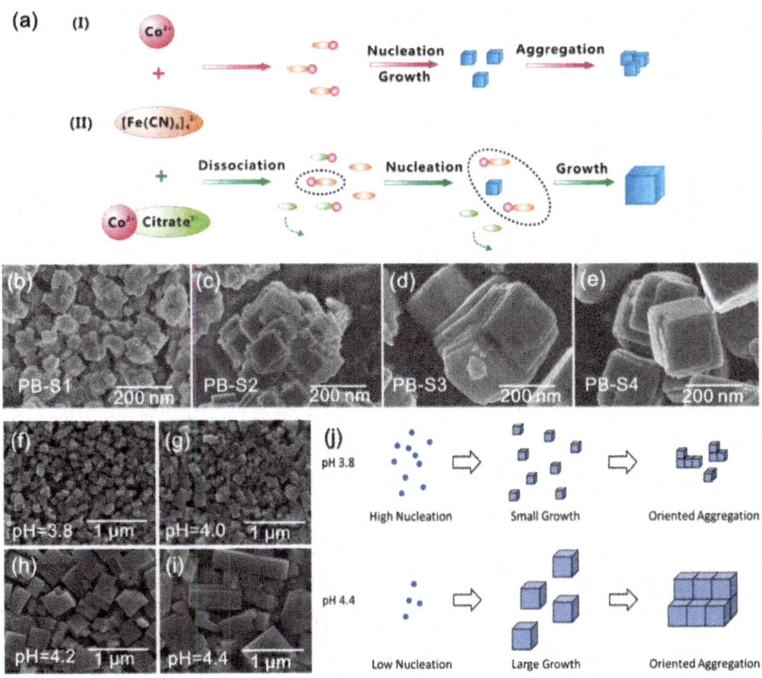

Figure 4.14 (a) Schematic illustration of (I) the non-citrate-assisted coprecipitation method and (II) the citrate-assisted controlled coprecipitation method. Reproduced with permission from Ref. [36]. Copyright ©2016, American Chemical Society. (b–e) SEM images of PB obtained by different synthesis parameters. Reproduced with permission from Ref. [59]. Copyright ©2020, Springer Nature. (f–i) SEM of PB (3.8), PB (4.0), PB (4.2), and PB (4.4). (j) Proposed mechanism of oriented aggregation between different synthesis pH. Reproduced with permission from Ref. [60]. Copyright ©2020, The Royal Society of Chemistry.

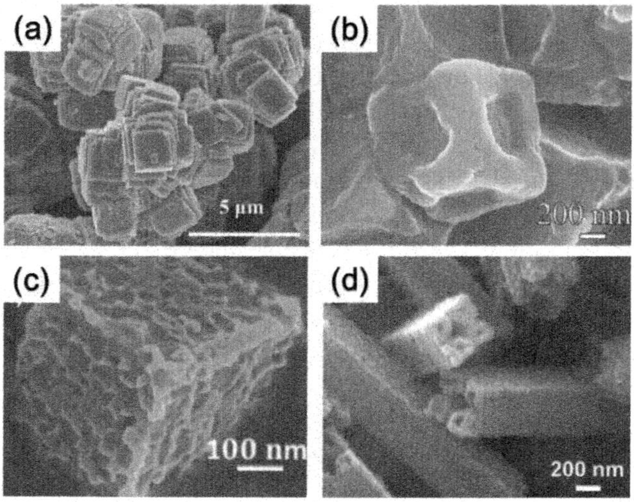

Figure 4.15 SEM images of (a) border-rich FeHCF. Reproduced with permission from Ref. [25]. Copyright ©2017, Elsevier. (b) Concave FeHCF. Reproduced with permission from Ref. [62]. Copyright ©2017, Wiley-VCH. (c) FeHCF with pore. Reproduced with permission from Ref. [63]. Copyright ©2016, American Chemical Society. (d) Prussian white hierarchical nanotubes. Reproduced with permission from Ref. [64]. Copyright ©2019, Wiley-VCH.

Morphology control can affect the kinetics of PBAs remarkably and may provide more active redox sites. Chen et al. [25] obtained border-rich FeHCF at a low temperature of 0°C, high concentration of Na_3CA, and PVP surfactant. The low reaction temperature is beneficial to two-dimensional (2D) nucleation, forming a border-rich morphology. Besides, Na_3CA decreases the supersaturation of the solution, strongly inhibiting the expansion rate of each layer. Liu et al. [62] found that PVP as a protective layer can be formed on the FeHCF surface, which leads to the adjustment of the size of distribution and the slowed etching effect of acids. In addition to size and homogeneity control, PVP, as a useful surfactant, can also affect the orientation of PBAs. Chen et al. [63] reported FeHCF with a pore-rich structure. In this study, PVP was found to play three important roles. First, PVP controls the micro-morphology of the particles because of its surfactant effect. Second, PVP can reduce Fe^{3+} to Fe^{2+}, resulting in better crystallinity with less defects. Finally, PVP protects the surface of particles from serious etching. Accordingly, the pore-rich FeHCF exhibited a capacity of 115.2/77.2 mAh/g at

50/400 mA/g and 99.6% retention after 200 cycles. Similar to PVP, polyallylamine hydrochloride (PAH) [65], cetyltrimethylammonium bromide (CTAB) [66], and sodium dodecyl benzenesulfonate (SDBS) [67] have also been reported to show similar effects. Self-assembly is another strategy to control the morphology of PBAs. Mai et al. [64] fabricated Prussian white ($Na_{3.1}Fe_4[Fe(CN)_6]_3$) hierarchical nanotubes with fully open configurations, which have enhanced Na^+ intercalation pseudocapacitance behavior and deliver a superior rate capability of 83 mAh/g at 50 C and superior outstanding cycling stability (65% retention over 10,000 cycles).

Compositing with carbon has been proved to be an efficient strategy no matter in LIBs or SIBs. Different from chemical or mechanical mixing, *in situ* compositing, can build more efficient electrical contact and improve the electrochemical performance of PBAs significantly. Our group [68] prepared FeHCF@KB with low vacancy and crystal water by using the single iron source method to grow FeHCF cubes on KB chains directly. The introduction of KB increased the nucleation sites in the synthesis process leading to a slower growth process. Therefore, more Fe^{LS} redox couples can be activated because of low $[Fe(CN)_6]$ vacancy content. Besides, fast kinetics (both electron transfer and ions diffusion) is assured because of the intimate contact of the FeHCF with the carbon and the small size of the precipitate. Hence, the FeHCF@KB achieved an unprecedented rate of performance and cyclability (77.5 mAh/g at 90 C, 90 mAh/g after 2000 cycles with 90% capacity retention at 20 C). Guo et al. [41] prepared a necklace-like FeHCF/CNT composite in which the CNT fibers pass through the cubic FeHCF nanocube. The obtained sample achieved high Coulombic efficiency of more than 99.4%, a discharge capacity of 142 mAh/g, an output specific energy density of 408 Wh/kg, with an 86% capacity retention over 1000 cycles. On account of the excellent kinetics, FeHCF/CNT also exhibits a notable performance at −25°C. FeHCFs composited with graphene oxide (GO) and reduced graphene oxide (RGO) have also been reported [66, 69–74]. Research shows that GO or RGO can not only provide excellent electronic conduction but also protect the FeHCF from the attack of electrolytes. Besides, the coordinated water can be further removed because of the existence of RGO [75]. The capacity of FeHCF@GO/RGO can even exceed the theoretical value due to the extra capacity contribution from GO/RGO.

4.4.3 Manganese Hexacyanoferrate (MnHCF)

MnHCF has a high theoretical specific capacity and high discharge voltage platforms at about 3.48~3.60 V (Mn^{HS}) and 3.26~3.40 V (Fe^{LS}). Besides, Mn^{2+} is stable during preparation in an aqueous solution leading to Na-rich MnHCF. Moritomo et al. [85] first reported $Na_{1.32}Mn[Fe(CN)_6]_{0.83}\cdot 3.5H_2O$ with cubic structure as a cathode for SIBs by the electrodeposition method, which exhibits a discharge capacity of 109 mAh/g at 0.5 C, as shown in Fig. 4.16. Removing interstitial H_2O from the MnHCF framework can induce the transition from monoclinic to rhombohedral. Song et al. [22] synthesized rhombohedral $Na_{2-\delta}$MnHFC by evacuating MnHCF to 15 mTorr at 100°C for 30 h, which delivers a high capacity of 123 mAh/g at 2860 mA/g and retains 75% capacity after 500 cycles. After dehydration, the plateaus merge two (3.17 Fe^{LS}; 3.49 Mn^{HS}) into a single flat plateau at 3.44 V. The same phenomenon is also reported by Jin et al. [28]. In order to explain the elusive redox behavior, standard ionization energy (SIE) and ligand field stabilization energy (LFSE) have been utilized [86]. The SIE of $Fe^{3+/2+}$ is lower than $Mn^{3+/2+}$ resulting in the lower redox potential of $Fe^{3+/2+}$. However, the LFSE of $Fe^{(LS)3+/2+}$ is higher than the LFSE of $Mn^{(HS)3+/2+}$ because of stronger Fe–C coordination than Mn–N, leading to the increase in the $Fe^{3+/2+}$ redox potential. When the interstitial water is removed, the competing effects of SIE and LFSE will balance each other so that $Fe^{(LS)2+}$ and $Mn^{(HS)2+}$ redox potentials would overlap and two plateaus merge into a single flat one.

However, MnHCFs suffer from a more severe Jahn–Teller effect (JTE) and Mn^{3+}dissolution than that of FeHCFs and NiHCFs, leading to poor cycling stability and low CE. Sottmann et al. [21] researched the effect of crystal water on the electrochemical performance of MnHCF through *operando* XRD and XAS. The results indicate that the Mn dissolution in the form of $NaMnCl_3$ caused the degradation of capacity, which will be aggravated at higher water content in the crystal of MnHCF. Furthermore, large volume changes between the phases of different voltage states may also result in poor cycling stability. A lot of studies have been conducted to suppress JTE, including coating and doping (elemental substitution). Carbon coating such as conductive polymer and GO is an effective strategy to promote the electrochemical performance and protect the active

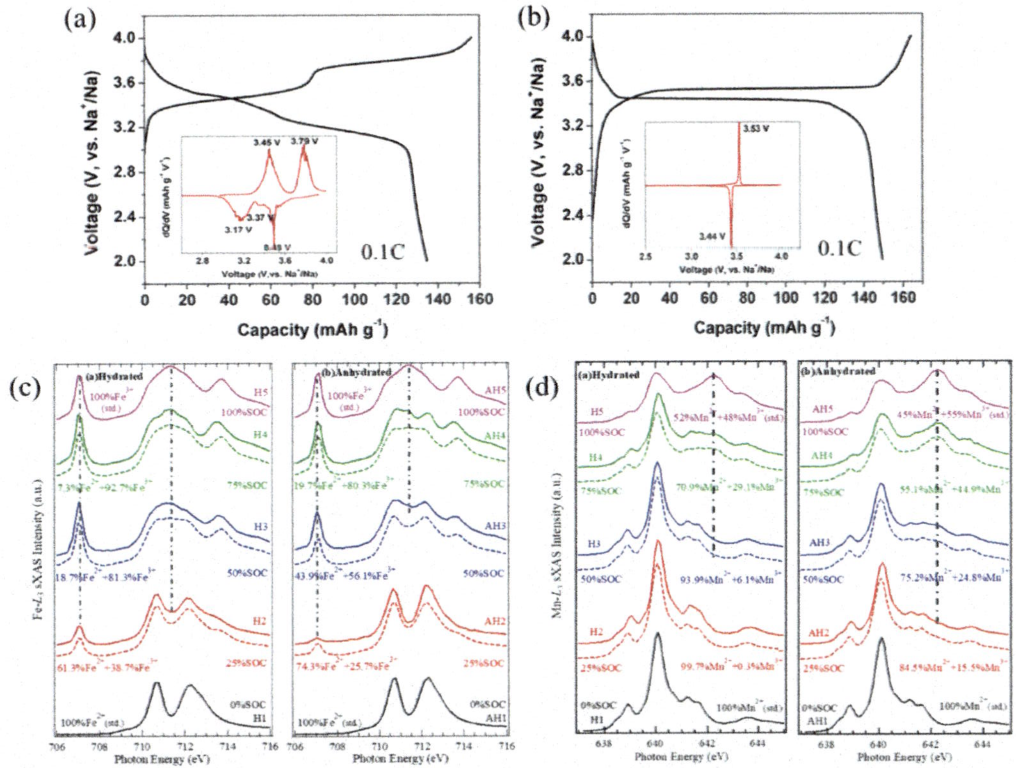

Figure 4.16 Galvanostatic initial charge and discharge profiles of (a) air-dried and (b) vacuum-dried $Na_{2-\delta}$MnHCF. Reproduced with permission from Ref. [22]. Copyright ©2015, American Chemical Society. L3-edge sXAS spectra of (c) Fe and (d) Mn. Reproduced with permission from Ref. [86]. Copyright ©2017, American Chemical Society.

Figure 4.17 SEM and TEM images of (a, b) glucose-treated MnHCF. Reproduced with permission from Ref. [87]. Copyright © 2015, MDPI. (c, d) MnHCF@PPY. Reproduced with permission from Ref. [88]. Copyright © 2017, WILEY-VCH. (e, f) MnHCF@PEDOT. Reproduced with permission from Ref. [89]. Copyright © 2020, American Chemical Society. (g, h) MnHCF@Graphene. Reproduced with permission from Ref. [90]. Copyright © 2018, American Chemical Society. (i, j) MnHCF@3DNC. Reproduced with permission from Ref. [91]. Copyright © 2019, Elsevier. (k, l) MnHCF@$Na_3(VOPO_4)_2F$. Reproduced with permission from Ref. [92]. Copyright © 2019, American Chemical Society.

material from reacting with the electrolyte. Moritomo et al. [87] reported glucose-treated MnHCF. Through heating, the mixture of MnHCF and glucose at 350°C, the thermal decomposition of the glucose modifies the surface of MnHCF, which enhances the electronic conductivity and improves the rate property. However, high-temperature treatment will damage the structure of MnHCF and bring severe loss of capacity.

Chou et al. [93] and Jiang et al. [88] reported polypyrrole (PPy)-coated MnHCF in order of time. $Na_{1.72}MnFe(CN)_6$@PPy exhibits a capacity of 120 mAh/g (at 12 mA/g) and 69 mAh/g (at 2400 mA/g) and retains 67% capacity after 200 cycles at 240 mA/g [93]. PPy plays important roles: It can increase the electronic conductivity and reduce the dissolution of Mn. PPy doped with ClO_4^- can increase capacity, and the polymerization of PPy can lead to additional Na insertion into MnHCF to improve reversible capacity. In addition to PPy, our group [89] reported MnHCF@PEDOT. The formula of MnHCF@PEDOT can be calculated as $Na_{1.71}Mn[Fe(CN)_6]_{0.94}\cdot 1.66H_2O$, which delivers a high capacity of 148 mAh/g at 12 mA/g, and as the current rate increases from 12 to 3000 mA/g, the reversible specific capacity decreases to 90.1 mAh/g. Because of the PEDOT coating, MnHCF@PEDOT shows a long lifetime with a capacity of 74.4 mAh/g (78.2% retained) at 10 C after 1000 cycles and has excellent performance even at a low temperature of −10°C. PEDOT surface layer on MnHCF can boost the electronic conduction, suppress phase transition and side reactions, and inhibit the dissolution of transition metal. GO coating is a normal way to enhance electrochemical performance both in LIBs and SIBs. Xie et al. [90] synthesized MnHCF/G using the precipitation method and vibration milling, which delivers an excellent rate performance.

Several new coating forms have been reported in recent years. Mao et al. [91] fabricated NaK-MnHCF@3DNC. First, 3D N-doped ultra-thin hierarchical porous carbon networks (3DNC) were synthesized by freeze-drying and high-heat treatment. Then NaK-MnHCF was grown *in situ* on 3DNC. The sodium storage mechanism of NaK-MnHCF@3DNC is both diffusion-controlled and capacitive. The composite delivers a capacity of 220 mAh/g at 20 mA/g and 99 mAh/g at 500 mA/g for over 1000 cycles. Peng et al. [92] reported a novel $Na_3(VOPO_4)_2F$ coating for MnHCF. $Na_3(VOPO_4)_2F$ is a kind of low specific capacity cathode material of SIBs with NASICON

Table 4.7 Representative performance of reported MnHCF in SIBs

Active Material	Water [wt.%]	Structure	Size and Morphology	V_{disch}	Rate Capability [mAh/g @ mA/g]	Voltage Window	Cycle Life [Retention, Cycles, Rate]	Electrolyte	Ref.
$Na_{1.80}Mn[Fe(CN)_6]_{0.98}$	10	M	Mixed particle/cube	3.7	146/87@15/1500	2–4.0	73.4%, 780 @75	1 M NP in EC:DEC = 1:1 with 5% FEC	[95]
$K_{1.80}Mn(Fe(CN)_6)_{0.94}$	4.7	M	300 nm agglomerate	3.28/3.43/3.57	124/93@120/2400	2–4.2	80%, 1000 @120	1 M NC in PC with 3% FEC	[96]
$Na_{1.66}Mn[Fe(CN)_6]_{0.94}$	10	M	10–30 nm particle	3.7/3.8	143/90@15/3000	2–4.0	90.6%, 500 @150	1 M NP in EC:DEC = 1:1 with 5% FEC	[90]
$Na_{2-\delta}MnFe(CN)_6$	2	R	200–400 nm irregular	3.47	150/123@14/2860	2–4.0	75%, 500 @100	1 M NC in EC:DEC = 1:1 with 10%FEC	[22]
$Na_{1.84}MnFe(CN)_6$	0.2	R	3–5 μm agglomerates	3.47	150.1@31	2–4.0	88.03%, 50 @155	1 M NP in PC:PP = 9.5:0.5 with 1%FEC	[28]

$Na_{1.80}Mn[Fe(CN)_6]_{0.98}$	10	M	200–600 nm particle	3.4/3.8	144/116/87@15/150/1500	2–4.0	72.7%, 2100 @150	1 M NP in EC:DEC = 1:1 with 5% FEC	[95]
$Na_{1.38}Mn[Fe(CN)6]_{0.92}$	14	C	2–4 μm cube	3.45/3.6	124/74@25/600	2–4.2	70%, 500 @200	1 M NP in PC with 10 wt% FEC	[43]
$Na_{1.92}Mn[Fe(CN)6]_{0.98}$	10.9	M	3–5 μm cube	3.45	153/110@10/1000	2–4.0	82%, 500 @100	1 M NP in PC:EMC:FEC = 50:48:2	[61]
$Na_{1.6}Mn_{0.75}[Fe(CN)_6]$	10.9	M	500–700 nm hexapod-like	3.3/3.7	137/105@25/500	2–4.2	72.3%, 2700 @500	1 M NP in EC:PC = 1:1 with 2% FEC	[94]
$Na_{1.60}Mn[Fe(CN)_6]_{0.9}$ @NVOPF	N/A	R	~500 nm irregular	3.25/3.34/3.7	113/91@10/1000	2–4.0	84.3%, 500 @100	1 M NP in PC:EMC:FEC = 50:48:2	[92]
$Na_{1.72}MnFe(CN)_6$@PPY	14	R	300 nm irregular	3.55	120/69@12/2400	2–4.2	67%, 200 @240	1 M NC in EC:DEC = 1:1	[93]

(Continued)

Table 4.7 (*Continued*)

$Na_{1.89}Mn[Fe(CN)_6]_{0.85}$ @PPY	9.3	C	500–600 nm particle	3.4/3.6	118/65@10/200	2–4.2	107/118, 150@10	1 M NC in EC:DEC = 1:1 with 2%FEC	[86]
$Na_{1.71}Mn[Fe(CN)_6]_{0.94}$ @PEDOT	4.2	M	0.3–1.2 μm particle	3.3/3.4	148/95@15/1500	2–4.0	78.2%, 1000@1500	1 M NP in EC:DEC = 1:1 with 5% FEC	[59]
$Na_{1.66}Mn[Fe(CN)_6]_{0.94}$ @GO	10	M	10–20 nm irregular	3.7/3.8	142.7/89.7@15/3000	2–4.2	86%, 2000 @1500	1 M NP in EC:DEC = 1:1 with 5% FEC	[90]
$Na_{1.73}K_{0.13}Mn$ $[Fe(CN)_6]_{0.97}$@3DNC	6.2	C	20–100 nm particle	3.1/3.55	220/110@20/500	2–4.2	90%, 1000 @500	1 M NC in EC:DEC = 1:1	[91]

Note: M = monoclinic, C = cubic, R = rhombohedral; NP = $NaPF_6$, NC = $NaClO_4$

structure, exhibiting promoted rate capability and long-term cyclability. The NVOPF-coating layer can protect MnHCF from the decay of HF generated by side reactions during cycling and suppress the transition metal dissolution, thus contributing to the structural integrity. The detailed performance is shown in Table 4.7. It is a similar strategy to NiHCF coating, using a material with high stability and rate capability to coat a material with high capacity and a high-voltage platform.

Introducing Mn vacancy artificially is another effective strategy to suppress JTE, which can substantially change the charge redistribution and mitigate structural deformation. Due to the strong chelation between EDTA and Mn atoms, Shang et al. [94] introduced Mn vacancies into MnHCF to form $Na_{1.6}Mn_{0.75}(\gamma Mn)_{0.25}[Fe(CN)_6]\cdot 1.57H_2O$. The active material displays excellent cyclability with a capacity retention of 72.3% after 2700 cycles at 500 mA/g while the Coulombic efficiency is nearly 100%. As shown in Fig. 4.18, during the second Na^+ extraction, four Mn–N bonds are slightly lengthened along the direction of $V_{(Mn)}$ in EDTA-NaMnHCF while the Mn–N bonds are severely lengthened in the traditional MnHCF with $V_{(FeCN)}$. Besides, introducing $V_{(Mn)}$ can hinder the generation of $V_{(FeCN)}$ to make sure the Fe–C_6 octahedra fixed skeletons are complete, which can stabilize the stretching of Mn–N bonds to suppress the JTE distortion. Doping (elemental substitution) is another effective way to improve the electrochemical performance of MnHCF, which will be discussed in Section 4.5.

4.4.4 Other Prussian Blue Analog Compounds

CoHCF exhibits two high-voltage platforms corresponding to Co^{HS} and Fe^{LS} and high specific capacity, the electrochemical properties of which are similar to MnHCF [10, 29, 35]. However, CoHCF is also suffering JTE, and the cost of cobalt restricts its research and application. Yang et al. [36] developed a facile citrate-assisted controlled crystallization method to synthesize CoHCF with low defect, which exhibits promoted Na^+-storage-specific capacity and cyclability. The capacity reached 153 mAh/g at 10 mA/g and decreased slightly from 128 to 114 mAh/g after 200 cycles at 100 mA/g, corresponding to a capacity retention of 90%. The cycling

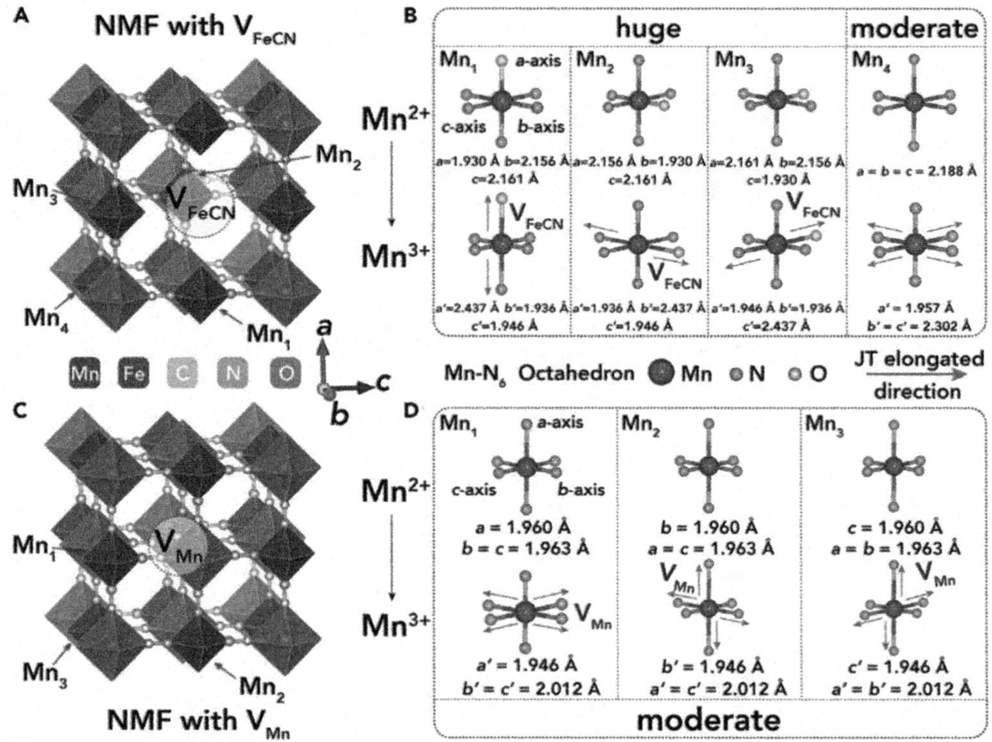

Figure 4.18 Theoretical calculation of the variations of Mn–N bond during the second Na^+ extraction, the cubic phase of (A) CIT-NaMnHCF with $[Fe(CN)_6]^{4-}$ vacancy. (C) EDTA-NaMnHCF with Mn vacancy; variations of Mn–N bond length (B) in CIT-NaMnHCF. (D) in EDTA-NaMnHCF during the second Na^+ extraction. Reproduced with permission from Ref. [94]. Copyright © 2020, Elsevier.

lifespan is remarkably longer than the sample obtained via the non-citrate-assisted coprecipitation method. This material also exhibited outstanding electrochemical performance in aqueous electrolytes (1 M Na_2SO_4 in H_2O), which delivered a capacity of 130 mAh/g, high rate performance up to 20 C, and excellent cycling stability with 90% capacity retention after 800 cycles. Wang et al. [35] synthesized high initial sodium content CoHCF ($Na_{1.87}Co[Fe(CN)_6]_{0.98}\cdot 2.2H_2O$ by a modified coprecipitation method, which is favorable for full cell fabrication.

Even though CuHCF was the first known PBA used for SIBs, there are fewer studies on the application of CuHCF in aprotic electrolytes. CuHCF is single-electron redox PBAs, exhibiting the high working potential of Fe^{LS} redox but inferior stability relative to NiHCF in organic electrolytes. As reported by Jiao et al. [97], CuHCF only exhibits a capacity of 44 mAh/g and short cycle life (57.1% retention over 50 cycles). The poor electrochemical performance in the organic system may be due to the poor crystallinity in the defect-rich framework. Instead, CuHCF was more intensively investigated in aqueous SIBs, exhibiting the discharge voltage platform close to the decomposition voltage of H_2O, which can make full use of the potential window of the aqueous electrolyte. Several full aqueous batteries with CuHCF cathodes were reported with a high rate, long life, and high power density [98–100]. Our previous work [101] demonstrated a unique intercalation pseudocapacitance in low-vacancy CuHCF, achieving outstanding rate capability and low-temperature electrochemical performance.

The general chemical formula of ZnHCF is different from most PBAs, represented as $A_xZn_3[Fe(CN)_6]_2\cdot zH_2O$, because Zn tends to form a tetrahedron (ZnN_4) with N. Moreover, rhombohedral ZnHCFs can be easy to synthesize after being hydrated. ZnHCF is also reported to be a promising candidate for aqueous batteries, especially A–Zn hybrid ion batteries [102, 103].

$Na_2Mn[Mn(CN)_6]$ (MnHCM) was presented first by Cui et al. [104] as a cathode electrode for SIBs. The MnHCM can deliver 209 mAh/g at 0.2 C, which is close to its theoretical capacity of 257.7 mAh/g. The MnHCM shows three voltage plateaus because of the redox between Mn^{I} and Mn^{III} during Na extraction/insertion. MnHCM was found to host an unprecedented 50% more Na cations than normal HCFs during the cycling. As shown in Fig. 4.19, the sodium storage mechanism of MnHCM is explained. Four of the eight interstitial sites in each unit cell are occupied by two Na^+ ions on average. However, the potential plateau at 1.8 V corresponding to one-third of the whole capacity is not high enough for cathode application. $Mn_3[Co(CN)_6]$ (MnHCC) was synthesized by Zhao et al. [105], which is used for supercapacitor. The as-prepared MnHCC delivers a power density of 177.1 mW/cm^3 at 20 mA/cm^2 and a volumetric energy density of 4.69 mWh/cm^3, but the electrochemical performances in SIBs have not been investigated. TiHCF [106] and VHCF [107] are other novel HCF materials. The theoretical capacity of TiHCF is calculated to be about 170 mAh/g, while the actual specific capacity of 92 mAh/g and two discharge voltage platforms at 2.6 V and 3.2 V. VHCF is reported as a cathode for aqueous SIBs, which achieves a capacity of 91 mAh/g at 110 mA/g. These works broadened the categories of PBAs and gave new ways to improve electrochemical performance.

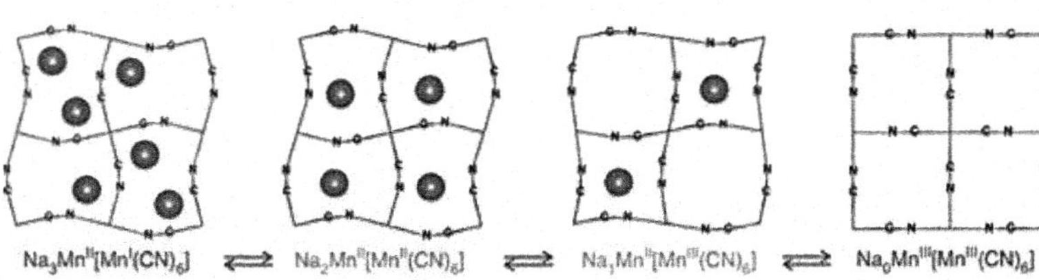

Figure 4.19 Sodium storage mechanism of MnHCM. Reproduced with permission from [104]. Copyright © 2014, Springer Nature.

4.4.5 Multi-Metal Hexacyanoferrate

Heteroatom doping or partial substitution of high-spin transition metal ions is an efficient way to provide an extra degree of freedom for the adjustment of the electrochemical properties of PBAs, such as specific capacity, output voltage, lifetime, rate performance. Heteroatom doping is especially suited for two-electron redox PBAs such as FeHCF and MnHCF, which have high specific capacities but poor cyclability and insufficient rate performance. Many thought-provoking works have been reported in recent years, as shown in Table 4.8.

Table 4.8 Representative performance of reported multi-metal PBAs in SIBs

Active Material	Water [wt.%]	Structure	Morphology and Size	V_{disch}	Rate Capability [mAh/g @ mA/g]	Voltage Window	Cycle Life [Retention, Cycles, Rate]	Electrolyte	Ref.
$Na_{1.76}Ni_{0.12}Mn_{0.88}[Fe(CN)_6]_{0.98}$	12.4	C	Agglomerates	3.0–3.3	123/50@10/600	2–4.0	83.8 %, 800 @100	1 M NC in EC:DMC = 1:1	[112]
$Na_{1.87}Ni_{0.05}Mn_{0.95}[Fe(CN)_6]_{0.98}$	11.3	C	500 nm cube	3.07/ 3.49	120/78@50/1600	2–4.0	96.7 %, 100 @50	1 M NP in EC:DEC = 1:1	[116]
$Na_{1.38}Ni_{0.07}Mn_{0.93}[Fe(CN)_6]_{0.82}$	9	C	150–300 nm cube	3.16/ 3.51	123/52@50/3200	2–4.2	82.9 %, 600 @50	1 M NP in EC:DEC = 1:1	[121]
$Na_{1.59}Mn_{0.17}Co_{0.18}Ni_{0.04}Fe_{0.61}[Fe(CN)_6]_{0.92}$	16.9	C	0.5–1 μm cube	2.9	111/75@170/1530	2–4.0	78.7 %, 1500 @170	1 M NC in EC:DEC = 1:1	[118]
$HQNi_{0.3}Co_{0.7}[Fe(CN)_6]$	10.95	M	50–200 nm	3.86/ 3.73 V	145/125.6@5C/ 0.1C	2.0–4.2	90 %, 600 @5C	1 M NP in DMC:EC = 1:1	[122]
$Na_{1.44}Mn_{0.72}Fe_{0.28}[Fe(CN)_6]_{0.864}$	N/A	C	21–92 nm	3.5–3.6	119/29@0.1C/50C	2–4.2	96 %, 50 @0.3C	1 M NC in PC	[116]

(*Continued*)

Table 4.8 (*Continued*)

$Na_{1.88}Mn_{0.75}Co_{0.25}$ $[Fe(CN)_6]_{0.973}$	N/A	C	11–84 nm	3.5–3.6	134@0.1C	2–4.2	92 %, 50 @0.3C	1 M NC in PC	[116]
$Na_{1.64}Mn_{0.74}Ni_{0.26}$ $[Fe(CN)_6]_{0.913}$	N/A	C	13–86 nm	3.5–3.6	139@0.1C	2–4.2	90 %, 50 @0.3C	1 M NC in PC	[116]
HNGPB	N/A	C	200–300 nm	2.77/ 3.3	114/@100/1000	2–4.0	80 %, 1000 @1000	1 M NC in EC:DEC:PC = 2:2:1	[119]
$Na_{1.37}(Ni_{0.28}Mn_{0.72})$ $[Fe(CN)_6]_{0.84}$	11	C	1–1.5 μm cube	3.2/3.5	110/72@50/1000	2–4.1	93 %, 1000 @500	1 M NP in EC:DMC = 1:1 with 5% FEC	[120]

Note: M = monoclinic, C = cubic, R = rhombohedral, T = trigonal; NP = $NaPF_6$, NC = $NaClO_4$

Nickel is intensively used to promote FeHCFs [108–110]/ MnHCFs [111–113] cycling and rate performance due to the "zero-stain" character and high rate capability of NiHCF. The Ni-substituted FeHCF shows lower capacity than that of the bare one, but higher than the physical mixture of FeHCF and NiHCF. Our group [108] reported that FeHCF with 20 at% nickel substitution exhibited excellent capacity retention of 96% over 100 cycles and superior rate capability to that of bare FeHCF. Similarly, Yang et al. reported that the cycling stability of $NaNi_{0.12}Mn_{0.88}HCF$ was significantly improved because of Ni substitution [112]. Ni substitution has been demonstrated to activate Fe^{LS} redox sites. Cao et al. [110] studied that 3% of nickel substitution can activate Fe^{LS} sites and contribute more capacity by Fe^{LS} redox (28% higher than pure FeHCF). The substitution of Ni ions lowered the bond force constant of Fe–C≡N. Besides, the substitution of Fe^{HS} can weaken the crystal field to promote high-spin states and increase the probability of high-spin–low-spin transitions, causing electrons to be transferred to and from the e_g orbitals instead of the t_{2g} orbitals of Fe^{LS}, which results in the activation of Fe^{LS} redox sites [114]. Ni substitution can also suppress the Jahn–Teller distortion in PBAs, especially in MnHCF. EXAFS results indicate that the distance of the Mn–N bond changed by 10% during cycles [115]. The substitution of Ni for Mn remarkably affects the lattice response to the Na^+ insertion/extraction. The difference (Δa) of the lattice constant was reduced from –0.06 Å to +0.01 Å [116]. Ma et al. [112] reported that the Ni sites were electrochemically inert and could balance the small structural distortion caused by the redox reactions of Mn sites.

In addition, iron substitution and cobalt substitution have a similar effect without lowering capacity [113, 114, 116]. Jiang et al. [117] reported that the structural stability and electronic conductivity of MnHCF were improved after Co doping. Moritomo et al. [116] investigated the performances of $NaM_{0.25}Mn_{0.75}HCF$ (M = Fe, Co, and Ni) in SIBs. The partial substitution of Co^{2+}, Fe^{2+}, and Ni^{2+} activates the Mn reduction reaction and suppresses the cooperative JTE distortion. The results suggested that Fe substitution is best for improving capacity, while Ni substitution is best for promoting rate capability and stability. Similar to the mechanism of Ni substitution, Fe substitution and Co substitution decrease the axial distance in MnN_6 octahedron, which means the covalent character of Mn≡N bonds is improved. The effect of the electrostatic interaction on the d_z2 orbital is weakened because of the covalent character. The modification of electron delocalization in the Mn–NC–Fe–CN–M (M = Co, Fe, and Ni) chain decreases the charge density around the Mn–N bond for the polarizing power of the external metal (M). Based on the preceding analysis, electrons are only removed from or to the e_g orbitals, and the kinetics of Mn^{HS} redox reaction can be modulated [113]. Besides single-element substitution, multi-element substitution was reported by Xie et al. [118]. The as-prepared sample, $Na_{1.59}Mn_{0.17}Co_{0.18}Ni_{0.04}Fe_{0.61}[Fe(CN)_6]_{0.92}$, delivers 117 mAh/g at 0.1 C, excellent cyclability of 78% retention after 1500 cycles at 1 C.

Gradient substitution was first applied to improve the performance of PBAs by our group [119], which is an efficient strategy in terms of coating and substitution. Through the unique gradient nickel substitution, the Ni-rich outer layer protects the Fe^{LS} from effectively reacting with the electrolyte, consequently resulting in superior cycling stability. In addition, the Ni ions across the entire grain of FeHCF play a key role in activating the Fe^{LS} redox couple and improving the electrochemical kinetics of the electrode. The strategy was also reported to enhance the performance of MnHCF [120].

It is to note that heteroatom substitution will affect the morphology and reduce the particle size of PBAs on account of the variation of nuclei composition, nucleation, and growth rate. Multi-metal PBAs with irregular morphology and small size can not only deliver excellent performance but also using chelating agents to slow the rate of precipitate and reducing interstitial water can improve the performance remarkably [109, 111]. In addition, substitution cannot suppress the dissolution of Mn completely, especially in an aqueous SIB system; more methods need to be considered to solve the problem.

4.5 Others

Introducing an additive into the electrolyte can improve the performance of PBAs, especially lifetime, and suppress the dissolution of transition metal. Xie et al. [118] added 1 wt% $AlCl_3$ into the electrolyte (1 M $NaClO_4$ in EC/DEC with 8 vol% FEC) to improve the cyclability of HQ-MnCoNiFeHCF. It is a novel way to *in situ* capture residual coordinated water during the Na^+ intercalation/deintercalation process, due to the Lewis-acid property of $AlCl_3$. On the one hand, the water in $Na(OH_2)^+$ was removed to further relieve the lattice deformation caused by the intercalation/deintercalation of large-size hydrated Na^+ ions. The $Al_2O_3{\cdot}3H_2O$ layer formed on the cathode electrode can sufficiently protect the active material from possible HF attack, thus improving the cycling stability. The reaction mechanism of the $AlCl_3$ additive can be expressed as the following equation:

$$6Na(OH_2)^+ + 2Al^{3+} = Al_2O_3{\cdot}3H_2O + 6Na^+ + 6H^+$$

A new electrolyte additive that can expand to other PBA materials is needed to be explored to improve the electrochemical performance.

4.6 Summary and Outlook

In summary, a comprehensive review of PBAs as cathode materials for SIBs is given. The structures and electric energy storage mechanisms of PBAs are systematically discussed, including detailed Na^+ diffusion paths and phase transition. The open framework structure and flexible stoichiometry give PBAs adjustable performance, such as electrochemical activity, kinetics, and stability. In terms of facile synthesis, low cost, and attractive electrochemical performance, PBAs, especially MnHCFs and FeHCFs, are emerging as promising candidates for the practical applications of SIBs. However, there are still several challenges to overcome, such as poor electrochemical stability, limited rate performance, and low initial Coulombic efficiency. Several strategies have been proposed to overcome the aforementioned issues and mainly focus on the improvement in the crystallinity, morphology, and conductivity of PBAs. By employing a reducing agent and an inert atmosphere, as well as using chelating agents and optimizing the synthesis parameters, PBAs with high crystallinity are obtained, which contain low defects and have ideal stoichiometry. Through morphology control, PBAs with unique morphology can be fabricated, which have excellent rate capability in general. Through *in situ* or *ex situ* carbon materials compositing, the conductivity of PBAs improves remarkably. In addition, dehydration, coating, and doping can further enhance electrochemical performance. However, there are still some issues that need to be paid more attention to promote the practical applications of PBAs toward SIBs.

1. **Water removal:** It does not matter whether interstitial water or coordinated water will strongly affect the performance of cells with organic electrolytes. However, it is hard to dehydrate PBAs completely since the synthesis process is usually carried out in an aqueous solution. Besides, water can be introduced even using the ball-milling method on account of the existence of interstitial water in precursors. Therefore, developing novel synthesis methods that can avoid the introduction of water and exploring efficient dehydration methods are necessary. In the theoretical aspect, the investigation of how water molecules affect the electrochemical behavior and performances of PBAs is limited. Specifically, the role of water in conduction is poorly understood.
2. **Electrolyte modification:** The electrolyte currently used is based on the ones in LIBs. The sodium salt, especially $NaPF_6$, can be decomposed at high voltage, and the decomposition products may attack the PBA cubes. In addition, the dissolution of TM ions, Mn^{2+} in particular, is severe. Finding appropriate electrolytes is critically important, which can significantly improve cyclability and rate capability.
3. **Choice of chelating agents:** Coprecipitation is the most promising synthesis method for large scale. However, the PBAs obtained by coprecipitation without adding chelating agents show severe defects. The current chelating agent, such as Na_3CA, EDTA, etc., can produce PBAs with ideal stoichiometry and more new chelating agents should be explored in coprecipitation.

Acknowledgment

This work was supported by the National Key Research and Development Program (2019YFE0111200).

References

1. Qian J, et al. Prussian blue cathode materials for sodium-ion batteries and other ion batteries. *Advanced Energy Materials*, 2018, **8**(17): 1702619.
2. Yabuuchi N, et al. Research development on sodium-ion batteries. *Chemical Reviews*, 2014, **114**(23): 11636–11682.
3. Xu Y, et al. Prussian blue and its derivatives as electrode materials for electrochemical energy storage. *Energy Storage Materials*, 2017, **9**: 11–30.
4. Karyakin AA, et al. Prussian blue-based first-generation biosensor. A sensitive amperometric electrode for glucose. *Analytical Chemistry*, 1995, **67**(14): 2419–2423.
5. Jia X, et al. Perfluoropentane-encapsulated hollow mesoporous prussian blue nanocubes for activated ultrasound imaging and photothermal therapy of cancer. *ACS Applied Materials & Interfaces*, 2015, **7**(8): 4579–4588.
6. Zong Y, et al. Preparation of a novel microsphere adsorbent of prussian blue capsulated in carboxymethyl cellulose sodium for Cs(I) removal from contaminated water. *Journal of Radioanalytical and Nuclear Chemistry*, 2016, **311**(3): 1577–1591.
7. Gütlich P, et al. Photoswitchable coordination compounds. *Coordination Chemistry Reviews*, 2001, **219–221**: 839–879.
8. Wessells CD, et al. Nickel hexacyanoferrate nanoparticle electrodes for aqueous sodium and potassium ion batteries. *Nano Letters*, 2011, **11**(12): 5421–5425.
9. Wessells CD, et al. Copper hexacyanoferrate battery electrodes with long cycle life and high power. *Nature Communications*, 2011, **2**: 550.
10. Wang B, et al. Prussian blue analogs for rechargeable batteries. *iScience*, 2018, **3**: 110–133.
11. Fernández-Ropero A, et al. Electrochemical characterization of $NaFe_2(CN)_6$ Prussian Blue as positive electrode for aqueous sodium-ion batteries. *Electrochimica Acta*, 2016, **210**: 352–357.
12. Kumar A, et al. Structural and magnetic properties of $Fe[Fe(CN)_6]\cdot 4H_2O$. *Physical Review B*, 2005, **71**(5): 054414.
13. Lu Y, et al. Prussian blue: a new framework of electrode materials for sodium batteries. *Chemical Communications*, 2012, **48**(52): 6544–6546.
14. Widmann A, et al. Structure, insertion electrochemistry, and magnetic properties of a new type of substitutional solid solutions of copper,

nickel, and iron hexacyanoferrates/hexacyanocobaltates. *Inorganic Chemistry*, 2002, **41**(22): 5706–5715.

15. Sun H, et al. $Fe_4[Fe(CN)_6]_3$: a cathode material for sodium-ion batteries. *RSC Advances*, 2014, **4**(81): 42991–42995.
16. Wu X, et al. Single-crystal $FeFe(CN)_6$ nanoparticles: A high capacity and high rate cathode for Na-ion batteries. *Journal of Materials Chemistry A*, 2013, **1**(35): 10130–10134.
17. Pramudita JC, et al. Sodium uptake in cell construction and subsequent in operando electrode behaviour of Prussian blue analogues, $Fe[Fe(CN)_6]_{1-x}\cdot yH_2O$ and $FeCo(CN)_6$. *Physical Chemistry Chemical Physics*, 2014, **16**(44): 24178–24187.
18. Yang Q, et al. Investigation of iron hexacyanoferrate as a high rate cathode for aqueous batteries: Sodium-ion batteries and lithium-ion batteries. *Electrochimica Acta*, 2018, **270**: 96–103.
19. You Y, et al. High-quality Prussian blue crystals as superior cathode materials for room-temperature sodium-ion batteries. *Energy & Environmental Science*, 2014, **7**(5): 1643–1647.
20. Kawamoto T, et al. Ab initio calculations on the mechanism of charge transfer in Co-Fe Prussian-blue compounds. *Physical Review B*, 1999, **60**(18): 12990.
21. Sottmann J, et al. In operando synchrotron XRD/XAS investigation of sodium insertion into the Prussian blue analogue cathode material $Na_{1.32}Mn[Fe(CN)_6]_{0.83}\cdot zH_2O$. *Electrochimica Acta*, 2016, **200**: 305–313.
22. Song J, et al. Removal of interstitial H_2O in hexacyanometallates for a superior cathode of a sodium-ion battery. *Journal of the American Chemical Society*, 2015, **137**(7): 2658–2664.
23. Brant WR, et al. Selective control of composition in Prussian white for enhanced material properties. *Chemistry of Materials*, 2019, **31**(18): 7203–7211.
24. Zhou A, et al. Hexacyanoferrate-type prussian blue analogs: principles and advances toward high-performance sodium and potassium ion batteries. *Advanced Energy Materials*, 2020, **11**(2): 2000943.
25. Huang Y, et al. A novel border-rich Prussian blue synthetized by inhibitor control as cathode for sodium ion batteries. *Nano Energy*, 2017, **39**: 273–283.
26. Liu Y, et al. Sodium storage in Na-rich $Na_xFeFe(CN)_6$ nanocubes. *Nano Energy*, 2015, **12**: 386–393.
27. You Y, et al. Sodium iron hexacyanoferrate with high Na content as a Na-rich cathode material for Na-ion batteries. *Nano Research*, 2015, **8**(1): 117–128.

28. Jo I-H, et al. Electrochemical properties of $NaMnFe(CN)_6 \cdot zH_2O$ synthesized in a Taylor-Couette reactor as a Na-ion battery cathode material. *Journal of Alloys and Compounds*, 2017, **729**: 590–596.

29. Yuan Y, et al. $Na_2Co_3[Fe(CN)_6]_2$: a promising cathode material for lithium-ion and sodium-ion batteries. *Journal of Alloys and Compounds*, 2016, **685**: 344–349.

30. Bie X, et al. Synthesis and electrochemical properties of Na-rich Prussian blue analogues containing Mn, Fe, Co, and Fe for Na-ion batteries. *Journal of Power Sources*, 2018, **378**: 322–330.

31. Xu Y, et al. Structure distortion induced monoclinic nickel hexacyanoferrate as high-performance cathode for Na-ion batteries. *Advanced Energy Materials*, 2019, **9**(4): 1803158.

32. Xiao P, et al. Theoretical study of the structural evolution of a $Na_2FeMn(CN)_6$ cathode upon Na intercalation. *Chemistry of Materials*, 2015, **27**(10): 3763–3768.

33. Ji Z, et al. On the mechanism of the improved operation voltage of rhombohedral nickel hexacyanoferrate as cathodes for sodium-ion batteries. *ACS Applied Materials & Interfaces*, 2016, **8**(49): 33619–33625.

34. Xie B, et al. Understanding the structural evolution and lattice water movement for rhombohedral nickel hexacyanoferrate upon sodium migration. *ACS Applied Materials & Interfaces*, 2019, **11**(50): 46705–46713.

35. Li C, et al. High crystalline prussian white nanocubes as a promising cathode for sodium-ion batteries. *Chemistry-An Asian Journal*, 2018, **13**(3): 342–349.

36. Wu X, et al. Highly crystallized $Na_2CoFe(CN)_6$ with suppressed lattice defects as superior cathode material for sodium-ion batteries. *ACS Applied Materials and Interfaces*, 2016, **8**(8): 5393–5399.

37. Xu Y, et al. Dual redox-active copper hexacyanoferrate nanosheets as cathode materials for advanced sodium-ion batteries. *Energy Storage Materials*, 2020, **33**: 432–441.

38. Muñoz P, et al. Prussian blue based batteries. *SpringerBriefs in Applied Sciences and Technology*, 2018.

39. Ling C, et al. First-principles study of alkali and alkaline earth ion intercalation in iron hexacyanoferrate: the important role of ionic radius. *Journal of Physical Chemistry C*, 2013, **117**(41): 21158–21165.

40. Wang L, et al. Rhombohedral Prussian white as cathode for rechargeable sodium-ion batteries. *Journal of the American Chemical Society*, 2015, **137**(7): 2548–2554.

41. You Y, et al. Subzero-temperature cathode for a sodium-ion battery. *Advanced Materials*, 2016, **28**(33): 7243–7248.

42. Wang W, et al. Understanding rhombohedral iron hexacyanoferrate with three different sodium positions for high power and long stability sodium-ion battery. *Energy Storage Materials*, 2020, **30**: 42–51.

43. Tang Y, et al. High-performance manganese hexacyanoferrate with cubic structure as superior cathode material for sodium-ion batteries. *Advanced Functional Materials*, 2020, **30**(10): 1908754.

44. Yin J, et al. In situ self-assembly of core-shell multimetal Prussian blue analogues for high-performance sodium-ion batteries. *ChemSusChem*, 2019, **12**(21): 4786–4790.

45. Gao H, et al. Stabilizing a high-energy-density rechargeable sodium battery with a solid electrolyte. *Chem*, 2018, **4**(4): 833–844.

46. Zhu Y-H, et al. High-energy-density flexible potassium-ion battery based on patterned electrodes. *Joule*, 2018, **2**(4): 736–746.

47. You Y, et al. A zero-strain insertion cathode material of nickel ferricyanide for sodium-ion batteries. *Journal of Materials Chemistry A*, 2013, **1**(45): 14061–14065.

48. Yue Y, et al. Mesoporous Prussian blue analogues: template-free synthesis and sodium-ion battery applications. *Angewandte Chemie International edition in English*, 2014, **53**(12): 3134–3137.

49. Rehman R, et al. Highly crystalline nickel hexacyanoferrate as a long-life cathode material for sodium-ion batteries. *RSC Advances*, 2020, **10**(45): 27033–27041.

50. Sun H, et al. Prussian blue analogue mesoframes for enhanced aqueous sodium-ion storage. *Crystals*, 2018, **8**(1): 23.

51. Ren W, et al. Activation of sodium storage sites in prussian blue analogues via surface etching. *Nano Letters*, 2017, **17**(8): 4713–4718.

52. Wang Z, et al. A string of nickel hexacyanoferrate nanocubes coaxially grown on a CNT@bipolar conducting polymer as a high-performance cathode material for sodium-ion batteries. *Nanoscale*, 2017, **9**(2): 823–831.

53. Xu Y, et al. In situ FTIR-Assisted synthesis of nickel hexacyanoferrate cathodes for long-Life sodium-ion batteries. *ACS Applied Materials & Interfaces*, 2019, **11**(33): 29985–29992.

54. Rudola A, et al. Monoclinic sodium iron hexacyanoferrate cathode and non-flammable glyme-based electrolyte for inexpensive sodium-ion batteries. *Journal of the Electrochemical Society*, 2017, **164**(6): A1098.

55. Yang Y, et al. Influence of structural imperfection on electrochemical behavior of prussian blue cathode materials for sodium ion batteries. *Journal of the Electrochemical Society*, 2016, **163**(9): A2117-A2123.

56. Wang L, et al. A superior low-cost cathode for a Na-ion battery. *Angewandte Chemie*, 2013, **125**(7): 2018–2021.

57. Wu X, et al. Low-defect Prussian blue nanocubes as high capacity and long life cathodes for aqueous Na-ion batteries. *Nano Energy*, 2015, **13**: 117–123.

58. Yan C, et al. A low-defect and Na-enriched Prussian blue lattice with ultralong cycle life for sodium-ion battery cathode. *Electrochimica Acta*, 2020, **332**: 135533.

59. Wang W, et al. Reversible structural evolution of sodium-rich rhombohedral Prussian blue for sodium-ion batteries. *Nature Communications*, 2020, **11**(1): 980.

60. Neale ZG, et al. Effect of synthesis pH and EDTA on iron hexacyanoferrate for sodium-ion batteries. *Sustainable Energy & Fuels*, 2020, **4**(6): 2884–2891.

61. Peng F, et al. Highly crystalline sodium manganese ferrocyanide microcubes for advanced sodium ion battery cathodes. *Journal of Materials Chemistry A*, 2019, **7**(39): 22248–22256.

62. Liu Y, et al. Role of acid in tailoring Prussian blue as cathode for high-performance sodium-ion battery. *Chemistry-A European Journal*, 2017, **23**(63): 15991–15996.

63. Chen R, et al. Preparation of Prussian blue submicron particles with a pore structure by two-step optimization for Na-ion battery cathodes. *ACS Applied Materials & Interfaces*, 2016, **8**(25): 16078–16086.

64. Ren W, et al. Prussian white hierarchical nanotubes with surface-controlled charge storage for sodium-ion batteries. *Advanced Functional Materials*, 2019, **29**(15): 1806405.

65. Meng Q, et al. Mesocrystalline coordination polymer as a promising cathode for sodium-ion batteries. *Chemical Communications*, 2016, **52**(9): 1957–1960.

66. Wang H, et al. Crystallographic-plane tuned Prussian-blue wrapped with RGO: a high-capacity, long-life cathode for sodium-ion batteries. *Journal of Materials Chemistry A*, 2017, **5**(7): 3569–3577.

67. Li L, et al. Novel acetic acid induced Na-rich Prussian blue nanocubes with iron defects as cathodes for sodium ion batteries. *Journal of Materials Chemistry A*, 2019, **7**(19): 12134–12144.

68. Jiang Y, et al. Prussian blue@C composite as an ultrahigh-rate and long-life sodium-ion battery cathode. *Advanced Functional Materials*, 2016, **26**(29): 5315–5321.

69. Wang H, et al. Reduced graphene oxide-anchored manganese hexacyanoferrate with low interstitial H_2O for superior sodium-ion batteries. *ACS Applied Materials & Interfaces*, 2018, **10**(40): 34222–34229.

70. Bu F, et al. One versatile route to three-dimensional graphene wrapped metal cyanide aerogels for enhanced sodium ion storage. *Chemistry-A European Journal*, 2017, **23**(35): 8358–8363.

71. Luo J, et al. Graphene-roll-wrapped Prussian blue nanospheres as a high-performance binder-free cathode for sodium-ion batteries. *ACS Applied Materials & Interfaces*, 2017, **9**(30): 25317–25322.

72. Prabakar SJR, et al. Highly crystalline prussian blue/graphene composites for high-rate performance cathodes in Na-ion batteries. *RSC Advances*, 2015, **5**(47): 37545–37552.

73. Chu H, et al. A 3D graphene current collector boosts ultrahigh specific capacity in a highly uniform Prussian blue@ graphene composite as a freestanding cathode for sodium ion batteries. *Nanoscale*, 2018, **10**(30): 14697–14704.

74. Qiu X, et al. Reverse microemulsion synthesis of nickel-cobalt hexacyanoferrate/reduced graphene oxide nanocomposites for high-performance supercapacitors and sodium ion batteries. *Applied Surface Science*, 2018, **434**: 1285–1292.

75. Yang D, et al. Prussian blue without coordinated water as a superior cathode for sodium-ion batteries. *Chemical Communications*, 2015, **51**(38): 8181–8184.

76. Li WJ, et al. Facile method to synthesize na-enriched $Na_{1+x}FeFe(CN)_6$ frameworks as cathode with superior electrochemical performance for sodium-ion batteries. *Chemistry of Materials*, 2015, **27**(6): 1997–2003.

77. Hu H, et al. Yolk-shell Prussian blue nanoparticles with fast ion diffusion for sodium-ion battery. *Materials Letters*, 2019, **249**: 206–209.

78. Zhang Q, et al. Surface engineering induced core-shell Prussian blue@ polyaniline nanocubes as a high-rate and long-life sodium-ion battery cathode. *Journal of Power Sources*, 2018, **395**: 305–313.

79. Kim DS, et al. Boosting the sodium storage capability of Prussian blue nanocubes by overlaying PEDOT:PSS layer. *Journal of Alloys and Compounds*, 2019, **791**: 385–390.

80. Qiao Y, et al. Prussian blue coupling with zinc oxide as a protective layer: an efficient cathode for high-rate sodium-ion batteries. *Chemical Communications*, 2019, **55**(4): 549–552.

81. Wan M, et al. Core-shell hexacyanoferrate for superior Na-ion batteries. *Journal of Power Sources*, 2016, **329**: 290–296.

82. Tang Y, et al. Polypyrrole-promoted superior cyclability and rate capability of NaxFe[Fe(CN)6] cathodes for sodium-ion batteries. *Journal of Materials Chemistry A*, 2016, **4**(16): 6036–6041.

83. Liu Y, et al. A heterostructure coupling of bioinspired, adhesive polydopamine, and porous Prussian blue nanocubics as cathode for high-performance sodium-ion battery. *Small*, 2020, **16**(11): 1906946.

84. Gong W, et al. Ultrafine Prussian blue as a high-rate and long-life sodium-ion battery cathode. *Energy Technology*, 2019, **7**(7): 1900108.

85. Matsuda T, et al. A sodium manganese ferrocyanide thin film for Na-ion batteries. *Chemical Communications*, 2013, **49**(27): 2750–2752.

86. Wu J, et al. Modification of transition-metal redox by interstitial water in hexacyanometalate electrodes for sodium-ion batteries. *Journal of the American Chemical Society*, 2017, **139**(50): 18358–18364.

87. Moritomo Y, et al. Glucose-treated manganese hexacyanoferrate for sodium-ion secondary battery. *Energies*, 2015, **8**(9): 9486–9494.

88. Hu F, et al. Hierarchical octahedral $Na_2MnFe(CN)_6$ and $Na_2MnFe(CN)_6$@ Ppy as cathode materials for sodium-ion batteries. *Chinese Journal of Chemistry*, 2017, **35**(4): 415–419.

89. Wang X, et al. Manganese hexacyanoferrate reinforced by PEDOT coating towards high-rate and long-life sodium-ion battery cathode. *Journal of Materials Chemistry A*, 2020, **8**(6): 3222–3227.

90. Shen Z, et al Manganese hexacyanoferrate/graphene cathodes for sodium-ion batteries with superior rate capability and ultralong cycle life. *Inorganic Chemistry Frontiers*, 2018, **5**(11): 2914–2920.

91. Mao Y, et al. Capacitance controlled, hierarchical porous 3D ultra-thin carbon networks reinforced prussian blue for high performance Na-ion battery cathode. *Nano Energy*, 2019, **58**: 192–201.

92. Peng F, et al. Enhanced electrochemical performance of sodium manganese ferrocyanide by $Na_3(VOPO_4)_2F$ coating for sodium-ion batteries. *ACS Applied Materials Interfaces*, 2019, **11**(41): 37685–37692.

93. Li W-J, et al. Multifunctional conducing polymer coated $Na_1^+MnFe(CN)_6$ cathode for sodium-ion batteries with superior performance via

a facile and one-step chemistry approach. *Nano Energy*, 2015, **13**: 200–207.

94. Shang Y, et al. Unconventional Mn vacancies in Mn-Fe Prussian blue analogs: suppressing Jahn-Teller distortion for ultrastable sodium storage. *Chem*, 2020, **6**(7): 1804–1818.

95. Shen Z, et al. Na-rich Prussian white cathodes for long-life sodium-ion batteries. *ACS Sustainable Chemistry & Engineering*, 2018, **6**(12): 16121–16129.

96. Zhou A, et al. Size-, water-, and defect-regulated potassium manganese hexacyanoferrate with superior cycling stability and rate capability for low-cost sodium-ion batteries. *Small*, 2019, **15**(42): 1902420.

97. Jiao S, et al. The electrochemical performance of $Cu_3[Fe(CN)_6]_2$ as a cathode material for sodium-ion batteries. *Materials Research Bulletin*, 2017, **86**: 194–200.

98. Jia Z, et al. Electrochemical sodium storage of copper hexacyanoferrate with a well-defined open framework for sodium ion batteries. *RSC Advances*, 2014, **4**(43): 22768–22774.

99. Wessells CD, et al. Tunable reaction potentials in open framework nanoparticle battery electrodes for grid-scale energy storage. *ACS Nano*, 2012, **6**(2): 1688–1694.

100. Jiang P, et al. Ion-selective copper hexacyanoferrate with an open-framework structure enables high-voltage aqueous mixed-ion batteries. *Journal of Materials Chemistry A*, 2017, **5**(32): 16740–16747.

101. Wang B, et al. Intercalation pseudocapacitance boosting ultrafast sodium storage in Prussian blue analogs. *ChemSusChem*, 2019, **12**: 2415 -2420.

102. Oliver-Tolentino MA, et al. Materials for aqueous sodium-ion batteries: cation mobility in a zinc hexacyanoferrate electrode. *RSC Advances*, 2016, **6**(110): 108627–108634.

103. Huang M, et al. Ultrafast cation insertion-selected zinc hexacyanoferrate for 1.9 VK-Zn hybrid aqueous batteries. *Journal of Materials Chemistry A*, 2020, **8**(14): 6631–6637.

104. Lee H-W, et al. Manganese hexacyanomanganate open framework as a high-capacity positive electrode material for sodium-ion batteries. *Nature Communications*, 2014, **5**(1): 5280–5285.

105. Zhao Q, et al. Facile synthesis of $Mn_3[Co(CN)_6]_2{\cdot}nH_2O$ nanocrystals for high-performance electrochemical energy storage devices. *Inorganic Chemistry Frontiers*, 2017, **4**(3): 442–449.

106. Xie M, et al. Sodium titanium hexacyanoferrate as an environmentally friendly and low-cost cathode material for sodium-ion batteries. *Journal of Power Sources*, 2016, **302**: 7–12.

107. Lee JH, et al. Metal-organic framework cathodes based on a vanadium hexacyanoferrate Prussian blue analogue for high-performance aqueous rechargeable batteries. *Advanced Energy Materials*, 2017, **7**(2): 1601491.

108. Yu S, et al. A promising cathode material of sodium iron-nickel hexacyanoferrate for sodium ion batteries. *Journal of Power Sources*, 2015, **275**: 45–49.

109. Wang J, et al. Sodium-rich iron hexacyanoferrate with nickel doping as a high performance cathode for aqueous sodium ion batteries. *Journal of Electroanalytical Chemistry*, 2018, **818**: 10–18.

110. Fu H, et al. Enhanced storage of sodium ions in Prussian blue cathode material through nickel doping. *Journal of Materials Chemistry A*, 2017, **5**(20): 9604–9610.

111. Chen R, et al. Chemical inhibition method to synthesize highly crystalline Prussian blue analogs for sodium-ion battery cathodes. *ACS Applied Materials & Interfaces*, 2016, **8**(46): 31669–31676.

112. Yang D, et al. Structure optimization of Prussian blue analogue cathode materials for advanced sodium ion batteries. *Chemical Communications*, 2014, **50**(87): 13377–13380.

113. Oliver-Tolentino M, et al. Electronic density distribution of Mn-N bonds by a tuning effect through partial replacement of Mn by Co or Ni in a sodium-rich hexacyanoferrate and its influence on the stability as a cathode for Na-ion batteries. *Dalton Transactions*, 2018, **47**(46): 16492–16501.

114. Pasta M, et al. Manganese-cobalt hexacyanoferrate cathodes for sodium-ion batteries. *Journal of Materials Chemistry A*, 2016, **4**(11): 4211–4223.

115. Mullaliu A, et al. Highlighting the reversible manganese electroactivity in na-rich manganese hexacyanoferrate material for Li- and Na- ion storage. *Small Methods*, 2019, **4**(1): 1900529.

116. Moritomo Y, et al. Enhanced battery performance in manganese hexacyanoferrate by partial substitution. *Electrochimica Acta*, 2016, **210**: 963–969.

117. Jiang X, et al. Hierarchical mesoporous octahedral $K_2Mn_{1-x}Co_xFe(CN)_6$ as a superior cathode material for sodium-ion batteries. *Journal of Materials Chemistry A*, 2016, **4**(41): 16205–16212.

118. Xie B, et al. Achieving long-life Prussian blue analogue cathode for Na-ion batteries via triple-cation lattice substitution and coordinated water capture. *Nano Energy*, 2019, **61**: 201–210.

119. Wang B, et al. Gradient substitution: an intrinsic strategy towards high performance sodium storage in Prussian blue-based cathodes. *Journal of Materials Chemistry A*, 2018, **6**(19): 8947–8954.

120. Hu P, et al. Concentration-gradient prussian blue cathodes for Na-ion batteries. *ACS Energy Letters*, 2019, **5**(1): 100–108.

121. Huang Y, et al. A chemical precipitation method preparing hollow-core-shell heterostructures based on the prussian blue analogs as cathode for sodium-ion batteries. *Small*, 2018, **14**(28): 1801246.

122. Peng J, et al. A dual-insertion type sodium-ion full cell based on high-quality ternary-metal prussian blue analogs. *Advanced Energy Materials*, 2018, **8**(11): 1702856.

Chapter 5

Polymer Electrodes for Sodium-Ion Batteries

Qinglan Zhao and Minhua Shao
Department of Chemical and Biological Engineering, The Hong Kong University of Science and Technology, Clear Water Bay, Kowloon, Hong Kong, China
keqzhao@ust.hk, qinglan.zhao@uqconnect.edu.au

5.1 Introduction

Current electrode materials for sodium-ion batteries (NIBs) include inorganic compounds (e.g., transition metal oxides, phosphates, fluorides, sulfides, phosphides, alloys, Prussian blue analogues, and carbonaceous materials) [1–6] and organic materials [7, 8]. Many inorganic compounds contain toxic metals, causing further concerns over resources and environmental contamination. In addition, the inorganic compounds undergo the insertion-type sodium storage mechanism. During the intercalation and deintercalation of sodium ions, large volume changes and irreversible phase transitions may occur, leading to a low reversible capacity and poor cycling performance [9].

Handbook of Sodium-Ion Batteries: Materials and Characterization
Edited by Rohit R. Gaddam and X. S. (George) Zhao

ISBN 978-981-4968-15-7 (Hardcover), 978-1-003-30874-4 (eBook)
www.jennystanford.com

Redox-active organic materials are gaining increasing interests due to a number of intrinsic advantages [10]. First, organic materials contain no heavy metals, an increasingly important consideration for development of green energy storage devices [11]. Second, the cost can be reduced because they can be produced from renewable natural resources [12]. Third, a range of polymer structures with different surface functionalities can be designed at the molecular level [13]. Fourth, organic materials exploit an alkali ion storage involving reactions of chemical bonds, making it less sensitive to the large sodium ion [14]. Finally, organic materials are inherently flexible, making it possible to realize commercial flexible energy storage devices [15].

Organic electrode materials include small organic molecules and polymers. Most small organic compounds suffer from rapid dissolution into organic electrolytes, leading to a short cycle life. Polymerization is considered a simple and efficient way to mitigate this problem [16]. Compared with small organic molecules, polymers with large extended backbones of repeating units have lower if not negligible solubility in electrolytes and thus better cycling performance. The advantages and disadvantages of electrode materials for NIB are listed in Table 5.1.

Table 5.1 Comparison of electrode materials for sodium-ion batteries

Electrode Materials		Advantages	Disadvantages
Inorganic materials		Good conductivity	Potential environmental contamination; Large volume changes; Irreversible phase transitions.
Organic materials	Small organic molecules	Environmental friendliness; low cost; structural diversity; tenability; flexibility	High solubility in electrolyte; poor conductivity
	Polymers	The same advantages of small organic molecules with lower solubility	Potential solubility in electrolyte; poor conductivity except conducting polymers

The reported polymer electrode materials for NIBs can be mainly divided into three categories according to the redox reactions they undergo. The first involves reactions at the C=O groups and includes polyimides and polyquinones. The second is based on reactions of C=N groups, and the typical representatives are the Schiff base polymers. The third type encompasses conducting polymers exploiting doping reactions, such as conjugated conductive polymers and non-conjugated conductive radical-containing polymers. However, apart from conducting polymers, most of the polymers have poor conductivity, and this remains one of the main challenges to wider exploitation of polymeric electrode materials.

This chapter provides a summary of recent research and development of polymer electrode materials for NIBs. The evolution of polymeric electrode materials from the initial work on polyimides, their electrochemical performance, sodium storage mechanisms, and the related characterization techniques will be discussed. The current grand challenges and strategies for improving polymer electrode materials for NIBs are identified and analyzed. Future research perspectives in this important field are projected.

5.2 Polymer Electrode Materials for NIBs

5.2.1 Polymers with Carbonyl Functional Groups

5.2.1.1 Polyimides

In the classical approach, aromatic polyimides are synthesized in two steps [17]. Initially, a dianhydride and a diamine are reacted in a polar aprotic solvent to generate a polyamic acid. In the second step, heating or treatment with chemical dehydrating agents is required to cyclodehydrate the polyamic acid to form the polyimide. The polyimide can be also obtained by a one-step solvothermal synthesis, in which the polycondensation and imidization occur spontaneously [17, 18].

The structure of the dianhydride and diamine precursors determines the nature of the aromatic backbone of the polyimides. Therefore, various molecular structures of polyimides (Fig. 5.1) can be realized through judicious choice of the dianhydrides and diamines used in the synthesis.

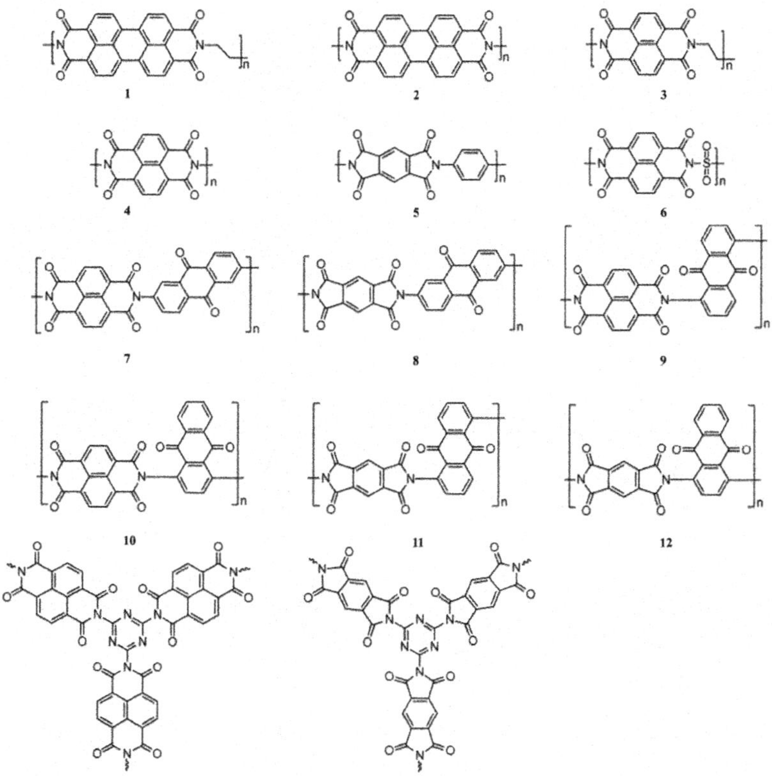

Figure 5.1 The structure of typical polyimides suitable for use in NIBs.

Aromatic polyimides are redox-active polymers (Fig. 5.1) and hence are very promising energy storage electrode materials [19]. The carbonyls of the polyimide provide the active sites for the redox reaction. The carbonyls can interact with sodium ions via enolization, involving two one-electron reduction steps to yield sequentially the anion radical and the dianion (Scheme 5.1).

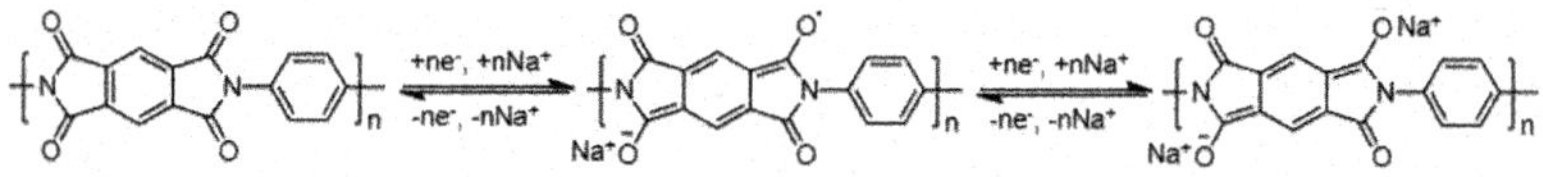

Scheme 5.1 Schematic representation of the redox reaction of aromatic polyimides.

The polyimides currently used in NIBs are usually derived from 3,4,9,10-perylenetetracarboxylic dianhydride (PTCDA) (**1, 2**) [20, 21], 1,4,5,8-naphthalenetetracarboxylic dianhydride (NTCDA) (**3, 4**) [22, 23], and pyromellitic dianhydride (PMDA) (**5**) [24]. The average discharge voltage of polyimides becomes progressively lower when the aromatic core changes from PTCDA (1.94 V) to NTCDA (1.89 V) and to PMDA (1.73 V) [20]. This is due to the increased energy of the lowest unoccupied molecular orbital (LUMO) and thus the decreased average discharge voltage. Therefore, PTCDA-derived polyimides are most commonly investigated as cathode materials for NIBs [20, 21], while the NTCDA- and PMDA-derived polyimides are studied as anode materials [22–24]. Improved cycling stability and increased gravimetric capacity can be achieved by employing larger conjugated aromatic cores and reducing the molecular weight of inactive chains, respectively.

The working voltage of the polyimides can be manipulated by introducing an electron-withdrawing group. For example, compared with the polyimide containing an ethylene connecting unit, polyimide (**6**) with a connecting sulfonyl group has a 0.14 V higher charge–discharge voltage, resulting from the decreased electron density of the redox-active carbonyl by inductive effects [25].

A very attractive advantage of polyimides as electrode materials for NIBs is their insolubility in electrolytes, which results in good cycling stability. However, only half of the carbonyls can be involved in the redox reactions, and as a result the capacity of polyimides is usually less than 140 mAh/g (Table 5.2). Further development of polyimide electrode materials requires this bottleneck to be overcome. Copolymerization of the imide with a high-capacity quinone is a promising strategy to obtain polyimides with enhanced capacity. Xu et al. prepared two imide-quinone copolymers with high reversible capacities of 192 mAh/g (**7**) and 165 (**8**) mAh/g [26]. A number of other imide-quinone copolymers (**9, 10, 11, 12**) have also displayed improved capacities [27]. In addition, introduction of electron-withdrawing groups into the polyimide chain can increase the capacity of the polyimides. Li et al. reported that the introduction of the triazine ring can increase the sodium-ion storage capability (**13, 14**), through the synergistic effects of electron-withdrawing amide groups and triazine rings [28].

Table 5.2 Summary of electrochemical properties of some polymer electrode materials

Number	Cathodic/ Anodic Peak Potential (V vs. Na/Na$^+$)	The 1st and 2nd Discharge Capacity (mAh/g)	Cycling Performance: Discharge Capacity Retention/Cycle/ Current Density (mA/g)	Rate Capability: Discharge Capacity (mAh/g)/Current Density (mA/g)	Conductive Additive (wt. %)	Electrolyte*
1 [20]	1.90/2.20	112, NA	89%/600/25	137.1/50, 38.9/20000	30	1M $NaPF_6$ in EC/DMC (1/1)
2 [21]	1.86/1.97, 2.45/2.75	126, 121	90%/50/100	120/25, 90/800	30	1 M $NaPF_6$ in PC
3 [22]	1.80/2.10, 2.25/2.4	140, NA	90%/500/140	140/140, 84/2520	50	1 M $NaClO_4$ in EC/DEC (1/1)
4 [23]	2.07/2.22, 2.27/2.47	184, 170	74%/20/50	NA	30	5 M $NaNO_3$ in H_2O
5 [24]	0.75/1~2, 1.25/1~2	NA	NA	125/25, 43/2000	30	1 M $NaClO_4$ in EC/PC (1/1) plus 0.3 wt% FEC
6 [25]	1.90/2.30, 2.10/2.80	122, NA	70%/100/50	NA	30	Saturated $NaPF_6$ in DME/DOL (1/1)
7 [26]	1.45/1.70, 1.80/2.10, 2.10/2.45	202, NA	95%/150/50	NA	40	Saturated $NaPF_6$ in DME/DOL (1/1)
8 [26]	1.30/1.75, 1.75/2.25	179, NA	92%/150/50	NA	40	Saturated $NaPF_6$ in DME/DOL (1/1)

Number	Cathodic/ Anodic Peak Potential (V vs. Na/Na^+)	The 1st and 2nd Discharge Capacity (mAh/g)	Cycling Performance: Discharge Capacity Retention/Cycle/ Current Density (mA/g)	Rate Capability: Discharge Capacity (mAh/g)/Current Density (mA/g)	Conductive Additive (wt. %)	Electrolyte*
9 [27]	1.42/1.78, 1.65/2.0, 2.00/2.25, 2.12/2.40	211, NA	57%/150/50	NA	40	Saturated $NaPF_6$ in DME/DOL (1/1)
10 [27]	1.52/1.68, 1.85/2.0, 2.15/2.30	234, NA	82%/150/50	200/50, 65/1000	40	Saturated $NaPF_6$ in DME/DOL (1/1)
11 [27]	1.35/1.65, 1.80/2.0, 1.90/2.25	242, NA	62%/150/50	NA	40	Saturated $NaPF_6$ in DME/DOL (1/1)
12 [27]	1.42/1.58, 1.85/2.15	248, NA	75%/150/50	NA	40	Saturated $NaPF_6$ in DME/DOL (1/1)
13 [28]	1.00/1.30, 1.76/2.20	1161.2, NA	15%/100/500	330.8/100, 102.3/5000	30	1 M $NaClO_4$ in EC/DEC (1/1)
14 [28]	0.90/1.30, 1.40/1.90	NA	NA	137.0/100, 69.4/5000	30	1 M $NaClO_4$ in EC/DEC (1/1)
15 [29]	1.46/1.8, 1.85/2.18	185, NA	100%/200/1600	210/400, 160/6000	50	Saturated $NaPF_6$ in DME/DOL (1/1)

(Continued)

Table 5.2 (*Continued*)

Number	Cathodic/ Anodic Peak Potential (V vs. Na/Na^+)	The 1st and 2nd Discharge Capacity (mAh/g)	Cycling Performance: Discharge Capacity Retention/Cycle/ Current Density (mA/g)	Rate Capability: Discharge Capacity (mAh/g)/Current Density (mA/g)	Conductive Additive (wt. %)	Electrolyte*
16 [30]	2.08	247, NA	68%/100/50	NA	30	1 M $NaN(CF_3SO_2)_2$ in DME/DOL (1/1)
17 [31]	NA	1081, NA	47%/1024/50	433/100, 122/3200	20	1 M $NaPF_6$ in EC/DEC (1/1)
18 [33]	1.02/1.11, 1.38/1.44	309, 224.9	45%/500/500	225/100, 131/1000	30	1 M $NaClO_4$ in EC/PC (1/1)
19 [35]	0.47/0.85	150, NA	NA	NA	20	1 M $NaN(SO_2F)_2$ in Me-THF
20 [35]	0.37/0.79	NA	NA	NA	20	1 M $NaN(SO_2F)_2$ in Me-THF
21 [35]	0.75/0.95	360, NA	53%/75/26	NA	50	1 M $NaN(SO_2F)_2$ in Me-THF
22 [35]	0.65/0.79	NA	NA	NA	20	1 M $NaN(SO_2F)_2$ in Me-THF
23 [35]	0.34/0.87	NA	NA	NA	20	1 M $NaN(SO_2F)_2$ in Me-THF

Number	Cathodic/ Anodic Peak Potential (V vs. Na/Na^+)	The 1st and 2nd Discharge Capacity (mAh/g)	Cycling Performance: Discharge Capacity Retention/Cycle/ Current Density (mA/g)	Rate Capability: Discharge Capacity (mAh/g)/Current Density (mA/g)	Conductive Additive (wt. %)	Electrolyte*
24 [35]	0.60/0.87	NA	NA	NA	20	1 M $NaN(SO_2F)_2$ in Me-THF
25 [36]	0.66, 1.04	206, NA	89%/25/19.7	NA	20	1 M $NaN(SO_2F)_2$ in Me-THF
28 [29]	3.4/3.7	91, NA	97%/200/500	96/200, 88/2000	50	Saturated $NaPF_6$ in DME/DOL (1/1)
29	NA	120, NA	94%/20/50	NA	20	5 mol% $NaClO_4$ in SCN[40]
	2.65/3.15, 3.5/3.75	181, NA	96%/50/50	180/50, 165/200	20	1 M $NaPF_6$ in EC/DMC/ DEC (1/1/1)[45]
30 [41]	2.8/3.0, 3.4/3.6	122, NA	97%/200/100	123/100, 76/800	10	1 M $NaPF_6$ in EC/DEC (1/1)
31 [42]	3.3/3.5, 3.6/3.7	103, NA	70%/100/50	92/100, 43/400	40	1 M $NaPF_6$ in EC/ DEC (1/1)
32 [43]	3.0/3.3	75, NA	100%/100/40	NA	30	1 M $NaPF_6$ in EC/ DEC (1/1)

(*Continued*)

Table 5.2 (*Continued*)

Number	Cathodic/ Anodic Peak Potential (V vs. Na/Na^+)	The 1st and 2nd Discharge Capacity (mAh/g)	Cycling Performance: Discharge Capacity Retention/Cycle/ Current Density (mA/g)	Rate Capability: Discharge Capacity (mAh/g)/Current Density (mA/g)	Conductive Additive (wt. %)	Electrolyte*
33 [46]	2.5, 3.7	115, NA	82%/50/50	105/100, 85/800	10	1 M $NaPF_6$ in EC/ DEC (1/1)
34 [47]	2.35/2.75	120, NA	92%/100/50	135/50, 75/1600	10	1 M $NaPF_6$ in EC/ DEC (1/1)
	NA	471.2, NA	40%/100/200	159.2/1800, 137.7/7200	20	1 M $NaClO_4$ in PC[48]
35	2.5/2.8	70, NA	79%/1000/400	100/20, 69/320	10	1 M $NaClO_4$ in EC/PC (1/1)[49]
	2.5/2.7	83, NA	100%/120/300	107/100, 75/300	15	1 M $NaPF_6$ in DEGDME[50]
36 [51]	3.54/3.78	121, NA	79%/10/20	60/100, 10/500	10	1 M $NaClO_4$ in PC
37 [53]	3.38/3.42	75, 73	65%/50/50	NA	60	1 M $NaClO_4$ in EC/DMC (1/1)
38 [54]	2.1/2.51, 3.36/3.51	217, NA	92%/100/112.5	222/22.5, 190/1125	30	1 M $NaClO_4$ in EC/DEC (1/1)

Number	Cathodic/ Anodic Peak Potential (V vs. Na/Na^+)	The 1st and 2nd Discharge Capacity (mAh/g)	Cycling Performance: Discharge Capacity Retention/Cycle/ Current Density (mA/g)	Rate Capability: Discharge Capacity (mAh/g)/Current Density (mA/g)	Conductive Additive (wt. %)	Electrolyte*
39 [58]	NA	~1030, 340	NA	182/300, 97/1000	20	1 M $NaClO_4$ in PC/FEC (98/2)
	NA	~850, 320	NA	208/300, 124/1000	20	1 M $NaClO_4$ in PC/FEC (98/2)
40 [58]	NA	~850, NA	NA	NA	20	1 M $NaClO_4$ in PC/FEC (98/2)
	NA	~850, NA	NA	NA	20	1 M $NaClO_4$ in PC/FEC (98/2)
41 [58]	NA	~850, NA	NA	NA	20	1 M $NaClO_4$ in PC/FEC (98/2)
	NA	~950, NA	NA	NA	20	1 M $NaClO_4$ in PC/FEC (98/2)
42 [59]	NA	NA	80%/7000/1000	55/10, 35/1000	20	1 M $NaClO_4$ in PC
	NA	NA	80%/7000/1000	185/10, 60/1000	20	1 M $NaClO_4$ in PC
43 [61]	NA	NA	NA	119/50, 99/100	30	1 M $NaClO_4$ in EC/DMC (1/1) plus 2 wt% FEC
44 [62]	NA	NA	95%/1100/1000	312/100, 137/10000	20	1 M $NaPF_6$ in EC/ DMC (1/1)

*EC: ethylene carbonate, DMC: dimethyl carbonate, PC: propylene carbonate, DEC: diethyl carbonate, FEC: fluoroethylene carbonate, DME: dimethoxy ethane, DOL: dioxolane, Me-THF: methyltetrahydrofuran, SCN: succinonitrile, DEGDME: diethylene glycol dimethyl ether. 1/1 and 1/1/1: volume ratio.

5.2.1.2 Polyquinones

The polyquinones employed as electrode materials for NIBs are usually sulfur-containing polyquinones, obtained by the polycondensation reaction of quinone monomers and sulfides [29–32]. Due to the reductive capacity of the quinonyl groups, a side redox reaction between the quinonyl groups and sulfide anions may occur during the formation of the polyquinone main chains. Therefore, the synthesis is usually followed by subsequent purification or oxidation steps [30, 32].

As shown in Fig. 5.2, the structure of the polyquinone chains is determined by the molecular structure of quinone monomers. The polyquinones usually contain a chain of quinone rings separated by heteroatoms with an unshared electron pair. Though the heteroatoms are not the active sites and do not contribute to the sodium storage, the heteroatoms can affect the discharge voltage [31].

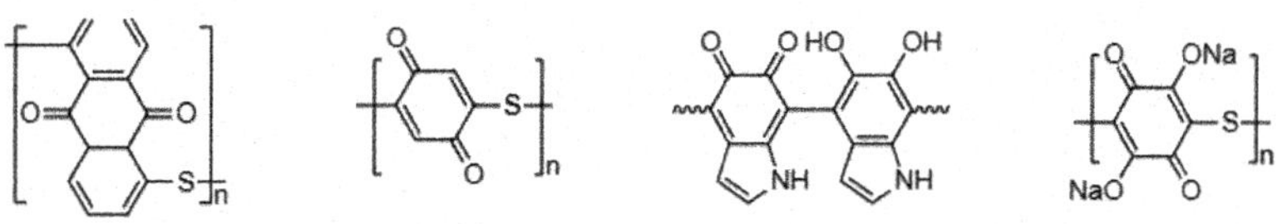

Figure 5.2 The structure of several polyquinones suitable for NIBs.

Similar to polyimides, polyquinones usually undergo two-electron redox reactions through reversible electrochemical association of sodium ions with their redox-active quinonyl groups (Scheme 5.2).

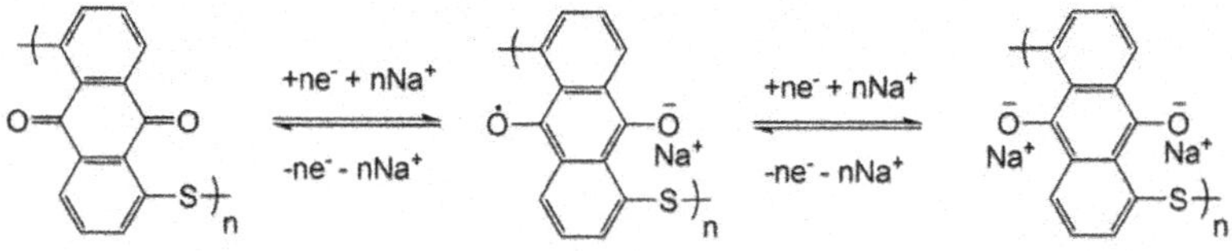

Scheme 5.2 Schematic representation of the redox reaction of polyquinones.

The first polyquinone used in NIBs was poly(anthraquinonyl sulfide) (15), reported by Yang et al. in 2013 [29]. The poly(anthraquinonyl sulfide) displayed two pairs of symmetric redox peaks at 1.85/2.18 V and at 1.46/1.8 V (Fig. 5.3a), implying a two-step redox reaction of the anthraquinonyl groups. Consistent with this, the material displayed two plateaus at 1.5 and 2.0 V with a capacity of 220 mAh/g (Fig. 5.3b), corresponding to 98% theoretical capacity of storage of two sodium ions. The polyquinone showed a good rate capability with the capacity of 160 mAh/g at a high current density of 6400 mA/g and a stable cyclability with 85% capacity retention at the 500th cycle at 1600 mA/g (Figs. 5.3c,d).

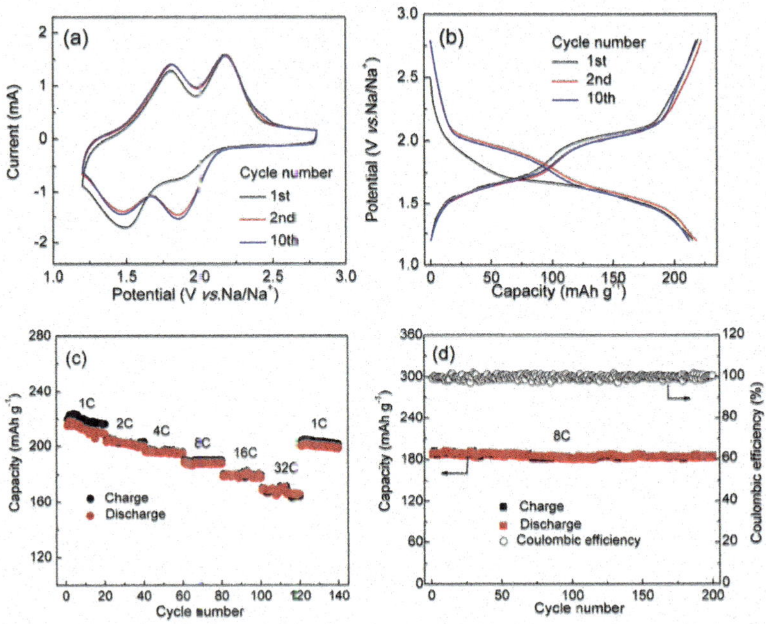

Figure 5.3 Electrochemical performance of polymer 15 as candidate anode material for sodium-ion batteries: (a) cyclic voltammograms (CV) curves at a scan rate of 5 mV/s; (b) charge and discharge profiles at a current density of 180 mA/g; (c) rate performance (1 C = 200 mA/g); (d) cycling performance at 8 C rate (1600 mA/g). Adapted from Ref. [29], with permission from Springer Nature, Copyright 2013, and acknowledgements to the authors.

In 2015, poly(benzoquinonyl sulfide) (**16**) was reported by Zhou et al. [30]. This polymer was synthesized by an oxidative polymerization technique due to the high reactivity of the benzoquinone monomers. The poly(benzoquinonyl sulfide) delivered a high capacity of 268 mAh/g and a high energy density of 557 Wh/kg. However, the capacity decayed at a rate of 0.88 mAh/g per cycle after the second cycle and remained only 68% after 100 cycles. The capacity was also unstable at current densities above

200 mA/g. The unsatisfactory cycling stability and rate performance were ascribed to the poor compatibility between the sodium and the 1 mol/L $NaN(CF_3SO_2)_2$/(dioxolane + dimethoxy ethane) electrolyte.

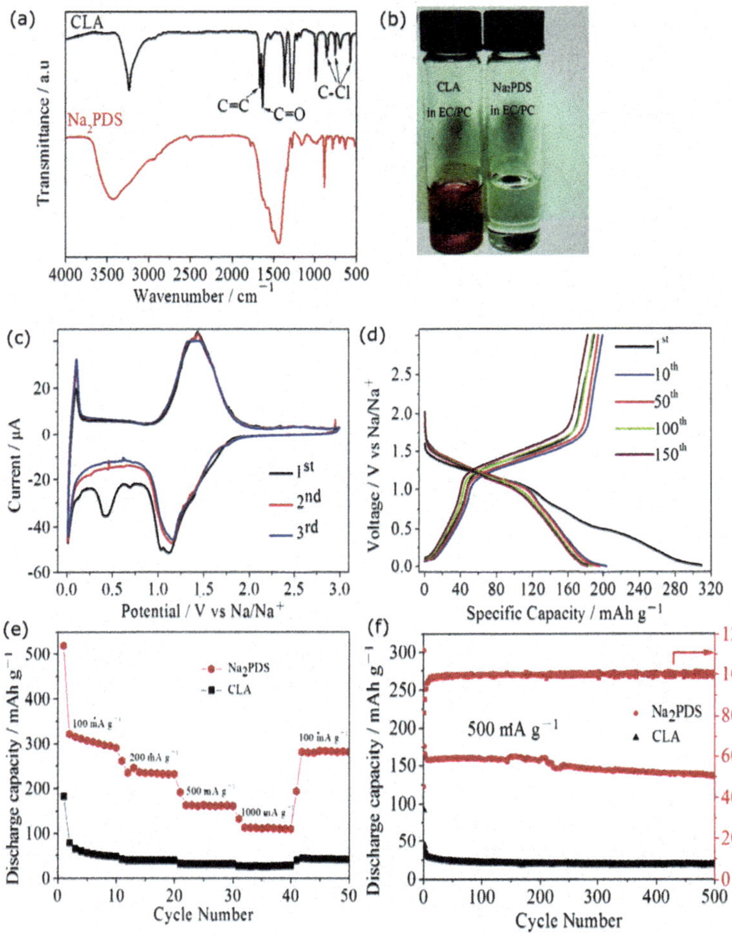

Figure 5.4 (a) Fourier transform infrared spectra of 2,5-dichloro-3,6-dihydroxy-*p*-quinone (CLA) and sodium salt of poly(2,5-dihydroxy-*p*-benzoquinonyl sulfide) (Na_2PDS); (b) photograph of the CLA powder and the Na_2PDS electrode in the electrolyte; (c) CV curves of the Na_2PDS electrode at a scan rate of 0.1 mV/s; (d) discharge and charge profiles of the Na_2PDS electrode at a current density of 100 mA/g; (e) rate performance; and (f) cycling performance of CLA and Na_2PDS as candidate anode materials for sodium-ion batteries. Adapted from Ref. [33], with permission from the Royal Society of Chemistry provided by Copyright Clearance Center.

Besides the above synthetic polymers, biomass-derived polydopamine (**17**) is an ideal redox-active biodegradable electrode material. The carbonyl groups of the *o*-benzoquinone group act as the active sites for coordination of sodium ions. The polydopamine exhibited excellent electrochemical performance in NIBs, with a high capacity of 500 mAh/g and a capacity retention of 100 % at the 1024th cycle at a current density of 50 mA/g [31].

The emergence of the polyquinones resulted from the high tendency for dissolution of quinones in the organic electrolytes [32]. However, polymerization can only mitigate rather than totally prevent the dissolution. Wu et al. [33] reported a sodium salt of poly(2,5-dihydroxy-*p*-benzoquinonyl sulfide) (**18**) as a novel anode material for NIBs. As seen from Fig. 5.4, the intrinsically insoluble salt form of the polyquinone exhibited an excellent cyclability with a capacity of about 138 mAh/g at the 500th cycle, much better than the original quinone (19 mAh/g). Thus, formation of the salt can be a feasible solution to the problem of solubilization, although further efforts are required in this direction.

5.2.2 Schiff Base Polymer Electrode Materials

Schiff base polymers can be synthesized by a simple polycondensation reaction of diamines and dialdehydes or diketones [34]. Schiff base polymers contain a system of conjugated –C=C– and –C=N– bonds. The structure of diamine and dialdehyde or diketone precursors determines the diversity of the Schiff base polymers (Fig. 5.5).

Schiff base polymers are increasingly being explored as organic electrode materials for NIBs. The active center of Schiff base polymers for the sodium-ion storage is the C=N of conjugated –N=CH–Ar–CH=N– (Ar = aromatic ring) (Scheme 5.3). Planarity and conjugation are the key properties required for electrochemical activity of Schiff base polymers [35]. The "reverse" configuration –C=NH–Ar–NH=C–, though isoelectronic, is not electrochemically active.

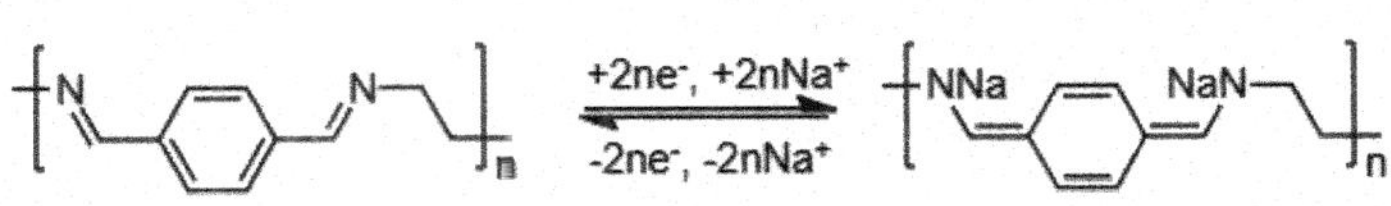

Scheme 5.3 Schematic representation of sodium-ion storage in Schiff base polymers.

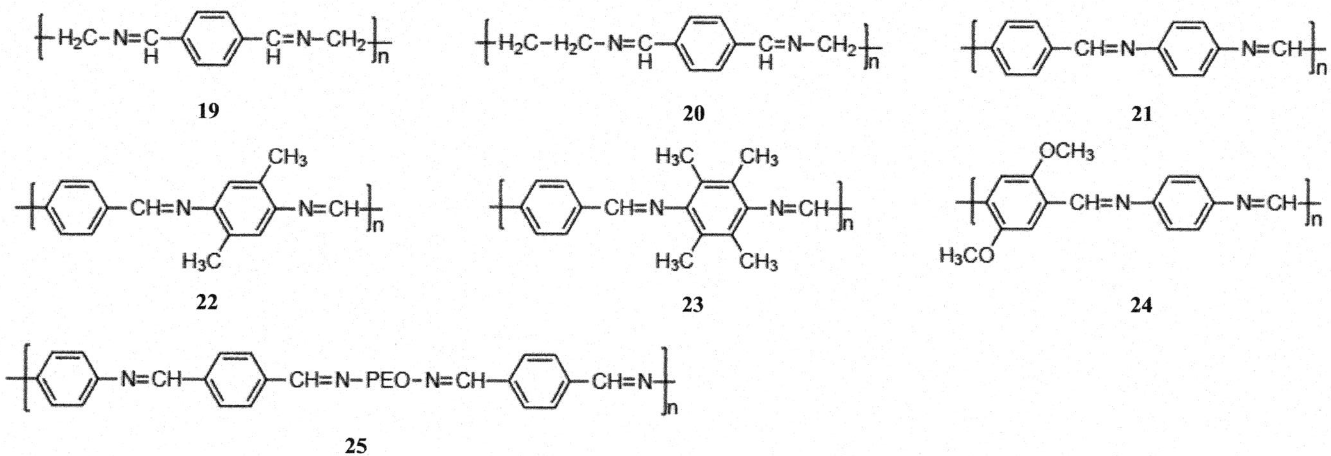

Figure 5.5 Structure of Schiff base polymers.

The first study on Schiff base polymers (**19–24**) as anode materials for NIBs was reported by Armand et al. in 2014 [35]. It was found that the reduction reaction takes place at potentials below 1.5 V versus Na/Na^+, indicating that Schiff base polymers are very promising anode materials for NIBs. The redox potential can be adjusted by the addition of appropriate substituents without compromising the polymer chain planarity and extent of conjugation. A maximum capacity of 350 mAh/g was achieved on the addition of 50 wt% of Ketjen Black, corresponding to 2.8 sodium ions per monomer unit.

Recently, Armand et al. reported a poly-Schiff-polyether terpolymer (**25**) with an electrochemical activity at low redox potential versus Na/Na^+ and self-binding property, allowing the preparation of laminated electrodes without binder [36]. At the same carbon content, the laminate electrode delivered higher capacities than the powder counterpart due to superior cohesive inter-particle contact. A stable reversible capacity of ~185 mAh/g could be obtained for the binder-free laminated electrode. Due to its intriguing adhesive properties, this poly-Schiff-polyether terpolymer was also tested as a binder for hard carbon, representing a proof of concept in the field of redox-active binders.

The low solubility of Schiff base polymers in most organic solvents [37] is an advantage for electrode materials; however, this makes the materials difficult to process. The balance between the solubility and processability can be manipulated by modifying the polymer chain using appropriate linkers between the azomethine units [35–37]. Furthermore, the electrical conductivity of Schiff base polymers can be increased up to the level of semiconductors by doping with iodine [38]. This important property makes Schiff base polymers very promising electrode materials for NIBs.

5.2.3 Conducting Polymer Electrode Materials

5.2.3.1 Conjugated conducting polymers

Conjugated conducting polymers (Fig. 5.6) can undergo p-doping and/or n-doping reactions (Scheme 5.4). In p-doping reactions, the polymer changes to the positively charged state and reacts with anions such as PF_6^-. In n-doping, the polymer changes to the negatively charged state and reacts with the sodium cations.

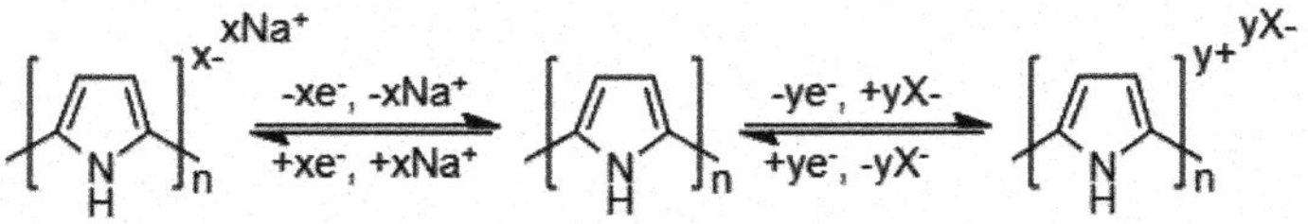

Scheme 5.4 Schematic representation of p-doping and/or n-doping reactions of conjugated conductive polymers.

The first study of conjugated conductive polymers for NIBs was reported by Shacklette et al. in 1985 [39]. It was found that polyacetylene (PAc) (**26**) and polyparaphenylene (PPP) (**27**) were possible anode materials. PAc and PPP are bipolar polymers, capable of undergoing both p-doping and n-doping redox reactions [10]. The bipolar polymers have potential in all-organic batteries since they can act as both cathode and anode.

Most of the well-studied conjugated conductive polymers are p-type polymers, such as polyaniline (PANi) and Polypyrrole (PPy). P-type conducting polymers (**28, 29**) can be used in the cathode to construct all-organic batteries with n-doping polymers as the anode [29, 40]. However, the p-doping mechanism can be modified by introducing electron-withdrawing groups. For example, by grafting the electron-withdrawing $-SO_3Na$ group onto the polyaniline chain, the sulfonated polyaniline/Ketjen black composite (**30**) mainly experienced an insertion/extraction mechanism of sodium ions rather than p-doping reactions [41]. A similar strategy was implemented for poly(diphenylaminesulfonic acid sodium) (**31**) [42] and poly(pyrrole-co-(sodium-3-(pyrrol-lyl) propanesulphonate)) (**32**) [43].

The advantage of conducting polymers over other polymers is their high electric conductivity comparable to semiconductors or even metals [44]. However, their capacity is low and limited by the degree of doping. Incorporation of a high-capacity redox-active unit can be used to overcome this drawback, as extensively studied by Yang et al. [45–47]. By grafting o-nitroaniline groups onto polyaniline chains, polymer (**29**) achieved a high reversible capacity of 180 mAh/g and retained at 173 mAh/g at the 50th cycle [45]. Similarly, by doping with diphenylamine sulfonate anions (**33**) [46] and ferrocyanides (**34**) [47], the anion-doped polypyrroles demonstrated increased capacities due to the contribution from the doped redox-active units. Thus, to develop high-performance conjugated conductive polymer electrode materials, it is feasible and promising to focus on the incorporation of other redox-active units onto the polymer chain.

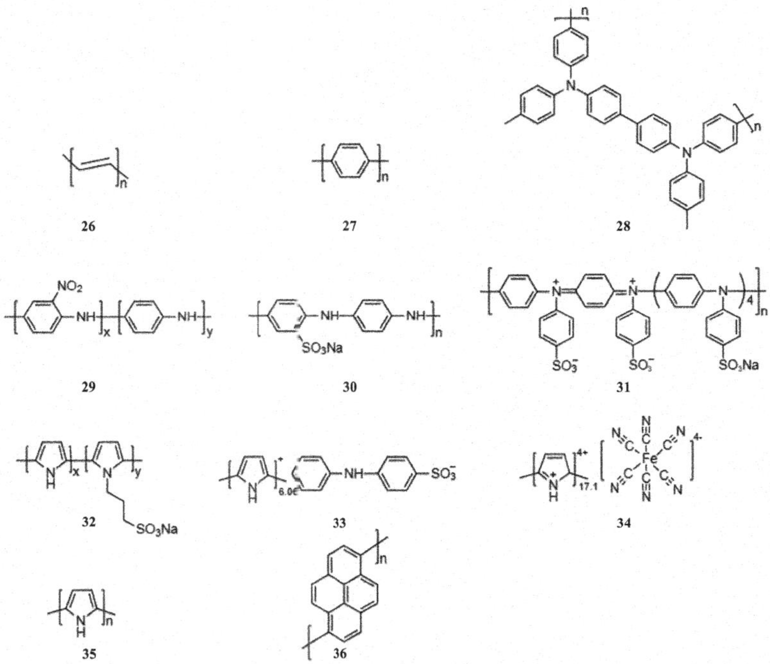

Figure 5.6 The structure of conjugated conductive polymers suitable for NIBs.

It has been reported that the particle size, morphology, crystalline structure of conjugated conductive polymers may affect their electrochemical performance. Wang et al. [48] found that submicron polypyrrole (**35**) with fluffy structure and chain-like morphology showed a higher capacity (183 mAh/g) than bulk polypyrrole (34.8 mAh/g). The excellent electrochemical performance is due to the unique structure, which increased the electrical contact between the polypyrrole particles and promoted the penetration of the electrolyte into the material. In addition, hollow nanospheres [49] and mesoporous nanosheets [50] of polypyrrole were prepared through the use of soft templates. The unique morphology enabled stable cycling performance and superior rate capability of the polypyrrole electrodes. Han et al. [51] reported crystalline and

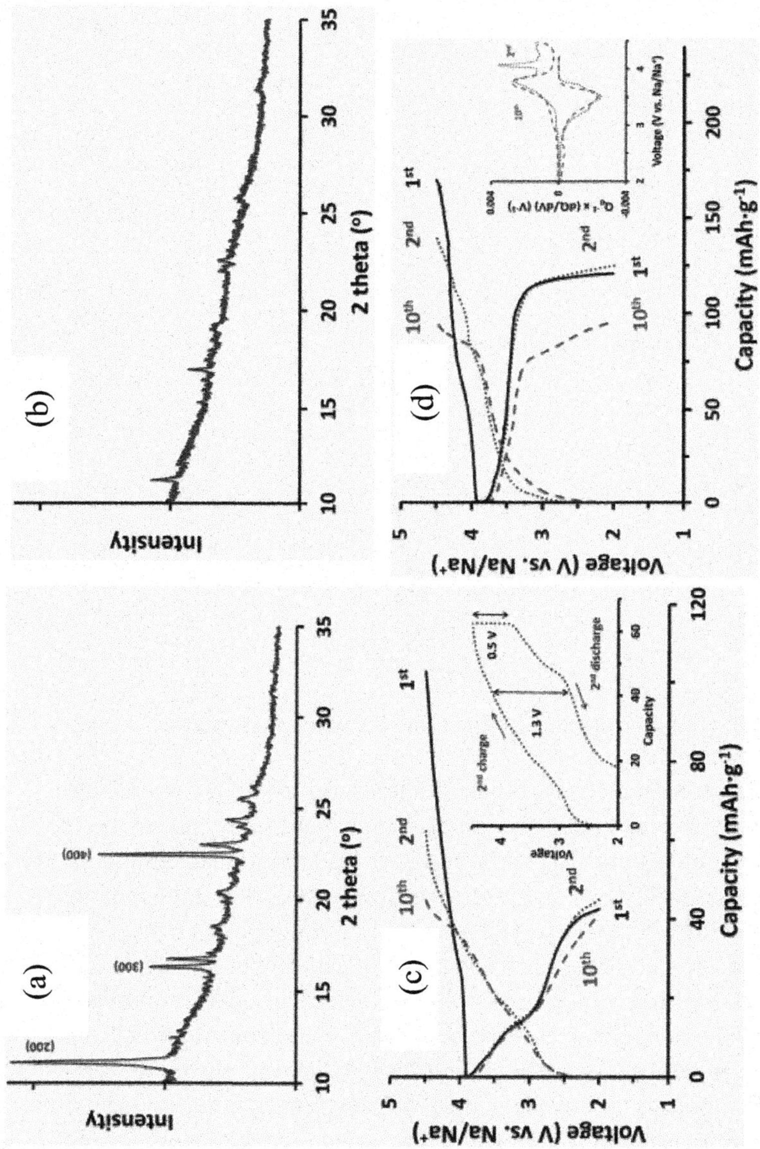

Figure 5.7 (a, b) XRD patterns and (c, d) charge and discharge profiles at a current density of 20 mA/g of crystalline oligopyrene (a, c) and amorphous oligopyrene (b, d) as candidate cathode materials for sodium-ion batteries; inset of (c) shows the second charge and discharge profile to indicate a large overpotential; inset of (d) shows the differential capacity (dQ/dV) curves for the second and 10th cycles. Adapted from Ref. [51], with permission from Elsevier provided by Copyright Clearance Center.

amorphous oligopyrenes (**36**) as high-voltage cathodes for NIBs. The crystalline oligopyrene with a layered structure (Fig. 5.7a) exhibited sloping charge and discharge curves with a discharge capacity of 42.5 mAh/g and an average potential of 2.9 V versus Na/Na^+ at a current density of 20 mA/g (Fig. 5.7c). The inset in Fig. 5.7c shows a large overpotential, which was ascribed to the slow ClO_4^- diffusion through the crystalline phase. The amorphous oligopyrene showed substantially reduced overpotential with a higher capacity of ~120 mAh/g and a discharge plateau at 3.5 V versus Na/Na^+ (Fig. 5.7d).

5.2.3.2 Non-conjugated conductive radical polymers

The most widely explored organic radical polymers for batteries are nitroxide radical polymers (Fig. 5.8). The nitroxide radical polymers can be not only reversibly n-doped to aminoxy anions in cathodic reactions at relatively high voltage but also p-doped to oxoammonium cations in anodic reactions at relative low voltage (Scheme 5.5).

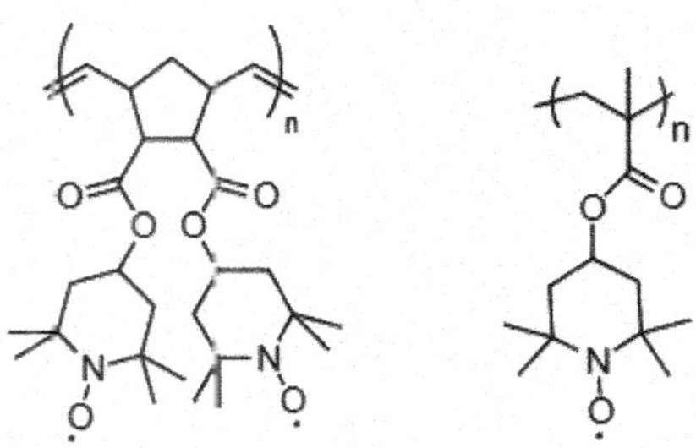

37 **38**

Figure 5.8 Molecular structures of two non-conjugated conductive radical polymers.

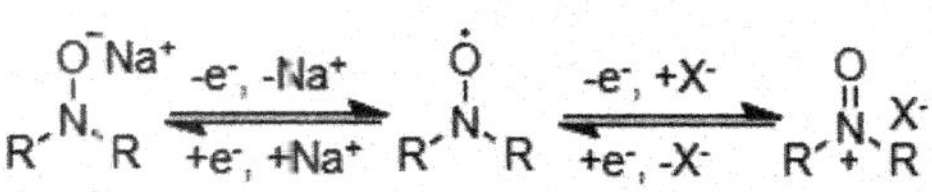

Scheme 5.5 N- and p-doping mechanism of nitroxide radical polymers.

Nakahara et al. first reported a stable nitroxyl polyradical for rechargeable lithium-ion batteries in 2002 and opened a new field of use for plastic batteries [52]. The first application in NIBs reported in 2010 by Dai et al. was a polynorbornene derivative radical polymer, poly[norbornene-2,3-endo, exo-(COO-4-TEMPO)$_2$] (**37**) as a cathode active material [53]. The radical polymer **37** delivered an initial discharge capacity of 75 mAh/g at a current density of 50 mA/g and retained a discharge capacity of 64.5% at the 50th cycle.

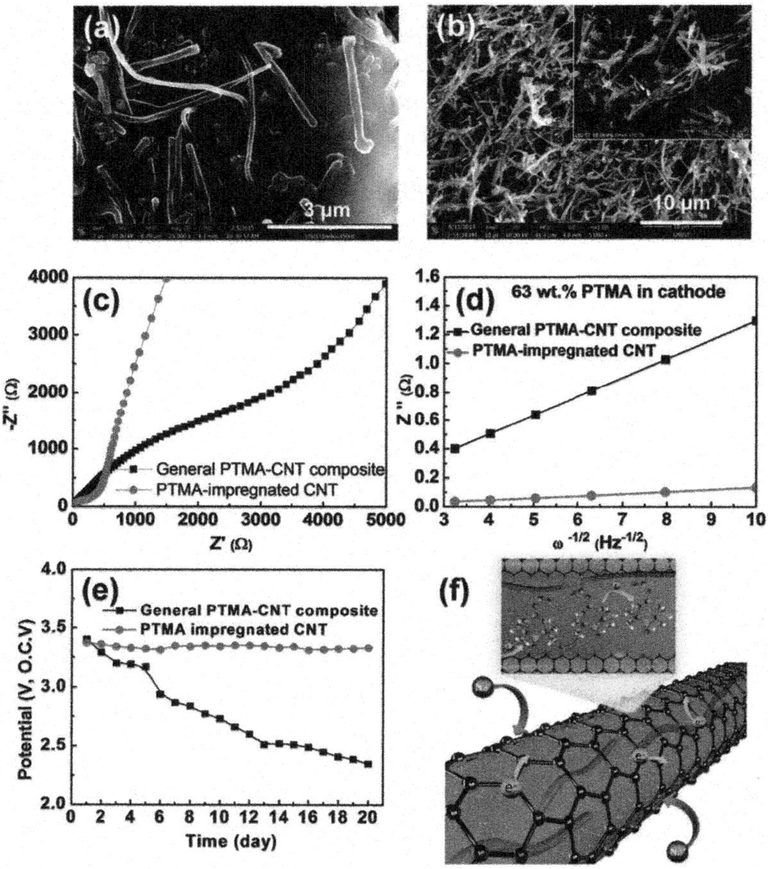

Figure 5.9 Scanning electronic microscopy images of (a) PTMA-CNT composite and (b) PTMA-impregnated CNT electrodes; (c) impedance spectra and (d) the relationship between the imaginary resistance and the square root of frequency in the low-frequency region, (e) open-circuit voltage during 12 days of PTMA-CNT composite and PTMA-impregnated CNT electrodes as candidate cathode materials for sodium-ion batteries; (f) schematic illustration of the PTMA-impregnated CNT structure. Adapted from Ref. [54], with permission from Royal Society of Chemistry provided by Copyright Clearance Center.

Kim et al. designed a unique organic electrode encapsulating the nitroxide radical polymer, poly(2,2,6,6-tetramethylpiperidinyloxy-4-vinylmethacrylate) (PTMA, **38**) into carbon nanotubes (CNTs) to form an electrode with a high polymer content [54]. To investigate the PTMA-impregnated CNT electrode, the PTMA–CNT composite electrode was compared as a control sample. The non-conducting PTMA layer on the PTMA–CNT composite electrode (Fig. 5.9a) blocked the free passage of ions at the electrode/electrolyte interface, leading to a larger charge-transfer resistance of 3716 Ω (Fig. 5.9c). However, the direct contact between CNTs of the PTMA-impregnated CNT electrode (Fig. 5.9b) improved the electron transfer leading to a decreased resistance of 283 Ω. In addition, the low slope of the relationship between the imaginary resistance and the square root of frequency in the low-frequency region suggested good ion kinetics in the PTMA-impregnated CNT electrode. The self-discharge caused by the dissolution of organic active material can also be minimized effectively by trapping PTMA in CNTs (Fig. 5.9e), leading to improved discharge capacity, cycling performance, and rate capability.

The advantages of the radical polymers include fast kinetics due to the small structural change during the oxidation process, stable cell voltage due to the stable structure, and good processability [55]. However, spontaneous self-discharging is one of the biggest challenges for radical polymers due to the dissolution into electrolyte, which functions as a redox shuttle [56]. To overcome this problem, strategies for reducing the dissolution of radical polymers in electrolytes must be developed.

It should be noted that the charge-storage mechanism of the radical polymers involves reversible one-electron redox reaction per repeating unit, and therefore the theoretical capacity of these polymers is strictly limited by the one-electron reaction and the molar mass of the repeat unit. To obtain high-capacity radical polymer electrode materials, it is essential to carefully design the molecular structure of the polymer chain.

5.2.4 Covalent Organic Frameworks

Covalent organic frameworks (COFs) are a kind of conjugated microporous polymers (Fig. 5.10). The well-defined crystalline porous structures as well as tailored functionalities make COFs promising in diverse applications, such as energy storage, gas storage and separation, optoelectricity, and catalysis [57]. COFs have been studied as electrode materials for organic NIBs due to their high specific surface area and microporous structure benefiting the rapid migration of sodium ions as well as the large amount of redox-active centers delivering relatively high theoretical capacity [58–62]. The capacity of COFs is generated based on the p-doping and/or n-doping mechanism. The charge-carrier conductivity tailored by polymer backbone planarity and accessible specific surface area also have large impacts on the performance of NIBs. Most COFs can be synthesized by condensation reactions via the formation of easily hydrolyzable species, such as imines, hydrazones, and borosilicates [57]. By varying the monomer combinations (COFs **39–41**), it is able to adjust the macromolecular ring diameter, porosity, network connectivity, and conductivity, which can affect the charge carrier and ion mobility and thus the performance [58].

The bipolar porous organic electrode (BPOE, **42**) is a COF consisting of benzene rings and triazine rings with two-dimensional (2D) porous-honeycomb structure [59]. The BPOE shows a p-dopable potential range of 4.1–2.8 V versus Na/Na^+ and an n-dopable potential range of 2.8–1.3 V versus Na/Na^+. Such a bipolar feature makes it suitable as both cathode and anode for all-organic battery, which enabled a large working window of 4.1–1.3 V versus Na/Na^+ and a high energy density of 34 Wh/kg [60]. Another conjugated microporous polymer 4,7-dicarbazyl-[2,1,3]-benzothiadiazole (PDCzBT, **43**) has a high specific surface area of 1166 m^2/g and a relatively uniform ultramicropore width at 0.52 and 0.86 nm, promoting the transport of sodium ions [61]. Ultrathin few-layer structure and an in-plane π-conjugated structure of 2D polymers (2DP, **44**) not only benefit the transport of sodium ions but also enable fast electron transport [62]. The 2DP delivered a high sodium storage capacity of 312 mAh/g at 0.1 A/g, a good rate capability and excellent cycling stability with a capacity retention of 95% at the 1100th cycle.

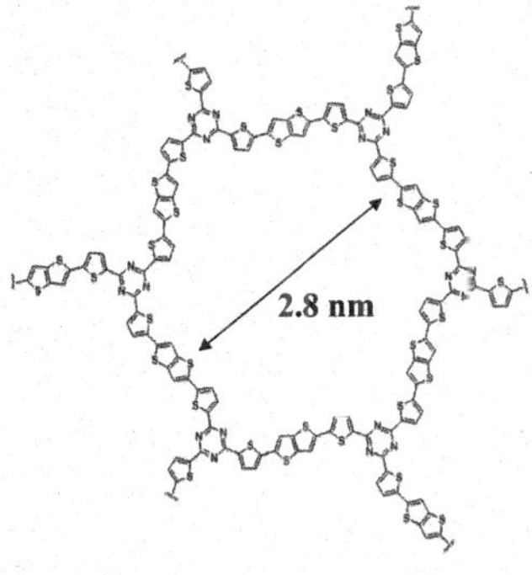

39

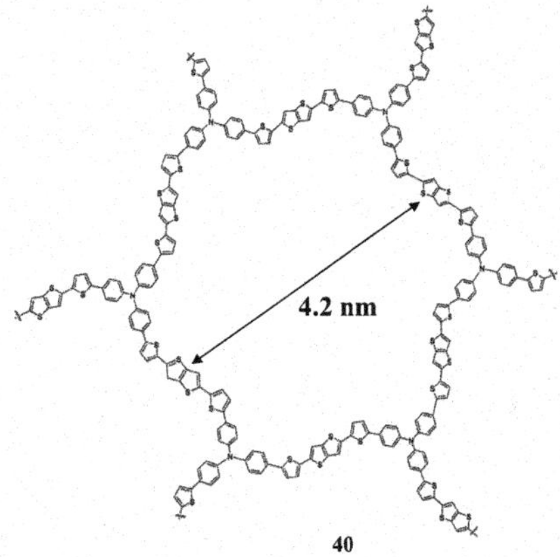

40

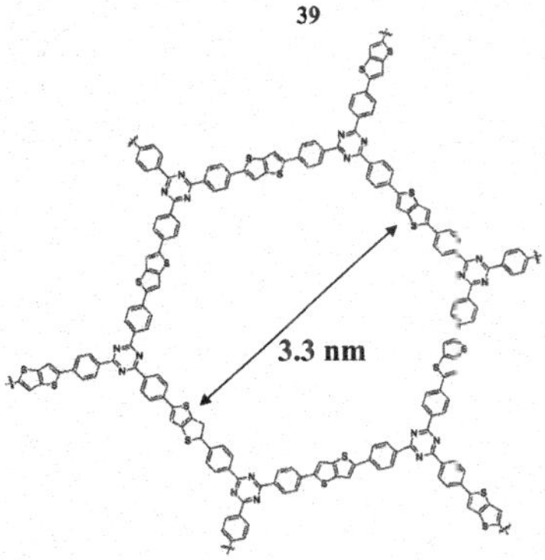

41

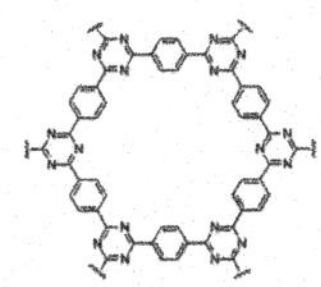

42

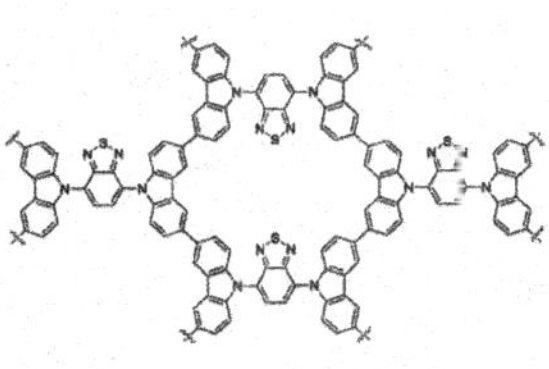

43

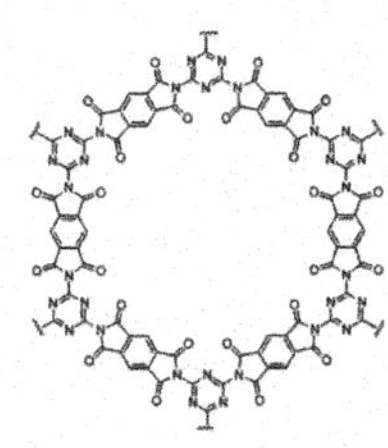

44

Figure 5.10 Molecular structures of COFs.

5.3 Characterizations for Polymer Electrode Materials

Due to the amorphous nature, which precludes the use of developed characterization techniques in battery research such as X-ray diffraction (XRD) and transmission electron microscopy (TEM), mechanistic studies of polymer electrode materials are hampered [63]. In this regard, solid-state nuclear magnetic resonance (ssNMR) spectroscopy and Fourier-transform infrared (FTIR) spectroscopy as common techniques to characterize polymers have emerged to provide insightful information of their electrochemical behaviors in batteries.

5.3.1 Solid-State NMR

^{13}C and ^{23}Na ssNMR spectroscopies are common techniques to characterize organic electrode materials in NIBs. Through the analyses of ssNMR spectroscopy, the local structure and ionic dynamics of electrode materials can be probed by extracting information of the chemical shift, peak shape, relaxation time, and coupling constant of nuclei [64–68].

Table 5.3 Comparisons of *ex situ* and *in situ* NMR characterizations in battery research

	Test Condition	Cell	Resolution	Timeliness
Ex situ NMR	Magic Angle Spinning	Disassembled	Relatively High	Delayed
In situ NMR	Static	Intact	Relatively Low	Real-Time

According to the condition of measurements, ssNMR in NIB research can be divided into two categories: *ex situ* and *in situ* NMR. For *ex situ* NMR characterization, electrode samples are collected from disassembled cells after testing, while *in situ* NMR characterization is conducted on functioning cells without disassembly. Such *in situ* NMR can monitor real-time information of dynamics and structural changes during operation [69]. However, *in situ* NMR needs to be performed on specific cell under the static condition, which is restricted by the circuit connecting cell and external potentiostat [70]. No sample spinning limits resolution of *in situ* NMR spectroscopy, due to anisotropic NMR interaction and the resulted signal broadening [71]. In addition, components of the cell and current of the circuit may also interfere NMR signals and detectability [72]. Comparisons of *ex situ* and *in situ* NMR characterizations are provided in Table 5.3. The procedures and setups of *ex situ* and *in situ* NMR for battery research as well as the ways to improve the quality of corresponding NMR spectra are introduced in the following sections.

5.3.1.1 *Ex situ* NMR

The general flow diagram of procedures for *ex situ* NMR is shown in Fig. 5.11. First of all, electrode samples are collected from a disassembled cell, which has been stopped at a certain state of charge. Then the electrode samples are ground into fine powders. Next, they are packed tightly into a rotor, which is equipped for ssNMR measurements. It should be noted that preparation of air-sensitive electrode samples needs to be conducted in glovebox under inert gas.

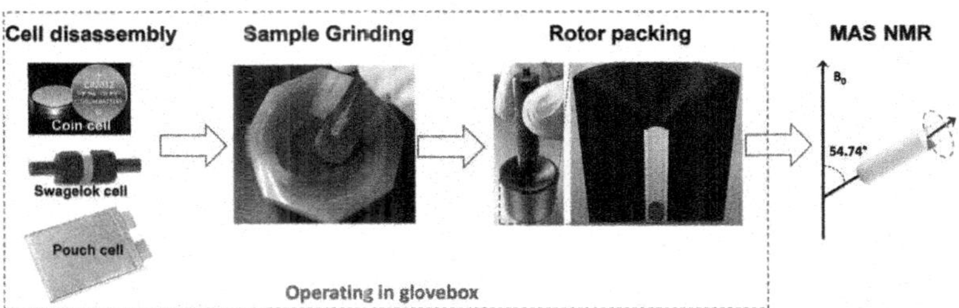

Figure 5.11 General flow diagram of procedures for *ex situ* NMR.

Sufficient amounts of samples are needed for ssNMR to fill the rotor and get NMR spectra with acceptable signal-to-noise (S/N) ratio. For example, 10–50 mg samples are usually needed for a high-resolution ^{13}C NMR spectrum. Therefore, commonly used 2032 coin cells for lab-scale study can hardly meet the requirement due to the limited loading amount of electrode materials per coin cell. Multiple disassemblies of coin cells are usually needed to obtain sufficient samples for ssNMR, though which can be more easily realized by replacing with Swagelok cells. In this case, pouch cells with larger sample loadings are applied to solve this problem [63, 73].

At this stage, all *ex situ* NMR experiments employ magic angle spinning (MAS), which is a powerful technique to obtain high-resolution NMR spectra. MAS introduces artificial motion by placing the axis of the sample rotor at the magic angle of 54.74° with respect to the direction of the magnetic field B_0. The MAS mimics the rapid molecular motion in solution NMR to average out the orientation-dependent interactions in ssNMR.

Many spinning sidebands usually occur at multiples of the MAS rate when the spinning rate is less than the magnitude of the anisotropic interaction [74]. As shown in Fig. 5.12 (marked rectangles), improper spinning rate (8 kHz) led to overlapping of sidebands and isotropic center bands of the target sample [75]. To avoid such overlap, spinning rate needs to be adjusted to separate the sidebands from the isotropic center bands [75, 76]. Another method is to suppress the sidebands to obtain sideband-free spectra by employing pulse sequences [77]. It should be noted that the sidebands are sometimes useful to provide important information on local electronic structure for analysis [78].

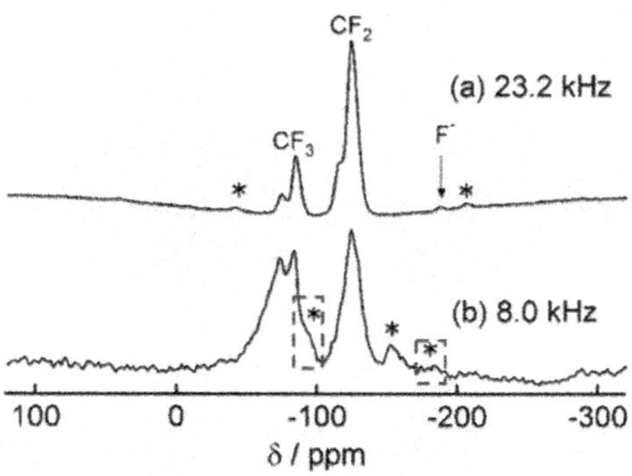

Figure 5.12 ^{19}F MAS NMR spectra of perfluorooctanesulfonic acid-doped graphite intercalation compounds at the spinning rate of (a) 23.2 kHz and (b) 8.0 kHz (* indicates spinning sidebands). Adapted from Ref. [75], with permission from Elsevier provided by Copyright Clearance Center.

The resolution of MAS NMR spectroscopy can be further improved by optimizing the experimental operation [70]. For example, smaller rotor can be used to achieve a higher MAS speed, resulting in an increased resolution. In the case of conductive samples, smaller sample particles and/or diluted sample with unreactive and diamagnetic powders can eliminate the eddy currents and possible signal interference caused by intrinsic magnetic moment at high magnetic fields. Cross polarization (CP) can be combined with MAS technique to improve the S/N ratio, by transferring polarization from abundant nuclei (e.g., ^{1}H and ^{19}F) to dilute rare nuclei such as ^{13}C and ^{15}N [74].

5.3.1.2 *In situ* NMR

The *in situ* NMR equipment for battery research is developed based on the development of two key components: the NMR probe and the electrochemical cell. Figure 5.13 shows a brief overview of historical development on the *in situ* NMR apparatus.

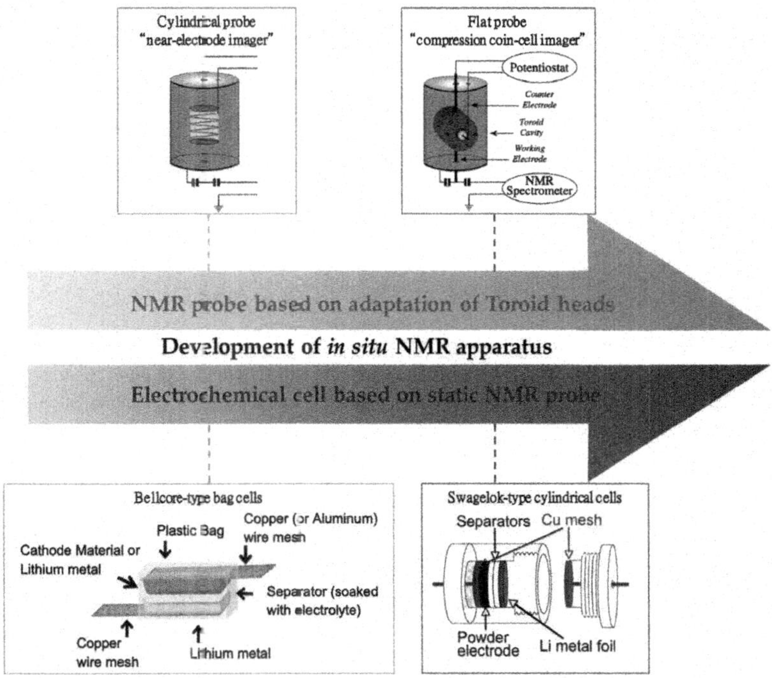

Figure 5.13 A brief overview of historical development of *in situ* NMR apparatus.

The design of NMR probe is focused on the adaptation of Toroid heads. In this area, Gerald et al. made great efforts in designing *in situ* NMR probes based on Toroid heads since the first *in situ* NMR investigation [79–81]. The initial setup is a "near-electrode imager" probe with a cylindrical electrochemical cell and a Toroid coil, as shown in Fig. 5.14a. A central copper rod was used as both a current collector of the electrochemical cell and a conductor of the toroid cavity NMR detector (Fig. 5.14b).

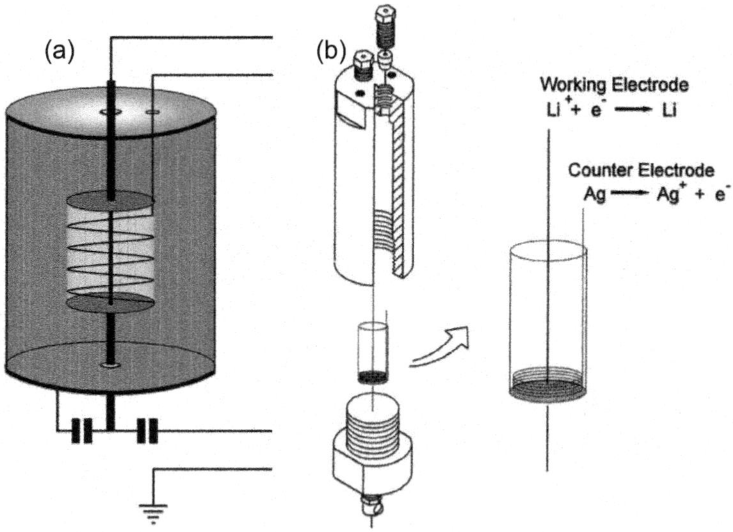

Figure 5.14 Schematic illustrations of (a) the cylindrical "near-electrode imager" probe and (b) the configuration of the cylindrical cell in NMR detector. Adapted from Refs. [79, 81], with permission from Elsevier and IOP Publishing provided by Copyright Clearance Center.

However, this non-standard cell design makes the preparation of electrodes complex and difficult, accelerating the design of flat "compression coin-cell imager" probe by the same group (Fig. 5.15a) [81]. In this probe (Fig. 5.15b), a solid copper plate replaced the central copper rod in the near-electrode imager. The electrochemical coin cell was installed in a toroid cavity, which allowed radio-frequency currents but shielded out extra signals. However, the S/N ratio of the Toroid cavity detector is still low, and the analysis of spectra is complicated resulting from nonlinear radio-frequency excitation and detection [82].

The other key component, the electrochemical cell, is designed based on the use of conventional static NMR probes. Such cells can be divided into two types: Bellcore-type bag cells and Swagelok-type cylindrical cells. Both these cells employ plastic container to eliminate the use of metallic components in battery assembly, which can reduce the undesired effects from the intrinsic magnetic moment of paramagnetic metallic components at high magnetic fields. Figure 5.16 shows a typical Bellcore-type bag cell, which is inside the coil of a conventional static NMR probe and cycled in NMR magnet. The flat shape may cause some susceptibility effects, inducing a shift in resonances in the NMR spectra [82]. In addition, the flat cells are also limited by the loading of electrode materials and electrolyte leakage due to the fragile edges.

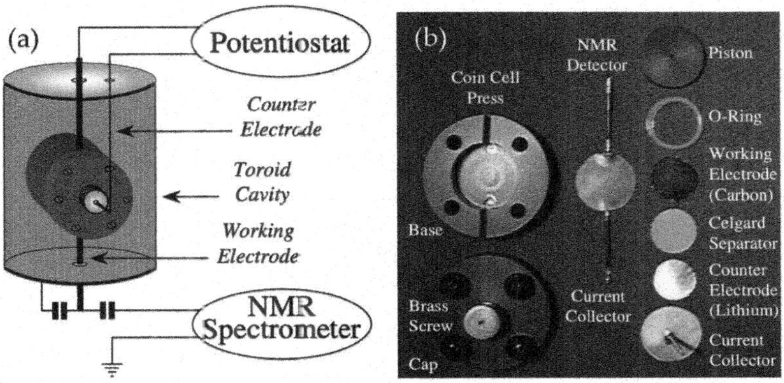

Figure 5.15 (a) Schematic illustration of the compression coin-cell battery imager. (b) Image of the components of the compression coin cell. Adapted from Ref. [81], with permission from IOP Publishing provided by Copyright Clearance Center.

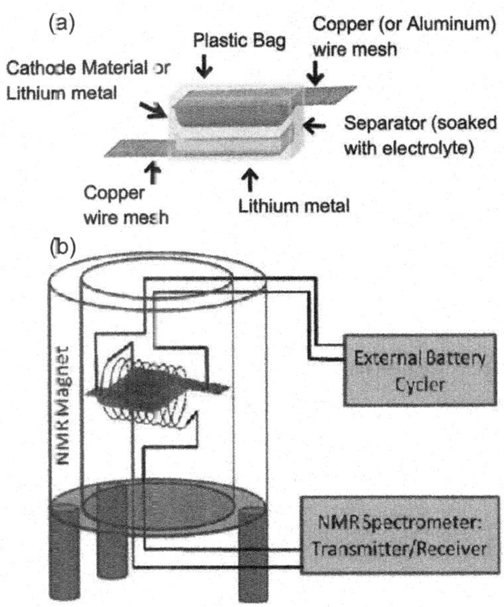

Figure 5.16 Schematic illustrations of (a) a typical Bellcore-type plastic bag cell and (b) the plastic bag cell in the coil of an NMR probe. Adapted from Refs. [82, 83], with permission from American Chemical Society and Elsevier provided by Copyright Clearance Center.

Figure 5.17 shows the configuration of cylindrical cells. Such cylindrical design enables higher loadings of electrode materials, diminishes the electrolyte leakage, and reduces some susceptibility effects in flat bag cells [84]. The Swagelok-type cylindrical cell with a spring can further enhance the electrical contact between electrodes and current collectors, assuring good cycling.

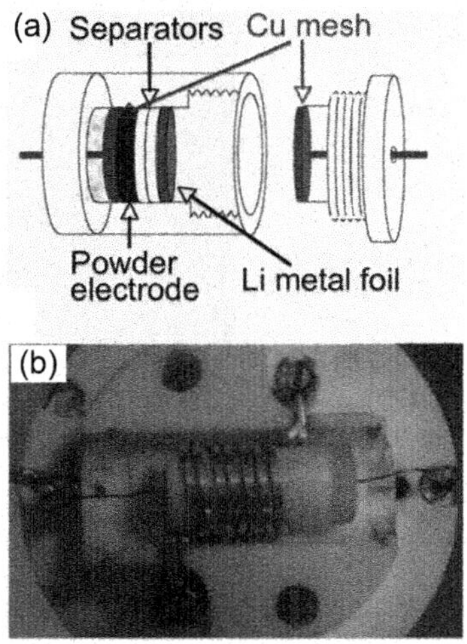

Figure 5.17 (a) The cylindrical cell design for *in situ* NMR experiments. (b) The cylindrical cell inside an NMR probe. Adapted from Ref. [84], with permission from Elsevier provided by Copyright Clearance Center.

The biggest advantage of *in situ* NMR characterization is to carry out real-time detection during cell functioning. However, this also generates a risk of interferences between the direct current of the functioning electrochemical cell circuits and the alternating current of the NMR. To reduce this risk, Grey et al. connected an automatic tuning matching cycler (ATMC) to filter the radio-frequency noise of the environment in the *in situ* NMR system [85]. As shown in Fig. 5.18, the ATMC *in situ* NMR probe system consists of a variable temperature unit (a1), an external automatic tuning/matching (eATM) controller (a2), radio frequency and electrochemical cell connectors (a3), Polyflon tuning/matching capacitors (a4), and a piezoelectric engine for automatic tuning/matching (a5). This ATMC probe enables an "on-the-fly" recalibration of the resonance circuit as well as a straightforward and highly shielded connection of electrochemical cells. Therefore, better NMR detectability and higher S/N ratios can be achieved by employing the ATMC.

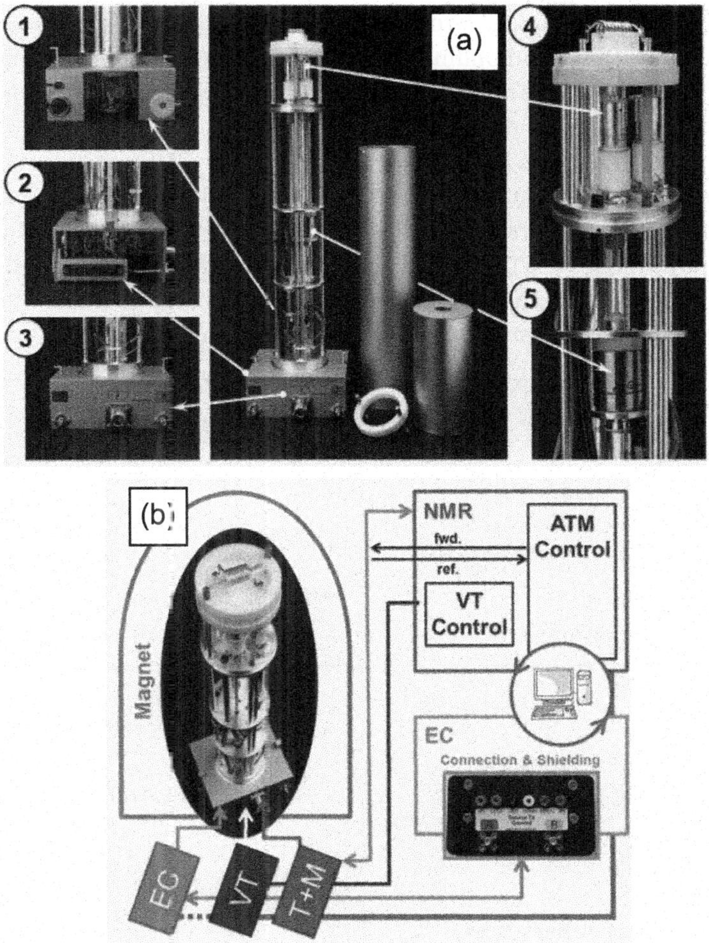

Figure 5.18 (a) Image of the ATMC *in situ* NMR probe system and components. (b) Schematic illustration of the ATMC NMR probe system. Adapted from Ref. [85], with permission from Elsevier provided by Copyright Clearance Center.

5.3.2 FTIR

FTIR is a technique based on the Fourier transform of the infrared (IR) spectrum [86]. When an IR beam passes through a sample, some IR radiation with specific frequencies of energy will be absorbed. The detected signal can form a unique IR spectrum, representing a molecular fingerprint of the sample. Therefore, the FTIR spectroscopy can provide specific information on molecular structure, chemical bonding, and molecular environment to probe the chemical nature of electrode materials and electrode/electrolyte interface species in the battery research [87].

In general, there are four working modes for FTIR spectroscopy: transmission, attenuated total reflectance (ATR), diffuse reflectance, and true specular reflectance/reflection absorption [88]. The four working modes of FTIR spectroscopy are schematically illustrated in Fig. 5.19. In the transmission FTIR mode, the energy of transmitted IR radiation is measured to obtain the spectrum [89].

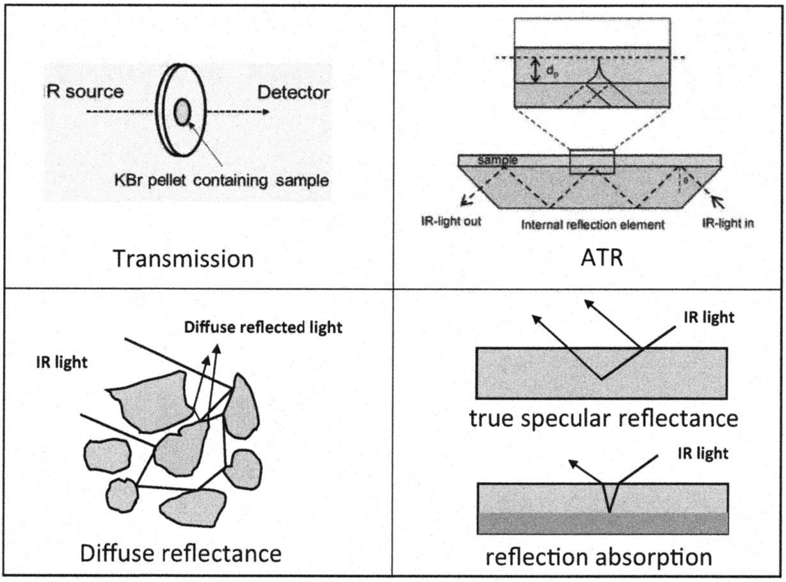

Figure 5.19 Schematic illustrations of four modes: transmission FTIR, ATR-FTIR, diffuse reflectance FTIR, and true specular reflectance/reflection absorption FTIR. Adapted from Refs. [89, 90], with permission from Multidisciplinary Digital Publishing Institute (MDPI) and Royal Society of Chemistry provided by Copyright Clearance Center.

In the ATR-FTIR mode, the electrode material is coated on the surface of an ATR crystal (internal reflection element) with high refractive index [90]. The IR beam enters the ATR crystal at an angle higher than the critical angle so that the total internal reflection takes place. After multiple internal reflectance, the IR beam exiting at the opposite end of the ATR crystal will be attenuated due to the absorbed evanescent wave by the sample. The intensity of the attenuated IR beam is recorded to generate the IR spectrum. The ATR-FTIR eliminates the problems arising from saturation effects in transmission mode, thus providing higher signal intensity and S/N ratio.

In the diffuse reflectance mode, the IR radiation comes to the sample and reflects off the particle surfaces, generating diffusion or scattering. The scattered energy of the altered IR radiation is recorded to produce the IR spectrum.

The true specular reflectance/reflection absorption FTIR works on the principle of reflective efficiencies. It is well known to note that every material has its own unique refractive index that varies with the frequency of light to which it is exposed. The biggest difference of this mode from others is that it does not measure the energy passing through the sample but reflected off the surface of a sample or its refractive index.

5.3.2.1 *Ex situ* FTIR

In *ex situ* FTIR spectroscopy experiments, electrode samples are analyzed from a disassembled cell which has been stopped at a certain state of charge. Since only the electrode materials are involved in the measurement, it eliminates the possible interference of the electrolyte. *Ex situ* FTIR measurement needs to be conducted in oxygen- and moisture-free atmosphere. The common methods are to take the measurements in glovebox or to seal the samples in specific cell. *Ex situ* FTIR spectroscopy can be realized by the preceding four modes. The FTIR modes can be selected based on the form of electrode materials. For the samples prepared in the form of KBr tablet, the transmission mode is usually used. For the electrode samples in the powder form, diffuse reflectance is a common mode. For the electrode samples coated on the current collector, reflection absorption mode can be used.

5.3.2.2 *In situ* FTIR

Compared with *ex situ* FTIR characterization, *in situ* FTIR measurement running during cell operation provides real-time probing of electrochemical processes. Same as *ex situ* FTIR, *in situ* FTIR measurement also needs to be conducted in oxygen- and moisture-free atmosphere. The first *in situ* FTIR characterization in the electrochemical processes was realized by Bewick and Kunimatsu et al. in the early 1980s [91, 92]. Figure 5.20 shows a typical *in situ* FTIR apparatus.

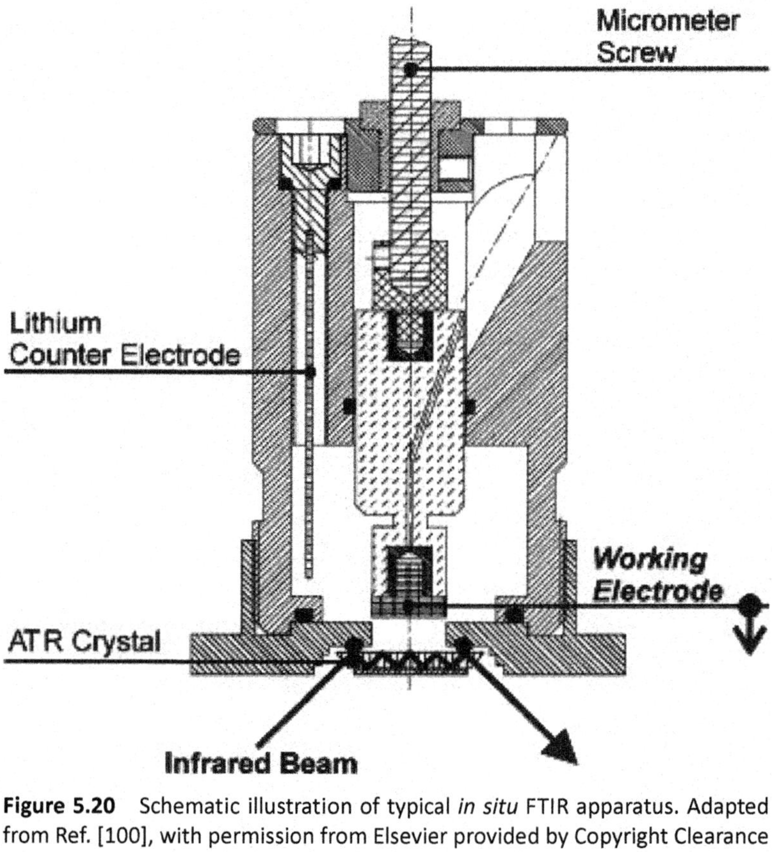

Figure 5.20 Schematic illustration of typical *in situ* FTIR apparatus. Adapted from Ref. [100], with permission from Elsevier provided by Copyright Clearance Center.

Most of the *in situ* FTIR characterizations employ diffuse reflectance or ATR mode [93–95]. However, diffuse reflectance or ATR mode usually suffers from low reflectance intensities and/or broad overlapping bands. In the *in situ* FTIR measurements, a thin-film electrode is usually needed to obtain enough reflection of the infrared beam, the preparation of which is still challenging. The thin-film electrode needs to face the window, though the signal from the bulk electrolyte can be minimized since the IR beam contacts the electrode and its surface species with little interaction with the electrolyte [96], this causes difficulties in fast charge and discharge process [97]. In the *in situ* ATR-FTIR spectroscopy, a Ge crystal is used to provide a high S/N ratio due to its high refractive index [98]. However, Ge may react with the alkaline metal counter to form alloy, causing damage to the ATR crystal [99].

Compared with *in situ* diffuse reflectance or ATR-FTIR spectroscopy, *in situ* transmission FTIR spectroscopy is seldomly reported. In the *in situ* transmission FTIR measurements, the interference of the signals from the electrolyte or other components is one of the main challenges [87, 100]. Therefore, the design of specific cell is the key. The principle of design is to make the path length of infrared radiation through the cell as short as possible. Burba et al. designed an *in situ* transmission FTIR spectroelectrochemical cell with the anode removed from the path of the infrared beam, as shown in Fig. 5.21 [101]. In this way, a separator does not need to completely cover the cathode, allowing the infrared radiation beaming through the cathode and electrolyte.

At this stage, a lot of space remains for improvement in the *in situ* FTIR technique. In practical *in situ* FTIR experiments, there are still some problems. For example, the growth of dendritic alkaline metal anode may damage the KBr window and short the *in situ* cell under charging. To alleviate this problem, it is proposed to replace the alkaline metal with an intercalation anode, such as graphite anode [101].

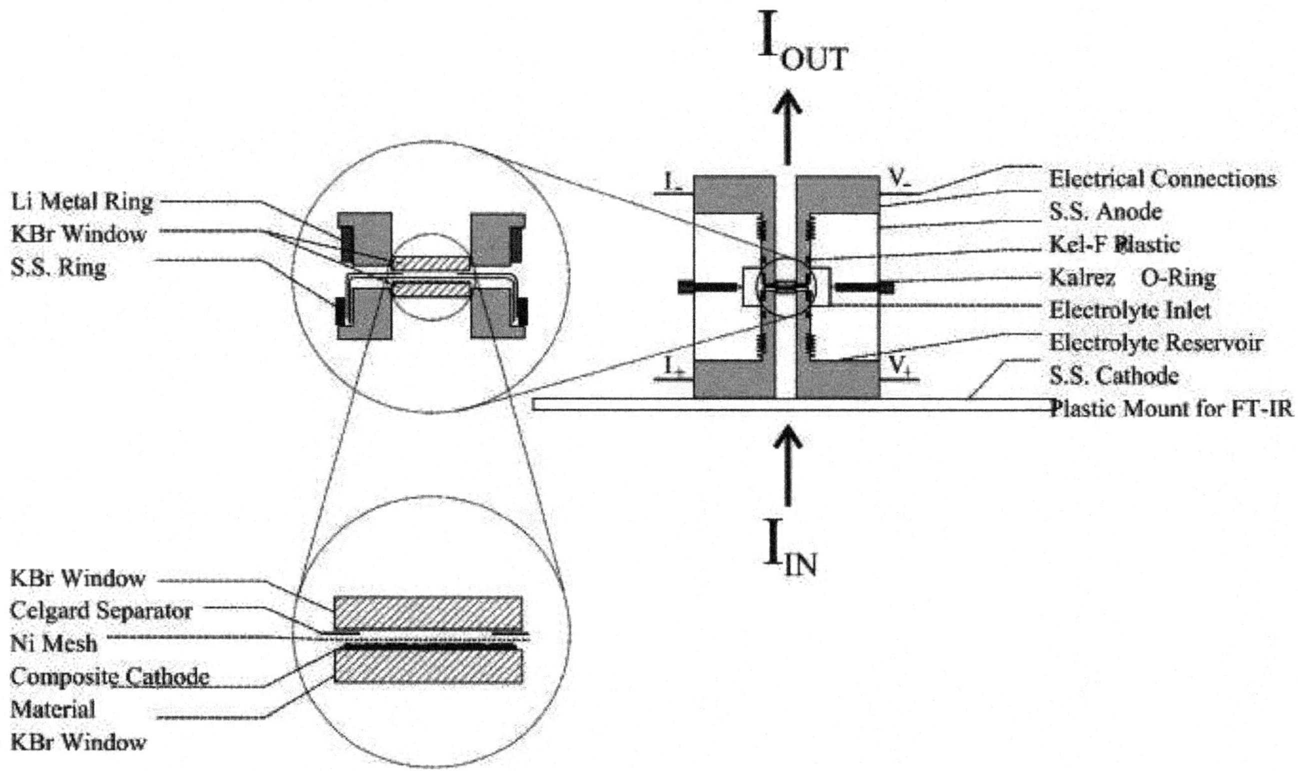

Figure 5.21 Schematic illustration of an *in situ* transmission FTIR apparatus. Adapted from Ref. [101], with permission from Elsevier provided by Copyright Clearance Center.

5.4 Summary and Outlook

The urgency of developing greener, cheaper, and more efficient energy storage systems has been widely recognized. NIBs based on polymer electrode materials have exhibited great potential due to the abundance of sodium sources, environmental friendliness, structure diversity, composition tenability, and multi-functionality of polymer materials. While it is believed that polymer NIBs represent a kind of promising future energy storage technology, the research still remains in its infancy.

On the basis of the current understanding of sodium-ion storage mechanisms in polymer electrode materials, carbonyl groups and/or heteroatoms in conjugated system and radicals can be the active sites for interaction with sodium ions. More mechanistic studies on the polymer electrode materials in the NIBs are still needed. The understanding of the sodium-ion storage mechanism can serve as the guiding principle to design and explore new polymer electrode materials for high-performance NIBs (Fig. 5.22).

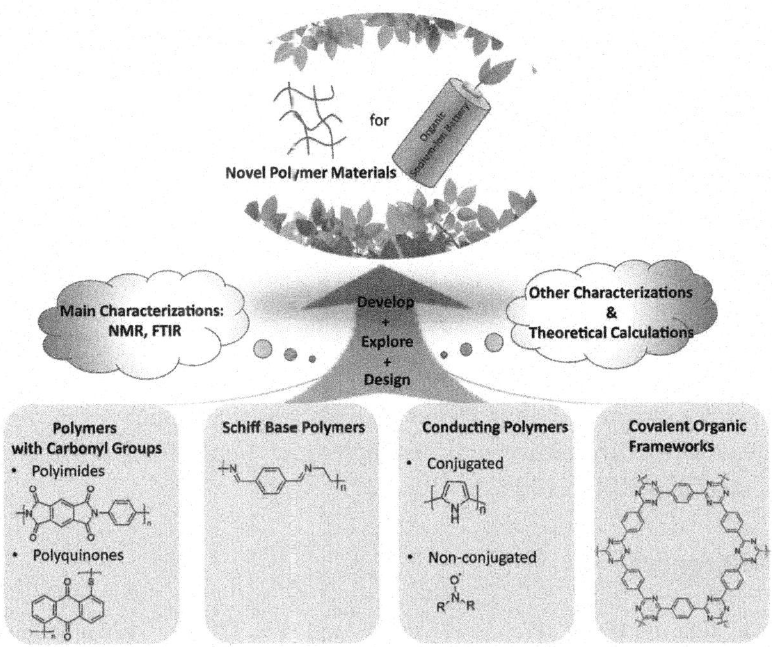

Figure 5.22 Summary and outlook of polymer electrode materials for sodium-ion batteries.

At this stage, some advanced characterization techniques have been applied in battery research to assist the mechanistic study and have made significant contributions. These characterizations, such as ssNMR and FTIR, are critical tools to understand the chemical bonding information and structural changes of electrode materials during charge and discharge. In particular, the *in situ* ssNMR and FTIR characterizations play a critical role in unveiling the redox reaction and true sodium-ion storage mechanism behind the electrochemical performance of the polymer batteries. However, there is still great space for improvement in the matching of the equipment and the *in situ* electrochemical cell, which can be operated under condition as closest to the practical battery. At the same time, it is also highly desired to explore and develop *in situ* techniques with increased sensitivity, higher resolution, and faster response, which is extremely important for ultrafast dynamic process. In addition, a combination of different characterizations and theoretical calculations is also needed to fully understand the complicated mechanism.

With the assistance of characterizations in the mechanistic study, the process of exploring new polymer electrode materials will be promoted. The development of rechargeable polymer batteries will certainly attract more and more research interest. It is very likely that emerging new polymer electrode materials with innovative structures and properties will pave the way toward future sustainable energy storage devices

Acknowledgments

Qinglan Zhao thanks the Hong Kong Postdoctoral Fellowship Scheme (HKUST PDFS2021-4S12) and acknowledges the integration of review paper published in materials (DOI: 10.3390/ma11122567) into this chapter. Minhua Shao acknowledges the support from the Department of Science and Technology of Guangdong Province, China (2019A050510043).

References

1. Hou, H., Qiu, X., Wei, W., Zhang, Y., and Ji, X. (2017). Carbon anode materials for advanced sodium-ion batteries, *Adv. Energy Mater.*, **7**, pp. 1602898.

2. Wang, L. P., Yu, L., Wang, X., Srinivasan, M., and Xu, Z. J. (2015). Recent developments in electrode materials for sodium-ion batteries, *J. Mater. Chem. A*, **3**, pp. 9353–9378.
3. Han, M. H., Gonzalo, E., Singh, G., and Rojo, T. (2015). A comprehensive review of sodium layered oxides: Powerful cathodes for Na-ion batteries, *Energy Environ. Sci.*, **8**, pp. 81–102.
4. Kim, H., Kim, H., Ding, Z., Lee, M. H., Lim, K., Yoon, G., and Kang, K. (2016). Recent progress in electrode materials for sodium-ion batteries, *Adv. Energy Mater.*, **6**, pp. 1600943.
5. Kim, S. W., Seo, D. H., Ma, X., Ceder, G., and Kang, K. (2012). Electrode materials for rechargeable sodium-ion batteries: Potential alternatives to current lithium-ion batteries, *Adv. Energy Mater.*, **2**, pp. 710–721.
6. Wang, P. F., You, Y., Yin, Y. X., and Guo, Y. G. (2018). Layered oxide cathodes for sodium-ion batteries: Phase transition, air stability, and performance, *Adv. Energy Mater.*, **8**, pp. 1701912.
7. Zhao, Q., Lu, Y., and Chen, J. (2017). Advanced organic electrode materials for rechargeable sodium-ion batteries, *Adv. Energy Mater.*, **7**, pp. 1601792.
8. Luo, W., Shen, F., Bommier, C., Zhu, H., Ji, X., and Hu, L. (2016). Na-ion battery anodes: Materials and electrochemistry, *Acc. Chem. Res.*, **49**, pp. 231–240.
9. Wang, L., Bi, X., and Yang S. (2016). Partially single-crystalline mesoporous Nb_2O_5 nanosheets in between graphene for ultrafast sodium storage, *Adv. Mater.*, **28**, pp. 7672–7679.
10. Song, Z. and Zhou, H. (2013). Towards sustainable and versatile energy storage devices: An overview of organic electrode materials, *Energy Environ. Sci.*, **6**, pp. 2280–2301.
11. Abouimrane, A., Weng, W., Eltaye, H., Cui, Y., Niklas, J., Poluektov, O., and Amine, K. (2012). Sodium insertion in carboxylate based materials and their application in 3.6 V full sodium cells, *Energy Environ. Sci.*, **5**, pp. 9632–9638.
12. Emanuelsson, R., Sterby, M., Strømme, M., and Sjödin, M. (2017). An all-organic proton battery, *J. Am. Chem. Soc.*, **139**, pp. 4828–4834.
13. Tarascon, J. M. (2010). Key challenges in future Li-battery research, *Phil. Trans. R. Soc. A*, **368**, pp. 3227–3241.
14. Wang, L., Wang, D., Zhang, F., and Jin, J. (2013). Interface chemistry guided long-cycle-life Li-S battery, *Nano Lett.*, **13**, pp. 4206–4211.
15. Wu, H., Meng, Q., Yang, Q., Zhang, M., Lu, K., and Wei, Z. (2015). Large-area polyimide/SWCNT nanocable cathode for flexible lithium-ion batteries, *Adv. Mater.*, **27**, pp. 6504–6510.

16. Zhang, Y., Wang, J., and Riduan, S. N. (2016). Strategies toward improving the performance of organic electrodes in rechargeable lithium (sodium) batteries, *J. Mater. Chem. A*, **4**, pp. 14902–14914.
17. Wilson, D., Stenzenberger, H. D., and Hergenrother, P. M. (1990) *Polyimides*, 1st ed. (Blackie and Son: Glasgow, Scotland).
18. Baumgartner, B., Bojdys, M. J., and Unterlass, M. M. (2014). Geomimetics for green polymer synthesis: Highly ordered polyimides via hydrothermal techniques, *Polym. Chem.*, **5**, pp. 3771–3776.
19. Song, Z., Zhan, H., and Zhou, Y. (2010). Polyimides: Promising energy-storage materials, *Angew. Chem.*, **122**, pp. 8622–8626.
20. Wang, H. G., Yuan, S., Ma, D. L., Huang, X. L., Meng, F. L., and Zhang, X. B. (2014). Tailored aromatic carbonyl derivative polyimides for high-power and long-cycle sodium-organic batteries, *Adv. Energy Mater.*, **4**, pp. 1301651.
21. Banda, H., Damien, D., Nagarajan, K., Hariharan, M., and Shaijumon, M. M. (2015). A polyimide based all-organic sodium ion battery, *J. Mater. Chem. A*, **3**, pp. 10453–10458.
22. Chen, L., Li, W., Wang, Y., Wang, C., and Xia, Y. (2014). Polyimide as anode electrode material for rechargeable sodium batteries, *RSC Adv.*, **4**, pp. 25369–25373.
23. Qin, H., Song, Z. P., Zhan, H., and Zhou, Y. H. (2014). Aqueous rechargeable alkali-ion batteries with polyimide anode, *J. Power Sources*, **249**, pp. 367–372.
24. Zhao, Q., Gaddam, R. R., Yang, D., Strounina, E., Whittaker, A. K., and Zhao, X. S. (2018). Pyromellitic dianhydride-based polyimide anodes for sodium-ion batteries, *Electrochim. Acta*, **265**, pp. 702–708.
25. Xu, F., Xia, J., Shi, W., and Cao, S. (2016). Sulfonyl-based polyimide cathode for lithium and sodium secondary batteries: Enhancing the cycling performance by the electrolyte, *Mater. Chem. Phys.*, **169**, pp. 192–197.
26. Xu, F., Xia, J., and Shi, W. (2015). Anthraquinone-based polyimide cathodes for sodium secondary batteries, *Electrochem. Commun.*, **60**, pp. 117–120.
27. Xu, F., Wang, H., Lin, J., Luo, X., Cao, S., and Yang, H. (2016). Poly(anthraquinonyl imide) as a high capacity organic cathode material for Na-ion batteries, *J. Mater. Chem. A*, **4**, pp. 11491–11497.
28. Li, Z., Zhou, J., Xu, R., Liu, S., Wang, Y., Li, P., Wu, W., and Wu, M. (2016). Synthesis of three-dimensional extended conjugated polyimide and application as sodium-ion battery anode, *Chem. Eng. J.*, **287**, pp. 516–522.

29. Deng, W., Liang, X., Wu, X., Qian, J., Cao, Y., Ai, X., Feng, J., and Yang, H. (2013). A low cost, all-organic Na-ion battery based on polymeric cathode and anode, *Sci. Rep.*, **3**, pp. 2671.

30. Song, Z., Qian, Y., Zhang, T., Otani, M., and Zhou, H. (2015). Poly(benzoquinonyl sulfide) as a high-energy organic cathode for rechargeable Li and Na batteries, *Adv. Sci.*, **2**, pp. 1500124.

31. Sun, T., Li, Z. J., Wang, H. G., Bao, D., Meng, F. L., and Zhang, X. B. (2016). A biodegradable polydopamine-derived electrode material for high-capacity and long-life lithium-ion and sodium-ion batteries, *Angew. Chem. Int. Ed.*, **55**, pp. 10662–10666.

32. Wu, Y., Zeng, R., Nan, J., Shu, D., Qiu, Y., and Chou, S. L. (2017). Quinone electrode materials for rechargeable lithium/sodium ion batteries, *Adv. Energy Mater.*, **7**, pp. 1700278.

33. Wu, D., Huang, Y., and Hu, X. (2016). A sulfurization-based oligomeric sodium salt as a high-performance organic anode for sodium ion batteries, *Chem. Commun.*, **52**, pp. 11207–11210.

34. Tunçel, M., Özbülbül, A., and Serı, S. (2008). Synthesis and characterization of thermally stable schiff base polymers and their copper(II), cobalt(II) and nickel(II) complexes, *React. Funct. Polym.*, **68**, pp. 292–306.

35. Castillo-Martínez, E., Carretero-González, J., and Armand, M. (2014). Polymeric Schiff bases as low-voltage redox centers for sodium-ion batteries, *Angew. Chem Int. Ed.*, **53**, pp. 5341–5345.

36. Fernández, N., Sánchez-Fontecoba, P., Castillo-Martínez, E., Carretero-González, J., Rojo, T., and Armand, M. (2018). Polymeric redox-active electrodes for sodium-ion batteries, *ChemSusChem*, **11**, pp. 311–319.

37. Gutch, P. K., Banerjee, S., Gupta, D. C., and Jaiswal, D. K. (2001). Poly-Schiff bases. V. Synthesis and characterization of novel soluble fluorine-containing polyether azomethines, *J. Polym. Sci. Poly. Chem.*, **39**, pp. 383–388.

38. Li, X., Jiao, Y., and Li, S. (1991). The syntheses, properties and application of new conducting polymers, *Eur. Polym. J.*, **27**, pp. 1345–1351.

39. Shacklette, L. W., Toth, T. E., Murthy, N. S., and Baughman, R. H. (1985). Polyacetylene and polyphenylene as anode materials for nonaqueous secondary batteries, *J. Electrochem. Soc.*, **132**, pp. 1529–1535.

40. Zhu, X., Zhao, R., Deng, W., Ai, X., Yang, H., and Cao, Y. (2015). An all-solid-state and all-organic sodium-ion battery based on redox-active polymers and plastic crystal electrolyte, *Electrochim. Acta*, **178**, pp. 55–59.

41. Zhou, M., Li, W., Gu, T., Wang, K., Cheng, S., and Jiang, K. (2015). A sulfonated polyaniline with high density and high rate na-storage performances as a flexible organic cathode for sodium ion batteries, *Chem. Commun.*, **51**, pp. 14354–14356.

42. Shen, Y. F., Yuan, D. D., Ai, X. P., Yang, H. X., and Zhou, M. (2014). Poly(diphenylaminesulfonic acid sodium) as a cation-exchanging organic cathode for sodium batteries, *Electrochem. Commun.*, **49**, pp. 5–8.

43. Zhu, L., Shen, Y., Sun, M., Qian, J., Cao, Y., Ai, X., and Yang, H. (2013). Self-doped polypyrrole with ionizable sodium sulfonate as a renewable cathode material for sodium ion batteries, *Chem. Commun.*, **49**, pp. 11370–11372.

44. Shirakawa, H., Louis, E. J., MacDiarmid, A. G., Chiang, C. K., and Heeger, A. J. (1977). Synthesis of electrically conducting organic polymers: halogen derivatives of polyacetylene, $(CH)_x$, *J. Chem. Soc. Chem. Commun.*, pp. 578–580.

45. Zhao, R., Zhu, L., Cao, Y., Ai, X., and Yang, H. X. (2012). An aniline-nitroaniline copolymer as a high capacity cathode for Na-ion batteries, *Electrochem. Commun.*, **21**, pp. 36–38.

46. Zhou, M., Xiong, Y., Cao, Y., Ai, X., and Yang, H. (2013). Electroactive organic anion-doped polypyrrole as a low cost and renewable cathode for sodium-ion batteries, *J. Polym. Sci. Poly. Phys.*, **51**, pp. 114–118.

47. Zhou, M., Zhu, L., Cao, Y., Zhao, R., Qian, J., Ai, X., and Yang, H. (2012). $Fe(CN)_6^{-4}$-doped polypyrrole: A high-capacity and high-rate cathode material for sodium-ion batteries, *RSC Adv.*, **2**, pp. 5495–5498.

48. Chen, X., Liu, L., Yan, Z., Huang, Z., Zhou, Q., Guo, G., and Wang, X. (2016). The excellent cycling stability and superior rate capability of polypyrrole as the anode material for rechargeable sodium ion batteries, *RSC Adv.*, **6**, pp. 2345–2351.

49. Su, D., Zhang, J., Dou, S., and Wang, G. (2015). Polypyrrole hollow nanospheres: Stable cathode materials for sodium-ion batteries, *Chem. Commun.*, **51**, pp. 16092–16095.

50. Liu, S., Wang, F., Dong, R., Zhang, T., Zhang, J., Zhuang, X., Mai, Y., and Feng, X. (2016). Dual-template synthesis of 2D mesoporous polypyrrole nanosheets with controlled pore size, *Adv. Mater.*, **28**, pp. 8365–8370.

51. Han, S. C., Bae, E. G., Lim, H., and Pyo, M. (2014). Non-crystalline oligopyrene as a cathode material with a high-voltage plateau for sodium ion batteries, *J. Power Sources*, **254**, pp. 73–79.

52. Nakahara, K., Iwasa, S., Satoh, M., Morioka, Y., Iriyama, J., Suguro, M., and Hasegawa, E. (2002). Rechargeable batteries with organic radical cathodes, *Chem. Phys. Lett.*, **359**, pp. 351–354.
53. Dai, Y., Zhang, Y., Gao, L., Xu, G., and Xie, J. (2010). A sodium ion based organic radical battery, *Electrochem. Solid St.*, **13**, pp. A22–A24.
54. Kim, J. K., Kim, Y., Park, S., Ko, H., and Kim, Y. (2016). Encapsulation of organic active materials in carbon nanotubes for application to high-electrochemical-performance sodium batteries, *Energy Environ. Sci.*, **9**, pp. 1264–1269.
55. Janoschka, T., Hager, M. D., and Schubert, U. S. (2012). Powering up the future: Radical polymers for battery applications, *Adv. Mater.*, **24**, pp. 6397–6409.
56. Bugnon, L., Morton, C. J. H., Novak, P., Vetter, J., and Nesvadba, P. (2007). Synthesis of poly(4-methacryloyloxy-TEMPO) via group-transfer polymerization and its evaluation in organic radical battery, *Chem. Mater.*, **19**, pp. 2910–2914.
57. Ding, S. Y. and Wang, W. (2013). Covalent organic frameworks (COFs): From design to applications, *Chem. Soc. Rev.*, **42**, pp. 548–568.
58. Kim, M. S., Lee, W. J., Paek, S. M., and Park, J. K. (2018). Covalent organic nanosheets as effective sodium-ion storage materials, *ACS Appl. Mater. Interfaces*, **10**, pp. 32102–32111.
59. Sakaushi, K., Hosono, E., Nickerl, G., Gemming, T., Zhou, H., Kaskel, S., and Eckert, J. (2013). Aromatic porous-honeycomb electrodes for a sodium-organic energy storage device, *Nat. Commun.*, **4**, pp. 1–7.
60. Sakaushi, K., Hosono, E., Nickerl, G., Zhou, H., Kaskel, S., and Eckert, J. (2014). Bipolar porous polymeric frameworks for low-cost, high-power, long-life all-organic energy storage devices, *J. Power Sources*, **245**, pp. 553–556.
61. Zhang, S., Huang, W., Hu, P., Huang, C., Shang, C., Zhang, C., Yang, R., and Cui, G. (2015). Conjugated microporous polymers with excellent electrochemical performance for lithium and sodium storage, *J. Mater. Chem. A*, **3**, pp. 1896–1901.
62. Duan, H., Lyu, P., Liu, J., Zhao, Y., and Xu, Y. (2019). Semiconducting crystalline two-dimensional polyimide nanosheets with superior sodium storage properties, *ACS Nano*, **13**, pp. 2473–2480.
63. Peng, C., Ning, G. H., Su, J., Zhong, G., Tang, W., Tian, B., Su, C., Yu, D., Zu, L., Yang, J., Ng, M. F., Hu, Y. S., Yang, Y., Armand, M., and Loh, K. P. (2017). Reversible multi-electron redox chemistry of π-conjugated N-containing heteroaromatic molecule-based organic cathodes, *Nature Energy*, **2**, pp. 17074.

64. Clément, R. J., Billaud, J., Armstrong, A. R., Singh, G., Rojo, T., Bruce, P. G., and Grey, C. P. (2016). Structurally stable Mg-doped P2-$Na_{2/3}Mn_{1-y}Mg_yO_2$ sodium-ion battery cathodes with high rate performance: Insights from electrochemical, NMR and diffraction studies, *Energy Environ. Sci.*, **9**, pp. 3240–3251.

65. Han, M. H., Gonzalo, E., Sharma, N., López del Amo, J. M., Armand, M., Avdeev, M., Garitaonandia, J. J. S., and Rojo, T. (2016). High-performance P2-phase $Na_{2/3}Mn_{0.8}Fe_{0.1}Ti_{0.1}O_2$ cathode material for ambient-temperature sodium-ion batteries, *Chem. Mater.*, 28, pp. 106–116.

66. Liu, Z., Hu, Y. Y., Dunstan, M. T., Huo H., Hao, X., Zou, H., Zhong, G., Yang, Y., and Grey, C. P. (2014). Local structure and dynamics in the Na ion battery positive electrode material $Na_3V_2(PO_4)_2F_3$, *Chem. Mater.*, **26**, pp. 2513–2521.

67. Shadike, Z., Zhao, E., Zhou, Y. N., Yu, X., Yang, Y., Hu, E., Bak, S., Gu L., and Yang, X. Q. (2018). Advanced characterization techniques for sodium-ion battery studies, *Adv. Energy Mater.*, **8**, pp. 1702588.

68. Billaud, J., Clément, R. J., Armstrong, A. R., Canales-Vázquez, J., Rozier, P., Grey, C. P., and Bruce, P. G. (2014). β-$NaMnO_2$: A high-performance cathode for sodium-ion batteries, *J. Am. Chem. Soc.*, **136**, pp. 17243–17248.

69. Blanc, F., Leskes, M., and Grey, C. P. (2013). *In situ* solid-state NMR spectroscopy of electrochemical cells: Batteries, supercapacitors, and fuel cells, *Acc. Chem. Res.*, **46**, pp. 1952–1963.

70. Pecher, O., Carretero-González, J., Griffith, K. J., and Grey, C. P. (2017). Materials' methods: NMR in battery research, *Chem. Mater.*, **29**, pp. 213–242.

71. Harks, P. P. R. M. L., Mulder, F. M., and Notten, P. H. L. (2015). *In situ* methods for Li-ion battery research: A review of recent developments, *J. Power Sources*, **288**, 92–105.

72. Magusin, P. C. M. M., Seymour, I. D., Pecher, O., and Grey, C. P. (2018). NMR studies of electrochemical storage materials. In ed. Hodgkinson, P., *New Developments in NMR*, vol. 15 (Royal Society of Chemistry), pp. 322–355.

73. Zhao, Q., Zhao, W., Zhang, C., Wu, Y., Yuan, Q., Whittaker, A. K., and Zhao, X. S. (2020). Sodium-ion storage mechanism in triquinoxalinylene and a strategy for improving electrode stability, *Energy Fuels*, **34**, pp. 5099–5105.

74. Grimmer, A. R. and Blümich, B. (1994). Introduction to solid-state NMR. In ed. Blümich, B., *Solid-State NMR I Methods*, vol. 30 (Springer, Berlin, Heidelberg), pp. 1–62.

75. Gotoh, K., Takeda, K., Lerner, M. M., Sueishi, Y., Maruyama, S., Goto, A., Tansho, M, Ohki, S., Hashi, K., Shimizu, T., and Ishida, H. (2011). Analysis of bis(trifluoromethylsulfonyl) imide-doped paramagnetic graphite intercalation compound using 19F very fast magic angle spinning nuclear magnetic resonance, *Carbon*, **49**, pp. 4064–4066.

76. Li, Q., Liu, Z., Zheng, F., Liu, R., Lee, J., Xu, G. L., Zhong, G., Hou, X., Fu, R., Chen, Z., Amine, K., Mi, J., Wu, S., Grey, C. P., and Yang, Y. (2018). Identifying the structural evolution of the sodium ion battery Na_2FePO_4F cathode, *Angewandte Chemie*, **130**, pp. 12094–12099.

77. Song, Z., Antzutkin, O. N., Feng, X., and Levitt, M. H. (1993). Sideband suppression in magic-angle-spinning NMR by a sequence of 5 π pulses, *Solid State Nucl. Magn. Reson.*, **2**, pp. 143–146.

78. Lee, Y. J. and Grey, C. P. (2002). Determining the lithium local environments in the lithium manganates $LiZn_{0.5}Mn_{1.5}O_4$ and Li_2MnO_3 by analysis of the ^{6}Li MAS NMR spinning sideband manifolds, *J. Phys. Chem. B*, **106**, pp. 3576–3582.

79. Rathke, J. W., Klingler, R. J., Gerald II, R. E., Kramarz, K. W., and Woelk, K. (1997). Toroids in NMR Spectroscopy, *Progr. NMR Spectr.*, **30**, pp. 209–253.

80. Rathke, J. W., Klingler, R. J., Woelk, K., and Gerald II, R. E. (2000). Near electrode imager, *US Patent* 6046592, issued 4 April.

81. Gerald, R., Sanchez, J., Johnson, C. S., Klingler, R. J., and Rathke, J. W. (2001). *In situ* nuclear magnetic resonance investigations of lithium ions in carbon electrode materials using a novel detector, *J. Phys.: Condens. Matter.*, **13**, pp. 8269–8285.

82. Trease, N. M., Zhou, L., Chang, H. J., Zhu, B. Y., and Grey, C. P. (2012). *In situ* NMR of lithium ion batteries-bulk susceptibility effects and practical considerations, *Solid State Nucl. Magn. Reson.*, **42**, pp. 62–70.

83. Key, B., Bhattacharyya, R., Morcrette, M., Seznec, V., Tarascon, J. M., and Grey, C. P. (2009). Real-time NMR investigations of structural changes in silicon electrodes for lithium-ion batteries, *J. Am. Chem. Soc.*, **131**, pp. 9239–9249.

84. Poli, F., Kshetrimayum, J. S., Monconduit, L., and Letellier, M. (2011). New cell design for in-situ NMR studies of lithium-ion batteries, *Electrochem. Commun.*, **13**, pp. 1293–1295.

85. Pecher, O., Bayley, P. M., Liu, H., Liu, Z., Trease, N. M., and Grey, C. P. (2016). Automatic tuning matching cycler (ATMC) *in situ* NMR spectroscopy as a novel approach for real-time investigations of Li-and Na-ion batteries, *J. Magn. Reson.*, **265**, pp. 200–209.

86. Gerwert, K. and Kötting, C. (2010). *Encyclopedia of Life Sciences* (ELS), Fourier Transform Infrared (FTIR) Spectroscopy (John Wiley & Sons, New York) DOI: 10.1002/9780470015902.a0003112.pub2.

87. Deng, Y., Dong, S., Li, Z., Jiang, H., Zhang, X., and Ji, X. (2018). Applications of conventional vibrational spectroscopic methods for batteries beyond Li-ion, *Small Methods*, **2**, pp. 1700332.

88. Mohamed, M. A., Jaafar, J., Ismail, A. F., Othman, M. H. D., and Rahman, M. A. (2017). Fourier transform infrared (FTIR) spectroscopy. In eds. Hilal, N., Ismail, A. F., Matsuura, T., and Oatley-Radcliffe, D., *Membrane Characterization*, vol. 1 (Elsevier), pp. 3–29.

89. Chen, Y., Zou, C., Mastalerz, M., Hu, S., Gasaway, C., and Tao, X. (2015). Applications of micro-fourier transform infrared spectroscopy (FTIR) in the geological sciences: A review, *Int. J. Mol. Sci.*, **16**, pp. 30223–30250.

90. Mojet, B. L., Ebbesen, S. D., and Lefferts, L. (2010). Light at the interface: The potential of attenuated total reflection infrared spectroscopy for understanding heterogeneous catalysis in water, *Chem. Soc. Rev.*, **39**, pp. 4643–4655.

91. Bewick, A. and Kunimatsu, K. (1980). Infrared spectroscopy of the electrode–electrolyte interphase, *Surf. Sci.*, **101**, pp. 131–138.

92. Bewick, A., Kunimatsu, K., and Pons, B. S. (1980). Infrared spectroscopy of the electrode–electrolyte interphase, *Electrochim. Acta*, **25**, pp. 465–468.

93. Pérez-Villar, S., Lanz, P., Schneider, H., and Novák, P. (2013). Characterization of a model solid electrolyte interphase/carbon interface by combined *in situ* Raman/Fourier transform infrared microscopy, *Electrochim. Acta*, **106**, pp. 506–515.

94. Zhang, Y., Katayama, Y., Tatara, R., Giordano, L., Yu, Y., Fraggedakis, D., Sun, J. G., Maglia, F., Jung, R., Bazant, M. Z., and Yang, S. H. (2020). Revealing electrolyte oxidation via carbonate dehydrogenation on Ni-based oxides in Li-ion batteries by *in situ* fourier transform infrared spectroscopy. *Energy Environ. Sci.*, **13**, pp. 183–199.

95. Haregewoin, A. M., Shie, T. D., Lin, S. D., Hwang, B. J., and Wang, F. M. (2013). An effective *in situ* drifts analysis of the solid electrolyte interface in lithium-ion battery, *ECS Trans.*, **53**, pp. 23–32.

96. Goren, E., Chusid, O., and Aurbach, D. (1991). The application of *in situ* FTIR spectroscopy to the study of surface films formed on lithium and noble metals at low potentials in Li battery electrolytes, *J. Electrochem. Soc.*, **138**, pp. L6.

97. Matsui, M., Deguchi, S., Kuwata, H., and Imanishi, N. (2015). In-operando FTIR spectroscopy for composite electrodes of lithium-ion batteries, *Electrochem.*, **83**, pp. 874–878.

98. Matsushita, T., Dokko, K., and Kanamura, K. (2005). *In situ* FTIR measurement for electrochemical oxidation of electrolyte with ethylene carbonate and diethyl carbonate on cathode active material used in rechargeable lithium batteries, *Journal of Power Sources*, **146**, pp. 360–364.

99. Köhler, T., Hanzig, J., and Koroteev, V. (2018). Optical spectroscopy as a tool for battery research, *Phys. Sci. Rev.*, **4**, pp. 20170154.

100. Novák, P., Goers, D., Hardwick, L., Holzapfel, M., Scheifele, W., Ufheil, J., and Würsig, A. (2005). Advanced *in situ* characterization methods applied to carbonaceous materials, *Journal of Power Sources*, **146**, pp. 15–20.

101. Burba, C. M. and Frech, R. (2006). *In situ* transmission FTIR spectroelectrochemistry: A new technique for studying lithium batteries, *Electrochim. Acta*, **52**, pp. 780–785.

Chapter 6

Transition Metal Dichalcogenides as Active Anode Materials for Sodium-Ion Batteries

Davi Marcelo Soares, Santanu Mukherjee, and Gurpreet Singh
Mechanical and Nuclear Engineering, Kansas State University, 3002 Rathbone Hall, 1701B Platt St., Manhattan, KS 66506, USA
soares@ksu.edu

6.1 Introduction

In the annual meeting of the American Physical Society in 1959, Richard Feynman gave a remarkable speech titled "There's plenty of room at the bottom" in which he emphasized the interest in investigating materials in their layered forms—in his words the "*right layers*" [1]. Four years later, in an attempt to obtain monolayer molybdenum disulfide (MoS_2), R. F. Frindt reported the preparation of single crystals of MoS_2 less than 100 Å thick [2]. Subsequently, in

Handbook of Sodium-Ion Batteries: Materials and Characterization
Edited by Rohit R. Gaddam and X. S. (George) Zhao

ISBN 978-981-4968-15-7 (Hardcover), 978-1-003-30874-4 (eBook)
www.jennystanford.com

1986, R. F. Frindt synthesized a single-layer MoS_2 by intercalation with lithium [3]. Nonetheless, these publications remained unobserved for decades.

It was only after 2004 when Novoselov et al. reported the isolation of a single graphite layer stable under ambient conditions—yielding graphene—that the research community started looking more closely into two-dimensional (2D) materials [4, 5]. Soon after, the researchers discovered that, similar to graphene, 2D materials exhibit unique properties usually not found in their bulk (three-dimensional, 3D) counterparts. Some examples of these properties are band gap dependence upon the number of layers, 2D superconductivity, and higher mechanical strength [6]. Arising from such interest, layered materials are currently an ever-growing family, whose produced compounds range from elemental 2D materials (e.g., graphene, silicene, and borophene) to transition metal dichalcogenides (TMDs) (e.g., MoS_2, VSe_2, and WS_2) [7–9]. This massive interest is illustrated in Fig. 6.1, which depicts the recent research efforts, i.e., articles, books, and open literature, to study molybdenum disulfide (MoS_2) [2].

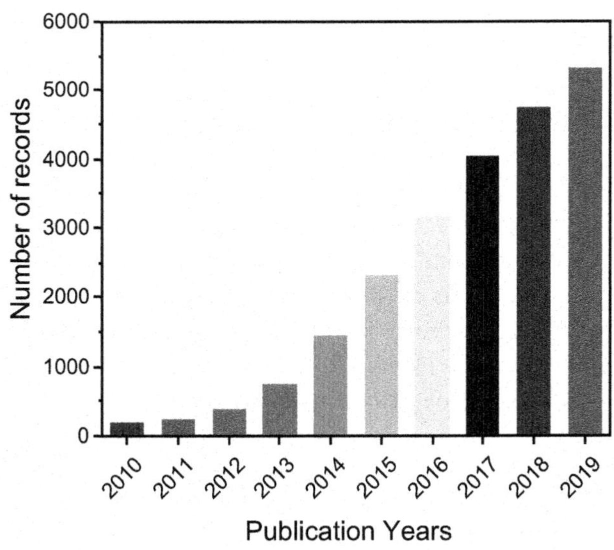

Figure 6.1 Number of published studies in the literature from 2010 until 2019 using the keyword "MoS_2," the most studied TMD species. Web of Science, Jul. 14, 2020.

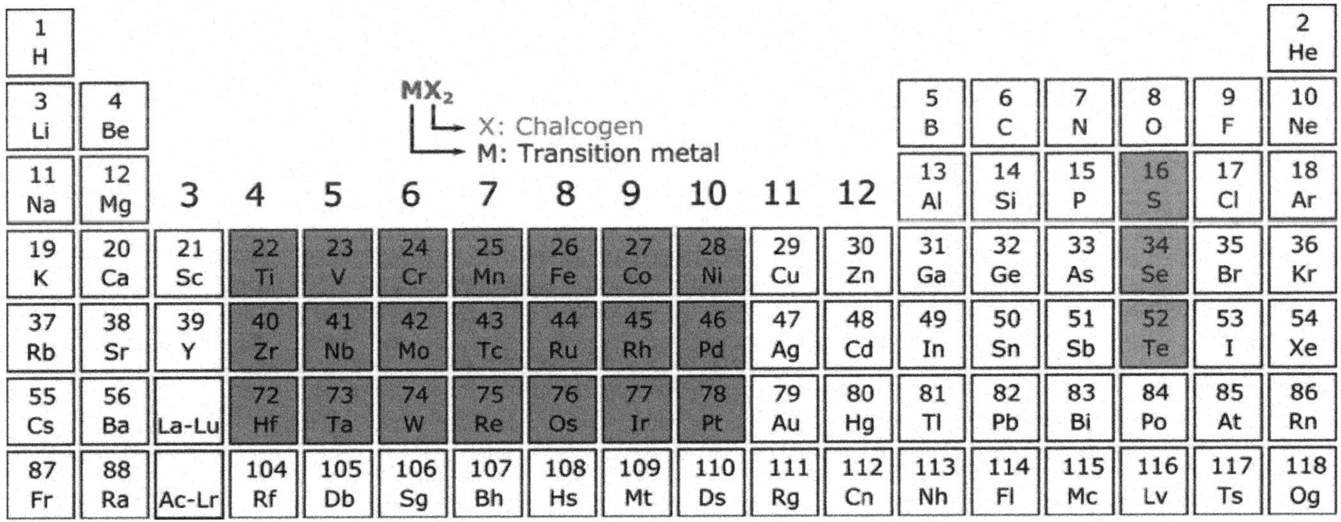

Figure 6.2 Periodic table showing the elements of TMDs: transition metal elements (highlighted in blue) and chalcogens (highlighted in red).

TMDs are materials identified by the formula MX_2, where M is a transition metal, from groups 4–10 in the periodic table, and X a chalcogen atom, both identified in Fig. 6.2. In addition, from the 60 TMDs identified, about 40 of them present layered structure. In particular, the non-layered TMDs commonly present the transition metal atom from groups 8–10 in the periodic table (Fig. 6.2) [10, 11].

Concerning the layered structures, TMDs present the transition metal atom sandwiched between chalcogen atoms layers, where the M-X atoms are bonded together by strong covalent interactions, yet interactions between the sandwiched layers—formed by X-M-X—are weakly bond together by weak van der Waals forces. Since van der Waals forces are interactions that require less energy to be broken, mechanical cleavage along the surface of a TMD crystal can be used to produce few-layers TMDs [2].

In addition, another property that makes TMDs singular materials is their polymorphism [12]. TMDs are commonly found in 1T, 2H, and 3R phases, where the numbers are ascribed to the number of X-M-X arrangements per repeat unit and the letters correspond to coordination of the transition metal atom, standing for tetragonal, hexagonal, and rhombohedral, respectively [13]. It is also worth noting that 1T' and T_d phases are also found in the literature and are assigned to monoclinic centrosymmetric unit cell and orthorhombic non-centrosymmetric unit cell, respectively (see Section 6.2 for further information regarding phases and phase-engineering approach) [14].

With respect to properties and TMD phases, MoS_2 is found in nature in the 2H phase, which is semiconducting [15]. On the other hand, 1T-MoS_2—not found in nature—presents metallic conductivity up to 10^7 times higher than 2H-MoS_2 and is hydrophilic, thus requiring phase transition methods from 2H-MoS_2 for its synthesis [16, 17].

As a consequence, synthetic routes are employed to exploit TMDs thermodynamically not preferred phases and to produce high phase-purity TMD crystals [16, 18]. These synthetic routes are divided into two major categories: *top-down* (e.g., chemical exfoliation or mechanical cleavage of bulk crystals) and *bottom-up* (e.g., chemical vapor deposition [CVD], and atomic layer deposition [ALD]) [19], as presented in Section 6.4. Based on this plethora of synthesis methods, properties, and chemistry, TMDs are expected to go beyond graphene in applications such as sensing, catalysis, and energy storage [11].

Regarding energy storage, in 1991 Sony introduced the lithium-ion full-cell battery to the market. The device consisted of a carbon-based anode and $LiCoO_2$ cathode [20]. Nowadays, this configuration is still being used in mobile devices and power tools, therefore making it an established technology [21]. However, since 2000, the prices of lithium (a commodity product) have been increasing, an inevitable result owing to the limited world resources of the metal, estimated in 14 million tons [22]. Moreover, meager resources of lithium may also lead to geopolitical issues because almost 90% of the reserves are concentrated in only five countries (Australia, Argentina, Bolivia, Chile, and China), thus raising economic concerns for multi-billion-dollar sectors, such as electronics and automobile.

In this regard, sodium-ion batteries (SIBs) are a feasible alternative to LIBs because sodium can be obtained from low-cost and abundant salts such as NaCl, Na_2S04, and Na_2CO_3 [23]. Moreover, SIB anode electrodes may use aluminum current collector, hence with the potential to provide cost competitive and light-weight batteries [23]. Nevertheless, graphite exhibits negligible capacity toward sodium (e.g., NaC_{70}); in comparison, graphite presents theoretical specific capacity of 372 mAh/g for Li^+ (e.g., LiC_6) [24]. Therefore, alternative layered materials are currently studied, such as elemental 2D materials (e.g., phosphorene and silicene) and TMDs.

In addition to having several species and a plethora of properties to be exploited, some layered TMDs present diffusion barrier energies for Na^+ lower than graphite presents for Li^+. As an example, 1H-VS_2 Na^+ diffusion barrier is approximately 0.085 eV, whereas graphite is approximately 0.2 eV [25]. Likewise, some TMD species present adequate OCP (open-circuit potential), within 0 and 1 V versus Na/Na^+, a relevant aspect for achieving both high energy densities and prevention of dendrites formation [23, 25, 26].

The lamellar structure of TMDs is also another attractive feature, because it may provide plenty of sites for electrochemical activity, along with an interlayer spacing that may efficiently intercalate the larger Na^+, unlike graphite. Additionally, TMDs may accommodate structural changes upon insertion and extraction of ions in their structure [27–29].

Nevertheless, given that several TMD thermodynamically stable phases are semiconductive in nature, one of the research challenges involving TMDs for electrochemical energy storage lies in improving their electronic conductivity through a scalable phase transition method, approach discussed in Section 6.2.

To summarize, the present chapter is organized in sections describing the structure and electronic properties of TMDs (Section 6.3), preparation methods of TMDs (Section 1.4), and TMDs as anode electrodes for SIBs (Section 1.5).

6.2 Structure of Transition Metal Dichalcogenides

Bulk TMDs present ordered monolayers stacked by weak van der Waals interactions, similar to graphene [30]. Another important feature exhibited by TMDs is polymorphism. In this regard, the coordination of the transition metal atom of TMDs plays an important role as TMDs may crystalize in different phases, for instance, 3R, 2H, 1H, 1T, 1T', and T_d, where the number in the nomenclature accounts for the number of X-M-X arrangements in the unit cell. For clarification, Figure 6.3 illustrates the mentioned phases [31].

First, the 3R phase presents rhombohedral symmetry with trigonal prismatic metal coordination, where three layers of X-M-X atoms per unit cell are found (Fig. 6.3a,b). Analogously, the stacking sequence is AbA CaC BcB, where the upper- and lower-case letters represent the chalcogen and metal atoms, respectively [11]. Likewise, for 2H phase the metal coordination is trigonal prismatic with hexagonal symmetry (D_{3h} point group), as shown in Fig. 6.3c, yet the stacking sequence is given by AbA BaB, as presented in Fig. 6.3d. It is worth noting that in accordance with Fig. 6.3c,e, for a single-layer TMD with trigonal prismatic phase, 1H is the reported structure [32]. As an example of the relevance of the phases, MoS_2 phases 3R and 2H are semiconducting, whereas 1T and 1T' are semimetallic [33]. More information about electronic properties is presented in Section 6.3.

The 1T phase displays tetragonal symmetry (D_{3d} point group) and octahedral coordination of metal atoms, with stacking sequence AbC, as exhibited by Fig. 6.3f and Fig. 6.3g, respectively. It is worth noting that compared with 1H- and 2H-, the tetragonal symmetry phase exhibits a shift of 60° in the chalcogen atoms in the arrangement X-M-X. Further, as presented In Fig. 6.3h and Fig. 6.3i, the 1T' phase is ascribed to distorted octahedral metal coordination. Examples of TMDs that present 1T' phases are telluride-based TMDs, e.g., $MoTe_2$ and WTe_2 [34]. Additionally, phases 1T' and T_d are fairly similar; however, they present centrosymmetric monoclinic and non-centrosymmetric orthorhombic unit cells, respectively. Figure 6.3h–k illustrates the differences of them.

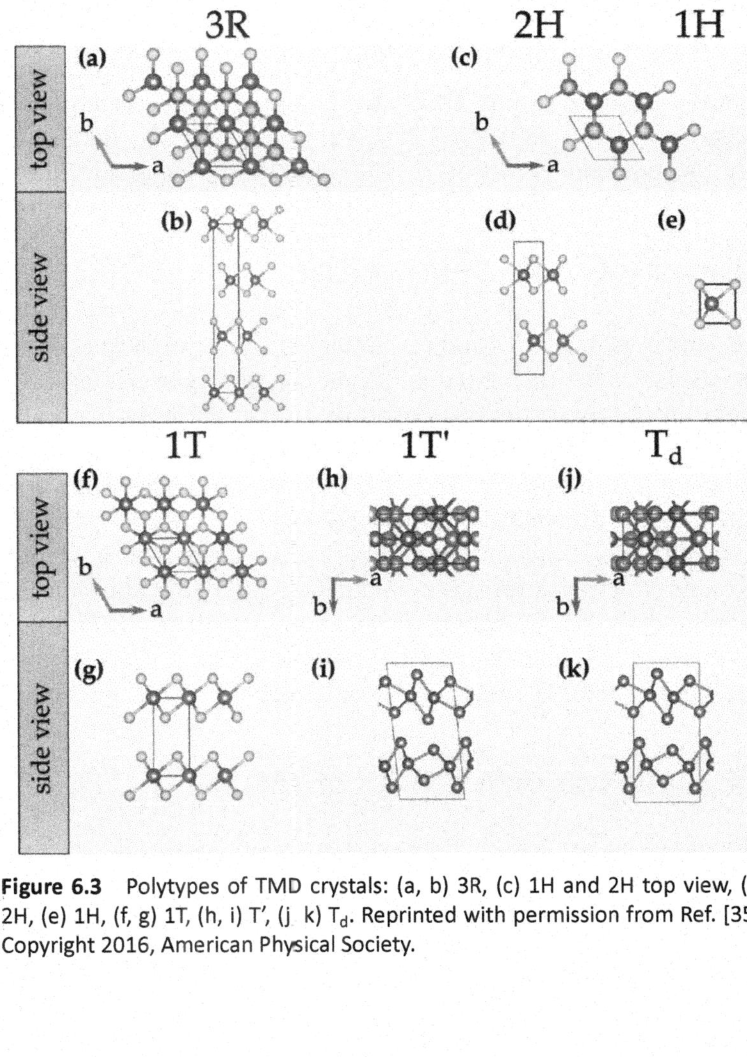

Figure 6.3 Polytypes of TMD crystals: (a, b) 3R, (c) 1H and 2H top view, (d) 2H, (e) 1H, (f, g) 1T, (h, i) T', (j k) T_d. Reprinted with permission from Ref. [35]. Copyright 2016, American Physical Society.

6.3 Electronic Properties of TMDs

The electronic properties of TMDs are dependent upon transition metal coordination—an aspect that also has implications in the phase—and the d-orbital filling [11]. For the trigonal prismatic metal coordination (i.e., 3R, 2H, and 1H), three components of the d-orbital of the transition metal (degenerate states) can be seen, i.e., d_{z^2}, $d_{x^2-y^2,xy}$, and $d_{xz,yz}$—from lower to higher energy levels, respectively [32, 36]. Hence, because the orbitals are entirely filled, a semiconducting behavior is observed [32]. On the other hand, the 1T (tetragonal symmetry, D_{3d} space group) exhibits the d metal orbitals in two energy levels, i.e., $d_{xy,yz,zx}$ (three degenerated orbitals) and $d_{x^2-y^2,z^2}$ (two degenerated orbitals), which can accommodate up to six electrons for group 10, shown in Fig. 6.4. As a result, group 10 presents telluride-based TMDs with metallic conductivities, as well as superconductivity (i.e., $PdTe_2$) [37]. Similarly, group 5—another D_{3d} group depicted by Fig. 6.4—presents narrow band gap, semimetals, or superconducting species. Further properties are introduced by Wilson and Yoffe in Ref. [37].

Regarding TMD electrodes for SIBs, although there are metallic and semimetallic phases and species of TMDs, potential candidates for SIB electrodes are located in D_{3d} and D_{3h} point groups (Fig. 6.4). These mentioned groups have species whose thermodynamically stable phase is semiconducting and consequently represents a drawback for their use in SIB electrodes [11, 32]. Considering that high electrical conductivity TMD may improve the charge transfer of Na^+ in SIBs and eliminate the necessity of conducting agents, phase-engineering strategies are studied [39, 40]. Additionally, the synthesis of heterostructures is another approach to improve the conductivity, volume expansion, and mitigate the capacity loss usually present in the first cycles [41].

6.4 Preparation Methods of TMDs

The synthesis process of TMDs is of great significance, as the method of preparation of these materials strongly influences their ultimate morphology, properties (both structural and electronic), and consequent electrochemical performance [42]. The layers of

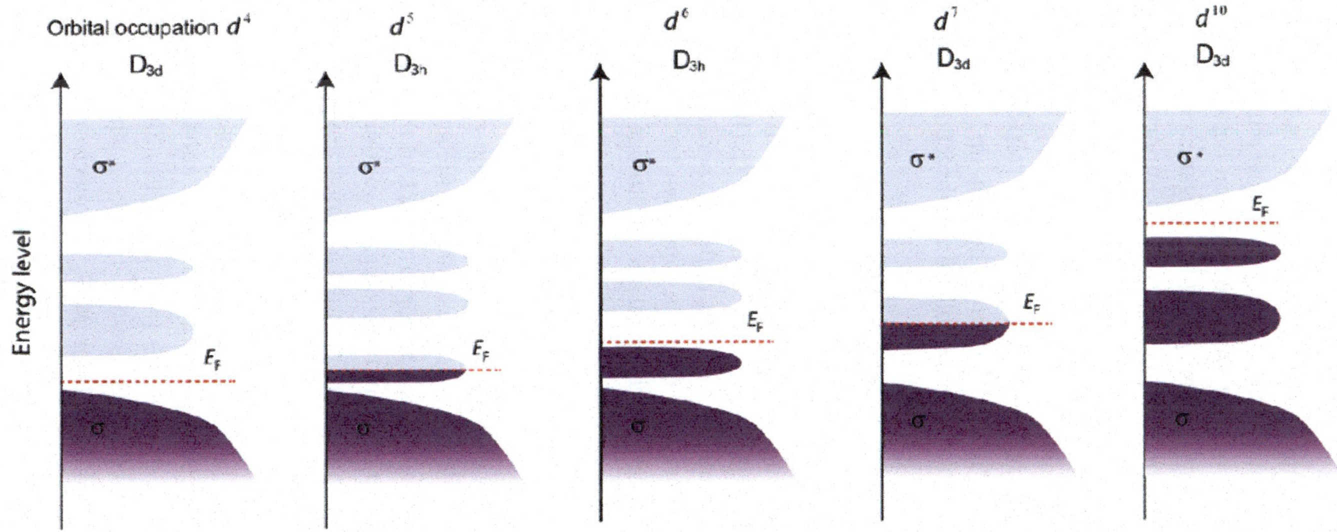

Figure 6.4 Schematic of the d-orbital filling in TMDs in groups 4, 5, 6, 7, and 10, respectively. σ and $\sigma*$ represent bonding and anti-bonding states, respectively. Reprinted with permission from Ref. [38]. Copyright 2015, American Chemical Society.

the TMDs in the bulk form are held together by weak van der Waals forces of attraction and synthesis processes, especially from the bulk material, tend to aim at using techniques to overcome these weak forces [43]. This absence of strong interaction between the layers of the material results in considerable reorganization of majority charge carriers and band gaps [44]. It is also important to note that the synthesis process must be thoroughly perfected so as to optimize the yield of the end product, as a sufficient quantity of end products is needed for electrochemical applications [43, 45].

The preparation/synthesis techniques of TMDs are, therefore, primarily twofold, "top down" and "bottom up," which will be discussed in greater detail in the following subsections [7, 43].

6.4.1 Top-Down Synthesis Techniques

Top-down fabrication techniques, as the name suggests, involve stripping down from the bulk material to achieve the desired layered phase, usually sometimes atomically thin [46, 47]. These approaches have the added advantage of being simple, inexpensive, and scalable [46, 48]. Also, importantly, only the top-down techniques provide enough material for electrochemical energy storage applications.

1. One of the oldest, yet simplest, top-down synthesis techniques that has been extensively applied is the "scotch-tape" or the mechanical exfoliation technique [7, 48]. This technique has the advantage of being able to provide high-purity individual layers; however, the yield is usually insufficient for large-scale applications, which is a major drawback of this process [46–48]. An important work in this regard was done by Budania et al. where the authors performed a comparative study of MoS_2 flakes exfoliated by the simple scotch-tape process alone as against a scotch-tape technique followed by a gel-assisted technique [49]. Results have indicated mechanical exfoliation assisted by gel-based process provided enhanced yield with larger lateral sized flakes (>20 μm) [49]. Similar work has been performed by Radisavljevic et al., in which the authors have obtained monolayer MoS_2 for electronic applications [50].

2. Keeping the limitations of mechanical exfoliation techniques in mind, extensive research has been done to improve and optimize the exfoliation process using other methodologies. Chemical exfoliation is an important technique that has, therefore, been developed to improve yield and obtain high-quality TMD nanosheets [51, 52]. The simplest way this is done by ultrasonicating bulk TMDs in appropriate solvents, where the energy of ultrasonication helps to overcome the weak van der Waals interactions between the sheets of the bulk material [53]. Ultrasonic energy is usually transferred by using an ultrasonic horn or bath. Selection of the solvent is one of the most critical factors in the ultrasonic-assisted liquid exfoliation process. Usually solvents are so chosen that they can efficiently transport the ultrasonic energy being generated, without any attenuation, to the bulk material in solution, so that energy loss is minimized and the yield is optimum [54]. Surface tension is another important parameter, as solvents with comparable surface tension are able to overcome the weak van der Waals forces faster than those with incomparable surface tension values. The simplest solvent that has been tried is water; however, the surface tension of water is 72 mJ/m^2 and is considered to be too high for exfoliation [55]. On the other hand, solvents whose surface tensions are ~40 mJ/m^2 are considered to be the best for exfoliation, as the surface tensions of these solvents are more comparable to those of TMD layers [55]. Considerable work by Coleman et al. has demonstrated that N-methyl-pyrrolidone (NMP) and dimethylformamide (DMF) are the best solvents for simple liquid-based TMD exfoliation [56, 57]. Kirubasankar et al. used a similar facile, sonochemical technique to synthesize a few layers of $MoSe_2$, which were then used for asymmetric supercapacitors and the subsequent electrodes were able to retain 88% of its initial capacitance even after 3000 cycles of operation [58].

The process of solvent-based exfoliation can be further refined/modified if ions are added to the solvent to facilitate the process further. An important advantage of this process is that it facilitates greater yield of exfoliated material. Li, Na, and K atoms are the most frequently used intercalating agents, and these form a metastable intermediate ion-intercalated phase with the TMDs in solution, followed by their sonication and subsequent delamination [53, 54]. Sometimes organolithium compounds are used as agents to provide Li ions for intercalation, such as t-butyllithium, to ultrasonically exfoliate MoS_2 as has been demonstrated by Fan et al. [59]. This method of using intercalating agents in solution to enhance the yield of the exfoliation process has been used quite extensively for a variety of TMDs such as MoS_2, WSe_2, WS_2, and $MoSe_2$ [5, 54].

A technique that stands out among these was developed by Singh et al., which exfoliated TMDs, including MoS_2, $MoSe_2$, WS_2, and WSe_2, using a very strong acid, such as chlorosulfonic acid in aqueous solution [60]. Protons were the intercalating agents in this case, and extreme care had to be taken to perform the exfoliation process due to the nature of the reaction involved [60].

These various liquid- and intercalant-liquid-based exfoliation techniques tend to provide large yields necessary for electrochemical applications as well as products that are phase pure. However, challenges still exist; for example, it has been observed that liquid-based exfoliation processes tend to modify the lattice structure of the final material resulting in the introduction of unwanted structural defects. This is especially important when monolayers are being sought after, and these layers present with lattice defects. Post-exfoliation treatments such as annealing are sometimes necessary to correct/mitigate the structural defects produced during the exfoliation process. Another problem that is commonly encountered is the physical breakage of the exfoliated layers that occurs due the high energy generated by the ultrasonication process itself, so care must be taken to select optimum ultrasonication energies.

6.4.2 Bottom-Up Synthesis Techniques

Bottom-up synthesis procedures essentially involve building the single/few layers of the TMDs one by one from the atomic level [61]. Therefore, these techniques involve a chemical interaction/reaction between the stoichiometric amounts of individual transition metal (M^{n+}) and chalcogenide atom (nX^{2-}) to form individual atomic-scale layers. The most important factor that influences the formation of the individual layers is the reaction kinetics and the energetics involving the sheet formation [43, 61]. Usually, individual MX_2 layers exhibit a significant amount of anisotropy, and this facilitates the growth of these layers, when grown using appropriate techniques.

Among the bottom-up synthesis techniques, the gas phase techniques have been used predominantly [62]. An advantage of the gas phase synthesis technique is that the crystal lattice can be tailored with a high degree of precision and highly phase pure end products are obtained [62]. Chemical vapor deposition (CVD) is an important gas phase technique, in which the chalcogen atom and the transition metal combine in the gas phase to form the TMD sheet [63]. Transition metal precursors are usually MoO_3, WO_3, and this technique usually provides nanosheets with large surface areas. CVD techniques have been utilized in synthesizing MoS_2, $MoSe_2$, WS_2, WSe_2, etc. [63, 64]. For the synthesis of atomically thin sheets, the choice of the substrate is equally important. The substrate must present an inert and almost atomically flat (or thin) surface and must not have any dangling bonds, which might interfere in the process of building up of TMD layers [64].

Liquid phase reactions have also been developed and studied for synthesizing thin-film TMDs. In these processes, the decomposition of the precursor mixture occurs in a solution, either thermally or due to the injection of a hot fluid. Finally, the TMDs that are obtained in solution are then filtered and removed as colloidal solutions from these solvents [65]. These techniques have yielded freestanding TMDs; however, they have been limited to multi-layers. Liquid phase techniques using CS_2 as the precursor of sulfur have been used to synthesize a wide spectrum of TMDs, including TaS_2, TiS_2, $TiSe_2$, WS_2, etc. [66, 67].

Even though bottom-up techniques provide high-purity and lattice-matched TMD nanosheets, the yield is not sufficient enough for applications in electrochemical energy storage systems [43, 46]. Therefore, top-down systems have attained greater applicability when it comes to synthesis for battery and supercapacitor applications [46].

6.5 TMDs as Anode Electrodes for SIBs

SIBs are candidates for large-scale grid storage systems [68, 69]. As an alternative monovalent ion to the consolidated lithium-ion technology, elemental sodium is abundant and environmentally friendly, making SIBs attractive due to the projected low cost per kWh [70]. Reported initially in the literature as a supplementary technology to LIB [23], new materials have showed SIBs with high energy densities, outperforming LIBs [70, 71].

This chapter presents some of the state-of-the-art layered TMD anode materials synthesized. As an intrinsic factor of every material, challenges are also presented. Some of these are (1) control of charge storage mechanisms for longer life cycle; (2) development of heterostructures for enhancing the electrochemical performance while avoiding 2D nanosheets to restack; and (3) mitigation of pulverization and aggregation of TMD nanosheets upon sodiation and desodiation processes. Further, the employed strategies to address such existing challenges on TMDs are presented.

6.5.1 Sulfide-Based TMDs

Arising from the promising findings on molybdenum disulfide (MoS_2), sulfide-based TMDs, or transition metal disulfides, are species also expected to share properties such as high theoretical capacity through conversion reactions and rich electrochemistry [72]. Compared with transition metal oxides (TMOs), the metal–sulfur bonds are weaker than metal–oxygen bonds in TMOs, and consequently metal–sulfur bonds are more favorable for reactions with sodium ions [73, 74].

In addition to conversion reaction, which allows the transfer of four electrons compared to a single electron in intercalation mechanism, the high theoretical capacity of 2H-MoS_2 can also be ascribed to interlayer space of 6.3 Å [75]. Therefore, MoS_2 allows the diffusion of larger ions, compared to other lamellar host materials (e.g., graphite) [76]. With respect to the charge storage mechanisms of MoS_2, both intercalation and conversion processes are seen. Overall, the proposed model for Na^+ storage in MoS_2 consists of intercalation (above 0.4 V) and conversion (below 0.4 V) processes, as presented by Eqs. (6.1) and (6.2), respectively [72, 77].

$$MoS_2 + xNa^+ + xe^- \rightarrow Na_xMoS_2 \qquad (6.1)$$

$$Na_xMoS_2 + (4 - x)Na^+ \rightarrow Mo + 2Na_2S \qquad (6.2)$$

As the Na^+ storage process on MoS_2 is voltage controlled, a determined Na^+ storage mechanism can be chosen by setting a specific cut-off voltage [78]. Although conversion reactions may provide up to six electrons per transition metal, they also lead to pulverization of the active material, resulting in capacity loss and poor cyclability performance [79]. On the other hand, although intercalation process yields lower theoretical specific capacity, it also provides higher cyclability performance.

Xu et al. prepared MoS_2 nanosheets anchored on nitrogen-doped graphene as exhibited by Fig. 6.5a [78]. Aiming to achieve electrochemistry based on intercalation, the potential was set from 0.4 to 3 V, as one can see in Fig. 6.5b. As depicted in Fig. 6.5c, US-MoS_2@NG presented the highest specific capacity and stability. According to the authors, the graphene scaffold provides mechanical stability and higher electronic conductivity. Also, the MoS_2 ultrathin nanosheets have lower energy barrier for Na^+ diffusion and improved transport path. The expanded nanosheets of MoS_2 reduce the charge transfer, again improving the insertion and extraction of Na^+. These factors combined made this composite material superior to those in previous reports, as shown in Fig. 6.5d. With respect to the intercalation-based mechanism of Na^+ storage, the stability up to 1000 cycles of US-MoS_2@NG is evidenced in Fig. 6.5e [78].

Using a similar strategy, Geng et al. prepared a 3D electrode of MoS_2– graphene–1T MoS_2 [80]. As regards electrochemical results, the high conductive freestanding material presented both high rate capability and reversibility, i.e., 313 mAh/g at 50 mA/g at the 200th cycle. Furthermore, the sandwich electrode assisted in mitigating the aggregation and pulverization of MoS_2.

As another strategy to safeguard MoS_2 upon sodiation and desodiation, Soares et al. functionalized MoS_2 nanosheets in polymer-derived silicon oxycarbide (SiOC) [7]. The high-temperature glass–ceramic material (SiOC) was used to protect MoS_2 species from polysulfide dissolution upon sodiation and desodiation, yet without affecting the electrochemical performance of MoS_2. Results indicate that the MoS_2–SiOC composite, at 200 mA/g, retains 86.5% capacity after 100 cycles. In addition, from electrochemical impedance spectroscopy (EIS) studies, the authors confirmed that the free carbon domains from SiOC improve the electronic conductivity of the material. In sum, this work may open new avenues with respect to polysulfide mitigation in sulfur-based anode electrodes for SIBs.

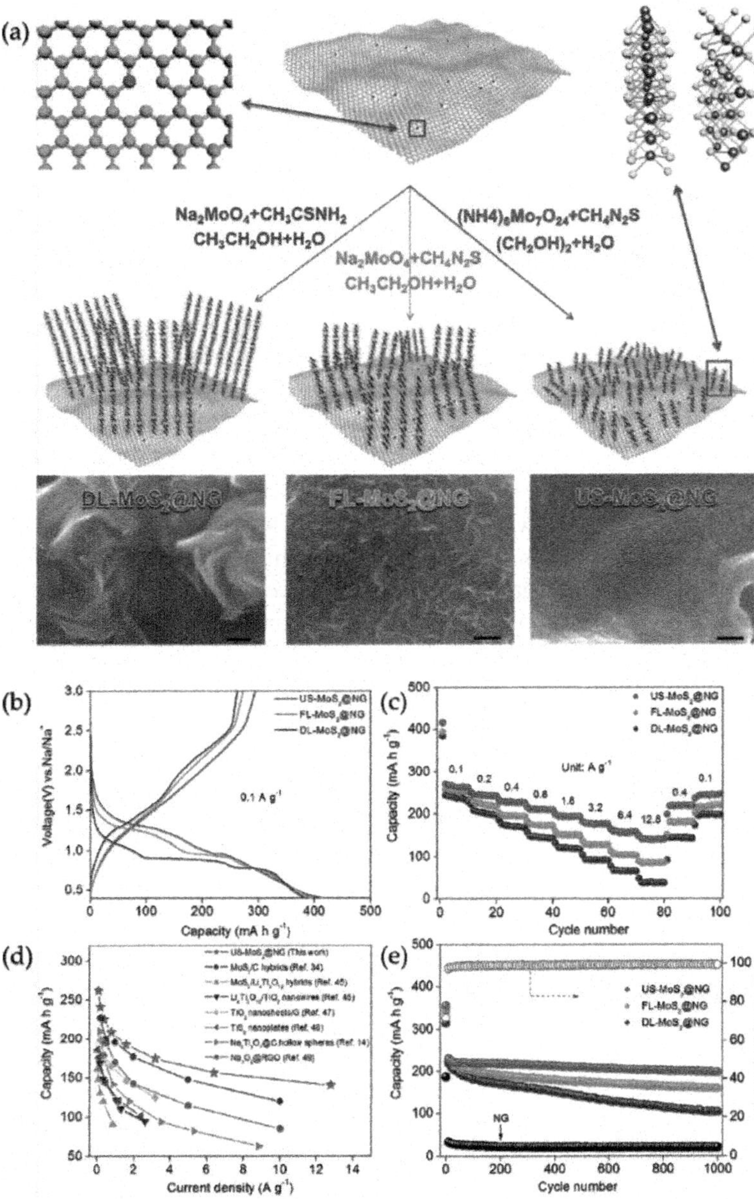

Figure 6.5 (a) Synthesis process of dozen-layered (DL), few-layered (FL), and ultrasmall (US) MoS_2 nanosheets in nitrogen-doped graphene (NG). Scale bars, 100 nm. (b) Voltage profiles in the first charge and discharge cycles at 100 mA/g. (c) Rate capability test of MoS_2@NG. (d) Rate performance comparison with other materials reported in the literature. (e) Cycling performance of MoS_2@NG at 1 A/g and Coulombic efficiency of US-MoS_2@NG. Reprinted with permission from Ref. [78]. Copyright 2018, John Wiley and Sons.

Similar to MoS_2, pyrite (FeS_2) is a TMD that also allows to control the charge storage mechanism upon setting a suitable cut-off voltage. In addition, FeS_2 is a potential TMD for negative electrodes of SIB due to low cost, nontoxic nature, and theoretical specific capacity of 894 mAh/g [81]. Studied within the potential range of 0.8–3 V, FeS_2 SIBs showed cycling stability up to 20,000 cycles at 1 A/g, with capacity retention of approximately 90% [82]. In the same study, Hu et al. also stated the importance of selecting a proper electrolyte. For FeS_2 SIBs, diglyme (DGM) was the best performing electrolyte.

Presenting higher conductivity than MoS_2, one of the first reports of tungsten disulfide (WS_2) as SIB anode appeared in 2014 [83, 84]. Since then, this TMD species has gained more attention. In the work published by Wen et al., WS_2 nanosheets with interlayer spacing expanded to 9.2 Å, in a nitrogen-doped carbon matrix, yielded a 3D structure, named H-WS_2@NC. The material provided several paths for Na^+ diffusion, consequently showing lower Na^+ diffusion barrier energy [85]. Evaluated in a half-cell configuration, H-WS_2@NC yielded over 99% capacity after 200 cycles at 100 mA/g. Additionally, at 1 A/g, the material presented cycling stability up 1000 cycles. In full-cell configuration, where $Na_3V_2(PO_4)_3$/rGO was the chosen cathode material, H-WS_2@NC exhibited reversible specific capacity of 296 mAh/g at 500 mA/g, therefore indicating potential for practical applications [85]. Nevertheless, challenges regarding scalability and cost are opportunities for further research.

6.5.2 Selenide-Based TMDs

Just like its sulfide counterparts, selenides have also found applications as negative electrode materials for SIBs. Selenides have the advantage of demonstrating high energy densities with respect to sulfides and also being relatively environmentally benign.

Among the layered TMDs, $MoSe_2$ has been studied the most due to its lamellar structure, and the ability to host intercalating ions in between these layers [86]. It has also demonstrated higher electronic and ionic conductivity as compared to MoS_2 [87]. In this regard, one of the earliest works have been performed by Morales et al. [88]. The authors have demonstrated that intercalation of Na ions into a primitive 2H $MoSe_2$ lattice results in a structural change (distortion), which results in a 20% capacity loss after the first charge–discharge cycle [88]. Wang et al. used a pyrolysis-based process to synthesize $MoSe_2$ using $(NH_4)_6Mo_7O_{24}·4H_2O$, Se, NH_4OH, and $N_2H_4·H_2O$ as the precursors [89]. Using it versus Na/Na^+ system demonstrated discharge and charge capacities of 513 and 440 mAh/g, respectively. Further investigation by the authors have indicated that a favorable sodium-ion hopping mechanism from octahedra to tetrahedra in the $MoSe_2$ structure helps to deliver high capacities [89]. Wang et al. have synthesized $MoSe_2$ via a colloidal route, thereby demonstrating highly ordered nanospherical morphology [90]. As an anode in SIB systems, these $MoSe_2$ nanospheres delivered an initial discharge capacity of 520 mAh/g at a current rate of 0.1 C within a potential window of 0.1–3 V and a capacity retention of 80% over 200 cycles. The authors have attributed the superior performance to the large surface area of the nanospheres, as obtained from first principles calculations [90]. Mukherjee et al. studied $MoSe_2$ flakes synthesized/exfoliated by a novel chlorosulfonic acid technique and were able to report 399.1 mAh/g as the first cycle discharge capacity [91]. Per the authors, the exfoliation technique prevented the flakes from aggregating, which resulted in the improved capacities [91]. Tang et al. were able to expand the interlayer spacing of $MoSe_2$ by stabilizing with carbon, using an acid-functionalized carbonization (from polydopamine) process [92]. The resultant sample $MoSe_2$@C was able to provide a reversible discharge capacity of 445 mAh/g at 1 A/g current density and good stability over 100 cycles of operation [92]. The authors have attributed the elevated performance to stabilization of the $MoSe_2$ lattice and increase in its interlayer spacing, both using carbon [92]. The structural characterization and electrochemical performance results of the $MoSe_2$@C sample are shown in Fig. 6.6a–h.

WSe_2 is another important layered selenide that has been studied as an electrode material for SIB systems. Share et al. were able to obtain a specific capacity as high as 200 mAh/g at a current density of 20 mA/g with a very stable capacity retention [93]. Further probing indicated that WSe_2 underwent a conversion reaction, with the formation of isolated W and Na_xSe islands, which re-formed back to WSe_2 once the sodium was extracted out [93]. WSe_2 hexagonal nanoplates as anodes in SIB systems have been studied by Yang et al. [94]. An initial discharge capacity of 762.4 mAh/g was obtained at a high current density of 400 mA/g and Coulombic efficiency

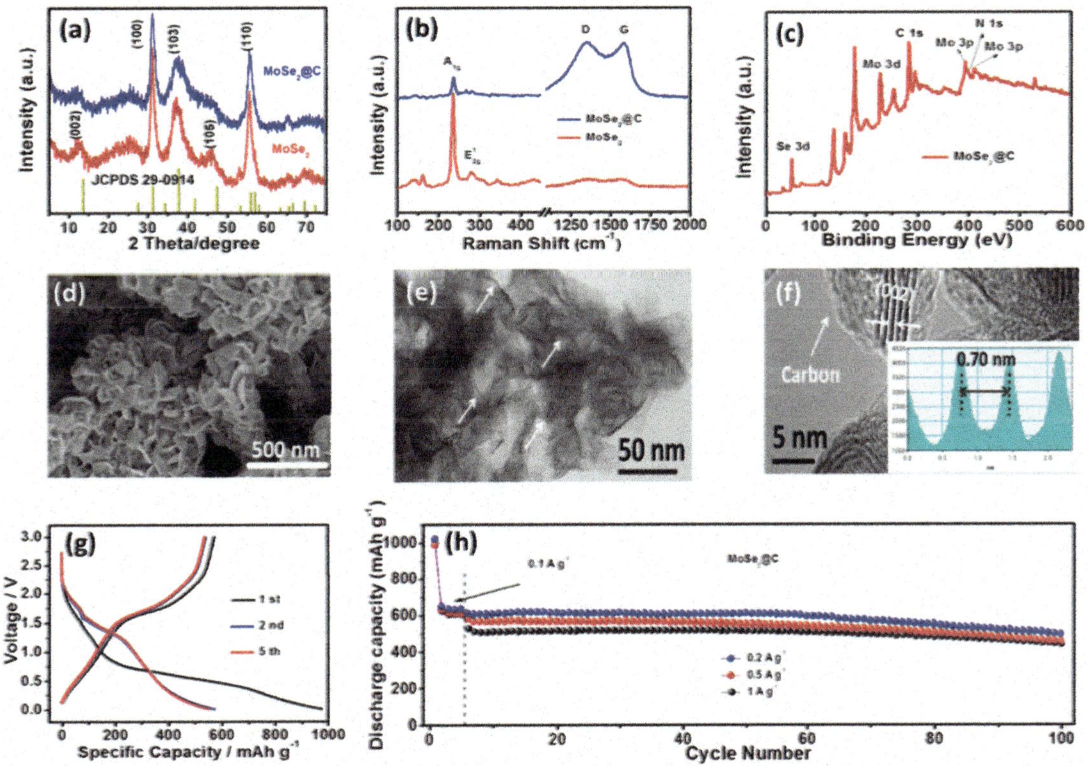

Figure 6.6 Structural characterization and electrochemical performance of $MoSe_2$@C nanosheets. (a, b) XRD patterns and Raman spectra of as-made $MoSe_2$ and $MoSe_2$@C samples showing the characteristic phases. (c) XPS spectra of the $MoSe_2$@C sample demonstrating the peaks corresponding to the different energy levels and bonds. (d) SEM images of the $MoSe_2$@C sample. (e, f) HRTEM images of the $MoSe_2$@C sample showing the characteristic lattice fringes. (g) Galvanostatic charge–discharge curves of the $MoSe_2$@C sample within a range of 0.01–3.0 V. (h) Cycling stability and capacity retention of the sample over 100 cycles at varying current rates. Reprinted with permission from Ref. [92]. Copyright 2016, American Chemical Society.

was almost 100% after the first cycle. The hexagonal nanoplates and their large surface area are necessary for providing the high capacities [94]. WSe_2 onions synthesized using a simple CVD-based technique have been studied by Wang et al. [95]. This CVD resulted in the formation of a spherical microstructure with temperature being controlled very carefully during the process. A specific capacity of 205 mAh/g at a current density of 10 A/g versus Na/Na^+ was obtained along with an energy density of 123.1 Wh/kg. The superior performance by the WSe_2 onions has been attributed to the morphology and high surface area [95]. Chun et al. synthesized WSe_2 nanosheets by a simple mechanical exfoliation followed by exfoliation in organic solvents [96]. They subsequently irradiated the nanosheets by a proton beam, with electrochemical results indicating proton beam irradiation enhancing performance (270 mAh/g discharge capacity and 63% capacity retention) as compared to bulk material (44% capacity retention) [96].

However, challenges such as loss of structural order due to the conversion reaction process are a major hurdle and future research is aimed at overcoming those hurdles.

6.6 Conclusion and Outlook

Solving the energy crisis is one of the most important issues currently facing the world. One possible solution to mitigate this crisis involves the development and implementation of large-area and cost-effective energy storage systems that can be seamlessly coupled with renewable energy generation plants and the electrical grid to supply energy to the grid during peak and outage times. To this end, among various ESS systems, SIBs have elicited promise and a means to overcome the shortcomings of other commercially available ESS such as the LIB systems in a more sustainable way. However, for the SIB systems to be commercially viable, the development of novel electrode materials is necessary. Layered electrode materials, especially TMDs, offer a way forward by virtue of their large surface areas, tunable properties, and enhanced electrochemical kinetics. A large variety of TMDs exist with a spectrum of electronic properties and structures—important advantages due to their chemical versatility. Nonetheless, the use of TMD-based electrodes for SIBs is still a relatively new area of research, and considerable work is ongoing to optimize TMD synthesis aiming to provide scalability in a cost-effective manner. Furthermore, certain shortcomings still exist and are not yet fully understood, these include, the mechanism of reaggregation and pulverization of exfoliated nanosheets that lowers long-term cycle life of electrodes. Voltage hysteresis and poor electrical conductivity of some TMDs are other issues adversely affecting the performance as SIB battery electrodes. Synthesis of state-of-the-art heterostructures and control of charge storage mechanisms on SIBs are some techniques being investigated to overcome some of these issues.

References

1. Feynman R P. There's plenty of room at the bottom. *Engineering and Science*, 1960; **23**(5): 22–36.
2. Frindt R F, Yoffe A D, and Bowden F P. Physical properties of layer structures : Optical properties and photoconductivity of thin crystals of molybdenum disulphide. *Proceedings of the Royal Society of London Series A Mathematical and Physical Sciences*, 1963; **273**(1352): 69–83.
3. Joensen P, Frindt R F, and Morrison S R. Single-layer MoS_2. *Materials Research Bulletin*, 1986; **21**(4): 457–461.
4. Novoselov K S, Geim A K, Morozov S V, Jiang D, Zhang Y, Dubonos S V, et al. Electric field effect in atomically thin carbon films. *Science*, 2004; **306**(5696): 666.
5. Lv R, Robinson J A, Schaak R E, Sun D, Sun Y, Mallouk T E, et al. Transition metal dichalcogenides and beyond: Synthesis, properties, and applications of single- and few-layer nanosheets. *Accounts of Chemical Research*, 2015; **48**(1): 56–64.
6. Wang L, Zou J, Chen S, Zhou G, Bai J, Gao P, et al. TiS_2 as a high performance potassium ion battery cathode in ether-based electrolyte. *Energy Storage Materials*, 2018; **12**: 216–222.
7. Soares D M and Singh G. SiOC functionalization of MoS_2 as a means to improve stability as sodium-ion battery anode. *Nanotechnology*, 2020; **31**(14): 145403.
8. Zhang R, Bao J, Pan Y, and Sun C-F. Highly reversible potassium-ion intercalation in tungsten disulfide. *Chemical Science*, 2019; **10**(9): 2604–2612.

9. Mannix A J, Kiraly B, Hersam M C, and Guisinger N P. Synthesis and chemistry of elemental 2D materials. *Nature Reviews Chemistry*, 2017; **1**(2): 0014.
10. Manzeli S, Ovchinnikov D, Pasquier D, Yazyev O V, and Kis A. 2D transition metal dichalcogenides. *Nature Reviews Materials*, 2017; **2**(8): 17033.
11. Chhowalla M, Shin H S, Eda G, Li L-J, Loh K P, and Zhang H. The chemistry of two-dimensional layered transition metal dichalcogenide nanosheets. *Nature Chemistry*, 2013; **5**(4): 263–275.
12. Lei Z, Zhan J, Tang L, Zhang Y, and Wang Y. Recent development of metallic (1T) phase of molybdenum disulfide for energy conversion and storage. *Advanced Energy Materials*, 2018; **8**(19): 1703482.
13. Novoselov K S, Mishchenko A, Carvalho A, and Castro Neto A H. 2D materials and van der Waals heterostructures. *Science*, 2016; **353**(6298): aac9439.
14. Sie E J, Nyby C M, Pemmaraju C D, Park S J, Shen X, Yang J, et al. An ultrafast symmetry switch in a Weyl semimetal. *Nature*, 2019; **565**(7737): 61–66.
15. Chua C K, Loo A H, and Pumera M. Top-down and bottom-up approaches in engineering 1T phase molybdenum disulfide (MoS_2): Towards highly catalytically active materials. *Chemistry: A European Journal*, 2016; **22**(40): 14336–14341.
16. Acerce M, Voiry D, and Chhowalla M. Metallic 1T phase MoS_2 nanosheets as supercapacitor electrode materials. *Nature Nanotechnology*, 2015; **10**(4): 313–318.
17. Lin Y-C, Dumcenco D O, Huang Y-S, and Suenaga K. Atomic mechanism of the semiconducting-to-metallic phase transition in single-layered MoS_2. *Nature Nanotechnology*, 2014; **9**(5): 391–396.
18. Chen X, Wang Z, Wei Y, Zhang X, Zhang Q, Gu L, et al. High phase-purity 1T-MoS_2 ultrathin nanosheets by a spatially confined template. *Angewandte Chemie International Edition*, 2019; **58**(49): 17621–17624.
19. Mane A U, Letourneau S, Mandia D J, Liu J, Libera J A, Lei Y, et al. Atomic layer deposition of molybdenum disulfide films using MoF_6 and H_2S. *Journal of Vacuum Science & Technology A*, 2017; **36**(1): 01A125.
20. Nishi Y. The development of lithium ion secondary batteries. *The Chemical Record*, 2001; **1**(5): 406–413.
21. Lyu Y, Wu X, Wang K, Feng Z, Cheng T, Liu Y, et al. An overview on the advances of $LiCoO_2$ cathodes for lithium-ion batteries. *Advanced Energy Materials*, 2020; 2000982.

22. Kim H, Kim J C, Bianchini M, Seo D-H, Rodriguez-Garcia J, and Ceder G. Recent progress and perspective in electrode materials for K-ion batteries. *Advanced Energy Materials*, 2018; **8**(9): 1702384.

23. Vaalma C, Buchholz D, Weil M, and Passerini S. A cost and resource analysis of sodium-ion batteries. *Nature Reviews Materials*, 2018; **3**(4): 18013.

24. Ryu W-H, Jung J-W, Park K, Kim S-J, and Kim I-D. Vine-like MoS_2 anode materials self-assembled from 1-D nanofibers for high capacity sodium rechargeable batteries. *Nanoscale*, 2014; **6**(19): 10975–10981.

25. Lin L, Lei W, Zhang S, Liu Y, Wallace G G, and Chen J. Two-dimensional transition metal dichalcogenides in supercapacitors and secondary batteries. *Energy Storage Materials*, 2019; **19**: 408–423.

26. Yang E, Ji H, and Jung Y. Two-dimensional transition metal dichalcogenide monolayers as promising sodium ion battery anodes. *The Journal of Physical Chemistry C*, 2015; **119**(47): 26374–26380.

27. Soares D M, Mukherjee S, and Singh G. TMDs beyond MoS_2 for electrochemical energy storage. *Chemistry: A European Journal*, 2020; **26**(29): 6320–6341.

28. Liu Y, Fan L-Z, and Jiao L. Graphene intercalated in graphene-like MoS_2: A promising cathode for rechargeable Mg batteries. *Journal of Power Sources*, 2017; **340**: 104–110.

29. Wen Y, He K, Zhu Y, Han F, Xu Y, Matsuda I, et al. Expanded graphite as superior anode for sodium-ion batteries. *Nature Communications*, 2014; **5**(1): 4033.

30. Yan C, Gong C, Wangyang P, Chu J, Hu K, Li C, et al. 2D group IVB transition metal dichalcogenides. *Advanced Functional Materials*, 2018; **28**(39): 1803305.

31. Tang S, Zhang C, Wong D, Pedramrazi Z, Tsai H-Z, Jia C, et al. Quantum spin Hall state in monolayer 1T'-WTe_2. *Nature Physics*, 2017; **13**(7): 683–687.

32. Voiry D, Mohite A, and Chhowalla M. Phase engineering of transition metal dichalcogenides. *Chemical Society Reviews*, 2015; **44**(9): 2702–2712.

33. Wang R, Yu Y, Zhou S, Li H, Wong H, Luo Z, et al. Strategies on phase control in transition metal dichalcogenides. *Advanced Functional Materials*, 2018; **28**(47): 1802473.

34. Yu Y, Nam G-H, He Q, Wu X-J, Zhang K, Yang Z, et al. High phase-purity 1T′-MoS_2- and 1T′-$MoSe_2$-layered crystals. *Nature Chemistry*, 2018; **10**(6): 638–643.

35. Lu P, Kim J-S, Yang J, Gao H, Wu J, Shao D, et al. Origin of superconductivity in the Weyl semimetal WTe_2 under pressure. *Physical Review B*, 2016; **94**(22): 224512.
36. Yang H, Kim S W, Chhowalla M, and Lee Y H. Structural and quantum-state phase transitions in van der Waals layered materials. *Nature Physics*, 2017; **13**(10): 931–937.
37. Wilson J A and Yoffe A D. The transition metal dichalcogenides discussion and interpretation of the observed optical, electrical and structural properties. *Advances in Physics*, 1969; **18**(73): 193–335.
38. Yuan H, Wang H, and Cui Y. Two-dimensional layered chalcogenides: From rational synthesis to property control via orbital occupation and electron filling. *Accounts of Chemical Research*, 2015; **48**(1): 81–90.
39. Sun D, Huang D, Wang H, Xu G-L, Zhang X, Zhang R, et al. 1T MoS_2 nanosheets with extraordinary sodium storage properties via thermal-driven ion intercalation assisted exfoliation of bulky MoS_2. *Nano Energy*, 2019; **61**: 361–369.
40. Xiong P, Ma R, Sakai N, Nurdiwijayanto L, and Sasaki T. Unilamellar metallic MoS_2/graphene superlattice for efficient sodium storage and hydrogen evolution. *ACS Energy Letters*, 2018; **3**(4): 997–1005.
41. Pomerantseva E and Gogotsi Y. Two-dimensional heterostructures for energy storage. *Nature Energy*, 2017; **2**(7): 17089.
42. Chen B, Chao D, Liu E, Jaroniec M, Zhao N, and Qiao S-Z. Transition metal dichalcogenides for alkali metal ion batteries: Engineering strategies at the atomic level. *Energy & Environmental Science*, 2020; **13**(4): 1096–1131.
43. Mukherjee S and Singh G. Two-dimensional anode materials for non-lithium metal-ion batteries. *ACS Applied Energy Materials*, 2019; **2**(2): 932–955.
44. Mak K F, Lee C, Hone J, Shan J, and Heinz T F. Atomically thin MoS_2: A new direct-gap semiconductor. *Physical Review Letters*, 2010; **105**(13): 136805.
45. Wang Y-H, Huang K-J, and Wu X. Recent advances in transition-metal dichalcogenides based electrochemical biosensors: A review. *Biosensors and Bioelectronics*, 2017; **97**: 305–316.
46. Mukherjee S, Ren Z, and Singh G. Beyond graphene anode materials for emerging metal ion batteries and supercapacitors. *Nano-Micro Letters*, 2018; **10**(4): 70.
47. Liu F, Wu W, Bai Y, Chae SH, Li Q, Wang J, et al. Disassembling 2D van der Waals crystals into macroscopic monolayers and reassembling into artificial lattices. *Science*, 2020; **367**(6480): 903.

48. Peng L, Zhu Y, Chen D, Ruoff R S, and Yu G. Two-dimensional materials for beyond-lithium-ion batteries. *Advanced Energy Materials*, 2016; **6**(11): 1600025.

49. Budania P, Baine P T, Montgomery J H, McNeill D W, Mitchell S J N, Modreanu M, et al. Comparison between Scotch tape and gel-assisted mechanical exfoliation techniques for preparation of 2D transition metal dichalcogenide flakes. *Micro & Nano Letters*, 2017; **12**(12): 970–973.

50. Radisavljevic B, Radenovic A, Brivio J, Giacometti V, and Kis A. Single-layer MoS_2 transistors. *Nature Nanotechnology*, 2011; **6**(3): 147–150.

51. Whittingham M S and Gamble F R. The lithium intercalates of the transition metal dichalcogenides. *Materials Research Bulletin*, 1975; **10**(5): 363–371.

52. Qiao W, Yan S, Song X, Zhang X, He X, Zhong W, et al. Luminescent monolayer MoS_2 quantum dots produced by multi-exfoliation based on lithium intercalation. *Applied Surface Science*, 2015; **359**: 130–136.

53. Chhowalla M, Liu Z, and Zhang H. Two-dimensional transition metal dichalcogenide (TMD) nanosheets. *Chemical Society Reviews*, 2015; **44**(9): 2584–2586.

54. Cunningham G, Lotya M, Cucinotta C S, Sanvito S, Bergin S D, Menzel R, et al. Solvent exfoliation of transition metal dichalcogenides: Dispersibility of exfoliated nanosheets varies only weakly between compounds. *ACS Nano*, 2012; **6**(4): 3468–3480.

55. Smith R J, King P J, Lotya M, Wirtz C, Khan U, De S, et al. Large-scale exfoliation of inorganic layered compounds in aqueous surfactant solutions. *Advanced Materials*, 2011; **23**(34): 3944–3948.

56. Nicolosi V, Chhowalla M, Kanatzidis M G, Strano M S, and Coleman J N. Liquid exfoliation of layered materials. *Science*, 2013; **340**(6139): 1226419.

57. Coleman J N, Lotya M, O'Neill A, Bergin S D, King P J, Khan U, et al. Two-dimensional nanosheets produced by liquid exfoliation of layered materials. *Science*, 2011; **331**(6017): 568–571.

58. Kirubasankar B, Vijayan S, and Angaiah S. Sonochemical synthesis of a 2D–2D $MoSe_2$/graphene nanohybrid electrode material for asymmetric supercapacitors. *Sustainable Energy & Fuels*, 2019; **3**(2): 467–477.

59. Fan X, Xu P, Zhou D, Sun Y, Li Y C, Nguyen M A T, et al. Fast and efficient preparation of exfoliated 2H MoS_2 nanosheets by sonication-assisted lithium intercalation and infrared laser-induced 1T to 2H phase reversion. *Nano Letters*, 2015; **15**(9): 5956–5960.

60. Mukherjee S, Turnley J, Mansfield E, Holm J, Soares D, David L, et al. Exfoliated transition metal dichalcogenide nanosheets for supercapacitor and sodium ion battery applications. *Royal Society Open Science*, 2019; **6**.
61. Brent J R, Savjani N, and O'Brien P. Synthetic approaches to two-dimensional transition metal dichalcogenide nanosheets. *Progress in Materials Science*, 2017; **89**: 411–478.
62. Jawaid A, Nepal D, Park K, Jespersen M, Qualley A, Mirau P, et al. Mechanism for liquid phase exfoliation of MoS_2. *Chemistry of Materials*, 2016; **28**(1): 337–348.
63. Ubaldini A and Giannini E. Improved chemical vapor transport growth of transition metal dichalcogenides. *Journal of Crystal Growth*, 2014; **401**: 878–882.
64. Ubaldini A, Jacimovic J, Ubrig N, and Giannini E. Chloride-driven chemical vapor transport method for crystal growth of transition metal dichalcogenides. *Crystal Growth & Design*, 2013; **13**(10): 4453–4459.
65. Shmeliov A, Shannon M, Wang P, Kim J S, Okunishi E, Nellist P D, et al. Unusual stacking variations in liquid-phase exfoliated transition metal dichalcogenides. *ACS Nano*, 2014; **8**(4): 3690–3699.
66. Najafi L, Bellani S, Oropesa-Nuñez R, Martín-García B, Prato M, Pasquale L, et al. TaS_2, $TaSe_2$, and their heterogeneous films as catalysts for the hydrogen evolution reaction. *ACS Catalysis*, 2020; **10**(5): 3313–3325.
67. Yun Q, Li L, Hu Z, Lu Q, Chen B, and Zhang H. Layered transition metal dichalcogenide-based nanomaterials for electrochemical energy storage. *Advanced Materials*, 2020; **32**(1): 1903826.
68. Weng G-M, Kong J, Wang H, Karpovich C, Lipton J, Antonio F, et al. A highly efficient perovskite photovoltaic-aqueous Li/Na-ion battery system. *Energy Storage Materials*, 2020; **24**: 557–564.
69. Miorandi D, Sicari S, De Pellegrini F, and Chlamtac I. Internet of things: Vision, applications and research challenges. *Ad Hoc Networks*, 2012; **10**(7): 1497–1516.
70. Hirsh H S, Li Y, Tan D H S, Zhang M, Zhao E, and Meng Y S. Sodium-ion batteries paving the way for grid energy storage. *Advanced Energy Materials*, 2020; 2001274.
71. Thangavel R, Samuthira Pandian A, Ramasamy H V, and Lee Y-S. Rapidly synthesized, few-layered pseudocapacitive SnS_2 anode for high-power sodium ion batteries. *ACS Applied Materials & Interfaces*, 2017; **9**(46): 40187–40196.

72. David L, Bhandavat R, and Singh G. MoS_2/graphene composite paper for sodium-ion battery electrodes. *ACS Nano*, 2014; **8**(2): 1759–1770.
73. Zhou Q, Liu L, Huang Z, Yi L, Wang X, and Cao G. Co_3S_4@polyaniline nanotubes as high-performance anode materials for sodium ion batteries. *Journal of Materials Chemistry A*, 2016; **4**(15): 5505–5516.
74. Hwang J-Y, Myung S-T, and Sun Y-K. Sodium-ion batteries: Present and future. *Chemical Society Reviews*, 2017; **46**(12): 3529–3614.
75. Rasamani K D, Alimohammadi F, and Sun Y. Interlayer-expanded MoS_2. *Materials Today*, 2017; **20**(2): 83–91.
76. Grey C P and Tarascon J M. Sustainability and in situ monitoring in battery development. *Nature Materials*, 2017; **16**(1): 45–56.
77. Hu Z, Wang L, Zhang K, Wang J, Cheng F, Tao Z, et al. MoS_2 nanoflowers with expanded interlayers as high-performance anodes for sodium-ion batteries. *Angewandte Chemie International Edition*, 2014; **53**(47): 12794–12798.
78. Xu X, Zhao R, Ai W, Chen B, Du H, Wu L, et al. Controllable design of MoS_2 nanosheets anchored on nitrogen-doped graphene: Toward fast sodium storage by tunable pseudocapacitance. *Advanced Materials*, 2018; **30**(27): 1800658.
79. Wang S, Cao F, Li Y, Zhang Z, Zhou D, Yang Y, et al. MoS_2-coupled carbon nanosheets encapsulated on sodium titanate nanowires as super-durable anode material for sodium-ion batteries. *Advanced Science*, 2019; **6**(10): 1900028.
80. Geng X, Jiao Y, Han Y, Mukhopadhyay A, Yang L, and Zhu H. Freestanding metallic 1T MoS_2 with dual ion diffusion paths as high rate anode for sodium-ion batteries. *Advanced Functional Materials*, 2017; **27**(40): 1702998.
81. Kim T B, Choi J W, Ryu H S, Cho G B, Kim K W, Ahn J H, et al. Electrochemical properties of sodium/pyrite battery at room temperature. *Journal of Power Sources*, 2007; **174**(2): 1275–1278.
82. Hu Z, Zhu Z, Cheng F, Zhang K, Wang J, Chen C, et al. Pyrite FeS_2 for high-rate and long-life rechargeable sodium batteries. *Energy & Environmental Science*, 2015; **8**(4): 1309–1316.
83. Su D, Dou S, and Wang G. WS_2@graphene nanocomposites as anode materials for Na-ion batteries with enhanced electrochemical performances. *Chemical Communications*, 2014; **50**(32): 4192–4195.
84. Wang Y, Kong D, Shi W, Liu B, Sim G J, Ge Q, et al. Ice templated freestanding hierarchically WS_2/CNT-rGO aerogel for high-performance rechargeable lithium and sodium ion batteries. *Advanced Energy Materials*, 2016; **6**(21): 1601057.

85. Hu X, Liu Y, Li J, Wang G, Chen J, Zhong G, et al. Self-assembling of conductive interlayer-expanded WS_2 nanosheets into 3D hollow hierarchical microflower bud hybrids for fast and stable sodium storage. *Advanced Functional Materials*, 2020; **30**(5): 1907677.

86. Shi Y, Hua C, Li B, Fang X, Yao C, Zhang Y, et al. Highly ordered mesoporous crystalline $MoSe_2$ material with efficient visible-light-driven photocatalytic activity and enhanced lithium storage performance. *Advanced Functional Materials*, 2013; **23**(14): 1832–1838.

87. Gobrecht J. Performance of synthetical n-$MoSe_2$ in electrochemical solar cells. *Journal of The Electrochemical Society*, 1978; **125**(12): 2085.

88. Morales J, Santos J, and Tirado J L. Electrochemical studies of lithium and sodium intercalation in $MoSe_2$. *Solid State Ionics*, 1996; **83**(1): 57–64.

89. Wang H, Lan X, Jiang D, Zhang Y, Zhong H, Zhang Z, et al. Sodium storage and transport properties in pyrolysis synthesized $MoSe_2$ nanoplates for high performance sodium-ion batteries. *Journal of Power Sources*, 2015; **283**: 187–194.

90. Wang H, Wang L, Wang X, Quan J, Mi L, Yuan L, et al. High quality $MoSe_2$ nanospheres with superior electrochemical properties for sodium batteries. *Journal of The Electrochemical Society*, 2016; **163**(8): A1627–A1632.

91. Mukherjee S, Turnley J, Mansfield E, Holm J, Soares D, David L, et al. Exfoliated transition metal dichalcogenide nanosheets for supercapacitor and sodium ion battery applications. *Royal Society Open Science*, **6**(8): 190437.

92. Tang Y, Zhao Z, Wang Y, Dong Y, Liu Y, Wang X, et al. Carbon-stabilized interlayer-expanded few-layer $MoSe_2$ nanosheets for sodium ion batteries with enhanced rate capability and cycling performance. *ACS Applied Materials & Interfaces*, 2016; **8**(47): 32324–32332.

93. Share K, Lewis J, Oakes L, Carter R E, Cohn A P, and Pint C L. Tungsten diselenide (WSe_2) as a high capacity, low overpotential conversion electrode for sodium ion batteries. *RSC Advances*, 2015; **5**(123): 101262-7.

94. Yang W, Wang J, Si C, Peng Z, and Zhang Z. Tungsten diselenide nanoplates as advanced lithium/sodium ion electrode materials with different storage mechanisms. *Nano Research*, 2017; **10**(8): 2584–2598.

95. Wang Y, Zhang X, Xiong P, Yin F, Xu Y, Wan B, et al. Insight into the intercalation mechanism of WSe_2 onions toward metal ion capacitors: Sodium rivals lithium. *Journal of Materials Chemistry A*, 2018; **6**(43): 21605–21617.

96. Chun Y-G, Lee W-J, Lee M, and Paek S-M. Effect of long-range and local order of exfoliated and proton-beam-irradiated WSe_2 nanosheets for sodium ion battery application. *Bulletin of the Korean Chemical Society*, 2018; **39**(5): 665–670.

Chapter 7

Effect of Polymeric Binders on the Sodium-Ion Storage Performance of Positive and Negative Electrode Materials

Ramaprabhu S. and Ajay Piriya V. S.
Department of Physics, Indian Institute of Technology Madras, Chennai, Tamil Nadu 600 036, India
ramp@iitm.ac.in

After several years, research progress on sodium-ion batteries has reached an advanced level pertaining to electrodes. But the progress in the advancement of sodium-ion batteries is also accompanied by the other external factors of electrodes such as binder, electrolyte, and conducting additives. The performance of the battery is affected not only by the electrode materials but also by the choice of binder. Though binder occupies a minor portion in the electrode material, the parameters like cyclic stability and rate capability are affected by the presence of binder. This chapter focuses on the effect of various binders affecting the performance of anode and cathode materials.

Handbook of Sodium-Ion Batteries: Materials and Characterization
Edited by Rohit R. Gaddam and X. S. (George) Zhao

ISBN 978-981-4968-15-7 (Hardcover), 978-1-003-30874-4 (eBook)
www.jennystanford.com

7.1 Introduction

The performance of the battery depends largely on the type and the amount of active electrode materials used in the battery design. The greater the amount of active electrode materials used in a restricted volume, the greater the amount of energy stored. In the electrode design, the active materials are used along with the conducting additives and binders in order to improve the electrical conductivity and mechanical stability of the electrode (Fig. 7.1). These inactive materials should be used as minimum as possible. The role of the binder in the electrode material is to adhere the active material to the current collector. Using a higher amount of binder in the electrode reduces the active material usage, and at the same time, not using a binder can lead to disintegration of the active material from the current collector. Generally, in a commercial battery electrode configuration, the binder contributes to about 2–5 wt.% of the total mass (active material, conductive agent, and binder). Certain desirable properties are required for a material to be used as a binder: (1) excellent adhesion to the metal electrode even in the case of volume expansion, (2) electrochemical stability over a potential range, (3) high melting points, (4) low swelling rate and insolubility into the electrolyte, (5) blocked ion dissolution and side reactions with the electrolyte, (6) facile electrode processing, and (7) environmentally friendly and low cost. Though the major contribution of a binder in the electrode material is to bind the active material, the performance of the electrode material is affected by the interaction of functional groups in the binder with the electrode material and the subsequent formation of the solid electrolyte interphase (SEI) layer [1]. Moreover, the choice of the binder also varies for the positive and negative electrodes. The binders that satisfy most of the requisite characteristics are polyvinylidene fluoride (PVDF) and styrene-butadiene copolymer (SBR). The PVDF binder is observed to be suitable for both positive and negative electrode materials. PVDF is oxidation and reduction resistant during the operation of the battery. SBR provides better properties compared to PVDF, but it is prone to oxidation at higher voltages and cannot be used for positive electrode materials. Therefore, SBR is used widely for negative electrode materials for its better binding ability and environmentally friendly due to the use of an aqueous solvent. In addition, other kinds of water-soluble binders like polyacrylonitrile are also used.

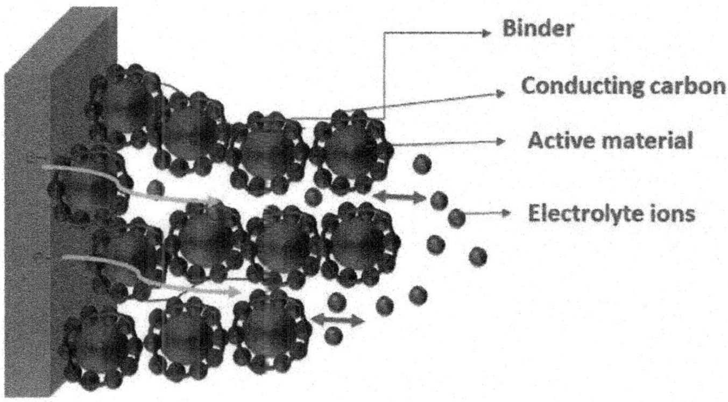

Figure 7.1 Schematic of an electrode with the dispersion of active material, binder, and conducting carbon.

7.2 Binders for Electrode Materials in Battery

Similar to the graphite anode in a lithium-ion battery, hard carbon is a widely accepted anode material for sodium-ion batteries. Initially, the PVDF binder is used along with the electroactive hard-carbon anode. But the use of the PVDF binder exhibits poor reversibility when compared to sodium carboxy methyl cellulose (CMC). Also the water-soluble binders provide potential benefits like low electrode processing cost and faster drying of electrodes. In fact, the cost reduction in using aqueous binder accounts for almost half of the current collector processing cost. Along with the use of binders soluble in environmentally benign solvents with the focus on the environmental concern, the chemical composition of a binder also decides the environmental friendliness of binders in the electrode. The presence of fluorine and ammonium lauryl sulfate in the binder is considered hazardous, since these polymers are highly difficult to dispose at the end of battery life. Binders like PVDF and polytetrafluoroethylene (PTFE), which are soluble in N-methyl-2-pyrrolidone (NMP) and water, respectively, are considered toxic due to the presence of fluorine in their chemical composition. The different categories of binders used in the electrode preparation are given in Fig. 7.2.

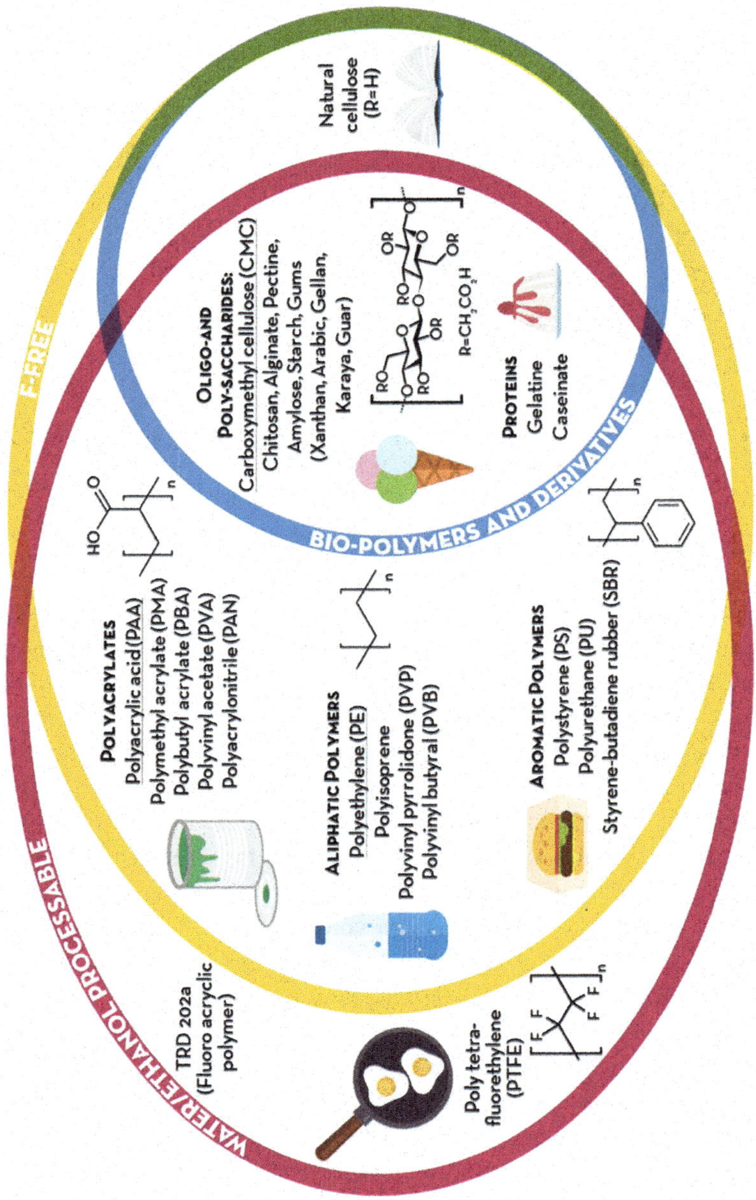

Figure 7.2 Classification of various types of binders (reproduced from Ref. [2]).

7.2.1 Physical Properties of PVDF Binder

PVDF is a type of fluoropolymer, also known as Kynar, with the chemical formula $-(C_2H_2F_2)_n$ and appears like a white crystal, which is insoluble in water. PVDF is a semicrystalline thermoplastic polymer with high mechanical strength and chemical resistance. The insoluble nature of PVDF polymer results from the presence of alternating CH_2 and CF_2 groups on the polymer chain (Fig. 7.3).

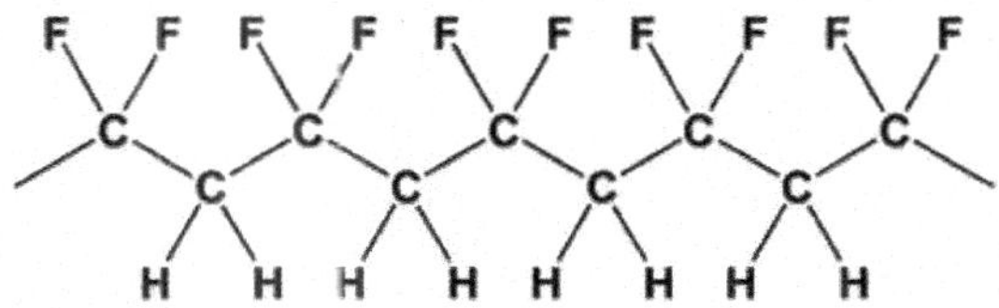

Figure 7.3 Molecular structure of polyvinylidene fluoride (PVDF).

The PVDF polymers are generally synthesized by the polymerization of 1,1-difluoroethylene at the temperature range of 10–150°C and a pressure of 10–300 atm (Fig. 7.4).

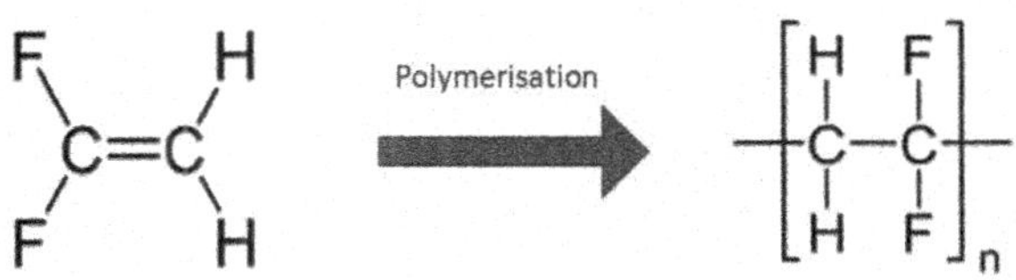

Figure 7.4 Polymerization of 1,1-difluoroethylene to form PVDF polymer.

PVDF polymer shows five crystalline phases α, β, γ, δ, and ε depending on the molecular chain conformation formed during processing methods. The α and ε phases of PVDF are nonpolar, and the other phases of PVDF polymer are polar, which exhibit both ferroelectric and piezoelectric properties. The α-phase of PVDF is the most commonly obtained polymorph of PVDF polymer and is obtained from the melt by the process of crystallization.

7.2.2 Polytetrafluoroethylene

Polytetrafluoroethylene is also a type of fluoropolymer formed by the polymerization of tetrafluoroethene molecules, which is also known as Teflon. The polytetrafluoroethylene is also chemically and thermally stable and insoluble in most of the solvents due to the stability of fluorine to carbon. The PTFE polymer consists of a chain of carbon atoms with two fluorine atoms bonded to each other (Fig. 7.5). The polymer is formed by the suspending monomers in water and undergoes polymerization under high pressure in the presence of free-radical initiators.

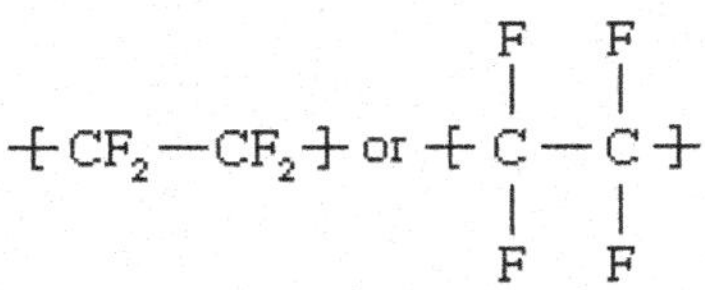

Figure 7.5 Chemical structure of polytetrafluoroethylene.

The PTFE polymer has the melting point of 327°C and is highly insoluble in most of the solvents.

7.2.3 Sodium Carboxymethyl Cellulose

Sodium carboxymethyl cellulose is a cellulose-derived polymer with carboxymethyl groups bonded to hydroxyl groups, which make the cellulose backbone. The sodium carboxymethyl cellulose is a water-soluble binder at all temperatures due to the presence of the (-CH_2-COOH-) group along with the presence of cellulose (Fig. 7.6).

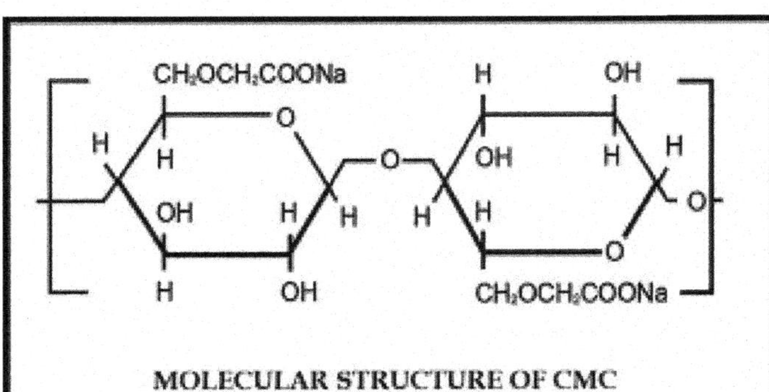

Figure 7.6 Chemical structure of sodium carboxymethyl cellulose.

The sodium carboxymethyl cellulose is synthesized by the catalytic reaction of cellulose with chloroacetic acid. Sodium carboxymethyl cellulose has adhesion, thickening, emulsifying, and suspending properties, which make them suitable to be used as the binder for electrode materials in a battery. CMC with a larger molecular weight has longer chain and aids in the strong binding of the active materials to the current collector. In addition, the degree of substitution (presence of positive ions such as sodium, potassium ions) in the CMC provides more active sites for binding the electrode materials. At the same time, the smaller degree of substitution results in the curling of the polymer and instability of the electrode. Hence, for the highly durable electrodes, CMC with a high degree of substitution is observed to be superior to CMC with a low degree of substitution.

7.2.4 Sodium Alginate

Sodium alginate is a linear polysaccharide derivative of alginic acid, which is distributed widely in the cell wall of marine brown algae and contains about 30 to 60% alginic acid. This alginic acid is converted into sodium alginate by reaction with sodium bicarbonate. The presence of a number of hydroxyl and carboxyl groups along the alginate backbone helps in the solubility of the binder in water (Fig. 7.7).

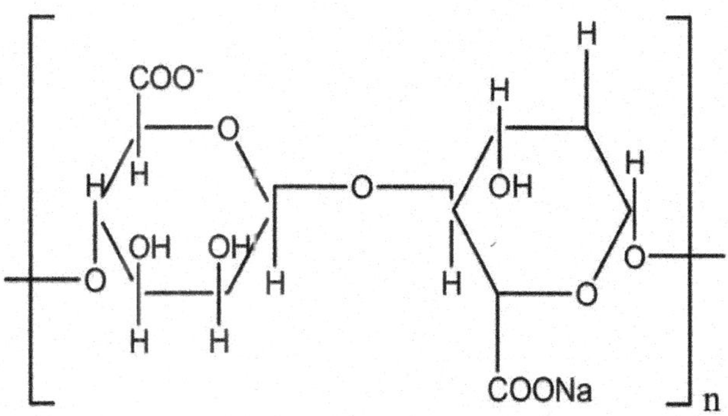

Figure 7.7 Chemical structure of sodium alginate.

7.2.5 Polyacrylic Acid

Polyacrylic acid is a synthetic polymer obtained by the free-radical polymerization of acrylic acid. Polyacrylic acid is soluble in aqueous media due to the presence of carboxyl chains on every two carbon atoms at neutral pH (Fig. 7.8). These are considered nontoxic and can be used as a good dispersing agent.

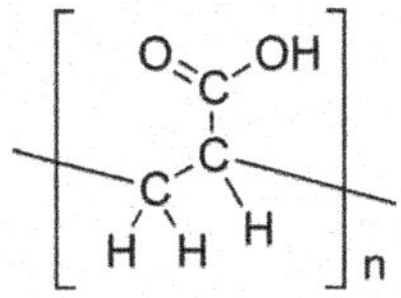

Figure 7.8 Chemical structure of polyacrylic acid.

7.3 Method of Making Slurry for Electrode Materials

The major component in the slurry preparation process includes the active material along with the formulating agents such as binder, conducting additive, and solvent to form the viscous slurry. The preparation of slurry at the desired viscosity is required for the fabrication of electrode with uniform thickness. The uniform distribution of active materials in the electrode is crucial as it decides the utilization of active materials and subsequently increases the specific capacity of the battery. The concentration of binder in the electrode varies between 3% and 15%, and the use of a higher concentration of binder improves the adhesion of the active material to the current collector but at the same time degrades the performance of the battery. The uniform mixing of slurry is important for the coating and drying process, which eventually affects the performance of the battery. Along with the choice of a suitable binder, the proper selection of coating process is critical for achieving uniformly coated electrodes. The dispersion of powder consisting of particle clusters suspended in the solvent forms the slurry. The slurry consists of larger agglomerates of particles, which in turn consist of smaller aggregates of particles. The larger agglomerates of particles are assembled by the weak van der Waals force, and the smaller aggregates of particles are bonded by electrostatic bonds (Fig. 7.9). The agglomerates of the particles are classified as soft agglomerates and hard agglomerates.

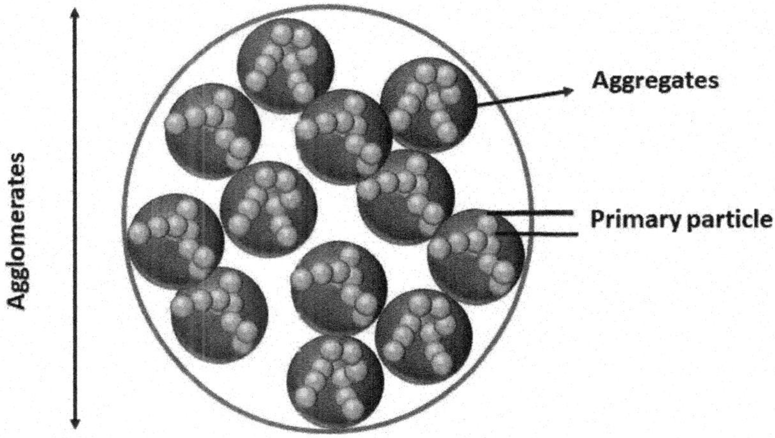

Figure 7.9 Structure of slurry with agglomerate/aggregate of particles.

The transformation of powdered substances to form a stable dispersion is the most significant step in the process of preparation of the slurry. The process of formation of a stable dispersion includes mixing of solids, wetting of the mixture, and formation of a uniform suspension.

In the laboratory, various kinds of slurry mixers are used such as ball milling, magnetic stirrers, and manual mixing. But for the preparation of an electrode slurry for a large-scale device, planetary mixers, homogenizers, universal type mixers, and static mixers are generally used (Fig. 7.10). The process/techniques employed for the fragmentation of the solid particle clusters can also be used for the preparation of the slurry. Ultrasonic mixers and shear fluid flow mixers are widely used for slurry preparation process. Based on the method of applying stress to the clusters, equipment is classified into two types: one applying shear stress through the liquid media and another applying stress through the solid media. Among the two-mixing equipment, shear stress applied through the solid particle offers better dispersion efficiency when compared to the technique of applying stress through the liquid media. The type of mixers used for making the slurry preparation also depends on the size of the cluster particles.

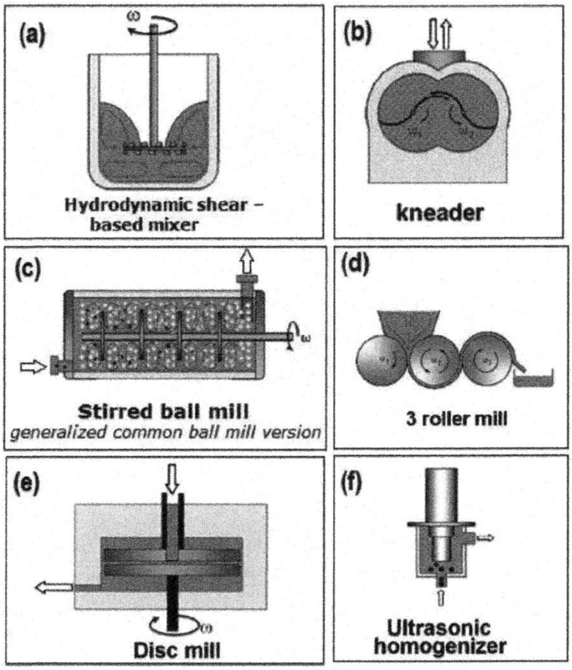

Figure 7.10 Different slurry coating methods. Reprinted with permission from Ref. [19], Copyright 2018, Elsevier.

7.3.1 Hydrodynamic Shear Mixing

Hydrodynamic shear mixing for the process of making slurry involves hydrodynamic forces, and the force is controlled by flow shear rate, viscosity of the fluid, and the cluster area. The process of hydrodynamic shear mixing involves the breakup of cluster particles and reassociation and restructuring of the suspended particles. The cluster breakage involves three major steps: erosion, rupture, and scattering. A weak force exists between the particles, and the process of erosion involves the fragmentation of particles using low energy. During the process, the small particles shear off from the larger cluster. By this method of rearrangement of clusters after the mixing process, the density of the slurry increases, thereby reducing the porosity of the electrode. The performance of the electrode strongly depends on the porosity of the active material and conductive additive with the maintenance of the internal structure of the electrode active material. Hydrodynamic shear mixing is the convenient method of slurry preparation and is considered more suitable for primary particles of size greater than 100 nm (submicron sized particles). Generally, the conducting carbon black is in the nanometer range, and the commercial cathode and anode materials are in the micrometer range with the primary cluster of particles in the size of nanometer. Since the slurry involves a mixture of nanometer- and micron-sized particles, the cluster particle size will be greater than 100 nm. This method of making electrode slurry can be used after gaining knowledge of the particle size of the anode and cathode materials. The technique is used for the preparation of anode and cathode materials for lithium-ion batteries and can be applied to sodium-ion batteries.

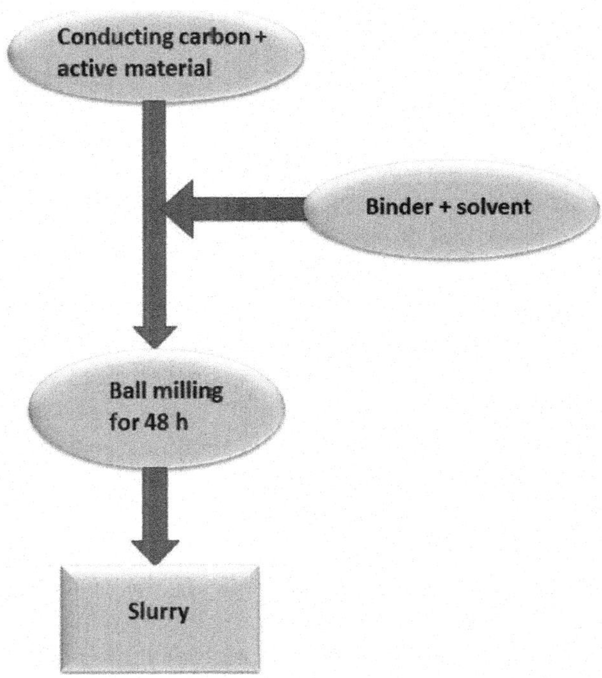

Figure 7.11 Process of preparation of anode slurry using the ball-milling technique.

7.3.2 Ball Milling

Among the various mixers used, ball milling is considered an effective mixer, which improves the mixing efficiency, although the process of mixing is long. So a mixer with a shorter mixing time along with better efficiency in mixing will be considered a suitable device for large-scale applications. During the process of ball milling, the active particles undergo structural and volumetric changes, which result in mechanochemical transformation of the material. Also the ball collisions and shear turbulence can lead to scission of the binder.

This method of slurry preparation is recommended for particles wherein the morphology of the conductive carbon and active material is not considered an essential factor in obtaining the performance of the battery.

7.4 Effect of Slurry Preparation Process on Electrode Morphology

The performance of the electrode depends on the size and morphology of the active materials and conductive additives that are added to the preparation of the electrode slurry. The use of bulk sized particles leads to the agglomeration of particles in the slurry, leading to the creation of pits in the electrode after the process of drying. The size of the agglomerates should be less than the size of the thickness of electrode coating. The presence of larger agglomerates in the slurry results in the creation of defects in the electrode-making process. So it is better to have smaller sized particles to operate the electrode at a higher current density due to the smaller diffusion path lengths of lithium ions. In addition, the downsizing of the particles also improves the coulombic efficiency of the battery. Along with the size of the active material, the optimum ratio of the active material and conducting additive is essential for the slurry preparation process. The electrode laminate thickness is another significant factor wherein higher thickness result in inhomogeneity of the distribution of the electrode materials and the creation of pin-hole defects. After the process of controlling the slurry parameters and the coating process, it is essential to characterize the uniformity of the coating by a simple approach that can be applied for large-scale production. The electrode coating thickness can be evaluated using the X-ray radiography technique, which is a nondestructive analytical technique. The technique provides information on the quality of electrode coating for a larger area with the resolution of

20 μm. This technique can be applied for quality checking of a large area electrode for the pilot plant. The technique can be applied for thick electrodes as well.

7.5 Binders for Anode Materials for Sodium-Ion Batteries

7.5.1 Binders for Carbon-Based Anode Materials

The commercial graphite used in lithium-ion batteries cannot be used as an anode material for sodium-ion batteries due to unfavorable thermodynamic kinetics. Therefore, research has been focused on carbonaceous materials with larger interlayer spacing like petroleum cokes, carbon black, carbon nanofiber, and soft and hard carbon. Among these carbon-based materials, hard carbon is an attractive anode material due to the better intercalation of sodium ions leading to higher specific capacity. On the other hand, the rate capability and cycle life of hard-carbon anode are found to be poor. The performance of the hard-carbon anode gets affected not only by its structure and morphology but also by the use of binders along with the active material in the electrode. Conventionally, PVDF is used as the binder for hard-carbon anode in sodium-ion batteries. Although PVDF is observed to have good binding capability that can adhere the hard carbon to the current collector, the cyclic stability of hard-carbon anodes using PVDF binder decays over cycles [3]. A synergy exists between the binder and the electrolyte additive on the performance of the hard-carbon anode. Dahbi et al. have analyzed the performance of hard-carbon anodes with PVDF and CMC binders along with the effect of the FEC electrolyte additive on performance. The synergistic effect of electrolyte and binder affects the passivation layer formed on the surface of hard-carbon anodes, which subsequently affects the performance of the anode. This is revealed from the initial high coulombic efficiency of the hard anode with CMC binder than PVDF binder.

The reversibility of sodium-ion storage in hard-carbon anodes using CMC binder is observed to be excellent when compared to hard-carbon anodes with PVDF binder. Pol et al. have used PVDF as the binder for the interconnected carbon network as the anode for sodium-ion batteries. The anode delivered reasonably good performance with the cyclic stability of 100 cycles, and the capacity retention at the end of 100 cycles was about 91.3%. The initial coulombic efficiency is observed to be very low (39.7%). The reason behind the low coulombic efficiency in this work is not studied in detail [4]. But the use of PVDF binder might be one of the reasons for the low coulombic efficiency of the carbon anode. The similar behavior of low initial coulombic efficiency for the hard-carbon anode using PVDF binder has been shown by Liu et al., which lies in the range of 29–34% [5]. It has been suggested in this study that the performance of the hard-carbon anode can be increased by changing the binder as well as the modification of the surface of hard-carbon anodes. The biomass-derived hard carbon has been widely studied as the anode material for sodium-ion batteries. The performance of the hard-carbon anode is optimized by tuning the structure of hard carbon by controlling the preparation conditions. The initial coulombic efficiency of biomass-derived hard-carbon anode is also observed to be less than the industrialized standard. Shao et al. derived hard carbon from pine cone with a high initial coulombic efficiency of 85.4% [6]. The electrode was fabricated using water-soluble sodium CMC binder. The high initial coulombic efficiency is due to the surface morphological control by optimizing the reaction temperature along with the use of CMC binder rather than the use of conventional PVDF-based binder. In addition to the hard-carbon-based anode for sodium-ion batteries, one-dimensional as well as two-dimensional carbon nanostructures are also used as the anode materials for sodium-ion batteries. Nitrogen-doped multiwalled carbon nanotubes with turbostratic structure have been employed as the anode material, and the effects of binders like PVDF and sodium polyacrylic acid (Na-PAA) on the performance of sodium-ion batteries have been analyzed. The use of water-soluble binders like polyacrylic acid (Na-PAA) enhances the specific capacity, coulombic efficiency, rate performance, and cyclic stability of the anode, when compared to organic-solvent-soluble PVDF binder.

Creating defects in the carbon structure increases the formation of irreversible reactions at the anode. It has been observed that nitrogen doping in any carbon structure increases the defects and leads to poor initial coulombic efficiency. The initial coulombic efficiency of nitrogen-doped carbon nanostructure is less than 50%,

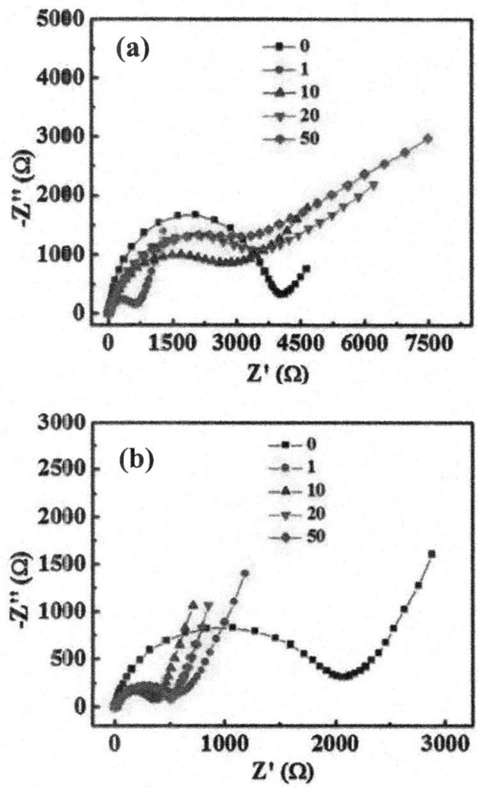

Figure 7.12 Electrochemical impedance spectra of N-CNT using (a) PVDF binder and (b) Na-PAA binder [7].

which is far behind the practical needs of commercial batteries. So the initial coulombic efficiency of sodium-ion battery anodes should be increased to couple with the suitable cathode material for the development of full cell sodium-ion batteries. Huang et al. have improved the initial coulombic efficiency of nitrogen-doped multiwalled carbon nanotubes with the use of Na-PAA binder. The use of PVDF binder shows an initial coulombic efficiency of 45.3%, and with the use of Na-PAA binder, the initial coulombic efficiency has enhanced to 61.2%, which is considered superior when compared with other reported carbon-based materials [7]. The rate capability of the electrode material with the use of Na-PAA is observed to be higher and is enhanced by ~25 times at a higher current density of 5 A/g. The reason behind the better performance of nitrogen-doped multiwalled carbon nanotubes with Na-PAA binder when compared to PVDF binder has been evaluated using the electrochemical impedance spectroscopy technique, as shown in Fig. 7.12.

The charge transfer resistance of the electrode with PVDF binder increases drastically before 20 cycles, but with the use of Na-PAA binder, the charge transfer resistance increases slightly during the initial cycles and stabilizes after 20 cycles. The increase in charge transfer resistance with the use of PVDF binder is due to the detachment of the electrode material from the current collector over the cycles. The nitrogen-doped carbon fibers have been evaluated as the anode material for sodium-ion batteries. For the fabrication of electrode, PVDF binder is used and the initial coulombic efficiency value is observed to be 46% [8]. The low initial coulombic efficiency is due to the use of PVDF binder, which is also evident from this study. So from these studies it is evident that the binder plays a critical role in improving the performance, like initial coulombic efficiency and rate capability of the carbon-based anode materials. So the selective choice of binder is crucial along with the design of electrode materials in order to obtain high-performance sodium-ion batteries.

7.5.2 Binders for Conversion-Based Anode Materials

Metal oxides have gained much attention as an anode material when compared to carbon-based materials due to their high theoretical capacity. Although the benefits of high theoretical capacity exist in conversion-based metal oxides, there are problems associated with them due to the phase change that occurs during the conversion process. The mechanism of sodium-ion storage in conversion-based materials is as follows:

$$MO_x + 2x\mathrm{Na}^+ \leftrightarrow M + x\mathrm{Na}_2\mathrm{O}$$

During the interaction of metal oxides with sodium ions, the metal oxides get converted into metal, and sodium oxide is formed. During the reverse process, the metal is converted back into metal oxide. These conversion reactions create phases in the electrode, and thereby volume changes occur at the electrode, which subsequently leads to pulverization of the electrode and results in rapid capacity fading. Though the use of conducting carbon polymer alleviates the problem of volume expansion, the significant role of binding the electrode material after pulverization is played by the binder. During this process, the adherence of the electrode material to the current collector is crucial to prevent the capacity fading. As the size of sodium ions is more when compared to lithium ions, the problem of pulverization exacerbates and hence the choice of binder in preparing the electrode material is crucial. The various metal oxides as well as metal sulfides such as CuO, Fe_2O_3, MnO_2, NiO, TiO_2, and FeS_2 are widely explored along with a few other metal oxides. Chen et al. demonstrated that CuO nanosheets coated using CMC binder showed enhanced performance when compared to PVDF binder. The CuO nanosheets using CMC binder show better structural integrity to the electrode aiding in accommodating the volume expansion during the cycling. This results in high specific capacity of 627 mAh/g at 50 mA/g and a capacity retention of about 73.4% at the end of 100 cycles. In the case of metal-oxide-based electrodes, added to the cycling and rate performance of the battery, the initial coulombic efficiency is also affected by the choice of binder. The reported initial coulombic efficiencies of CuO-nanosheets-based anode using PVDF and CMC binders are observed to be 19.6% and 42%, respectively. The use of PVDF binder results in more irreversible reaction in the initial cycle and results in the formation of thicker SEI layer. This signifies that the binder plays a significant role in controlling the irreversible capacity loss in the initial cycles. This is also evident from the surface morphological image of the electrode in the fresh state and cycled after 100 cycles. The surface morphology of the electrode with PVDF binder is observed to have cracks on the surface, while the electrode with CMC binder does not have any cracks and pits, as the CMC binder binds closely toward the current collector. The binder helps in maintaining the structural integrity of the electrode by controlling the volume expansion of the electrode materials. Moreover, the effects of binder on the electrochemical performance have been analyzed using the electrochemical impedance spectroscopy technique. The CuO nanosheets with CMC binder offer lower charge transfer resistance when compared to the electrode using PVDF binder.

There are reports wherein CuO nanoparticles confined to ordered mesoporous carbon are used as anode material for sodium-ion batteries, with the use of PVDF binder. The presence of ultrasmall 4 nm CuO nanoparticles confined to mesoporous carbon helps to accommodate the volume expansion of nanosized particles. The initial coulombic efficiency was in the range of 50–55% for the different sized CuO nanoparticles. The use of smaller sized particles as the active material with the use of PVDF binder does not create much volume expansion when compared to the larger sized CuO particles [9]. Chen et al. have loaded 56% nanosized CuO (10 nm) into the carbon spheres and the CuO particles are homogenously distributed in the carbon matrix. Even at the high loading of CuO in the carbon matrix, nanosized CuO embedded in carbon matrix exhibits a good cyclic stability of 600 cycles. In this study, PVDF is used as the binder and the effect of the binder on the electrochemical performance of the electrode has not been analyzed in this study [10]. So the effect of the binder on different sized particles will be a good approach for analyzing the role of the binder in affecting the performance of different sized particles. Iron oxide is another type of cost-effective anode material for sodium-ion batteries. For iron oxide as the anode material, different binders such as PVDF, CMC, and sodium alginate have been used in the literature [11–15]. MoS_2 is a 2D layered material similar to graphite, which facilitates the intercalation of sodium ions as well as conversion reactions. In addition, MoS_2 has high theoretical specific capacity for the intercalation of sodium ions. The intercalation of sodium ions into the layered structured MoS_2 results in restacking of the layers, leading to capacity fade over the cycles. Various binders such as PVDF, CMC, SBR, and poly(amide imide) have been used for the fabrication of anode. But the effect of the binder on the performance of sodium-ion battery anode has not been studied in detail. Although the SEI layer formation depends on the surface of the material, the formation of irreversible compounds in the first cycle also depends on the binder used. The pure MoS_2 shows initial coulombic efficiency of about 70.3% with the use of PVDF binder [16]. MoS_2@ rGO delivers initial coulombic efficiency of 74.6% when CMC is used as the binder for the preparation of electrode [17]. The use of different binders does not significantly affect the initial coulombic efficiency of MoS_2-based anode materials for sodium-ion batteries.

7.6 Binders for Cathode Materials for Sodium-Ion Batteries

Binders for cathode materials should have properties such as electrochemical stability under an oxidative environment as the cathode material will be operated at higher voltages. PVDF is the most commonly used binder for the preparation of cathodes, but PVDF suffers from swelling due to the interaction of the binder with the electrolyte, especially with ester-based electrolytes. In addition, it also imposes environmental threats due to the use of N-methyl-2-pyrrolidone (NMP) as the solvent for the binder. So moving toward an aqueous binder seems to be environmentally friendly, but the use of an aqueous binder for the hygroscopic cathode materials poses several challenges. So the impact of using water-based binders on the performance of sodium-ion batteries needs further investigation.

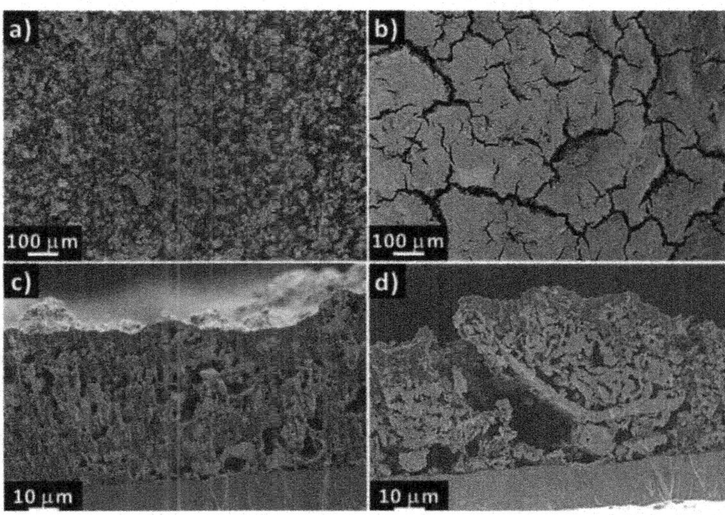

Figure 7.13 SEM image of $Na_{0.67}Mn_{0.6}Fe_{0.25}Co_{0.15}O_2$ using (a) PVDF binder and (b) CMC binder; cross-sectional image of $Na_{0.67}Mn_{0.6}Fe_{0.25}Co_{0.15}O_2$ electrode with (c) PVDF binder and (d) CMC binder (reproduced from Ref. [18]).

The effect of using water-based binders on the performance of P2-layered metal oxide $Na_{0.67}Mn_{0.6}Fe_{0.25}Co_{0.15}O_2$ has been investigated by Villevieille et al. [18]. The morphology of the electrode with the use of CMC binder creates cracks on the surface of the electrode, and the performance of P2-layered metal oxide deteriorates with the use of CMC binder. The use of CMC-based binder causes agglomeration of the conducting carbon on the surface of the electrode without properly embedding the active material in the electrode (Fig. 7.13). The inhomogeneous distribution of the active material leads to poor electrochemical performance, especially at high currents. The use of water as the solvent results in the reaction of moisture-sensitive cathode material with the water, leading to the formation of inactive phases on the surface of the electrode.

7.7 Conclusion

This chapter illustrates the use of different binders in the electrode preparation of sodium-ion batteries. The effects of using different binders on the electrochemical performance of anode and cathode materials have been elaborately discussed, and an outline on the binder's effects on obtaining the performance of the battery in terms of storage capacity and durability of the battery has been provided. Although much focus is on electrode material properties, exploration of new water-soluble environmentally friendly binders is highly essential in the development of sodium-ion battery technology. The role of binders in cathode materials needs deeper investigation, which will further help in the advancement of performance and the manufacturing process of electrodes for sodium-ion batteries. The use of water-soluble binders in the electrode-manufacturing process can significantly reduce the cost of commercial sodium-ion batteries.

References

1. Bommier, C. & Ji, X. Electrolytes, SEI formation, and binders: A review of nonelectrode factors for sodium-ion battery anodes. *Small* **14**, 1–20 (2018).
2. Bresser, D., Buchholz, D., Moretti, A., Varzi, A. & Passerini, S. Alternative binders for sustainable electrochemical energy storage-the transition to aqueous electrode processing and bio-derived polymers. *Energy Environ. Sci.* **11**, 3096–3127 (2018).

3. Dahbi, M. *et al.* Sodium carboxymethyl cellulose as a potential binder for hard-carbon negative electrodes in sodium-ion batteries. *Electrochem. commun.* **44**, 66–69 (2014).
4. Hong, S.-M. *et al.* Enhanced lithium- and sodium-ion storage in an interconnected carbon network comprising electronegative fluorine. *ACS Appl. Mater. Interfaces* acsami.7b03456 (2017) doi:10.1021/acsami.7b03456.
5. Wang, H. *et al.* Biomass derived hierarchical porous carbons as high-performance anodes for sodium-ion batteries. *Electrochim. Acta* **188**, 103–110 (2016).
6. Zhang, T. *et al.* Pinecone biomass-derived hard carbon anodes for high-performance sodium-ion batteries. *RSC Adv.* **7**, 41504–41511 (2017).
7. Fan, Q. *et al.* Effects of binders on electrochemical performance of nitrogen-doped carbon nanotube anode in sodium-ion battery. *Electrochim. Acta* **174**, 970–977 (2015).
8. Fu, L. *et al.* Nitrogen doped porous carbon fibres as anode materials for sodium ion batteries with excellent rate performance. *Nanoscale* **6**, 1384–1389 (2014).
9. Chandra Rath, P. *et al.* Highly enhanced electrochemical performance of ultrafine CuO nanoparticles confined in ordered mesoporous carbons as anode materials for sodium-ion batteries. *J. Mater. Chem. A* **4**, 14222–14233 (2016).
10. Lu, Y., Zhang, N., Zhao, Q., Liang, J. & Chen, J. Micro-nanostructured CuO/C spheres as high-performance anode materials for Na-ion batteries. *Nanoscale* **7**, 2770–2776 (2015).
11. Li, D., Zhou, J., Chen, X. & Song, H. Amorphous Fe_2O_3/Graphene composite nanosheets with enhanced electrochemical performance for sodium-ion battery. *ACS Appl. Mater. Interfaces* **8**, 30899–30907 (2016).
12. Li, M. *et al.* Well-ordered mesoporous Fe_2O_3/C composites as high performance anode materials for sodium-ion batteries. *Dalt. Trans.* **46**, 5025–5032 (2017).
13. Wang, Q. *et al.* Plasma enabled Fe_2O_3/Fe_3O_4 nano-aggregates anchored on nitrogen-doped graphene as anode for sodium-ion batteries. *Nanomaterials* **10**, 1–12 (2020).
14. Park, G. D., Cho, J. S., Lee, J. K. & Kang, Y. C. Na-ion storage performances of $FeSe_x$ and Fe_2O_3 Hollow nanoparticles-decorated reduced graphene oxide balls prepared by nanoscale kirkendall diffusion process. *Sci. Rep.* **6**, 1–10 (2016).

15. Liang, M., Zou, J., Zeng, X. & Ding, W. Nanostructured Fe_2O_3 based composites prepared through arc plasma method as anode materials in the lithium-ion battery. *J. Nanomater.* **2016**, (2016).

16. Xu, Z. *et al.* Sulfur nanodots as MoS_2 antiblocking agent for stable sodium ion battery anodes. *J. Mater. Chem. A* **6**, 10535–10542 (2018).

17. Wang, R. *et al.* MoS_2@rGO Nanoflakes as high performance anode materials in sodium ion batteries. *Sci. Rep.* **7**, 1–9 (2017).

18. Marino, C., Marelli, E., Park, S. & Villevieille, C. Impact of water-based binder on the electrochemical performance of P2-$Na_{0.67}Mn_{0.6}Fe_{0.25}Co_{0.15}O_2$ electrodes in Na-ion batteries. *Batteries* **4**, 1–12 (2018).

19. Schilde, C., Mages-Sauter, C., Kwade, A. & Schuchmann, H. P. Efficiency of different dispersing devices for dispersing nanosized silica and alumina. *Powder Tech.* **207**, 353–361 (2018).

Chapter 8

Organic Liquid Electrolytes for Sodium-Ion Batteries

Qingbing Xia[a] and X. S. (George) Zhao[a,b]
[a]*School of Chemical Engineering, The University of Queensland, St Lucia, Brisbane, QLD 4072, Australia*
[b]*Institute of Materials for Energy and Environment, Qingdao University, Shandong 266071, China*
george.zhao@uq.edu.au

8.1 Introduction

Lithium-ion batteries (LIBs) have been dominating the battery market for decades [1–3]. However, the rapidly increasing demand for LIBs has led to great concerns over lithium and cobalt resources [1, 4–7]. As a result, alternative rechargeable battery technologies have received great interests. Sodium-ion batteries (NIBs) that have a working principle similar to that of LIBs have proven to be a promising alternative to LIBs [5–11]. Sodium has a redox potential of −2.71 V versus the standard hydrogen electrode (SHE) and similar chemical and physical properties with that of lithium [7, 8, 12, 13].

Handbook of Sodium-Ion Batteries: Materials and Characterization
Edited by Rohit R. Gaddam and X. S. (George) Zhao

ISBN 978-981-4968-15-7 (Hardcover), 978-1-003-30874-4 (eBook)
www.jennystanford.com

Besides, sodium ranks the fourth most abundant element on earth. Its elemental properties of high abundance and low cost make NIB chemistry an attractive potential complementary to LIBs, especially in large-scale energy storage applications [7–9].

In the past few years, many electrode materials have been investigated to enhance the energy density and cycling stability of NIBs. Transition metal layered oxides Na_xMO_2 (M = Co, Mn, Fe Ni, etc.) [14–18], Na-based polyanionic phases (like $NaFePO_4$, $Na_2Fe_2(SO_4)_3$, $NaVPO_4F$, etc.) [19–22], and Prussian blue phases $A_xMM'(CN)_6$ (A = Na, K; M and M' = Fe, Co, Ni, Mn) [23–25] have been extensively studied as promising cathode candidates for NIBs. Anode materials including carbon (e.g., hard carbon, graphite and soft carbon) [26–28], metal and metal oxides (like Sn, Sb, Bi, Ge, SnO_2, Sb_2O_3, TiO_2, etc.) [29–31], metal sulfides and selenides (MoS_2, $MoSe_2$, NiS_2, FeS, etc.) [32–35], metal phosphides (FeP, CoP_2, NiP_4, Cu_3P, etc.) [12, 36–38], and phosphorus (red P and black P) [39–42] have been exploited. Hard carbon is regarded the most promising anode material for NIBs due to its comparable electrochemical performance to that of graphite anode in LIBs [27].

In an NIB battery configuration, the electrolyte contacts with both the cathode, anode and separator. The electrolyte can profoundly determine the cell performance and properties, including safety, coulombic efficiency, rate capability, cycling and thermal stability [6, 9, 13, 43]. Besides, electrolytes have the ability to form a kinetically and structurally favorable interphase at the electrode/electrolyte interface, which is crucial in regulating the reaction kinetics and maintaining the structural integrity of electrodes.

Electrolytes can be generally classified into five categories, as illustrated in Fig. 8.1, namely organic liquid electrolytes, ionic electrolytes, aqueous liquid electrolytes, solid polymer electrolytes, and inorganic solid electrolytes [43–46]. In organic liquid electrolytes, the transport of Na ions is driven by the solvation effect and the electric filed, whereas segment and vacancy migration are the driving forces in polymer and solid electrolytes [43, 45]. As a result, the Na-ion conductivity in organic liquid electrolyte is much higher than that in solid electrolytes due to the superior fluidity. Solid electrolytes, however, show merits of good thermal stability and wide voltage windows, which are much safer than the widely used organic liquid electrolytes [47, 48]. Besides, good mechanical property of solid electrolytes shows potential in suppressing the aggressive growth of sodium metal dendrites. Despite the advantages of solid electrolytes, their high manufacturing cost and poor interfacial compatibility with the electrode materials are critical issues. Aqueous electrolytes can provide high ionic conductivity, excellent safety as well as low cost, but their narrow electrochemical voltage window results in a low cell-level energy density [49].

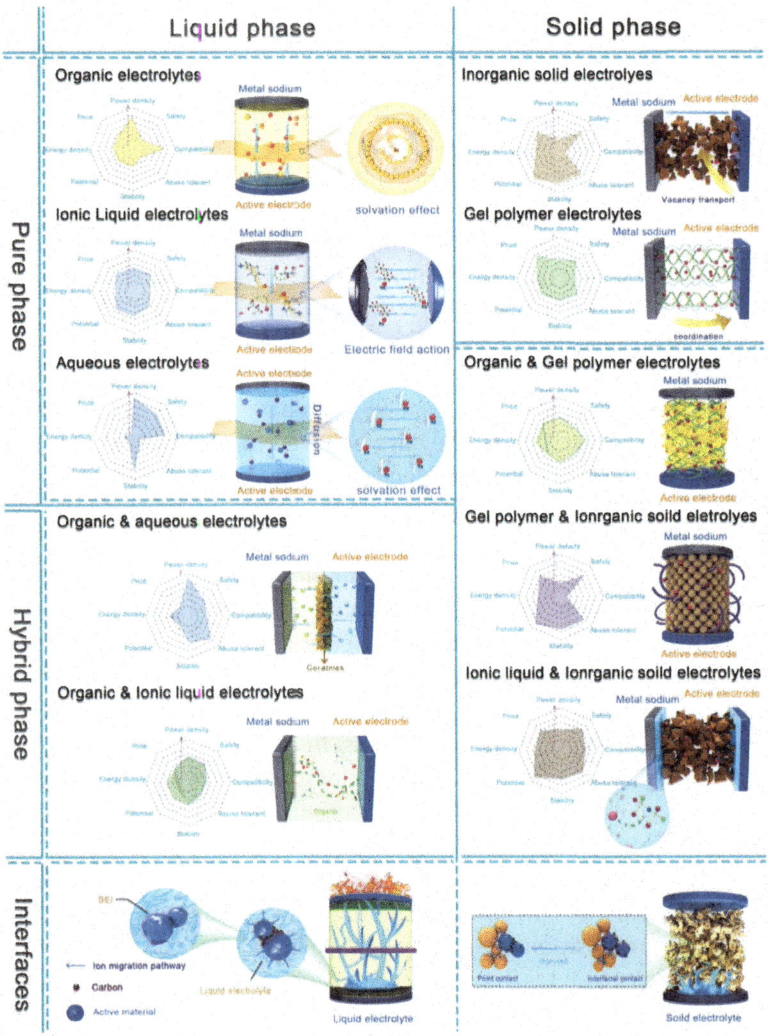

Figure 8.1 Typical electrolytes for NIBs [52].

Non-aqueous electrolytes for NIBs, especially organic liquid electrolytes, have been studied in the past few years [50, 51]. The non-aqueous electrolytes are capable of sufficient dissolving capacity of sodium salts to meet the requirement of high ionic conductivity. Besides, organic liquids ensure good compatibility with a wide range of electrode materials. For future grid energy storage, NIBs with high energy density, long cycling stability, low production cost, and good compatibility under extreme conditions are highly demanded. In current NIB chemistry, organic liquid electrolytes are wildly used and have been proven to be preferential to other electrolytes in achieving excellent battery performance, owing to their merits of high ionic conductivity and excellent wettability.

In battery cells with an organic liquid electrolyte, a passivation solid electrolyte interphase (SEI) film, which is a mixture of inorganic and organic Na^+-containing species, forms at the electrode/electrolyte interface, especially on the anode side [43, 45]. Such an SEI film is beneficial to improve the interface compatibility and stability of the electrode. However, the formation of SEI film by decomposition of the electrolytes will sacrifice the CE and irreversible capacity to a certain extent. The formation mechanism, chemical components, and structure of the SEI film have not been completely understood. Nevertheless, they are believed to be closely related to the chemical properties of the electrolyte and the electrode materials [53–55]. Besides, for some electrode materials, the electrolyte has a profound impact on sodium-ion storage capacity and mechanism of the electrode [53, 56, 57]. For example, graphite does not show significant sodium ion storage activity in carbonate ester-based electrolytes but exhibits excellent electrochemical performance in ether-based electrolytes through the Na^+-solvent co-intercalation process [56, 57]. Thus, the choice of electrolytes is important to improve the cell performance of NIBs. In addition, battery safety is closely associated with electrolytes. Flammable solvents are easy to catch fire when thermal runaway occurs inside the cell, resulting in serious safety hazards [58].

In this chapter, we summarise the recent development of organic liquid electrolytes in NIBs, including their designing principles, chemical compositions (e.g., solvents, sodium salts, and additives), the role of electrolytes in tuning cell performance, and the interfacial chemistry between electrolytes and electrode materials.

8.2 Characteristics of Organic Liquid Electrolytes

Electrolyte is one of the key elements in an NIB cell to achieve high energy/power density, long lifespan, and good safety. Meanwhile, its chemical and physical properties significantly govern the cell performance. The design of advanced electrolytes to achieve excellent battery performance plays increasingly significant role. Generally, the selection of organic liquid electrolytes should meet the following requirements (Fig. 8.2).

8.2.1 Ionic and Electronic Conductivity

In an NIB, the electrolyte is the Na-ion transport carrier between the cathode and anode separated by the separator. An ideal organic liquid electrolyte should have characteristics of high ionic (Na^+) conductivity and negligible electron conductivity [50, 52, 59, 60]. High ionic conductivity requires that the organic solvents should be capable of dissolving a large amount of sodium salts and minimising the resistance of the Na-ion transport. The requirement for zero electron conductivity can minimise cell self-discharge to the maximum extent, particularly under the fully charged state. Aqueous liquid electrolytes tend to have high ionic conductivity due to the low dynamic viscosity, strong dissolving ability, and high salt concentration in water. For organic liquid electrolytes, the ionic conductivity is determined by factors such as the viscosity and dielectric of organic solvents, the concentration of sodium salts, and the dissociation degree of the sodium salts [43].

The concentration of the sodium salt determines the Na-ion transport number in the electrolyte, which relies on the dissolving capacity of sodium salt in the organic solvent. Generally, carbonate ester solvents can dissolve more sodium salts than ether solvents. The ionic conductivity increases with the increase in sodium salt concentration. However, it tends to decrease under a high sodium salt concentration, rising from the electrostatic interference effect, which reduces the number of free ions [61, 62]. Besides, the interaction between the solvents and the cations/anions also influences the conductivity of electrolytes. Ponrouch et al. and Bhide et al. compared the conductivity of different sodium salts ($NaClO_4$, $NaPF_6$, NATFSI, and NaOTf) in PC solvent and EC/EMC mixture

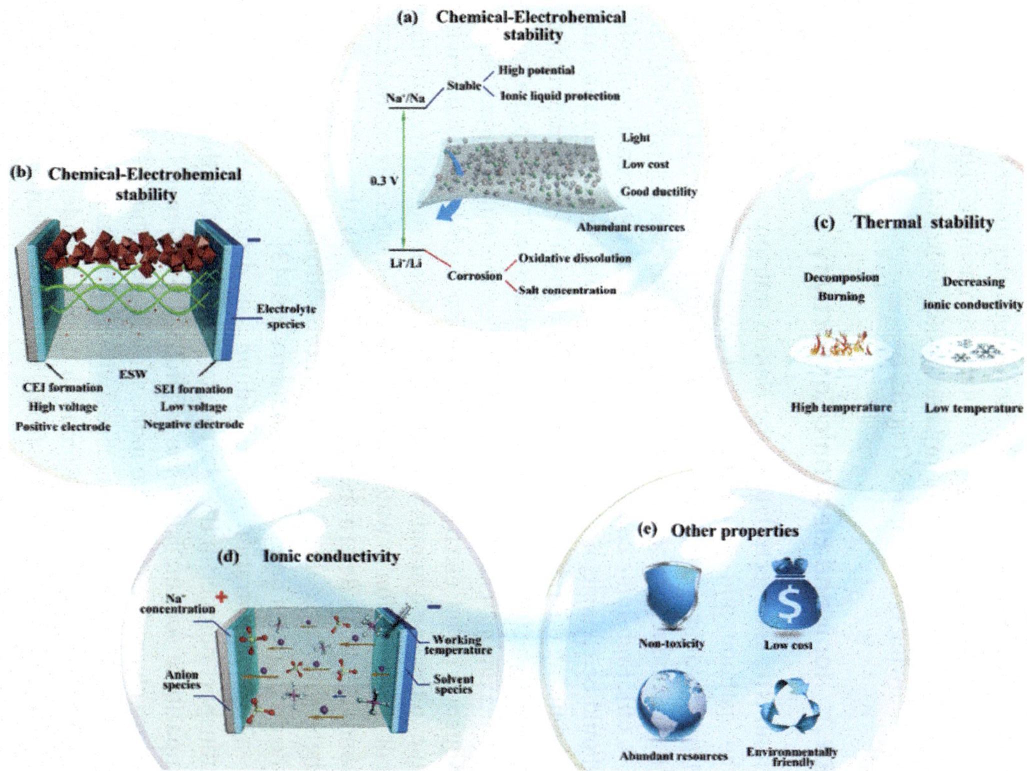

Figure 8.2 Summary of the main properties of electrolytes for NIBs [52].

solvent and found that $NaPF_6$ achieves the highest conductivity than other sodium salts in both solvents, attributing to the low polarization of the PF_6^- anions, which is helpful in improving sodium salt dissolution and ionic mobility [45, 60, 61].

The solvation capability is another intrinsic factor that impacts the ionic conductivity of organic liquid electrolytes. The Na-ion transport is defined as the process of turning a solvated Na ion in the electrolyte into Na in the electrode by accepting an electron from the electrode. The de-solvation of solvated Na ion at the electrolyte/electrode interface is the *rate-limiting step* of the sodiation process, i.e., intercalation of Na ions in the active materials and the following electrochemical reactions. A lower de-solvation energy provides a faster ion-transport rate and a higher electrochemical reaction activity [63]. The solvation effect between the solvent molecule and Na ion is also impacted by the sodium salt concentration and the type of solvents [64]. The chemical structure, electronic structure, and molecular chain segments of the solvents exhibit prominent effect on the de-solvation energy.

Besides, viscosity of the electrolyte also influences the ionic conductivity. In carbonate ester-based electrolytes, if the same sodium salt is used, the viscosities follow the order of DMC < EC/DMC < EC/PC < PC. Generally, binary-solvent-based electrolytes containing a solvent with high dielectric constant (like EC) and a solvent with low viscosity (such as DME and DMC) display a high ionic conductivity. On the other hand, sodium salts $NaClO_4$, $NaPF_6$, NaTFSI, and NaFSI in the same solvent show negligible effect on the viscosity of the electrolytes. However, their concentration has significant influence on viscosity. High concentration provides large Na-ion transfer numbers, nevertheless, bringing high viscosity which may rise adverse impact to the electrolyte ionic conductivity.

8.2.2 Electrochemical Stability

Electrochemical stability in a wide range of organic liquid electrolytes is an important factor for developing high-performance NIBs with high energy density, long lifetime, and safety. The electrolyte should be electrochemically inert toward both the positive and negative materials in the operation voltage window. The term of electrochemical stability window (ESW) represents the upper and lower limits of the redox reactions of an electrolyte [60]. An electrolyte with wide electrochemically stable window ensures a high cell voltage and contributes to reduce the side reactions with the electrode materials. Besides, it is also beneficial to the development and application of cathode materials with high working potential, thus achieving high energy density in cell level. For organic liquid electrolytes, the electrochemical window is highly associated with the species of solvent and the concentration of the sodium salt [52, 60]. Generally, carbonate ester-based electrolytes exhibit a higher ESW than ether-based electrolytes. Ponrouch et al. investigated the electrochemical stability of a series of electrolytes with 1 M $NaClO_4$ sodium salt. It was found that the electrochemical stability follows DEC > PC > triglyme > DME > DMC > THF and EC/PC > EC/DMC > EC/DME > EC/DEC > EC/triglyme [60].

The ESW is also related to the interfacial compatibility between the electrolyte and the electrode. SEI film-formation additives, such as fluoroethylene carbonate (FEC) and vinylene carbonate (VC), are helpful in promoting the formation of a stable SEI film as a protective layer on the surface of the electrode, thus contributing to improve the electrochemical stability of the organic liquid electrolytes [52].

8.2.3 Thermal Stability

During the charge/discharge processes, electrochemical side reactions will occur on the surface of electrodes, which results in the decomposition of organic liquid electrolytes. Severe electrolyte decomposition will lead to heat release and rise in temperature in the cell, bringing in safety concerns. For practical applications, especially for large-scale energy storage, thermal stability is crucial [45]. Generally, the battery cell should be capable of working in a wide temperature range from low temperature (below 0 °C) to high temperature (more than 60 °C), for which the freezing point and the melting point of the electrolytes must be outside the range of the operating temperature. Ponrouch et al. investigated the thermal stability of sodium salts using the DSC technique (Fig. 8.3b) and found the following trend: $NaClO_4$ (310 °C) > $NaPF_6$ (280 °C) > NaTFSI (250 °C) [60]. From the aspect of organic solvents, linear carbonate ester-based solvents, such as DMC and DEC, are highly reactive and may result in poor thermal stability. The strategy of adding EC and

PC solvents can dramatically improve the thermal stability of liquid organic electrolytes, as shown in Fig. 8.3a [52, 60].

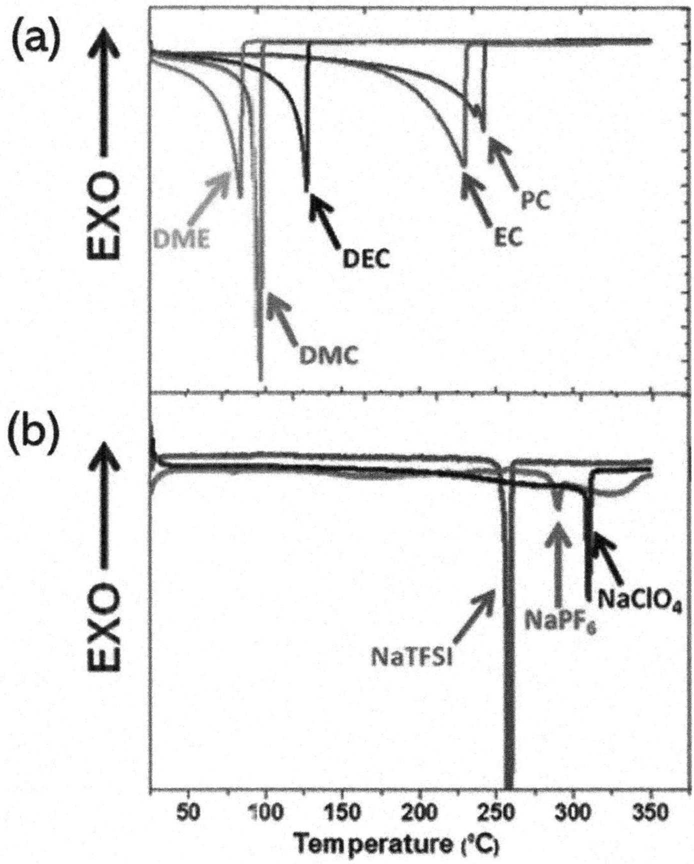

Figure 8.3 DSC curves up to 350 °C of (a) solvents, (b) 1 M sodium salts in PC solvent [60].

8.3 Chemical Compositions of Organic Liquid Electrolytes

Figure 8.4 summarises the widely used sodium salts, organic solvents, and additives in organic liquid electrolytes for NIBs. $NaPF_6$, $NaClO_4$, Na bis(trifluoromethane) sulfonimide (NaTFSI), and Na bis(fluorosulfonyl) imide (NaFSI) are the typical sodium salts for the liquid electrolytes. The frequently used solvents can be divided into two categories: carbonate ester-based and ether-based solvents. The carbonate ester-based solvents include ethylene carbonate (EC), propylene carbonate (PC), dimethyl carbonate (DMC), diethyl carbonate (DEC), and methy-ethyl carbonate (EMC). 1,2-Dimethoxyethane (DME), diethylene glycol dimethyl ether (DEGDME), 1,3-dioxolane (DOL), etc., are the frequently used ether-based solvents. Apart from the aforementioned sodium salts and solvents, some additives such as prop-1-ene-1,3-sultone (PES), fluoroethylene carbonate (FEC), and succinic anhydride (SA) are widely introduced into the organic liquid electrolytes to optimize the chemical and physical properties of the SEI film to enhance the battery performance of the NIBs.

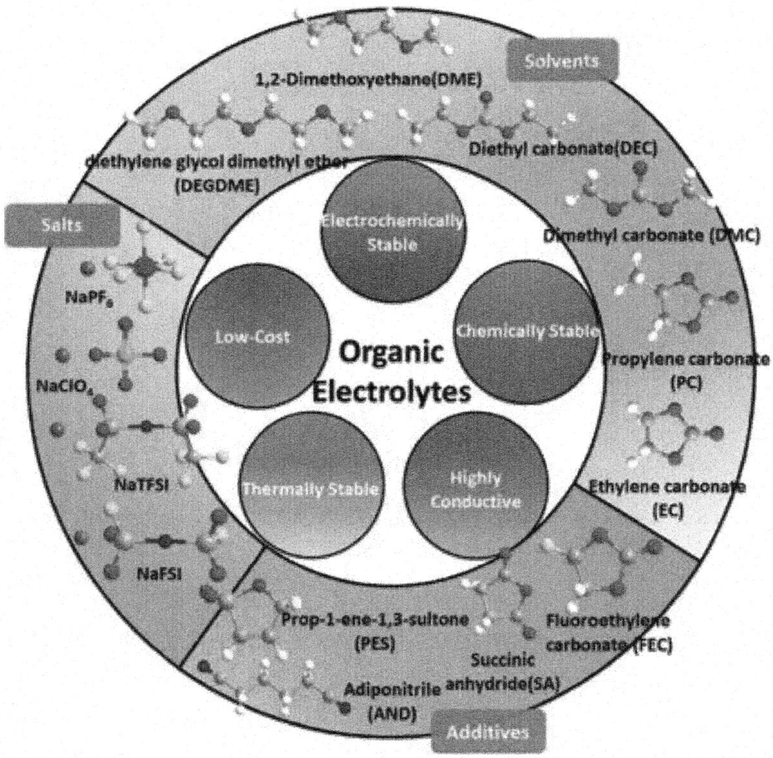

Figure 8.4 Summary of the sodium salts, organic solvents, additives for organic liquid electrolytes for NIBs [46].

8.3.1 Sodium Salts

Sodium salts, composed of sodium and anion ions, are the core component of the electrolyte and have profound influence on the battery performance of NIBs. Ideal sodium salts for organic liquid electrolytes should meet the following requirements: (i) high solubility in the organic solvent [65]; (ii) dissociated Na^+ can be well solvated by the organic solvent and can diffuse inside the organic solvents without energy and kinetic barriers theoretically [46, 65]; (iii) high electrochemical stability in a wide electrochemical window. The oxidation and reduction potentials of sodium salts directly determine the ESW of the organic liquid electrolytes [60]; (iv) chemically and electrochemically inert to other battery components except for active materials, like separators, current collectors, binders, conductive carbon additives, packing materials, etc. [54, 66]; (v) good thermal stability. The sodium salts should have the ability to work in extreme environmental (e.g., high and low temperature) while maintaining their chemical and physical properties, thus ensuring the safety of the batteries [58]; and (vi) low price, nontoxicity, and environment friendly.

Table 8.1 Physical and chemical properties of the widely reported sodium salts for organic liquid electrolytes in NIBs

Salt	Molecular Weight (g/mol)	Decomposition Temperature (°C)	Ionic Conductivity (mS/cm)	Ref.
$NaPF_6$	167.9	302	7.98	[52, 60, 67]
$NaClO_4$	122.4	472	6.4	[52, 60, 67]
NaFSI	203.3	122	—	[52, 65, 67]
NaTFSI	303.1	263	6.2	[52, 60, 67]
NaFTFSI	253.13	160	—	[52, 67]
$NaBF_4$	109.8	384	—	[52, 65, 67]

Note: Adopted from Refs. [46, 52].

Table 8.1 summarises and compares the chemical and physical properties of sodium salts in organic liquid electrolytes for NIBs reported in the literature. Compared with lithium salts for LIBs, there are more choices in sodium salts for NIBs. Usually, the anionic groups of the sodium salt play a critical role in determining the properties of the electrolytes, such as solubility and conductivity. In terms of the thermal stability of sodium salts, they follow the order of $NaClO_4$ > NaTFSI > $NaPF_6$ > NaFTFSI > NaFSI (Fig. 8.5a) [67]. $NaClO_4$ is currently the most popular sodium salt for organic liquid electrolytes, especially in the carbonate ester-based electrolytes. $NaClO_4$ outperforms other salts in terms of ionic transport, solubility, and compatibility with solvents, as well as low cost [45, 46, 52, 60]. Nevertheless, similar to the lithium counterpart in LIBs, the potential risk for $NaClO_4$ is its instinct explosive nature, which results from the highly oxidisability.

$LiPF_6$ is the most widely used lithium salt for commercialized LIBs. However, its sodium-containing counterpart $NaPF_6$ is not popular in NIBs, despite having higher thermal and electrochemical stability than those of $LiPF_6$, and the highest Na^+ conductivity when compared to other salts in PC-based electrolytes (Fig. 8.5b) [54, 60, 67]. One potential issue of $NaPF_6$ is its moisture-sensitive nature, which may release highly toxic and corrosive HF. The HF can dramatically damage the SEI film by etching the alkaline components, resulting in a fragile SEI and battery performance decay. Besides, $NaPF_6$ has a solubility of 1.4 M in EC solvent, but much lower values in solvents of PC (1.0 M), DEC (0.8 M), and DMC (0.6 M). Sodium salts that contain large anionic groups of $TFSI^-$, FSI^-, $FTFSI^-$, BF^{4-}, etc., like NaTFI and NaFSI, can provide high ionic conductivity and are also reported for NIBs. These salts are nontoxic and display good Na-ion storage electrochemical stability. However, they are highly corrosive to the aluminum current collector in the carbonate ester-based electrolytes [67].

Apart from the aforementioned sodium salts with solvent-dependent performances, some unconventional salts, such as sodium-difluoro(oxalato)borate (NaDFOB) [68], sodium-4,5-dicyano-2-(trifluoromethyl) imidazolate (NaTDI) [66], and sodium-4,5-dicyano-2-(pentafluoroethyl)imidazolate (NaPDI) [66], have been adopted in the organic liquid electrolytes for NIBs. As for NaDFOB, the F element in the molecular structure presents strong electronegativity, which provides more delocalized charges and weak interactions with the Na ions, contributing to high conductivity. Meanwhile, NaDFOB displays excellent compatibility with carbonate ester solvents of EC, PC, DEC, DMC, etc. and provides good Na-ion storage performance in $Na/Na_{0.44}MnO_2$ half-cells, which forms sharp contrast to $NaClO_4$ and $NaPF_6$ [68]. NaTDI and NaPDI are two typical imidazole fluorine-derived sodium salts with F atom in their heterocyclic ring structures, providing good thermal stability over 300°C and an electrochemical stability window up to 4.5 V for NaTDI and 4.2 V for NaPDI versus Na/Na^+, respectively. Meanwhile, even at a low salt concentration of 0.5 M, the NaTDI-PC-based electrolyte can achieve a good ionic conductivity of 4 mS/cm [66].

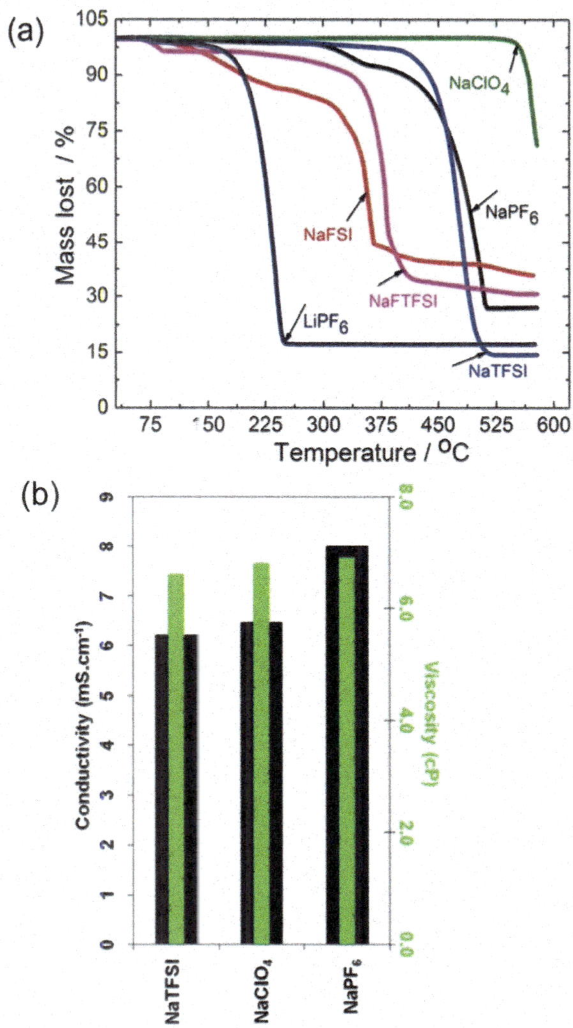

Figure 8.5 (a) TGA profiles of sodium salts and lithium salt ($LiPF_6$) [67]. (b) Conductivity (black bars and left-hand side *y*-axis) and viscosity (green bars and right-hand side *y*-axis) values of 1 M sodium salts in PC-based electrolyte [60].

8.3.2 Solvents

For organic liquid electrolytes, solvents are vital in determining physicochemical and electrochemical properties, like the ability to dissolve sodium salts, thermal safety, and electrochemical reaction kinetics. The organic solvents used for liquid electrolytes are mainly based on two categories: carbonate esters and glyme-based ethers. Table 8.2 summarises the physical and chemical properties of wildly used organic solvents in NIBs. According to the attributes of the solvents, organic liquid electrolytes are divided into carbonate ester-based electrolytes and ether-based electrolytes.

Table 8.2 Physical and chemical properties of carbonate esters and ethers for organic liquid electrolytes in NIBs [52, 69]

Name	Chemical Structure	T_m/°C	T_b/°C	T_f/ °C	η (25°C) [cP]	ε (25°C)
EC	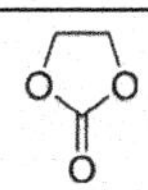	36.4	248	160	2.1	89.78
PC	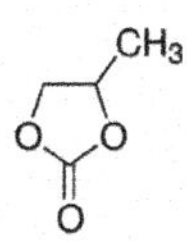	-48.8	242	132	2.53	64.92
DEC	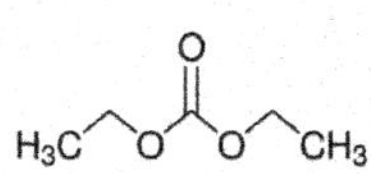	-74.3	126	31	0.75	2.81
DMC	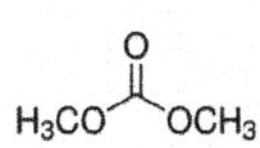	4.6	91	18	0.59	3.11
EMC	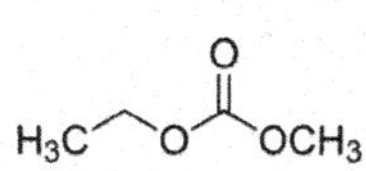	-53	110	23.9	0.65	2.96
DME	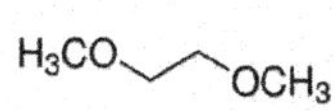	-58	84	0	0.46	7.18
DEGDME	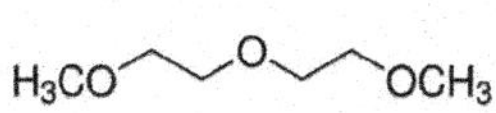	-64	162	57	1.06	7.4
TEGDME	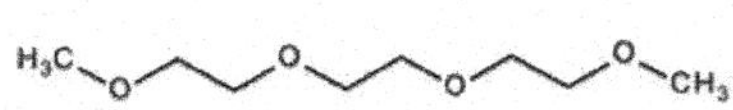	-46	216	111	3.39	7.53

8.3.2.1 Carbonate ester-based electrolytes

Carbonate ester-based electrolytes have been widely developed and successfully commercialized in LIBs over the pass decades.

Analogous to LIBs, the early research activities on organic liquid electrolytes for NIBs originated from carbonate esters, such as EC, PC, DEC, DMC, and EMC, which are the primarily used alkyl carbonate solvents, thanks to their good electrochemical stability [45]. These solvents can be further divided into two types according to their molecular structure: cyclic carbonates (including EC, PC, and VC) and linear carbonates (such as DMC, DEC, and EMC) [45, 46, 52]. Usually, the organic liquid electrolytes for NIBs employ binary or ternary solvents rather than a single solvent to meet the requirements for achieving high battery performance. A statistical report reveals that the EC/DEC mixture is the most used solvent for organic liquid electrolytes in NIBs, which is followed by EC/PC, PC, and EC/DMC solvents, respectively [70]. As for the sodium salts, $NaClO_4$ and $NaFP_6$ are the frequently reported salts for the carbonate ester-based electrolytes.

The cyclic carbonate esters of EC and PC have been intensively studied as solvent components for organic liquid electrolytes due to their high dielectric constant, stable chemical and electrochemical properties. Among them, the EC solvent displays a high dielectric constant value of 89.78, which provides a capability to dissolve higher concentration sodium salts. The EC solvent shows a high viscosity and has good ionic conductivity, high thermal stability, and wide electrochemically stable window. Meanwhile, the EC solvent can interact with sodium salts by the formation of strong dipole–dipole intermolecular forces, thus promoting to dissolve more sodium slats. Besides, the EC solvent can be electrochemically decomposed to form a chemically stable protective film (SEI) on the surface of electrode materials, especially in the graphite anode, to hinder the electrolyte decomposition [60, 71]. However, EC has a high melting point of 36°C and is solid in nature at room temperature, which means that single EC is not suitable for serving as a solvent for electrolyte. Usually, the EC is used as a co-solvent with other carbonate esters for an optimized electrolyte, in which the addition of EC solvent can dramatically improve the solubility and ionic conductivity of the mixture solvents. Similarly, PC, with a desirable wide liquid temperature range but a slightly lower dielectric constant value (64.92) than EC, is another attractive carbonate ester solvent for organic liquid electrolytes [60]. In LIBs, the utilization of PC solvent results in the PC molecule co-intercalation with the graphite interlayers, leading to graphene layer exfoliation [72]. In contrast, it is well compatible with hard-carbon anodes for NIBs, in which the disordered structure consisted of curved graphene layers that prevented the possibility of solvent molecules co-intercalation. However, the continuous decomposition of PC solvent and formation of SEI during cycling result in significant capacity fading in the electrolytes when single PC is used as solvent. At this point, it is not a good choice to use PC as the sole solvent for organic liquid electrolytes for NIBs. But it can be used as an SEI film-forming solvent to combine with other solvents for practical application [60].

Linear carbonate esters of DMC, DEC, EMC, etc. show lower viscosity and melting point than cyclic carbonate easter solvents. These linear carbonate esters containing electrolytes usually exhibit different ionic conductivity, thermal stability, and viscosity, which have further impact on the battery performance [45, 60]. For practical use, the linear carbonate esters are rarely used individually and commonly used in compound mode of binary or ternary, or combined with cyclic carbonates, such as EC/PC, EC/DEC, EC/DMC, and EC/DEC/PC, to obtain an optimized property.

8.3.2.1.1 Interaction behavior of Na ions with carbonate ester solvents

Compared with lithium salts, sodium salts display about 15–20% lower ion-pair dissociation energy than lithium salts and smaller de-solvation energy when interacting with carbonate ester solvents due to the weaker Lewis acidity of Na^+ [73, 74]. Shakourian-Fard et al. studied the Na-ion solvation behavior with carbonate ester solvents by using molecular dynamics simulations (Fig. 8.6). It was found that the HOMO–LUMO energy gap increases when interacting with Na^+. But it decreases in binary solvents by adding EC into PC, DMC, and EMC carbonates [71]. The interactions between Na^+ ions and carbonate solvent complexes can be classified as a closed-shell (electrostatic) interaction. Among the complexes of Na^+ and solvents, the highest number of EC molecules exists in the first solvent sphere followed by PC molecules, and then the linear carbonates, in which the DMC molecules concentrate at the reduced interface owing to their weaker interaction with Na^+ ions. There are four kinds of interaction forces between Na^+ ions and carbonate ester solvents: electrostatic energy (ΔE_{ele}), exchange energy (ΔE_{ex}), polarization energy (Δ_{Epol}), and dispersion energy (ΔE_{disp}). The contribution of these interaction forces to the final interaction energy follows the order of $\Delta E_{ele} > \Delta E_{pol} > \Delta E_{disp} > \Delta_{Eex}$ [52, 74].

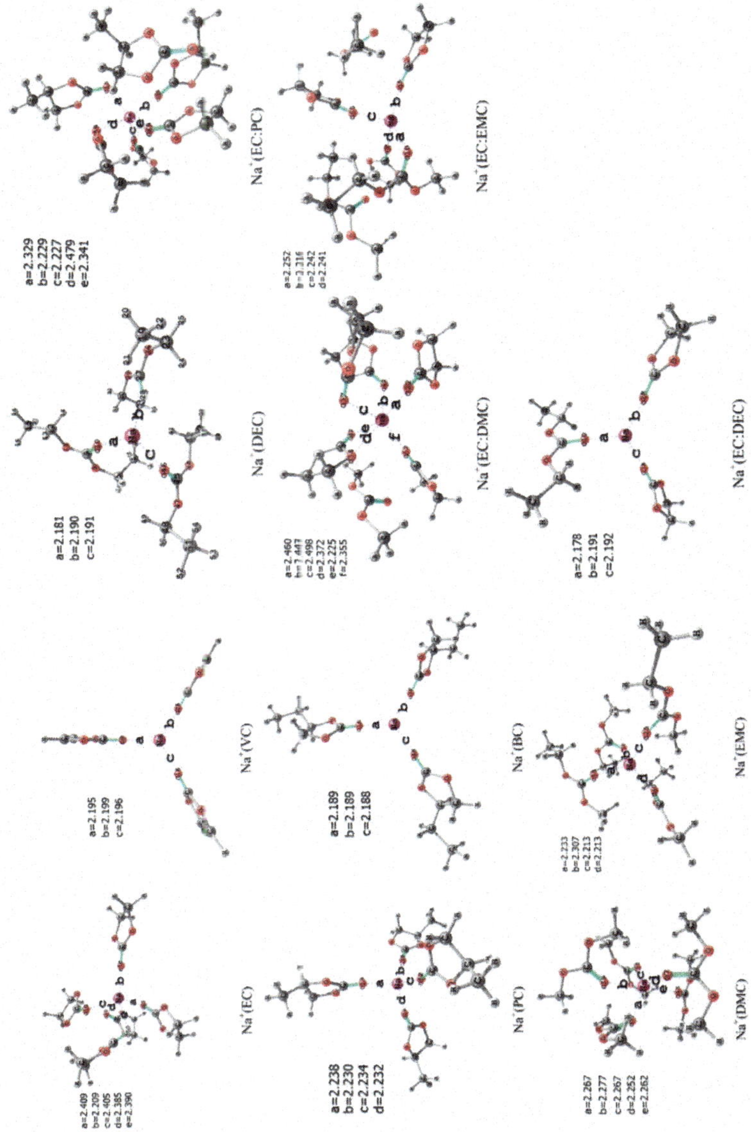

Figure 8.6 Molecule models of the Na-ion interactions with carbonate esters [71].

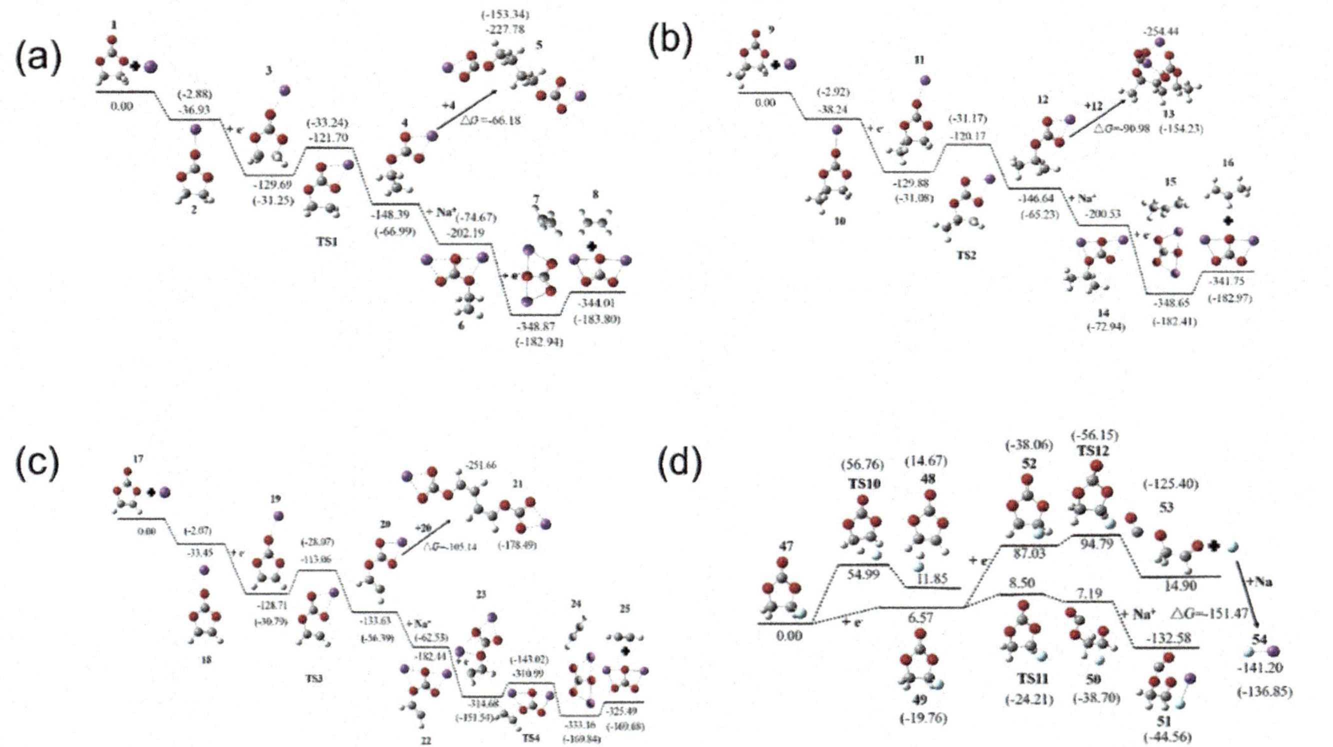

Figure 8.7 Potential energy diagram for the reduction of (a) EC, (b) PC, (c) VC, and (d) FEC. Dark gray, white, red, violet, and blue spheres denote C, H, O, Na, and F atoms/ions, respectively [75].

8.3.2.1.2 Reduction of the carbonate ester-based electrolytes

The formation of SEI is associated with electrolyte solvent reduction. The reduction mechanism is crucial in understanding the chemical composition of the SEI film. DFT calculations show the potential energy diagram for the reduction of EC and PC (Fig. 8.7). Solvents EC and PC, and additive VC can decompose into $(CH_2CH_2CO_3Na)_2$, $(CH_3CH_2CH_2CO_3Na)_2$, and $(CHCHCO_3Na)_2$ organic SEI components via a one-electron reduction process. The reductive decomposition to form these species follows the order of VC > PC > EC. Apart from the organic components, inorganic Na_2CO_3 is also formed through a two-electron reduction process but undergoes through a totally different order of EC > PC > VC. In the case of FEC, DFT calculation results suggest that reduction of FEC intends to produce inorganic NaF through both one-electron and two-electron reduction routes [75]. The research into the reduction mechanisms of solvents and additives will be beneficial to deeply understand the formation mechanism and composition of the SEI film, then providing new insights for designing new electrolytes.

8.3.2.1.3 Electrochemical compatibility with electrodes

The compatibility between the electrode and electrolyte is associated with the chemical composition of the electrode and electrolyte materials, surface chemistry of the electrode, and the contact interface area. Previous studies have revealed that linear carbonate ester-based electrolytes exhibit good wettability to hydrophobic polyethylene (PE) separators. However, PC (cyclic carbonate ester) solvent is not compatible with the PE separator [76]. In laboratory-scale research, glass fibers that can be well compatible with the PC-containing electrolytes are widely used as separator to evaluate the performance of electrode materials. However, there are still challenges for the practical utilization of glass fiber separators because of their disadvantages of fragility, high cost, inflammability, sensitivity to moisture, etc.

The investigation of carbonate ester-based electrolytes in representative anode and cathode materials for NIBs has been widely reported. Ponrouch et al. systematically evaluated the effects of sodium salts ($NaClO_4$, $NaPF_6$, and NaTFSI) and solvents (EC, PC, DMC, DME, DEC, THF, triglyme, EC/PC, EC/DMC, and EC/DME) on the electrochemical performance of hard-carbon anodes [59, 60]. Remarkable electrochemical performance difference was observed depending on the components of electrolytes. Single PC solvent-based electrolytes achieve the highest columbic efficiency, whereas the electrolytes consisting of the EC/DMC and EC/DME solvents exhibit poor performance attributing to DMC decomposition below 1 V and large overpotential, respectively. The electrolytes with binary solvents of EC/PC with $NaClO_4$ or $NaPF_6$ are promising as optimal electrolytes in terms of capacity retention and columbic efficiency owing to their capability of forming a stable and ionic conducting SEI film on hard-carbon electrodes. In contrast, in PC-, EC/DMC-, EC/DEC-, and EC/DME-based electrolytes, the cells present different levels of capacity fading. Nevertheless, the continuous growth of SEI film caused by the continuous electrolyte decomposition occurred in PC-based electrolyte, leading to significant increase in charge-transfer resistance. Besides, the impact of sodium salts on electrochemical performance was also investigated. Very similar electrochemical response is observed for both 1 M $NaClO_4$ EC/PC and 1 M $NaPF_6$ EC/PC electrolytes. But in the case of 1 M $NaPF_6$ EC/PC electrolyte, the cell shows slightly higher overpotential because of the formation of less conductive SEI on hard-carbon electrode. Stable cycling performance can be achieved by both electrolytes, even without the addition of any additives. However, Zhao et al. reported a completely opposite result that the hard-carbon material synthesized by thermal carbonization of organic polymer with an aromatic ring under 1600°C can be well compatible with $NaClO_4$/EC/DMC electrolyte and achieves better electrochemical performance and thermal stability than in $NaClO_4$/PC electrolyte [77]. These results suggest that the difference in solvent species may result in big impact on the Na-ion storage mechanism of hard-carbon materials. Komaba et al. studied the electrochemical performance of hard-carbon materials in carbonate ester-based electrolytes, including single solvent-based (EC, PC, and BC) and binary solvent-based (EC + X and PC + X, where X = DMC, EMC, and DEC) [78, 79]. It is found that the single PC-based and EC/DEC-based electrolytes, which contain 1 M $NaClO_4$, were well compatible with the hard-carbon electrodes and form a passivation layer mainly consisting of inorganic compounds on the surface of hard-carbon electrode. In other electrolytes, significant performance degradation was observed due mainly to solvent decomposition. However, the full cell by using hard carbon as anode and $NaNi_{0.5}Mn_{0.5}O_2$ failed to work in the case of 1 M $NaClO_4$/PC electrolyte. Later, the authors evaluated the effect of sodium salts with PC solvent and found that the hard-carbon electrodes in PC-based electrolytes with $NaPF_6$ or NaTFSI salts displayed better cycling performance than that of $NaClO_4$.

Apart from the compatibility with the anode electrode, the formulation optimization of electrolytes should also consider their influence on cathode electrodes regarding the upper limit of charging voltage, fast charge/discharge kinetics, reversible capacity, capacity retention during cycling, etc. For instance, the electrochemical influence of solvent species on polyanion cathode electrodes was reported. The electrolytes of single PC solvent and EC/DEC mixture solvent containing $NaPF_6$ salt were selected to evaluate $Na_{1.8}FePO_4F$ electrodes [80]. Using single PC as solvent, electrochemical instability was observed at lower-end voltages; whereas for EC/DEC mixture solvent-based electrolyte, a similar phenomenon occurred at the higher-end voltages. Besides, the upper limit of the voltage window strongly affects the oxidation state of iron and the discharge capacity when using PC solvent. The capacity and capacity retention can be improved by using the electrolyte of $NaPF_6$ in EC/DEC mixture solvent. Jang et al. studied the electrochemical performance of $Na_4Fe_3(PO_4)_2(P_2O_7)$ cathode by using electrolytes of EC/PC and EC/DEC containing 1 M $NaClO_4$ [81]. The EC/PC-based electrolyte exhibits superior oxidation durability at the surface of cathode, which does not undergo electrochemical decomposition when the cells are charged up to 4.2 V and is also highly stable toward Na-metal electrode. The electrochemical performance of $Na_3V_2(PO_4)_2F_3$/HC full cells in $NaPF_6$-based electrolytes containing either single linear carbonate of DMC, EMC, DEC or single cyclic carbonate of PC and EC was investigated by using *in situ* UV and CV analytical techniques [82]. The investigation demonstrated that the partial decomposition of linear carbonates (DMC and EMC) forms soluble species, which fails to provide a stable SEI film on the surface of the hard-carbon electrodes, leading to poor battery performance. In contrast, the cyclic carbonate ester-based electrolytes exhibit better cycle capability.

Similarly, the carbonate ester-based electrolytes greatly affect the electrochemical performance of the PB cathodes. The effects of solvents (EC/PC and EC/DMC) and sodium salts ($NaClO_4$ and $NaPF_6$) on the $Na_{0.75}Fe_{2.08}(CN)_6 \cdot 3.4H_2O$ were investigated [83]. With the addition of 2 vol% FEC additive, the electrolyte of 1 M $NaPF_6$ in EC/PC/FEC mixture solvent shows the most promise for PB electrodes in the voltage range of 2.4–4.2 V, delivering high reversible specific capacity of 130 mAh/g with 99.5% of columbic efficiency and 87% of capacity retention after 40 cycles of C-rate capability test. Lee et al. compared the cycling stability of $Na_2Zn_3[Fe\ (CN)_6]_2 \cdot 9H_2O$ in three different electrolytes of 1 M $NaClO_4$ in PC or EC/DMC, and 1 M $NaPF_6$ in EC/DMC [84]. The results show that replacing the solvent of PC with EC/DMC mixture solvent substantially improves capacity retention from 85.2% to 90.1% after 50 cycles, which might associate with smoother shell exchange of the solvated Na-ion clusters at the electrode–electrolyte interface. By changing the sodium salt from $NaClO_4$ to $NaPF_6$ in EC/DMC solvent, a further cycling retention improvement from 90.1% to 94.6% was achieved.

The carbonate ester-based electrolytes are also crucial for layered transition metal oxide cathodes to achieve high energy density and enhance the cycling performance. For example, the studies on the effect of salt concentration of $NaPF_6$, $NaClO_4$, and $NaCF_3SO_3$ in EC/DMC mixture solvent (weight ratio 3:7) show that the best conductivities can be achieved in 0.6 M $NaPF_6$ (6.8 mS/cm), 1.0 M $NaClO_4$ (5.0 mS/cm), and 0.8 M $NaCF_3SO_3$ [61]. When $Na_{0.7}CoO_2$ was chosen as the cathode electrode, an electrochemically stable SEI film can be formed on its surface in $NaPF_6$-based electrolytes, achieving favor kinetics. Apart from the aforementioned conventional sodium salts, sodium-difluoro(oxalato)borate (NaDFOB) was proposed as sodium salt for NIBs, which presented excellent compatibility with EC, PC, DMC, and DEC solvents [68]. The electrolytes containing NaDFOB sodium salt endow the $Na/Na_{0.44}MnO_2$ half-cells with significantly enhanced reversible capacities and high rate capability than those containing $NaClO_4$ or $NaPF_6$ salts, which show solvent-dependent performance.

8.3.2.2 Ether-based electrolytes

For LIBs applications, ether-based solvents are rarely used because of their narrow ESW (less than 4 V Li/Li^+) and the formation of inferior passivation layer on anodes [85]. However, for NIB applications, ether-based solvents have proven to be promising owing to their nonflammability and capability to realise the Na^+-solvent co-intercalation mechanism and fast transport of Na ions, and to form thin SEI film [46, 86]. Dimethyl ether (DME), diethylene glycol dimethyl ether (DEGDME), and tetraethylene glycol dimethyl ether (TEGDME) are the frequently used ethers for the ether-based electrolytes.

Graphite, the most used anode material in commercialized LIBs, is electrochemically inert to Na^+ in carbonate ester-based electrolytes because of the high energy barrier to form stable Na-carbon intercalation compounds [87, 88]. The ether-based electrolytes make the Na-ion storage in graphite a reality through the Na^+-solvent co-intercalation mechanism. Jache et al. first reported the co-intercalation phenomenon in a diglyme-based electrolyte in graphite anode, achieving a superior cycle performance with reversible capacities close to 100 mAh/g for 1000 cycles and Coulomb efficiencies >99.87%. The Na-ion storage mechanism involves the formation of ternary graphite intercalation compounds (*t*-GICs) $Na(diglyme)_2C_{20}$ through the co-intercalation of solvated sodium ion into graphite: $C_n + e^- + Na^+ + y\,solvent \rightarrow Na^+(solvent)_yC_n^-$) [56]. Later, the same group investigated the correlation between the glyme molecular weight with the intercalation potential and the rate capability. The authors comprehensively studied the formation of the *t*-GICs by evaluating a series of glymes with different chain lengths (mono-, di-, tri-, and tetraglyme) and their derivatives as well as cyclic ethers (tetrahydrofuran (THF)) in graphite anodes [89]. A staged redox behavior is observed, i.e., the redox potentials are determined by the chain lengths of the ether molecules. The redox potentials systematically shift to higher potential with increase in the chain length. But in the case of triglyme, ill-defined voltammogram and much lower capacity were observed, which were proved to be poor match between the sizes of the Na^+ and the triglyme molecule, impeding a geometrically favorable coordination. Other solvents of glyme derivatives and THF displayed poor co-intercalation redox potential and redox activity. However, despite the co-intercalation mechanism enabling Na-ion storage in graphite anode, the co-intercalation leads to a 230–255% expansion in the interlayer distance.

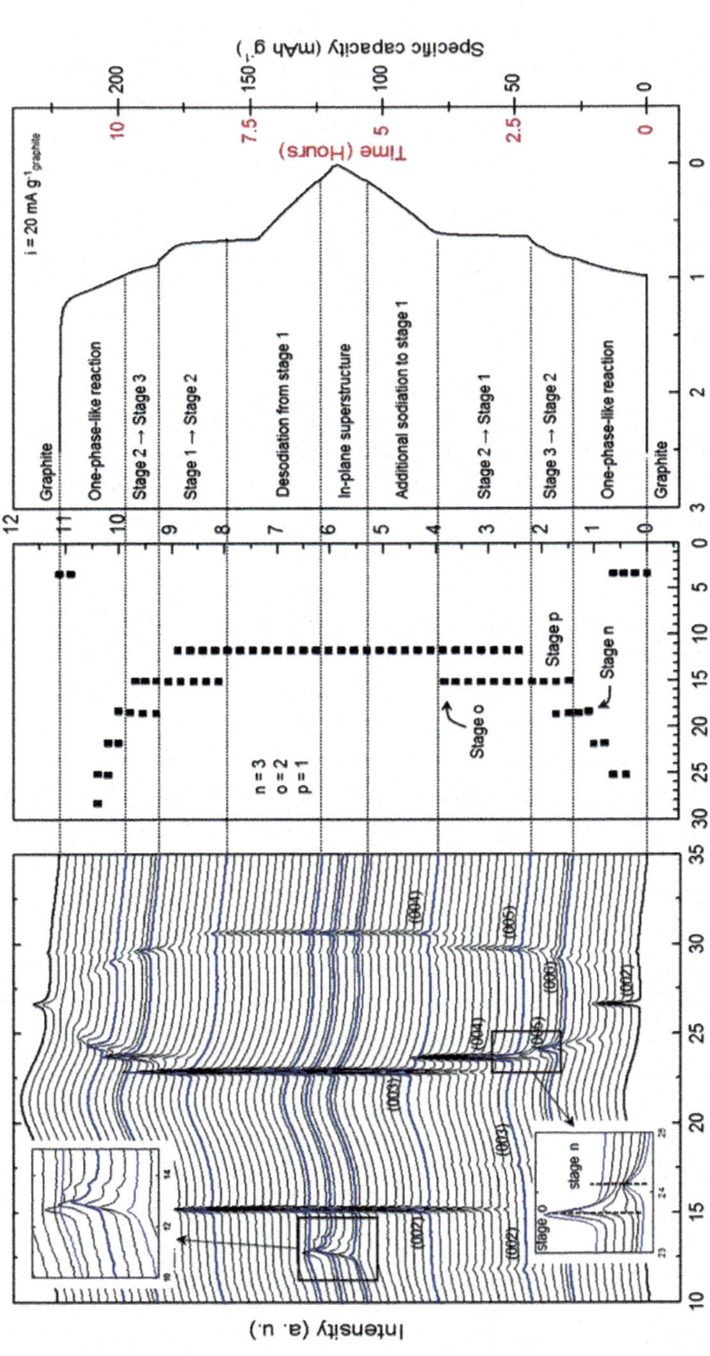

Figure 8.8 *In operando* synchrotron X-ray diffraction patterns of solvated Na^+ intercalation and de-intercalation into/out of graphite [90].

Kang's group also fully investigated the Na-ion storage behavior of natural graphite in ether-based electrolytes consisting of different solvents of TEGDME, DEGDME, DME and different salts of $NaPF_6$, $NaClO_4$, $NaCF_3SO_3$ [91]. The results indicate that both the sodium salts and the solvent show negligible influence on the electrochemical performance, while the chain length of solvents determines not only the plateau redox potential, which increases proportionally from 0.60 V (DME) to 0.78 V (TEGDME), but also the rate capability. Solvents with longer chain lengths can generate energetically more stable GICs and exhibit a higher Na-ion storage potential. Meanwhile, the Na^+-solvent co-intercalation occurs combined with partial pseudocapacitive behavior. Without any chemical modification or treatment, the natural graphite with a particle size distribution around 100 micrometers could deliver an excellent cycle stability up to 2500 cycles and rate capability (75 mAh/g at 10 A/g) through the Na^+-solvent co-intercalation mechanism. A full cell fabricated with $Na_{1.5}VPO_{4.8}F_{0.7}$ cathode and graphite anode in $NaPF_6$ DEGDME electrolyte delivered an energy of 120 Wh/kg while maintaining 70% of its initial capacity after 250 cycles, highlighting the feasibility of practical utilization of solvent co-intercalation graphite anode. *In operando* XRD coupled with electrochemical titration and DFT calculations were used to investigate the Na^+-solvent co-intercalation mechanism in graphite [90]. It is found that the Na-ion intercalation occurs through multiple staging reactions and finally forms first-stage Na-GICs within a wide range of Na/C from 1/28 to 1/21 with excellent reversibility (Fig. 8.8). The intercalated Na ions and ether solvents are in the form of $[Na\text{-}ether]^+$ complexes double stacked in parallel with graphene layers. The DFT calculations reveal that the redox potential shifts to a high value along with increase in the solvent chain length. Such a redshift in redox potential was attributed to the stronger screening effects of longer solvent molecules on the repulsion between positively charged Na ions in the discharge products.

DFT calculations demonstrate that the unfavorable local Na–graphene interaction primarily leads to the formation of unstable Na-GICs, but their chemical compositions can be effectively controlled by optimizing solvent molecules (Fig. 8.9) [92]. Moreover, the reversible Na-ion intercalation into graphite anode occurs only under specific conditions, i.e., a solvation energy $E_s > 1.75$ eV is required. Among the carbonate esters and ethers, only the DME, DEGDME, TEGTME, and PC are suitable for realizing the co-intercalation. Besides, in order to retain the chemical stability of the Na-GICs, the LUMO levels of the $[\text{Na-solvent}]^+$ molecules must be higher than the Fermi level of graphite. Such an energy barrier makes the injection of electrons into the Na-solvent complexes difficult and is less likely to initiate the decomposition reaction and subsequent degradation of graphite. Gotoh et al. investigated the coordination structure and dynamic behavior of Na-Diglyme-d_{14} by using ^{2}H solid-state NMR and revealed that two diglyme molecules coordinate to each sodium ion rigidly, except for rotation of the methyl groups at low temperatures below 233 K [93]. At room temperature, diglyme weakly coordinates to Na ion through oxygen atom in the ligand and rotates around the O–Na axis. The active motion of sodium–diglyme complexes is favorable for Na-ion diffusion between graphene layers. *Ab initio* study found that the $Na(digl)_2C_n$ has the lowest negative intercalation energy at $n \approx 21$, which enables relatively fast the diffusion for $Na(digl)_2$ complex in the interlayer space; and the electronic conductance of graphite was enhanced upon the co-intercalation of Na^+ and solvents. Meanwhile, both the diglyme molecule and Na atom will donate electrons to the graphene layer, resulting an ionic bond between the graphene layer and the diglyme molecule [94].

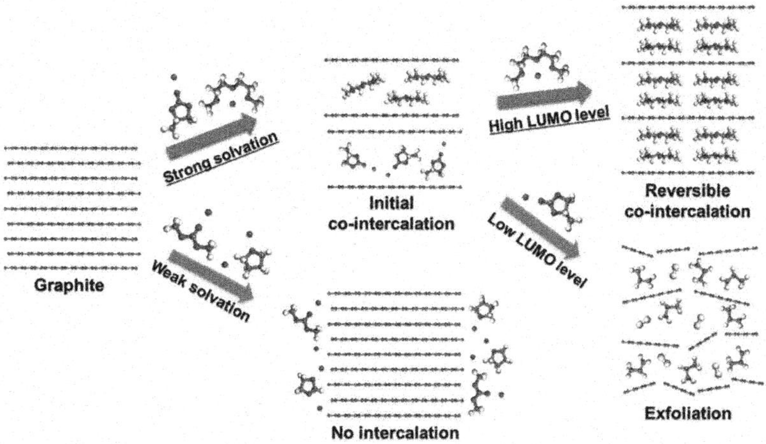

Figure 8.9 Schematic of solvent dependent of the Na^+-solvent co-intercalation in graphite anode for NIBs [92].

The application of graphite based on Na^+-solvent co-intercalation mechanism in full cells has been reported. Zhu et al. fabricated a full cell consisting of graphite anode and $Na_3V_2(PO_4)_3$@C cathode in 1 M $NaCF_3SO_3$/diglyme electrolyte, delivering a reversible capacity of ~100 mAh/g at a working voltage of 2 V and good rate capability up to 2 A/g [95]. Hasa et al. studied the performance of graphite//$Na_{0.7}CoO_2$ full cells in the electrolyte of 1 M $NaClO_4$ TEGDME, providing a capacity of 80 mAh/g and good cycling life for 1250 cycles [86]. Kang et al. reported full cells by coupling the graphite anode with $Na_{1.5}VPO_{4.8}F_{0.7}$ cathode in 2 M $NaPF_6$ DEGDME electrolyte; the cells presented a voltage of 3.1 V, a high power density of 3863 W/kg in both electrodes, and an extremely low-capacity fading rate of 0.007% per cycle over 1000 cycles, showing a promising choice for large-scale energy storage systems [57].

Apart from the graphite, ether-based electrolytes are also effective with other anodes for NIBs, including graphene [96], hard carbon [97–99], metal oxides [53, 100], and chalcogenides [101–103]. For instance, Yun et al. investigated the Na-ion storage behaviors of hard-carbon materials thermally treated at different temperatures from 1600 to 2800°C in both carbonate ester-based and ether-based electrolytes [98]. The high-temperature treated samples present superior Na-ion storage performance with much increased capacity delivered in the plateau region in ether-based electrolyte than those in ester-based electrolyte. The carbonate ester-based electrolyte tends to decompose on the hard-carbon electrode surface to form a thick SEI film with poor electrochemical stability. In contrast, the glyme–sodium ion complex can be delivered without de-solvation, and its motion in the graphitic carbon structures is much faster. Hence, the complex charge carriers improve the charge delivery kinetics in both surface region and bulk region of the hard-carbon electrodes. *In situ* EIS characterization results reveal significantly lower surface film resistance (R_f) and charge-transfer resistance (R_{ct}) values in ether-based electrolyte, providing more rapid and effective charge delivery kinetics. Xu et al. developed a "foreign SEI" from ether-based electrolyte to realize the "bulk ion transport" and "interphasial" requirements, which can electrochemically stabilize the hard-carbon electrodes to achieve significantly enhanced electrochemical performance [99]. At a high rate of 500 mA/g, a capacity of 200 mAh/g was retained for over

1000 cycles without detectable capacity fading. Kang et al. studied the pseudocapacitive Na-ion storage behavior of anatase TiO_2 anode and found that pseudocapacitive sodium-ion storage performance is sensitive to the nature of SEI film [100]. The ether-based electrolytes enable the formation of thin (~2.5 nm) and robust SEI film during cycling, which is in sharp contrast to the thick (~10 nm) and growing SEI in carbonated ester-based electrolytes. Such a difference can be attributed to the higher LUMO energies of the ether solvents/ion complexes, which are further confirmed by first principle calculations. Consequently, the anatase TiO_2 with the engineering SEI film achieved an exceptional cyclic stability and high rate capability. Later, Li et al. investigated the electrochemical interface evolution in anatase TiO_2 anode with both ether-based and ester-based electrolytes [53]. The TiO_2 anode coupled with the diglyme-based electrolyte shows a reversible capacity of 257.9 mAh/g at 100 mA/g and more than 100 mAh/g at 2000 mA/g, which are much higher than those in EC/DEC-based electrolyte. Such an electrolyte-dependent performance correlates with the different structural evolution induced by a varied sodiation depth. *In operando* Raman spectra and XRD techniques confirm that the ether-based electrolyte facilitates a depth sodiation-induced structural transition: $TiO_2 + Na^{+} + e_{-} \rightarrow Na_x(TiO_2) + Ti + NaO_2$, while an incompletely sodiation was observed in EC/DEC-based electrolyte. Meanwhile, a more uniform and thinner SEI layer consisting of organic (RCH_2ONa) and inorganic (Na_2CO_3 and NaF) components was observed in the diglyme-based electrolyte, in which the sodium alkoxide (RCH_2ONa) is beneficial to reduce the energy barriers for Na-ion diffusion, resulting in favorable sodiation dynamics. The charge-transfer energy barrier of the interface formed in the diglyme-based electrolyte is only 172 meV, which is 1.4 times lower than that in EC/DEC-based electrolyte, providing faster charge transfer across the electrolyte/electrode interface.

8.3.3 Additives

Additives have proven to be helpful in improving battery performance. Generally, additives are added into the electrolyte in a small quantity, less than 10% by weight or volume, to realize some special functions. Fluoroethylene carbonate (FEC) [104–106], vinylene carbonate (VC) [107], tris(trimethylsilyl)phosphite (TMSP) [108], prop-1-ene-1,3-sultone (PES) [109, 110], ethylene sulfite (ES) [109] etc. are the frequently used additives for organic liquid electrolytes. According to their functions, there are four types of additives: (i) film-formation additives; (ii) overcharge protection additives; (iii) flame-retardant additives; and (iv) other types to realize some special requirements, such as salt stabilizer, wetting agents, and current collector corrosion inhibitors [46, 58, 111]. Currently, the studies on additives in NIBs are mainly focused on their ability in regulating the SEI film, enhancing the safety, and overcharge protection.

8.3.3.1 Film-formation additives

During electrochemical cycling, an SEI film is formed on the surface of anode electrode, which results from the decomposition of sodium salts and solvents. The film-formation additives have a direct impact on the morphology, chemical compositions, thickness, and mechanical properties of the SEI film, further influencing the battery performance. The SEI films formed in traditional carbonate ester-based electrolytes, such as 1 M $NaClO_4$/PC, are usually not homogeneous in thickness and chemical compositions, and not mechanically stable to afford repeated Na-ion intercalation/de-intercalation [78, 112], especially for those electrode materials suffering from large volume expansion [39, 113, 114]. Thus, introduction of film-formation additives to regulate the properties of the SEI is much more necessary. FEC is the most widely used film-formation additive in both LIBs and NIBs for facilitating the formation of an SEI film with high chemical and mechanical stability [104–106]. Komaba et al. studied the impact of FEC on the electrochemical performance of hard carbon and $NaNi_{1/2}Mn_{1/2}O_2$ in 1 M $NaClO_4$/PC electrolyte. The results show that the FEC additive can efficiently enhance the reversibility of the electrochemical Na-ion intercalation/de-intercalation. The formed passivated SEI film is useful in suppressing the side reactions between Na metal and PC solvent, ensuring electrochemical deposition/dissolution of metallic Na with higher reversibility [104]. Later, the same group investigated Na-ion insertion behavior in hard-carbon electrode in cyclic alkylene carbonate-based electrolytes with $NaClO_4$ or $NaPF_6$ [115]. The $NaPF_6$-containing electrolyte provides superior reversibility and cyclability of Na-ion insertion into hard-carbon electrode than that of $NaClO_4$-based electrolyte. With the addition of FEC additive, a thinner passivation film was formed on the surface of hard-carbon electrode, resulting an enhanced capacity retention. The combination of $NaPF_6$ and FEC achieves a synergetic effect on passivation for the hard-carbon electrodes, especially when PVDF is used as binder for preparing the electrodes, presenting enhanced cycling performance.

Additionally, the FEC additive has also been proved to be effective for stabilizing electrodes that suffer from huge volume expansion during Na-ion insertion, such as Sn [116], Sb [117, 118], P [39], and metal phosphides [119], through the formation of a mechanically stable SEI film. Sn_4P_3 is an attractive anode material for NIBs because of its high theoretical capacity (1132 mAh/g) but suffers from large volumetric variation after complete sodiation [12]. With the presence of an FEC additive, a robust stable SEI film can be formed, which results in a much improved cycling stability. For instance, by adding 5 wt.% FEC into the 1 M $NaClO_4$ EC/DEC electrolyte, the Sn_4P_3 electrode presented an excellent cycling stability with negligible capacity fading over 100 cycles [119]. In contrast, the Sn_4P_3 electrode without addition of FEC suffers from huge volume expansion and finally fails to work with no capacity delivered after about 50 cycles. Cao et al. studied the impact of the FEC additive on the SiC-Sb-C electrode in carbonate ester-based electrolyte [120]. The results indicate that the FEC additive participates in the formation process of SEI film, which can minimize the reductive decomposition of the electrolyte and stabilize the surface structure of the electrode, achieving an improved capacity retention. Despite that a compact protective film can be formed on the surface of the electrode, the presence of HF, a byproduct resulting from the decomposition of the FEC additive, remains a risk for maintaining stable battery performance. Jang et al. reported the utilization of binary additives of FEC and TMSP for Sn_4P_3 electrode, in which the TMSP additive can help to scavenge of the HF generated from the decomposition of FEC [108]. Benefiting from the synergistic advantages of binary additives, a unique protective surface film was constructed on the Sn_4P_3 electrode to protect against unwanted electrolyte decomposition and to prevent the formation of the $Na_{15}Sn_4$ phase, thus alleviating the large volume expansion. Other additives, such as adiponitrile (AND)

and N-propyl-N-methyl pyrrolidinium bis(trifluoromethanesulfony) imide, were also proven to be effective in promoting the formation of uniform and compact SEI film to achieve stable cycling performance [121, 122].

Despite achieving significantly enhanced performance by adding FEC additives, it also leads to increased charge-transfer resistance and then decreases the reversible specific capacity, resulting in a low initial Columbic efficiency, particularly when excess FEC is used [104, 119]. Besides, the thermal stability and chemical stability of the additive-derived SEI films for NIBs, especially under elevated temperature conditions, have been rarely studied.

8.3.3.2 Flame-retardant additives

The increase in the energy density of NIBs is good to promote the practical application of NIBs but also triggers safety issues, such as thermal runaway, fire, and explosion. Non-flammable electrolytes show promise in addressing these issues; however, they usually have low ionic conductivity or cannot well protect the electrodes. Adding flame-retardant additives into the electrolytes would be an efficient strategy to low the flame risk while arousing minimum impact on the electrochemical performance [69]. Under elevated temperature, the decomposition of carbonate ester solvents generates a large amount of hydrogen radicals, which will further react with oxygen and catalyze the formation of highly active oxygen free radicals [58, 69]. The typical working principle of the flame-retardant additives is to chemically capture active free radicals to terminate radical chain reactions [58].

In NIBs, however, studies on flame-retardant additives have been rarely reported. Feng et al. adopted the ethoxy-pentafluoro-cyclotriphosphazene (EFPN) as flame-retardant additive in 1 M $NaPF_6$ EC/PC [123]. The authors found that the electrolyte's self-extinguishing time decreased from 58 s to 0 s upon increasing the percentage of EFPN from 0 to 5%. The flammable carbonate ester-based electrolyte flame becomes inhibiting when 5% EFPN was added. Meanwhile, the addition of EPFN did not sacrifice the ionic conductivity of the electrolyte with a value of 5.7 mS/cm achieved. Besides, the EFPN-containing electrolyte has also been proved stable to Na metal, ensuring enhanced cyclability of both the acetylene black and $Na_{0.44}MnO_2$ electrodes when tested in half-cells with Na-metal anode.

8.3.3.3 Overcharge protection additives

Overcharge is another potential safety issue for batteries, which can lead to the severe decomposition of electrolytes, generating flammable gases and eventually fire, even explosion [58]. The addition of overcharge protection additives into the electrolyte is effective in suppressing the uncontrollable voltage increase of the batteries under overcharge status [111]. The overcharge protection additives protect the batteries through two principles: (1) redox shuttle-type additives by continuously repeating the redox reaction between the electrodes; (2) shutdown-type additives, which form insulating polymer species on the electrode surface via electrochemical polymerization to prevent further deterioration reactions [124, 125]. The choice of overcharge protection additives is highly dependent on the operation voltage window and the work environment of the batteries.

Currently, there are only limited reports on the application of overcharge protection additives in organic liquid electrolytes for NIBs. Biphenyl (BP), a frequently used additive in LIBs, is also promising as an overcharge protection additive for NIBs. The BP additive can be electro-polymerized at 4.3 V on the surface of the $Na_{0.44}MnO_2$ electrode to prevent the batteries from voltage runway upon 800% overcharge capacity in carbonate ester-based electrolyte while bringing negligible impact on the cycling performance and reversible capacity [126].

8.4 Summary and Outlook

As a low-cost battery technology, NIBs have been widely studied over the past decade. Electrolytes play a significant role in coulombic efficiency, electrochemical window, rate capability, cycling stability, energy density, and safety. Carbonate ester- and ether-based organic liquid electrolytes hold great promise for commercial NIBs. In this chapter, we have reviewed the working principle, basic characteristics, chemical compositions (including sodium salts, solvents, and additives), and electrochemical compatibility with electrode materials of organic liquid electrolytes for NIBs. For future development of high-performance organic liquid electrolytes for NIBs, the following aspects deserve more attention.

1. The effect of composition on electrolyte performance. The composition of organic liquid electrolytes, including sodium salts, solvents, and additives, have profound influence on the performance of NIB cells. An in-depth understanding of the solvation mechanism is critical for optimising the physical and chemical properties of the electrolyte to improve the electrochemical performance of NIBs.
2. The chemical and electrochemical compatibility between electrode and electrolyte materials. Research has shown that ether-based electrolytes are promising for hard-carbon anode. However, the ether-based electrolyte is not well compatible with those cathode materials which have high redox potential. Development of electrolytes with compatibility toward both cathode and anode is needed. The match between the anode and cathode electrodes should be fully considered when evaluating the electrolytes. Besides, the compatibility mechanisms should be comprehensively studied both experimentally and theoretically.
3. The SEI films formed on the surface of electrodes are crucial in stabilising the structural integrity of electrodes and determining the sodiation kinetics and cycling stability. The chemical composition of the SEI film is directly determined by the electrolyte compositions and their electrochemical reduction mechanism. Usually, film-formation additives are introduced to promote the formation of a multifunctional SEI film. However, the formation mechanisms of SEI film in different electrolytes and at different electrode surfaces are poorly understood. Besides, strategies for modifying the chemical and physical properties of the SEI to achieve desirable functions are rarely available. Advanced *in situ* and *operando* characterization techniques should be developed to realise the physical and chemical processes of electrolyte decomposition on electrode surface for understanding the formation mechanism of SEI films.

Acknowledgment

The authors are grateful for the financial support from the Australian Research Council (project No. DP200102573). Professor X. S. Zhao thanks for the funding support from the Australian Research Council Australian Laureate Fellowship (No. FL170100101).

References

1. Tarascon, J.-M. Is lithium the new gold? *Nat. Chem.* 2010, **2**, 510.
2. Sun, Y., Liu, N., and Cui, Y. Promises and challenges of nanomaterials for lithium-based rechargeable batteries. *Nat. Energy* 2016, **1**, 16071.
3. Dunn, B., Kamath, H., and Tarascon, J. M. Electrical energy storage for the grid: A battery of choices. *Science* 2011, **334**, 928–935.
4. Palacín, M. and de Guibert, A. Why do batteries fail? *Science* 2016, **351**, 1253292.
5. Vaalma, C., Buchholz, D., Weil, M., and Passerini, S. A cost and resource analysis of sodium-ion batteries. *Nat. Rev. Mater.* 2018, **3**, 18013.
6. Kumar, N. P., Liangtao, Y., Wolfgang, B., and Philipp, A. From lithium-ion to sodium-ion batteries: Advantages, challenges, and surprises. *Angew. Chem. Int. Ed.* 2018, **57**, 102–120.
7. Tarascon, J.-M. Na-ion versus Li-ion batteries: Complementarity rather than competitiveness. *Joule* 2020, **4**, 1616–1620.
8. Yabuuchi, N., Kubota, K., Dahbi, M., and Komaba, S. Research development on sodium-ion batteries. *Chem Rev* 2014, **114**, 11636–11682.
9. Hwang, J. Y., Myung, S. T., and Sun, Y. K. Sodium-ion batteries: Present and future. *Chem. Soc. Rev.* 2017, **46**, 3529–3614.
10. Luo, W., Shen, F., Bommier, C., Zhu, H., Ji, X., and Hu, L. Na-ion battery anodes: Materials and electrochemistry. *Acc. Chem. Res.* 2016, **49**, 231–240.
11. Hu, Z., Liu, Q., Chou, S. L., and Dou, S. X. Advances and challenges in metal sulfides/selenides for next-generation rechargeable sodium-ion batteries. *Adv. Mater.* 2017, **29**, 1700606.
12. Xia, Q. B., Li, W. J., Miao, Z. C., Chou, S. L., and Liu, H. K. Phosphorus and phosphide nanomaterials for sodium-ion batteries. *Nano Res.* 2017, **10**, 4055–4081.
13. Kundu, D., Talaie, E., Duffort, V., and Nazar, L. F. The emerging chemistry of sodium ion batteries for electrochemical energy storage. *Angew. Chem. Int. Ed.* 2015, **54**, 3431–3448.
14. Cao, Y., Xiao, L., Wang, W., Choi, D., Nie, Z., Yu, J., Saraf, L. V., Yang, Z., and Liu, J. Reversible sodium ion insertion in single crystalline manganese oxide nanowires with long cycle life. *Adv. Mater.* 2011, **23**, 3155–3160.

15. Yu, C.-Y., Park, J.-S., Jung, H.-G., Chung, K.-Y., Aurbach, D., Sun, Y.-K., and Myung, S.-T. $NaCrO_2$ cathode for high-rate sodium-ion batteries. *Energy Environ. Sci.* 2015, **8**, 2019–2026.

16. Wang, Y., Xiao, R., Hu, Y.-S., Avdeev, M., and Chen, L. P2-$Na_{0.6}[Cr_{0.6}Ti_{0.4}]O_2$ cation-disordered electrode for high-rate symmetric rechargeable sodium-ion batteries. *Nat. Commun.* 2015, **6**.

17. Dose, W. M., Sharma, N., Pramudita, J. C., Brand, H. E. A., Gonzalo, E., and Rojo, T. Structure–electrochemical evolution of a Mn-rich P2 $Na_{2/3}Fe_{0.2}Mn_{0.8}O_2$ Na-ion battery cathode. *Chem. Mater.* 2017, **29**, 7416–7423.

18. Luo, C., Langrock, A., Fan, X., Liang, Y., and Wang, C. P2-type transition metal oxides for high performance Na-ion battery cathodes. *J. Mater. Chem. A* 2017, **5**, 18214–18220.

19. Kim, H., Yoon, G., Park, I., Park, K.-Y., Lee, B., Kim, J., Park, Y.-U., Jung, S.-K., Lim, H.-D., Ahn, D., Lee, S., and Kang, K. Anomalous Jahn–Teller behavior in a manganese-based mixed-phosphate cathode for sodium ion batteries. *Energy Environ. Sci.* 2015, **8**, 3325–3335.

20. Wongittharom, N., Lee, T.-C., Wang, C.-H., Wang, Y.-C., and Chang, J.-K. Electrochemical performance of Na/$NaFePO_4$ sodium-ion batteries with ionic liquid electrolytes. *J. Mater. Chem. A* 2014, **2**, 5655–5661.

21. Barpanda, P., Oyama, G., Nishimura, S.-I., Chung, S.-C., and Yamada, A. A 3.8-V earth-abundant sodium battery electrode. *Nat. Commun.* 2014, **5**, 4358.

22. Law, M. and Balaya, P. $NaVPO_4F$ with high cycling stability as a promising cathode for sodium-ion battery. *Energy Storage Mater.* 2018, **10**, 102–113.

23. Li, W., Han, C., Xia, Q., Zhang, K., Chou, S., Kang, Y.-M., Wang, J., Liu, H. K., and Dou, S. X. Remarkable enhancement in sodium-ion kinetics of $NaFe_2(CN)_6$ by chemical bonding with graphene. *Small Methods* 2018, **2**, 1700346.

24. Li, W., Han, C., Wang, W., Xia, Q., Chou, S., Gu, Q., Johannessen, B., Liu, H., and Dou, S. Stress distortion restraint to boost the sodium ion storage performance of a novel binary hexacyanoferrate. *Adv. Energy Mater.* 2020, **10**, 1903006.

25. Wang, W., Gang, Y., Hu, Z., Yan, Z., Li, W., Li, Y., Gu, Q.-F., Wang, Z., Chou, S.-L., Liu, H.-K., and Dou, S.-X. Reversible structural evolution of sodium-rich rhombohedral Prussian blue for sodium-ion batteries. *Nat. Commun.* 2020, **11**, 980.

26. Wen, Y., He, K., Zhu, Y. J., Han, F. D., Xu, Y. H., Matsuda, I., Ishii, Y., Cumings, J., and Wang, C. S. Expanded graphite as superior anode for sodium-ion batteries. *Nat. Commun.* 2014, **5**, 4033.
27. Chen, D., Zhang, W., Luo, K., Song, Y., Zhong, Y., Liu, Y., Wang, G., Zhong, B., Wu, Z., and Guo, X. Hard carbon for sodium storage: Mechanism and optimization strategies toward commercialization. *Energy Environ. Sci.* 2021, **14**, 2244–2262.
28. Hou, H. S., Qiu, X. Q., Wei, W. F., Zhang, Y., and Ji, X. B. Carbon anode materials for advanced sodium-ion batteries. *Adv. Energy Mater.* 2017, **7**, 1602898.
29. Xia, Q. B., Liang, Y. R., Lin, Z. H., Wang, S. W., Lai, W. H., Yuan, D., Dou, Y. H., Gu, Q. F., Wang, J. Z., Liu, H. K., Dou, S. X., Fang, S. M., and Chou, S. L. Confining ultrathin 2D superlattices in mesoporous hollow spheres renders ultrafast and high-capacity Na-ion storage. *Adv. Energy Mater.* 2020, **10**, 2001033.
30. Xia, Q., Huang, Y., Xiao, J., Wang, L., Lin, Z., Li, W., Liu, H., Gu, Q., Liu, H. K., and Chou, S. L. Phosphorus-modulation-triggered surface disorder in titanium dioxide nanocrystals enables exceptional sodium-storage performance. *Angew Chem Int Ed* 2019, **58**, 4022–4026.
31. Xia, Q., Lin, Z., Lai, W., Wang, Y., Ma, C., Yan, Z., Gu, Q., Wei, W., Wang, J. Z., Zhang, Z., Liu, H. K., Dou, S. X., and Chou, S. L. 2D titania-carbon superlattices vertically encapsulated in 3D hollow carbon nanospheres embedded with 0D TiO_2 quantum dots for exceptional sodium-ion storage. *Angew. Chem. Int. Ed.* 2019, **58**, 14125–14128.
32. Wang, X., Chen, Y., Fang, Y., Zhang, J., Gao, S., and Lou, X. W. D. Synthesis of cobalt sulfide multi-shelled nanoboxes with precisely controlled two to five shells for sodium-ion batteries. *Angew. Chem. Int. Ed.* 2019, **58**, 2675–2679.
33. Ru, J., He, T., Chen, B., Feng, Y., Zu, L., Wang, Z., Zhang, Q., Hao, T., Meng, R., Che, R., Zhang, C., and Yang, J. Covalent assembly of MoS_2 nanosheets with SnS nanodots as linkages for lithium/sodium-ion batteries. *Angew. Chem. Int. Ed.* 2020, **59**, 14621–14627.
34. Zhang, K., Park, M., Zhou, L., Lee, G. H., Shin, J., Hu, Z., Chou, S. L., Chen, J., and Kang, Y. M. Cobalt-doped FeS_2 nanospheres with complete solid solubility as a high-performance anode material for sodium-ion batteries. *Angew. Chem. Int. Ed.* 2016, **55**, 12822–12826.
35. Song, X., Li, X., Bai, Z., Yan, B., Li, D., and Sun, X. Morphology-dependent performance of nanostructured Ni_3S_2/Ni anode electrodes for high performance sodium ion batteries. *Nano Energy* 2016, **26**, 533–540.

36. Chen, S., Wu, F., Shen, L., Huang, Y., Sinha, S. K., Srot, V., van Aken, P. A., Maier, J., and Yu, Y. Cross-linking hollow carbon sheet encapsulated CuP_2 nanocomposites for high energy density sodium-ion batteries. *ACS Nano* 2018, **12**, 7018–7027.

37. Wang, X., Chen, K., Wang, G., Liu, X., Wang, H. Rational design of three-dimensional graphene encapsulated with hollow FeP@carbon nanocomposite as outstanding anode material for lithium ion and sodium ion batteries. *ACS Nano* 2017, **11**, 11602–11616.

38. Wu, C., Kopold, P., van Aken, P. A., Maier, J., and Yu, Y. High performance graphene/Ni_2P hybrid anodes for lithium and sodium storage through 3D yolk-shell-like nanostructural design. *Adv. Mater.* 2017, **29**, 1604015.

39. Kim, Y., Park, Y., Choi, A., Choi, N. S., Kim, J., Lee, J., Ryu, J. H., Oh, S. M., and Lee, K. T. An amorphous red phosphorus/carbon composite as a promising anode material for sodium ion batteries. *Adv. Mater.* 2013, **25**, 3045–3049.

40. Qian, J., Wu, X., Cao, Y., Ai, X., and Yang, H. High capacity and rate capability of amorphous phosphorus for sodium ion batteries. *Angew. Chem. Int. Ed.* 2013, **52**, 4633–4636.

41. Sun, J., Lee, H. W., Pasta, M., Yuan, H. T., Zheng, G. Y., Sun, Y. M., Li, Y. Z., and Cui, Y. A phosphorene-graphene hybrid material as a high-capacity anode for sodium-ion batteries. *Nat. Nanotechnol.* 2015, **10**, 980-U184.

42. Liu, S., Xu, H., Bian, X., Feng, J., Liu, J., Yang, Y., Yuan, C., An, Y., Fan, R., and Ci, L. Nanoporous red phosphorus on reduced graphene oxide as superior anode for sodium-ion batteries. *ACS Nano* 2018, **12**, 7380–7387.

43. Huang, Y., Zhao, L., Li, L., Xie, M., Wu, F., and Chen, R. Electrolytes and electrolyte/electrode interfaces in sodium-ion batteries: From scientific research to practical application. *Adv. Mater.* 2019, **31**, e1808393.

44. Sun, Y., Shi, P., Xiang, H., Liang, X., and Yu, Y. High-safety nonaqueous electrolytes and interphases for sodium-ion batteries. *Small* 2019, **15**, 1805479.

45. Che, H., Chen, S., Xie, Y., Wang, H., Amine, K., Liao, X.-Z., and Ma, Z.-F. Electrolyte design strategies and research progress for room-temperature sodium-ion batteries. *Energy Environ. Sci.* 2017, **10**, 1075–1101.

46. Lin, Z., Xia, Q., Wang, W., Li, W., and Chou, S. Recent research progresses in ether- and ester-based electrolytes for sodium-ion batteries. *InfoMat* 2019, **1**, 376–389.

47. Wu, E. A., Banerjee, S., Tang, H., Richardson, P. M., Doux, J.-M., Qi, J., Zhu, Z., Grenier, A., Li, Y., Zhao, E., Deysher, G., Sebti, E., Nguyen, H., Stephens, R., Verbist, G., Chapman, K. W., Clément, R. J., Banerjee, A., Meng, Y. S., and Ong, S. P. A stable cathode-solid electrolyte composite for high-voltage, long-cycle-life solid-state sodium-ion batteries. *Nat. Commun.* 2021, **12**, 1256.

48. Li, Z., Liu, P., Zhu, K., Zhang, Z., Si, Y., Wang, Y., and Jiao, L. Solid-state electrolytes for sodium metal batteries. *Energy Fuels* 2021, **35**, 9063–9079.

49. Pahari, D. and Puravankara, S. Greener, safer, and sustainable batteries: An insight into aqueous electrolytes for sodium-ion batteries. *ACS Sustain. Chem. Eng.* 2020, **8**, 10613–10625.

50. Ponrouch, A., Monti, D., Boschin, A., Steen, B., Johansson, P., and Palacín, M. R. Non-aqueous electrolytes for sodium-ion batteries. *J. Mater. Chem. A* 2015, **3**, 22–42.

51. Hijazi, H., Desai, P., and Mariyappan, S. Non-aqueous electrolytes for sodium-ion batteries: Challenges and prospects towards commercialization. *Batteries Supercaps* 2021, **4**, 881–896.

52. Huang, Y., Zhao, L., Li, L., Xie, M., Wu, F., and Chen, R. Electrolytes and electrolyte/electrode interfaces in sodium-ion batteries: From scientific research to practical application. *Adv. Mater.* 2019, **31**, 1808393.

53. Li, K., Zhang, J., Lin, D., Wang, D.-W., Li, B., Lv, W., Sun, S., He, Y.-B., Kang, F., Yang, Q.-H., Zhou, L., and Zhang, T.-Y. Evolution of the electrochemical interface in sodium ion batteries with ether electrolytes. *Nat. Commun.* 2019, **10**, 725.

54. Eshetu, G. G., Diemant, T., Hekmatfar, M., Grugeon, S., Behm, R. J., Laruelle, S., Armand, M., and Passerini, S. Impact of the electrolyte salt anion on the solid electrolyte interphase formation in sodium ion batteries. *Nano Energy* 2019, **55**, 327–340.

55. Zhang, J., Wang, D.-W., Lv, W., Zhang, S., Liang, Q., Zheng, D., Kang, F., and Yang, Q.-H. Achieving superb sodium storage performance on carbon anodes through an ether-derived solid electrolyte interphase. *Energy Environ. Sci.* 2017, **10**, 370–376.

56. Jache, B. and Adelhelm, P. Use of graphite as a highly reversible electrode with superior cycle life for sodium-ion batteries by making use of co-intercalation phenomena. *Angew. Chem. Int. Ed.* 2014, **53**, 10169–10173.

57. Xu, Z. L., Yoon, G., Park, K. Y., Park, H., Tamwattana, O., Joo Kim, S., Seong, W. M., and Kang, K. Tailoring sodium intercalation in graphite for high energy and power sodium ion batteries. *Nat. Commun.* 2019, **10**, 2598.

58. Sun, Y., Shi, P., Xiang, H., Liang, X., and Yu, Y. High-safety nonaqueous electrolytes and interphases for sodium-ion batteries. *Small* 2019, **15**, 1805479.

59. Ponrouch, A., Dedryvère, R., Monti, D., Demet, A. E., Ateba Mba, J. M., Croguennec, L., Masquelier, C., Johansson, P., and Palacín, M. R. Towards high energy density sodium ion batteries through electrolyte optimization. *Energy Environ. Sci.* 2013, **6**, 2361–2369.

60. Ponrouch, A., Marchante, E., Courty, M., Tarascon, J.-M., and Palacín, M. R. In search of an optimized electrolyte for Na-ion batteries. *Energy Environ. Sci.* 2012, **5**, 8572–8583.

61. Bhide, A., Hofmann, J., Katharina Dürr, A., Janek, J., and Adelhelm, P. Electrochemical stability of non-aqueous electrolytes for sodium-ion batteries and their compatibility with $Na_{0.7}CoO_2$. *Physical Chemistry Chemical Physics* 2014, **16**, 1987–1998.

62. Feng, J., Zhang, Z., Li, L., Yang, J., Xiong, S., and Qian, Y. Ether-based nonflammable electrolyte for room temperature sodium battery. *J. Power Sources* 2015, **284**, 222–226.

63. Pham, T. A., Kweon, K. E., Samanta, A., Lordi, V., and Pask, J. E. Solvation and dynamics of sodium and potassium in ethylene carbonate from ab initio molecular dynamics simulations. *J. Phys. Chem. C* 2017, **121**, 21913–21920.

64. Kuratani, K., Uemura, N., Senoh, H., Takeshita, H. T., and Kiyobayashi, T. Conductivity, viscosity and density of $MClO_4$ (M = Li and Na) dissolved in propylene carbonate and γ-butyrolactone at high concentrations. *J. Power Sources* 2013, **223**, 175–182.

65. Chen, S., Ishii, J., Horiuchi, S., Yoshizawa-Fujita, M., and Izgorodina, E. I. Difference in chemical bonding between lithium and sodium salts: Influence of covalency on their solubility. *Physical Chemistry Chemical Physics* 2017, **19**, 17366–17372.

66. Plewa-Marczewska, A., Trzeciak, T., Bitner, A., Niedzicki, L., Dranka, M., Żukowska, G. Z., Marcinek, M., and Wieczorek, W. New tailored sodium salts for battery applications. *Chem. Mater.* 2014, **26**, 4908–4914.

67. Eshetu, G. G., Grugeon, S., Kim, H., Jeong, S., Wu, L., Gachot, G., Laruelle, S., Armand, M., and Passerini, S. Comprehensive insights into the reactivity of electrolytes based on sodium ions. *ChemSusChem* 2016, **9**, 462–471.

68. Chen, J., Huang, Z., Wang, C., Porter, S., Wang, B., Lie, W., and Liu, H. K. Sodium-difluoro(oxalato)borate (NaDFOB): A new electrolyte salt for Na-ion batteries. *Chem. Commun.* 2015, **51**, 9809–9812.
69. Eshetu, G. G., Elia, G. A., Armand, M., Forsyth, M., Komaba, S., Rojo, T., and Passerini, S. Electrolytes and interphases in sodium-based rechargeable batteries: Recent advances and perspectives. *Adv. Energy Mater.* 2020, **10**, 2000093.
70. Bommier, C. and Ji, X. Electrolytes, SEI formation, and binders: A review of nonelectrode factors for sodium-ion battery anodes. *Small* 2018, **14**, 1703576.
71. Shakourian-Fard, M., Kamath, G., Smith, K., Xiong, H., Sankaranarayanan, S. K. R. S. Trends in Na-ion solvation with alkyl-carbonate electrolytes for sodium-ion batteries: Insights from first-principles calculations. *J. Phys. Chem. C* 2015, **119**, 22747–22759.
72. Tasaki, K., Goldberg, A., and Winter, M. On the difference in cycling behaviors of lithium-ion battery cell between the ethylene carbonate- and propylene carbonate-based electrolytes. *Electrochim. Acta* 2011, **56**, 10424–10435.
73. Jónsson, E. and Johansson, P. Modern battery electrolytes: Ion–ion interactions in Li+/Na+ conductors from DFT calculations. *Phys. Chem. Chem. Phys.* 2012, **14**, 10774–10779.
74. Okoshi, M., Yamada, Y., Yamada, A., and Nakai, H. Theoretical analysis on de-solvation of lithium, sodium, and magnesium cations to organic electrolyte solvents. *J. Electrochem. Soc.* 2013, **160**, A2160–A2165.
75. Liu, Q., Mu, D., Wu, B., Wang, L., Gai, L., and Wu, F. Density functional theory research into the reduction mechanism for the solvent/additive in a sodium-ion battery. *ChemSusChem* 2017, **10**, 786–796.
76. Lee, Y., Lee, J., Kim, H., Kang, K., and Choi, N.-S. Highly stable linear carbonate-containing electrolytes with fluoroethylene carbonate for high-performance cathodes in sodium-ion batteries. *J. Power Sources* 2016, **320**, 49–58.
77. Zhao, J., Zhao, L., Chihara, K., Okada, S., Yamaki, J.-I., Matsumoto, S., Kuze, S., and Nakane, K. Electrochemical and thermal properties of hard carbon-type anodes for Na-ion batteries. *J. Power Sources* 2013, **244**, 752–757.
78. Komaba, S., Murata, W., Ishikawa, T., Yabuuchi, N., Ozeki, T., Nakayama, T., Ogata, A., Gotoh, K., and Fujiwara, K. Electrochemical Na insertion and solid electrolyte interphase for hard-carbon electrodes and application to Na-ion batteries. *Adv. Funct. Mater.* 2011, **21**, 3859–3867.

79. Dahbi, M., Yabuuchi, N., Kubota, K., Tokiwa, K., and Komaba, S. Negative electrodes for Na-ion batteries. *Phys. Chem. Chem. Phys.* 2014, **16**, 15007–15028.

80. Vidal-Abarca, C., Lavela, P., Tirado, J. L., Chadwick, A. V., Alfredsson, M., and Kelder, E. Improving the cyclability of sodium-ion cathodes by selection of electrolyte solvent. *J. Power Sources* 2012, **197**, 314–318.

81. Jang, J. Y., Kim, H., Lee, Y., Lee, K. T., Kang, K., and Choi, N.-S. Cyclic carbonate based-electrolytes enhancing the electrochemical performance of $Na_4Fe_3(PO_4)_2(P_2O_7)$ cathodes for sodium-ion batteries. *Electrochem. Commun.* 2014, **44**, 74–77.

82. Yan, G., Alves-Dalla-Corte, D., Yin, W., Madern, N., Gachot, G., and Tarascon, J.-M. Assessment of the electrochemical stability of carbonate-based electrolytes in Na-ion batteries. *J. Electrochem. Soc.* 2018, **165**, A1222–A1230.

83. Piernas-Muñoz, M. J., Castillo-Martínez, E., Gómez-Cámer, J. L., and Rojo, T. Optimizing the electrolyte and binder composition for sodium Prussian blue, $Na_{1-x}Fe_{x+(1/3)}(CN)_6 \cdot yH_2O$, as cathode in sodium ion batteries. *Electrochim. Acta* 2016, **200**, 123–130.

84. Lee, H., Kim, Y.-I., Park, J.-K., and Choi, J. W. Sodium zinc hexacyanoferrate with a well-defined open framework as a positive electrode for sodium ion batteries. *Chem. Commun.* 2012, **48**, 8416–8418.

85. Xu, K. Electrolytes and interphases in Li-ion batteries and beyond. *Chem. Rev.* 2014, **114**, 11503–11618.

86. Hasa, I., Dou, X., Buchholz, D., Shao-Horn, Y., Hassoun, J., Passerini, S., and Scrosati, B. A sodium-ion battery exploiting layered oxide cathode, graphite anode and glyme-based electrolyte. *J. Power Sources* 2016, **310**, 26–31.

87. Stevens, D. A. and Dahn, J. R. High-capacity anode materials for rechargeable sodium-ion batteries. *J. Electrochem. Soc.* 2000, **147**, 1271.

88. Thomas, P. and Billaud, D. Electrochemical insertion of sodium into hard carbons. *Electrochim. Acta* 2002, **47**, 3303–3307.

89. Jache, B., Binder, J. O., Abe, T., and Adelhelm, P. A comparative study on the impact of different glymes and their derivatives as electrolyte solvents for graphite co-intercalation electrodes in lithium-ion and sodium-ion batteries. *Phys. Chem. Chem. Phys.* 2016, **18**, 14299–14316.

90. Kim, H., Hong, J., Yoon, G., Kim, H., Park, K.-Y., Park, M.-S., Yoon, W.-S., and Kang, K. Sodium intercalation chemistry in graphite. *Energy Environ. Sci.* 2015, **8**, 2963–2969.

91. Kim, H., Hong, J., Park, Y.-U., Kim, J., Hwang, I., and Kang, K. Sodium storage behavior in natural graphite using ether-based electrolyte systems. *Adv. Funct. Mater.* 2015, **25**, 534–541.

92. Yoon, G., Kim, H., Park, I., and Kang, K. Conditions for reversible Na intercalation in graphite: Theoretical studies on the interplay among guest ions, solvent, and graphite host. *Adv. Energy Mater.* 2017, **7**, 1601519.

93. Gotoh, K., Maruyama, H., Miyatou, T., Mizuno, M., Urita, K., and Ishida, H. Structure and dynamic behavior of sodium–diglyme complex in the graphite anode of sodium ion battery by 2H nuclear magnetic resonance. *J. Phys. Chem. C* 2016, **120**, 28152–28156.

94. Yu, C.-J., Ri, S.-B., Choe, S.-H., Ri, G.-C., Kye, Y.-H., and Kim, S.-C. Ab initio study of sodium co-intercalation with diglyme molecule into graphite. *Electrochim. Acta* 2017, **253**, 589–598.

95. Zhu, Z., Cheng, F., Hu, Z., Niu, Z., and Chen, J. Highly stable and ultrafast electrode reaction of graphite for sodium ion batteries. *J. Power Sources* 2015, **293**, 626–634.

96. Cohn, A. P., Share, K., Carter, R., Oakes, L., and Pint, C. L. Ultrafast solvent-assisted sodium ion intercalation into highly crystalline few-layered graphene. *Nano Lett.* 2016, **16**, 543–548.

97. Zhu, Y.-E., Yang, L., Zhou, X., Li, F., Wei, J., and Zhou, Z. Boosting the rate capability of hard carbon with an ether-based electrolyte for sodium ion batteries. *J. Mater. Chem. A* 2017, **5**, 9528–9532.

98. Lee, M. E., Lee, S. M., Choi, J., Jang, D., Lee, S., Jin, H. J., and Yun, Y. S. Electrolyte-dependent sodium ion transport behaviors in hard carbon anode. *Small* 2020, **16**, e2001053.

99. Bai, P., He, Y., Xiong, P., Zhao, X., Xu, K., and Xu, Y. Long cycle life and high rate sodium-ion chemistry for hard carbon anodes. *Energy Storage Mater.* 2018, **13**, 274–282.

100. Xu, Z. L., Lim, K., Park, K. Y., Yoon, G., Seong, W. M., and Kang, K. Engineering solid electrolyte interphase for pseudocapacitive anatase TiO_2 anodes in sodium-ion batteries. *Adv. Funct. Mater.* 2018, **28**, 1802099.

101. Zhang, K., Park, M. H., Zhou, L. M., Lee, G. H., Li, W. J., Kang, Y. M., and Chen, J. Urchin-like $CoSe_2$ as a high-performance anode material for sodium-ion batteries. *Adv. Funct. Mater.* 2016, **26**, 6728–6735.

102. Guo, Q., Ma, Y., Chen, T., Xia, Q., Yang, M., Xia, H., and Yu, Y. Cobalt sulfide quantum dot embedded N/S-doped carbon nanosheets with superior reversibility and rate capability for sodium-ion batteries. *ACS Nano* 2017, **11**, 12658–12667.

103. Su, D., Kretschmer, K., and Wang, G. Improved electrochemical performance of Na-ion batteries in ether-based electrolytes: A case study of ZnS nanospheres. *Adv. Energy Mater.* 2016, **6**, 1501785.

104. Komaba, S., Ishikawa, T. Yabuuchi, N., Murata, W., Ito, A., and Ohsawa, Y. Fluorinated ethylene carbonate as electrolyte additive for rechargeable Na batteries. *ACS Appl. Mater. Interfaces* 2011, **3**, 4165–4168.

105. Markevich, E., Salitra, G., and Aurbach, D. Fluoroethylene carbonate as an important component for the formation of an effective solid electrolyte interphase on anodes and cathodes for advanced Li-ion batteries. *ACS Energy Lett.* 2017, **2**, 1337–1345.

106. Huang, Y., Xie, M., Zhang, J., Wang, Z., Jiang, Y., Xiao, G., Li, S., Li, L., Wu, F., and Chen, R. A novel border-rich Prussian blue synthetized by inhibitor control as cathode for sodium ion batteries. *Nano Energy* 2017, **39**, 273–283.

107. Hwang, J.-Y., Myung, S.-T., and Sun, Y.-K. Sodium-ion batteries: Present and future. *Chem. Soc. Rev.* 2017, **46**, 3529–3614.

108. Jang, J. Y., Lee, Y., Kim, Y., Lee, J., Lee, S.-M., Lee, K. T., and Choi, N.-S. Interfacial architectures based on a binary additive combination for high-performance Sn_4P_3 anodes in sodium-ion batteries. *J. Mater. Chem. A* 2015, **3**, 8332–8338.

109. Mogensen, R., Colbin, S., and Younesi, R. An attempt to formulate non-carbonate electrolytes for sodium-ion batteries. *Batteries Supercaps* 2021, **4**, 791–814.

110. Self, J., Hall, D. S., Madec, L., and Dahn, J. R. The role of prop-1-ene-1,3-sultone as an additive in lithium-ion cells. *J. Power Sources* 2015, **298**, 369–378.

111. Zhang, S. S. A review on electrolyte additives for lithium-ion batteries. *J. Power Sources* 2006, **162**, 1379–1394.

112. Song, J., Xiao, B., Lin, Y., Xu, K., and Li, X. Interphases in sodium-ion batteries. *Adv. Energy Mater.* 2018, **8**, 1703082.

113. Wu, C., Maier, J., and Yu, Y. Sn-based nanoparticles encapsulated in a porous 3D graphene network: Advanced anodes for high-rate and long life Li-ion batteries. *Adv. Funct. Mater.* 2015, **25**, 3488–3496.

114. Liu, J., Kopold, P., Wu, C., van Aken, P. A., Maier, J., and Yu, Y. Uniform yolk-shell Sn_4P_3@C nanospheres as high-capacity and cycle-stable anode materials for sodium-ion batteries. *Energy Environ. Sci.* 2015, **8**, 3531–3538.

115. Dahbi, M., Nakano, T., Yabuuchi, N., Fujimura, S., Chihara, K., Kubota, K., Son, J.-Y., Cui, Y.-T., Oji, H., and Komaba, S. Effect of hexafluorophosphate

and fluoroethylene carbonate on electrochemical performance and the surface layer of hard carbon for sodium-ion batteries. *Chemelectrochem* 2016, **3**, 1856–1867.

116. Wang, J. W., Liu, X. H., Mao, S. X., and Huang, J. Y. Microstructural evolution of tin nanoparticles during in situ sodium insertion and extraction. *Nano Lett.* 2012, **12**, 5897–5902.

117. Wu, L., Hu, X., Qian, J., Pei, F., Wu, F., Mao, R., Ai, X., Yang, H., and Cao, Y. Sb–C nanofibers with long cycle life as an anode material for high-performance sodium-ion batteries. *Energy Environ. Sci.* 2014, **7**, 323–328.

118. Liu, S., Feng, J. K., Bian, X. F., Liu, J., and Xu, H. The morphology-controlled synthesis of a nanoporous-antimony anode for high-performance sodium-ion batteries. *Energy Environ. Sci.* 2016, **9**, 1229–1236.

119. Kim, Y., Kim, Y., Choi, A., Woo, S., Mok, D., Choi, N. S., Jung, Y. S., Ryu, J. H., Oh, S. M., and Lee, K. T. Tin phosphide as a promising anode material for Na-ion batteries. *Adv. Mater.* 2014, **26**, 4139–4144.

120. Lu, H., Wu, L., Xiao, L., Ai, X., Yang, H., and Cao, Y. Investigation of the effect of fluoroethylene carbonate additive on electrochemical performance of Sb-based anode for sodium-ion batteries. *Electrochim. Acta* 2016, **190**, 402–408.

121. Song, X., Meng, T., Deng, Y., Gao, A., Nan, J., Shu, D., and Yi, F. The effects of the functional electrolyte additive on the cathode material $Na_{0.76}Ni_{0.3}Fe_{0.4}Mn_{0.3}O_2$ for sodium-ion batteries. *Electrochim. Acta* 2018, **281**, 370–377.

122. Manohar, C. V., Forsyth, M., MacFarlane, D. R., and Mitra, S. Role of N-propyl-N-methyl pyrrolidinium bis(trifluoromethanesulfonyl)imide as an electrolyte additive in sodium battery electrochemistry. *Energy Technology* 2018, **6**, 2232–2237.

123. Feng, J., An, Y., Ci, L., and Xiong, S. Nonflammable electrolyte for safer non-aqueous sodium batteries. *J. Mater. Chem. A* 2015, **3**, 14539–14544.

124. Odom, S. A., Ergun, S., Poudel, P. P., and Parkin, S. R. A fast, inexpensive method for predicting overcharge performance in lithium-ion batteries. *Energy Environ. Sci.* 2014, **7**, 760–767.

125. Lee, H., Kim, S., Jeon, J., and Cho, J.-J. Proton and hydrogen formation by cyclohexyl benzene during overcharge of Li-ion batteries. *J. Power Sources* 2007, **173**, 972–978.

126. Feng, J., Ci, L., and Xiong, S. Biphenyl as overcharge protection additive for nonaqueous sodium batteries. *RSC Adv.* 2015, **5**, 96649–96652.

Chapter 9

Cycling Stability of Sodium-Ion Batteries in Analogy to Lithium-Ion Batteries

Yverick Rangom,[a] Timothy T. Duignan,[a] Xin Fan,[a] and X. S. (George) Zhao[a,b]
[a]*School of Chemical Engineering, The University of Queensland, Brisbane, 4072, Australia*
[b]*Institute of Materials for Energy and Environment, Qingdao University, Shandong 266071, China*
george.zhao@uq.edu.au

Lithium-ion batteries (LIBs) have been widely used to power portable electronic devices. Recently, LIBs are being adopted for large-scale energy-storage applications. However, a major concern with the use of LIBs is safety. Another concern about LIBs is the geographically limited resources of lithium and cobalt. Sodium-ion batteries (NIBs) have come back after being silent for over two decades and represent the future of energy-storage technology, especially for large-scale energy-storage applications. The successful adoption of this sustainable electrochemical energy-storage technology by end users requires fundamental understandings of charge storage

Handbook of Sodium-Ion Batteries: Materials and Characterization
Edited by Rohit R. Gaddam and X. S. (George) Zhao

ISBN 978-981-4968-15-7 (Hardcover), 978-1-003-30874-4 (eBook)
www.jennystanford.com

behavior in electrode materials and cell-failure mechanisms. This review focuses on discussion of current research efforts aiming at preventing/minimizing NIB failures that leverage lessons learned from LIBs. After comparison of the similarity and difference of electrode materials, electrolytes, and configurations between LIBs and NIBs, the knowledge on LIB cell-failure mechanisms has been adopted to interpret the causes leading to the failure of NIBs. Appropriate research approaches and future research and development directions toward increasing the lifetime of NIBs have been suggested.

9.1 Introduction

The high energy density and low weight of LIBs compared with alternative technologies such as lead acid batteries have significantly improved our living standards in terms of portable electronic devices such as cell phones. LIBs are also the current technology of choice for electric vehicles (EVs) and for solving the intermittency issue of renewable energy sources such as solar and wind. The rapid growth in EVs and the needs of the renewable energy industry has led to increasing demands for LIBs, leading to price increases for both lithium and cobalt minerals. In addition, these minerals are unevenly distributed in the earth's crust. These situations have resulted in huge concerns of the sustainability and reliability of the LIB technology.

In the past decade, research into alternative electrochemical energy-storage technologies has accelerated substantially [1–3]. A technology similar to LIBs is the NIBs, which use sodium ions as charge carriers instead of lithium ions. Because of the abundance of sodium compared with lithium, NIBs are positioned to be a sustainable and viable alternative to LIBs, especially in large-scale energy-storage applications. For these applications, the battery must operate for many years with little to no servicing. Therefore, these batteries need to have a long shelf life. It is, therefore, critical to understand all aspects of the failure of NIBs. Many reviews discussing anode, cathode, electrolyte materials as well as challenges for NIBs have been published [4–11]. However, no review of their failure mechanisms is available in the literature.

Because of the similarities in the cell configuration and underlying principle between LIBs and NIBs, in addition to the vast breadth of knowledge readily available for LIBs, the general approach to studying NIBs is largely carried over from LIBs. This includes design of electrode materials and solvent for electrolytes, choice of separators and current collector, characterization methods and approaches, and infrastructure used for cell fabrication. Therefore, many explanations of the failures of NIBs explicitly, or implicitly, refer to the failure mechanisms of LIBs. It is true that the common failure mechanisms are largely shared between the two systems, ranging from decomposition of electrolytes, metal plating, electrode pulverization and delamination of anodes, to various structural degradations at the cathode. However, the sodium chemistry does differ significantly from the lithium chemistry in several important aspects, so this "copy-and-paste" approach enjoys only a limited level of success, and the limitations must be kept in mind.

In this review, the phenomena that cause sodium-ion-based electrochemical energy-storage systems to fail to store charge reversibly, namely failure mechanisms, are discussed, along with possible solutions reported in the literature. Failure mechanisms for NIBs will be discussed and analyzed referencing to the lessons learned from the LIB system. Significant differences between the two systems will be highlighted wherever applicable.

9.2 Electrolytes

The environment in which a battery electrolyte operates is both highly reducing on the anode side and highly oxidizing on the cathode side. As energy stored is directly proportional to the voltage difference (ΔV) between the two electrodes ($Q = C \times \Delta V$), anode and cathode materials are selected to have the maximum voltage difference that is allowed by the electrochemical stability window (ESW) of the electrolyte. The ESW is defined as the potential range where neither the solvent nor the salt of the electrolyte undergoes irreversible reactions. However, in practice, electrolytes still decompose on the surface of both electrodes, but in a controlled and limited manner. The products of the electrolyte decompositions form a passivating layer on the electrode surface, which prevents the electrolyte from further decomposition. The passivating layer essentially increases the practical ESW of the electrolyte. These layers are, therefore, formed during the first few cycles. The phenomenon depletes the electrolyte; therefore, it will eventually cause cell failure if the growth of these layers continues uncontrolled after these first few cycles.

9.2.1 Electrolyte Degradation at Anode (Negative Electrode)

The reduction of electrolyte solvents at the anode occurs in both NIB and LIB systems. It is one of the major causes leading to cell failure. It is also the main mechanism for the formation of a solid electrolyte interphase (SEI), which is a passivating layer that has a tremendous impact on the battery performance [12, 13]. The solvents employed for NIBs are mostly carried over from LIBs. Blends of cyclic carbonates, such as propylene carbonate (PC) and ethylene carbonate (EC), dimethyl carbonate (DMC), diethyl carbonate (DEC), and linear ether dimetoxyethane (DME), are the commonly used solvents for NIB studies. As elucidated by Dey and Sullivan [14], the reduction of cyclic carbonates is driven by reactions with solvated cations where the C–O bonds break via a two-electron transfer reaction mechanism. For example, the reduction of PC ($CH_3C_2H_3O_2CO$) to form propylene (C_3H_6) and carbonate ion (CO_3^{2-}) proceeds as follows:

$$CH_3C_2H_3O_2CO + 2e^- \rightarrow C_3H_6 + CO_3^{2-}$$

Cyclic carbonates are opened to form highly reactive radicals first. Then, the organic and carbonate compounds making up the SEI are produced along with gases, such as CO, CO_2, and C_2H_4 [15, 16]. Figure 9.1 shows the energy levels associated with the successive reactions that are believed to occur during the reduction of EC, vinylene carbonate (VC), and fluoroethylene carbonate (FEC) in an NIB [17]. In the case of EC, it first combines with one Na^+ (1) to form an EC–Na^+ complex (2), which has a lower potential energy than the EC molecule and Na^+ with an energy difference of about 36.93 kcal/mol. Then, an electron is gained to form complex (3), which pulls the Na cation closer to the other O atom, leading to the opening of the ring C–O bond to form transition state 1 (TS1) after a small energy barrier is passed. Afterward, TS1 can spontaneously transform to molecule (5), which is the combination of two (4) molecules. Alternatively, molecule (5) can convert to species (6) by combining with another Na^+ by gaining another electron, resulting in the formation of charge-balanced species (7). Finally, complex (7) can dissociate into species (8) to form sodium carbonate and ethylene gas. Similar reduction processes occur to solvents VC and FEC [17].

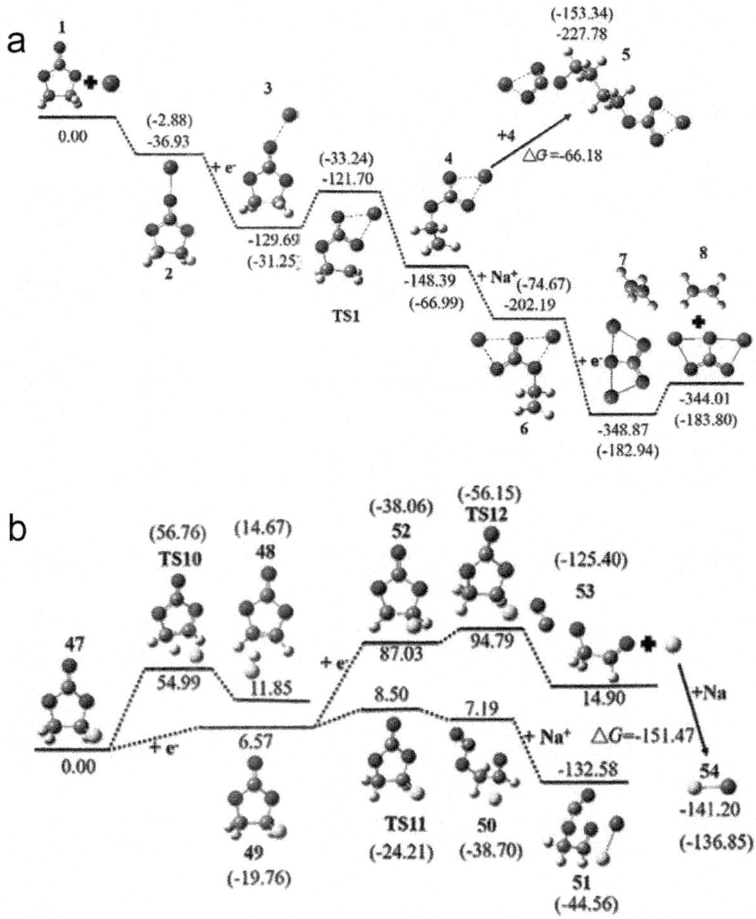

Figure 9.1 Potential energy diagram for the reduction of solvents (a) EC and (b) FEC. Gray, white, red, and violet spheres denote C, H, O, and Na atoms or ions, respectively. Unit for energy is kcal/mol. Data in parentheses were from C-PCM-B3LYP/6–311++G(3df,3dp) single point calculations. Reproduced with permission from Ref. [17]. Copyright (2017), Wiley-VCH.

The reduction process can differ between LIB and NIB cells as the cations in the electrolytes are different. One consequence of the cation changes is changes in the energy required for the two-electron reduction of EC. In LIB cells, there is an energy barrier of 2.3 kcal/mol barrier for the reduction of EC. While this energy barrier is small, it can prevent the electrolyte from further formation of SEI [18]. However, for NIB cells, there is no such energy barrier for this same reaction, making the decomposition of the electrolyte spontaneous. This spontaneity also explains why the SEI films in NIB cells are less passivating, i.e., further electrolyte degradation is not stopped by the previously formed SEI.

Using high-precision coulometry with LIB cells, Dahn et al. observed that the SEI layers continue to grow in each cycle. However, the SEI growth is largely subdued on later cycles with each subsequent cycle being exponentially smaller than the previous one [19, 20]. In general, the SEIs in LIBs form a layered structure with inorganic species inside and organic species outside of the layer, which renders the SEI stable [21, 22]. Philippe et al. confirmed that the SEI in NIBs forms spontaneously without cycling [23]. Rojo et al. also demonstrated that the SEI layer in NIBs formed from the decomposition of carbonate-based electrolytes and is thicker than that in LIBs. The SEI in NIBs contains a higher amount of brittle inorganic salts than that in LIBs, thus making the former less stable than the latter [24]. The presence of sodium carbonate species in the SEI was confirmed by Baggetto et al. [25].

Unlike ester-based solvents, ether-based solvents (e.g., diglyme and triglyme) have recently been reported to enable the storage of sodium ions in graphite via a co-intercalation mechanism [26–29]. It has also been shown that glymes form smaller amounts of SEI on the anode than ester-based solvents, therefore leading to a high columbic efficiency [30, 31]. Contrary to SEI layers made from the degradation of carbonate-based electrolyte, SEI layers from diglyme-based electrolyte were also demonstrated to be more stable [32]. Cyclic ether tetrahudrofuran (THF) has also been investigated as a solvent [22, 33–35]. However, these solvents are not often used in practical applications because they are not chemically stable at higher voltage.

To mitigate cell failure due to electrolyte solvent decomposition at the anode of NIBs, additives such as FEC are employed to stabilize the SEI as has been the case for the LIB system [36, 37]. The basic principle for SEI modification with the FEC additive in NIBs is similar to what occurs in LIBs: FEC decomposes at a higher potential versus Na/Na^+ than the organic solvent. DFT calculations predicted a 7.05 kcal/mol barrier for FEC to decompose versus 11.7 kcal/mol for PC [38]. Note these calculations are based on FEC decomposition following the one-electron reduction pathway instead of the more commonly accepted two-electron route [39–41].

Figure 9.2a compares the energy levels for the reduction of EC, PC, and VC and FEC additives. The upper path represents the opening of the EC and PC dimer rings, whereas the lower path represents a scenario where the additive decomposes first. The calculated energy levels show that the decomposition of the additive in the lower path is energetically more favorable. The reduction reactions shown in Fig. 9.2a are, therefore, less favorable. Both VC and FEC additives can lower the reduction energy of the EC molecule in the first step [42]. More importantly, these additives prevent the ring opening of the EC molecule. The pathways involving the reduction of additives lead to radicals with a greater likelihood of oligomerization and polymerization. When these polymeric compounds get incorporated into the SEI, they make it mechanically stronger [43]. In addition, decomposition of FEC releases fluoride and forms NaF, a known stable component for SEIs [44]. This may explain why FEC provides better results than VC in both NIBs and LIBs, in particular with hard-carbon anodes [43]. On the contrary, with phosphorous-based anode materials, VC is a better additive than FEC [45]. A time-of-flight secondary ion mass spectroscopy study on phosphorous anodes showed that the external part of the SEI layer formed with VC additive contains both organic and inorganic species whereas of the FEC-induced SEI layer is mostly inorganic species [45].

Nagaoka et al. developed a new and iterative computational simulation method (the Red Moon method) [46], which uses cycles of Monte Carlo and reactive molecular dynamic methods to study the layer-by-layer formation of the SEI on carbon anode. Results showed that a thinner (3.9 versus 5.7 nm) but denser SEI layer forms when FEC is present. The authors confirmed that inorganic salts (e.g., NaF and Na_2CO_3) tend to be located close to the electrode surface, while organic species such as NaPC reside mainly on the outside of the SEI layer. This compact and orderly morphology results in the formation of a more efficient SEI for minimizing the irreversible capacity loss. The authors also found that FEC molecules remain unreacted. This finding contradicts the commonly accepted mechanism that FEC is the first compound to decompose during the formation of the SEI. Figure 9.2b schematically illustrates the coordination of FEC

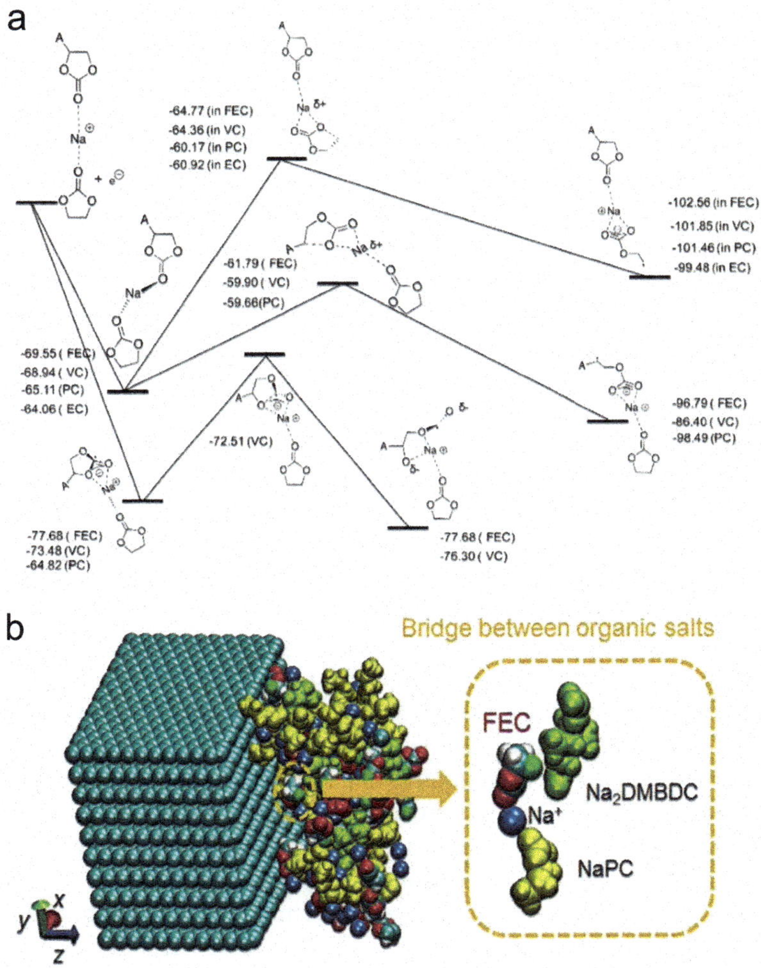

Figure 9.2 (a) Pathways for the reduction of EC, PC, VC, and FEC complexes. A marking next to molecules denotes additive molecules (A = EC, PC, VC, FEC). Reproduced with permission from Ref. [42]. Copyright (2016), American Chemical Society. (b) Representation of SEI film formation from a FEC-added electrolyte on the surface of electrode material (yellow, NaPC; green, Na_2DMBDC; red, Na_2CO_3; pink, NaF). Reproduced with permission from Ref. [46]. Copyright (2015), American Chemical Society.

molecules with organic salts such as sodium propylene carbonate (NaPC) and 2,3-dimethylsodium butylene dicarbonate (Na_2DMBDC) in the SEI, thanks to the strong electronegativity of fluoride. This explanation supplements the commonly accepted understanding that the formation of NaF is the main reason for the SEI layer to be stable.

Besides the reduction of solvents as described earlier, the salt component in the electrolyte can also potentially be reduced at the anode. However, salt reduction is less of an issue as salts such as $NaPF_6$ are very stable compared to organic solvents such as EC and DEC. Dahn et al. [47] showed that $NaPF_6$ does not decompose at elevated temperatures on the surface of sodiated carbon. In contrast, $LiPF_6$ decomposes to form LiF under the same conditions. Nevertheless, when these solvated salts dissociate into Li^+/Na^+ cations and PF_6^- anions, PF_6^- by itself will decompose and form various species (including NaF) as demonstrated by Dabhi et al. [48], who employed the hard X-ray photoelectron spectroscopy and time-of-flight secondary ion mass spectroscopy techniques to analyze the decomposition products. Similarly, the ClO_4^- anion also decomposes. The ClO_4^- decomposition does not explicitly contribute to a more stable SEI, and its continuous decomposition can become a potential failure mechanism of the cell [48]. Remarkably, with $NaPF_6$, the decomposition of the anion is limited and stops after the SEI starts forming on hard-carbon or sodium metal anodes. This means that $NaPF_6$ can contribute to a longer cell cycle life compared to other salts like $NaClO_4$.

9.2.2 Electrolyte Degradation at Cathode (Positive Electrode)

As in LIBs, oxidation of electrolyte solvents at the cathode occurs in NIBs. This, however, is of a lesser issue compared to the reduction at the anode. The SEI formed on the cathode is named cathode electrolyte interphase (CEI), which is thinner and more stable compared to the SEI formed at the anode [49]. Nonetheless, PC can be oxidized at the cathode to produce propylene oxide and CO_2 followed by further reactions as proven by using an *in situ* online mass spectroscopy technique [50]. Figure 9.3a depicts the successive reactions that occur resulting in H_2 and CO_2 formation. The gases generated can build up pressure inside the battery cell, potentially leading to explosions [51].

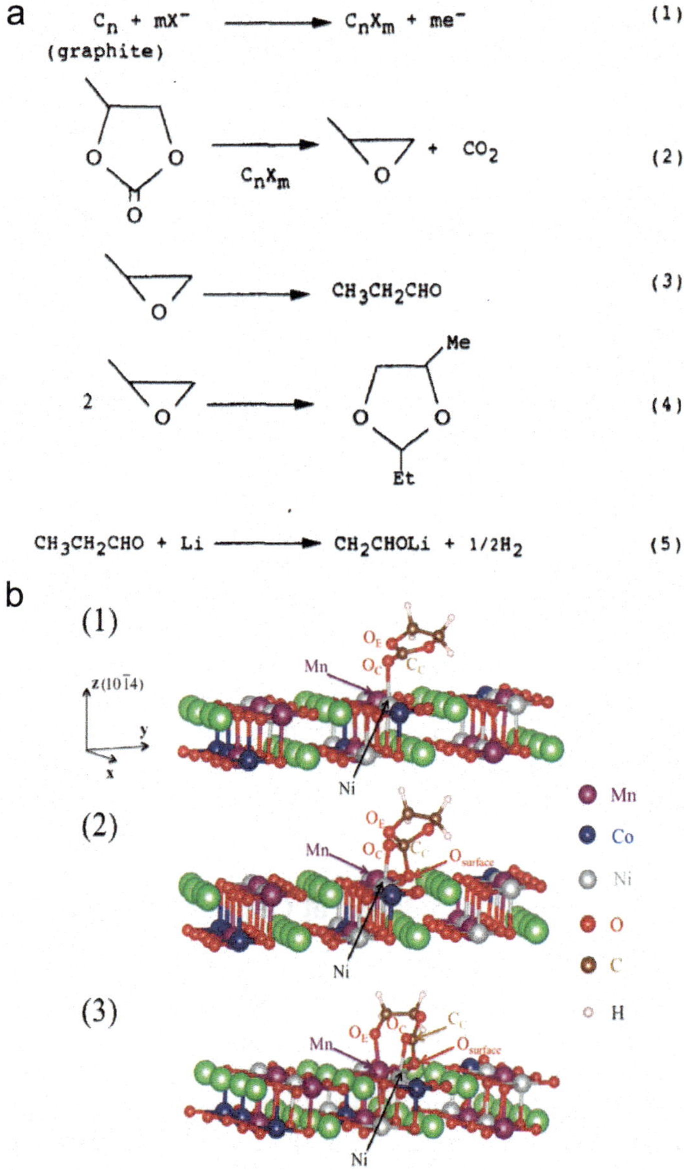

Figure 9.3 (a) Mechanism for the oxidation of propylene carbonate (X represents an anion such as ClO_4^-). Reproduced with permission from Ref. [50]. Copyright (1995), Elsevier. (b) Representation of the CC–OE bond breaking process on the $(10\bar{1}4)$ surface (*xy* plane) in three (1–3) steps. Reproduced with permission from Ref. [52]. Copyright (1998), The Electrochemical Society.

In LIBs, a recent *ab initio* study showed that the oxidation of EC can occur via the opening of its carbon ring caused by transition metal atoms on the surface of the cathode, as shown in Fig. 9.3b [52]. The oxidation rate of EC is, therefore, dependent on the electrolyte environment, i.e., coordination with the cation as well as surface termination groups (e.g., –OH and –F) on the cathode. However, the very high degradation rates reported strongly suggest that further reactions involving solvent molecules also take place, making the oxidation of electrolyte a highly complex phenomenon with many possible oxidation pathways. The most likely degradation products are organic salts and oligomers of EC and DMC with EC being the most easily oxidized solvent. Apart from the bis(trifluoromethane) sulfonimide lithium (TFSI) salt, solvents are much more oxidized than salts. Interestingly, a comprehensive study by Aurbach et al. [53] on the electrochemical stability of commonly used carbonate solvents and their mixtures, namely dimethyl carbonate (DMC), DMC+EC, and EC+DEC (diethylcarbonate), revealed that the products of degradation of these solvents stay suspended in solution and are not taking part in the CEI formation on some LIB cathodes [54].

Compared to LIBs, NIBs have sodium instead of lithium in both the electrolyte salt and the electrode materials. Because of this change in cation, oxidation of solvents in NIBs will most likely be different in terms of reaction rates compared to the LIBs. The general outcome shall nonetheless be similar. Oxidation of solvents at high potentials is a cumulative process taking place at every cycle, therefore eventually leading to cell failure. Currently, no information on oxidation of solvent at the cathode of the NIB system is available. Nevertheless, researchers frequently point toward electrolyte degradation at the cathode as the main cause for the low coulombic efficiency experienced by high-voltage cathode materials (> 4 V versus Na^+) [55–58]. In general, it is the solvent that dictates the ESW for an electrolyte except for the known case of aluminum corrosion by the TFSI anion [22]. The most frequently used carbonate-based electrolytes as well as triglym-based electrolytes are stable in a voltage range between 0 and 5 V versus Na/Na^+ when tested against aluminum. However, cathode materials can catalyze solvent degradation, which reduces the ESW [22, 59].

To mitigate electrolyte oxidation, researchers routinely use cycling voltage ranges to a smaller potential window than the stable window of the electrolyte solvent. This practice, on one hand, can prevent the electrolyte from degradation and, on the other hand, can prevent reversible redox storage mechanism of sodium ions from taking place. For example, manganese is oxidized at a higher potential than the tetravalent state in P2-$Na_xMn_{0.5}Fe_{0.5}O_2$; therefore, the operating potential will not allow for manganese to be oxidized, thereby limiting the amount of sodium stored in the electrode [60, 61]. According to Delmas et al. [62, 63], the "P," "O," and "T" designations refer to the alkali metal (Li or Na) prismatic, octahedral, or tetrahedral possible environments. The numeral that follows the letter (e.g., the "2" in P2) indicates the number of repeating transition metal stacked in the direction of the c-axis and the apostrophe denotes distortion in the unit cell.

The subject of salt oxidation at the cathode has not been investigated for NIBs. It is, nevertheless, safe to assume that the problem is most likely minor compared to solvent oxidation as what has been observed in LIBs [52]. Furthermore, the formation of CEI is a minor issue compared to decomposition at the anode site. Overall, it can be concluded that salt decomposition at the cathode is unlikely to become a failure mechanism in the NIB system.

9.2.3 Degradation of SEI Layer

Once formed, the SEI layer prevents further degradation of the electrolyte by blocking electron transport across it while still allowing ion transport. The SEI is, therefore, commonly accepted to be a functional part of the battery cell in its own right. The SEI acts as both an electrical insulator and an ionic conductor, thereby fulfilling the same role as the liquid electrolyte. It is, therefore, acceptable to describe an "electrolyte system" made of both the liquid phase and the solid phase (the SEI). As a result, the cyclic breaking-off of the SEI can be viewed as a failure mechanism of the "electrolyte system."

A study showed that the dissolution of the constituents of the SEI in NIBs is more severe than that in LIBs [64]. Using the electrochemical quartz crystal microbalance (EQCM) technique, the authors found that the SEI layer formed upon sodiation on the anode and dissolved upon desodiation. The severe instability of the SEI in NIBs was also observed by Zarrabeita et al. [65], who studied the SEI on the $Na_2Ti_3O_7$ electrode. They attributed this to the larger diameter of Na^+ compared to Li^+ because the latter can form salts with stronger bonds, making them less soluble than sodium salts. The solubility of the SEI layer as a whole in carbonate solvents in the LIBs was thoroughly studied by Jones et al. [66]. The atomic absorption spectrophotometry method was used to determine the solubility of lithium compounds commonly found in the SEI. The solvents examined were EC, PC, DMC, DEC, and VC as well as a ternary mixture of EC:PC:3DMC. Among all other factors including solubility of the ions, it was observed that colloidal agglomeration and particle size of the lithium compounds are the key factors determining the solubility as described by the Gibbs–Kelvin equation, which establishes the solubility (s) and particle size (D_p):

$$\ln\left[\frac{s}{s_0}\right] = \frac{4\gamma_{SL}}{\nu D_p}\frac{V_m}{RT}$$

where s_0 is the equilibrium solubility for particles with an infinite diameter (assuming spherical particles), γ_{SL} is solid–liquid surface tension, V_m is molar volume of particles, ν is the number of moles of ions, D_p is particle diameter, R is the gas constant, and T is the absolute temperature. It is clear that solubility is inversely proportional to particle size. For example, LiF is the most readily soluble compound in all alkylcarbonate solvents commonly used with a solubility one and two orders of magnitude higher compared to Li_2CO_3, LiOH, Li_2O, $LiOCH_3$, and $LiOC_2H_5$. The reason is that LiF particles formed are, on average, four times smaller than that of other salt particles [66].

It follows that the agglomeration of particles and their immediate physical environment in the SEI layer are paramount to determining the dissolution. A greater rate of dissolution of SEIs in NIBs compared to SEIs dissolution in LIBs is observed [43]. Assuming the mechanism for dissolution is the same between the two systems, the reason is most likely the relative lack of agglomeration through polymerizing and cross linking of the species present in SEIs in the NIB system [43]. The less tightly linked and more soluble particles that make up the SEI in NIBs pose a greater risk of failure than in the LIB system. Furthermore, the dissolution of SEI also promotes the formation of fresh SEI on subsequent cycles, irreversibly depleting the liquid electrolyte.

While SEI is largely the product of the electrolyte components, including solvent and salt, the chemical composition of the SEI layer formed on an electrode also depends upon the electrode [12]. For example, studies of the decomposition of electrolyte on carbon have shown that SEI layers formed on hard carbon are similar to SEI layers formed on the cross section of highly oriented pyrolytic graphite (HOPG). While SEI formed on soft carbon more closely resembles SEI formed on the basal plane of HOPG [12, 67, 68].

Mogensen et al. [69] carried out a comparative study on lithium- and sodium-based SEIs. They performed in-depth analysis of the SEI layers using synchrotron X-ray photoelectron spectroscopy. They attributed the generally higher self-discharge characteristic of sodium cells compared to lithium cells to a greater degree of dissolution of the SEI in the former. Furthermore, hard-carbon-based and "super P"-based cells cycled against lithium and sodium were given a 100 h rest period between the 10th and 11th cycles. During that 100 h rest period, the potential of the cells was continuously measured and it demonstrated that the self-discharge behavior of high-surface area "super P" carbon is much higher than that of low-surface area hard carbon for both the lithium and sodium cells, as shown in Fig. 9.4a. The self-discharge was also more severe for the sodium cells.

Another cause of SEI degradation is the physical breaking of the SEI layer due to the evolution of gas bubbles, as shown in Fig. 9.4b [70]. The formation of bubbles progressively dislodges the active material and disintegrates the electrode. This process will cause progressive electrode failure due to a gradual loss in electronic pathways connecting with the electrode particles.

To prevent SEI degradation and dissolution, one way is to limit the potential range at which cells are cycled. Komaba et al. [71] reported that setting the voltage lower limit to 0.70 V instead of 0.65 V versus Na/Na^+ significantly benefited capacity retention and coulombic efficiency for a tin/graphite composite electrode. The SEI formation and dissolution potentials appear as plateaus on the GCD curves at about 0.4 V versus Na/Na^+ for SEI formation and 0.68 V versus Na/Na^+ for SEI dissolution. XPS data confirmed the presence of a thicker SEI layer for a sample cycled up to 0.70 V versus Na/Na^+ compared to samples cycled up to 0.65 V versus Na/Na^+ (Fig. 9.5). The stronger intensity of the peaks associated with carbon species in the SEI is seen from the C1s XPS spectrum of the 0.70 V sample than from the 0.65 V sample (Fig. 9.5a), indicating a larger amount

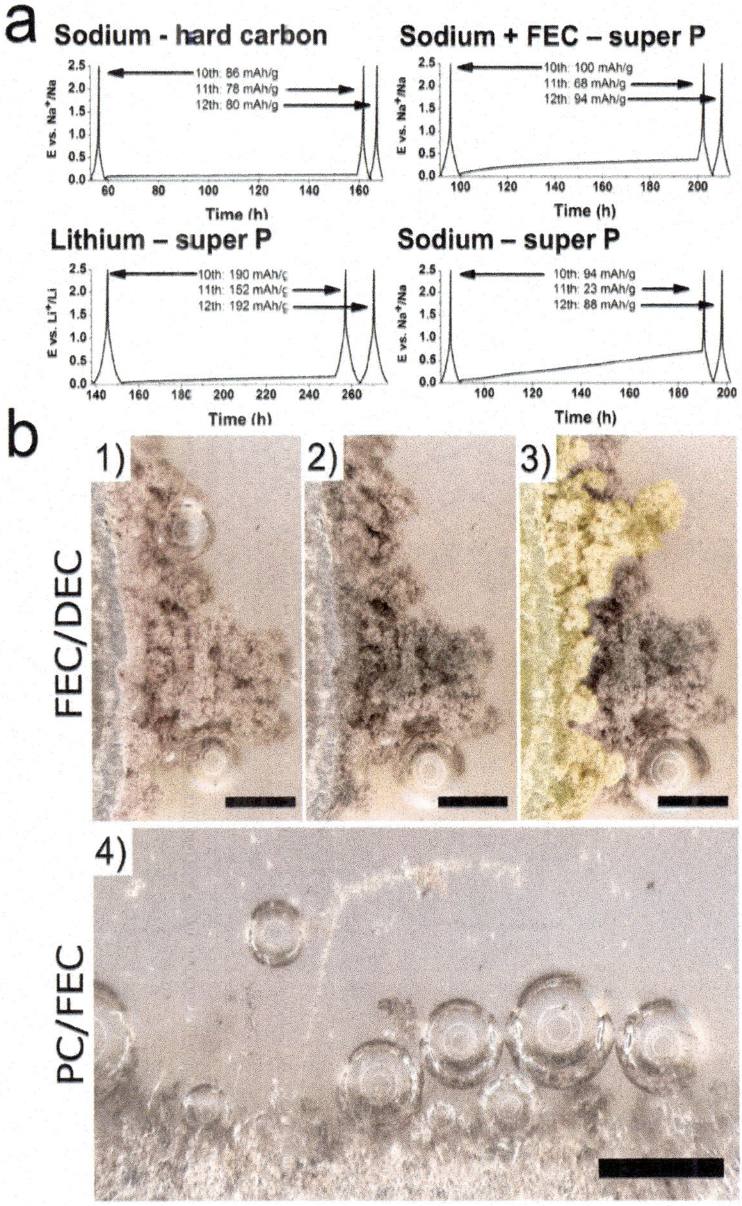

Figure 9.4 (a) GCD curves for 10th, 11th, and 12th cycles of super P and hard-carbon anodes cycled against lithium and sodium. There is a 100 h rest period between the 10th and 11th cycles. Self-discharge is shown by an increasing potential during the 100 h pause. Reproduced with permission from Ref. [69]. Copyright (2016), American Chemical Society. (b) Optical pictures of gassing in FEC/DEC (1–3) and PC/FEC on sodium metal (4). 500 μm scale bar. Reproduced with permission from Ref. [70]. Copyright (2017), American Chemical Society.

of carbonate species present in the SEI of the 0.70 V sample than that of the 0.65 V sample. In Fig. 9.5b, the Sn peak is smaller for the 0.70 V sample compared to the 0.65 V and pristine samples. As a thicker SEI prevents the XPS signal from penetrating down the Sn bulk electrode underneath the SEI, the smaller Sn peak is again indicative of a thicker SEI layer on the 0.70 V sample.

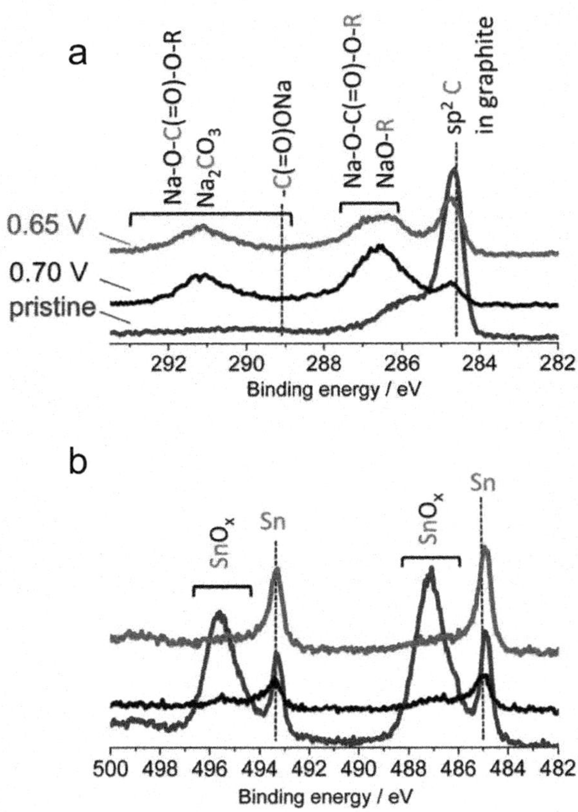

Figure 9.5 Hard XPS spectra of Sn composite electrodes cycled at 0.65 and 0.70 V versus Na/Na^+ for the (a) C1s and (b) Sn3d regions. Reproduced with permission from Ref. [71]. Copyright (2016), American Chemical Society.

9.3 Failure of Anode

The failure of the anode in NIBs can be due to pulverization, delamination, and dendrite formation. Dendrite formation causes catastrophic failure of the whole cell due to short circuit, resulting in thermal runaways and possible fires and explosions.

9.3.1 Pulverization

Pulverization, as schematically illustrated in Fig. 9.6, generally arises from repeated volume changes, generating mechanical stress to lead to cracking of the electrode during cycling. The phenomenon is most severe for alloying and conversion-type electrode materials, such as tin and red phosphorous, which can have volume expansion up to 400% [72, 73]. Due to the larger size of the sodium ion compared to the lithium ion, volume expansion is more severe in NIBs than in LIBs.

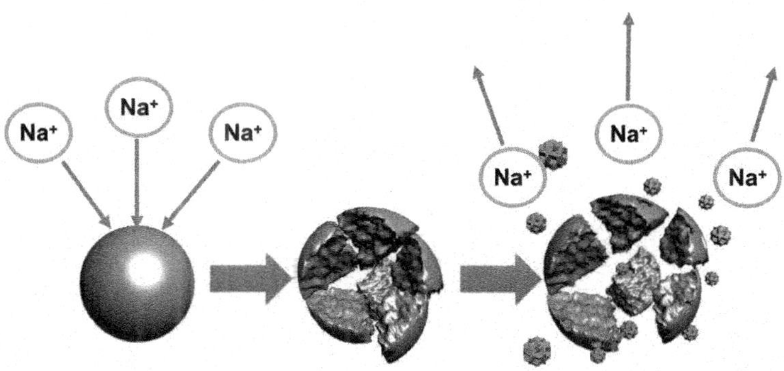

Figure 9.6 Schematic illustration of anode failure due to pulverization.

For alloy anode materials, Obrovac et al. [74] proposed a universal expansion equation, which describes the expansion of electrodes upon lithiation and is independent of the alloying host metal, as shown in the following equation:

$$\tilde{U} = \frac{F}{k} V_{\text{avg}} \left(\frac{\xi_f}{1+\xi_f} \right)$$

where $\tilde{U}$ = volumetric energy density $\left[\frac{\text{J}}{\text{m}^3}\right]$, F = Faraday's number $\left[\frac{\text{Ah}}{\text{mol}}\right]$, k = volume occupied by lithium ion [m^3], V_{avg} = average voltage of the cell [V], ξ_f = volume expansion of alloy at full lithiation (or sodiation) [%].

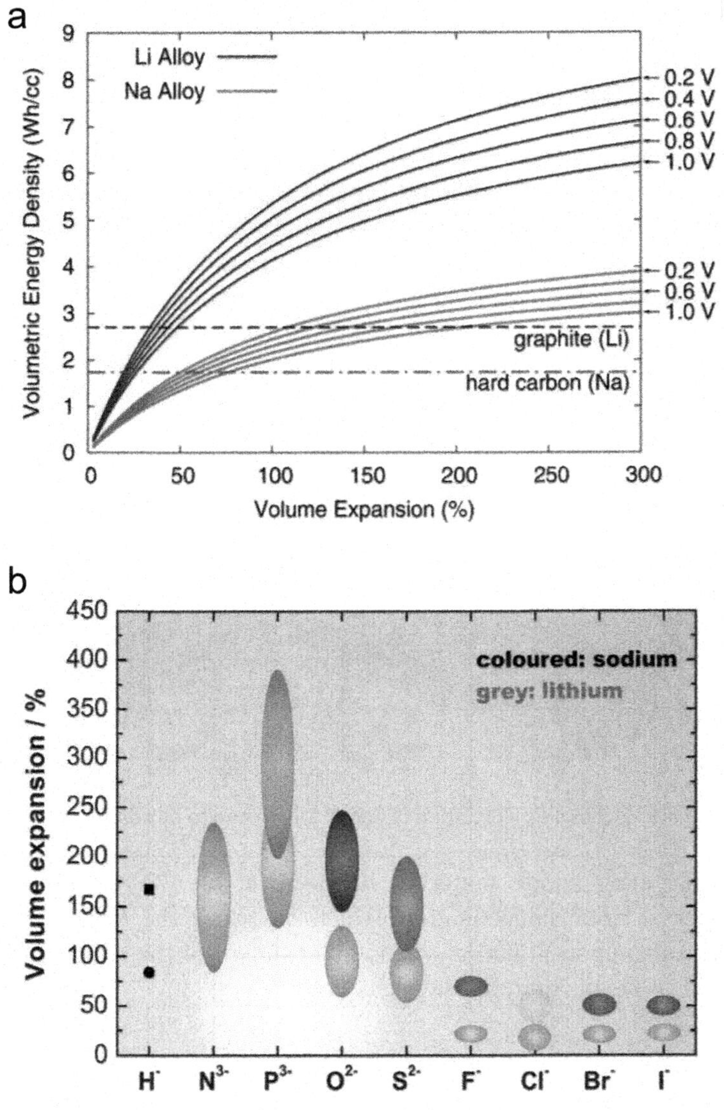

Figure 9.7 (a) Theoretical expansion curves of Li and Na alloys, along with fixed volumetric energy density of graphite (for Li) and hard carbon (for Na). Reproduced with permission from Ref. [6]. Copyright (2011), The Electrochemical Society. (b) Volumetric expansion of transition metal compounds $M_a^{n+}X_b^{m-}$ as a function of X in lithium-ion (grey) and sodium-ion storage (colored) via the conversion mechanism. Reproduced with permission from Refs. [72, 75]. Copyright (2013), Royal Society of Chemistry.

Chevrier and Ceder used the preceding equation to compare NIBs with LIBs, and the results are shown in Fig. 9.7a [6]. It can be seen that the Na alloy electrode undergoes a considerably larger volume expansion than the Li alloy electrode at the same level of volumetric energy density. Therefore, sodium-based alloys are expected to be more prone to pulverization than lithium-based alloys.

Transition metal compounds (TMCs) store sodium ions through the conversion mechanism as follows:

$$M_a^{n+}\mathrm{X}_b^{\mathrm{m}-} + (bc)\mathrm{Na}^+ \leftrightarrow a\mathrm{M} + b\mathrm{Na}_c^+\mathrm{X}$$

where *M* stands for the transition metal and *X* represents a non-metal element, such as oxygen (O), selenium (Se), and phosphorous (P). These materials also suffer from large volume change during cycling, especially the P-based materials, as illustrated in Fig. 9.7b [6, 75]. The issue of large volume expansion has been reported for TMCs for both LIB and NIB applications [76, 77].

Alloying materials are also very promising anode materials, particularly silicon for LIBs because of its very high theoretical lithium-ion storage capability of 4200 mAh/g corresponding to 4.4 lithium atom for every silicon atom [78]. However, its large volume expansion reaching 400% upon lithiation makes it prone to severe pulverization [79–81]. For sodium-ion storage, amorphous silicon can only store 0.76 sodium for every silicon, yielding a theoretical capacity of 725 mAh/g with a volume expansion of 114% [82, 83]. In practice, the capacity for sodium-ion storage is less than 400 mAh/g because of the poor electronic conductivity of silicon [84]. Nonetheless, silicon anodes in NIB applications still experience severe capacity fade after just a few tenths of cycles due to pulverization [85, 86]. Few studies have been published on silicon as anode material for NIBs, and its failure mechanisms are not well understood.

Tin is also prone to pulverization due to large volume expansion when electrochemically cycled against sodium. Wang et al. [72] conducted a thorough investigation of the sodiation of Sn using automated, full-field, X-ray microscopic tomography. This technique produces three-dimensional images of particles at sub-50 nm resolution. It was applied *in situ* to characterize the lithiation of tin nanoparticles over 10 cycles [87]. The study showed that pulverization occurs due to lattice mismatch at the interface between sodium-poor domain and sodium-rich domain. During the sodiation process, the volume increase of pristine particles starts moderately, and the overall concave shape of the particles is initially preserved. This step is referred as a "two-phase" sodiation because of the simultaneous presence of a non-sodiated tin phase and a sodiated tin phase [88]. Upon further sodiation, the electrode starts to develop cracks accompanied by apparition of concave features and a rapid increase in overall volume, as schematically illustrated in Fig. 9.8a. This second step is the single-phase sodiation as has been revealed by XRD characterization, which showed no crystalline tin was left [25]. However, the exact mechanism for the sodiation remains the subject of debate. Subsequent desodiation does not induce further physical damages. This is in remarkable contrast to the lithiation of tin where pulverization occurs during delithiation, as comparatively illustrated in Fig. 9.8a, and damage appears only at the end of delithiation. Compared to desodiation, where the damaged particles (on the right of Fig. 9.8a) develop concave features, the damaged delithiated particles display convex features. Particle size also plays a major role in pulverization, as shown in Fig. 9.8b, which illustrates that large particles break apart while small particles remain intact, both with volume expansion. Particles below 0.5 μm showed no fracture at all upon sodiation, whereas particles between 0.5 μm and 1.6 μm showed moderate fractures and particles bigger than 1.6 μm exhibited severe fracturing [72].

Another effective approach to mitigating this type of failure for materials with large volume expansion is to use alloys [25]. The core–shell structure with the charge storage metal as the core and an amorphous phase of another metal as the shell that acts as an effective electronic pathway and a volume-change accommodating matrix is a good design. Zhu and Deng demonstrated the advantage of this approach with a Co–Sn alloy [89]. This technique is also believed to be present in Sony's Nexellion™ LIB [89]. The study by Zhu and Deng revealed that using the Co–Sn alloy, electrochemical performance is significantly enhanced compared to crystalline tin. An alloy of tin and antimony also demonstrated very good performance with a stable capacity of 525 mAh/g over 125 cycles [90].

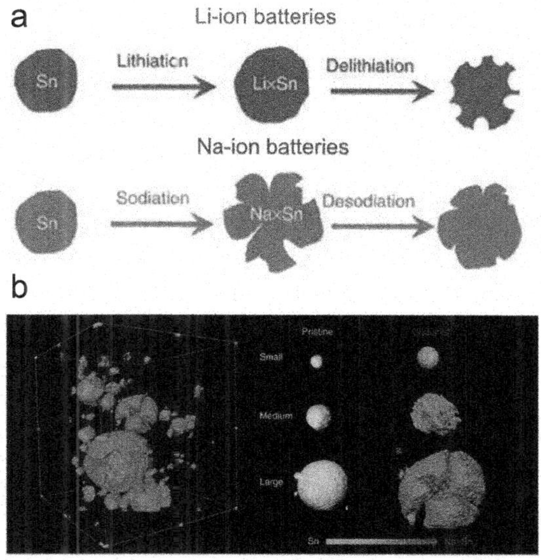

Figure 9.8 (a) Schematic of Sn microparticle distortion change upon sodiation and lithiation. (b) 3D morphological representation of the sodiated particles of three different sizes. Reproduced with permission from Ref. [72]. Copyright (2015), Nature Publishing Group.

Composite electrode materials with metal nanoparticles wrapped by carbon are another effective strategy of minimizing electrode pulverization failures. The carbon usually serves as a superstructure that can provide electronic pathways while also accommodating the strain from the metal particle expansions during charge/discharge. Lui et al. [91] described an anode material with Sn nanoparticles encapsulated in carbon spheres with a stable 415 mAh/g of Na^+ capacity for 500 cycles at 1 A/g. Wu et al. [92] produced nanospheres with antimony (Sb) embedded in carbon for 655 mAh/g of which 93% is preserved after 100 cycles. Composite architectures involving carbon fibers, nanotubes, or graphene sheets with metals have also been reported for mitigating volume expansion for alloying and conversion electrode materials for storage [93–95].

Combining two mitigation techniques such as using reduced particle size at the nano-level inside composite materials is also very effective at reducing pulverization as demonstrated by Guo et al. [96]. Red phosphorous nanoparticles of sizes between 10 and 20 nm embedded in graphene aerogel can deliver a storage capacity as high as 1095.5 mAh/g after 200 cycles at 1 C.

9.3.2 Delamination

Like the pulverization mechanism described earlier, failure through delamination arises from a loss of electronic contact. This occurs specifically at the current collector/active material interface where the active material becomes physically separated from the current collector. As with pulverization, delamination is caused by volume change of particles. Delamination is the principal cause of failure for alloying metal anodes. As the active material becomes detached, it is electrically cut off from the rest of electrode and can no longer take part in charge storage in subsequent cycles. Delamination can lead to rapid cell failure because large capacity reductions can occur after just a few tens of cycles [97].

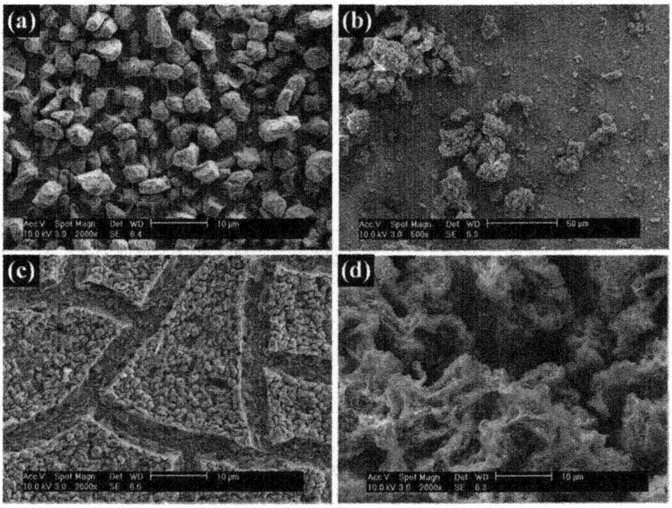

Figure 9.9 *Ex situ* SEM images of a P-type Sn electrode after (a) 1 cycle and (b) 20 cycles and an L-type Sn electrode after (c) 1 cycle and (d) 40 cycles. Reproduced with permission from Ref. [97]. Copyright (2016), Wiley-VCH.

One phenomenon that can cause delamination is when anisotropic volume expansion occurs during sodiation, followed by isotropic contraction during desodiation. Anisotropic volume expansion is a result of the material's crystalline orientation. Nam et al. [97] showed this by electrodepositing two thin films of tin on copper with different crystalline orientations resulting in significantly different cycle lives. The P-type films of crystalline Sn grew along the [101], [211], and [112] directions and exhibited visible film detachment from the copper substrate after 40 cycles, whereas the L-type films grew along the [200] direction and transformed into a porous structure firmly attached to the current collector. The film detachment can be clearly seen from the *ex situ* SEM images shown in Fig. 9.9 [97]. The underlying smooth current collector is visible from Fig. 9.9b. Detachment is directly correlated to large losses in reversible capacity.

Ultimately, delamination is explained by the failure of polymeric binders to maintain electrical pathways through the physical contact of active materials with the current collector. In that respect, the replacement of common polyvinylidene fluoride (PVDF) binder by cellulose binder has been shown to provide a significant improvement in electrode cyclability in LIBs [98–100]. This improvement is the result of cross linking, and the formation of networks that are similar to nets as opposed to a blanketing of the active material particles [101]. Hochgatterer et al. [102] demonstrated that strong cross linking between binder molecules and particles appears to be the key to reducing electrical contact loss and, therefore, better cycle life, as shown in Fig. 9.10a. PVDF-hexafluoropropylene binder performance is compared with stryrene-butadiene-rubber/sodium carboxymethylcellulose binder with varying types and amounts of substituted functional groups. Cycle life increased when functional groups are changed from C_2H_4CN (cyanothylcellulose) to C_2H_4OH (hydroxyethylcellulose) and finally to CH_2COONa (Na-CMC).

These results indicate that all cellulose-based binders present better cycle life than PVDF. Furthermore, the number of functional groups also plays a significant role on cycle life. This discovery demonstrates that the interaction of functional groups and active material is crucial to the prevention of the destruction of electrodes especially for materials with large volume changes. When Na-CMC is sprayed onto SiO_2, covalent bonding was detected through infrared spectroscopy, further indicating that these bonds are the main reason for the improved capacity retention. A possible bonding mechanism involving hydrolyzed CH_2COONa groups and SiO_2 layer on top of silicon substrate is speculated and described by Hochgatterer et al. [102].

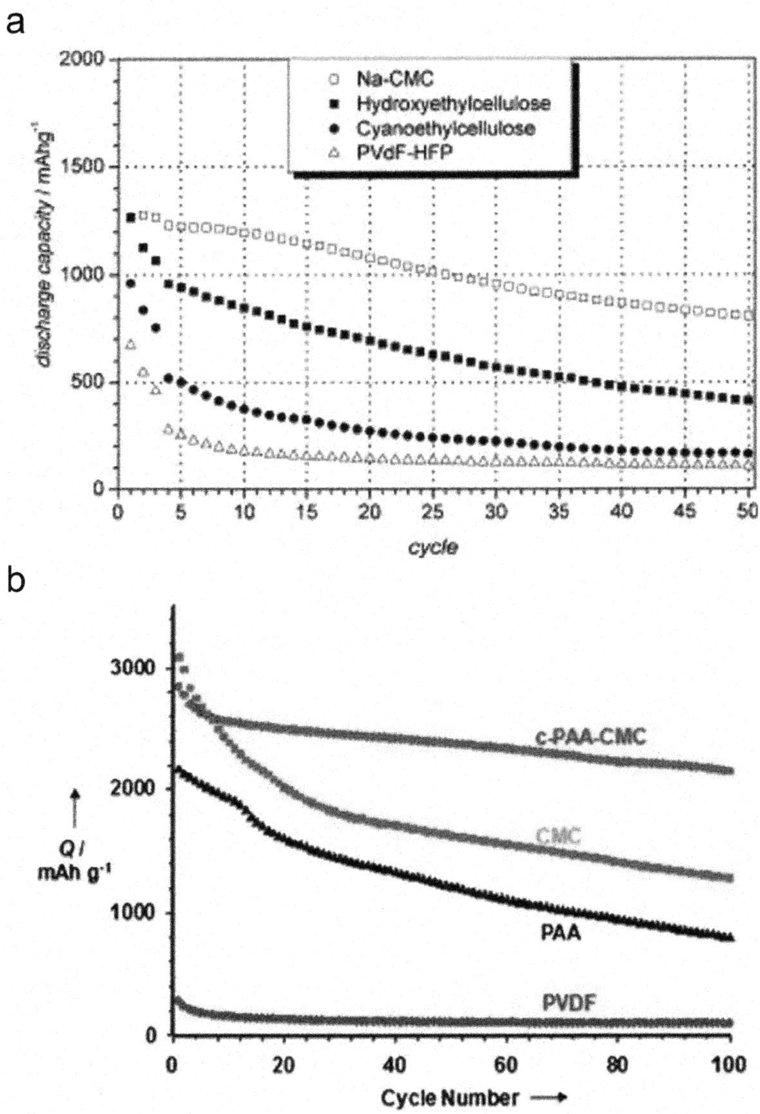

Figure 9.10 (a) Cycle life of anodes using different cellulose binders. Reproduced with permission from Ref. [102]. Copyright (2008), The Electrochemical Society. (b) Cycle life of silicon composite electrodes in LIB application compared for different binders. Reproduced with permission from Ref. [101]. Copyright (2012), Wiley-VCH.

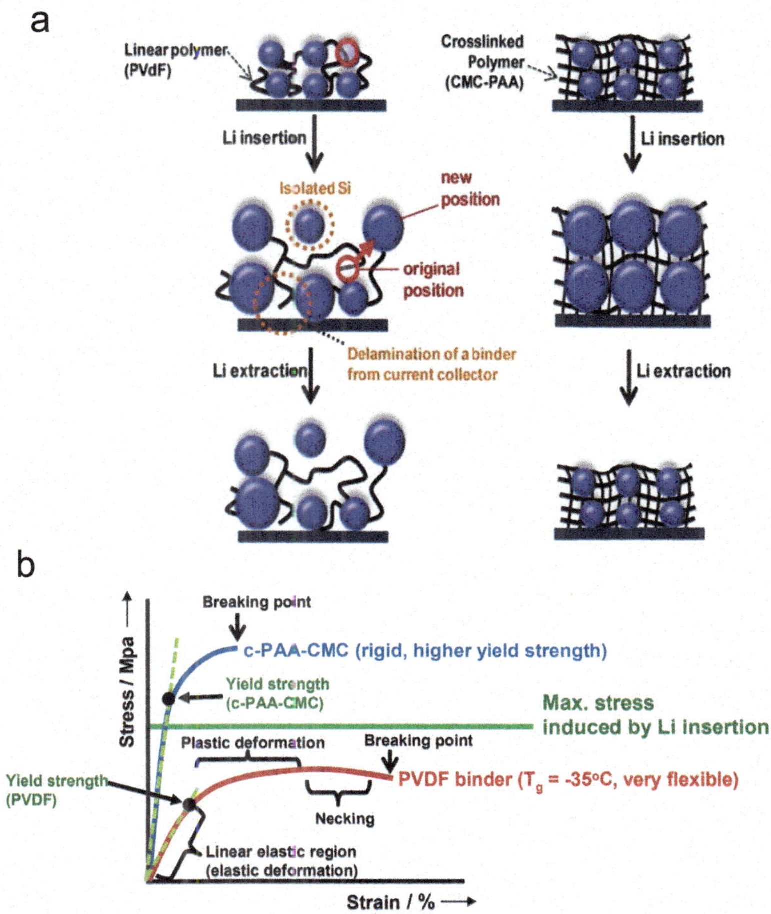

Figure 9.11 (a) Schematic depicting morphology changes of Si-based anodes during the first lithiation with PVDF and c-PAA-CMC as the binders. (b) Stress–strain curves for c-PAA-CMC and PVDF binders with max strain imparted by lithium insertion. Reproduced with permission from Ref. [101]. Copyright (2012), Wiley-VCH.

In another study, Koo et al. [101] compared cross-linked CMC-polyacrylic acid (c-CMC-PAA) binder and PVDF. They showed improved cycle stability for the former, as shown in Fig. 9.10b. The better performance of c-CMC-PAA is attributed to the two following factors. The first factor is that covalent ester bonds (cross linking) are formed between the hydroxyl groups on the thin oxidized layer on the surface of the silicon particles and the carboxylic acid from the PAA. The presence of these bonds is confirmed through magic angle spinning NMR characterization. The second factor is the stronger mechanical stress modulus of the c-CMC-PAA binder that retains the silicon particle in the same spot on the binder net, as shown in Fig. 9.11a. Strong mechanical stress also reduces the maximum volume expansion from 130% to 35% compared to PVDF (Fig. 9.11b) and returns the particles to their original size after cycling. These findings made in the context of LIBs directly apply to Na-ion applications.

Using an Sn–Co alloy as NIB anode, Yui et al. [103] did a systematic comparison of several binders, including VDF, PAA, sodium polyacrylate (PAANa), CMC, and polyimide (PI). The electrodes were tested against cycling stability, volumetric expansion, and adhesion strength to a copper current collector. Volumetric expansion was tested using *in situ* light microscopy, and the adhesion to the current collector was assessed via a mechanical surface and interfacial cutting analysis system, as shown in Fig. 9.12a. Quantitative results are presented in Fig. 9.12b. The results of these two experiments are given in Table 9.1. The adhesion to copper is attributed to hydrogen bonding through the oxide layer. The SEM results showed that binders PAANa, PVDF, and PI all led to larger cracks in the electrode films and larger particles after cycling compared with PAA [104].

Kim et al. [106] found that PAA binder improved the cycling stability of amorphous red phosphorous significantly compared to PVDF. The authors assumed a similar positive effect from cross linking with the binder, as described earlier. These researchers also showed that the amount of carbon additive was remarkably important for the rate capability. The increased electronic conductivity is assumed to be responsible for the observed improved performance.

Dai et al. [107] showed superior capacity and cycle stability of tin nanoparticles with poly(9,9-dioctylfluorene-co-fluorenone-co-methylbenzoic ester) (PFM) conductive polymer as the binder when compared to binders CMC and PVDF. The electrodes they used in the study were made without conductive additive to clearly compare binder influence on electrochemical performance. SEM images showed that PVDF leads to detachment from the tin electrode, while binders CMC and PFM did not show observable detachment.

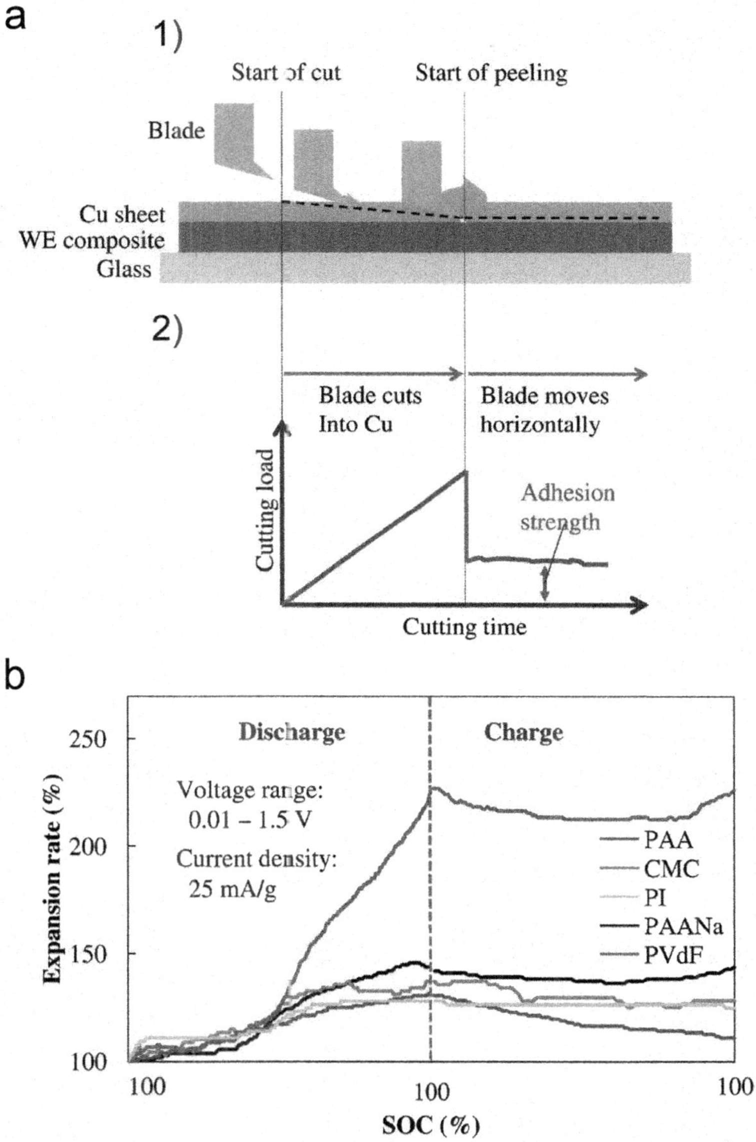

Figure 9.12 (a) Expansion rate of anodes prepared using different binders (PAA, CMC, PI, PAANa, and PVdF) at different states of charge (SOC) [103]. (b) (1) Schematic of adhesion strength experiment. (2) Cutting load versus cutting time diagram. Reproduced with permission from Ref. [103]. Copyright (2016), Elsevier.

Table 9.1 Average adhesion strength and the rate for expansion and contraction of WEs with PAA, CMC, PI, PAANa, and PVDF as binders [105]

	Adhesion Strength (kN/m)	Volume Expansion (%)	
		After Discharge	After Charge
PAA	0.72	130	111
CMC	0.62	137	125
PI	0.58	128	125
PVDF	0.60	227	198
PAANa	0.65	146	143

9.3.3 Sodium Plating and Dendrite Formation

While sodium plating and dendrite formation do not result in electrode breaking off, both phenomena can lead to severe battery failure. Sodium plating, as illustrated in Fig. 9.13a, refers to the deposition of metallic sodium on the anode of the battery. Plating can also occur on the anode materials that store sodium at potentials close to 0 V versus Na/Na^+. This is particularly true for hard carbons that are the most promising and currently the most used anode material for NIBs [108, 109]. Fundamentally, sodium plating and lithium plating are similar, and they arise under similar circumstances, particularly with a fast sodiation or lithiation rate. There are two fundamental causes for plating. The first is when interfacial overpotential between the active anode material and the electrolyte is close to 0 V versus Li/Li^+ or versus Na/Na^+. The second is when the Li^+ or Na^+ saturation concentration is reached in the electrolyte close to that interface. The saturation concentration reaches when the total current density driven by the external current exceeds the rate at which the anode material can uptake Li or Na [105]. The former cause can be described by the following equation, as reported by Arora et al. [110], who derived the equation based on the Doyle–Fuller–Newman model and Butler–Volmer equation [111]. This theory describes the overpotential (η) as a function of the potential in the electrolyte (ϕ_e) as well as the open-circuit potential (U_0) of the electrode [105]:

$$\eta = \phi_n - U_0$$

Here ϕ_n includes the potential in the electrode (ϕ_s) and the potential in the electrolyte (ϕ_e) as follows:

$$\phi_n = \phi_s - \phi_e - Fj_nR_{SEI}$$

where j_n is the pore wall flux, R_{SEI} is the SEI layer resistance, and F is the Faraday constant.

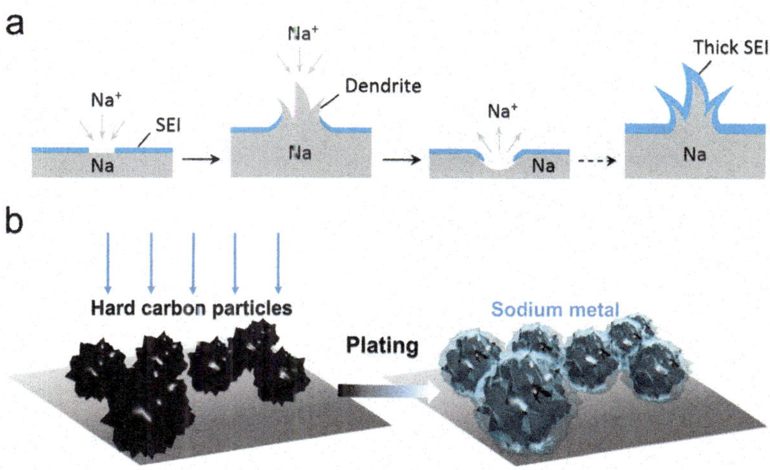

Figure 9.13 (a) Schematic illustration of plating of sodium on hard carbon particles. (b) Schematics of dendrite formation on sodium. Reproduced with permission from Ref. [112]. Copyright (2015), American Chemical Society.

Plating occurs when $\eta < 0$ in the preceding equation. Concentration gradients in the electrode determine the location where this last condition becomes true first. Particles near the electrode–separator tend to be the first to intake sodium, which pushes their potential close to 0 V [113]. Therefore, plating usually starts to occur on top of the anode close to the separator. While plating itself is theoretically reversible, the exposure of electrolyte to freshly plated sodium metal promotes the formation of fresh SEI that consumes alkali (lithium/sodium) ions irreversibly, which precipitates cell failure [112]. Capacity gradually declines toward failure of the cells. Catastrophic failure by short circuit may also occur due to the growth of the dendrites out of the plated sodium metal and puncturing the separator, contacting the cathode [114]. Stevens and Dahn reported that failure occurs more readily when sodium is purposely plated on metal substrates. Figures 9.14a–d show that plating on aluminum results in an increased potential drop, a less cycling stability, and a quicker short circuiting.

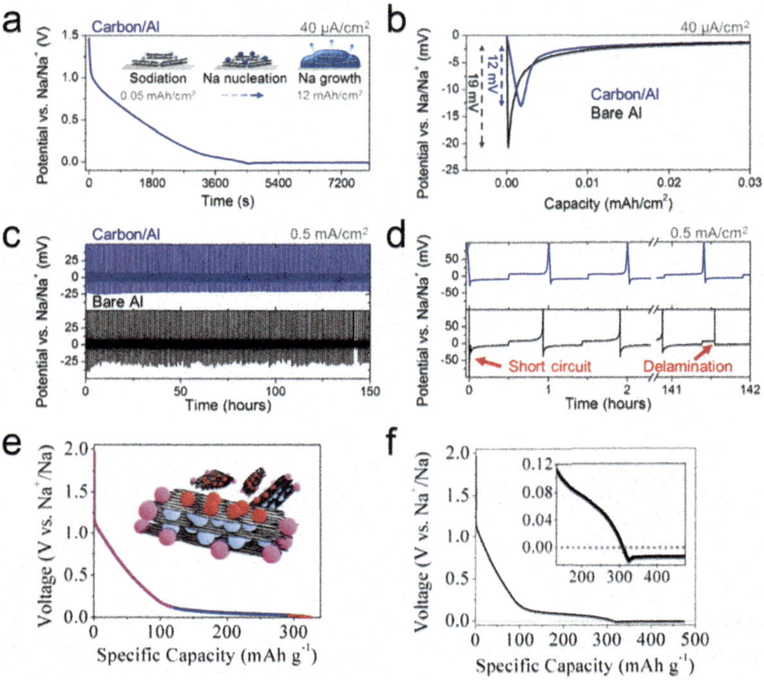

Figure 9.14 (a) Schematic representation of sodiation and plating for carbon/Al current collector at 40 μA/cm^2 with carbon loading of 400 μg/cm^2. (b) Comparison between bare Al and carbon/Al current collectors at 40 μA/cm^2 for sodium nucleation overpotential for (c) cycling of bare Al and carbon/Al current collectors at 0.5 mA/cm^2 with 30 min plating times with (d) enlarged voltage profiles showing short circuit and delamination. Reproduced with permission from Ref. [114]. Copyright (2017), American Chemical Society. (e) Color-coded potentiogram coupled with schematic of proposed three-part sodium storage mechanism. (f) Discharge curve of sodiation continued until Na-metal plating is induced. Reproduced with permission from Ref. [116]. Copyright (2015), American Chemical Society.

As stated earlier, plating can also occur on anode materials like hard carbon [114, 115]. Hard carbon (HC) can be especially susceptible to plating, as illustrated in Fig. 9.13b, because it exhibits a sodium-ion storage mechanism that operates at very low potential versus Na/Na$^+$ [115]. Figure 9.14e illustrates different sodium-ion storage mechanisms in hard carbon at different potentials. In the sloping region (pink color), sodium ions adsorb/interact with defective sites. The plateau region (blue color) is due to sodium-ion intercalation between graphene layers of graphite domains. The small low potential region (red color) is due to plating. This phenomenon has been attributed to the adsorption of sodium on sp^2-configured pore surfaces [116]. This is because these pores have been calculated to be the least energetically favorable binding sites for sodium compared to defects and storage in graphitic domains. Therefore, only sodium storage occurs at the lowest voltage close to 0 V versus Na/Na^+ on these surfaces favoring plating at these sites [116–118]. Bommier et al [116] found that plating starts occurring at –0.02 V versus Na/Na^+ and stabilizes at –0.015 V versus Na/Na^+, as shown in Fig. 9.14f.

Dendrite formed on anode can lead to short circuit of battery devices, leading to safety problems. A number of strategies have been shown for minimizing dendrite formation. Forsyth et al. [119] observed that high concentration of electrolytes can inhibit dendrite formation because of the elimination of the mass transport limitation. In contrast with regular electrolytes, highly concentrated electrolytes present no concentration gradients and consequently no polarization at the surface of the anode. Polarization of the electrolyte locally decreases the potential of the anode surface, creating conditions for metal plating to occur. Basile et al. [120] used an electrolyte with 45 mol% Na concentration of sodium bis(fluorosulfoniyl)imide in liquid-state tri(isobutyl)methylphosphonium nis(fluorosulfonyl) imide to study dendrite-free plating and stripping of sodium for a few tenths of cycles. A blend of inorganic and organic ionic liquid was used at 50°C in order to reduce viscosity. No dendrites were observed during cycling at this temperature. Surprisingly, water can also be added to the electrolyte blend to reduce viscosity and further decrease resistivity without hydrogen evolution [121].

While the protection of lithium metal by various coatings designed to prevent dendrite formation and to promote the use of lithium metal as anode for LIBs is currently a hot topic in academic research, comparatively few equivalent research efforts for Na-ion systems have been reported. However, as in the lithium case, prevention of dendrite formation on the sodium metal anode is provided by the creation of artificial SEI layers made either by direct coating or through electrolyte additives. Luo et al. [122] demonstrated that sodium metal can be protected by a thin layer of alumina (Al_2O_3) formed by atomic layer deposition. Alumina is an electronic insulator, which means it prevents the creation of further SEI. In addition, alumina is a better conductor for sodium ions compared to lithium ions [123]. The thin alumina layer, therefore, regulates the flow of ions uniformly across the whole surface of sodium metal, thereby preventing the metal electrode from dendrite formation. Huang et al. utilized $SnCl_2$ additive to the electrolyte to form an Na–Sn alloy layer on top of the sodium metal. The $SnCl_2$ additive also helped change the SEI composition by promoting the formation of NaCl salt inside the electrolyte. They demonstrated stable plating and stripping of 1 mAh/cm^2 over 500 h of operation at 0.5 mA/cm^2 [124]. Jiao et al. used $NaAsF_6$ as additive to promote the formation of greater amount of NaF as well as O-As-O polymer in the SEI. Four hundred stable cycles of plating-stripping 0.5 mAh/cm^2 at 0.1 mA/cm^2 were obtained [125].

Metal oxides generally do not have the metal plating issue because their sodium insertion/extraction voltages are well above the plating potential. One exception is $Na_2Ti_3O_7$, which has a sodium insertion voltage of only 0.3 V versus Na/Na^+ [126]. Therefore, $Na_2Ti_3O_7$ anode is prone to sodium plating [127].

9.4 Cathodes

Similar to anodes, cathodes also suffer from degradation during cycling. Transition metal oxides (TMOs) are the most common cathode materials for rechargeable alkali metal batteries [11, 128]. They store alkali metal ions at higher voltages relative to anodes. Therefore, cathodes degradation mechanisms are quite different compared with anodes. The mechanisms for cathode failures include material degradations, phase changes, and foreign molecule insertion. It is worth noting that TMO cathode materials can act as an accelerant if thermal runaway occurs because of the presence of oxygen in high concentration in the materials and can promote a catastrophic failure that mirrors the catastrophic failure from dendrite formation in the anode case.

9.4.1 Material Degradation

9.4.1.1 Dissolution of transition metals

Of all the degradation phenomena that affect cathode compounds, dissolution of transition metals (TMs) in the electrolyte is common [129, 130]. This has two consequences: first, loss of capacity due to leaching of the active material from the electrode; and second, the dissolved metal destabilizes the SEI layer at the anode surface [131–135].

Let us use the LIB system as an example to discuss. The mechanism for meal dissolution in a fluorine-containing electrolyte involves the continuous production of HF and protic species (alcohols) through a revolving chain reaction starting with the presence of trace water impurities in the electrolyte as follows [136–138]:

$$LiPF_6 \leftrightarrow LiF + PF_5 \quad (9.1)$$

$$PF_5 + H_2O \rightarrow OPF_3 + 2HF \quad (9.2)$$

$$OPF_3 + 2xLi^+ \, 2xe^- \rightarrow Li_xPF_{3-x}O + xLiF \quad (9.3)$$

$$OPF_3 + 3H_2O \rightarrow PO_4H_3 + 3HF \quad (9.4)$$

While the decomposition of $NaPF_6$ has not yet been reported, similar hydrolysis of the PF_6^- anion is expected in the NIB system, as proposed by Dahbi et al. [45]:

$$NaPF_6 \leftrightarrow NaF + PF_5 \quad (9.5)$$

$$PF_5 + nNa^+ + ne^- \rightarrow Na_xPF_y + NaF \quad (9.6)$$

$$NaPF_6 + H_2O \rightarrow NaF + POF_3 + 2HF \quad (9.7)$$

$$POF_3 + nNa^+ + ne^- \rightarrow Na_xPOF_y + NaF \quad (9.8)$$

In addition, a trace amount of alcohol is believed to be produced from the decomposition of solvents, such as methanol from dimethyl carbonate (DMC), ethanol from diethyl carbonate (DEC), ethylene glycol from ethylene carbonate (EC), and propylene glycol from propylene carbonate (PC). Various reactions are suggested where the production of aprotic solvent through the decomposition of the surrounding organic solvent perpetuates the formation of acid and water in the electrolyte [131]. HF can also be present as a common impurity [139]. And the acid reacts with oxide compounds on the cathode surfaces to release TM ions and water molecules as described by Hunter on an $LiMnO_4$ cathode [140]:

$$2LiMn_2O_4 + 4H^+ \rightarrow 2Li^+ + 3\lambda\text{-}MnO_2 + Mn^{2+} + 2H_2O \quad (9.9)$$

where manganese disproportionates as follows:

$$2Mn^{3+} \rightarrow Mn^{2+} + Mn^{4+}$$

The preceding migration of manganese to the particle surface as manganese oxide (MnO_2) forms a non-passivating layer that continues to grow, thereby depleting manganese [141]. The dissolution will continue through the migration of Mn^{3+} from the bulk to the surface of the compound where MnO_2 is formed. An analysis of K-XANES (X-ray absorption near-edge structure) spectra of an electrolyte solution in an Li-ion battery system revealed that Mn exists as Mn^{2+}, thereby supporting the conclusion that dissolution happens in the charged state [142]. The presence of Mn^{3+} is fundamental to the dissolution mechanism regardless of the crystalline structure as shown by a study of Choi and Manthiram [143]. Layered $LiMn_{0.8}Cr_{0.2}O_2$ with orthorhombic $LiMnO_2$ showed a similar behavior of manganese dissolution (2.6 to 3.2%) as compared with the spinel $LiMn_2O_4$ (3.2%). However, layered $LiNi_{0.5-y}Mn_{0.5-y}Co_{2y}O_2$ and spinel $LiMn_{1.5}Ni_{0.5}O_4$ both showed manganese dissolutions lower than 1%. Dissolution is particularly severe for manganese oxide compounds because of the Jahn–Teller distortion, which exacerbates stresses on the crystalline structure during cycling and phase changes, promoting even more loss of manganese. Continuous dissolution of transition metal from cathodes eventually prompts the failure of the cells.

In a study on cycling of a monoclinic $NaMnO_2$ cathode that contained Mn^{3+}, Ceder et al. [58] detected manganese dissolution in an $NaPF_6$ EC:DMC electrolyte using the inductive coupled plasma (ICP) technique on a water-dissolved sample. Figure 9.15a shows a voltage versus capacity graph for the $NaMnO_2$ electrode with noticeable capacity loss (>20%) after 20 cycles partly due to dissolution. In this study, the overall effects of the manganese loss at the cathode were marginal compared to the deleterious effects from increased SEI growth at the anode due to the dissolved manganese in the electrolyte [58].

Other transition metals such as cobalt, nickel, and iron also dissolve in organic electrolytes making transition metal dissolution arguably the most widespread failure mechanism for cathodes [142, 144–146].

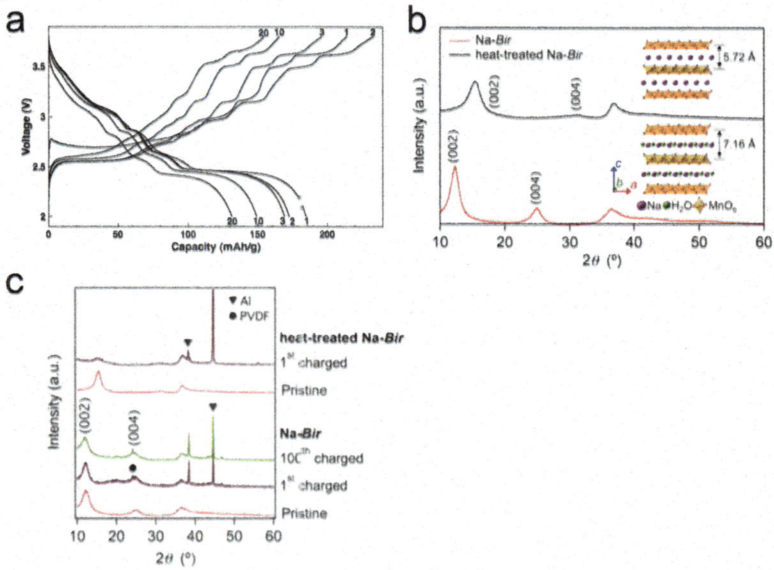

Figure 9.15 (a) Charge–discharge profiles of $NaMnO_2$ at C/10 until the 20th cycle. Reproduced with permission from Ref. [58]. Copyright (2011), The Electrochemical Society. (b) XRD spectra and schematic of crystal structures of Na-Bir with and without heat treatment. (c) XRD spectra and schematic of crystal structures of pristine Na-Bir, Na-Bir after first charge, pristine heat-treated Na-Bir, heat-treated Na-Bir after first charge, and heat-treated Na-Bir after 100th charge. Reproduced with permission from Ref. [147]. Copyright (2015), American Chemical Society.

A recent effort by the Aurbach group to curtail manganese dissolution was to use water, which is purposefully inserted inside the layered MnO_2 oxide structure [147]. The water molecules settle between the MnO_2 sheets, thus increasing the inter-sheet distance from 5.72 to 7.16 Å as seen from the XRD patterns depicted in Fig. 9.15b. The water-stabilized layered material retained far better crystallinity than pristine material after 100 cycles as confirmed by the XRD data in Fig. 9.15c. In Fig. 9.15c, the water-stabilized sample retained characteristic (002) and (004) crystalline peaks even after the 100th charge, whereas these peaks disappeared from heat-treated sample after the first charge. ICP characterization revealed an 11.7-fold decrease in the amount of dissolved Mn^{2+}, indicating a significant improvement on the electrode stability against dissolution. This is explained by the reduction in strain on MnO_6 layers during cycling enabled by the water molecules serving as a pillar between the sheets. Concurrently, the lower impedance observed and the similar capacity to pristine material indicate that the water does not occupy Na crystallographic sites nor impede transport of electrolyte ions.

9.4.1.2 Oxygen evolution at elevated temperatures

As discussed earlier, the transition metal component in cathodes has the problem of dissolution in a fluorine-containing electrolyte. Oxygen, the other main constituent of TMO cathodes, can also be dissociated from the structure. At elevated temperatures (between 150 and 500°C), cathode materials, in particular layered oxides, tend to undergo thermal runaway. The high temperatures release oxygen from the structure and contribute to accelerate oxidation, which, in turn, increases temperature further releasing more oxygen [148]. This phenomenon occurs because the decomposition of TMO materials releases oxygen gas. The evolved oxygen then contributes to the combustion of solvent molecules, resulting in a complete failure of the cell [149]. Therefore, oxygen evolution is not only of safety concern, but also one of the causes leading to cell failure [150, 151]. Large cells are particularly sensitive because heat dissipation is more difficult compared with smaller cells. The causes for initial heating can be as diverse as accidental over-discharge, puncture by nail-like objects, or external short circuiting [146]. As the Na-ion battery technology is being developed mainly for large-scale energy-storage applications, oxygen evolution is a particularly important issue that must be considered in electrode research.

Oxygen evolution for the sodium cathode is similar to that of the lithium counterpart. However, one notable difference separates the two systems: There is rapid defluorination of $NaPF_6$ and production of $NaCoF_3$. Because of the fast kinetics of the reaction, the authors suggest that this last decomposition may even occur during normal cycling conditions at low temperatures [150].

Dahn et al. [150] investigated oxygen evolution on P2-Na_xCoO_2 using an accelerated rate calorimetry technique. They revealed three successive exothermic reactions:

$$Na_{0.35}CoO_2 \rightarrow 1/2Na_{0.7}CoO_2 + 1/6Co_3O_4 + 1/6O_2 \qquad (9.10)$$

$$1/2Na_{0.7}CoO_2 \rightarrow 1/6Co_3O_4 + \text{Na-containing compounds} + 1/6O_2 \quad (9.11)$$

$$1/3Co_3O_4 \rightarrow \text{Co or CoO} \quad (9.12)$$

The products in the preceding three reactions were confirmed by the XRD data shown in Fig. 9.16a, which depicts the XRD patterns of the samples after P2-Na_xCoO_2 was heated to different temperatures from 180 to 300°C.

This research group further investigated $NaCrO_2$ but did not find oxygen evolution occurring [152]. Oxygen remained bonded to the chromium as described by the following disproportionation reaction: $Na_{0.5}CrO_2 \rightarrow 1/2NaCrO_2 + 1/2CrO_{2-\Delta}$. As a result, this cathode material exhibits an excellent resistance to self-heating.

Guo et al. [153] showed that desodiated $NaMnO_2$ decomposes at 255°C exothermically, releasing 181.2 J/g of heat as measured using the differential scanning calorimetry (DSC) method. Modification with titanium oxide lowered the heat released of some 101.8 J/g at a higher decomposition temperature of 283°C.

Zhao et al. [154] investigated $NaFeO_2$ and proposed the following exothermic reaction based on the decomposition of $LiCoO_2$:

$$Na_{0.58}FeO_2 \rightarrow 0.58NaFeO_2 + 0.21Fe_2O_3O_4 + 0.105O_2$$

DSC measurements showed the decomposition temperature of about 380°C. The authors also revealed a remarkable lowering of the exothermic reaction onset depending on the amount of electrolyte present relative to the cathode material mass.

$NaCoO_2$ shows remarkable stability in the $Na_{0.7}CoO_2$ discharged state with no further loss in mass at temperatures up to 1035°C as observed from the TG/DTA data shown in Fig. 9.16b [155]. However, at higher charge states, $NaCoO_2$ shows decomposition signs at much lower temperatures, as shown by Hwang et al. [156]. They monitored the crystal appearance of $Na_{0.12}CoO_2$ (4.3 V versus Na), $Na_{0.24}CoO_2$ (4.1 V versus Na), and $Na_{0.52}CoO_2$ (3.5 V versus Na) under an electronic microscope in a 2017 study. Figure 9.16c shows multiple microscope images of the material with noticeable changes to the surface morphology occurring as the temperature is increased. This performance is on par with other oxides. Dahn et al. [157] reported thermal instability for $Na_{1/3}CoO_2$ with partial loss of oxygen around 400°C and 960°C. At the time of this review, we are not aware of any thermal stability studies for $NaNiO_2$ or $NaVO_2$ layered oxides.

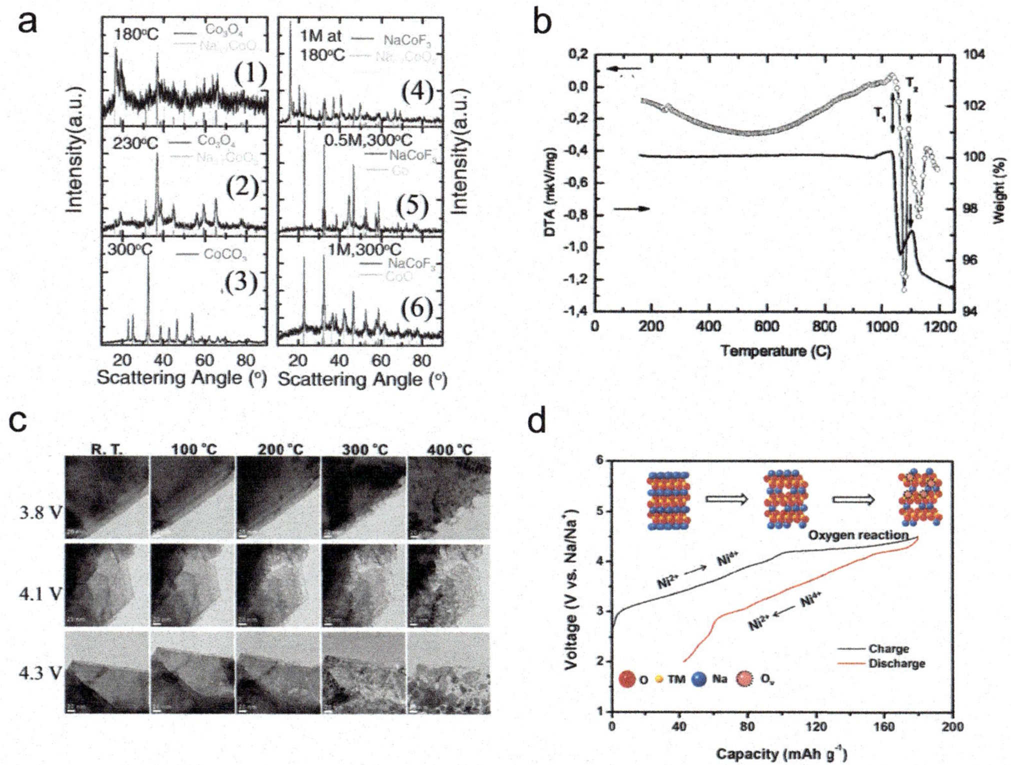

Figure 9.16 (a) XRD patterns of (1) 100 mg $Na_{0.35}CoO_2$ in 100 mg EC: DEC (1:2 v/v) at 300°C, (2) $Na_{0.65}CoO_2$ in 100 mg EC: DEC (1:2 v/v) heated at 230°C, (3) $Na_{0.65}CoO_2$ in 100 mg EC: DEC (1:2 v/v) at 300°C. (Right panel) XRD patterns of (4) $Na_{0.35}CoO_2$ in solvent at 300°C, (5) $Na_{0.65}CoO_2$ in

(*Continued*)

solvent heated to 230°C, (6) $Na_{0.65}CoO_2$ in solvent heated to 300°C. Reproduced with permission from Ref. [150]. Copyright (2012), The Electrochemical Society. (b) DTA-TG analysis of single crystalline $Na_{0.7}CoO_2$ heated at 7.5°C/min up to 1200–1250°C in oxygen atmosphere. T1 is 1035°C (melting temperature) and T2 is 1092°C with a weight-loss reversal. Reproduced with permission from Ref. [155]. Copyright (2004), American Physical Society. (c) Images of the surface of NCO cathode materials charged to 3.8 V, 4.1 V, and (3) 4.3 V at different temperatures. Reproduced with permission from Ref. [156]. Copyright (2017), American Chemical Society. (d) Schematic depicting charge–discharge mechanisms in the P2–$Na_{0.78}Ni_{0.23}Mn_{0.69}O_2$ cathode. Reproduced with permission from Ref. [151]. Copyright (2017), American Chemical Society.

P2-type $Na_{2/3}Fe_{1/3}Mn_{2/3}O_2$ shows good stability up to 379°C [158]. The amount of heat generated as well as the onset of the decomposition varies depending on the charge (desodiated state) of the compound [158]. Ma et al. [151] reported that a P2-$Na_{0.78}Ni_{0.23}Mn_{0.69}O_2$ cathode material exhibits an irreversible plateau above 4.1 V versus Na beyond oxidation of transition metal elements (Fig. 9.16d). This is an overcharging reaction. The additional capacity is a by-product of a surface oxygen reaction. The authors expect the oxygen reaction to cause the free evolution of the gas.

Hwang et al. [159, 160] studied the effect of transition metal (nickel, cobalt, manganese) substitution on the performance and thermal stability of layered $Na[Ni_xCo_yMn_z]O_2$ cathodes. Oxygen evolution was characterized via TGA and found a weight-loss onset from 240°C onward for bulk P3-$Na_{0.44}[Ni_{0.60}Co_{0.05}Mn_{0.35}]O_2$. High-temperature *in situ* XRD on an accelerator beamline showed that the P3 crystalline structure evolves progressively to a spinel structure as oxygen evaporates. Interestingly, the onset of the transformation is highly dependent on the surface area as samples with 6.41 m^2/g start evolving at 240°C while 0.77 m^2/g samples are mainly stable up to 400°C. The authors also report that increased nickel contents are detrimental to thermal stability while increased manganese contents are beneficial as characterized through DSC analysis [160]. The same group investigated the $Na[Fe_{0.75-x}Ni_xMn_{0.25}]O_2$ system for thermal stability [161]. They found that lowering the amount of iron (x) compared to nickel from 0.55 to 0.4 also lowered the oxygen evolution temperature and increased heat generation. For compounds charged to 3.9 V, oxygen evolution temperatures range from 267.5 to 289.8°C. Oh et al. [162], from the same group, investigated $Na_{0.38}[Ni_{0.25}Fe_{0.5}Mn_{0.25}]O_2$ and found the compound to decompose at 280.3°C with heat generation of 401.7 J/g. This result is characterized using DSC. They also investigated $Na_{0.28}[Ni_{0.25}Fe_{0.25}Mn_{0.5}]O_2$ and $Na_{0.28}[Li_{0.05}(Ni_{0.25}Fe_{0.25}Mn_{0.5})_{0.95}]O_2$ and found decomposition at 277.7°C and 297.6°C, with 633.7 J/g and 473.9 J/g of heat evolved, respectively [163]. TGA/mass spectroscopy analysis by Hasa et al. [164] shows that $Na_{0.5}[Ni_{0.23}Fe_{0.13}Mn_{0.63}]O_2$ starts releasing oxygen above 300°C.

9.4.2 Phase Change

9.4.2.1 Permanent phase changes due to over- or under-charging

Cathode materials may also experience phase changes during cycling. $Na_xM_yO_z$ materials require some alkali atoms to remain inside the structure in order to avoid structural collapses [165]. Therefore, these materials have a limited sodium extraction range ($\Delta < x < 1$) that defines reversible capacity. If that reversible range is exceeded and too many sodium atoms are removed during charging, then the materials undergo irreversible phase changes, even forming an amorphous phase [166]. When too many sodium atoms are removed, a drastic reduction in reversible capacity follows, which will ultimately result in cell failure. Cedar et al. [167] conducted a computational study of cathode materials $Na_xM_yO_z$ and $Li_xM_yO_z$ (with M = Ti, V, Cr, Mn. Fe, Co, Ni) using the first principles method. Interestingly, sodium- and lithium-layered compounds show fundamentally different pathways to structural changes with the sodium material being generally more stable. Higher energy barriers are observed for the migration of transition metals in layered sodium materials (see Table 9.2). Also, for the sodium materials at half depletion (i.e., $Na_{x=0.5}M_yO_z$), there is generally a lack of a thermodynamic driving force for a transformation from layered to spinel structure with the exception of $NaTiO_2$ (Fig. 9.17a). Sodium compounds with manganese, cobalt, and nickel thermodynamically favor P3 over O3 (Figs. 9.17b,c) configurations. The more stable behavior of sodium compounds is ultimately driven by the lower number of sites suitable for sodium versus lithium. The sodium atom is larger in size, and it can only be six-atom coordinated (octahedral sites only) while lithium can be either six- or four-atom coordinated (octahedral and tetrahedral sites). Note that the phase changes of the $Na_xM_yO_z$ compounds are irreversible, thus resulting in immediate failure of the battery cell.

Table 9.2 Computed migration barriers in eV for transition metal ions in layered O3 $A_{0.5}MO_2$ [167]

	Ti	V	Cr	Mn	Fe	Co	Ni
$Na_{0.5}MO_2$	1.45	1.50	2.01	0.82	1.10	2.45	1.61
$Li_{0.5}MO_2$	0.98	0.78	1.23	0.44	0.55	1.62	1.10

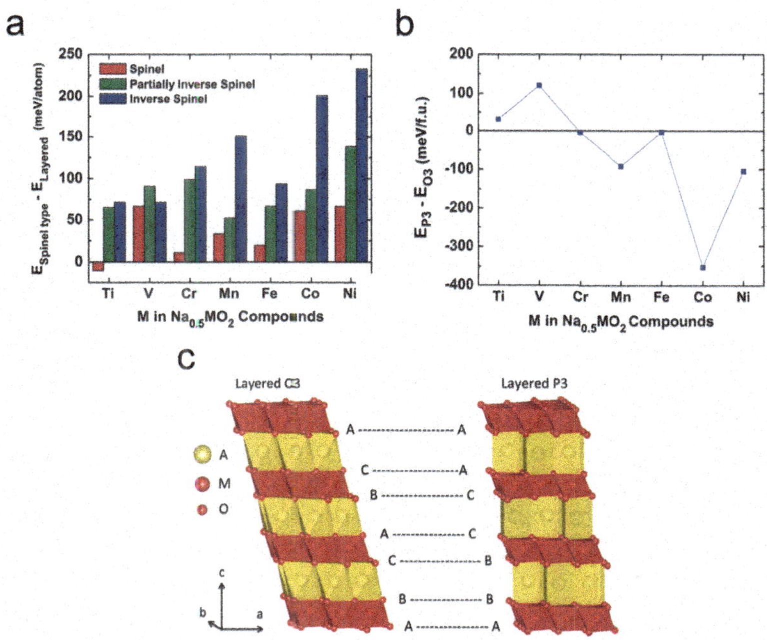

Figure 9.17 (a) Energy per atom differences between NaM_2O_4 spinel-type structures and the layered O3 $Na_{0.5}MO_2$ structure [167]. (b) Energy per formula unit and per transition metal atom between layered P3 $Na_{0.5}MO_2$ and layered O3 $Na_{0.5}MO_2$. (c) Crystalline structure representation of layered O3 and P3 structures, with positions of the oxygen planes labeled as A, B, and C. Reproduced with permission from Ref. [167]. Copyright (2012), Royal Society of Chemistry.

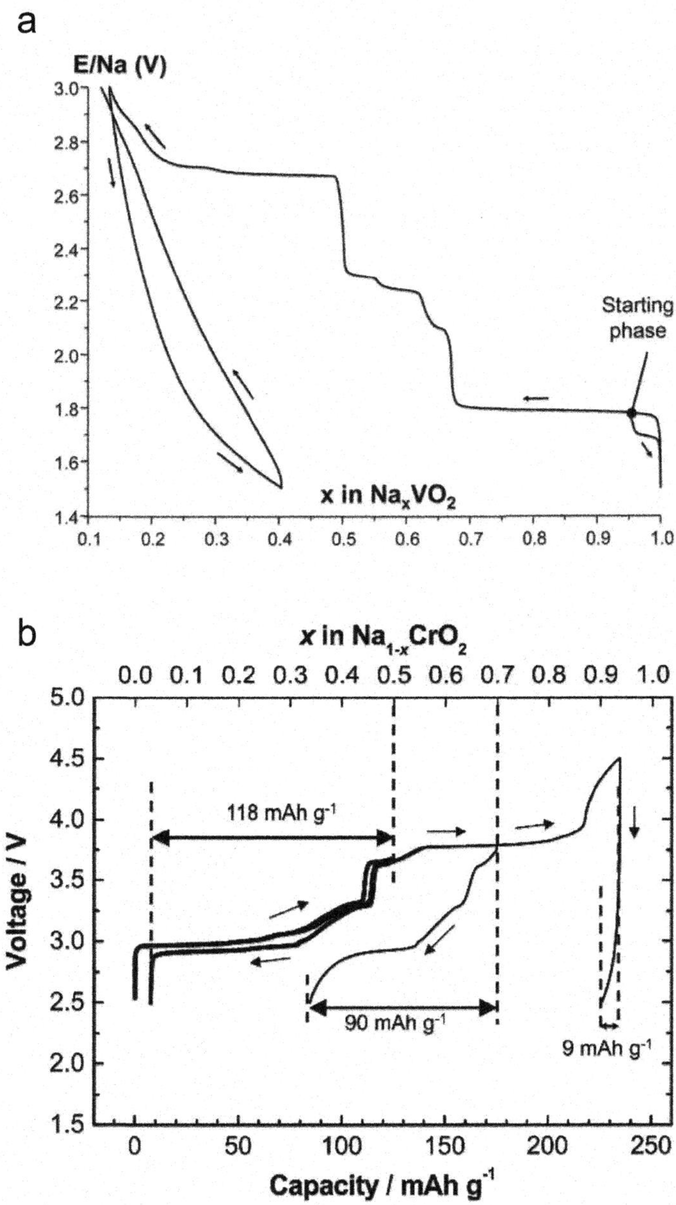

Figure 9.18 (a) Galvanostatic cycling of Na_xVO_2 at C/100 rate. Reproduced with permission from Ref. [168]. Copyright (2011), The Electrochemical Society. (b) First galvanostatic curves of Na//$NaCrO_2$ cells at a rate of 1/20 C between $0 \leq x \leq 0.5$ and between $0 \leq x \leq 0.7$ in Na_xCrO_2. Reproduced with permission from Ref. [4]. Copyright (2015), American Chemical Society.

Liu et al. [166] studied overcharging of a cathode material, $Na_xV_6O_{15}$. During overcharging, the $Na_xV_6O_{15}$ structure is emptied of most of its sodium content. This causes structural collapses of the crystal to become an amorphous phase. The charge storage sites are destroyed, and the material loses the ability to store charges, resulting in the failure of the electrode.

Na_xVO_2 can be charged up to $x = 0.5$. At this point, the potential of $Na_{0.5}VO_2$ is about 2.5 V versus Na/Na^+ according to Didier et al. [168]. Further charging destroys the reversibility of the process. Subsequent charge profiles are also very different from the original profile: the bi-phasic plateaus expected from intercalation compounds disappear and they are replaced by a capacitive-like sloping behavior, as shown in Fig. 9.18a. Such a drastic change in charge profile indicates that crystalline changes have occurred. XRD characterization results reveal an increase in inter-sheet distance and a decrease in vanadium-to-vanadium distance due to the oxidation of vanadium from V^{3+} to V^{4+} during desodiation. The authors explained the irreversibility of sodium cycling and decrease in sodium-ion storage capacity to be due to the migration of vanadium into the inter-sheet space.

Komaba et al. [4, 55, 169] showed that the reversible capacity for Na_xCrO_2 is limited to half of the theoretical capacity. Na_xCrO_2 can reversibly store 118 mAh/g of Na^+ at a maximum voltage of 3.7 V versus Na/Na^+. At the charge potentials of 3.8 V and 4.5 V, the reversible capacities dropped to 90 and 9 mAh/g, respectively, as shown in Fig. 9.18b. The XRD characterization data showed that sodium extraction beyond 3.8 V triggers a strong reduction in inter-sheet distance with a permanent phase change from P'3- to O3'-type according to Delmas et al. [62, 63]. This phase change is also supported by *ab initio* calculations of Kubota et al. [170]. Chromium ions also migrate from their original sheet location to intra-sheet locations, henceforth drastically reducing sodium storage in this space, as depicted in Fig. 9.19a.

As predicted by Cedar et al. [171], Na_xTiO_2 cycles reversibly only up to $x > 0.7$, and then it suffers from irreversible structural change beyond that limit. Maazaz et al. [171] performed an XRD analysis on the layered compound and revealed a decrease in intra-sheet distance and an increase in inter-sheet distance due to the higher oxidation state of the titanium atoms (Table 9.2). A new phase

$(Na_xTi_yO_{1-x-y})(Ti_{1-y}O_y)O_2$ appeared at the charge states with x smaller than 0.67. This study confirmed that during overcharging, migration of titanium atoms to the original sodium sites occurred.

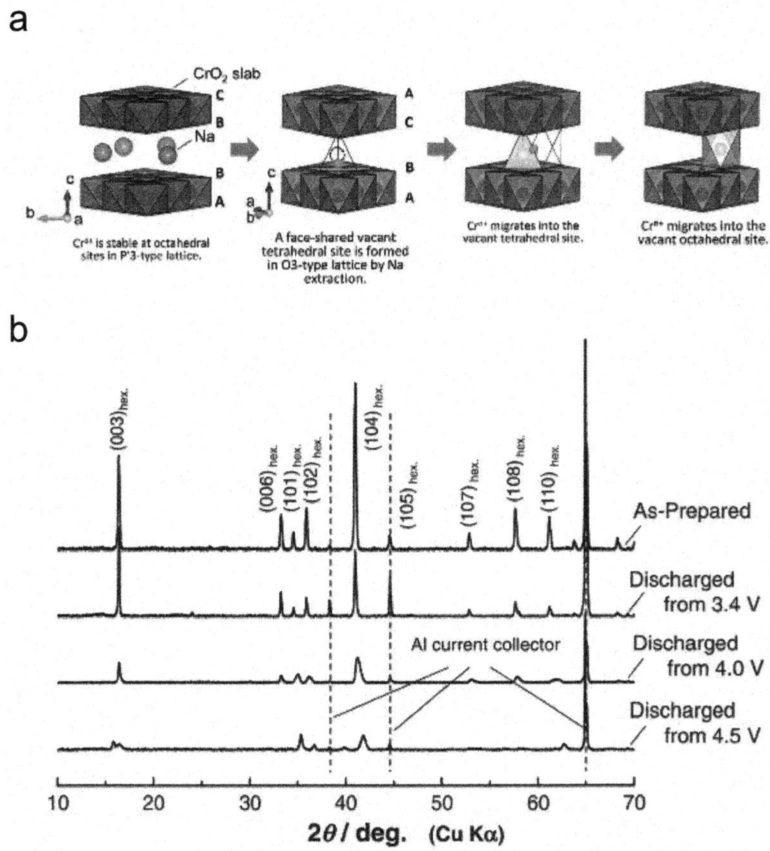

Figure 9.19 (a) Proposed mechanism for the migration of chromium in Na_xCrO_2. Reproduced with permission from Ref. [4]. Copyright (2015), American Chemical Society. (b) XRD patterns of $Na_{1-x}FeO_2$ for the as-prepared electrodes and after initial discharge at 2.5 V versus Na/Na^+ at 12.1 mA/g from different cut-off voltages (3.4, 4.0, 4.5 V versus Na/Na^+). Reproduced with permission from Ref. [172]. Copyright (2012), The Electrochemical Society of Japan.

Yu et al. [173] reported a loss of reversibility of compound $Na_xTi_{0.5}Ni_{0.5}O_2$. When charged to 170 mAh/g or 4.7 V versus Na/Na^+, the discharge capacity was only 121 mAh/g. Namba et al. [174] tested this compound as well and observed phase transformation

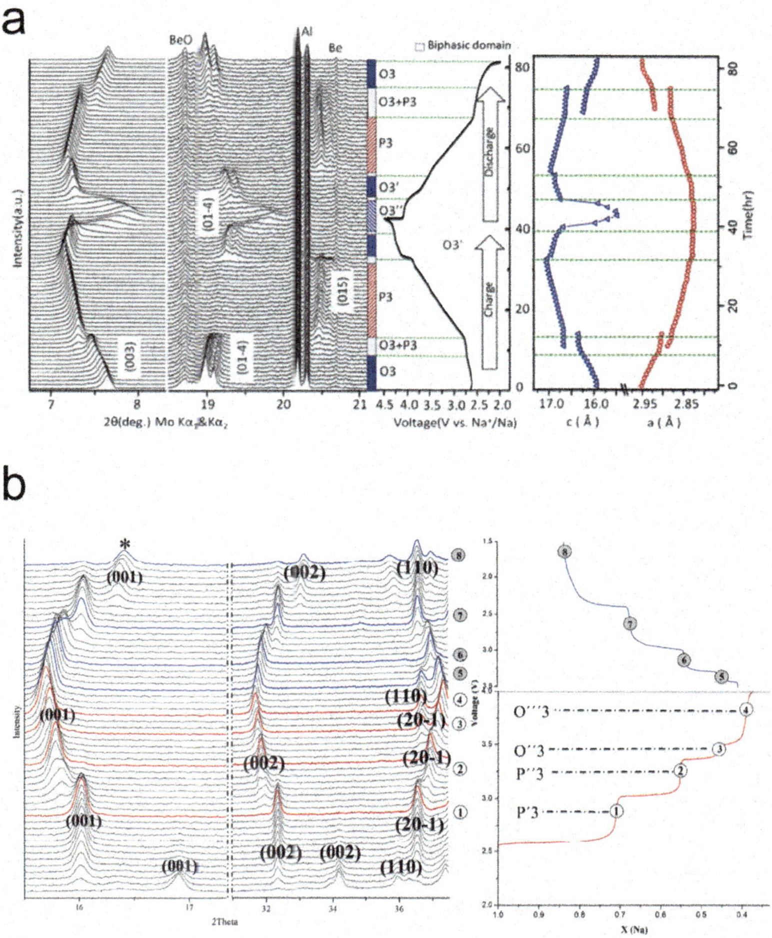

Figure 9.20 (a) *In situ* XRD pattern during galvanostatic charge–discharge cycling at C/50 rate (left) with corresponding lattice parameter evolution (right). Reproduced with permission from Ref. [176]. Copyright (2014), Elsevier Science. (b) *In situ* XRD pattern of Na_xNiO_2 electrode during the first charge/discharge with corresponding phase transitions (right). Reproduced with permission from Ref. [177]. Copyright (2014), Elsevier.

from O3 to P3 on charging to $x = 0.5$. Yabuuchi et al. [172] did a thorough investigation of the Na_xFeO_2 system. They found that this cathode material suffers from severe irreversible behavior for $x > 0.5$. The change from reversible to irreversible regimes seems to happen after charging over 4 V versus Na/Na^+. From the XRD results presented in Fig 9.20b, researchers postulated an iron migration from the 3a sites to the 6c tetrahedral sites or 3b octahedral site [172]. Such a migration is expected to obstruct sodium-conducting paths, hence reducing the reversibility and sodium storage capacity of the material.

In 2013, Ceder et al. [175, 176] investigated Na_xFeO_2-derived cathode materials prepared by partial substitution of Fe with Ni, CO, and Mn, namely ternary material $Na_xNi_{1/3}Co_{1/3}Fe_{1/3}O_2$ and quaternary material $Na_xMn_{0.25}Fe_{0.25}Co_{0.25}Ni_{0.25}O_2$. The ternary compound supports delivered a reversible capacity of 165 mAh/g at 4.2 V versus Na [175]. Such high voltage corresponds to a fully desodiated state according to the first principles calculations. The quaternary phase could deliver a reversible capacity between 160 and 180 mAh/g at 4.2 V versus Na [176]. The authors used *in situ* XRD technique to characterize a smooth transition from O3 to P3 to O3' and further to O3'', as shown in Fig. 9.20a. It is clear that during charging, the material exhibited an O3 symmetry at 2.6 V versus Na/Na^+. Then the structure adopted the P3 symmetry at 2.8 V and finally the O3' and O3'' symmetries at 3.7 and 4.1 V, respectively. The authors attributed the excellent reversibility to a monoclinic distortion. They also noted that reversibility was compromised above 4.25 V in the O3'' phase.

Nevertheless, doping with other transition metals can further stabilize prospective cathode structures that have to go through extensive crystalline changes during cycling. For example, substituting cobalt in layered oxides enhances structure stability through shortening of TM–O and O–O distances and widening of the d-spacing enlargements according to Liu et al. [170].

Na_xNiO_2 can be cycled reversibly between 1.25 and 3.75 V versus Na/Na^+, providing 118 mAh/g of sodium-ion storage according to Vassilaras et al. [57]. The first cycle has a coulombic efficiency of 83.8%. If Na_xNiO_2 is cycled outside 1.25 to 3.75 V versus Na/Na^+, cycling becomes irreversible. As predicted by the first principles calculations, this compound undergoes O3 to P3 phase transformations during cycling as characterized by *in situ* XRD shown in Fig. 9.20b [177]. Thanks to *in operando* high-energy XRD using synchrotron light, Wang et al. [178] identified irreversible structural change occurring during the first cycle at potentials below 3 V and above 4 V versus Na/Na^+. Above 4 V, there is a discrepancy in the inter-planar distance (in the c-direction) between the material during charging compared with discharging even though the amount of charge stored is the same. This is due to the gliding of the NiO_2 slabs, resulting in an irreversible transformation, and structural distortion increases the overpotential. A material cycled over 4 V versus Na/Na^+ also forms an irreversible core in the core–shell reaction mechanism that does not disappear below 3 V versus Na/Na^+.

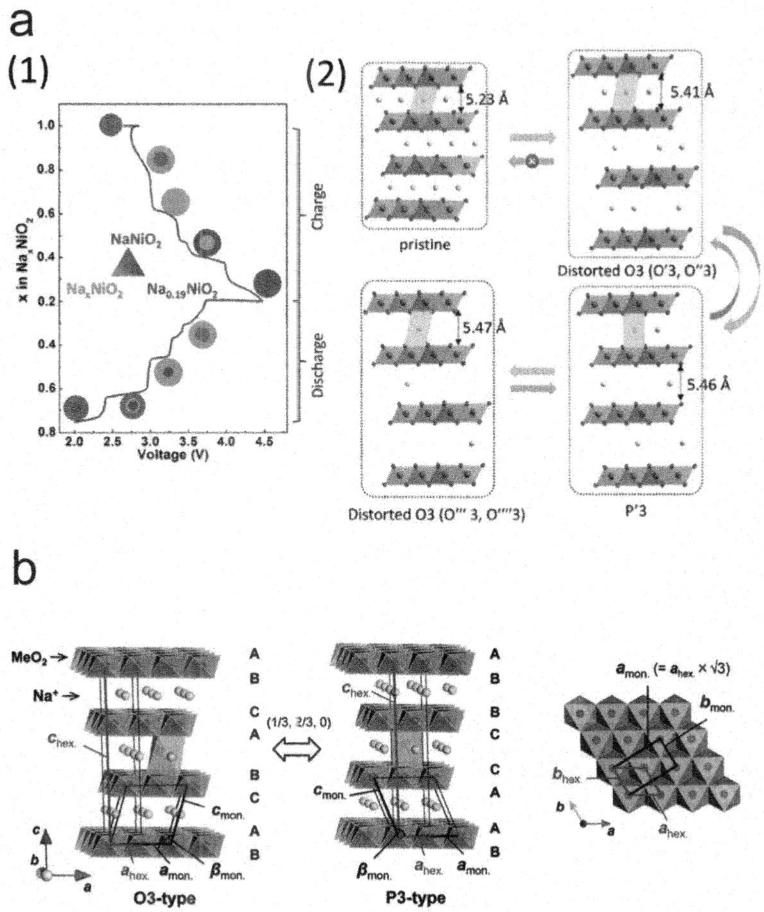

Figure 9.21 (a) Illustration of phase transition of $NaNiO_2$ material during electrochemical cycling (1) and schematic of its structure during intercalation and deintercalation processes (2). Reproduced with permission from Ref. [178]. Copyright (2017), Elsevier. (b) Schematic illustrations of crystalline structure of $Na_{1-x}Ni_{0.5}Mn_{0.5}O_2$ for O3 and P3 configurations; top view on the right. Reproduced with permission from Ref. [179]. Copyright (2012), American Chemical Society.

Komaba et al. [180] introduced materials similar to Na_xFeO_2 discussed earlier but with half of the iron atoms substituted with manganese atoms. The resulting capacity was more than doubled going from 80 mAh/g to 190 mAh/g for a P2-type material. The extraordinary improvement is due to a reversible crystalline structural change from P2 to OP4. Systematic work from this group on this system continued with the investigation of $NaFeO_2$-$NaNi_{0.5}Mn_{0.5}O_2$ [181]. The addition of $NaNi_{1/2}Mn_{1/2}O_2$ stabilized cycling compared to Na_xFeO_2. Specific capacities between 130 and 135 mAh/g were extracted from the compound.

$Na_xNi_{0.5}Mn_{0.5}O_2$ delivers up to 125 mAh/g of reversible capacity thanks to progressive structural change (O3 → O'3 → P3 → P'3 → P3'') revealed by X-ray absorption spectroscopy (XAS) in Fig. 9.21b according to Komaba et al. [179]. The research group then reported a cobalt-substituted $Na_xFe_{0.5}Co_{0.5}O_2$ material. This solid solution intermediate compound between Na_xFeO_2 and Na_xCoO_2 shows reversible capacity of about 160 mAh/g, which is higher than either Na_xFeO_2 or Na_xCoO_2 [182].

Nazar et al. [183] used pair distribution function (PDF) analysis on XRD data to characterize charged P2-$Na_{0.66\text{-}z}[Mn_{0.5}Fe_{0.5}]O_2$ after its crystalline structure collapsed to an amorphous state. They observed that P2-$Na_{0.66\text{-}z}[Mn_{0.5}Fe_{0.5}]O_2$ undergoes two distinct crystalline transformations, as shown in Fig. 9.22a. First, there is a transition from hexagonal P2 to orthorhombic P'2. Then the crystal goes to an amorphous phase called the "Z" phase. PDF analysis reveals a shrinkage of inter-sheet distances explained by the migration of transition metal atoms into the alkali layer. In contrast with O3-Na_xCrO_2 and O3-Na_xFeO_2, the transition metal migration associated with the transformation to the Z-phase is reversible especially for the Ni-substituted compound $Na_x[Mn_{0.65}Ni_{0.15}Fe_{0.2}]O_2$ [4, 184]. The study also confirmed that substituting Ni, Fe, and Mn in layered oxides prevents long-range ordering of Na vacancies. The ordering of vacancies directly promotes the structural transitions at the onset of voltage steps in the charge/discharge curves. Substituting atoms produces smoother solid solution-like discharge curves, i.e., curves without plateaus denoting phase changes. Capacity for these compounds is about 200 mAh/g at the first cycle; capacity is derived from galvanostatic tests shown in Fig. 9.22b.

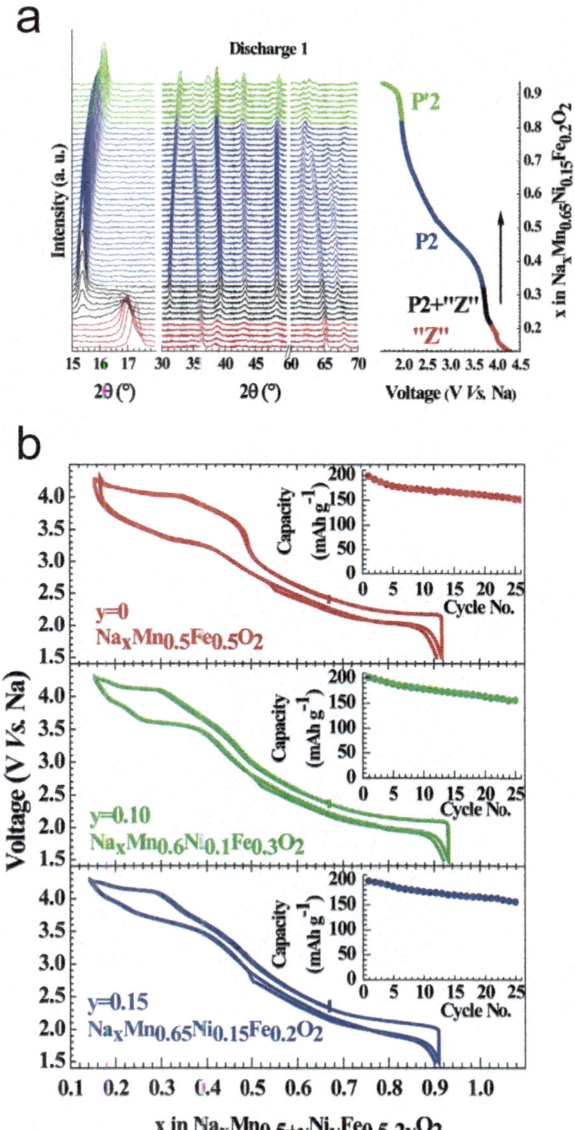

Figure 9.22 (a) *Operando* XRD pattern from galvanostatic cycling of $Na_{0.67}[Mn_{0.65}Ni_{0.15}Fe_{0.2}]O_2$ at the rate of C/20 (left) with illustration of corresponding color-coded phase transitions for the first discharge (right). (b) Charge/discharge cycles at C/20 rate with specific capacity insets for P2-$Na_{0.67}[Mn_{0.5+y}Ni_yFe_{0.52y}]O_2$ (y = 0, 0.1, 0.15). Reproduced with permission from Ref. [183]. Copyright (2015), Royal Society of Chemistry.

9.4.2.2 Permanent phase change due to anisotropic stress

In phase changes that cathode materials undergo, it is found that the stresses associated with the phase change can be anisotropic in nature, meaning that upon sodiation and desodiation, the h, k, and l Miller indices of host crystals do not change proportionally with each other. Intermediate crystalline structures form during cycling to mitigate stresses. However, the occurrence of these structures is not symmetrical between sodiation and desodiation as exemplified by the study of the $NaFePO_4$ system [185]. Because sodium is bigger than lithium, the insertion of sodium into a crystal leads to a stronger stress on the crystal, leading to the formation of more intermediate structures in order to mitigate the stress. While the stress normally only entails reversible crystalline modifications, it may cause permanent collapses of the cathode structure, resulting in electrode failure.

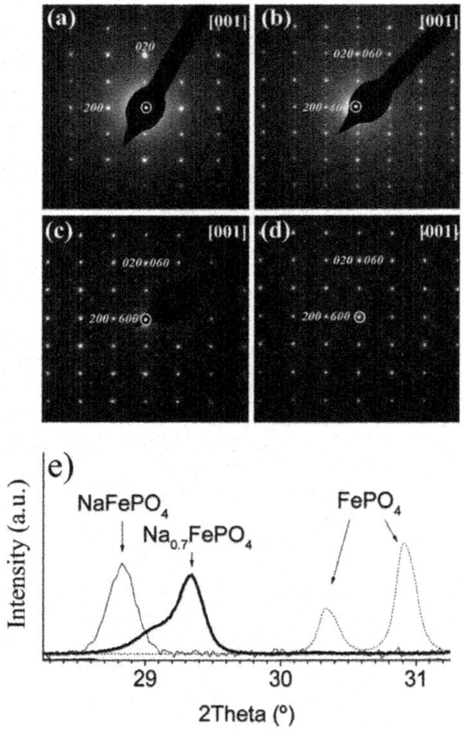

Figure 9.23 Electron diffraction patterns of [001] zone axis of (a) $FePO_4$ and (b–d) samples of $Na_{\sim 0.7}FePO_4$ material. (e) XRD patterns of $NaFePO_4$, $Na_{\sim 0.7}FePO_4$, and $FePO_4$ at room temperature. Reproduced with permission from Ref. [185]. Copyright (2012), Royal Society of Chemistry.

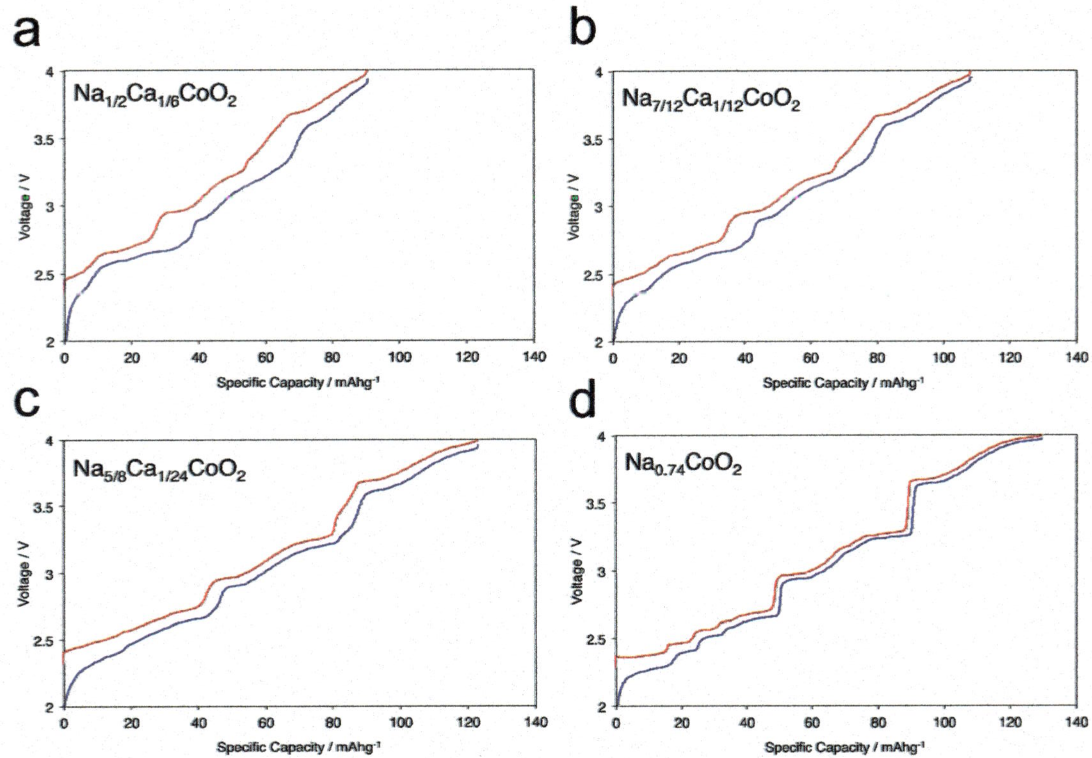

Figure 9.24 Galvanostatic charge–discharge data for (a) $Na_{1/2}Ca_{1/6}CoO_2$, (b) $Na_{7/12}Ca_{1/12}CoO_2$, (c) $Na_{5/8}Ca_{1/24}CoO_2$, and (d) $Na_{0.74}CoO_2$. Reproduced with permission from Ref. [187]. Copyright (2015), Elsevier.

Rojo et al. found that the Na_xFePO_4 system with x between 1 and 0 became an intermediate structure at $x = 0.7$. $Na_{0.7}FePO_4$ contains at least four different crystal structures as detected by selected area diffraction patterns depicted in Figs. 23a–d [185]. The Miller indices for these phases are double or triple that of $FePO_4$. In addition, there is a 17.58% crystal size mismatch between $FePO_4$ and $Na_{0.7}FePO_4$ where $Na_{0.7}FePO_4$ is larger. The crystal size mismatch is only 3.62% between $Na_{0.7}FePO_4$ and $NaFePO_4$ where $NaFePO_4$ is larger as measured by XRD characterization shown in Fig. 9.23e. This explains why the phase changes occur simultaneously on discharge and consequently on charge. The lithium counterpart does not form an intermediate phase between $LiFePO_4$ and $FePO_4$ owing to the smaller size of the lithium ion.

Cycling stresses due to bi-phasic changes can be reduced by using atom substitution [186]. Calcium atom substitution in Na_xCoO_2 reduces crystalline mismatches according to Matsui et al. [187]. Calcium is wedged between MO_2 planes preventing the organization of phases. The effect is an expansion of solid solution behavior, which reduces the voltage steps on the charge/discharge curves. Figures 9.24a–d shows more prominent voltage steps in each profile progressing from a to d as the Ca content in the $Na_xCa_yCoO_2$ compound decreases. As the Ca content decreases from 1/6, 1/12, 1/24 to 0, reversible capacity grows from about 85, 115, 122, and 125 mAh/g, as shown in Figs. 9.24a–d, respectively. As expected, the Ca content decreases and the GCD curves display more pronounced voltage steps.

9.4.2.3 Amorphization of crystalline structures

The cycling stability of a layered $Na_{0.6}MnO_2$ cathode material was studied by Caballero et al. [188]. The material delivered capacities of 140 mAh/g when cycled between 2.0 and 3.8 V versus Na/Na^+, and 100 mAh/g when cycled between 2.0 and 3.0 V versus Na/Na^+. However, a steep decline in performance was observed to lead to a complete failure of the electrode after eight cycles, as shown in Fig. 9.25a. The XRD data reveal that crystallinity is lost upon sodiation and desodiation (Fig. 9.25b). The authors attribute the structure collapses to the distortion of the Mn–O layers. Insertion of water molecules in the material after all the sodium atoms were removed from the structure was also observed. This hydration is believed to contribute to the irreversible amorphization of the layered material $Na_{0.6}MnO_2$.

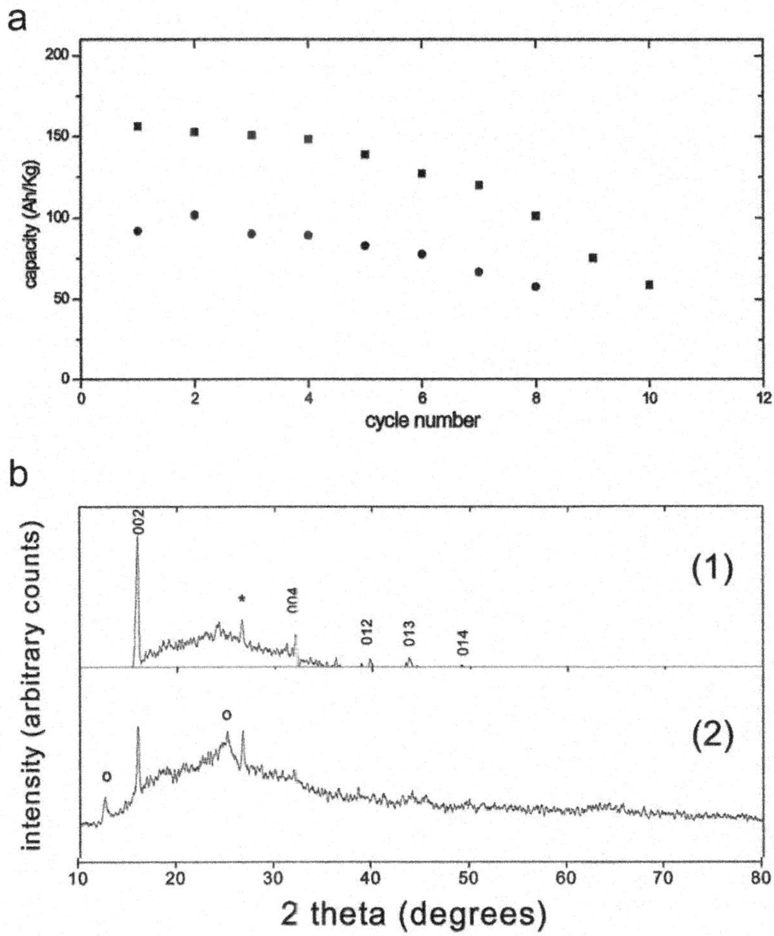

Figure 9.25 (a) Specific capacity versus cycle number cycled between 2.0 and 3.8 V versus Na/Na^+ (square dots) and 2.0 and 3.0 V versus Na/Na^+ (circle dots). (b) XRD data of pristine $Na_{0.6}MnO_2$ (1) and cycled $Na_{0.6}MnO_2$ (2) with peaks representing hydrated phase identified by circles. Reproduced with permission from Ref. [181]. Copyright (2002), Royal Society of Chemistry.

9.4.3 Foreign Molecule Inclusion/Intercalation

The inclusion of a foreign molecule can cause a failure of NIBs [189–192]. The larger size of sodium atoms compared to lithium atoms makes it possible for the insertion of foreign atoms and molecules in sodium sites, while this phenomenon does not occur in LIBs. Foreign molecules get trapped when electrode materials are in their desodiated states where sodium sites are vacant. The absence of the charged Na^+ cations in these sites increases the repulsion between negatively charged oxide layers creating bigger gaps between these layers as described by Passerini et al. [193]. Despite the lack of structure destruction or permanent collapse, the inclusion of foreign molecules is a major problem for NIB cathode materials. These foreign molecules originally come from precursor impurities as well as exposure to air during production of the cathode materials [193]. The presence of these molecules reduces the capacity and cyclability of the cathode materials and in severe cases can prompt the complete failure of cells. Foreign molecules capable of intercalation in these compounds include water and carbon dioxide, which are two very common molecules presented in the air [191]. This air sensitivity dramatically raises manufacturing, transportation, and storage costs.

As explained earlier, desodiated oxides leave empty sites in the crystal that can be filled with water molecules. For instance, in sodium coboltate (Na_xCoO_2), sodium atoms are determined via neutron diffraction to occupy sites labeled Na1 and Na2 [194]. Atoms are stored in site Na1 only for $x = 1/3$ and Na1 and Na2 sites equally for $x = 1/2$, as illustrated in Fig. 9.26a [192]. Takada et al. [195] showed that hydration of this oxide can cause superconductivity when water molecules are in certain sites. Polar water molecules in bi-hydrate $Na_xCoO_{2-y}H_2O$ intercalate in between sodium atoms and CoO_2 layers giving the material its superconductivity at 5 K. It is speculated by Zhang et al. [196] that the screening of Na charges inhibits the formation of a charge ordering pattern in the CoO_2 layer as the water molecules intercalate within the Na^+ plane explaining the overall good performance of the electrode.

Lu and Dahn [197] performed neutron scattering characterization of P2, T2, and O2 structures of $A_z[Co_xNi_{0.33-x}Mn_{0.66}]O_2$ where A represents Li or Na. They found experimentally that water would only intercalate in the sodium-containing compound as long as there are cobalt atoms ($x = 0.33$) in the structure. However, the lithium-containing compound cannot be hydrated. Water molecules are presumably located in the 2c sites. The presence of these water

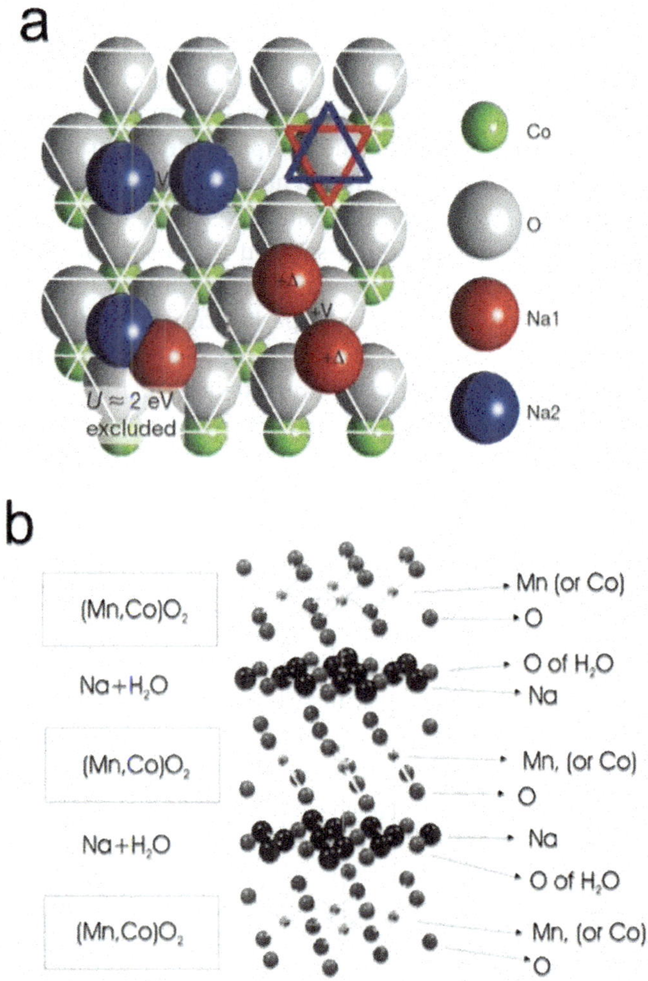

Figure 9.26 (a) Schematic representation of crystalline structure of Na_xCoO_2. Reproduced with permission from Ref. [192]. Copyright (2017), American Chemical Society. (b) Illustration of the crystalline structure of phase I P2-$Na_{2/3}(H_2O)_{2/3}[Co_{1/3}\text{-}Mn_{2/3}]O_2$ with Na positions statistically half filled. Reproduced with permission from Ref. [196]. Copyright (2005), American Physical Society.

molecules, as identified by their O atom in Fig. 9.26b, leads to an increase in the inter-slab distance from 5.65 to 7.05 Å. Hydration of the sodium-containing compound is obtained merely by leaving the material in humid air. Water molecules intercalate in free sites in the sodium layer as a way to mitigate the strong repulsion between oxygen atoms via the creation of hydrogen bonds. Further hydration occurs when the material is immersed in water, but the phase becomes unstable. After removing the immersed phase from water and exposing it to air, the sample hydration reverts to the hydration phase acquired in humid air. Finally, all hydration water can be removed by treating the sample at 200°C for 2 days. In an earlier work, the same authors showed that incorporating cobalt in $Ni_xMn_yO_2$-layered material removes ordering of the transition metals in the oxide layers (i.e., superlattice ordering) [198]. The cobalt-substituted structure becomes the equivalent of a 2D gas where atoms are not bound to a crystalline position. To explain why only the cobalt-substituted material can be hydrated, Lu and Dahn [198] speculated that superlattice ordering induces stronger interlayer attraction, preventing the intercalation of water.

Fully sodiated $NaNi_{1/3}Mn_{1/3}Co_{1/3}O_2$ (NaNMC) may also react with atmospheric moisture and carbon dioxide to form some sodium hydroxide and carbonate as reported by Tarascon et al. and Prakash et al. [199]. The groups point out the open-circuit potential of the compound against sodium to be lower than that of water against sodium (2.4 versus 3.94 V) as cause for the spontaneous reaction in air. Unsurprisingly, $Na_{0.5}Ni_{0.33}Mn_{0.33}Co_{0.33}O_2$ can be intercalated with water by simple exposure, as demonstrated by XRD characterization in Fig. 9.27a where an extra peak appears around 30°.

Buchholz et al. [193] studied a P2/P3-$Na_xNi_{0.22}Co_{0.11}Mn_{0.66}O_2$ cathode material that exhibited a tendency to allow for water intercalation. They observed a similar increase in the inter-sheet distances to Lu and Dahn's material at low sodium content (Fig. 9.27b). The material was characterized by XRD, and it was shown that for $x < 0.34$, water intercalates in a well-ordered fashion with defined peaks (Fig. 9.27c). A shift in the positions of the (002)' and (003)' peaks occurs for the P2 and P3 phase, respectively, as shown in Fig. 9.27d.

Alpha-$NaFeO_2$ can intercalate both water and CO_2 from the atmosphere as characterized through TGA and energy-dispersive X-ray spectroscopy, respectively, by Monyoncho et al. [200]. Water can be extracted from the structure by heating at 100°C or above, and CO_2 by heating between 500 and 600°C.

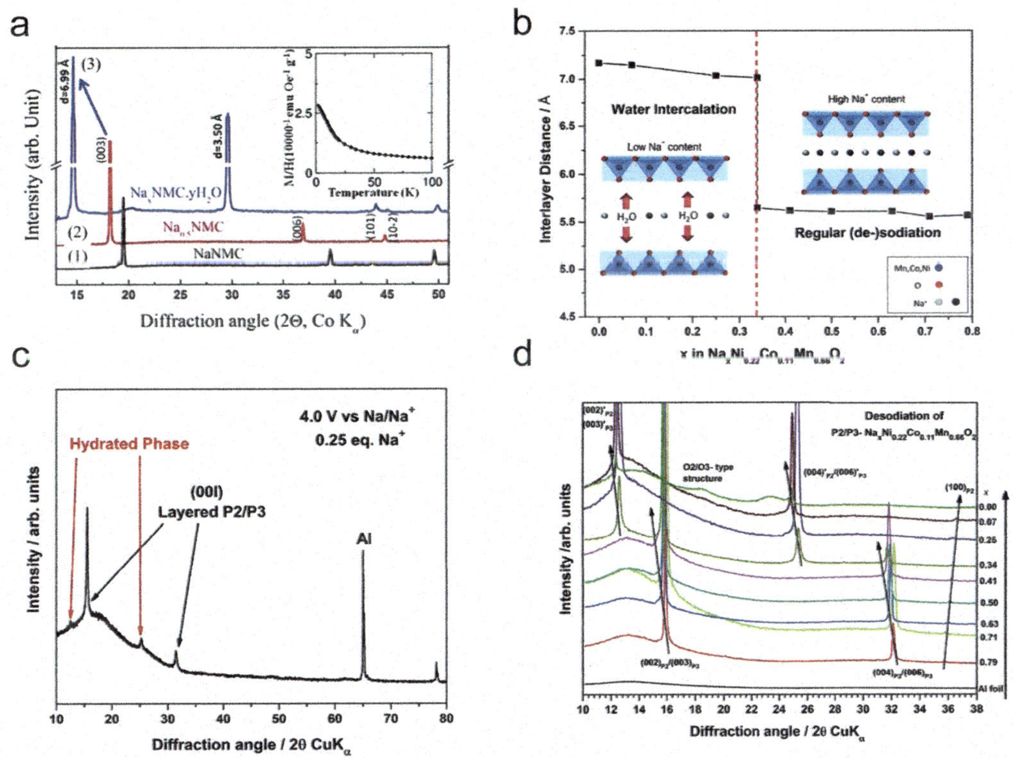

Figure 9.27 (a) X-ray diffraction patterns for (1) NaNMC, (2) Na_xNMC charged to 3.75 V ($x = 0.5$), and (3) after water-treated $Na_xNMC \cdot yH_2O$. Magnetization data are shown in the inset. Reproduced with permission from Ref. [197]. Copyright (2000), American Chemical Society. (b) Schematic representation and calculated interlayer distance with regard to sodium content x in P2/P3-$Na_xNi_{0.22}Co_{0.11}Mn_{0.66}O_2$ electrodes. (c) XRD pattern data for sealed P2/P3-$Na_xNi_{0.22}Co_{0.11}Mn_{0.66}O_2$ electrodes. (d) XRD pattern data for P2/P3-$Na_xNi_{0.22}Co_{0.11}Mn_{0.66}O_2$ electrodes at different charge states. Reproduced with permission from Ref. [199]. Copyright (2012), American Chemical Society.

Nazar et al. [191] demonstrated the inclusion and adsorption of water, carbon dioxide, and oxygen within layered P2-$Na_{0.67}Mn_{0.5}Fe_{0.5}O_2$. The presence of water, carbon dioxide, and oxygen was detected with thermogravimetric analysis coupled with mass spectroscopy (TGA-MS). However, the authors believe that it is mainly carbonate that incorporates into the crystal. Rietveld refinement of neutron time-of-flight diffraction data demonstrated that exposure to carbon dioxide and oxygen from the air promotes the insertion of CO_3^{2-} ions within the tetrahedral sites of the transition metal layer. The final formula for air-exposed P2-$Na_{0.67}Mn_{0.5}Fe_{0.5}O_2$ is

$$Na^{I}_{2/3}Fe^{III}_{1/2}Mn^{III}_{1/6}Mn^{IV}_{1/3}O_2^{-II} + \frac{1}{12}CO_3^{2-} \rightarrow \frac{9}{8}\left(Na^{I}_{16/27}Fe^{III}_{4/9}Mn^{IV}_{4/9}C^{IV}_{2/27}O_2^{-II}\right)$$

Charge balancing of the carbonate ions is carried out by the further oxidation of the manganese Mn^{3+} and Mn^{4+} according to the preceding equation. The carbonate ions reduce the capacity by 30% and increase the polarization compared to the pristine phase. Nazar et al. also showed that progressive substitution of manganese by nickel helps retain electrochemical performance. Nickel mechanically increases the mean oxidation of manganese, therefore preventing carbonate incorporation.

Research efforts have shifted toward making air-resistant compounds for Na-ion applications. In their 2015 publication, Mu et. al. [201] presented an O3-type material, O3-$Na_{0.9}[Cu_{0.22}Fe_{0.30}Mn_{0.48}]O_2$, which is resistant to air exposure as proven by XRD characterization. In the same paper, the authors show that P2-$Na_{2/3}Ni_{1/3}Mn_{2/3}O_2$ and P2-$Na_{2/3}Co_{2/3}Mn_{1/3}O_2$ are stable against water, while O3-$NaNi_{1/2}Mn_{1/2}O_2$ and O3-$NaNi_{1/3}Co_{1/3}Mn_{1/3}O_2$ are very unstable.

In 2017, Yao et al. [192] published a systematic approach to reduce air sensitivity in $NaNi_{0.5}Mn_{0.5}O_2$ via co-doping with an atom of similar electronegativity and another sporting a different Fermi level; here copper and titanium atoms. The difference in Fermi levels between Ti^{4+} and Mn^{4+} prevents charge ordering, resulting in an expansion of the transition metal layer, contraction of interlayer spacing and decreased interaction of the transition metal with oxygen atoms [202]. The result is remarkable stability of the $NaNi_{0.45}Cu_{0.05}Mn_{0.4}Ti_{0.1}O_2$ compound when exposed to air or water, as demonstrated by XRD and XANES characterization. The co-doping also had the effect of smoothing the charge–discharge profiles. Undoped compound charge–discharge curves are presented in Fig. 9.28a, whereas doped compound curves are shown in Fig. 9.28b. This effect is due to electronic delocalization, making the material behave more like a solid solution.

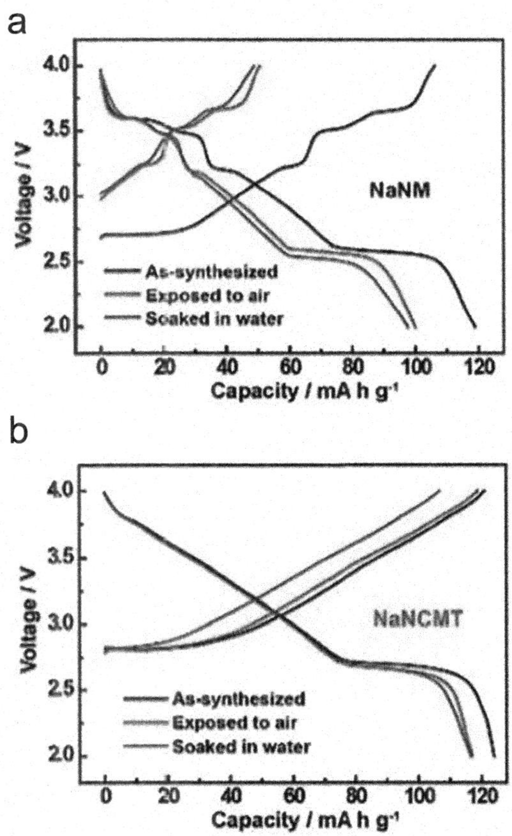

Figure 9.28 First charge–discharge profiles for the as-synthesized, aged, and soaked-in-water (a) NaNM samples and (b) NaNCMT samples. Reproduced with permission from Ref. [192]. Copyright (2017), American Chemical Society.

9.5 Concluding Remarks

Sodium-ion batteries are closely related to Li-ion batteries in terms of both material design and cell-failure mechanisms. However, degradations upon cycling are generally more severe for the former than the latter due to the bigger size and the more active chemistry of sodium ions than that of lithium ions. Mitigation strategies are currently mostly just carried over from the Li-ion systems even though these two systems have fundamental differences. It follows that NIBs will always be more prone to failures than LIBs unless new approaches specifically designed for sodium applications are developed. With the renewed interests in developing NIB technology, more efforts are being focused on increasing the cycle life of Na-ion systems using tailor-made approaches. One example is the use of the fluoroethylene carbonate additive in the electrolyte. It is likely that this trend will continue predominantly on the anode side because graphite is not a practical anode for NIBs, so materials such as hard carbons are used instead, which have different storage mechanisms. On the cathode side, the development of novel materials that can cope with repeated insertion of the larger cation will be needed to increase the cycle life of NIBs.

References

1. Fang, Y.; Yu, X.-Y.; and Lou, X. W. D., Nanostructured electrode materials for advanced sodium-ion batteries. *Matter* 2019, **1**(1), 90–114.
2. Xu, Y.-S.; Duan, S.-Y.; Sun, Y.-G.; Bin, D.-S.; Tao, X.-S.; Zhang, D.; Liu, Y.; Cao, A.-M.; and Wan, L.-J., Recent developments in electrode materials for potassium-ion batteries. *J. Mater. Chem. A* 2019, **7**(9), 4334–4352.
3. Fu, J.; Liang, R.; Liu, G.; Yu, A.; Bai, Z.; Yang, L.; and Chen, Z., Recent progress in electrically rechargeable zinc–air batteries. *Adv. Mater.* 2019, **31**(31), 1805230.
4. Kubota, K.; Ikeuchi, I.; Nakayama, T.; Takei, C.; Yabuuchi, N.; Shiiba, H.; Nakayama, M.; and Komaba, S., New insight into structural evolution in layered $NaCrO_2$ during electrochemical sodium extraction. *J. Phys. Chem. C* 2014, **119**(1), 166–175.
5. Dahbi, M.; Yabuuchi, N.; Kubota, K.; Tokiwa, K.; and Komaba, S., Negative electrodes for Na-ion batteries. *Phys. Chem. Chem. Phys.* 2014, **16**(29), 15007–15028.

6. Chevrier, V. and Ceder, G., Challenges for Na-ion negative electrodes. *J. Electrochem. Soc.* 2011, **158**(9), A1011–A1014.
7. Slater, M. D.; Kim, D.; Lee, E.; and Johnson, C. S., Sodium-ion batteries. *Adv. Funct. Mater.* 2013, **23**(8), 947–958.
8. Luo, W.; Shen, F.; Bommier, C.; Zhu, H.; Ji, X.; and Hu, L., Na-ion battery anodes: materials and electrochemistry. *Acc. Chem. Res.* 2016, **49**(2), 231–240.
9. Hwang, J.-Y.; Myung, S.-T.; and Sun, Y.-K., Sodium-ion batteries: Present and future. *Chem. Soc. Rev.* 2017, **46**(12), 3529–3614.
10. Palomares, V.; Serras, P.; Villaluenga, I.; Hueso, K. B.; Carretero-González, J.; and Rojo, T., Na-ion batteries, recent advances and present challenges to become low cost energy storage systems. *Energy Environ. Sci.* 2012, **5**(3), 5884–5901.
11. Han, M. H.; Gonzalo, E.; Singh, G.; and Rojo, T., A comprehensive review of sodium layered oxides: powerful cathodes for Na-ion batteries. *Energy Environ. Sci.* 2015, **8**(1), 81–102.
12. Peled, E. and Menkin, S., SEI: Past, present and future. *J. Electrochem. Soc.* 2017, **164**(7), A1703–A1719.
13. Petibon, R.; Aiken, C.; Sinha, N.; Burns, J.; Ye, H.; VanElzen, C. M.; Jain, G.; Trussler, S.; and Dahn, J., Study of electrolyte additives using electrochemical impedance spectroscopy on symmetric cells. *J. Electrochem. Soc.* 2013, **160**(1), A117–A124.
14. Dey, A. N. and Sullivan, B. P., The electrochemical decomposition of propylene carbonate on graphite. *J. Electrochem. Soc.* 1970, **117**(2), 222–224.
15. Martinez de la Hoz, J. M.; Leung, K.; and Balbuena, P. B., Reduction mechanisms of ethylene carbonate on Si anodes of lithium-ion batteries: Effects of degree of lithiation and nature of exposed surface. *ACS Appl. Mater. Interfaces* 2013, **5**(24), 13457–13465.
16. Hamidah, N. L.; Wang, F. M.; and Nugroho, G., The understanding of solid electrolyte interface (SEI) formation and mechanism as the effect of flouro-o-phenylenedimaleimaide (F-MI) additive on lithium-ion battery. *Surface Interface Anal.* 2019, **51**(3), 345–352.
17. Liu, Q.; Mu, D.; Wu, B.; Wang, L.; Gai, L.; and Wu, F., Density functional theory research into the reduction mechanism for the solvent/additive in a sodium-ion battery. *ChemSusChem* 2017, **10**(4), 786–796.
18. Leung, K., Two-electron reduction of ethylene carbonate: A quantum chemistry re-examination of mechanisms. *Chem. Phys. Lett.* 2013, **568**, 1–8.

19. Smith, A.; Burns, J. C.; Zhao, X.; Xiong, D.; and Dahn, J., A high precision coulometry study of the SEI growth in Li/graphite cells. *J. Electrochem. Soc.* 2011, **158**(5), A447–A452.

20. Smith, A.; Burns, J.; Trussler, S.; and Dahn, J., Precision measurements of the coulombic efficiency of lithium-ion batteries and of electrode materials for lithium-ion batteries. *J. Electrochem. Soc.* 2010, **157**(2), A196–A202.

21. Komaba, S.; Murata, W.; Ishikawa, T.; Yabuuchi, N.; Ozeki, T.; Nakayama, T.; Ogata, A.; Gotoh, K.; and Fujiwara, K., Electrochemical Na insertion and solid electrolyte interphase for hard-carbon electrodes and application to Na-ion batteries. *Adv. Funct. Mater.* 2011, **21**(20), 3859–3867.

22. Ponrouch, A.; Marchante, E.; Courty, M.; Tarascon, J.-M.; and Palacin, M. R., In search of an optimized electrolyte for Na-ion batteries. *Energy Environ. Sci.* 2012, **5**(9), 8572–8583.

23. Philippe, B.; Valvo, M.; Lindgren, F.; Rensmo, H. K.; and Edström, K., Investigation of the electrode/electrolyte interface of Fe_2O_3 composite electrodes: Li vs Na batteries. *Chem. Mater.* 2014, **26**(17), 5028–5041.

24. Muñoz-Márquez, M. A.; Zarrabeitia, M.; Castillo-Martínez, E.; Eguía-Barrio, A.; Rojo, T. F.; and Casas-Cabanas, M., Interfaces, composition and evolution of the solid-electrolyte interphase in $Na_2Ti_3O_7$ electrodes for Na-ion batteries: XPS and Auger parameter analysis. *ACS Appl. Mater.* 2015, **7**(14), 7801–7808.

25. Baggetto, L.; Ganesh, P.; Meisner, R. P.; Unocic, R. R.; Jumas, J.-C.; Bridges, C. A.; and Veith, G. M., Characterization of sodium ion electrochemical reaction with tin anodes: Experiment and theory. *J. Power Sources* 2013, **234**, 48–59.

26. Jache, B.; Binder, J. O.; Abe, T.; and Adelhelm, P., A comparative study on the impact of different glymes and their derivatives as electrolyte solvents for graphite co-intercalation electrodes in lithium-ion and sodium-ion batteries. *Phys. Chem. Chem. Phys.* 2016, **18**(21), 14299–14316.

27. Yoon, G.; Kim, H.; Park, I.; and Kang, K., Conditions for reversible Na intercalation in graphite: Theoretical studies on the interplay among guest ions, solvent, and graphite host. *Adv. Energy Mater.* 2017, **7**(2), 1601519.

28. Jache, B. and Adelhelm, P., Use of graphite as a highly reversible electrode with superior cycle life for sodium-ion batteries by making use of Co-intercalation phenomena. *Angew. Chem.* 2014, **126**(38), 10333–10337.

29. Cohn, A. P.; Share, K.; Carter, R.; Oakes, L.; ad Pint, C. L., Ultrafast solvent-assisted sodium ion intercalation into highly crystalline few-layered graphene. *Nano Lett.* 2015, **16**(1), 543–548.

30. Zhang, J.; Wang, D.-W.; Lv, W.; Zhang, S.; Liang, Q.; Zheng, D.; Kang, F.; and Yang, Q.-H., Achieving superb sodium storage performance on carbon anodes through an ether-derived solid electrolyte interphase. *Energy Environ. Sci.* 2017, **10**(1), 370–376.

31. Zhang, B.; Rousse, G.; Foix, D.; Dugas, R.; Corte, D. A. D.; and Tarascon, J. M., Microsized Sn as advanced anodes in glyme-based electrolyte for Na-ion batteries. *Adv. Mater.* 2016, **28**(44), 9824–9830.

32. Tang, S.; Zhang, X.-G.; Cui, X.-Y.; Xu, H.-Y.; Wu, D.-Y.; Yan, J.-W.; Zheng, M.-S.; Dong, Q.-F. and Mao, B.-W., Mitigating concentration polarization for highly reversible plating/stripping electrochemistry: Li versus Na. *J. Mater. Chem. A* 2019, **7**(40), 23216–23224.

33. Ponrouch, A.; Monti, D.; Boschin, A.; Steen, B.; Johansson, P.; and Palacin, M., Non-aqueous electrolytes for sodium-ion batteries. *J. Mater. Chem. A* 2015, **3**(1), 22–42.

34. Kubota, K. and Komaba, S., Practical issues and future perspective for Na-ion batteries. *J. Electrochem. Soc.* 2015, **162**(14), A2538–A2550.

35. Ponrouch, A.; Dedryvère, R.; Monti, D.; Demet, A. E.; Mba, J. M. A.; Croguennec, L.; Masquelier, C.; Johansson, P.; and Palacín, M. R., Towards high energy density sodium ion batteries through electrolyte optimization. *Energy Environ. Sci.* 2013, **6**(8), 2361–2369.

36. Haregewoin, A. M.; Wotango, A. S.; and Hwang, B.-J., Electrolyte additives for lithium ion battery electrodes: Progress and perspectives. *Energy Environ. Sci.* 2016, **9**(6), 1955–1988.

37. Ma, L.; Ellis, L.; Glazier, S.; Ma, X.; and Dahn, J., Combinations of $LiPO_2F_2$ and other electrolyte additives in Li $[Ni_{0.5}Mn_{0.3}Co_{0.2}]$ O_2/graphite pouch cells. *J. Electrochem. Soc.* 2018, **165**(9), A1718–A1724.

38. Purushotham, U.; Takenaka, N.; and Nagaoka, M., Additive effect of fluoroethylene and difluoroethylene carbonates for the solid electrolyte interphase film formation in sodium-ion batteries: A quantum chemical study. *RSC Adv.* 2016, **6**(69), 65232–65242.

39. Evans, D. H., One-electron and two-electron transfers in electrochemistry and homogeneous solution reactions. *Chem. Rev.* 2008, **108**(7), 2113–2144.

40. Shkrob, I. A.; Zhu, Y.; Marin, T. W.; and Abraham, D., Reduction of carbonate electrolytes and the formation of solid-electrolyte interface (SEI) in lithium-ion batteries. 1. Spectroscopic observations of radical

intermediates generated in one-electron reduction of carbonates. *J. Phys. Chem. C* 2013, **117**(38), 19255–19269.

41. Shkrob, I. A.; Zhu, Y.; Marin, T. W.; and Abraham, D., Reduction of carbonate electrolytes and the formation of solid-electrolyte interface (SEI) in Lithium-ion batteries. 2. Radiolytically induced polymerization of ethylene carbonate. *J. Phys. Chem. C* 2013, **117**(38), 19270–19279.
42. Kumar, H.; Detsi, E.; Abraham, D. P.; and Shenoy, V. B., Fundamental mechanisms of solvent decomposition involved in solid-electrolyte interphase formation in sodium ion batteries. *Chem. Mater.* 2016, **28**(24), 8930–8941.
43. Madec, L.; Petibon, R.; Tasaki, K.; Xia, J.; Sun, J.-P.; Hill, I.; and Dahn, J., Mechanism of action of ethylene sulfite and vinylene carbonate electrolyte additives in $LiNi_{1/3}Mn_{1/3}Co_{1/3}O_2$/graphite pouch cells: Electrochemical, GC-MS and XPS analysis. *Phys. Chem. Chem. Phys.* 2015, **17**(40), 27062–27076.
44. Yildirim, H.; Kinaci, A.; Chan, M. K.; and Greeley, J. P., First-principles analysis of defect thermodynamics and ion transport in inorganic SEI compounds: LiF and NaF. *ACS Appl. Mater. Interfaces* 2015, **7**(34), 18985–18996.
45. Dahbi, M.; Yabuuchi, N.; Fukunishi, M.; Kubota, K.; Chihara, K.; Tokiwa, K.; Yu, X.-f.; Ushiyama, H.; Yamashita, K.; and Son, J.-Y., Black phosphorus as a high-capacity, high-capability negative electrode for sodium-ion batteries: Investigation of the electrode/electrolyte interface. *Chem. Mater.* 2016, **28**(6), 1625–1635.
46. Takenaka, N.; Sakai, H.; Suzuki, Y.; Uppula, P.; and Nagaoka, M., A computational chemical insight into microscopic additive effect on solid electrolyte interphase film formation in sodium-ion batteries: Suppression of unstable film growth by intact fluoroethylene carbonate. *J. Phys. Chem. C* 2015, **119**(32), 18046–18055.
47. Xia, X.; Obrovac, M.; and Dahn, J., Comparison of the reactivity of Na_xC_6 and Li_xC_6 with non-aqueous solvents and electrolytes. *Electrochem. Solid State Lett.* 2011, **14**(9), A130–A133.
48. Dahbi, M.; Nakano, T.; Yabuuchi, N.; Fujimura, S.; Chihara, K.; Kubota, K.; Son, J. Y.; Cui, Y. T.; Oji, H.; and Komaba, S., Effect of hexafluorophosphate and fluoroethylene carbonate on electrochemical performance and the surface layer of hard carbon for sodium-ion batteries. *ChemElectroChem* 2016, **3**(11), 1856–1867.
49. Zhang, J.-N.; Li, Q.; Wang, Y.; Zheng, J.; Yu, X.; and Li, H., Dynamic evolution of cathode electrolyte interphase (CEI) on high voltage $LiCoO_2$ cathode and its interaction with Li anode. *Energy Storage Mater.* 2018, **14**, 1–7.

50. Arakawa, M. and Yamaki, J.-I., Anodic oxidation of propylene carbonate and ethylene carbonate on graphite electrodes. *J. Power Sources* 1995, **54**(2), 250–254.
51. Kanamura, K.; Toriyama, S.; Shiraishi, S.; and Takehara, Z. I., Electrochemical oxidation processes on Ni electrodes in propylene carbonate containing various electrolyte salts. *J. Electrochem. Soc.* 1996, **143**(8), 2548–2558.
52. Xu, S.; Luo, G.; Jacobs, R.; Fang, S.; Mahanthappa, M. K.; Hamers, R. J.; and Morgan, D., Ab initio modeling of electrolyte molecule ethylene carbonate decomposition reaction on Li (Ni, Mn, Co) O_2 cathode surface. *ACS Appl. Mater. Interfaces* 2017, **9**(24), 20545–20553.
53. Moshkovich, M.; Cojocaru, M.; Gottlieb, H.; and Aurbach, D., The study of the anodic stability of alkyl carbonate solutions by in situ FTIR spectroscopy, EQCM, NMR and MS. *J. Electroanal. Chem.* 2001, **497**(1–2), 84–96.
54. Aurbach, D. and Moshkovich, M., A study of lithium deposition-dissolution processes in a few selected electrolyte solutions by electrochemical quartz crystal microbalance. *J. Electrochem. Soc.* 1998, **145**(8), 2629–2639.
55. Komaba, S.; Nakayama, T.; Ogata, A.; Shimizu, T.; Takei, C.; Takada, S.; Hokura, A.; and Nakai, I., Electrochemically reversible sodium intercalation of layered $NaNi_{0.5}Mn_{0.5}O_2$ and $NaCrO_2$. *ECS Trans.* 2009, **16**(42), 43–55.
56. Serras, P.; Palomares, V.; Goñi, A.; de Muro, I. G.; Kubiak, P.; Lezama, L.; and Rojo, T., High voltage cathode materials for Na-ion batteries of general formula $Na_3V_2O_{2x}(PO_4)_2F_{3-2x}$. *J. Mater. Chem.* 2012, **22**(41), 22301–22308.
57. Vassilaras, P.; Ma, X.; Li, X.; and Ceder, G., Electrochemical properties of monoclinic $NaNiO_2$. *J. Electrochem. Soc.* 2013, **160**(2), A207–A211.
58. Ma, X.; Chen, H.; and Ceder, G., Electrochemical properties of monoclinic $NaMnO_2$. *J. Electrochem. Soc.* 2011, **158**(12), A1307–A1312.
59. Xu, K.; Ding, S. P.; and Jow, T. R., Toward reliable values of electrochemical stability limits for electrolytes. *J. Electrochem. Soc.* 1999, **146**(11), 4172–4178.
60. Hirayama, M.; Tomita, H.; Kubota, K.; and Kanno, R., Structure and electrode reactions of layered rocksalt $LiFeO_2$ nanoparticles for lithium battery cathode. *J. Power Sources* 2011, **196**(16), 6809–6814.
61. Kundu, D.; Talaie, E.; Duffort, V.; and Nazar, L. F., The emerging chemistry of sodium ion batteries for electrochemical energy storage. *Angew. Chem.* 2015, **54**(11), 3431–3448.

62. Delmas, C.; Fouassier, C.; and Hagenmuller, P., Structural classification and properties of the layered oxides. *Physica B+C* 1980, **99**(1–4), 81–85.

63. Szajwaj, O.; Gaudin, E.; Weill, F.; Darriet, J.; and Delmas, C., Investigation of the new P′3-$Na_{0.60}VO_2$ phase: Structural and physical properties. *Inorg. Chem.* 2009, **48**(19), 9147–9154.

64. Moshkovich, M.; Gofer, Y.; and Aurbach, D., Investigation of the electrochemical windows of aprotic alkali metal (Li, Na, K) salt solutions. *J. Electrochem. Soc.* 2001, **148**(4), E155-E167.

65. Zarrabeitia, M.; Nobili, F.; Muñoz-Márquez, M. Á.; Rojo, T.; and Casas-Cabanas, M., Direct observation of electronic conductivity transitions and solid electrolyte interphase stability of $Na_2Ti_3O_7$ electrodes for Na-ion batteries. *J. Power Sources* 2016, **330**, 78–83.

66. Jones, J.; Anouti, M.; Caillon-Caravanier, M.; Willmann, P.; Sizaret, P.-Y.; and Lemordant, D., Solubilization of SEI lithium salts in alkylcarbonate solvents. *Fluid Ph. Equilibria* 2011, **305**(2), 121–126.

67. Peled, E.; Tow, D. B.; Merson, A.; Gladkich, A.; Burstein, L.; and Golodnitsky, D., Composition, depth profiles and lateral distribution of materials in the SEI built on HOPG-TOF SIMS and XPS studies. *J. Power Sources* 2001, **97**, 52–57.

68. Peled, E.; Golodnitsky, D.; and Penciner, J., The anode/electrolyte interface. In *Handbook of Battery Materials* 1998, 419–456, Wiley-VCH Verlag GmbH & Co. KGaA.

69. Mogensen, R.; Brandell, D.; and Younesi, R., Solubility of the solid electrolyte interphase (SEI) in sodium ion batteries. *ACS Energy Lett.* 2016, **1**(6), 1173–1178.

70. Rodriguez, R.; Loeffler, K. E.; Nathan, S. S.; Sheavly, J. K.; Dolocan, A.; Heller, A.; and Mullins, C. B., In situ optical imaging of sodium electrodeposition: Effects of fluoroethylene carbonate. *ACS Energy Lett.* 2017, **2**(9), 2051–2057.

71. Fukunishi, M.; Yabuuchi, N.; Dahbi, M.; Son, J.-Y.; Cui, Y.; Oji, H.; and Komaba, S., Impact of the cut-off voltage on cyclability and passive interphase of Sn-polyacrylate composite electrodes for sodium-ion batteries. *J. Phys. Chem. C* 2016, **120**(28), 15017–15026.

72. Wang, J.; Eng, C.; Chen-Wiegart, Y.-C. K.; and Wang, J., Probing three-dimensional sodiation–desodiation equilibrium in sodium-ion batteries by in situ hard X-ray nanotomography. *Nat. Commun.* 2015, **6**, 7496.

73. Lao, M.; Zhang, Y.; Luo, W.; Yan, Q.; Sun, W.; and Dou, S. X., Alloy-based anode materials toward advanced sodium-ion batteries. *Adv. Mater.* 2017, **29**(48), 1700622.

74. Obrovac, M.; Christensen, L.; Le, D. B.; and Dahn, J. R., Alloy design for lithium-ion battery anodes. *J. Electrochem. Soc.* 2007, **154**(9), A849–A855.

75. Klein, F.; Jache, B.; Bhide, A.; and Adelhelm, P., Conversion reactions for sodium-ion batteries. *Phys. Chem. Chem. Phys.* 2013, **15**(38), 15876–15887.

76. Fang, S.; Bresser, D.; and Passerini, S., Transition metal oxide anodes for electrochemical energy storage in lithium- and sodium-ion batteries. *Adv. Energy Mater.* 2020, **10**(1), 1902485.

77. Wang, W.; Li, W.; Wang, S.; Miao, Z.; Liu, H. K.; and Chou, S., Structural design of anode materials for sodium-ion batteries. *J. Mater. Chem. A* 2018, **6**(15), 6183–6205.

78. Chen, X.; Li, H.; Yan, Z.; Cheng, F.; and Chen, J., Structure design and mechanism analysis of silicon anode for lithium-ion batteries. *Sci. China Mater.* **2019**, 1–22.

79. Beaulieu, L.; Eberman, K.; Turner, R.; Krause, L.; and Dahn, J., Colossal reversible volume changes in lithium alloys. *Electrochem. Solid-State Lett.* 2001, **4**(9), A137–A140.

80. McDowell, M. T.; Lee, S. W.; Nix, W. D.; and Cui, Y., 25th anniversary article: Understanding the lithiation of silicon and other alloying anodes for lithium-ion batteries. *Adv. Mater.* 2013, **25**(36), 4966–4985.

81. Chan, C. K.; Peng, H.; Liu, G.; McIlwrath, K.; Zhang, X. F.; Huggins, R. A.; and Cui, Y., High-performance lithium battery anodes using silicon nanowires. *Nat. Nanotechnol.* 2008, **3**(1), 31.

82. Jung, S. C.; Jung, D. S.; Choi, J. W.; and Han, Y.-K., Atom-level understanding of the sodiation process in silicon anode material. *J. Phys. Chem. Lett.* 2014, **5**(7), 1283–1288

83. Legrain, F.; Malyi, O. I.; and Manzhos, S., Comparative computational study of the energetics of Li, Na, and Mg storage in amorphous and crystalline silicon. *Comput. Mater. Sci.* 2014, **94**, 214–217.

84. Lim, C.-H.; Huang, T.-Y.; Shao, P.-S.; Chien, J.-H.; Weng, Y.-T.; Huang, H.-F.; Hwang, B. J.; and Wu, N.-L., Experimental study on sodiation of amorphous silicon for use as sodium-ion battery anode. *Electrochim. Acta* 2016, **211**, 265–272.

85. Qiu, D.-F.; Ma, X.; Zhang, J.-D.; Lin, Z.-X.; and Zhao, B., Mesoporous silicon microspheres produced from in situ magnesiothermic reduction of

silicon oxide for high-performance anode material in sodium-ion batteries. *Nanoscale Res. Lett.* 2018, **13**(1), 275.

86. Xu, Y.; Swaans, E.; Basak, S.; Zandbergen, H. W.; Borsa, D. M.; and Mulder, F. M., Reversible Na-ion uptake in Si nanoparticles. *Adv. Energy Mater.* 2016, **6**(2), 1501436.
87. Wang, J.; Karen Chen, Y.-C.; Yuan, Q.; Tkachuk, A.; Erdonmez, C.; Hornberger, B.; and Feser, M., Automated markerless full field hard x-ray microscopic tomography at sub-50 nm 3-dimension spatial resolution. *Appl. Phys. Lett.* 2012, **100**(14), 143107.
88. Wang, J. W.; Liu, X. H.; Mao, S. X.; and Huang, J. Y., Microstructural evolution of tin nanoparticles during in situ sodium insertion and extraction. *Nano Lett.* 2012, **12**(11), 5897–5902.
89. Zhu, J. and Deng, D., Amorphous bimetallic Co_3Sn_2 nanoalloys are better than crystalline counterparts for sodium storage. *J. Phys. Chem. C* 2015, **119**(37), 21323–21328.
90. Darwiche, A.; Sougrati, M. T.; Fraisse, B.; Stievano, L.; and Monconduit, L., Facile synthesis and long cycle life of SnSb as negative electrode material for Na-ion batteries. *Electrochem. Commun.* 2013, **32**, 18–21.
91. Liu, Y.; Zhang, N.; Jiao, L.; Tao, Z.; and Chen, J., Ultrasmall Sn nanoparticles embedded in carbon as high-performance anode for sodium-ion batteries. *Adv. Funct. Mater.* 2015, **25**(2), 214–220.
92. Wu, L.; Lu, H.; Xiao, L.; Ai, X.; Yang, H.; and Cao, Y., Electrochemical properties and morphological evolution of pitaya-like Sb@ C microspheres as high-performance anode for sodium ion batteries. *J. Mater. Chem. A* 2015, **3**(10), 5708–5713.
93. Dirican, M.; Lu, Y.; Ge, Y.; Yildiz, O.; and Zhang, X., Carbon-confined SnO_2-electrodeposited porous carbon nanofiber composite as high-capacity sodium-ion battery anode material. *ACS Appl. Mater. Interfaces* 2015, **7**(33), 18387–18396.
94. Hou, H.; Jing, M.; Yang, Y.; Zhang, Y.; Song, W.; Yang, X.; Chen, J.; Chen, Q.; and Ji, X., Antimony nanoparticles anchored on interconnected carbon nanofibers networks as advanced anode material for sodium-ion batteries. *J. Power Sources* 2015, **284**, 227–235.
95. Zhang, Y.; Wang, H.; Luo, Z.; Tan, H. T.; Li, B.; Sun, S.; Li, Z.; Zong, Y.; Xu, Z. J.; and Yang, Y., An air-stable densely packed phosphorene–graphene composite toward advanced lithium storage properties. *Adv. Energy Mater.* 2016, **6**(12), 1600453.
96. Gao, H.; Zhou, T.; Zheng, Y.; Liu, Y.; Chen, J.; Liu, H.; and Guo, Z., Integrated carbon/red phosphorus/graphene aerogel 3D architecture via

advanced vapor redistribution for high-energy sodium-ion batteries. *Adv. Energy Mater.* 2016, **6**(21), 1601037.

97. Nam, D.-H.; Hong, K.-S.; Lim, S.-J.; Kim, T.-H.; and Kwon, H.-S., Electrochemical properties of electrodeposited Sn anodes for Na-ion batteries. *J. Phys. Chem. C* 2014, **118**(35), 20086–20093.

98. Wachtler, M.; Wagner, M. R.; Schmied, M.; Winter, M.; and Besenhard, J. O., The effect of the binder morphology on the cycling stability of Li–alloy composite electrodes. *J. Electroanal. Chem.* 2001, **510**(1–2), 12–19.

99. Drofenik, J.; Gaberscek, M.; Dominko, R.; Poulsen, F. W.; Mogensen, M.; Pejovnik, S.; and Jamnik, J., Cellulose as a binding material in graphitic anodes for Li ion batteries: a performance and degradation study. *Electrochim. Acta* 2003, **48**(7), 883–889.

100. Liu, W.-R.; Yang, M.-H.; Wu, H.-C.; Chiao, S.; and Wu, N.-L., Enhanced cycle life of Si anode for Li-ion batteries by using modified elastomeric binder. *Electrochem. Solid State Lett.* 2005, **8**(2), A100–A103.

101. Koo, B.; Kim, H.; Cho, Y.; Lee, K. T.; Choi, N. S.; and Cho, J., A highly cross-linked polymeric binder for high-performance silicon negative electrodes in lithium ion batteries. *Angew. Chem.* 2012, **124**(35), 8892–8897.

102. Hochgatterer, N. S.; Schweiger, M. R.; Koller, S.; Raimann, P. R.; Wöhrle, T.; Wurm, C.; and Winter, M., Silicon/graphite composite electrodes for high-capacity anodes: influence of binder chemistry on cycling stability. *Electrochem. Solid State Lett.* 2008, **11**(5), A76–A80.

103. Yui, Y.; Hayashi, M.; Hayashi, K.; and Nakamura, J., Electrochemical properties of Sn-Co electrode with various kinds of binder materials for sodium ion batteries. *Solid State Ionics* 2016, **288**, 219–223.

104. Magasinski, A.; Zdyrko, B.; Kovalenko, I.; Hertzberg, B.; Burtovyy, R.; Huebner, C. F.; Fuller, T. F.; Luzinov, I.; and Yushin, G., Toward efficient binders for Li-ion battery Si-based anodes: Polyacrylic acid. *ACS Appl. Mater. Interfaces* 2010, **2**(11), 3004–3010.

105. Li, Z.; Huang, J.; Liaw, B. Y.; Metzler, V.; and Zhang, J., A review of lithium deposition in lithium-ion and lithium metal secondary batteries. *J. Power Sources* 2014, **254**, 168–182.

106. Kim, Y.; Park, Y.; Choi, A.; Choi, N. S.; Kim, J.; Lee, J.; Ryu, J. H.; Oh, S. M.; and Lee, K. T., An amorphous red phosphorus/carbon composite as a promising anode material for sodium ion batteries. *Adv. Mater.* 2013, **25**(22), 3045–3049.

107. Dai, K.; Zhao, H.; Wang, Z.; Song, X.; Battaglia, V.; and Liu, G., Toward high specific capacity and high cycling stability of pure tin nanoparticles with conductive polymer binder for sodium ion batteries. *J. Power Sources* 2014, **263**, 276–279.

108. Irisarri, E.; Ponrouch, A.; and Palacin, M., Hard carbon negative electrode materials for sodium-ion batteries. *J. Electrochem. Soc.* 2015, **162**(14), A2476.

109. Balogun, M.-S.; Luo, Y.; Qiu, W.; Liu, P.; and Tong, Y., A review of carbon materials and their composites with alloy metals for sodium ion battery anodes. *Carbon* 2016, **98**, 162–178.

110. Arora, P.; Doyle, M.; and White, R. E., Mathematical modeling of the lithium deposition overcharge reaction in lithium-ion batteries using carbon-based negative electrodes. *J. Electrochem. Soc.* 1999, **146**(10), 3543–3553.

111. Doyle, M.; Fuller, T. F.; and Newman, J., Modeling of galvanostatic charge and discharge of the lithium/polymer/insertion cell. *J. Electrochem. Soc.* 1993, **140**(6), 1526–1533.

112. Seh, Z. W.; Sun, J.; Sun, Y.; and Cui, Y., A highly reversible room-temperature sodium metal anode. *ACS Cent. Sci.* 2015, **1**(8), 449–455.

113. Gallagher, K. G.; Trask, S. E.; Bauer, C.; Woehrle, T.; Lux, S. F.; Tschech, M.; Lamp, P.; Polzin, B. J.; Ha, S.; and Long, B., Optimizing areal capacities through understanding the limitations of lithium-ion electrodes. *J. Electrochem. Soc.* 2016, **163**(2), A138–A149.

114. Cohn, A. P.; Muralidharan, N.; Carter, R.; Share, K.; and Pint, C. L., Anode-free sodium battery through in situ plating of sodium metal. *Nano Lett.* 2017, **17**(2), 1296–1301.

115. Irisarri, E.; Ponrouch, A.; and Palacin, M., Hard carbon negative electrode materials for sodium-ion batteries. *J. Electrochem. Soc.* 2015, **162**(14), A2476–A2482.

116. Bommier, C.; Surta, T. W.; Dolgos, M.; and Ji, X., New mechanistic insights on Na-ion storage in nongraphitizable carbon. *Nano Lett.* 2015, **15**(9), 5888–5892.

117. Tsai, P.-C.; Chung, S.-C.; Lin, S.-k.; and Yamada, A., Ab initio study of sodium intercalation into disordered carbon. *J. Mater. Chem. A* 2015, **3**(18), 9763–9768.

118. Malyi, O. I.; Sopiha, K.; Kulish, V. V.; Tan, T. L.; Manzhos, S.; and Persson, C., A computational study of Na behavior on graphene. *Appl. Surf. Sci.* 2015, **333**, 235–243.

119. Forsyth, M.; Girard, G.; Basile, A.; Hilder, M.; MacFarlane, D.; Chen, F.; and Howlett, P., Inorganic-organic ionic liquid electrolytes enabling high energy-density metal electrodes for energy storage. *Electrochim. Acta* 2016, **220**, 609–617.

120. Basile, A.; Makhlooghiazad, F.; Yunis, R.; MacFarlane, D. R.; Forsyth, M.; and Howlett, P. C., Extensive sodium metal plating and stripping in a highly concentrated inorganic–organic ionic liquid electrolyte through surface pretreatment. *ChemElectroChem* 2017, **4**(5), 986–991.

121. Basile, A.; Ferdousi, S. A.; Makhlooghiazad, F.; Yunis, R.; Hilder, M.; Forsyth, M.; and Howlett, P. C., Beneficial effect of added water on sodium metal cycling in super concentrated ionic liquid sodium electrolytes. *J. Power Sources* 2018, **379**, 344–349.

122. Luo, W.; Lin, C. F.; Zhao, O.; Noked, M.; Zhang, Y.; Rubloff, G. W.; and Hu, L., Ultrathin surface coating enables the stable sodium metal anode. *Adv. Energy Mater.* 2017, **7**(2), 1601526.

123. Jung, S. C.; Kim, H.-J.; Choi, J. W.; and Han, Y.-K., Sodium ion diffusion in Al2O3: A distinct perspective compared with lithium ion diffusion. *Nano Lett.* 2014, **14**(11), 6559–6563.

124. Zheng, X.; Fu, H.; Hu, C.; Xu, H.; Huang, Y.; Wen, J.; Sun, H.; Luo, W.; and Huang, Y., Toward a stable sodium metal anode in carbonate electrolyte: a compact, inorganic alloy interface. *J. Phys. Chem. Lett.* 2019, **10**(4), 707–714.

125. Wang, S.; Cai, W.; Sun, Z.; Huang, F.; Jie, Y.; Liu, Y.; Chen, Y.; Peng, B.; Cao, R.; and Zhang, G., Stable cycling of Na metal anodes in a carbonate electrolyte. *Chem. Commun.* 2019, **55**(95), 14375–14378.

126. Senguttuvan, P.; Rousse, G.; Seznec, V.; Tarascon, J.-M.; and Palacin, M. R., $Na_2Ti_3O_7$: Lowest voltage ever reported oxide insertion electrode for sodium ion batteries. *Chem. Mater.* 2011, **23**(18), 4109–4111.

127. Wang, Y.; Yu, X.; Xu, S.; Bai, J.; Xiao, R.; Hu, Y.-S.; Li, H.; Yang, X.-Q.; Chen, L.; and Huang, X., A zero-strain layered metal oxide as the negative electrode for long-life sodium-ion batteries. *Nat. Commun.* 2013, **4**, 2365.

128. Jin, T.; Li, H.; Zhu, K.; Wang, P.-F.; Liu, P.; and Jiao, L., Polyanion-type cathode materials for sodium-ion batteries. *Chem. Soc. Rev.* 2020, **49**(8), 2342–2377.

129. Jang, D. H.; Shin, Y. J.; and Oh, S. M., Dissolution of spinel oxides and capacity losses in 4 V li/li$_x$Mn$_2$O$_4$ cells. *J. Electrochem. Soc.* 1996, **143**(7), 2204–2211.

130. Komaba, S.; Kumagai, N.; and Kataoka, Y., Influence of manganese (II), cobalt (II), and nickel (II) additives in electrolyte on performance of graphite anode for lithium-ion batteries. *Electrochim. Acta* 2002, **47**(8), 1229–1239.

131. Pieczonka, N. P.; Liu, Z.; Lu, P.; Olson, K. L.; Moote, J.; Powell, B. R.; and Kim, J.-H., Understanding transition-metal dissolution behavior in $LiNi_{0.5}Mn_{1.5}O_4$ high-voltage spinel for lithium ion batteries. *J. Phys. Chem. C* 2013, **117**(31), 15947–15957.

132. Delacourt, C.; Kwong, A.; Liu, X.; Qiao, R.; Yang, W.; Lu, P.; Harris, S.; and Srinivasan, V., Effect of manganese contamination on the solid-electrolyte-interphase properties in Li-ion batteries. *J. Electrochem. Soc.* 2013, **160**(8), A1099–A1107.

133. Ochida, M.; Domi, Y.; Doi, T.; Tsubouchi, S.; Nakagawa, H.; Yamanaka, T.; Abe, T.; and Ogumi, Z., Influence of manganese dissolution on the degradation of surface films on edge plane graphite negative-electrodes in lithium-ion batteries. *J. Electrochem. Soc.* 2012, **159**(7), A961-A966.

134. Tsunekawa, H.; Tanimoto, S.; Marubayashi, R.; Fujita, M.; Kifune, K.; and Sano, M., Capacity fading of graphite electrodes due to the deposition of manganese ions on them in Li-ion batteries. *J. Electrochem. Soc.* 2002, **149**(10), A1326–A1331.

135. Shin, H.; Park, J.; Sastry, A. M.; and Lu, W., Degradation of the solid electrolyte interphase induced by the deposition of manganese ions. *J. Power Sources* 2015, **284**, 416–427.

136. Aurbach, D.; Markovsky, B.; Shechter, A.; Ein-Eli, Y.; and Cohen, H., A comparative study of synthetic graphite and Li electrodes in electrolyte solutions based on ethylene carbonate-dimethyl carbonate mixtures. *J. Electrochem. Soc.* 1996, **143**(12), 3809–3820.

137. Balbuena, P. B. and Wang, Y., *Lithium-Ion Batteries: Solid-Electrolyte Interphase*. Imperial College Press: 2004.

138. Sloop, S. E.; Kerr, J. B.; and Kinoshita, K., The role of Li-ion battery electrolyte reactivity in performance decline and self-discharge. *J. Power Sources* 2003, **119**, 330–337.

139. Bhide, A.; Hofmann, J.; Dürr, A. K.; Janek, J.; and Adelhelm, P., Electrochemical stability of non-aqueous electrolytes for sodium-ion batteries and their compatibility with $Na_{0.7}CoO_2$. *Phys. Chem. Chem. Phys.* 2014, **16**(5), 1987–1998.

140. Hunter, J. C., Preparation of a new crystal form of manganese dioxide: λ-MnO_2. *J. Solid State Chem.* 1981, **39**(2), 142–147.

141. Billy, E.; Laucournet, R.; Joulie, M.; Meyer, D.; Boulineau, A.; and De Vito, E. In *Dissolution Mechanism of $LiNi_{1/3}Mn_{1/3}Co_{1/3}O_2$ Positive Electrode Material from Used Lithium-Ion Batteries*, Meeting Abstracts, The Electrochemical Society: 2016; pp 1229–1229.

142. Terada, Y.; Nishiwaki, Y.; Nakai, I.; and Nishikawa, F., Study of Mn dissolution from $LiMn_2O_4$ spinel electrodes using in situ total reflection X-ray fluorescence analysis and fluorescence XAFS technique. *J. Power Sources* 2001, **97**, 420–422.

143. Choi, W. and Manthiram, A., Comparison of metal ion dissolutions from lithium ion battery cathodes. *J. Electrochem. Soc.* 2006, **153**(9), A1760–A1764.

144. Du Pasquier, A.; Blyr, A.; Courjal, P.; Larcher, D.; Amatucci, G.; Gerand, B.; and Tarascon, J. M., Mechanism for Limited 55 C Storage Performance of $Li_{1.05}Mn_{1.95}O_4$ Electrodes. *J. Electrochem. Soc.* 1999, **146**(2), 428–436.

145. Amatucci, G.; Schmutz, C.; Blyr, A.; Sigala, C.; Gozdz, A.; Larcher, D.; and Tarascon, J., Materials' effects on the elevated and room temperature performance of $CLiMn_2O_4$ Li-ion batteries. *J. Power Sources* 1997, **69**(1–2), 11–25.

146. Lu, J.; Zhan, C.; Wu, T.; Wen, J.; Lei, Y.; Kropf, A. J.; Wu, H.; Miller, D. J.; Elam, J. W.; and Sun, Y.-K., Effectively suppressing dissolution of manganese from spinel lithium manganate via a nanoscale surface-doping approach. *Nat. Commun.* 2014, **5**, 5693.

147. Nam, K. W.; Kim, S.; Yang, E.; Jung, Y.; Levi, E.; Aurbach, D.; and Choi, J. W., Critical role of crystal water for a layered cathode material in sodium ion batteries. *Chem. Mater.* 2015, **27**(10), 3721–3725.

148. Biensan, P.; Simon, B.; Peres, J.; De Guibert, A.; Broussely, M.; Bodet, J.; and Perton, F., On safety of lithium-ion cells. *J. Power Sources* 1999, **81**, 906–912.

149. MacNeil, D. and Dahn, J. R., The reactions of $Li_{0.5}CoO_2$ with nonaqueous solvents at elevated temperatures. *J. Electrochem. Soc.* 2002, **149**(7), A912–A919.

150. Xia, X. and Dahn, J., A study of the reactivity of de-intercalated P2-Na_xCoO_2 with non-aqueous solvent and electrolyte by accelerating rate calorimetry. *J. Electrochem. Soc.* 2012, **159**(5), A647–A650.

151. Ma, C.; Alvarado, J.; Xu, J.; Clément, R. J.; Kodur, M.; Tong, W.; Grey, C. P.; and Meng, Y. S., Exploring oxygen activity in the high energy P2-type $Na_{0.78}Ni_{0.23}Mn_{0.69}O_2$ cathode material for Na-ion batteries. *J. Am. Chem. Soc.* 2017, **139**(13), 4835–4845.

152. Xia, X. and Dahn, J., $NaCrO_2$ is a fundamentally safe positive electrode material for sodium-ion batteries with liquid electrolytes. *Electrochem. Solid State Lett.* 2011, **15**(1), A1–A4.

153. Guo, S.; Li, Q.; Liu, P.; Chen, M.; and Zhou, H., Environmentally stable interface of layered oxide cathodes for sodium-ion batteries. *Nat. Commun.* 2017, **8**(1), 135.

154. Zhao, J.; Zhao, L.; Dimov, N.; Okada, S.; and Nishida, T., Electrochemical and thermal properties of α-$NaFeO_2$ cathode for Na-ion batteries. *J. Electrochem. Soc.* 2013, **160**(5), A3077–A3081.

155. Chen, D.; Chen, H.; Maljuk, A.; Kulakov, A.; Zhang, H.; Lemmens, P.; and Lin, C., Single-crystal growth and investigation of Na_xCoO_2 and $Na_xCoO_2 \cdot yH_2O$. *Phys. Rev. B* 2004, **70**(2), 024506.

156. Hwang, S.; Lee, Y.; Jo, E.; Chung, K. Y.; Choi, W.; Kim, S. M.; and Chang, W., Investigation of thermal stability of P2–Na_xCoO_2 cathode materials for sodium ion batteries using real-time electron microscopy. *ACS Appl. Mater. Interfaces* 2017, **9**(22), 18883–18888.

157. Paulsen, J. and Dahn, J., Studies of the layered manganese bronzes, $Na_{2/3}$ $[Mn_{1-x}M_x]$ O_2 with M = Co, Ni, Li, and $Li_{2/3}$ $[Mn_{1-x}M_x]$ O_2 prepared by ion-exchange. *Solid State Ionics* 1999, **126**(1–2), 3–24.

158. Zhao, J.; Xu, J.; Lee, D. H.; Dimov, N.; Meng, Y. S.; and Okada, S., Electrochemical and thermal properties of P2-type $Na_{2/3}Fe_{1/3}Mn_{2/3}O_2$ for Na-ion batteries. *J. Power Sources* 2014, **264**, 235–239.

159. Hwang, J.-Y.; Oh, S.-M.; Myung, S.-T.; Chung, K. Y.; Belharouak, I.; and Sun, Y.-K., Radially aligned hierarchical columnar structure as a cathode material for high energy density sodium-ion batteries. *Nat. Commun.* 2015, **6**, 6865.

160. Hwang, J.-Y.; Yoon, C. S.; Belharouak, I.; and Sun, Y.-K., A comprehensive study of the role of transition metals in O3-type layered Na $[Ni_xCo_yMn_z]$ O2 (x= 1/3, 0.5, 0.6, and 0.8) cathodes for sodium-ion batteries. *J. Mater. Chem. A* 2016, **4**(46), 17952–17959.

161. Hwang, J.-Y.; Myung, S.-T.; Aurbach, D.; and Sun, Y.-K., Effect of nickel and iron on structural and electrochemical properties of O3 type layer cathode materials for sodium-ion batteries. *J. Power Sources* 2016, **324**, 106–112.

162. Oh, S.-M.; Myung, S.-T.; Yoon, C. S.; Lu, J.; Hassoun, J.; Scrosati, B.; Amine, K.; and Sun, Y.-K., Advanced Na $[Ni_{0.25}Fe_{0.5}Mn_{0.25}]$ O_2/C–Fe_3O_4 sodium-ion batteries using EMS electrolyte for energy storage. *Nano Lett.* 2014, **14**(3), 1620–1626.

163. Oh, S.-M.; Myung, S.-T.; Hwang, J.-Y.; Scrosati, B.; Amine, K.; and Sun, Y.-K., High capacity O3-type Na $[Li_{0.05}(Ni_{0.25}Fe_{0.25}Mn_{0.5})_{0.95}]O_2$ cathode for sodium ion batteries. *Chem. Mater.* 2014, **26**(21), 6165–6171.

164. Hasa, I.; Buchholz, D.; Passerini, S.; Scrosati, B.; and Hassoun, J., High performance $Na_{0.5}[Ni_{0.23}Fe_{0.13}Mn_{0.63}]O_2$ cathode for sodium-ion batteries. *Adv. Energy Mater.* 2014, **4**(15), 1400083.

165. Gao, R.-M.; Zheng, Z.-J.; Wang, P.-F.; Wang, C.-Y.; Ye, H.; and Cao, F.-F., Recent advances and prospects of layered transition metal oxide cathodes for sodium-ion batteries. *Energy Storage Mater.* 2020, **30**, 9–26.

166. Liu, H.; Zhou, H.; Chen, L.; Tang, Z.; and Yang, W., Electrochemical insertion/deinsertion of sodium on NaV_6O_{15} nanorods as cathode material of rechargeable sodium-based batteries. *J. Power Sources* 2011, **196**(2), 814–819.

167. Kim, S.; Ma, X.; Ong, S. P.; and Ceder, G., A comparison of destabilization mechanisms of the layered Na_xMO_2 and Li_xMO_2 compounds upon alkali de-intercalation. *Phys. Chem. Chem. Phys.* 2012, **14**(44), 15571–15578.

168. Didier, C.; Guignard, M.; Denage, C.; Szajwaj, O.; Ito, S.; Saadoune, I.; Darriet, J.; and Delmas, C., Electrochemical Na-deintercalation from $NaVO_2$. *Electrochem. Solid State Lett.* 2011, **14**(5), A75–A78.

169. Komaba, S.; Takei, C.; Nakayama, T.; Ogata, A.; and Yabuuchi, N., Electrochemical intercalation activity of layered $NaCrO_2$ vs. $LiCrO_2$. *Electrochem. Commun.* 2010, **12**(3), 355–358.

170. Li, Z.-Y.; Zhang, J.; Gao, R.; Zhang, H.; Hu, Z.; and Liu, X., Unveiling the role of Co in improving the high-rate capability and cycling performance of layered $Na_{0.7}Mn_{0.7}Ni_{0.3-x}Co_xO_2$ cathode materials for sodium-ion batteries. *ACS Appl. Mater. Interfaces* 2016, **8**(24), 15439–15448.

171. Maazaz, A.; Delmas, C.; and Hagenmuller, P., A study of the Na_xTiO_2 system by electrochemical deintercalation. *J. Inclusion Phenomena* 1983, **1**(1), 45–51.

172. Yabuuchi, N.; Yoshida, H.; and Komaba, S., Crystal structures and electrode performance of alpha-$NaFeO_2$ for rechargeable sodium batteries. *Electrochem.* 2012, **80**(10), 716–719.

173. Yu, H.; Guo, S.; Zhu, Y.; Ishida, M.; and Zhou, H., Novel titanium-based O3-type $NaTi_{0.5}Ni_{0.5}O_2$ as a cathode material for sodium ion batteries. *Chem. Commun.* 2014, **50**(4), 457–459.

174. Nanba, Y.; Iwao, T.; Boisse, B. M.; Zhao, W.; Hosono, E.; Asakura, D.; Niwa, H.; Kiuchi, H.; Miyawaki, J.; and Harada, Y., Redox potential paradox in Na_xMO_2 for sodium-ion battery cathodes. *Chem. Mater.* 2016, **28**(4), 1058–1065.

175. Vassilaras, P.; Toumar, A. J.; and Ceder, G., Electrochemical properties of $NaNi_{1/3}Co_{1/3}Fe_{1/3}O_2$ as a cathode material for Na-ion batteries. *Electrochem. Commun.* 2014, **38**, 79–81.

176. Li, X.; Wu, D.; Zhou, Y.-N.; Liu, L.; Yang, X.-Q.; and Ceder, G., O3-type Na $(Mn_{0.25}Fe_{0.25}Co_{0.25}Ni_{0.25})$ O_2: A quaternary layered cathode compound for rechargeable Na ion batteries. *Electrochem. Commun.* 2014, **49**, 51–54.

177. Han, M. H.; Gonzalo, E.; Casas-Cabanas, M.; and Rojo, T., Structural evolution and electrochemistry of monoclinic $NaNiO_2$ upon the first cycling process. *J. Power Sources* 2014, **258**, 266–271.

178. Wang, L.; Wang, J.; Zhang, X.; Ren, Y.; Zuo, P.; Yin, G.; and Wang, J., Unravelling the origin of irreversible capacity loss in $NaNiO_2$ for high voltage sodium ion batteries. *Nano Energy* 2017, **34**, 215–223.

179. Komaba, S.; Yabuuchi, N.; Nakayama, T.; Ogata, A.; Ishikawa, T.; and Nakai, I., Study on the reversible electrode reaction of $Na_{1-x}Ni_{0.5}Mn_{0.5}O_2$ for a rechargeable sodium-ion battery. *Inorg. Chem.* 2012, **51**(11), 6211–6220.

180. Yabuuchi, N.; Kajiyama, M.; Iwatate, J.; Nishikawa, H.; Hitomi, S.; Okuyama, R.; Usui, R.; Yamada, Y.; and Komaba, S., P2-type $Na_x[Fe_{1/2}Mn_{1/2}]$ O_2 made from earth-abundant elements for rechargeable Na batteries. *Nat. Mater.* 2012, **11**(6), 512.

181. Yabuuchi, N.; Yano, M.; Yoshida, H.; Kuze, S.; and Komaba, S., Synthesis and electrode performance of O3-type $NaFeO_2$-$NaNi_{1/2}Mn_{1/2}O_2$ solid solution for rechargeable sodium batteries. *J. Electrochem. Soc.* 2013, **160**(5), A3131–A3137.

182. Yoshida, H.; Yabuuchi, N.; and Komaba, S., $NaFe_{0.5}Co_{0.5}O_2$ as high energy and power positive electrode for Na-ion batteries. *Electrochem. Commun.* 2013, **34**, 60–63.

183. Talaie, E.; Duffort, V.; Smith, H. L.; Fultz, B.; and Nazar, L. F., Structure of the high voltage phase of layered P2-$Na_{2/3-z}[Mn_{1/2}Fe_{1/2}]$ O_2 and the positive effect of Ni substitution on its stability. *Energy Environ. Sci.* 2015, **8**(8), 2512–2523.

184. Yabuuchi, N. and Komaba, S., Recent research progress on iron-and manganese-based positive electrode materials for rechargeable sodium batteries. *Sci. Technol. Adv. Mater.* 2014, **15**(4), 043501.

185. Casas-Cabanas, M.; Roddatis, V. V.; Saurel, D.; Kubiak, P.; Carretero-González, J.; Palomares, V.; Serras, P.; and Rojo, T., Crystal chemistry of Na insertion/deinsertion in $FePO_4$–$NaFePO_4$. *J. Mater. Chem.* 2012, **22**(34), 17421–17423.

186. Su, J.; Pei, Y.; Yang, Z.; and Wang, X., First-principles investigation on the structural, electronic properties and diffusion barriers of Mg/Al doped $NaCoO_2$ as the cathode material of rechargeable sodium batteries. *RSC Adv.* 2015, **5**(35), 27229–27234.

187. Matsui, M.; Mizukoshi, F.; and Imanishi, N., Improved cycling performance of P2-type layered sodium cobalt oxide by calcium substitution. *J. Power Sources* 2015, **280**, 205–209.

188. Caballero, A.; Hernan, L.; Morales, J.; Sanchez, L.; Pena, J. S.; and Aranda, M., Synthesis and characterization of high-temperature hexagonal P2-$Na_{0.6}MnO_2$ and its electrochemical behaviour as cathode in sodium cells. *J. Mater. Chem.* 2002, **12**(4), 1142–1147.

189. Lu, Z. and Dahn, J., Intercalation of water in P2, T2 and O2 structure $A_z[Co_xNi_{1/3-x}Mn_{2/3}]O_2$. *Chem. Mater.* 2001, **13**(4), 1252–1257.

190. Buchholz, D.; Chagas, L. G.; Vaalma, C.; Wu, L.; and Passerini, S., Water sensitivity of layered P2/P3-$Na_xNi_{0.22}Co_{0.11}Mn_{0.66}O_2$ cathode material. *J. Mater. Chem. A* 2014, **2**(33), 13415–13421.

191. Duffort, V.; Talaie, E.; Black, R.; and Nazar, L. F., Uptake of CO_2 in layered P2-$Na_{0.67}Mn_{0.5}Fe_{0.5}O_2$: Insertion of carbonate anions. *Chem. Mater.* 2015, **27**(7), 2515–2524.

192. Yao, H.-R.; Wang, P.-F.; Gong, Y.; Zhang, J.; Yu, X.; Gu, L.; OuYang, C.; Yin, Y.-X.; Hu, E.; and Yang, X.-Q., Designing air-stable O3-type cathode materials by combined structure modulation for Na-ion batteries. *J. Am. Chem. Soc.* 2017, **139**(25), 8440–8443.

193. Buchholz, D.; Chagas, L. G.; Vaalma, C.; Wu, L.; and Passerini, S., Water sensitivity of layered P2/P3-$Na_xNi_{0.22}Co_{0.11}Mn_{0.66}O_2$ cathode material. *J. Mater. Chem. A* 2014, **2**(33), 13415–13421.

194. Roger, M.; Morris, D.; Tennant, D.; Gutmann, M.; Goff, J.; Hoffmann, J.-U.; Feyerherm, R.; Dudzik, E.; Prabhakaran, D.; and Boothroyd, A., Patterning of sodium ions and the control of electrons in sodium cobaltate. *Nature* 2007, **445**(7128), 631.

195. Takada, K.; Sakurai, H.; Takayama-Muromachi, E.; Izumi, F.; Dilanian, R. A.; and Sasaki, T., Superconductivity in two-dimensional CoO_2 layers. *Nature* 2003, **422**(6927), 53.

196. Zhang, P.; Capaz, R. B.; Cohen, M. L.; and Louie, S. G., Theory of sodium ordering in Na_xCoO_2. *Phys. Rev. B* 2005, **71**(15), 153102.

197. Lu, Z. and Dahn, J., Intercalation of water in P2, T2 and O2 structure $A_z[Co_xNi_{1/3-x}Mn_{2/3}]O_2$. *Chem. Mater.* 2001, **13**(4), 1252–1257.

198. Lu, Z.; Donaberger, R.; and Dahn, J., Superlattice ordering of Mn, Ni, and Co in layered alkali transition metal oxides with P2, P3, and O3 structures. *Chem. Mater.* 2000, **12**(12), 3583–3590.

199. Sathiya, M.; Hemalatha, K.; Ramesha, K.; Tarascon, J.-M.; and Prakash, A., Synthesis, structure, and electrochemical properties of the layered sodium insertion cathode material: $NaNi_{1/3}Mn_{1/3}Co_{1/3}O_2$. *Chem. Mater.* 2012, **24**(10), 1846–1853.

200. Monyoncho, E. and Bissessur, R., Unique properties of α-$NaFeO_2$: De-intercalation of sodium via hydrolysis and the intercalation of guest molecules into the extract solution. *Mater. Res. Bull.* 2013, **48**(7), 2678–2686.

201. Mu, L.; Xu, S.; Li, Y.; Hu, Y. S.; Li, H.; Chen, L.; and Huang, X., Prototype sodium-ion batteries using an air-stable and Co/Ni-Free O3-layered metal oxide cathode. *Adv. Mater.* 2015, **27**(43), 6928–6933.

202. Wang, Y.; Xiao, R.; Hu, Y.-S.; Avdeev, M.; and Chen, L., P2-$Na_{0.6}[Cr_{0.6}Ti_{0.4}]O_2$ cation-disordered electrode for high-rate symmetric rechargeable sodium-ion batteries. *Nat. Commun.* 2015, **6**, 6954.

Chapter 10

Atomistic Modeling and Analysis of Electrolyte Properties for Sodium-Ion Batteries

Argyrios V. Karatrantos,[a,b] Emilia Olsson,[a] and Qiong Cai[a]
[a]*Department of Chemical and Process Engineering, Faculty of Engineering and Physical Sciences, University of Surrey, Guildford GU2 7XH, United Kingdom*
[b]*Luxembourg Institute of Science and Technology, 5, avenue des Hauts-Fourneaux, L-4362 Esch-sur-Alzette, Luxembourg*
q.cai@surrey.ac.uk

10.1 Introduction

Due to the high abundance and low cost of sodium (Na), Na-ion batteries (NIBs) have emerged as a promising candidate for medium- and large-scale stationary energy storage [1–4]. To date, NIB research has focused on understanding and optimizing the electrode materials, electrolytes, and their interfaces [5–10]. The importance of electrolyte in battery performance (e.g., operation voltage, lifetime, practically accessible capacity, rate capability, safety) has

Handbook of Sodium-Ion Batteries: Materials and Characterization
Edited by Rohit R. Gaddam and X. S. (George) Zhao

ISBN 978-981-4968-15-7 (Hardcover), 978-1-003-30874-4 (eBook)
www.jennystanford.com

been gaining wide recognition. This has sparked increasing volume of research in searching for electrolyte systems that enable wider operation voltage, longer lifetime, higher accessible capacity, rate capability, and safety. In the search for optimal NIB electrolyte systems, atomic-scale computational studies are highly valuable, alongside the experimental research. NIB electrolyte systems include liquid, polymeric, and ceramic solid candidates. Due to the different nature of the candidate electrolyte systems, different modeling techniques must be applied, which are reviewed in Section 10.2.

The physical and chemical properties of the electrolyte systems are important measures as they determine the battery performance and, therefore, are discussed in this chapter. Both for liquid and solid electrolytes (SEs), these materials must have high ionic conductivity and stability [11–13]. For liquid electrolytes (LEs), salt dissociation, Na solvation, and electrochemical stability are key properties, whereas the solid ceramic electrolytes need to have high ionic conductivity, thermal stability, volume expansion compatible with both cathode and anode, and form stable solid–solid interfaces with the electrodes that do not limit Na interfacial diffusion [14, 15]. In this chapter, we review the NIB electrolyte simulation and computational modeling studies, aiming to bring fundamental insights into the current understanding of the electrolyte and their implications on battery design.

10.2 Computational Methods

In this chapter, we will focus on atomic-scale studies of NIB electrolytes. The different electrochemistry and properties of the NIB electrolytes require the employment of different computational methods. In common for all NIB electrolytes is the need to computationally evaluate and predict the ionic conductivity and Na^+ ion diffusion mechanisms. Two major molecular modeling techniques have been applied: quantum-mechanics-based density functional theory (DFT) methods and statistical-mechanics-based molecular dynamics (MD) methods. Recently, *ab initio* molecular dynamics (AIMD) has also been employed to model the Na^+ diffusion, combining DFT and MD [16–20]. The main differences between these methods are the complexity and system size that can be evaluated.

DFT simulations are useful for studying the solvation structure and electrochemical stability of LEs, as well as phase stability, defect formation, and dopant structures of SEs [21–25], typically employing small molecular models [15, 26–30]. MD and AIMD simulations are useful for modeling the dynamics, transport properties, and ionic diffusion of LEs and SEs, typically allowing for more complex models containing bigger number of atoms/molecules to be evaluated at longer timescales and higher temperatures [31–33]. In the next sections, we will first briefly introduce the different computational methods outlined earlier and then discuss what properties can be obtained. More detailed and comprehensive overviews of these methods are available elsewhere [9, 19, 23, 34–45].

10.2.1 DFT Simulations of Electrolyte Properties

At the smallest length and time scale, DFT has been used to solve the time-independent Schrödinger equation to obtain a complete quantum mechanical description of practical model systems. DFT simulations have been applied both to LEs and SEs. The total energy of a given system (E) in the Kohn–Sham DFT formalism can be calculated from Eq. (10.1) [9, 39–41].

$$E_i\psi_i(\boldsymbol{r}) = \left(-\nabla^2 + V_{\mathrm{H}}\left[n(\boldsymbol{r})\right] + V_{\mathrm{N}}(\boldsymbol{r}) + V_{\mathrm{XC}}\left[n(\boldsymbol{r})\right]\right)\psi_i(\boldsymbol{r}) \tag{10.1}$$

Here $-\nabla^2$ is the electronic kinetic energy, V_{H} is the Hartree term describing the Coulomb repulsion between electrons, V_{N} is the energy relating to all the nuclei, and V_{XC} is the exchange-correlation (XC) energy. V_{XC} is a correction to the kinetic energy and the non-classical interelectronic repulsion due to exchange and correlation and accounts for the Pauli exclusion and electron correlation effects [41, 43, 46]. The exact form of V_{XC} cannot be directly obtained, which has led to a number of approximations having been developed [34–37, 40–45]. The accuracy of DFT simulations is directly dependent on the choice of XC functionals, as the XC functionals introduce inherent inaccuracies that one must be aware of when conducting these calculations [28, 34–37, 40–45]. Typically, simulations using different functionals are validated against experimental data. A variety of different functionals (based on different methods to account for XC such as LDA, GGA, and hybrids) have to date been applied, with PBE, M06-2X, TPSSh, and B3LYP functionals being among the most popular [19, 28, 47–52]. For polymer SEs, Hartree–Fock methods have also been employed, which considers every electron of the studied system as interacting with a mean-field of all the other electrons. The consequence of this is that only the motion of electrons with parallel spins is correlated, making the total energy derived from Hartree–Fock simulations equal to the exact total energy of a system minus the correlation energy [41, 53, 54]. Furthermore, there are an inherent electron self-interaction error within DFT, which means that unphysical interactions between an electron and itself are not canceled. This is especially critical when modeling materials containing d- and/or f-electrons [47, 48]. Hence, for semiconductors and strongly correlated electronic systems used for inorganic SE materials, correction schemes need to be used in conjunction with GGA functionals. One of the most popular methods for doing so in ceramic materials is DFT+U (sometimes also referred to as GGA+U or PBE+U) [55, 56].

For LEs, DFT can be used to study the equilibrium solvation structure of the solvent molecules and salts in LEs, solvation energy, reduction and oxidation potentials, and the highest occupied molecular orbital (HOMO) and the lowest unoccupied molecular orbital (LUMO) states for the electrolyte constituents [15, 27–30].

The solvation strength (ΔE_b, also referred to as binding energy) between Na^+ and solvent molecules in LEs can be calculated from DFT simulations according to Eq. (10.2) [57].

$$\Delta E_{\text{b}} = E_{\text{complex}} - [nE_{\text{solvent}} + E_{\text{Na}^+}] \tag{10.2}$$

Here E_{complex} is the calculated energy of the solvent Na^+ complex, n is the number of solvent molecules in the complex, E_{solvent} is the energy of a solvent molecule, and E_{Na^+} is the energy of the Na ion. Following the convention of Eq. (10.2), negative ΔE_b indicates an energetically stable complex; the stronger the solvation of Na, the more negative it is.

To evaluate the electrochemical stability window, the Gibbs free energy of reduction ($\Delta G^s_{\text{redox}}$) and oxidation ($\Delta G^s_{\text{oxidation}}$) of a complex M in its solvated state can be calculated according to the following equations [57–61]:

$$\Delta G^s_{\text{redox}} = G[M^{n-}(s)] - G[M(s)] - nG[e^- s] \tag{10.3}$$

$$\Delta G^s_{\text{oxidation}} = G[M^{n+}(s)] + nG[e^- s] - G[M(s)] \tag{10.4}$$

It has been shown that the reduction potential of the organic molecules is correlated to DFT-calculated electron affinity, and that the oxidation potential, in turn, is correlated to their ionization potentials [60, 62, 63]. The HOMO and LUMO levels of an isolated single molecule of an electrolyte component (e.g., solvent) are often used to evaluate the electrochemical stability of the electrolyte component. To have more realistic and accurate evaluation of the electrochemical stability of an electrolyte system, the reduction and oxidation potentials of a molecular cluster containing all the electrolyte components (e.g., solvent molecules and salts) should be used.

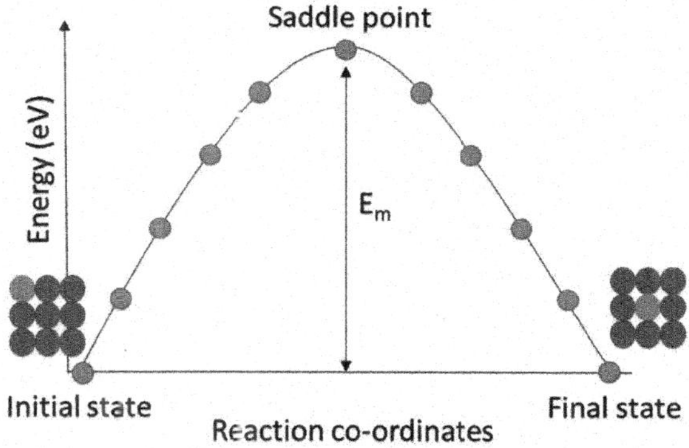

Figure 10.1 Schematic representation of the calculation of migration energy (E_m) from NEB simulations. E_m is obtained as the energy difference between the initial state total energy and the saddle point total energy.

For SEs, DFT can be used to derive migration barriers by comparing the total energy of the starting/end points with the total energy of the transition state (Fig. 10.1). The results have been shown to be comparable to experimentally measured Na migration barriers and diffusion coefficients, obtainable from, for example, galvanostatic intermittent titration technique (GITT), or muon spin rotation spectroscopy [10, 64–67]. The Na migration barriers from DFT simulations can be calculated using the nudged elastic band (NEB) method [68]. The NEB simulations are dependent on the knowledge and optimization of the start and end point structures, and initial guesses of migration paths between these points. From this initial setup, the NEB simulation can identify the migration energy barrier (E_m), transition state, energy profile of the pathway, and the lowest energy migration pathway. NEB-derived migration barriers being based on the difference in DFT total energies between the different states (Fig. 10.1) do not typically consider temperature or pressure [69]. Furthermore, while metal migration barriers from DFT simulations have been found reliable and in good agreement with experiment, they are computationally very expensive and challenging. Especially for complex migration mechanisms in liquids and disordered ceramics, a large number of migration pathways have to be sampled, and in consequence a large number of NEB images included.

10.2.2 MD Simulations of Electrolyte Properties

MD can be used to study the time evolution of interacting particles by numerically integrating the interacting particles equations of motion and has been employed to study structural and dynamical properties of both LEs and SEs [70, 71]. Classical MD methods are based on Newtonian dynamics and can be used to model ionic diffusion and does not account for electrons explicitly, making this a less computationally expensive method than DFT and AIMD [72]. Atomic interactions are instead treated by potential energy functions, which are dependent on force-field models and interatomic potentials to account for the intra- and inter-molecular interactions [72]. This leads to the possibility to model materials at longer time scales, larger systems, and at higher temperatures than DFT for less computational expense [9, 15, 73, 74]. The force fields or interatomic potentials are typically empirically fitted based on experimental or DFT data. Hence, the accuracy of the interatomic potential is reliant on high-quality data of properties such as structural properties (bond lengths, lattice, parameters, bond angles, dihedral angles), density, mechanical properties, or heat of vaporization. These models are, hence, for the most part nonreactive and do not simulate processes involving bond breaking. Interatomic potential methods are computationally inexpensive compared to DFT and allow for simulation cells consisting of between thousands and millions of atoms to be modeled.

When the modeling of reactivity (such as electronic structure affected migration, or salt dissociation) is required, AIMD, sometimes also referred to as first principles MD (FPMD), becomes more suitable, particularly for systems where prior knowledge of force fields is not available and reactions become important. For these simulations, the interatomic potentials in MD are replaced by DFT steps to calculate the interatomic interactions, and the diffusion coefficients are obtained using the same equations as detailed for MD earlier (essentially solving the electronic Schrödinger equation at each time step and propagating the nuclei as classical objects along the trajectory). In this method, the forces are computed on the fly using first principles electronic structure calculations, and as the electronic structure is treated explicitly, many-body forces, electronic polarization, or even bond-breaking events can be described within the accuracy of the electronic structure representation. This does, however, make AIMD simulations significantly more computationally expensive than MD simulations, and considerations of system size, trajectory length, and real-time calculation must be made. These simulations are more computationally expensive than the classical MD simulations but have been used to great success in the study and optimization of battery materials and are implemented in DFT codes [10, 15, 16, 75–82].

MD and AIMD simulations are used to model the dynamic properties of electrolytes such as diffusivity and ionic conductivity [19, 69, 83]. The dynamics (diffusion) of ions can be calculated from the mean square displacement (MSD) [70, 84] obtainable for MD and AIMD trajectories:

$$MSD = \langle r^2(t) \rangle = 6D_i t + B_i$$

$$D_i = \frac{1}{6t} \lim_{t \to \infty} \langle |r_i(t) - r_i(0)|^2 \rangle \quad (10.5)$$

where t is simulation time, D_i is the self-diffusion coefficient for species i, and B_i is a thermal factor associated with atomic vibrations. D_i can be obtained by plotting the MSD versus t and can consequently be used to calculate the diffusion activation energy (E_a, similar to the migration barrier obtained from DFT NEB simulations) from Eq. (10.6) [85–88].

$$D_i = D_0 \exp\left({}^{-E_a}/_{kT} \right) \Leftrightarrow \ln D_i = \ln D_0 - \frac{E_a}{kT} \quad (10.6)$$

Here, D_0 is a temperature-independent pre-exponential term, T is the temperature, and k is Boltzmann constant. E_a is commonly obtained by Arrhenius plots from the gradient of $\ln D_i$ versus $1/T$. Finally, the ionic conduction of species i (σ_i) can be obtained through the Nernst–Einstein relationship assuming that the ions move independently of each other and is related to the self-diffusion coefficients [89]:

$$\sigma_i = \frac{1}{6t} \lim_{t \to \infty} \frac{e^2}{6VK_BT} \sum_i^n z_i^2 \langle |r_i(t) - r_i(0)|^2 \rangle$$
$$= \frac{e^2}{VK_BT}\left(\eta^+ D_+ - \eta^- D_-\right) = \frac{Nq^2 D_i}{fk_BT} \quad (10.7)$$

where N is the total number of mobile i ions, q is the species charge, k_B is Boltzmann constant, f is the Haven ratio, and T is temperature [90]. However, in reality the motion of anions and cations is correlated, and the Nernst–Einstein relation provides an upper boundary prediction for conductivity calculation. Based on the diffusion simulations, one can also calculate the Na^+ transport number (t^+), comparing the diffusivity of the cation (D_+) and anion (D_-) salt constituents [89].

$$t^+ = \frac{D_+}{D_+ + D_-} \quad (10.8)$$

In order to obtain information about the structure, the radial distribution function (RDF) $g(r)$ is calculated. In general, the $g_{AB}(r)$ between two particles A and B is defined by [91]:

$$g_{AB}(r) = \frac{V}{4\pi r^2} \sum_i^{N_A} \sum_j^{N_B} P(r) \quad (10.9)$$

10.2.3 Available Software

As described earlier, the computational modeling field is diverse, which is further demonstrated in the range of simulation codes available. Popular DFT codes include VASP [92] (also suitable for AIMD simulations), CP2K [93] (also suitable for AIMD simulations), Gaussian [94], CRYSTAL [95], CASTEP [96], Wien2K [97], and Quantum Espresso [98], whereas MD simulations are commonly conducted using DL_POLY [99], GROMACS [100], NAMD [101], or LAMMPS [102]. The choice of code depends on the material being studied, required information, and code capabilities (such as functional, interatomic potential, periodicity, etc.), which can be obtained from each code's website as listed in references.

10.3 Liquid Electrolytes

The NIB LEs normally consist of three key components to achieve desirable properties and performances: salts, solvents, and additives. The choices of salt [103–105], solvent [106], and additives [11] are instrumental for the performance of NIBs and need to provide high ionic conductivity, long cycle life, high energy density, greatly reducing side reaction and improving the electrochemical stability [107]. Based on this, the NIB LEs should have (1) high polarity (a high dielectric constant is necessary for salt ion solvation and the ion pairing limit), (2) low viscosity to improve the Na^+ mobility, (3) chemical stability (to remain inert on charged surfaces of the cathode and the anode during cell operation), (4) the ability to remain liquid over a broad range of temperatures, and (5) safety (nontoxicity, and economical) [108–110].

NIB salt candidates include $NaPF_6$, $NaClO_4$, NaTFSI, NaFSI, and $NaCF_3SO_3$ [11]. Both $NaPF_6$ and $NaClO_4$ have been used extensively as salts in NIBs, even though $NaClO_4$ has intrinsic safety issues [108]. Regarding the solvent, either cyclic or linear solvent molecules have been used. These solvents are typically ethylene carbonate (EC), vinylene carbonate (VC), propylene carbonate (PC), butylene carbonate (BC), dimethyl carbonate (DMC), ethyl methyl carbonate (EMC), diethyl carbonate (DEC), monoglyme, diglyme, triglyme, or mixtures of these. EC and PC have been shown to be essential for LEs due to their low viscosity (1.9 mPas and 2.5 mPas, respectively), which leads to rapid ionic conduction, and large dielectric constants (89.6 for EC and 66.1 for PC), which is beneficial for facile sodium salt dissociation [27]. Additives including VC, and fluoroethylene carbonate (FEC) can be applied to overcome some shortcomings with solvents [16, 111]. In particular, the additive's task is to improve the chemical and electrochemical stability window or even to widen the operation window. These can modify the solid electrolyte interface (SEI) or act as flame retardants, viscosity reducers, or radical scavengers [11]. For example, the reduction mechanism of EC was shown to be dependent on the choice of additive molecule, in the DFT study by Kumar et al. [112]. This study showed that the additive VC molecules have lower decomposition barriers than EC and will, hence, decompose before EC if added to the LE [112]. Furthermore, both FEC and VC increased the EC decomposition barrier, changing the most energetically favorable reduction product of EC by forming a dimer with the additive molecules. Following these simulations, the choice of additive does directly affect the decomposition products of LEs, which can be utilized for SEI design [112]. However, the amount of additives should be kept low (less than 5 wt%) to limit side reactions. One issue of the LEs is safety, which can be compensated by the use of ionic liquids as plasticizers or (as discussed in Sections 10.4 and 10.5) SEs [106, 113].

10.3.1 Electrochemical Stability

For ideal LE stability under NIB operating conditions, the LE components should not react with the electrode surfaces, i.e., the electrochemical interfacial side reactions should be minimized. In practical terms, this means that the cell voltage needs to remain within the electrochemical window of which the LE is stable [60]. Should the cell voltage be outside this electrochemical window, reduction and oxidation of the LE can occur, which can lead to NIB breakdown. The DFT-calculated HOMO and LUMO states of the isolated solvent molecule and salt can be used to estimate electrolyte stability, both within a given solvent salt mixture and with the electrode materials [6, 24, 114]. DFT simulations of the reduction and oxidation potentials of the LE components give vital insight not only to the isolated electrolyte properties, but also the electrode–electrolyte interphases. For side reactions to be avoided at the electrode interfaces, the LUMO level of the LE should be lower than the anode reduction energy, and the LE HOMO level should be higher than the cathode oxidation energy [24]. DFT simulations using M06-2X hybrid functional in Gaussian have shown that the HOMO–LUMO energy gaps in EC, PC, BC, VC, DMC, EMC, and DEC increase when they are interacting with Na^+ ions (Fig. 10.2a), with the LUMO and HOMO levels shifted downward [62]. When EC is co-solvated with PC, DMC, or EMC, the HOMO–LUMO gap decreases due to the beneficial coordination effect of the EC molecule and Na^+

ions (which is further observed in the strong solvation energies for these systems as discussed in the following paragraphs) [27]. These results were also seen in DFT simulations (using the same M06-2X functional in Gaussian) of the reduction and oxidation potentials for both diglyme and carbonate solvents with Na^+, which showed that the Na^+–EC complex has a higher reduction potential (1.21 V) than pure EC solvent (1.16 V) without the cation [61]. All the investigated sodium complexes with EC, DMC, and diglyme, respectively, showed higher oxidation potentials than 4.7 V, indicating that no oxidative decomposition at the cathode is expected for these electrolytes [61]. Pure DMC had a reduction potential of 0.56, whereas when forming a complex with Na, the reduction potential increased to 1.85 V, indicating that DMC would start to decompose at higher voltages, and being less stable than EC [61]. The reduction potentials of diglyme complexes are much lower at −0.54 V for Na^+ with one diglyme molecule, −1.05 V for Na^+ with two diglyme molecules, and −1.07 V for two diglyme molecules with Na^+ and PF_6^- [61]. These negative reduction potentials show that diglyme is stable until voltages below 0 V, which confirms the experimental observations of less diglyme decomposition products in SEIs as compared to the carbonate based LE SEIs [61, 115]. In reality, the usefulness of these measures in practical NIB cells are still being discussed, due to the prevalence of side reactions at the LE electrode interphases and impurities present in the LE and at the electrode surfaces [116, 117].

10.3.2 Solvation Structures

The solvation energy combined with the cation–anion interaction energy is a powerful measure for the solvation and local cation coordination. Between the Na^+ ions and the solvent molecules, there are four types of interactions that have a direct influence on the electrolyte solvation energy: electrostatic, exchange, polarization, and dispersion [12, 57, 118]. The largest contribution to the cluster interaction energy comes from the electrostatic interaction, thereafter polarization, dispersion, and exchange [57]. Based on these, the solvation structures of a variety of solvent Na^+ structures have been evaluated from first principles simulations. DFT simulations showed that the entropy of LEs decreased when Na^+ was introduced to the carbonate solvent molecules, decreasing the

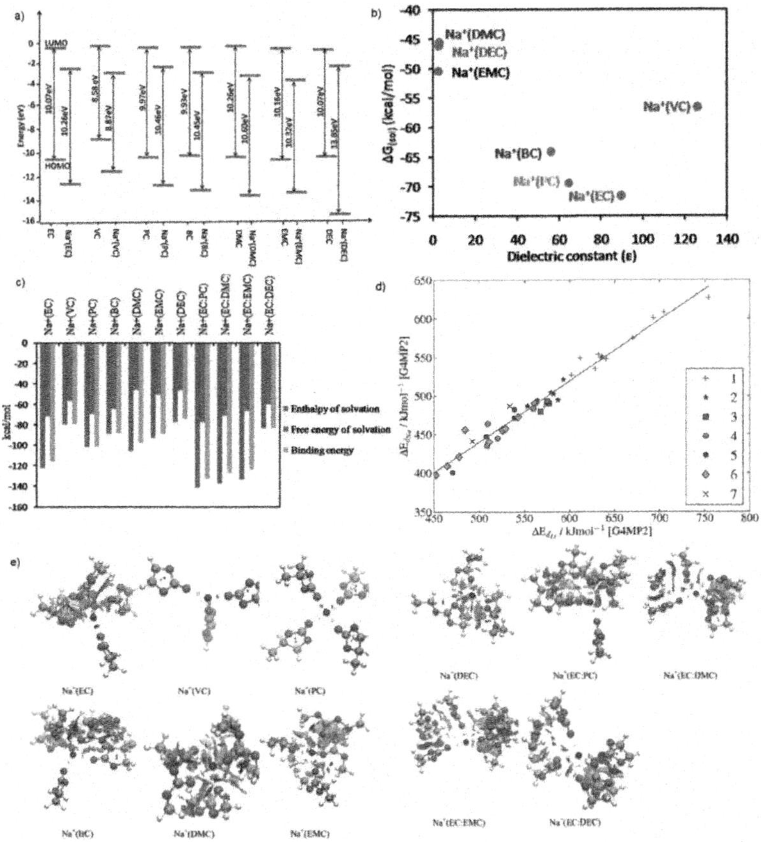

Figure 10.2 (a) DFT-calculated HOMO and LUMO energies of carbonate solvents, both in isolation and in Na^+ complexes. Figure from Ref. [62]. (b) shows the free solvation energy of Na^+ in carbonate solvents as a function of the dielectric constant of the solvent, with (c) showing the enthalpy of solvation (in blue), free energy of solvation (in red), and binding energy (in green) of the Na^+ solvent complexes from DFT simulations. Figures from Ref. [62] (d) $\Delta E_{d,Na^+}$ versus $\Delta E_{d,Li^+}$ calculated with G4MP2. The numbering corresponds to the grouping of the anions (seven groups, including standards for LIBs, such as PF_6^-, imides such as TFSI, boron-based anions, phosphorous anions with bidentate ligands such as TOP, 2 heterocyclic anions, other Huckel type, etc.). Figure from Ref. [119]. Finally, in (e) the optimized structures of the Na^+ solvent complexes from DFT simulations (M06-2X functional) are shown, with blue, green, and red illustrating the strong attractive, weak van der Waals, and strong repulsion interactions, respectively. Figure from Ref. [62].

translational degree of freedom for the solvent molecules [62]. In the same work, the highest Na solvent binding energies were found for EC-containing electrolytes (Fig. 10.2b,c) [62]. The calculations of the Gibbs free solvation energy further confirmed that EC forms more stable solvation structures with Na^+ than the other common LE solvents PC, BC, VC, EMC, DMC, and DEC. The solvation structures could be further improved by mixing EC with the other organic solvents, with the strongest binding and solvation energies obtained for the EC:PC mixture [62].

The Na^+ ion–anion interaction was investigated using DFT methods benchmarked to the accurate G4(MP2) calculations (Gaussian-4 theory series in the effects of basis set extension are obtained from calculations at second-order Møller–Plesset perturbation theory) for 53 different anions [119] (traditional salt anions such as TFSI and PF_6^- and recently developed anions, some designed for LIB electrolytes) in a mixture of EC and DMC solvents. Replacing Li^+ with Na^+ affects the ion-pair dissociation energy and decreases approximately 15–20% [119]. In particular, the ion-pair dissociation energy ($\Delta E_{d,Na^+}$) was calculated using the expression

$$\Delta E_{d,Na^+} = E_{Na^+} + E_{An^-} - E_{NaAn} \tag{10.10}$$

where $\Delta E_{d,Na^+}$ is the dissociation energy of Na^+ cation and An^- is the anion (Fig. 10.2d) [119]. The largest reduction in binding energy was observed for the simple F^-, although it is still strongly bound, and the second largest reduction for $B(CH_3)_4^-$. The N_5C10^- and TOP anions are the only anions with $\Delta E_{d,Na^+} < 400$ kJ/mol [119]. DFT simulations of the solvation structures further showed that the strongest attractive interactions were obtained for Na in EC mixtures (Fig. 10.2e) [62]. Thus, such anions would be interesting candidates to be used in NIB studies.

Experimentally, it was shown using nuclear magnetic resonance (NMR) that Na^+ ions have a higher tendency to form contact pairs in linear carbonate solvents (such as EMC, DMC, DEC) than Li^+ [120]. Using AIMD simulations, Pham et al. [121] could show that Na^+ exhibits a more disordered and flexible solvation structure (as a result of having weaker solvation energy) than Li^+ with EC molecules (Fig. 10.3). In particular, the coordination number of Na^+ was 5.7, which was comprised mainly from carbonyl oxygens but also involved a fraction of ether oxygens [121].

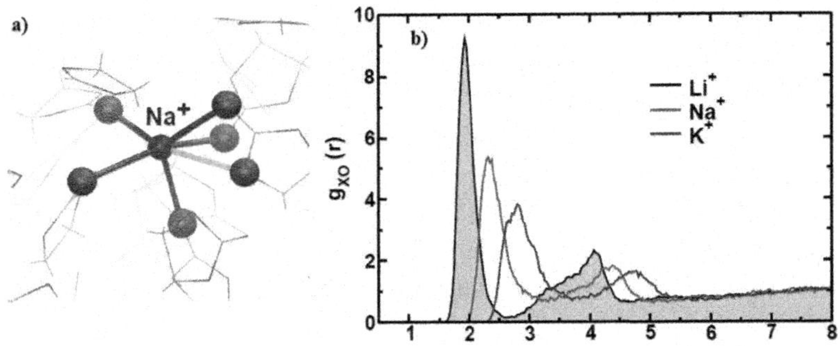

Figure 10.3 (a) The solvation structure of Na^+ can be described as tetrahedral and distorted trigonal bipyramid (or square pyramidal), (b) metal ion oxygen radial distribution functions, $g_{XO}(r)$, for solvated ions in EC, obtained from AIMD simulations. Black, red, and blue lines indicate results for X = Li^+, Na^+, K^+, respectively [121].

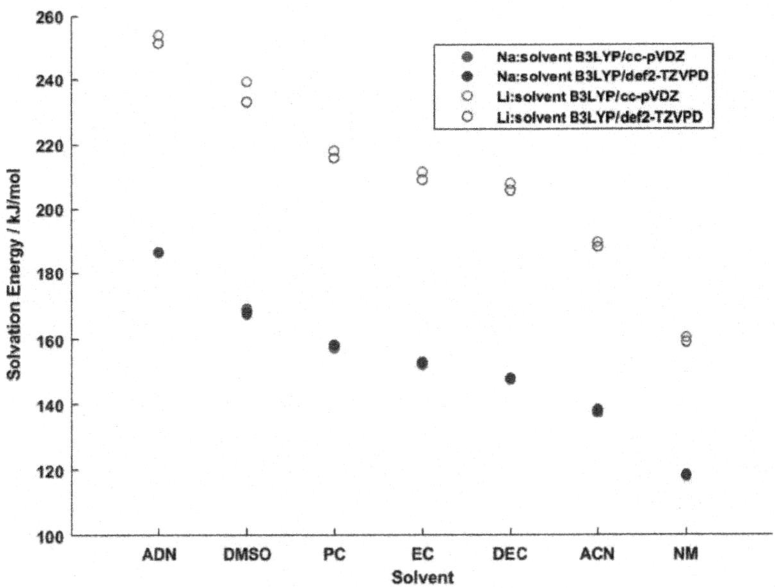

Figure 10.4 The solvation energies of adiponitrile (ADN), dimethyl sulfoxide (DMSO), PC, EC, diethyl carbonate (DEC), acetonitrile (ACN), and NM. Open circles denote Li^+ and filled circles Na^+ [123].

Moreover, it was shown by MD simulations that the Na^+–EC solvation was stronger than the Na^+–PC solvation as shown by the broader Na^+–PC distributions compared to the Na^+–EC distributions [29]. This was in agreement with the work of Kamath et al. [26], which used a combination of DFT and MD simulations. Solvation studies of Na^+ in 27 common solvents were performed using DFT by Okoshi et al. [122, 123]. It was found that cyclic carbonates have intermediate solvation energies, whereas nitro-based solvents, such as nitro methane (NM) and nitro ethane, were stronger coordinated; however, dinitrile-based solvents resulted in small solvation energies (Fig. 10.4). The substitution of Li^+ by Na^+ resulted in a reduction in solvation energies by 20%, which implied faster Na^+ kinetics as further confirmed by impedance spectroscopy experiments [124].

Kamath et al. studied the interaction energy and first solvation shell between organic solvents and $NaClO_4$ salt using MD [26] (Fig. 10.5a). The highest free solvation energy (ΔG_{sol}) values for the interaction of Na^+ ions with organic solvent such as EC, VC, PC, BC, DMC, EMC, and DEC solvents were recently computed. It was found that EC:DMC and EC:EMC, EC:PC solvent mixtures were the best electrolyte candidates, as they present the highest free solvation energy, where on the other hand EC:DEC was the worst choice. Especially, EC:DMC and EC:EMC present higher conductivity and diffusion [125]. In addition, for a single solvent electrolyte, EC is preferred and adding EC to any of the other solvents increases the Na^+-solvent interaction energy [62] as observed by DFT calculations. RDFs showed a low coordination number (~3) of Na^+ for solvents of low absolute values of Gibbs free energy, which implies an incomplete solvation shell [26]. An incomplete solvation shell is not able to screen the positive charge of Na^+, increasing the ion–ion correlation, and thus leading to lower conductivities. Specifically, several MD studies have investigated the coordination number of Na^+ and concluded that a coordination number between 5 and 7 is optimal, but with exceptions of linear carbonate solvents. By using a semiempirical (PM7) approach, the coordination number of Na^+ was measured to be 6.3, 5.7, 6.3 for dilute, concentrated, and highly concentrated, respectively, in PC solutions containing $NaPF_6$ salt [126]. In acetonitrile solutions, the Na^+ coordination number was reported as 4.8, 6.0, and 7 for dilute, concentrated, and highly concentrated solutions, respectively [126]. Thus, it was concluded, from such PM7 simulation approach, that the solvent influenced the Na^+ coordination number even at high salt concentrations. In addition, Na^+ appeared to have a higher coordination number variance and a stronger dependence on salt concentration as compared to Li^+ [126].

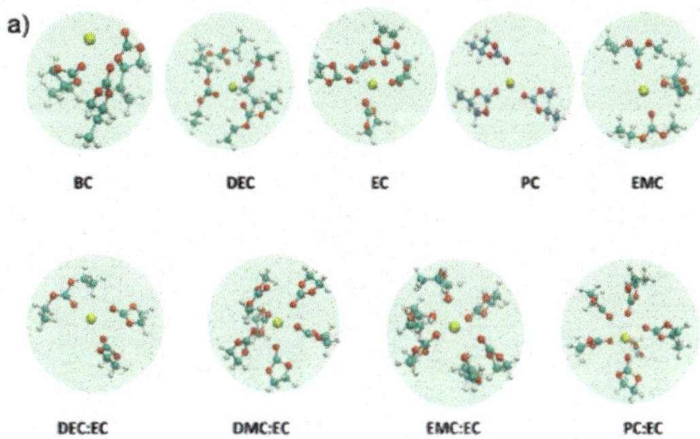

b)

Electrolyte	Dielectric constant	Viscosity (cP)	Transport Number*
EC	89.78	1.95	0.49
PC	64.92	6.8	0.44
DMC	3.11	0.5	0.44
DEC	2.81	0.3	0.34
VC	--	--	0.30
BC	--	--	0.49
EMC	2.96	0.65	0.44
EC:PC	--	5.3	0.51
EC:DMC	--	2.2	0.49
EC:DEC	--	2.7	0.41
EC:EMC	--	--	0.49

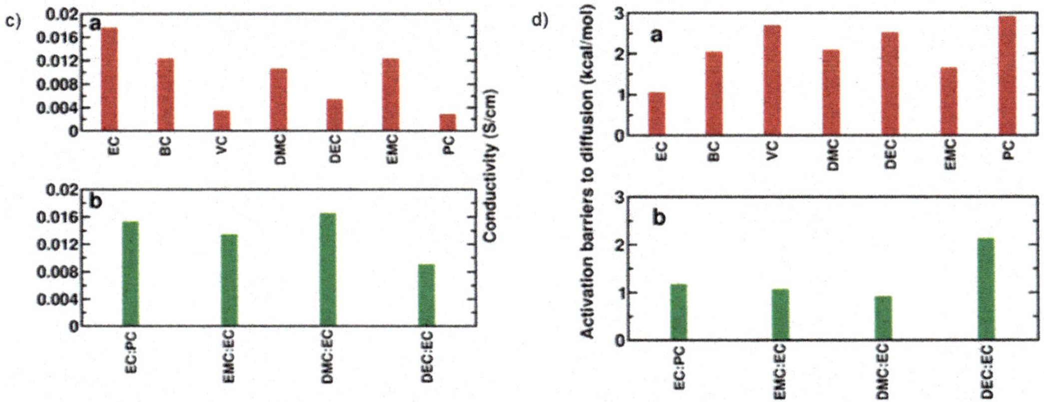

Figure 10.5 (a) Snapshots from MD simulations showing the local oxygen coordination of Na^+ ion within the first solvation shell for a selection of LE solvents. Top panel shows the pure solvents, whereas bottom panel shows snapshots for the mixed solvents, highlighting the difference in coordination number for Na^+ as a function of LE composition. (b) Lists the dielectric constants, viscosities, and transport number from MD simulations for the solvents with $NaClO_4$ salt, and the simulated ionic conductivity and corresponding activations barriers are presented in (c) and (d), respectively. Figure is compiled from Ref. [26].

10.3.3 Transport Properties

Apart from electrochemical stability, NIB performance is intricately dependent on Na^+ transference number and transport, which in turn is also dependent on the solvation structures presented in the previous sections, and LE composition [24]. Typically, the interactions between Na^+ and carbonate solvent molecules are mainly electrostatic, with EC being the solvent molecule with the largest number of solvent molecules in the first shell (coordination numbers between 5 and 6) [27]. Although there is a plethora of studies investigating the structure and solvation of Na^+ specifically in solvents that are used in batteries [127], there are not many experimental NMR studies and even fewer computational studies that focus on diffusion in such solvents. Pham et al. [121] showed using AIMD that the short-scale dynamics, up to 10 ps, of the Na^+ ion, in EC, is faster than that of Li^+ or K^+ ion due to the weaker solvation.

Viscosity (Fig. 10.5b) is also an important property for LE but has not been well studied. An ideal LE should have a relatively low viscosity so that ionic conduction resistance is minimized, and the LE remains in liquid form over a wide temperature range. To this end, the lower viscosity linear carbonate organic solvents (such as DMC with viscosity 0.58 mPas, DEC 0.75 mPas, EMC 0.65 mPas, and 1,2-dimethoxyethane (DME) 0.46 mPas [27] are commonly used in mixtures with the cyclic carbonates (viscosity of EC 1.9 mPas, PC 2.5 mPas, and BC 3.2 mPas) [27] in LEs to increase NIB performance [24]. Limited experimental studies show that the viscosity (η) of 1 M NaTFSI–PC solutions is lower than that of 1 M LiTFSI–PC solutions in the temperature range of 353–283 K [128]. The diffusion of TFSI metal anions in the preceding solutions was measured by a pulsed gradient spin-echo nuclear magnetic resonance (PGSE-NMR) [128]. It was observed that PC diffusivity is higher in the 1 M NaTFSI–PC solutions than in 1 M LiTFSI–PC solutions for the preceding temperature range. A higher discrepancy on PC diffusivity is observed for 2 M solutions. However, the TFSI diffusivity in 1 M NaTFSI–PC solution is higher than that in 1 M LiTFSI–PC solution, only in the 353–313 K range, and their diffusivities coincide at lower temperatures. The ionic conductivity, as measured experimentally, of 1 M NaTFSI–PC solutions is higher than that of 1 M LiTFSI–PC solutions in very dilute (≤0.25 M) or highly concentrated solutions (1.5 M, 2 M) [128].

MD simulations of $NaPF_6$–EC solutions disclosed that self-diffusion of ions increases with temperature following an Arrhenius plot (Eq. 10.6), and the electrolyte tends to form multi-ion aggregates at higher temperature and concentration [129]. From other MD simulations of $NaClO_4$ in various solvents, EC and BC were found to have the highest transport numbers ($t^+ = 0.49$, calculated according to Eq. (10.8)) for the single solvent component LEs, indicating that the cation and anion in the salt in these solvents diffuse at similar speeds in the solvent (Fig. 10.5b) [89]. $NaClO_4$ in PC, DMC, DEC, VC, and EMC had lower t^+ at 0.44, 0.44, 0.34, 0.30, and 0.44, respectively. For the mixed solvent LEs, t^+ remains close to 0.49 for EC:DMC (0.49), EC:PC (0.51), EC:DEC (0.41), and EC:EMC (0.49). t^+ closer to 0.4 than 0.5 indicates that the anion is more mobile than Na^+ in the solvent, and that the anion carries the majority of the charge. The variation in transport number with varying LE composition was further observed in the ionic conductivity simulations (Fig. 10.5c), with EC showing the highest ionic conductivity, and PC and VC the lowest. This is in agreement with the dielectric constant and viscosity criteria for high performing LEs (Fig. 10.5b) and is also reflected in the high Na migration barrier in PC and VC electrolytes (obtained from MD simulations), as compared to EC with the lowest migration barrier (Fig. 10.5d) [89]. Furthermore, MD simulations of NaTFSI-DME solutions, based on a polarized force field at 298 K, predicted the Na^+ self-diffusion coefficient in quantitatively agreement to diffusion-ordered spectroscopy (DOSY)-NMR experiments. The Na^+ transference number was calculated as 0.45–0.5 depending on the salt concentration, and the correlated Na^+ and $TFSI^-$ diffusion slightly increased with salt concentration. The Na^+ diffusion mechanism was less vehicular than that of Li^+ in DME [130].

10.4 Polymeric Solid Electrolytes

The organic solvent used in NIB LEs is usually flammable and can be substituted by a polymeric matrix, gel, or plasticized system, in order to enhance the safety. Polymeric solid electrolytes (PSEs), which are composed of only polyethylene oxide (PEO) matrices and sodium salts (sodium bis(fluorosulfonyl)imide), have shown good mechanical properties and decent ionic conductivities (4.1×10^{-4}

S/cm) at elevated temperatures (80°C), but their conductivities are lower than 10^{-4} S/cm at room temperature and cannot be compared to LEs [131]. However, the introduction of liquid plasticizer can improve the ionic conductivity of the electrolytes.

PEO has been extensively used since the 1970s [132, 133] both with Li^+ and Na^+ electrolytes. Recently, polycarbonates [134] and polyacrylonitrile [135] have also been used as polymer matrices. DFT calculations were implemented by Memboeuf et al. to oligomer (up to 18 repeating units) poly(ethylene glycol) (PEG) matrix with Na^+ cations (Fig. 10.6) [136]. The first solvation shell of Na^+ consisted of six ether oxygen atoms. It was shown that the stiffness of the PEG chains prevented the formation of a second solvation shell even for the longest oligomers. The binding energy ΔE_n of the Na^+ ion to C was lower than that of the Li^+ to the PEO oligomers.

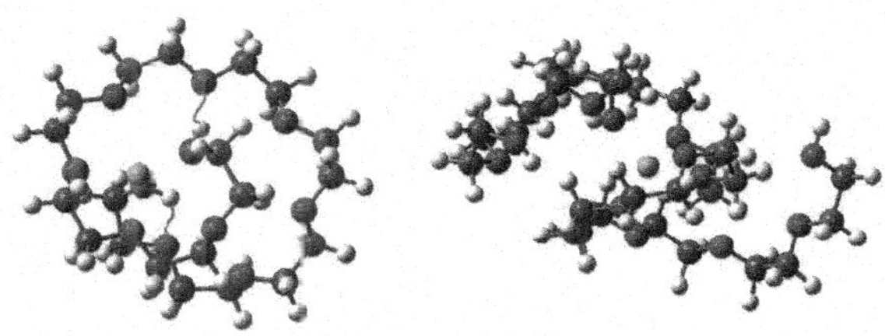

Figure 10.6 Structures of cation PEG_n with n = 10 (left) and 14 (right). The cation is denoted with blue color; O, C, and H atoms are red, green, and white, respectively [136].

Moreover, Hartree–Fock calculations of 1:1 complexes of $NaBF_4$, $NaClO_4$, NaSCN, $NaCF_4SO_3$ (NaTf), $Na(CF_3SO_2)_2N$, NaTFSI, $NaPF_6$, $NaAsF_6$, and $NaSbF_6$ as oligomer models (diglyme, triglyme, tetraglyme) [137] of PEO showed that the increasing number of ether oxygen atoms led the coordination number of Na^+ to increase from 5 to 7 [138–141]. In addition, AIMD simulations were implemented to explore the dynamics of the PEO_8:$NaAsF_6$ [142] system. The Na^+ coordination in PEO was found to be influenced, strongly, by the AsF_6^- anion migration, which tended to dominate the ion transport in this system, in particular at the high-temperature regime [143].

A lot of the atomic modeling studies for PSEs are on the understanding of the ion conduction mechanisms and calculation of transport properties, using MD methods. In a study by Laasonen and Klein, the ion diffusion both in a PEO crystal (Fig. 10.7) with a single inserted NaI ion pair (effectively $PEO_{224}NaI$) and in the same system rendered amorphous [144] was simulated. In contrast to the highly concentrated crystalline PEO_3NaI, where Na^+ resides inside single PEO helices, the cations in the dilute $PEO_{224}NaI$ crystal were found to coordinate to three neighboring polymer chains [144].

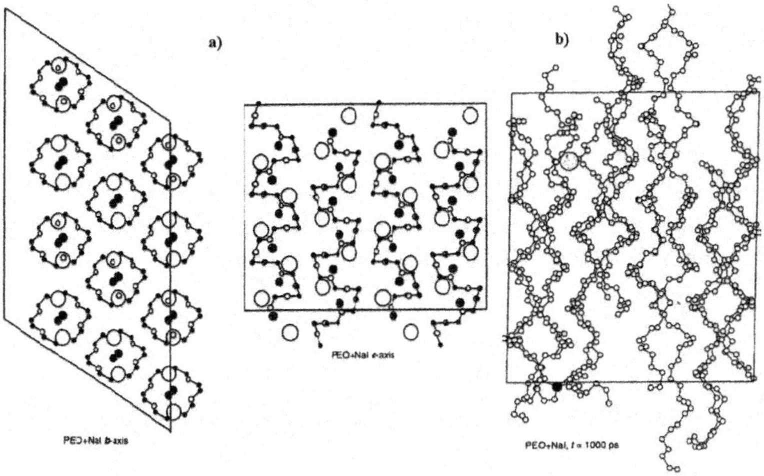

Figure 10.7 (a) Crystal structure of PEO–NaI (3:1), viewed from ***b*** (chain axis) and ***c*** directions. The full 2×3×3 simulation cell is shown. The large empty spheres are the I^-, and the grey spheres are the Na^+ ions, respectively. (b) Snapshot (t = 1000 ps) of the PEO–NaI (224:1) simulation viewed from the ***a*** direction. The large empty sphere is the I^- ion, and the smaller darker sphere is the Na^+ ion [144].

The PEO:NaI system was modeled by a simple Rouse model. Initially the Na^+-ether oxygen (of PEO) coordination was modeled equal to the distance of monomer chain segments; thus, permanent crosslinks between the polymer chains were formed. However, this approximation can help to show how the created crosslinks slow down the dynamics. On long time scales, cation dynamics follows the polymer network dynamics (Fig. 10.8a). Incorporating dynamic physical crosslinks in the model can lead to the calculation of coefficients and thus conductivity predictions, as can be seen in Fig. 10.8b [145].

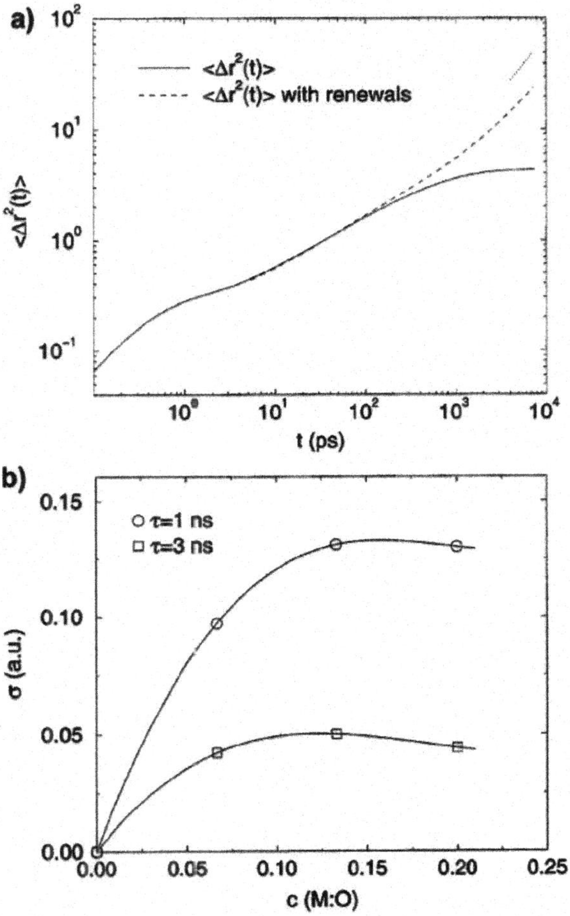

Figure 10.8 (a) MSD for cat ons in a Rouse model for polymer electrolytes. As time goes to infinity, the displacement is bounded in the model without renewal of crosslinks. (b) Ionic conductivity as a function of concentration for the Rouse model with dynamic crosslinks [145].

Diffusion of ions takes place through ion hopping with the polymer chains. The simulations further indicated a lower ion mobility in the amorphous regions of PEO [144]. Ion clustering, which can hinder ionic conductivity, was also studied by Mills and Catlow in PEO:NaI systems for different salt concentrations [146, 147]. In particular, not only ionic cluster formation was observed, in disagreement with previous simulation efforts [148] and experiments [149], but also the aggregation was found to decrease with temperature [146, 147]. The movement of cations between solvation sites on the PEO chains was found to depend on the availability of anions at these sites [147].

Payne et al. used DME and monoglyme with NaI [150–152] salt and found monoglyme to solvate Na^+ efficiently, leading to less ion clustering. The ion clustering was affected by the scaled ionic charges of the model used, and such models led to phase separation in dimethyl ether [151]. In a more recent study by Dong et al., tetraglyme with NaTf (sodium triflate) was modeled [153]. The use of the molecular Tf anion at a ratio of 10:1 concluded that the anions contribute with more oxygen atoms than the solvent to the first coordination shell of Na^+, showing that Na^+:Tf coordination is more favorable than Na^+:ether coordination [153].

Fully atomistic models of an amorphous PEO:NaI system and crystalline PEO_3NaI phase at different salt concentrations were developed by Neyertz et al. [154]. The Na–O coordination sphere of crystalline PEO_3NaI was reproduced and single PEO segments appeared to coil around Na^+ in a similar fashion as in crystalline PEO_3NaI regardless of concentration or temperature [154, 155]. There was a competition between ether oxygen atoms and anions for coordination. At high salt concentration, the cation–polymer interactions are reduced and ionic clustering was increased, and a single percolating cluster linked with the polymer chain was formed [154, 155]. An increase in ion aggregation was also observed by increasing the temperature from 500 to 1000 K for the lower concentration systems, with the formation of a permanent single cluster, confirmed by both simulations and experiments [149].

Different models have been used to predict the dynamics of PEO:NaI electrolytes as can be seen in Fig. 10.9a. By taking into account the polarizability of the ions and polymer matrix, a reduced interaction between Na^+ and I^+ and more coordination of Na^+ to the PEO chains are observed [156]. The polarizable and the springs model give the best prediction of relaxation times in good agreement with neutron scattering experiments (Fig. 10.9b).

Using a full atomistic MD model, Chen et al. simulated the effects of introducing either solvating ether or non-solvating styrene spacer groups in sodium poly[4-styrenesulfonyl(trifluoromethanesulfonyl) imide] (NaPSTFSI) single ion conductors (Fig. 10.10) [157].

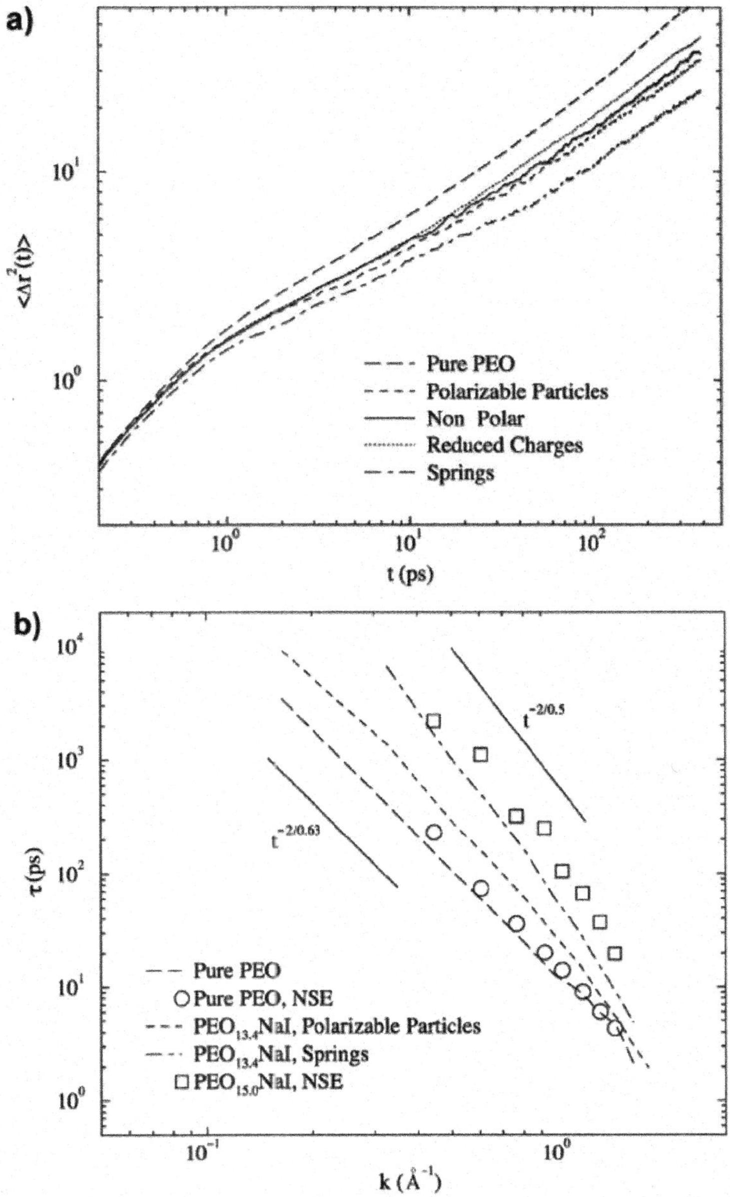

Figure 10.9 (a) MSD at T = 343 K for both pure PEO and PEO:NaI electrolytes from different models, show ng the slowing down of chain dynamics for the PEO:NaI electrolyte, (b) comparison of relaxation times from MD simulations with experimental data from neutron spin-echo measurements [156].

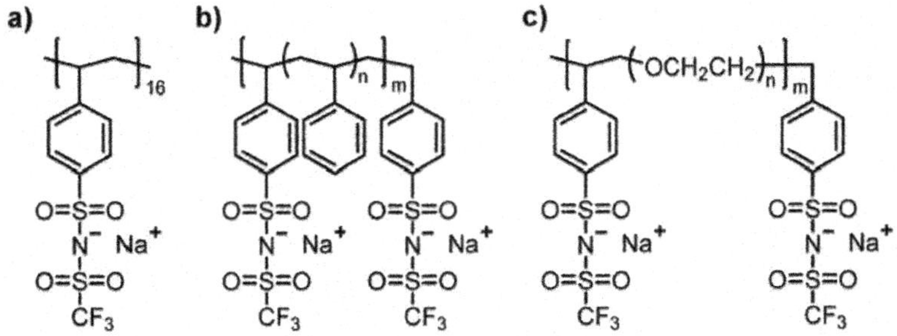

Figure 10.10 Structures of (a) NaPSTFSI homopolymer and its copolymer with (b) styrene ($n = 2$, $m = 5$ and $n = 4$, $m = 3$), (c) ethylene oxide ($n = 5$, $m = 7$ and $n = 13$, $m = 4$) spacer groups [157].

A scaled-charges model was used in order to weaken the interactions and enable ion hopping. In the NaPSTFSI homopolymer, fast ion dynamics is facilitated through a continuous ionic aggregated network, whereas the introduction of long spacer groups disrupts this network, breaking it into discrete clusters [157]. The ether segments coordinate to Na^+ and limit their association with the tethered anions. Ion transport can take place via segmental motions of the oligoether segments, similar to regular PEs, whereas Na^+ moves through decoupled hopping between anionic sites in the rigid styrene segments. The same type of ion solvation can be achieved by the addition of low-molecular-weight plasticizers and solvents. Recently, Chen et al. [32] investigated the addition of tetraglyme to dual-cation (Li^+/Na^+ in combination with methyl triethylammonium) ionomers based on poly(2-acrylamido-2-methyl-1-propanesulfonic acid) by means of MD simulations and revealed a higher tendency for Li^+ than Na^+ to coordinate to the tetraglyme plasticizer [147]. In contrast, the Na^+ system was still dominated by large clusters and appearing an enhanced ion hopping after adding the tetraglyme plasticizer, which explains the observation that ionic conductivity is decreased by tetraglyme addition in the Li^+ system whereas is increased for the Na^+ system [32].

Polyethylene oxide-based sulfonated Na^+ ionomers (Fig. 10.11) were also investigated using MD simulations by Maranas et al. [158–161]. In particular, by using a united-atom force field with scaled charges, it identified a transport mechanism that involved rearrangement of chain-like aggregates and showed that their formation was dependent on the stacking of the benzene rings of the polymer backbone [161]. At 343 K (1.3 Tg), intermediate-size aggregates (8.5–10.5 Å) were formed, which facilitated "superionic" conduction (charge diffusion exceeded the ion diffusion) through a mechanism that involved the movement of Na^+ at either chain ends [159].

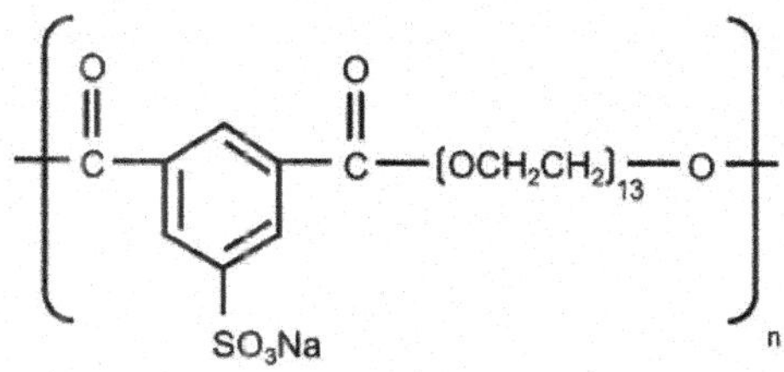

Figure 10.11 Chemical structure of the PEO-based sulfonated ionomer: PEO600-100% Na [158].

When PEO chains are anchored to the anionic group (sulfonated isophthalate), a mobility gradient in the chain is created [158] and all atoms in the free PEO chains are equally mobile [159]. The free oligoether segment presents a higher flexibility and enables adoption of a larger number of conformations that favor cation solvation. This can lead to more efficient solvation of cations and break down the larger ionic aggregates, which is similar to the disruptive effects observed on ionic aggregates by introducing oligoether segments in NaPSTFSI [157]. The higher mobility of free PEO gives higher ion mobility [159]. The dynamics of the oligoether segment and of the cations is dependent on the balance between cation–anion and cation–ether interactions [160]. By using MD models, where anionic charges are found to be either concentrated on the sulfonate group or distributed throughout the ionic isophthalate segments, PEO backbone dynamics is reduced due to either more free cations or more ionic aggregates that act as ionic crosslinks [160].

10.5 Ceramic Solid-State Electrolytes

Ceramic, or inorganic, solid-state electrolytes (CSEs) have the potential to improve battery lifetime, safety, and performance. A high-performance electrolyte needs to have high phase stability, sodium storage potential, and fast sodium migration. Furthermore, they should be electronically insulating, which makes DFT simulations excellently posed to screen, predict, and optimize these materials. One of the main attractions of CSEs is the possibility to pair these electrolytes with sodium metal anodes, something that from a safety perspective is impossible for liquid organic electrolytes [162, 163]. This would give the possibility to store a larger amount of energy per mass (or volume), leading to higher power density. Furthermore, CSEs could theoretically be coupled to a wider range of high-voltage cathode materials, further enhancing their power density [160, 161]. β-alumina has been used as a CSE for large-scale solid-state sodium batteries, but operates at 270°C [162, 163]. Furthermore, β-alumina needs to be synthesized at high temperatures (>1200°C) to lower grain boundary resistance and requires sintering at 1800°C to achieve good solid–solid contact with the electrodes, making these batteries uneconomical for most applications [162, 164]. CSEs that can be synthesized at lower temperatures, more cheaply and that have high ionic conductivities at room temperature are hence sought.

For a material to be a suitable CSE, it should have high ionic conductivity (1–6 mS/cm to be comparable with LEs) at ambient temperatures, and high mechanical strength to suppress dendrite growth [164, 165]. Currently, the development of CSEs for NIBs is limited by the high costs of these materials, complex processing, inadequate ionic conductivity, as well as challenges in finding suitable CSEs that can form stable and low-resistance interfaces with both cathode and anode materials [11]. The ionic conductivity of CSEs is dependent on composition and structure, but the exact relation between these remains to be fully understood [11]. To this end, atomic-scale studies are invaluable to further the technology and allow for the next frontier in commercial secondary batteries to be realized. Atomic-scale simulations are critical in understanding the underlying factors that control Na transport in CSEs. DFT and MD simulations can also be used to propose new structures for CSEs that can then be experimentally studied. In the next sections, we will review atomic-scale studies of ionic conductivity, mainly commenting on the simulated Na diffusion coefficients (D_{Na}), Na migration energy barrier (E_a), and Na-ion conductivity (σ_{Na}). These properties can be obtained using either DFT (through NEB simulations), MD, or AIMD (as detailed in the computational section). Due to the plethora of different CSEs, we will discuss each CSE type separately, starting with NASICONs in Section 10.5.1, anti-perovskites in Section 10.5.2, and finally thio-phosphates in Section 10.5.3. As opposed to the liquid and polymer electrolytes discussed in Sections 10.3 and 10.4, the performance of CSEs is heavily influenced by crystal structure, defects, and dopants. Hence, we will discuss the ionic conductivity and Na diffusion in relation to CSE type, and crystal, defect, and dopants structures.

10.5.1 NASICON

NASICON (sometimes stylized as NaSICON) is the short for NAtrium Superionic CONductor and is a family of solid-state ceramic electrolytes with the general formula $Na_xM_2(SiO_4)_z(PO_4)_{3-z}$ (M = transition metal, x = 0–4) [165, 166]. These materials have high conductivity, good thermal stability, low ductility, and excellent chemical stability [167]. As seen from the general formula of NASICON, this electrolyte material provides ample possibility for property tuning, something that has been widely exploited within the computational modeling community [14]. Substituting M with different valence transition state metals, one can tune both the ionic conduction and electrochemical properties of the NASICON CSE. The NASICON material $Na_3V_2(PO_4)_3$ is, for example, widely seen as a promising NIB cathode material in addition to its use as a CSE [168–170].

The NASICON family of materials have corner-sharing PO_4 tetrahedra, and MO_6 octahedra (an example is provided in Fig. 10.12a), with the high Na conduction occurring within the 3D framework between the Na(1) and Na(2) sites (Fig. 10.12a) [166, 171]. The Na conduction has been confirmed from MD simulations with a full interionic potential and agrees with experimental studies [166]. Determining the structure of these electrolytes has been challenging as the crystal structure can vary depending on the synthesis procedure and the thermal history of the material [171]. In the computational literature, three crystal structures are commonly used: α-, β-, and γ-NASICON. Monoclinic α-NASICON is the low-temperature (300 K) structure, whereas γ-NASICON with a rhombohedral structure is achieved at temperatures above 450 K. At temperatures between 300 and 450 K, the monoclinic β phase is found. Furthermore, the structure can also change from rhombohedral to monoclinic dependent on x (in the general chemical structure) [167].

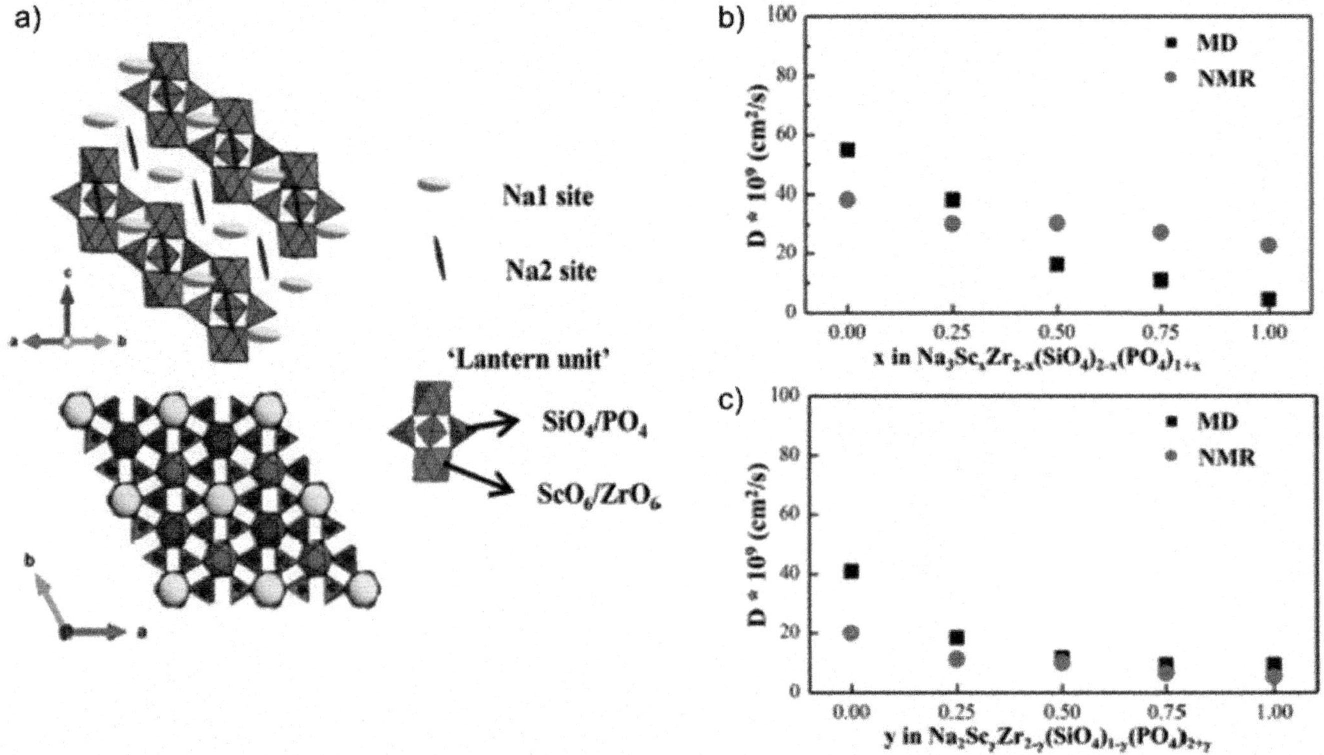

Figure 10.12 (a) A schematic view of the rhombohedral NASICON structure where ScO_6 and ZrO_6 polyhedra in grey, and SiO_4 and PO_4 polyhedra in red. The Na1 sites are shown in yellow and the Na2 sites in blue. The Na sites are shown as ellipsoids with different probability values of 95%. Simulated and experimental D_{Na} in (b) $Na_3Sc_xZr_{2-x}(SiO_4)_{2-x}(PO_4)_{1+x}$ and (c) $Na_2Sc_yZr_{2-y}(SiO_4)_{1-y}(PO_4)_{2+y}$. Figures have been adapted from Ref. [171].

Deng et al. used DFT and MD simulations, in conjunction with experiments, to investigate the Na conductivity in $Na_3Sc_xZr_{2-x}(SiO_4)_{2-x}(PO_4)_{1+x}$ and $Na_2Sc_yZr_{2-y}(SiO_4)_{1-y}(PO_4)_{2+y}$, where x and y are between 0 and 1 [171]. $Na_3Sc_xZr_{2-x}(SiO_4)_{2-x}(PO_4)_{1+x}$ undergoes a phase transition from monoclinic to rhombohedral between 295 and 473 K, whereas in the same temperature range, $Na_2Sc_yZr_{2-y}(SiO_4)_{1-y}(PO_4)_{2+x}$ is only found in the rhombohedral crystal structure. MD simulations confirmed the Na diffusivity from NMR experiments, showing that D_{Na} decreases with increasing Sc concentration (x, y) [171]. An additional Na site, Na5, was identified as a major contributor to the Na diffusion in $Na_{1+n}Zr_2Si_nP_{3-n}O_{12}$, which has also been proposed as a promising CSE. The enhanced ionic conductivity in this NASICON was from DFT simulations shown to be due to correlated migration at increased Na concentration [167]. The increasing Na concentration leads to increased Coulombic repulsion, consequently activating the correlated migration paths. These findings were confirmed experimentally and show that one optimization route for CSEs could, instead of framework expansion, be increasing the Na concentration [167]. This was investigated by introducing excess Na in $Na_3Zr_2Si_2PO_{12}$ [172]. Typical NASICON CSEs have a 75% Na site occupancy. DFT simulations showed that an Na excess of 16.67% (in between the experimentally prepared 10% and 20% excess) had negative formation energies, similar to those of the non-excess NASICON but were slightly less thermodynamically stable [172]. Comparing the measured activation energy for the excess and non-excess NASICON material, both in the monoclinic and rhombohedral structures, introducing excess Na resulted in lower Na migration barriers for both structures [172]. E_a in monoclinic and rhombohedral Na-excess NASICON is 0.24 eV and 0.18 eV, respectively, whereas for the non-excess materials, they are 0.28 eV, and 0.21 eV, respectively [172].

High Na mobility has also been found for molybdate CSEs. These materials can, except for the alkali cation and molybdenum oxide unit, contain a variety of di- and trivalent cations. DFT simulations have been employed to simulate the most probable Na conduction paths from NEB simulations in $Na_9Sc(MoO_4)_6$ (Fig. 10.13) [173]. As for the NASICON-type electrolytes, the ionic conduction takes place in a 3D network of migration pathways. The calculated migration barriers are between 0.1 and 0.8 eV, which would qualify this material as a high-ionic-conductivity CSE [173]. The diffusion paths consist of a combination of curved and linear paths, with the linear paths having the smaller migration barriers (0.12 eV, between Na2 and Na3 sites in Fig. 10.13), and the curved paths responsible for the higher barriers (0.59 eV for Na2 to Na2 and 0.53 eV for Na3 to Na3). The highest reported migration barriers (0.69 eV and 0.84 eV, respectively) correspond to the longer Na jumps (~3.7 Å) between Na1 and Na3 sites, and Na1 and Na2 sites, respectively [173]. These different migration paths, or Na jumping mechanisms, were found to be activated at different temperatures, explaining the different ionic conductivity regimes observed by NMR studies between 300 and 750 K [173].

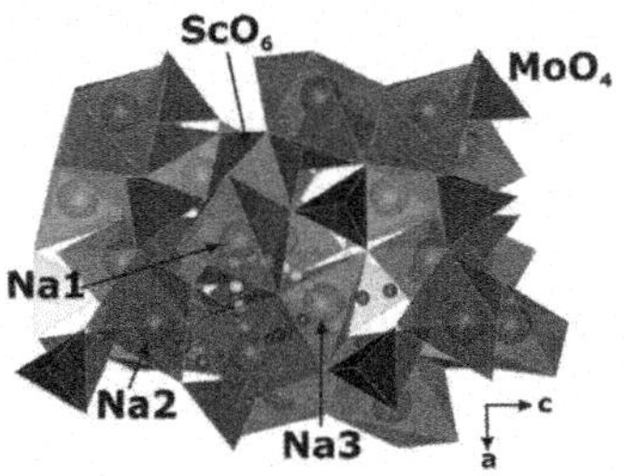

Figure 10.13 Polyhedral model of $Na_9Sc(MoO_4)_6$ showing the simulated diffusion paths for Na jumping mechanism between Na1 and Na2 or Na3 (indicated by yellow circles), between two Na2 sites and between two Na3 sites (shown in pink circles), and finally between Na2 and Na3 sites (black circles). The ScO_6 polyhedra are shown in purple, MoO_4 polyhedra are shown in brown, Na1 sits inside the green polyhedral, Na2 inside the blue, and Na3 inside the pink polyhedral. O sits at the corners of the polyhedra. The figure is taken from Ref. [173].

10.5.2 Anti-perovskites

Anti-perovskites have the general formula Na_3OX, where X is a halogen or mixture of halogens. An example of the cubic anti-perovskite structure is shown in Fig. 10.14, where Na occupies the perovskites traditional oxygen sites, and the two anions (O and X) in the traditional cation sites. These CSEs have high ionic conductivity, low Na migration barriers, negligible electronic conductivity, good cyclability and stability, wide electrochemical window, and can be synthesized from both cheap and environmentally friendly materials [174, 175]. Similar to the CSEs discussed in the previous sections, the anti-perovskites can be optimized through chemical and structural modification [174]. Defects in anti-perovskites that facilitate ionic conduction are typically thermally generated Frenkel or Schottky defects [31]. From both DFT and MD simulations, these defects are only expected to be present at low concentrations [31, 174]. MD simulations of $Na_3OCl_{0.5}Br_{0.5}$ with Na vacancy concentrations (δ) of 0.038 and 0.150 showed Na migration activation energies of 0.24 and 0.22 eV, respectively [174]. Although compared to Na_3OCl (E_a = 0.29 eV for δ = 0.038, and 0.26 eV for δ = 0.150) and Na_3OBr (E_a = 0.28 eV for δ = 0.038, and 0.26 eV for δ = 0.150), the activation energies for the mixed systems are slightly lower, no significant increase in the ionic conductivity could be identified from these simulations. Mixed Li and Na anti-perovskites were further investigated as CSEs but were found to have high metal migration activation energy barriers [174]. MD simulations of these materials showed that the activation energies are lower at Δ = 0.038 than Δ = 0.150. Both $LiNa_2OBr$ and $LiNa_2OCl$ have activation energies of 0.31 eV, whereas Li_2NaOCl and Li_2NaOBr have higher Na activation energies of 0.40 and 0.44 eV, respectively [174]. Hence, mixing Li and Na in these CSEs does not lead to any improved conductivities, but instead very poor CSEs. The introduction of divalent cations such as Ca^{2+} and Mg^{2+} in the Na_3OX CSE increases the Na vacancy concentration and the ionic conductivity due to the charge imbalance introduced by the divalent cations. This strategy has been shown to improve the conductivity from both experimental and computational studies and has been suggested to be a promising strategy if combined with mixed halide site anti-perovskites for CSEs [175].

As for perovskites, the ionic conductivity in anti-perovskites can also be increased by structural distortions [176, 177]. These distortions take the shape of polyhedra tilting and rotations, bond length variations, and space group changes from cubic to orthorhombic (Fig. 10.14b) [176]. A comprehensive DFT study of Na_3AX (where A = O, S, or Se, and X = F, Cl, Br, or I) showed that high ionic conductivity could be achieved by tuning the distortions of these lattices, but that the increased distortions could lead to poor thermodynamic stability [176]. Hence, a balancing act is required to fine tune these materials for fast ionic conduction. Figure 10.14c

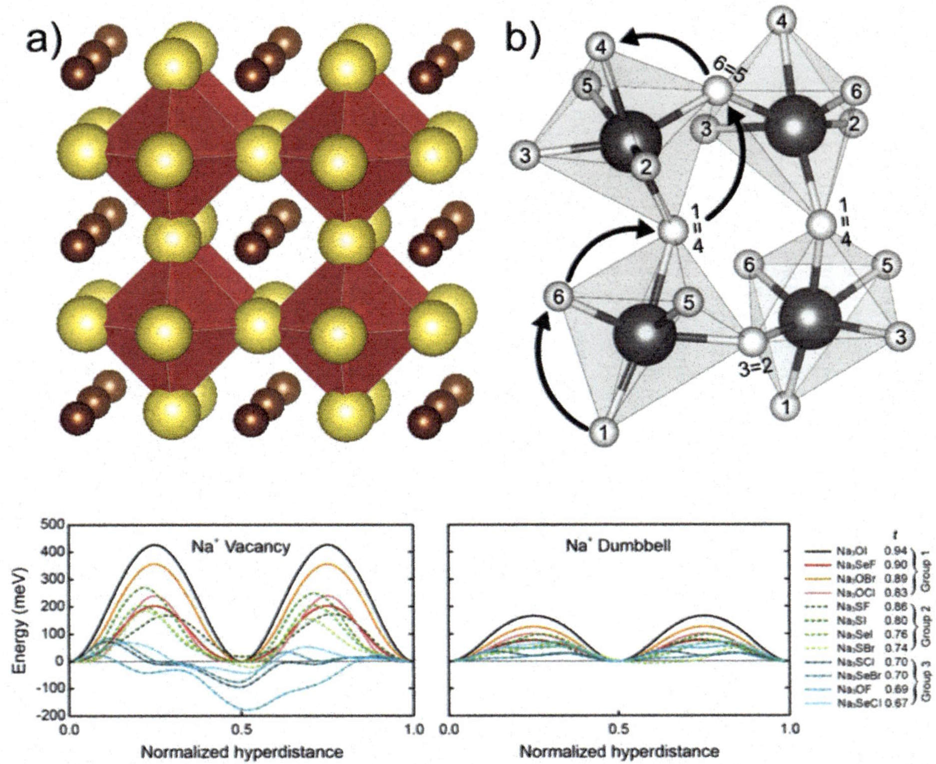

Figure 10.14 (a) Polyhedral representation of the Pm-3m anti-perovskite where yellow spheres are Na, brown halogens, and O sits in the middle of the red polyhedra. (b) Distortions in the polyhedral structure, which can be used to tune ionic conductivity. Numbers indicate the mobile Na ions, with same numbers showing equivalent sites, and arrows indicate typical migration paths. (c) Lowest energy Na migration pathways in anti-perovskites with Na following either a vacancy migration path or an interstitial dumbbell path. For all materials, the dumbbell path migration barriers are lower than the vacancy path. t is the Goldschmidt tolerance factor ($0.9 < t < 1$ indicates stable cubic structure, $0.75 < t < 0.9$ orthorhombic, and $t < 0.75$ hexagonal), and groups 1, 2, and 3 refer to materials with $t > 0.83$ (well-ordered octahedra and cubic), $0.74 < t < 0.81$ (moderate distortions, octahedral tilting, and quasi-cubic structure), and $t < 0.7$ (quasi-orthorhombic structure and significant octahedral distortions), respectively. Figures (b) and (c) are from Ref. [176].

shows the calculated Na migration energies in Na_3AX following either a vacancy or interstitial migration path; for all materials, the interstitial site was found to have lower migration barriers and higher ionic conductivity. Among the investigated compounds, Na_3SI was identified as having high ionic conductivity, while being expected to be thermodynamically unstable. Recently, experimental synthesis of Na_3SI was attempted using *in situ* synchrotron X-ray diffraction. The formation of Na_3SI was, however, found to be hindered, and hence further experimental studies are required for this CSE to become a practical candidate for solid-state batteries [178].

10.5.3 Thiophosphates

Sodium thiophosphate (Na_3PS_4) counts as one of the most promising CSEs for solid-state sodium batteries due to their high ionic conductivity and is inspired by the all-solid-state lithium battery electrolyte counterpart [179–182]. Na_3PS_4 crystallizes in two different structural polymorphs at room temperature: cubic or tetragonal (Fig. 10.15a,b). The room-temperature phase of this glass ceramic is 0.2 mS/cm, and higher ionic conductivities have been achieved by heteroatom doping [164]. Its cubic phase has been employed as a CSE in solid-state batteries at room temperature and was first reported in 1992 [179]. Apart from the cubic phase, Na_3PS_4 is tetragonal at room temperature where the PS_4^{3-} polyhedra arrange in a distorted bcc structure with Na^+ filling the octahedral and tetrahedral voids [179]. However, the ionic conductivity of this tetragonal room-temperature phase was found to be unsuitably low for battery applications (4×10^{-6} S/cm), and hence other phases, leading to the stabilization of the cubic phase in 2012, has been sought [179]. Experimental studies have reported that the cubic phase Na_3PS_4 (where all PS_4^{3-} polyhedra align, and the Na^+ only occupy tetrahedral sites) has higher ionic conductivity than the tetragonal phase, whereas from computational studies, no discernible difference in ionic conductivity between the two phases is found [182–184]. Using PDF, impedance, and XRD analysis, it was shown that the reason for this is that Na_3PS_4 prepared from ball milling (regardless of phase) has the highest ionic conductivity, whereas Na_3PS_4 prepared via high-temperature routes has an average tetragonal structure [183]. The ball-milling-prepared samples do typically have a cubic structure (with local tetragonal motifs), which may serve to explain the cubic phase being reported as the superior ionic conductor. Hence, the conclusion was drawn that, as suggested by DFT simulations, neither phase had the highest conductivity, but rather that ball-milling and high-temperature synthetic routes lead to different amounts of defects. This conclusion was previously vented by Bo et al. in 2016, where from AIMD and DFT simulations, the ionic conductivity in cubic and tetragonal Na_3PS_4 was found to be almost identical (Fig. 10.15c) [182]. From Na vacancy defect formation energy calculations, it was found that the cubic phase is more likely to contain Na vacancy defects (E_{def} = 0.34 eV) than the tetragonal phase (E_{def} = 0.57 eV) [182]. Hence, the inherently higher defect concentration in practical cubic Na_3PS_4 samples, as compared to the tetragonal phase, could also be responsible for the higher ionic conductivity in the cubic phase (Fig. 10.15c) [182]. This highlights the importance for computational studies to not only consider the pristine polymorphs but also account for heterogeneities in these electrolytes.

Until recently, Na_3PS_4 was thought to be unstable at high temperatures, with a melting point of 500°C, but impedance spectroscopy, diffraction, and *ab initio* studies could show that a high-temperature solid polymorph (γ-Na_3PS_4) exists [180]. This orthorhombic polymorph was found to have fast ion conduction, further allowing for this material to be considered for solid-state sodium batteries. γ-Na_3PS_4 has a low E_a of 0.1 eV, which indicates that it could be a promising CSE for continued study [180]. MD simulations identified that sodium conduction and PS_4^{3-} rotational motion both contribute to the molecular motion, leading to plastic characteristics [180].

The effect of grain boundaries on ionic conductivity in Na_3PS_4 was studied using a novel computational approach combining MD simulations with models generated from Atomsk and Voronoi tessellation [185]. Using this approach, cubic polycrystal models with different grain volumes could be constructed, and the grain boundary resistance evaluated (Fig. 10.16) [185]. Polycrystalline Na_3PS_4 had insignificant grain boundary resistance, whereas the ionic conduction in the oxide counterpart (Na_3PO_4) was found to be highly sensitive to grain boundaries, with decreasing grain volume leading to decreasing ionic conductivity (Fig. 10.16a,b) [185].

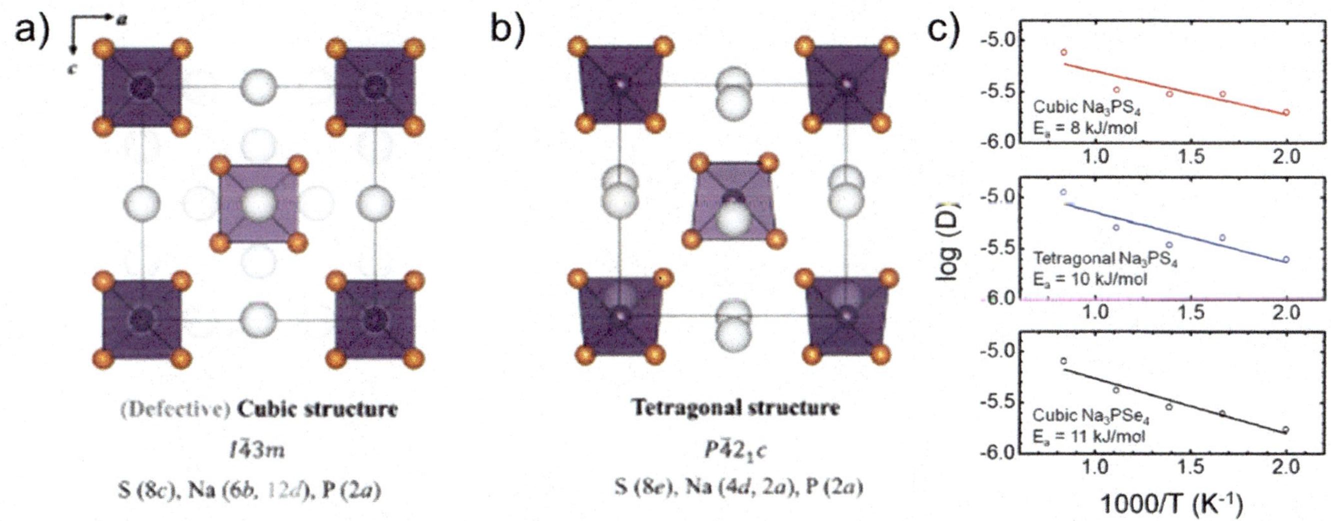

Figure 10.15 Crystal structure of (a) cubic and (b) tetragonal Na_3PS_4. Figure from Ref. [181]. (c) Na diffusion coefficients (D) for cubic and tetragonal Na_3PS_4 and cubic Na_3PSe_4 from AIMD simulations. Figure from Ref. [182].

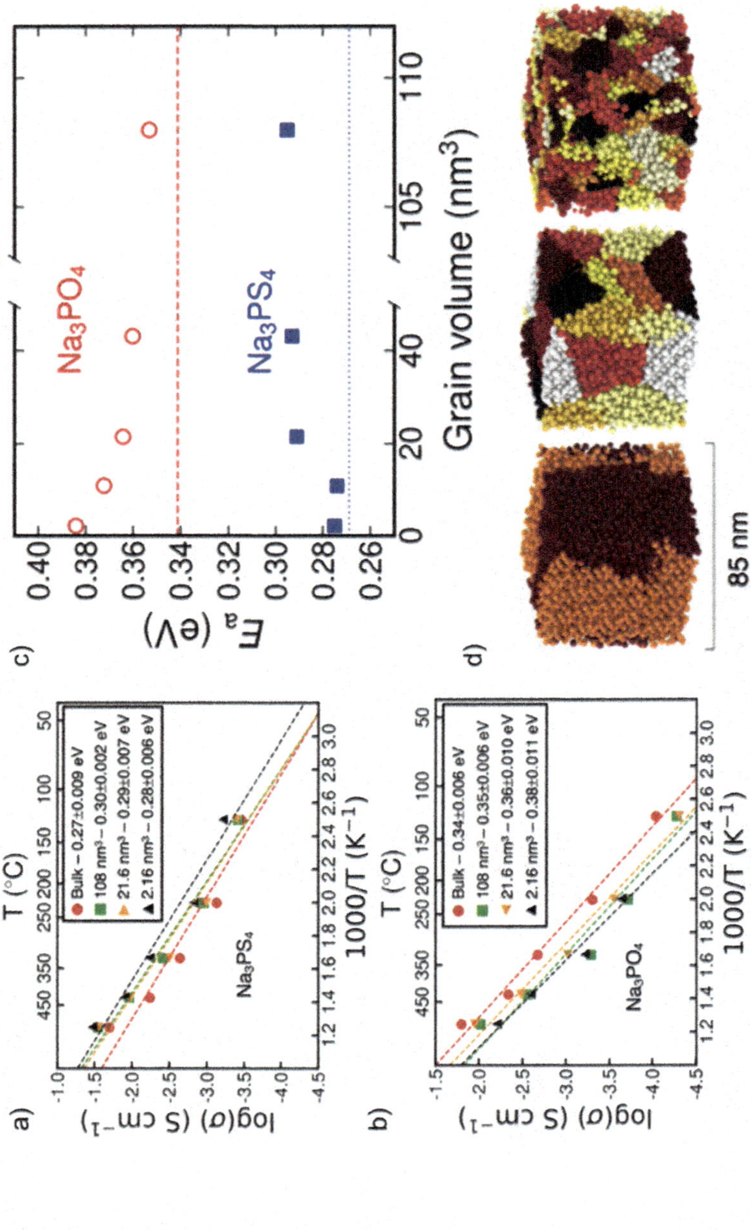

Figure 10.16 Arrhenius plots of Na ionic conduction in (a) Na_3PS_4 and (b) Na_3PO_4 for bulk (crystalline) and polycrystalline (green squares, yellow triangles, and black triangles) models. (c) Corresponding Na diffusion activation energies for bulk (blue and red dashed lines) and polycrystalline (circles and squares) Na_3PO_4 and Na_3PS_4. The polycrystalline models are shown in (d), with the first model containing 2, 10, and 100 grains (each color represents a separate grain), corresponding to 108, 21.6, and 2.16 nm^3 in (a) and (b). All figures are from Ref. [185].

Comparing the local structures of the sulfide and oxide CSE, Na_3PS_4 shows much less difference between its bulk and polycrystalline structure than Na_3PO_4 [185]. Hence, Na_3PS_4 is much less influenced by the structural changes brought about by the grains and its ionic conductivity is not sensitive to the structural changes induced by the grain boundaries (Fig. 10.16) [185].

Combining Na_3PS_4 with other compounds, such as NaI, and heteroatom doping can increase its ionic conductivity and improve its stability [186–188]. Yu et al. performed a combined DFT and experimental study to optimize the ionic conductivity in heteroatom-doped Na_3PS_4 and to also combat its inherent moisture instability [164]. This was achieved by doping on the phosphorous site with arsenic, giving the compound $Na_3P_{1-x}As_xS_4$, where $0 < x < 1$. The highest ionic conductivity for this material (1.46 mS/cm) was achieved at $x = 0.38$ [164]. These DFT simulations showed that the high conductivity was a direct result of the elongated Na–S bond in the transition state structures in the migration path. Furthermore, it was found that the ionic conduction follows a sodium jumping route (similar to NASICON CSEs), where the Na1 to Na2 site jump is more energetically favorable than the Na2 to Na2 site jump with around 30–40 meV [164]. Hence, the ionic conduction in $Na_3P_{1-x}As_xS_4$ occurs mainly through the a- and b-axis directions. The end members of this family of CSEs ($x = 0$ and 1) both had lower conductivity, and higher E_a (0.289 eV for $x = 0$ and 0.341 eV for $x = 1$, respectively) than $Na_3P_{0.62}As_{0.38}S_4$ ($E_a = 0.256$ eV) [164]. Sn-doping increases the ionic conductivity at 300 K to 2.4 mS/cm for $Na_{11}Sn_2PS_{12}$ and $Na_{11.125}Sn_2PS_{12}$, and 2.3 mS/cm for $Na_{10.875}Sn_2PS_{12}$ [189, 190]. Doping on the P site with other group XIV elements does not lead to the same increase in ionic conductivity. $Na_{11}Si_2PS_{12}$ and $Na_{11}Ge_2PS_{12}$ have simulated ionic conductivities of 0.3 and 0.6 mS/cm, respectively [189]. For the Sn-doped material, all sodium sites can participate in the sodium conduction as long as the sodium vacancies are well distributed within the lattice, and the enhanced ionic conductivity as compared to the Si- and Ge-doped samples was ascribed to the larger lattice expansion in this material. As opposed to other sulfide CSEs, $Na_{11}Sn_2PS_{12}$ was found to be stable at 0 K from thermodynamic stability calculations, but that it is only stable in the electrochemical window of 1.16–1.92 eV making the selection of electrode materials to be paired with this CSE challenging [189].

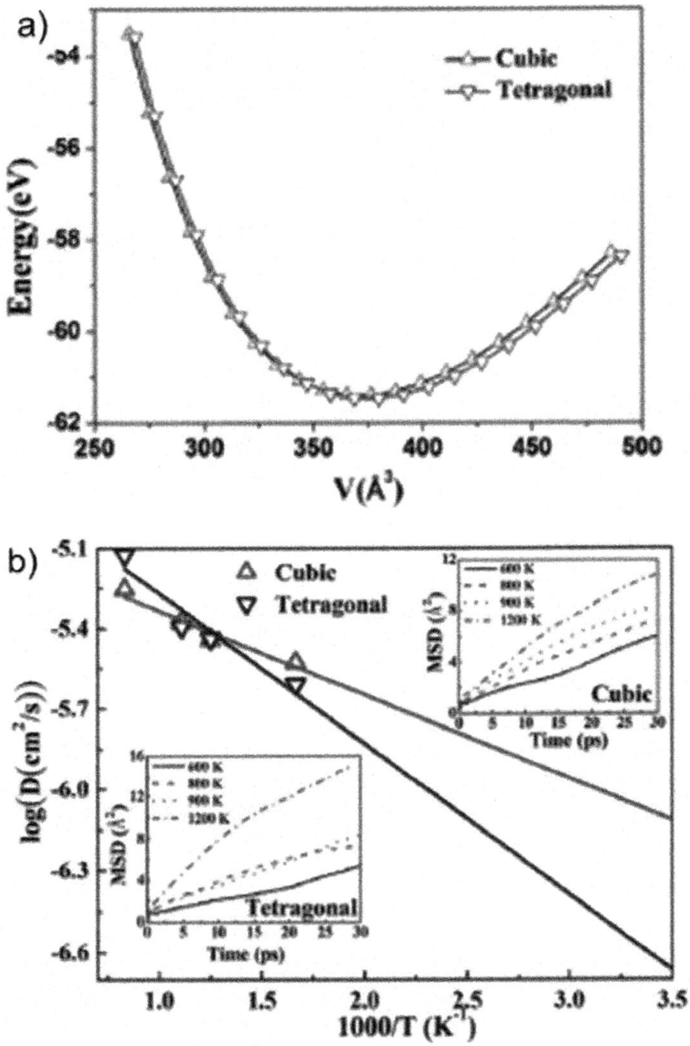

Figure 10.17 (a) DFT total energy as a function of cell volume fitted to the Murnaghan equation of state for cubic and tetragonal Na_3SbS_4. (b) Arrhenius plots of Na diffusion from AIMD simulations for the cubic and tetragonal Na_3SbS_4 with Na vacancy (concentration 2%). Insets in the figure are the MSD plots from which the diffusion coefficients were derived. Figures are from Ref. [191].

Sb substitution at the P site resulting in Na_3SbS_4 has also been investigated as a CSE [190]. As for Na_3PS_4, both cubic and tetragonal phases of this CSE are possible. From DFT simulations of the phase stability of both the cubic and tetragonal phases, it was shown that the two phases are both thermodynamically stable, and that the cubic phase was only 4.3 meV/atom less stable than the tetragonal phase (Fig. 10.17a) [190]. Hence, it was expected that the two phases would be interconvertible. The reversible phase transition from tetragonal to cubic, and back, was confirmed by *in situ* XRD [190]. However, from AIMD simulations, no significant ionic conductivity was found for either of the stoichiometric Na_3SbS_4 phases, in a similar vein to the Na_3PS_4 and Na_3PSe_4, and hence for the diffusion simulations, a 2% concentration of Na vacancies was included (Fig. 10.17b) [190]. From these simulations, it was found that at low temperatures (<800 K), the cubic phase showed the highest sodium diffusion (and hence highest ionic conductivity). The diffusivity increases with increasing temperatures, and at higher temperatures, the tetragonal phase shows higher sodium diffusivity [190]. E_a from AIMD simulations was similar to those found from experimental AC measurements, 0.06 eV for cubic (experimental 0 06 eV) and 0.1 eV (experimental 0.25 eV, 0.08 eV at higher temperatures when the cubic phase becomes more stable). Both the phases have promising sodium conductivity and could be promising CSEs [190]. One issue relating to all CSEs when implemented in a full battery cell setup is the reactivity between the CSE and the sodium metal anode [192]. DFT simulations have showed that the Na_3SbS_4 CSE can be protected by applying a hydrate coating [193]. This coating was also formed in experimental studies when exposing the CSE to air. NEB simulations also confirmed that sodium diffusion in the hydrate phase ($Na_3SbS_4{\cdot}8H_2O$) is in line with other CSEs (E_a = 502 meV) [193].

S-site doping with halogens improves CSE performance [194]. AIMD simulations of $Na_{3-y}PS_{4-x}Cl_x$ containing different concentrations of Na vacancies and Cl showed that at $y = 0.0625$ and $x = 0.125$, the ionic conductivity is increased to 9.03 mS/cm, with $E_a = 0.174$ eV. Filling the Na vacancies, while retaining the Cl doping, lowers the ionic conductivity to the same level as Na_3PS_4 [187]. MD simulations further showed that other halogens such as Br, F, and I can also increase the bulk conductivity of Na_3PS_4 [162]. In these simulations, a 2 × 2 × 2 cubic supercell was used with two S sites substituted with F, Cl, Br, or I and two Na vacancies introduced. This results in $Na_{2.88}PS_{3.88}X_{0.12}$ (X = Cl, F, Br, I) [1620]. At 525 K, the ionic conductivity of these compounds was all found to be similar to the undoped cubic phase (with the same vacancy concentration as the halogen-doped systems), with the ionic conductivity increasing from F to Cl to Br. For $Na_{2.88}PS_{3.88}I_{0.12}$, the ionic conductivity decreases due to the reduced I–P bond lengths as compared to the other X–P lengths [162]. The larger ionic radius of I^-, hence, reduces the diffusion window, and hence less Na jumping paths are open. From these simulations, Br was found to be the optimum halogen dopant [162].

Na_3PSe_4 is isostructural with cubic Na_3PS_4 and has been investigated as a CSE using the rationale that the larger ionic size of Se, as compared to S, will provide an enhanced structural framework for the relatively large Na^+ ion to diffuse through. Ionic conductivity in stoichiometric Na_3PSe_4 is negligible, making the inclusion of defects in modeling this CSE necessary [197]. To this end, a variety of defect structures were simulated, and their defect formation energies were calculated. These DFT simulations showed that the single Na vacancy is the most energetically probable defect in Na_3PSe_4 (E_f = 0.39 eV), with Na Frenkel (E_f = 1.14 eV) and Na interstitial defects (E_f = 1.78 eV) having much higher formation energies [179]. P deficiency in the lattice has much higher E_f (4.04 eV), and doping of the Na site with Al^{3+} ($Na_{45}AlP_{16}Se_{64}$) and Zr^{4+} ($Na_{44}ZrP_{16}Se_{64}$) was not found to be as energetically favorable as the formation of an Na vacancy [179]. From these simulations, it was postulated that the Na vacancy can be stabilized by either partial oxidation of Se^{2-} or by displacement of the Na ions neighboring the vacancy site, which would minimize the Coulombic repulsion between the remaining Na. NEB simulations of the Na vacancy diffusion revealed a migration barrier of 0.05 eV, suggesting facile sodium diffusion [179].

10.6 Summary

In this chapter, we have reviewed atomic-scale modeling of different types of NIB electrolytes, including LEs, PSEs, and CSEs. As summarized in Table 10.1, there are a set of key properties to characterize each electrolyte type, and atomic-scale computational methods (DFT, MD, and AIMD) are available to (1) understand and optimize the atomic/molecular structures, (2) calculate and predict key electrolyte properties, and (3) provide insights into the fundamental mechanisms such as Na-ion mobility and paths.

Table 10.1 Na-ion battery electrolytes, key properties, and computational methods

Na-Ion Battery (NIB) Electrolytes	Electrolyte Components	Key Electrolyte Properties	Computational Methods
Liquid electrolytes (LEs)	**Salts:** $NaPF_6$, $NaClO_4$, NaTFSI, NaFSI, $NaCF_3SO_3$ **Solvents:** EC, PC, BC, VC, EMC, DMC, DEC, monoglyme, diglyme, triglyme, and their mixtures **Additives:** VC, FEC	Electrochemical stability: HOMO/LUMO, reduction/oxidation potentials	DFT
		Solvation structure	DFT, MD, AIMD
		Transport properties: Transport number, viscosity, ionic conductivity/diffusivity	MD, AIMD
Polymeric solid electrolytes (PSEs)	**Salts:** $NaBF_4$, NaI, $NaClO_4$, NaSCN, $NaCF_4SO_3$ (NaTf), $Na(CF_3SO_2)_2N$, NaTFSI, $NaPF_6$, $NaAsF_6$, $NaSbF_6$ **Polymer matrices:** PEO, PEG	Structure optimization/Na-ion coordination structure	DFT, MD
		Ion mobility and paths, ionic conductivity/diffusivity	MD, AIMD
Ceramic solid electrolytes (CSEs)	NASICON materials, anti-perovskites, thiophosphates	Structure optimization, defects, dopants	DFT, MD, AIMD
		Ion migration barrier and paths	DFT
		Ionic conductivity/diffusivity	MD, AIMD

Although there is a considerable volume of research in atomic modeling of NIB electrolytes, some shortfalls are noted. For example, transport properties of NIB LEs are scarcely studied, probably due to the unavailability of robust force fields to describe atomic–ionic interactions for such systems. The evaluation of the electrochemical stability of LEs so far has been primarily based on the HOMO/LUMO of a single electrolyte component, which may not give accurate evaluation. As we discussed in Section 10.2, all the electrolyte components (salts, solvents, and additives) affect the reduction and oxidation potentials of the electrolyte systems; therefore, a more accurate and robust evaluation of an electrolyte's electrochemical stability should be based on the reduction and oxidation potentials of a model molecular cluster containing all the electrolyte components. Furthermore, defects and dopants are common in real CSEs, but their effects on Na-ion migration and ionic conductivity have not been well studied using atomic-scale modeling.

The advantage of atomistic modeling is the ability to examine and play with the atomic/molecular structures of electrolytes and link with electrolyte properties. To this end, atomic modeling could be used for designing next-generation NIB electrolytes with optimal properties. For this, systematic studies of NIB electrolytes using atomic modeling are desired, which require (2) generation and optimization of various possible electrolytes, (2) calculations of a wide range of electrolyte properties, and (3) selection and screening of optimal electrolytes. Such systematic studies, if combined with machine learning methods, could speed up the discovery and design of optimal NIB electrolytes.

Acknowledgments

Funding support from the Engineering and Physical Sciences Research Council (EPSRC) in the United Kingdom under grant number EP/M027066/1 and EP/R021554/1 is gratefully acknowledged.

References

1. J.-Y. Hwang, S.-T. Myung, and Y.-K. Sun, *Chem. Soc. Rev.*, 2017, **46**, 3529–3614.
2. K. Kubota and S. Komaba, *J. Electrochem. Soc.*, 2015, **162**, A2538.
3. C. Delmas, *Adv. Energy Mater.*, 2018, **8**, 1703137.
4. J. Tang, A. D. Dysart, and V. G. Pol, *Curr. Opin. Chem. Eng.*, 2015, **9**, 34–41.

5. N. Yabuuchi, K. Kubota, M. Dahbi, and S. Komaba, *Chem. Rev.*, 2014, **114**, 11636–11682.
6. J. Song, B. Xiao, Y. Lin, K. Xu, and X. Li, *Adv. Energy Mater.*, 2018, **8**, 1703082.
7. H. Au, H. Alptekin, A. C. S. Jensen, E. Olsson, C. A. O'Keefe, T. Smith, M. Crespo-Ribadeneyra, T. F. Headen, C. P. Grey, Q. Cai, A. J. Drew, and M-M. Titirici, *Energy Environ. Sci.*, 2020, **13**, 3469–3479.
8. E. Olsson, J. Cottom, H. Au, Z. Guo, A. C. S. Jensen, H. Alptekin, A. J. Drew, M.-M. Titirici, and Q. Cai, *Adv. Funct. Mater.*, 2020, **30**, 1908209.
9. M. S. Islam and C. A. J. Fisher, *Chem. Soc. Rev.*, 2014, **43**, 185–204.
10. Q. Bai, H. Chen, L. Yang, and Y. Mo, *Adv. Energy Mater.*, 2018, **8**, 1702998.
11. J. Zhang, X. Yao, R. K. Misra, Q. Cai, and Y. Zhao, *J. Mater. Sci. Technol.*, 2020, **44**, 237–257.
12. A. Wang, S. Kadam, H. Li, S. Shi, and Y. Qi, *npj Comput. Mater.*, 2018, **4**, 15.
13. Q. Zhao, S. Stalin, C. Z. Zhao, and L. A. Archer, *Nat. Rev. Mater.*, 2020, **5**, 229–252.
14. C. Zhou, S. Bag, and V. Thangadurai, *ACS Energy Lett.*, 2018, **3**, 2181–2198.
15. G. Åvall, J. Mindemark, D. Brandell, and P. Johansson, *Adv. Energy Mater.*, 2018, **8**, 1703036.
16. K. Leung, S. B. Rempe, M. E. Foster, Y. Ma, J. M. Martinez del la Hoz, N. Sai, and P. B. Balbuena, *J. Electrochem. Soc.*, 2014, **161**, A213–A221.
17. T. A. Pham, *Int. J. Quantum Chem.*, 2019, **119**, e25795.
18. T. A. Pham, K. E. Kweon, A. Samanta, V. Lordi, and J. E. Pask, *J. Phys. Chem. C*, 2017, **121**, 7b06457.
19. G. Åvall, J. Mindemark, D. Brandell, and P. Johansson, *Adv. Energy Mater.*, 2018, **1703036**, 1–22.
20. J. Wahlers, K. D. Fulfer, D. P. Harding, D. G. Kuroda, R. Kumar, and R. Jorn, *J. Phys. Chem. C*, 2016, **120**, 17949–17959.
21. A. Urban, D. H. Seo, and G. Ceder, *npj Comput. Mater.*, 2015, **2**, 16002.
22. G. Ceder, S. P. Ong, and Y. Wang, *MRS Bull.*, 2018, **43**, 782–788.
23. Y. Xiao, Y. Wang, S. H. Bo, J. C. Kim, L. J. Miara, and G. Ceder, *Nat. Rev. Mater.*, 2020, **5**, 105–126.
24. H. Che, S. Chen, Y. Xie, H. Wang, K. Amine, X. Z. Liao, and Z. F. Ma, *Energy Environ. Sci.*, 2017, **10**, 1075–1101.

25. Z. Lin, Q. Xia, W. Wang, W. Li, and S. Chou, *InfoMat*, 2019, **1**, 376–389.
26. G. Kamath, R. W. Cutler, S. A. Deshmukh, M. S. Fard, R. Parrish, J. Huether, D. P. Butt, H. Xiong, and S. K. R. S. Sankaranarayanan, *J. Phys. Chem. C*, 2014, **118**, 13406–13416.
27. Y. Huang, L. Zhao, L. Li, M. Xie, F. Wu, and R. Chen, *Adv. Mater.*, 2019, **31**, 1–41.
28. B. Narayanan, P. C. Redfern, R. S. Assary, and L. A. Curtiss, *Chem. Sci.*, 2019, **10**, 7449–7455.
29. A. Karatrantos, S. Khan, T. Ohba, and Q. Cai, *Phys. Chem. Chem. Phys.*, 2018, **20**, 6307–6315.
30. A. V. Karatrantos, T. Ohba, and Q. Cai, *J. Mol. Liq.*, 2020, **320**, 114351.
31. H. Nguyen, S. Hy, E. Wu, Z. Deng, M. Samiee, T. Yersak, J. Luo, S. P. Ong, and Y. S. Meng, *J. Electrochem. Soc.*, 2016, **163**, A2165–A2171.
32. X. Chen, F. Chen, and M. Forsyth, *Phys. Chem. Chem. Phys.*, 2017, **19**, 16426.
33. D. Morales, R. E. Ruther, J. Nanda, and S. Greenbaum, *Electrochim. Acta*, 2019, **304**, 239–245.
34. M. Ernzerhof and G. E. Scuseria, *J. Chem. Phys.*, 1999, **110**, 5029.
35. D. Fritsch, B. J. Morgan, and A. Walsh, *Nanoscale Res. Lett.*, 2017, **12**, 19.
36. C. Freysoldt, B. Grabowski, T. Hickel, J. Neugebauer, G. Kresse, A. Janotti, and C. G. Van De Walle, *Rev. Mod. Phys.*, 2014, **86**, 253–305.
37. N. D. M. Hine, K. Frensch, W. M. C. Foulkes, and M. W. Finnis, *Phys. Rev. B*, 2009, **79**, 024112.
38. D. Frenkel and B. Smit, *Understanding Molecular Simulation: From Algorithms to Applications*, Academic Press, 2nd edn., 2001.
39. D. G. Ladha, *Mater. Today Chem.*, 2019, **11**, 94–111.
40. W. Kohn and L. Sham, *Phys. Rev.*, 1965, **385**, 1133.
41. L. J. Bartolotti and K. Flurchick, *Rev. Comput. Chem.*, 2007, **7**, pp. 187–216.
42. A. Labdelli, A. Boukortt, S. Meskine, H. Abbassa, and A. Zaoui, *Int. J. Comput. Mater. Sci. Eng.*, 2018, **07**, 1850018.
43. J. F. Janak, *Phys. Rev. B*, 1978, **18**, 7165–7168.
44. K. Burke, F. G. Cruz, and K.-C. Lam, *J. Chem. Phys.*, 1998, **109**, 8161–8167.
45. B. Lee, C. K. Lee, C. S. Hwang, and S. Han, *Curr. Appl. Phys.*, 2011, **11**, S293–S296.

46. C. J. Cramer and D. G. Truhlar, *Phys. Chem. Chem. Phys.*, 2009, **11**, 10757–816.

47. J. Perdew and A. Zunger. *Phys. Rev. B.* 1981, **23**, 5048–5079.

48. S. Lany and A. Zunger, *Phys. Rev. B Condens. Matter Mater. Phys.*, 2008, **78**, 17–20.

49. A. Janotti and C. G. Van de Walle, LDA + U and hybrid functional calculations for defects in ZnO, SnO_2, and TiO_2. In *Advanced Calculations for Defects in Materials: Electronic Structure Methods*, Wiley-VCH, 2011, pp. 155–164.

50. Y.-L. Lee, J. Kleis, J. Rossmeisl, and D. Morgan, *Phys. Rev. B*, 2009, **80**, 224101.

51. S. F. Sousa, P. A. Fernandes, and M. J. Ramos, *J. Phys. Chem. A*, 2007, **111**, 10439–10452.

52. Q. He, B. Yu, Z. Li, and Y. Zhao, *Energy Environ. Mater.*, 2019, **2**, 264–279.

53. M. Rahm and R. Hoffmann, *J. Am. Chem. Soc.*, 2015, **137**, 10282–10291.

54. S. J. Clark and M. C. Payne, 2002, **14**, 2717–2744.

55. S. Dudarev, G. Botton, and S. Savrasov, *Phys. Rev. B*, 1998, **57**, 1505–1509.

56. A. V. Krukau, O. A. Vydrov, A. F. Izmaylov, and G. E. Scuseria, *J. Chem. Phys.*, 2006, **125**, 224106.

57. M. Shakourian-Fard, G Kamath, and S. K. R. S. Sankaranarayanan, *ChemPhysChem*, 2016, 2916–2930.

58. H. Kumar, E. Detsi, D. P. Abraham, and V. B. Shenoy, *Chem. Mater.*, 2016, **28**, 8930–8941.

59. A. Urban, D. H. Seo, and G. Ceder, *npj Comput. Mater.*, 2016, **2**, 16002.

60. O. Borodin, *Curr. Opin. Electrochem.*, 2019, **13**, 86–93.

61. K. Westman, R. Dugas, P. Jankowski, W. Wieczorek, G. Gachot, M. Morcrette, E. Irisarri, A Ponrouch, M. R. Palacín, J. M. Tarascon, and P. Johansson, *ACS Appl. Energy Mater.*, 2018, **1**, 2671–2680.

62. M. Shakourian-Fard, G Kamath, K. Smith, H. Xiong, and S. K. R. S. Sankaranarayanan, *J. Phys. Chem. C*, 2015, **119**, 22747–22759.

63. T. Binninger, A. Marcolongo, M. Mottet, V. Weber, and T. Laino, *J. Mater. Chem. A*, 2020, **8**, 1347–1359.

64. C. Bommier and X. Ji, *Isr. J. Chem.*, 2015, **55**, 486–507.

65. S. Alvin, D. Yoon, C. Chandra, H. S. Cahyadi, J. H. Park, W. Chang, K. Y. Chung, and J. Kim, *Carbon N. Y.*, 2019, **145**, 67–81.

66. A. C. S. Jensen, E. Olsson, H. Au, H. Alptekin, Z. Yang, S. Cottrell, K. Yokoyama, Q. Cai, M.-M. Titirici, and A. J. Drew, *J. Mater. Chem. A*, 2020, **8**, 743–749.

67. H. Alptekin, H. Au, A. C. Jensen, E. Olsson, M. Goktas, T. F. Headen, P. Adelhelm, Q. Cai, A. J. Drew, and M.-M. Titirici, *ACS Appl. Energy Mater.*, 2020, **3**, 9918–9927.

68. G. Henkelman and H. Jónsson, *J. Chem. Phys.*, 2000, **113**, 9978–9985.

69. S. H. Bo, Y. Wang, J. C. Kim, W. D. Richards, and G. Ceder, *Chem. Mater.*, 2016, **28**, 252–258.

70. M. P. Allen and D. J. Tildesley, *Computer Simulation of Liquids*, Oxford University Press, Oxford, 1989.

71. A. R. Leach, *Molecular Modelling: Principles and Applications*, Prentice Hall, Pearson, 2nd edn., 2001.

72. M. E. Tuckerman and G. J. Martyna, *J. Phys. Chem. B*, 2000, **104**, 159–178.

73. P. H. Hünenberger, *Adv. Polym. Sci.*, 2005, **173**, 105–147.

74. E. A. Carter, *Science (80-)*, 2008, **321**, 800.

75. P. Ganesh, D. E. Jiang, and P. R. C. Kent, *J. Phys. Chem. B*, 2011, **115**, 3085–3090.

76. J.-F. Wax, M. R. Johnson, L. E. Bove, and M. Mihalkovič, *Phys. Rev. B*, 2011, **83**, 144203.

77. K. Leung, *Phys. Chem. Chem. Phys.*, 2015, **17**, 1637–1643.

78. F. A. Soto, A. Marzouk, F. El-Mellouhi, and P. B. Balbuena, *Chem. Mater.*, 2018, **30**, 3315–3322.

79. V. Sharma, K. Ghatak, and D. Datta, *J. Mater. Sci.*, 2018, **53**, 14423–14434.

80. Z. Zhang, P.-N. Roy, H. Li, M. Avdeev, and L. F. Nazar, *J. Am. Chem. Soc*, 2019, **141**, 26.

81. C. Cazorla, *Thin Solid Films*, 2010, **518**, 6951–6961.

82. W. W. Tipton, C. R. Bealing, K. Mathew, and R. G. Hennig, *Phys. Rev. B Condens. Matter Mater. Phys.*, 2013, **87**, 184114.

83. M. J. Boyer and G. S. Hwang, *Electrochim. Acta*, 2018, **266**, 326–331.

84. A. Karatrantos, R. J. Composto, K. I. Winey, M. Kröger, and N. Clarke, *Polymers (Basel)*, 2019, **11**(5), 876.

85. E. Olsson, X. Aparicio-Anglès, and N. H. de Leeuw, *Phys. Chem. Chem. Phys.*, 2017, **19**, 13960–13969.

86. M. S. Islam, *Solid State Ionics*, 2002, **154–155**, 75–85.

87. M. S. Islam, M. Cherry, and C. R. A. Catlow, *J. Solid State Chem.*, 1996, **124**, 230–237.

88. F. Chen, H. Zhang, and Z. Hu, *ACTA Metall. Sin.*, 2004, **40**, 731–735.

89. G. Kamath, R. W. Cutler, S. A. Deshmukh, M. Shakourian-Fard, R. Parrish, J. Huether, D. P. Butt, H. Xiong, and S. K. R. S. Sankaranarayanan, *J. Phys. Chem. C*, 2014, **118**, 13406–13416.

90. T. C. Yeh, J. L. Routbort, and T. O. Mason, *Solid State Ionics*, 2013, **232**, 138–143.

91. M. S. Khan, A. V. Karatrantos, T. Ohba, and Q. Cai, *Phys. Chem. Chem. Phys.*, 2019, **21**, 22722–22731.

92. VASP, https://www.vasp.at.

93. CP2K, https://www.cp2k.org.

94. Gaussian, http://www.gaussian.com.

95. CRYSTAL, http://www.crystal.unito.it.

96. CASTEP, http://www.castep.org.

97. Wien2K, http://susi.theochem.tuwien.ac.at/index.html, (accessed 12 January 2021).

98. Quantum Espresso, https://www.quantum-espresso.org.

99. DL_POLY, https://www.scd.stfc.ac.uk/Pages/DL_POLY.aspx.

100. GROMACS, http://www.gromacs.org.

101. NAMD, https://www.ks.uiuc.edu/Research/namd.

102. LAMMPS, http://lammps.sandia.gov.

103. A. Mauger, C. M. Julien, A. Paolella, M. Armand, and K. Zaghib, *Mater. Sci. Eng. R*, 2018, **134**, 1–21.

104. M. Marcinek, J. Syzdek, M. Marczewksi, M. Piszcz, L. Niedzicki, M. Kalita, A. Plewa-Marczewska, A. Bitner, P. Wieczorek, and T. Trzeciak, *Solid State Ionics*, 2015, **276**, 107–126.

105. X. Qu, Y. Zhang, N. N. Rajput, A. Jain, K. A. Maginn, and E. Persson, *J. Phys. Chem. C*, 2017, **121**, 16126–16136.

106. M. Li, C. Wang, Z. Chen, K. Xu, and J. Lu, *Chem. Rev.*, 2020, **120**, 6783–6819.

107. G. Yan, D. Alves-Dalla-Corte, W. Yin, N. Madern, G. Gachot, and J. M. Tarascon, *J. Electrochem. Soc.*, 2018, **165**, A1222–A1230.

108. A. Ponrouch, D. Monti, A. Boschin, B. Steen, P. Johansson, and M. R. Palacín, *J. Mater. Chem. A*, 2015, **3**, 22–42.

109. A. Ponrouch, R. Dedryvere, D. Monti, A. E. Demet, J. M. Ateba Mba, L. Croguennec, C. Masquelier, P. Johansson, and M. R. Palacin, *Energy Environ. Sci.*, 2013, **6**, 2361.

110. K. Vignarooban, R. Kushagra, A. Elango, P. Badami, B. E. Mellander, X. Xu, T. G. Tucker, C. Nam, and A. M. Kannan, *J. Mater. Chem. A*, 2016, **41**, 2829–2846.

111. A. Basile, M. Hilder, F. Makhlooghiazad, C. Pozo-Gonzalo, D. R. Macfarlane, P. C. Howlett, and M. Forsyth, *Adv. Energy Mater.*, 2018, **1703491**, 1–20.

112. H. Kumar, E. Detsi, D. P. Abraham, and V. B. Shenoy, *Chem. Mater.*, 2016, **28**, 8930–8941.

113. M. Watanabe, M. L. Thomas, S. Zhang, K. Ueno, T. Yasuda, and K. Dokko, *Chem. Rev.*, 2017, **117**, 7190–7239.

114. D. Monti, E. Jónsson, A. Boschin, M. R. Palacín, A. Ponrouch, and P. Johansson, *Phys. Chem. Chem. Phys.*, 2020, **22**, 22768–22777.

115. M. Goktas, C. Bolli, J. Buchheim, E. J. Berg, P. Novák, F. Bonilla, T. Rojo, S. Komaba, K. Kubota, and P. Adelhelm, *ACS Appl. Mater. Interfaces*, 2019, **11**, 32844–32855.

116. G. Ramos-Sanchez, F. A. Soto, J. M. Martinez De La Hoz, Z. Liu, P. P. Mukherjee, F. El-Mellouhi, J. M. Seminario, and P. B. Balbuena, *J. Electrochem. Energy Convers. Storage*, 2016, **13**, 1–10.

117. R. Jorn and R. Kumar, *Electrochem. Soc. Interface*, 2017, **26**, 55–59.

118. O. Borodin, X. Ren, J. Vatamanu, A. Von Wald Cresce, J. Knap, and K. Xu, *Acc. Chem. Res.*, 2017, **50**, 2886–2894.

119. E. Jónsson and P. Johansson, *Phys. Chem. Chem. Phys.*, 2012, **14**, 10774–10779.

120. A. V. Cresce, S. M. Russell, O. Borodin, J. A. Allen, M. A. Schroeder, M. Dai, J. Peng, M. P. Gobet, S. G. Greenbaum, R. E. Rogers, and K. Xu, *Phys. Chem. Chem. Phys.*, 2017, **19**, 574–586.

121. T. Anh Pham, K. E. Kweon, A. Samanta, V. Lordi, and J. E. Pask, *J. Phys. Chem. C*, 2018, **121**, 21913–21920.

122. M. Okoshi, Y. Yamada, S. Komaba, A. Yamada, and H. Nakai, *J. Electrochem. Soc.*, 2016, **164**, A54–A60.

123. M. Okoshi, Y. Yamada, A. Yamada, and H. Nakai, *J. Electrochem. Soc.*, 2013, **160**, A2160–A2165.

124. K. L. Browning, R. L. Sacci, and G. M. Veith, *J. Electrochem. Soc.*, 2017, **164**, A580–A586.

125. M. Okoshi, C.-P. Chou, and H. Nakai, *J. Phys. Chem. B*, 2018, **122**, 2600–2609.

126. E. Flores, G. Åvall, S. Jeschke, and P. Johansson, *Electrochim. Acta*, 2017, **233**, 134–141.

127. H. Zhang, I. Hasa, and S. Passerini, *Adv. Energy Mater.*, 2018, **8**, 1870082.

128. S. Seki, K. Hayamizu, S. Tsuzuki, K. Takahashi, Y. Ishino, M. Kato, E. Nozaki, H. Watanabe, and Y. Umebayashi, *J. Electrochem. Soc.*, 2018, **165**, A542–A546.

129. B. Ravikumar, S. Kumari, M. Mynam, and R. Beena, *Proceedings of the 7th International Conference on Advances in Energy Research*, 2021, 1071-1081

130. T. P. Liyana-Arachchi, J. B. Haskins, C. M. Burke, K. M. Diederichsen, B. D. McCloskey and J. W. Lawson, *J. Phys. Chem. B*, 2018, **122**, 8548–8559.

131. L. Qiao, X. Judez, T. Rojo, M. Armand, and H. Zhang, *J. Electrochem. Soc.*, 2020, **167**, 070534.

132. D. E. Fenton, J. M. Parker, and P. V Wright, *Polymer (Guildf)*, 1973, **14**, 589.

133. P. V Wright, *Br. Polym. J.*, 1975, **7**, 319–327.

134. J. Mindemark, R. Mogensen, M. J. Smith, M. M. Silva, and D. Brandell, *Electrochem. Commun.*, 2017, **77**, 58–61.

135. K. Nazir, A. F. Aziz, M. Z. A. Yahya, and A. M. M. Ali, *AIP Conf. Proc.*, 2017, **1877**, 40003.

136. A. Memboeuf, K. Vékey, and G. Lendvay, *Eur. J. Mass Spectrom.*, 2011, **17**, 33–46.

137. K. Li, S. R. Galle Kankanamge, T. K. Weldeghiorghis, R. Jorn, D. G. Kuroda, and R. Kumar, *J. Phys. Chem. C*, 2018, **122**, 4747–4756.

138. T. V Kaulgud, N. R. Dhumal, and S. P. Gejji, *J. Phys. Chem. A*, 2006, **110**, 9231–9239.

139. N. R. Dhumal and S. P. Gejji, *J. Mol. Struct. Theochem*, 2008, **859**, 86–92.

140. N. R. Dhumal and S. P. Gejji, *J. Phys. Chem. A*, 2006, **110**, 219–227.

141. N. R. Dhumal and S. P. Gejji, *J. Mol. Struct. Theochem*, 2006, **758**, 233–240.

142. C. Zhang, S. Gamblel, D. Ainsworth, S. A. M. Z., Y. G. Andreev, and P. G. Bruce, *Nat. Mater.*, 2009, **8**, 580–584.

143. A. Liivat, *Electrochimica Acta*, 2011, **57**, pp. 244–249.

144. K. Laasonen and M. L. Klein, *J. Chem. Soc. Faraday Trans.*, 1995, **91**, 2633–2638.

145. A. Van Zon and S. W. De Leeuw, *Electrochim. Acta*, 2001, **46**, 1539–1544.

146. G. E. Mills and C. R. A. Catlow, *J. Chem. Soc. Chem. Commun.*, 1994, **18**, 2037-2039.

147. C. R. A. Catlow and G. E. Mills, *Electrochim. Acta*, 1995, **40**, 2057–2062.

148. M. Forsyth, V. A. Payne, M. A. Ratner, S. W. de Leeuw, and D. F. Shriver, *Solid State Ionics*, 1992, **53–56**, 1011–1026.

149. N. A. Stolwijk, M. Wiencierz, and S. Obeidi, *Faraday Discuss.*, 2007, **134**, 157–169.

150. V. A. Payne, M. C. Lonergan, M. Forsyth, M. A. Ratner, D. F. Shriver, S. W. de Leeuw, and J. W. Perram, *Solid State Ionics*, 1995, **81**, 171–181.

151. V. A. Payne, J. Xu, M. Forsyth, M. A. Ratner, D. F. Shriver, and S. W. de Leeuw, *J. Chem. Phys.*, 1995, **103**, 8734–8745.

152. V. A. Payne, J. Xu, M. Forsyth, M. A. Ratner, D. F. Shriver, and S. W. de Leeuw, *J. Chem. Phys.*, 1995, **103**, 8746–8755.

153. C. M. Burba, N. M. Rocher, R. Frech, and D. R. Powell, *J. Phys. Chem. B*, 2008, **112**, 2991–2995.

154. S. Neyertz, J. Thomas, and D. Brown, *Comput. Polym. Sci.*, 1995, **5**, 107–120.

155. S. Neyertz and D. Brown, *J. Chem. Phys.*, 1996, **104**, 3797–3809.

156. J. J. de Jonge, A. van Zon, and S. W. de Leeuw, *Solid State Ionics*, 2002, **147**, 349–359.

157. X. Chen, F. Chen, M. S. Liu, and M. Forsyth, *Solid State Ionics*, 2016, **288**, 271–276.

158. K. J. Lin and J. K. Maranas, *Macromolecules*, 2012, **45**, 6230–6240.

159. K.-J. Lin and J. K. Maranas, *Phys. Rev. E*, 2013, **88**, 52602.

160. K. J. Lin, K. Li, and J. K. Maranas, *RSC Adv.*, 2013, **3**, 1564–1571.

161. K. J. Lin and J. K. Maranas, *Phys. Chem. Chem. Phys.*, 2013, **15**, 16143-16151.

162. N. J. J. De Klerk and M. Wagemaker, *Chem. Mater.*, 2016, **28**, 3122–3130.

163. V. S. Kandagal, M. D. Bharadwaj, and U. V Waghmare, *J. Mater. Chem. A*, 2015, **3**, 12992–12999.

164. Z. Yu, S. L. Shang, J. H. Seo, D. Wang, X. Luo, Q. Huang, S. Chen, J. Lu, X. Li, Z. K. Liu, and D. Wang, *Adv. Mater.*, 2017, **29**, 0–6.

165. C. Zhao, L. Liu, X. Qi, Y. Lu, F. Wu, J. Zhao, Y. Yu, Y. S. Hu, and L. Chen, *Adv. Energy Mater.*, 2018, **8**, 14–16.

166. Z. Deng, Y. Mo, and S. P. Ong, *NPG Asia Mater.*, 2016, **8**, 254.

167. Z. Zhang, Z. Zou, K. Kaup, R. Xiao, S. Shi, M. Avdeev, Y. S. Hu, D. Wang, B. He, H. Li, X. Huang, L. F. Nazar, and L. Chen, *Adv. Energy Mater.*, 2019, **9**, 1–14.

168. P. A. Aparicio and N. H. De Leeuw, *Phys. Chem. Chem. Phys.*, 2020, **22**, 6653–6659.

169. P. A. Aparicio, J. A. Dawson, M. S. Islam, and N. H. de Leeuw, *J. Phys. Chem. C*, 2018, **122**, 25829–25836.

170. W. Ling, N. Fu, J. Yue, X. X. Zeng, Q. Ma, Q. Deng, Y. Xiao, L. J. Wan, Y. G. Guo, and X. W. Wu, *Adv. Energy Mater.*, 2020, **10**, 1903966.

171. Y. Deng, C. Eames, L. H. B. Nguyen, O. Pecher, K. J. Griffith, M. Courty, B. Fleutot, J. N. Chotard, C. P. Grey, M. S. Islam, and C. Masquelier, *Chem. Mater.*, 2018, **30**, 2618–2630.

172. H. Park, K. Jung, M. Nezafati, C. S. Kim, and B. Kang, *ACS Appl. Mater. Interfaces*, 2016, **8**, 27814–27824.

173. A. Buzlukov, I. Arapova, Y. V Baklanova, N. I. Medvedeva, T. A. Denisova, A. A. Savina, B. Lazoryak, E. G. Khaikina, and M. Bardet, *Phys. Chem. Chem. Phys.*, 2020, **22**, 144.

174. J. A. Dawson, H. Chen, and M. Saiful Islam, *J. Phys. Chem. C*, 2018, **122**, 23978–23984.

175. E. Ahiavi, J. A. Dawson, U. Kudu, M. Courty, M. S. Islam, O. Clemens, C. Masquelier, and T. Famprikis, *J. Power Sources*, 2020, **471**, 228489.

176. K. Kim and D. J. Siegel, *J Mater. Chem. A*, 2019, **7**, 3216–3227.

177. E. Olsson, J. Cottom, X. Aparicio-Anglès, and N. H. De Leeuw, *Phys. Chem. Chem. Phys.*, 2020, **22**, 692–699.

178. L. Yin, M. Murphy, K. Kim, L. Hu, J. Cabana, D. J. Siegel, and S. H. Lapidus, *Inorg. Chem.*, 2020, **59**, 11244–11247.

179. S. H. Bo, Y. Wang, J. C. Kim, W. D. Richards, and G. Ceder, *Chem. Mater.*, 2016, **28**, 252–258.

180. T. Famprikis, J. A. Dawson, F. Fauth, O. Clemens, E. Suard, B. Fleutot, M. Courty, J.-N. Chotard, M. S. Islam, and C. Masquelier, *ACS Mater. Lett.*, 2019, **15**, 641–646.

181. H. Tang, Z. Deng, Z. Lin, Z. Wang, I. H. Chu, C. Chen, Z. Zhu, C. Zheng, and S. P. Ong, *Chem. Mater.*, 2018, **30**, 163–173.

182. S. H. Bo, Y. Wang, and G. Ceder, *J. Mater. Chem. A*, 2016, **4**, 9044–9053.

183. T. Krauskopf, S. P. Culver, and W. G. Zeier, *Inorg. Chem.*, 2018, **57**, 4739–4744.

184. T. Krauskopf, C. Pompe, M. A. Kraft, and W. G. Zeier, *Chem. Mater.*, 2017, **29**, 8859–8869.

185. J. A. Dawson, P. Canepa, M. J. Clarke, T. Famprikis, D. Ghosh, and M. S. Islam, *Chem. Mater.*, 2019, **31**, 5296–5304.

186. Y. Hibi, N. Tanibata, A. Hayashi, and M. Tatsumisago, *Solid State Ionics*, 2015, **270**, 6–9.

187. X. Feng, P. H. Chien, Z. Zhu, I. H. Chu, P. Wang, M. Immediato-Scuotto, H. Arabzadeh, S. P. Ong, and Y. Y. Hu, *Adv. Funct. Mater.*, 2019, **29**, 1–9.

188. R. P. Rao, H. Chen, L. L. Wong, and S. Adams, *J. Mater. Chem. A*, 2017, **5**, 3377–3388.

189. K. Oh, D. Chang, I. Park, K. Yoon, and K. Kang, *Chem. Mater.*, 2019, **31**, 6066–6075.

190. Z. Zhang, E. Ramos, F. Lalère, A. Assoud, K. Kaup, P. Hartman, and L. F. Nazar, *Energy Environ. Sci.*, 2018, **11**, 87–93.

191. D. Zhang, X. Cao, D. Xu, N. Wang, C. Yu, W. Hu, X. Yan, J. Mi, B. Wen, L. Wang, and L. Zhang, *Electrochim. Acta*, 2018, **259**, 100–109.

192. P. Hu, Y. Zhang, X. Chi, K. Kumar Rao, F. Hao, H. Dong, F. Guo, Y. Ren, L. C. Grabow, and Y. Yao, *ACS Appl. Mater. Interfaces*, 2019, **11**, 9672–9678.

193. Y. Tian, Y. Sun, D. C. Hannah, Y. Xiao, H. Liu, K. W. Chapman, S. H. Bo, and G. Ceder, *Joule*, 2019, **3**, 1037–1050.

194. I.-H. Chu, C. S. Kompella, H. Nguyen, Z. Zhu, S. Hy, Z. Deng, Y. S. Meng, and S. P. Ong, *Sci. Rep.*, 2016, **6**, 33733.

Chapter 11

Production, Characteristic, and Development of Separators

Weihua Chen, Jiyu Zhang, Xinle Li, and Xiaoniu Guo
College of Chemistry, Zhengzhou University, Zhengzhou 450001, P. R. China
chenweih@zzu.edu.cn

11.1 Introduction

Separators are the critical component for sodium-ion batteries (SIBs), whose primary function is to prevent the direct contact between anode and cathode and provide channels for sodium-ion transmission. Indeed, the separator also acts as an electrolyte reservoir in liquid electrolyte systems to ensure the rapid transport of sodium ions inside batteries. Although the separator is not directly participating in the electrochemical reaction, its physical and chemical properties strongly affect the energy density, cycle life, and rate performance of batteries, such as pore structure, surface chemical groups and wettability, and so on.

Handbook of Sodium-Ion Batteries: Materials and Characterization
Edited by Rohit R. Gaddam and X. S. (George) Zhao

ISBN 978-981-4968-15-7 (Hardcover), 978-1-003-30874-4 (eBook)
www.jennystanford.com

Polyolefin separators have been widely used in traditional lithium-ion batteries (LIBs) because of their mature production process, high electrochemical and chemical stability, including polyethylene (PE), polypropylene (PP) monolayer separator, and PP/PE/PP tri-layers composite separator. However, the development of SIB separator has been ignored for a long time. Due to poor wettability to electrolytes and low electrolyte uptake, using a polyolefin separator often leads to long activation time and low discharge capacity, which cannot fully deliver electrode's charge and discharge capacity in SIBs, especially in high-viscosity electrolytes. Through modification and structural design of polyolefin separators, such as coating, surface grafting modification, the introduction of new functional materials, etc., the wettability and thermal/mechanical stability of the polyolefin separator can be effectively improved, facilitating their applications in SIBs.

Alternatively, glass fiber (GF) membranes are selected as separators for SIBs in lab-scale experiments due to their excellent wettability to electrolytes, including ethylene carbonate (EC) and propylene carbonate (PC). Despite this, it is still challenging to achieve a safe and stable battery during the Na^+ insertion/extraction process. Large pores in the GF separator make it invalid in preventing dendrite penetration during the long cycling process and shorten out the batteries. And insufficient mechanical property poses additional barriers during the process of manufacturing battery devices. Additionally, the large thickness of the GF separator further limits the enhancement of battery energy density. Therefore, it is critical to developing novel separators for achieving practical high-performance SIBs.

Ideally, there are some essential requirements for separators, such as good wettability and electrolyte uptake, excellent thermal/mechanical stability, and suitable pore structures. The separator should also be as thin as possible, and the process should be simple, where the energy density and application costs of SIBs are not compromised. However, there are only a limited number of separators available in the market for SIBs. In particular, the separator performance in SIBs is immature compared with LIBs, and there is plenty of room for improvement. This situation is because most of the separators currently used in SIBs are inherited from LIBs. They are not usually explicitly developed for SIBs and are not fully optimized for sodium-based systems [1]. In this sense, this lowers the costs of separators for mass production and the research workload, as well as encourages scientific workers to develop specialized sodium-ion separators via exploring various possible components and advanced technology, better serving the performance improvement and application of SIBs. Therefore, this chapter will mainly describe the following sections, including separator preparation, test technology, separator components, the improvement strategies for separator performance, and make a prospect for the future development of separators in SIBs.

11.2 Separator Preparation Technology

Recently, related studies on SIB separators emerged in an endless stream, one of the important reasons being the continuous development of various separator preparation technologies. With advanced preparation and modification technology, various physical structures of the separator can be designed effectively, thus greatly promoting the development of high-performance and functional separators.

11.2.1 Electrospinning and Electrostatic Spraying

The principles of electrospinning and electrostatic spraying are remarkably similar. Both are driven by charged solution through an electric field and collected in a collector. Their biggest difference is the preparation conditions and products. Typically, when the spinning liquid has a high molecular weight and high liquid viscosity, the hydrodynamic stability of the electrostatic jet is proportional to the viscosity, and the jet will generate nanofibers. This process is named as electrospinning. Unlike the above situation, when the precursor solution concentration is deficient, the jet is sprayed as droplets. The process is defined as electrostatic spray. Generally, the organic/inorganic composite separator obtained by the combination of the two methods and the inorganic nanoparticles are attached to the surface of nanofibers, which can directly contact the electrolyte and further improve the performance of the separator. For example, Yanilmaz et al. [2] used electrospinning and electrostatic spraying to spray nanofibers and nanoparticles on the receiving drum simultaneously and obtained a composite separator with SiO_2 nanoparticles attached to PVDF nanofibers (Fig. 11.1).

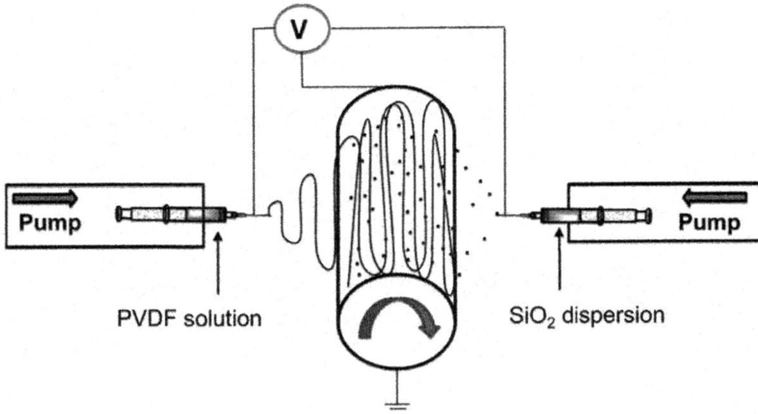

Figure 11.1 Preparation of composite separator by electrospinning and electrostatic spray. Reproduced with permission from Ref. [2]. Copyright 2014, Elsevier.

11.2.1.1 Principle

Electrospinning is a special fiber manufacturing process, and its equipment is mainly composed of four parts: driving device, liquid storage device, high-voltage power supply, and collecting device (Fig. 11.2). The high-voltage electric field is filled between the liquid storage device and the collecting device. The polymer solution is sprayed out by the driving device and passes through the high-voltage electric field and is filled with an electric charge. Due to the electric field force and surface tension, the spherical droplets ejected from the syringe become conical, called the "Taylor cone." With the gradual increase in the electric field force, the high-voltage electrostatic force will overcome the surface tension, air resistance, gravity, intermolecular force, etc., to form a charged jet. The formed electric jet is accelerated under the electric field and suffers from a series of fiber splits, volatilization, fiber coagulation, and finally collected in the receiving device to form a nonwoven fabric.

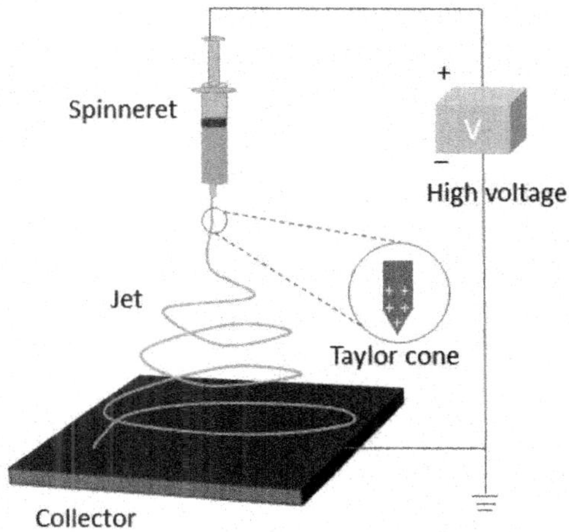

Figure 11.2 Schematic diagram of electrospinning principle.

11.2.1.2 Influencing factors

Polymer solution

Solvent: If the electrospinning precursor solvent is difficult to volatilize smoothly during the spinning process, the nanofibers will adhere to each other when they reach the collecting device, which will affect the morphology of the nonwoven separator. If the solvent evaporates too fast, the nanofibers cannot effectively split and stretch in the electric field during spinning, forming several beaded structures on the nanofibers, which will cause defects in the separator fiber structure and result in inferior mechanical properties.

Viscosity: High viscosity of the electrospinning precursor solution makes the polymer molecular chains entangled too much and too densely, leading to greater instability of the jet and the increasing difficulty in spinning process, resulting in the varying diameter and distribution of nanofibers. However, if the viscosity is too low, it will be difficult to form a normal jet, eventually, only forming droplets that fall from the spinning jet. Therefore, the viscosity of the precursor solution plays a decisive role in electrospinning. The factors that affect the viscosity mainly include the types of solvent, polymer's molecular weight, and the ratio of each component in the precursor solution. Therefore, to obtain the best effect, the preceding factors should consider simultaneously when preparing the spinning solutions, and plenty of experiments should be carried out to select the appropriate spinning solution concentration.

Conductivity: In the electrospinning process, the spinning solution is greatly affected by the electrostatic field, so the conductivity has a great influence on the electrospinning. The electrical conductivity affects the fiber diameter size. Therefore, if salts, ionic solutions, and some conductive metal ions are added to the precursor solution, the conductivity can be improved effectively and the charge density of solution can be increased. After the spinning solution enters the electric field, it will be stretched and transformed to nanofibers with a more acceptable diameter due to the stronger electric field force.

Voltage

The influence of voltage on the preparation of nanofibers by electrospinning is divided into the following two main aspects:

1. If voltage increases, the jet will be continuously stretched, so that nanofibers with small diameters can be prepared.
2. If the voltage continues to rise, the jetting speed will be too fast, which will make it difficult to achieve relatively enough volatilization of the solvent during the jetting process, and the stretching and splitting of small droplets will not be enough.

Receiving distance

The receiving distance will directly impact the electric field strength, the stretching and flight time, and the solvent volatilization. The relatively close receiving distance can prepare fiber separators with smaller diameters. But if the receiving distance is relatively small, the electric field strength will be increased to a large extent, so that the nanofiber cannot be formed stably. A short receiving distance will also shorten the stretching time and cause insufficient solvent volatilization, which will eventually lead to the uneven distribution of fiber on the collector. If the receiving distance is far and the electric field strength is small, the nanofibers cannot be stretched smoothly.

Humidity

During the spinning process, if the humidity of the surrounding environment is too low, polymer solution will volatilize faster and form a beaded structure on the nanofibers. If the ambient humidity is relatively high, the solvent will be difficult to volatilize, which will worsen the morphology of the electrospinning separator.

Temperature

Ambient temperature directly or indirectly affects the conductivity, viscosity, surface tension, and volatility of spinning solutions. As the temperature increases, the solution volatility increases, the viscosity and surface tension decrease, and the electrical conductivity and dielectric constant change. These parameters have a non-negligible effect on the electrospinning process. For example, the variation in ambient temperature can change the volatilization speed of the jet solvent and, in turn, affect the solution fluidity, which ultimately affects the morphology of fibers. The variation in ambient temperature will also indirectly affect the surface tension of solution, the dielectric constant, and the electrical conductivity of nozzle, etc., which have a certain impact on the electric field of the entire spinning equipment.

Feeding rate of spinning solution

When the pore diameter of the spinneret is fixed, the average velocity of fluid ejection is proportional to the diameter of fiber formed on the receiver, so it can regulate the fiber size by controlling the fluid ejection velocity.

Collector

If the collector is fixed, the formed fibers will have random and irregular shapes, and the thickness will be uneven. If a rotating collector is used, the nanofibers will be arranged in parallel and regular patterns.

11.2.1.3 Method evaluation

Electrospinning can easily prepare nanofibers with a few microns to a few hundred nanometers, and the nanofiber separators have large specific surface area, high porosity, and good lipophilicity. Its non-straight 3D hole structure has a high degree of tortuosity, resulting in better suppression to micro-short circuits. At the same time, the nanofiber size is easy to control, and the surface is comfortable to be functionalized (surface coating and modification, etc.).

In practical applications, electrospinning technology still has some problems:

1. The mechanical strength of the electrospinning separator is low, which is often improved by increasing the thickness.
2. Electrospinning can only prepare high molecular polymer nanofibers and has not penetrated deeply into the field of inorganic nanofibers. The inorganic nanofibers prepared by electrospinning technology have poor flexibility and continuity.
3. When applied to large-scale preparation, there are challenges in process parameter optimization, equipment development, etc.

11.2.2 Phase Inversion Method

11.2.2.1 Principle

The phase inversion method prepares a homogeneous polymer solution of a certain composition and then uses a specific physical way to make the casting liquid, which carries out the mass transfer exchange between the solvent and the non-solvent in the surrounding environment. It changes the thermodynamic state of the original stable solution and makes it apart from the homogeneous polymer solution, undergoing phase separation, transforming into a three-dimensional macromolecular network type gel structure, and finally solidifying into a separator.

11.2.2.2 Classify

According to the physical state of changing the thermodynamic state of solution, the preparation strategy can be divided into the following types: solvent evaporation phase inversion method, thermally induced phase inversion method, vapor precipitation phase inversion method, and solution phase inversion method.

The solution phase inversion method (Fig. 11.3) is also called the immersion gel phase inversion method. Among the preceding methods, the solution phase inversion method delivers a simple preparation process and flexible operation, making it possible to adjust the structure and performance of the separator according to the application scenarios. Nowadays, it has become the main method for preparing microporous separators.

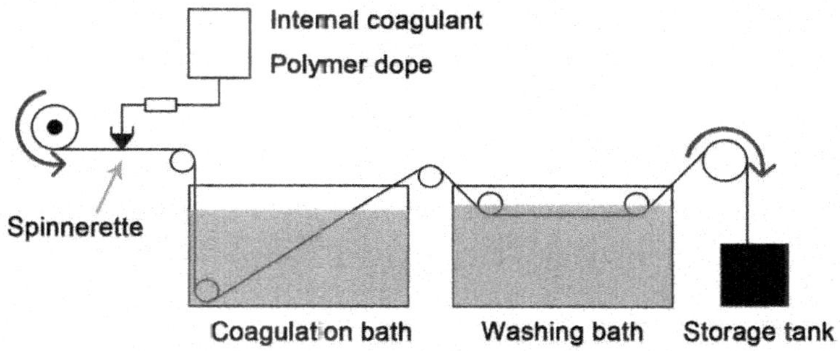

Figure 11.3 Schematic diagram of the phase conversion method.

11.2.2.3 Preparation process

Phase separation process: When the casting solution is immersed in the coagulation bath, the solvent and the non-solvent will diffuse mutually between the liquid separator and coagulation bath interface. When the exchange between the solvent and the non-solvent reaches a certain level, the casting solution becomes a thermodynamically unstable system, leading to phase separation of the casting solution. This stage is a key step in determining the pore structures of separator. The main research content is the thermodynamic properties and mass transfer dynamics of the separator system.

Phase inversion process: After the phase separation of the casting solution system, the solvent and non-solvent are further exchanged. Meanwhile, the separator pores condense, the flow between the phases, and the polymer-rich phase solidifies to form a separator. These stages have a great influence on the structure and morphology of the final polymer separator, but they are not the main factors for pore formation, which is controlled by the gelation kinetics process.

11.2.2.4 Method evaluation

Advantages: It can mix and separate one or more components simultaneously with low energy consumption and high separation efficiency in the process, which can be used to modify the surface coating of the separator and prepare a porous separator.

Disadvantages: The mechanism of the method is complex and various. The structure and distribution of pores are related to the mass transfer process between different phases, but it is difficult to derive a general multiphase mass transfer process mechanism [3].

11.2.3 Stretching Method

The stretching method is the leading way to produce polyolefin separators. According to the different pore-forming mechanisms, it can be divided into dry method and wet method. Based on the

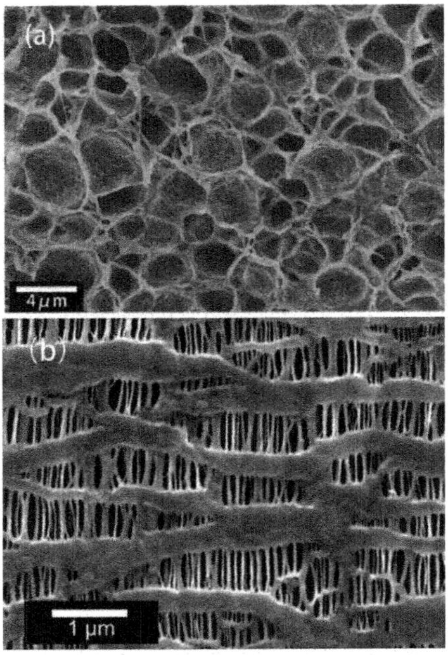

Figure 11.4 (a) Scanning electron microscope (SEM) image of PE separator prepared by wet stretching. Reproduced with permission from Ref. [4]. Copyright 2014, Elsevier. (b) SEM image of PP separator prepared by dry stretching. Reproduced with permission from Ref. [1]. Copyright 2004, American Chemical Society.

different stretching orientations, the dry method can be divided into dry unidirectional stretching and dry biaxial stretching method. Depending on whether the stretching direction is synchronous, the wet method can be further divided into biaxial synchronous stretching and biaxial asynchronous stretching. The microstructures of the separator prepared by the two methods are shown in Fig. 11.4, showing different surface morphologies, pore sizes, and size distributions.

11.2.3.1 Dry stretching method

Dry unidirectional stretching: The principle of dry unidirectional stretching is crystal stretching. The dry unidirectional stretching process prepares a high alignment polyolefin separator with low crystallinity by producing stiff elastic fibers and then uses high-temperature annealing to obtain an alignment separator with high crystallinity. This process extrudes the polyolefin melt from the mold, and then cools and crystallizes under the stress field of the high-speed drawing of the casting roll to obtain a casting base membrane with a parallel lamella structure perpendicular to the extrusion direction. The base membrane has good rigid and elastic properties. Then the base membrane is heat treated to eliminate defects in the crystal region, further improving the lamella structure and increasing the crystallinity of the membrane. Finally, the prepared rigid elastic polypropylene cast membrane is first stretched at low temperature to form micro-defects such as craze. Then the defects are pulled apart at high temperature to form micropores. The brief preparation process is shown in Fig. 11.5.

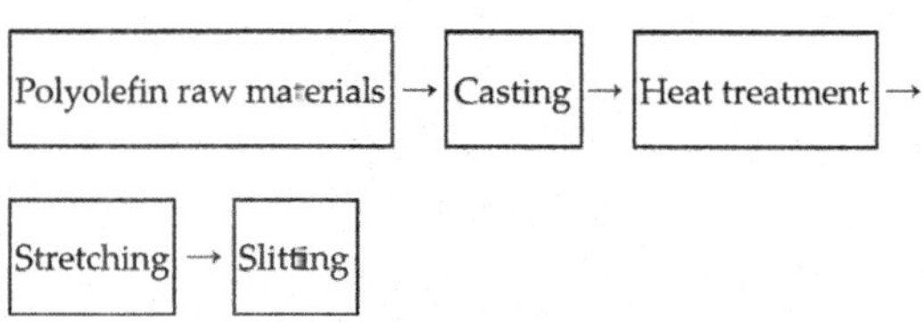

Figure 11.5 Process flow chart of unidirectional stretching.

1. **Feeding:** After pre-processing polyolefin and other raw materials according to the formula, they are transported to the extrusion system.

2. **Casting:** The pretreated raw material is melted and plasticized in the extrusion system; then the melt is extruded from the die and cast to form a base membrane with a specific crystalline structure.
3. **Heat treatment:** The base membrane is heat treated to obtain a hard-elastic membrane.
4. **Stretching:** Cold stretch and hot stretch are applied on the hard-elastic membrane to form a nanoporous separator.
5. **Slitting:** The nanoporous separator is cut according to the required specifications.

Dry biaxial stretching (β crystal method): The principle of dry biaxial stretching method is crystal conversion. The biaxial stretching process is an independent intellectual property developed by the Institute of Chemistry, Chinese Academy of Sciences, and a unique separator manufacturing process in China. The biaxial stretching process involves mainly adding β crystal into polyolefin, forming modifier with nucleation effect, and finally obtaining a polyolefin separator with high β crystal content. Because the β crystal form is a hexagonal crystal system, single crystal nucleation, loose wafer arrangement, it has a lamellar structure that grows in a divergent beam shape along the radial direction but does not have a complete spherulite structure. It will transform into a denser and more stable α crystal under the existence of heat and stress. Finally, the density difference between the polyolefin phases is used to form a microporous structure, and then a single-layer polyolefin separator is prepared. The brief preparation process is shown in Fig. 11.6.

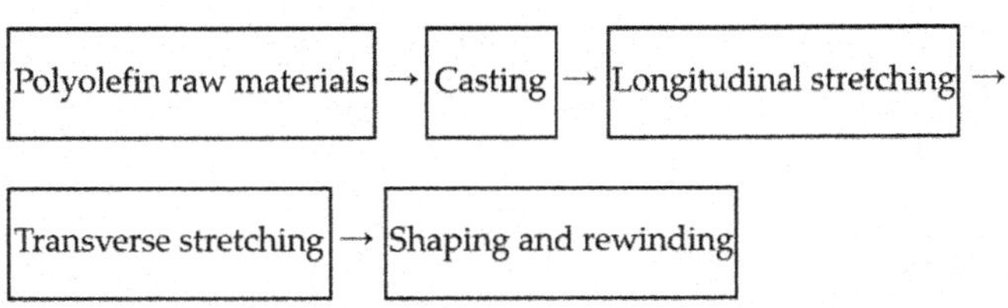

Figure 11.6 Process flow chart of biaxial stretching.

a. **Feeding:** The raw materials are pretreated according to the formula and then transported to the extrusion system.
b. **Casting:** High β crystal content and good β crystal morphological uniformity are achieved.

c. **Longitudinal stretching:** The cast piece is stretched longitudinally at a certain temperature, and the β crystal is easy to form pores under tensile stress.
d. **Transverse stretching:** The sample is stretched transversely at a higher temperature to expand the pores and improve the uniformity of pore size distribution.
e. **Shaping and rewinding:** Heat treatment on the separator at high temperature to reduce its thermal shrinkage and improve dimensional stability.

Characteristics of separators prepared by stretching method

a. **Unidirectional stretched separator:** The micropore size porosity is about 40%, and the longitudinal tensile strength is good, but the transverse tensile strength is poor. Because it can only be stretched in one direction, there is almost no thermal shrinkage in the transverse direction. The obturator temperature and the melting temperature are both higher. No solvent is used in the production process, which is environmentally friendly.
b. **Biaxially stretched separator:** It is based on the heterogeneous nucleation and forms micropores through the crystal transformation, which is a unique separator manufacturing process in China. The separator has uniform biaxial orientation, enabling it with stronger mechanical strength and better air permeability than the unidirectional stretch separator. Meantime, the separator' thickness can be adjusted for a wider range with a greater puncture strength.

11.2.3.2 Wet stretching method

The wet method principle involves thermally induced phase separation, mixing plasticizers (high-boiling hydrocarbon liquids or some relatively low molecular weight substances) with polyolefin resins, and using the phenomenon of solid–liquid or liquid–liquid phase separation during the cooling process of the molten mixture. Then the separator is compressed, heated to a temperature closed to the melting point, and stretched to make the molecular chain orientation consistent. After a certain period, volatile solvents (such as dichloromethane and trichloroethylene) are used to remove the plasticizer from the separator and then the interpenetrating submicron size microporous separator material is obtained. The brief preparation process is shown in Fig. 11.7.

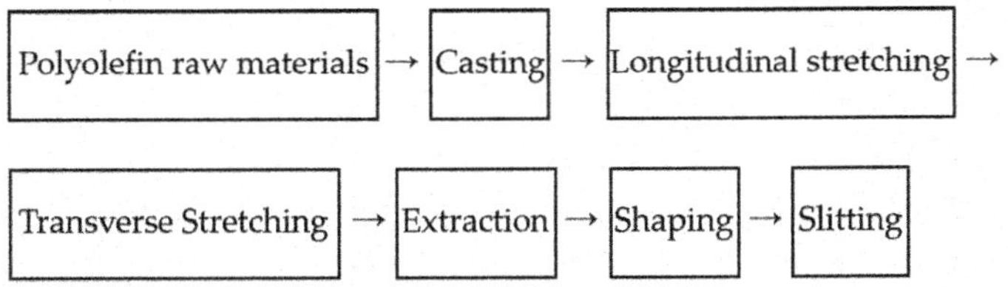

Figure 11.7 Process flow chart of wet asynchronous stretching.

a. **Feeding:** The raw materials such as polyolefin and pore former are preprocessed and transported to the extrusion system according to the formula.
b. **Casting:** The pretreated raw materials are melted and plasticized in a twin-screw extrusion system, and then the melt is extruded from the die, and the melt is cast to form a thick cast sheet containing a pore-forming agent.
c. **Longitudinal stretch:** The thick cast sheet is stretched longitudinally.
d. **Transverse stretch:** The thick cast sheet is stretched transversely to obtain a base separator containing a pore-forming agent.
e. **Extraction:** Solvent extraction of the base separator is carried out to form a base separator without a pore-forming agent.
f. **Shaping:** The base separator without a pore-forming agent is dried and shaped to obtain a nanoporous separator.
g. **Slitting:** The nanoporous separator is cut according to the required size.

Wet synchronous stretching technology: The process of wet synchronous stretching technology is the same as that of asynchronous stretching technology, except that it can be oriented in both the horizontal and vertical directions during stretching, eliminating the need for longitudinal stretching process and strengthening the separator thickness uniformity. However, the problem of synchronous stretching is, first, the slow speed and, second, the worse adjustability. Only the transverse stretch ratio is adjustable, while the longitudinal stretch ratio is fixed.

11.2.3.3 Comparison of dry and wet methods

Compared with the dry method, the wet method is more suitable for low-melting polymers such as polyethylene-based materials (melting point is 140°C), because of the high-temperature melting process, but this also means that the thermal stability of the separator prepared by the wet method is not as good as the dry method. The wet method involves biaxial stretching to orient polymer molecular chains. Synchronous stretching technology can be oriented in both horizontal and vertical directions simultaneously, thereby eliminating the difference in membrane thickness caused by unidirectional stretching and improving the uniformity of film thickness. Therefore, they can show high product transparency, no scratches, and excellent optical and surface properties. At the same time, air permeability and porosity are significantly better than dry-processed separators.

Besides, wet preparation can increase the simplification of coating process and improve the thermal stability of the separator. The coating-layer separator has obvious performance advantages and has a good development prospect in high energy density batteries in the future. However, the dry method is still superior to the wet method in terms of pore size distribution, heat resistance, high pressure resistance, and oxidation resistance, and the dry method can prepare three-layer separators. The dry method is lower than the wet method in terms of cost and energy consumption.

11.2.4 Solid Particle Sintering Method

The solid particle sintering method is the most used method for manufacturing all-ceramic separators. The all-ceramic separators' composition does not include organic components.

For example, Xiang et al. [5] prepared pure ceramic separator for LIBs by compressing nano-Al_2O_3 particles and calcining at high temperature. Although the thickness of the obtained separator reaches 200 μm (Fig. 11.8), the lithium-ion conductivity at room temperature shows 3.45 mS/cm, which is about eight times that of polyolefin separators and as high as 0.78 mS/cm at –20°C, offering superior low temperature resistance.

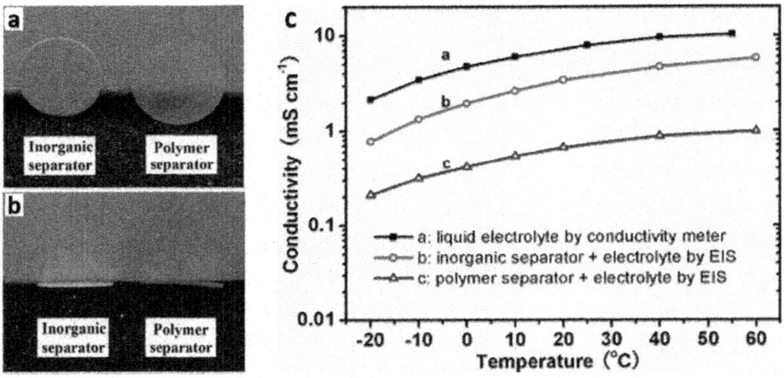

Figure 11.8 (a, b) The photos of Al_2O_3 inorganic separator and commercial polymer separator and (c) dependence of ion conductivity on temperature for the electrolyte-infiltrated inorganic separator and the polymer separator, and the electrolyte of 1 M $LiPF_6$/EC: diethyl carbonate (DEC). Reproduced with permission from Ref. [5]. Copyright 2011, Elsevier.

11.2.4.1 Principle

The solid ceramic particles form a stable suspension in the dispersion, and then the slurry method is used to develop a membrane on the porous ceramic support. When the support is immersed in the suspension, water will enter the support under the capillary force's action, while the solid ceramic particles accumulate on the surface of the support to form a membrane. The contact surface of the particles is partially sintered at high temperature, so that the separator has certain pores, pore size, and strength. In the separator production process, the pore size of the separator layer is mainly regulated by controlling the particle size of solid particles, and the thickness of the separator layer is controlled primarily by the concentration, viscosity, and dipping time of the slurry. Different pore diameters, separator thicknesses, and porous ceramic separators with porosity can be prepared. The brief preparation process is shown in Fig. 11.9.

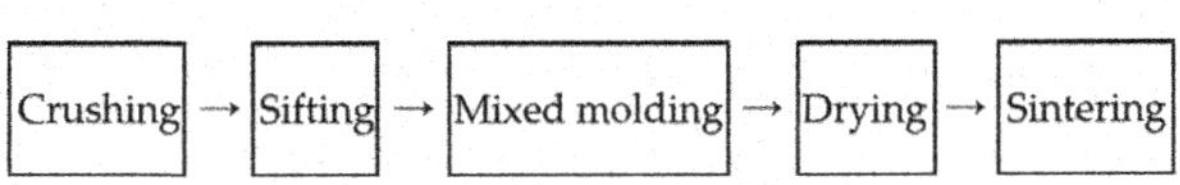

Figure 11.9 Process flow chart of solid particle sintering method.

a. **Crushing:** Crush inorganic materials into particles.
b. **Sifting:** Screen out the required particle size for preparation.
c. **Mixed molding:** Mix the sieved powder with a specific particle size with binder, surfactant, and water to form a molding material.
d. **Drying:** Dry the body to be molded.
e. **Sintering:** Sinter the dried body at high temperature to produce a microporous separator.

11.2.4.2 Influencing factors

Particle size of powder: Powder size and classification will affect the separator pore structure because voids are formed between the particles after the particles agglomerate. The pore structure is affected by the three-dimensional spatial geometric distribution or the geometric size of particles. Large particles will form large pores after agglomeration; small particles will form small pores. Therefore, the more uniform the particle size distribution of powder, the more uniform the formed pore sizes.

Molding methods: Molding methods are divided into three types: dry pressing, grouting, and extrusion. Dry pressing involves putting the powder in the metal mold and then pressing the hydraulic pressure to press the loose powder into a green body quickly. Producing the green body requires a certain amount of water content and a suitable binder. The obtained material has precise dimensions, smooth surface, compact and uniform structure, and a high mechanical strength.

Grouting involves injecting the slurry into the gypsum mold. After the slurry is adsorbed, the excess slurry is poured out, and then the green body is obtained by absorbing and shrinking the mold. The slurry should have a certain degree of fluidity and stability, and it requires pore formers and adhesives, so the process is difficult to control, and the green body is easy to crack and deform.

The extrusion molding method puts the plastic mud into the screw extruder and presses it to a certain shape when it passes through the mold of the extrusion head.

The comparison of the three molding methods is shown in Table 11.1.

Table 11.1 Comparison of three molding methods

Molding Methods	Pore Structure Uniformity	Porosity/%	Production Cycle
Dry pressing	Best	40.1	Shortest
Grouting	Poor	45	Longer
Extrusion molding	Better	42.3	Longest

Drying and sintering conditions: The sintering process is divided into three stages: initial stage, middle stage, and final stage.

a. **Initial sintering stage:** The particles touch each other to form a linked entirety without significantly changing the original structure. The separator has a certain mechanical strength and better pore structure.
b. **Middle stage of sintering:** The adhesion of particles further expands, the pore structure changes much, the pore performance is significantly decreased, and the mechanical strength increases.
c. **Final stage of sintering:** The pore structure of the separator completely disappears and becomes a dense separator.

Pore performance and mechanical strength are contradictory factors during the sintering process. The best sintering temperature must be selected and determined to obtain a good pore structure and a certain mechanical strength.

11.2.5 Melt-Blown Spinning Process

11.2.5.1 Principle

The melt-blown method is a nonwoven technology that directly uses polymer as raw material to prepare fine fibers or fiber net products. The primary process involves extruding the molten polymer from the spinneret, stretching it into short fibers under high-speed and high-temperature airflow, and then drawing the refined fibers to gather them on a roller or netting curtain, and finally consolidating into a net by thermal bonding or self-bonding. The principle is shown in Fig. 11.10.

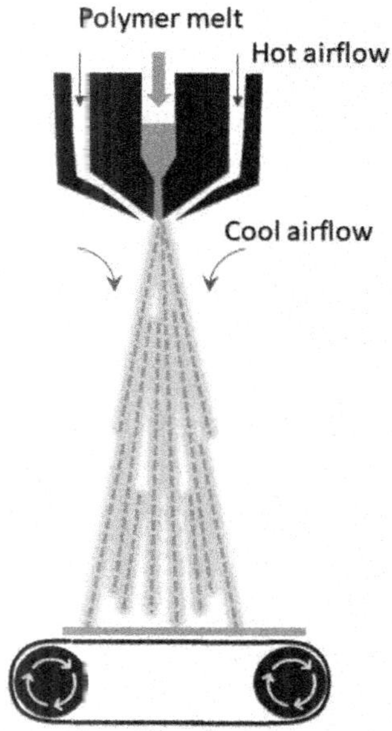

Figure 11.10 Schematic diag-am of the melt-blown spinning process.

11.2.5.2 Process

The brief preparation pro:ess is shown in Fig. 11.11.

a. **Feeding:** The polymer is generally made into small spherical or granular slices, poured into a barrel or hopper, and fed into the screw extruder.
b. **Melt extrusion:** The treated polymer is mixed with inorganic /organic additives, and then heated to the melting point of polymer and delivered to the syringe. Subsequently, the melt is sent to the spinneret and pump through the filter. In the melt-blown process, the general extruder also uses its shearing and thermal degradation effects to reduce of polymer's molecular weight.
c. **Fiber formation:** The filtered clean melt passes through the distribution system and then evenly feeds each group of spinnerets to make the extrusion volume of each spinneret uniform. The spinneret of melt-blown fiber is different from other spinning methods. The spinneret holes must be

arranged in a straight line, and the upper and lower sides have high-speed air jet holes.

d. **Fiber cooling:** A large amount of room-temperature air is simultaneously sucked in on both sides of the spinneret, mixed with the hot air stream containing ultra-fine fibers to cool it, and the molten ultra-fine fibers are cooled and solidified.

e. **Net formation:** In melt-blown fiber nonwoven fabric production, the spinneret can be placed horizontally or vertically. If placed horizontally, the superfine fibers are sprayed on a circular collecting drum to form a web. If placed vertically, the fibers fall onto a horizontally moving web developing curtain to condense into a net.

f. **Adhesion (fixed net):** The self-adhesive reinforcement mentioned earlier is enough for some applications of melt-blown cloth, such as relatively fluffy structure, reasonable air retention rate or voids rate, etc. For many other applications, self-adhesive reinforcement alone is not enough. Calendar bonding, ultrasonic bonding, or other reinforcement methods are also required.

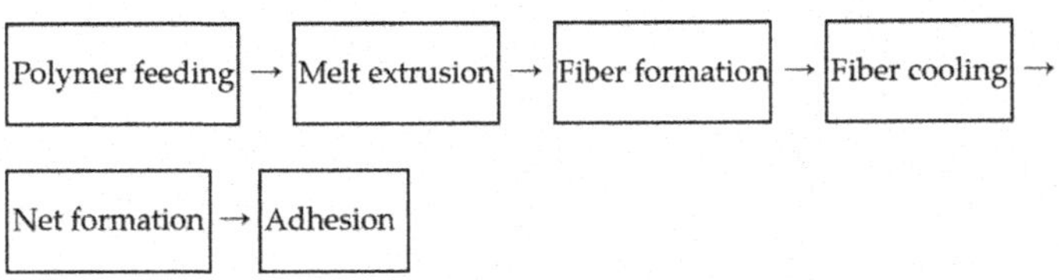

Figure 11.11 The flow chart of melt-blown spinning process.

11.2.5.3 Influencing factors

The formed superfine fiber nonwoven material has the advantages of good porosity, large specific surface area, safety, low cost, etc. The production process is short with production efficiency. However, there are also disadvantages of poor fiber orientation, low strength, and high energy consumption.

Melt flow index: The strength and elongation at break of melt-blown fiber cloth decrease as the melt flow index (MFI) of the raw material increases. Therefore, in order to get a better draft of the melt stream in the hot air blowing process, the MFI of raw material is required to be as high as possible.

Hot air velocity: Under the conditions of the same temperature, screw speed, and receiving distance, the hot air velocity increases, the fiber diameter decreases, the fiber entanglement increases, the fiber net becomes dense and smooth, and the strength somewhat increases. But if the air velocity is too large, it is easy to appear flying, which seriously affects the cloth's appearance.

Receiving distance: The receiving distance increases, the vertical and horizontal strengths decrease, and the bending strength decreases. With this technology's continuous progress, the mixed raw materials such as polypropylene/polyamide, polypropylene/polyethylene, and polypropylene/polystyrene can also be used in the melt-blown manufacturing of nonwoven fabrics. Due to the poor heat resistance of melt-blown nonwoven fabrics, battery separators prepared by the melt-blown method should not be used under conditions of excessively high temperature.

11.2.6 Coating

Coating is a surface-modified technology that needs to use other separators as the base membrane, such as polyolefin microporous separators, nonwoven, etc. Different kinds of ceramic particles and binders are generally used in the coating process. They always are mixed into slurry with dispersant, and then, is thrown on the base membrane. After coating, the ceramic and the base membrane are tightly bonded together. The ceramic material with good thermal stability, high mechanical strength, rich pore structure, and good electrolyte wettability is used to enhance the safety of separators and improve the battery's electrochemical performance. Coating can obtain single-layer composite, double-layer composite, and bulk composite separator materials, which can effectively improve the base separator's thermal stability and mechanical strength. It is widely used in the separator production process, but it is relatively complicated.

11.2.6.1 Coating method

Commonly used methods include atomic layer deposition, hot rolling, surface grafting, and other technologies.

Atomic layer deposition

Atomic layer deposition (ALD) is a method in which substances can be deposited layer by layer on the substrate's surface in the form of a monoatomic separator. In this process, the chemical reaction of the new atomic layer is directly related to the previous layer. In this way, only one layer of atoms is deposited per reaction. The specific method involves forming a deposited separator by passing the gas-phase precursor alternately into the reactor in pulses and chemically adsorbing and reacting on the deposition substrate.

For example, Jung et al. [6] deposited an ultra-thin Al_2O_3 functional layer with a thickness of about 6 nm on the surface of PP microporous separator by ALD technology and obtained a ceramic composite separator, which is different from conventional ceramic-coated inorganic/polymer/inorganic tri-layer separator. The ceramic layer improves the heat resistance and the lipophilicity of the original PP base separator. Under the 4C rate of charging and discharging conditions, the discharge capacity of the battery still maintains about 80% after 1000 cycles.

ALD allows precise control of separator thickness at the atomic level. Moreover, ALD can form multilayer structures with different materials relatively easily. Because of its excellent projection ability, it can be easily applied to powders and shapes with complex structures. But the ALD process is very time consuming. For example, the separator formation of Al_2O_3 is 0.11 nm per cycle, and the standard separator formation amount per hour is 100 to 300 nm.

Hot rolling

Hot rolling is a composite separator manufacturing process, which requires the presence of a base separator. The pressure is applied between two rollers, supplemented by a certain temperature, and then double or even three-layer base separator is bonded together by pressure to prepare the composite separator. Usually, the composite of ceramic and organic separators is studied to improve the thermal stability and mechanical strength of separators.

Cho et al. prepared a composite ceramic separator successfully with a sandwich structure through electrospinning and hot-rolling processes (Fig. 11.12). The ceramic particle layer is restricted between the double-layer polyacrylonitrile nonwoven [7]. The ceramic particles are effectively prevented from falling off, and the thermal stability and mechanical strength of separator are improved simultaneously.

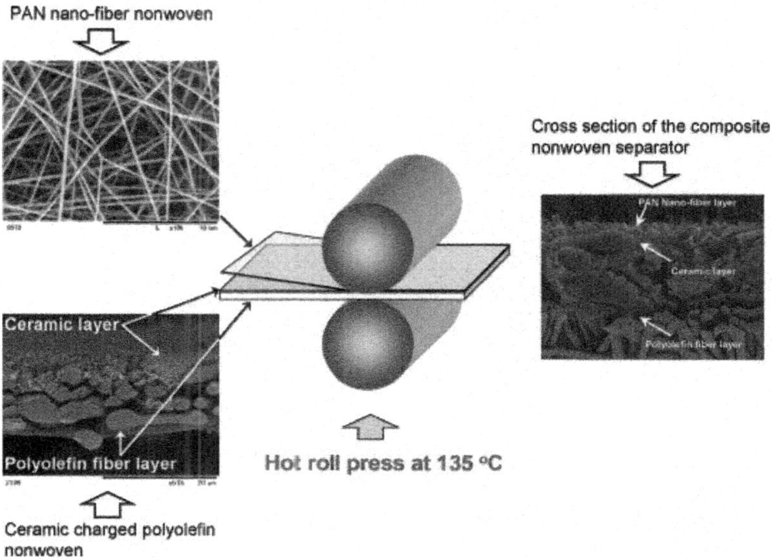

Figure 11.12 Preparation of composite ceramic separator with sandwich structure by the hot-rolling method. Reproduced with permission from Ref. [7]. Copyright 2010, Elsevier.

Surface grafting

In view of the shortcomings of the low surface energy of polyolefin separators, they can be modified by surface-active agent treatment and surface grafting of hydrophilic groups. Surfactant treatment can improve the surface wettability of separators to electrolytes but cannot improve the retention ability of electrolytes. And it is difficult to maintain the modification effect for a long time due to the repeated washing of the surfactant by electrolytes during battery operation. The grafting of hydrophilic groups on the surface is by plasma treatment to induce the opening of the C–C and C–H bonds on the polyolefin to generate polymer free radicals, and then the hydrophilic groups such as hydroxyl, carbonyl, carboxyl, amino, and imine groups are grafted onto the separator surface. After grafting, the liquid absorption and retention of separators are significantly improved, and the cycle and rate performance of battery are greatly enhanced.

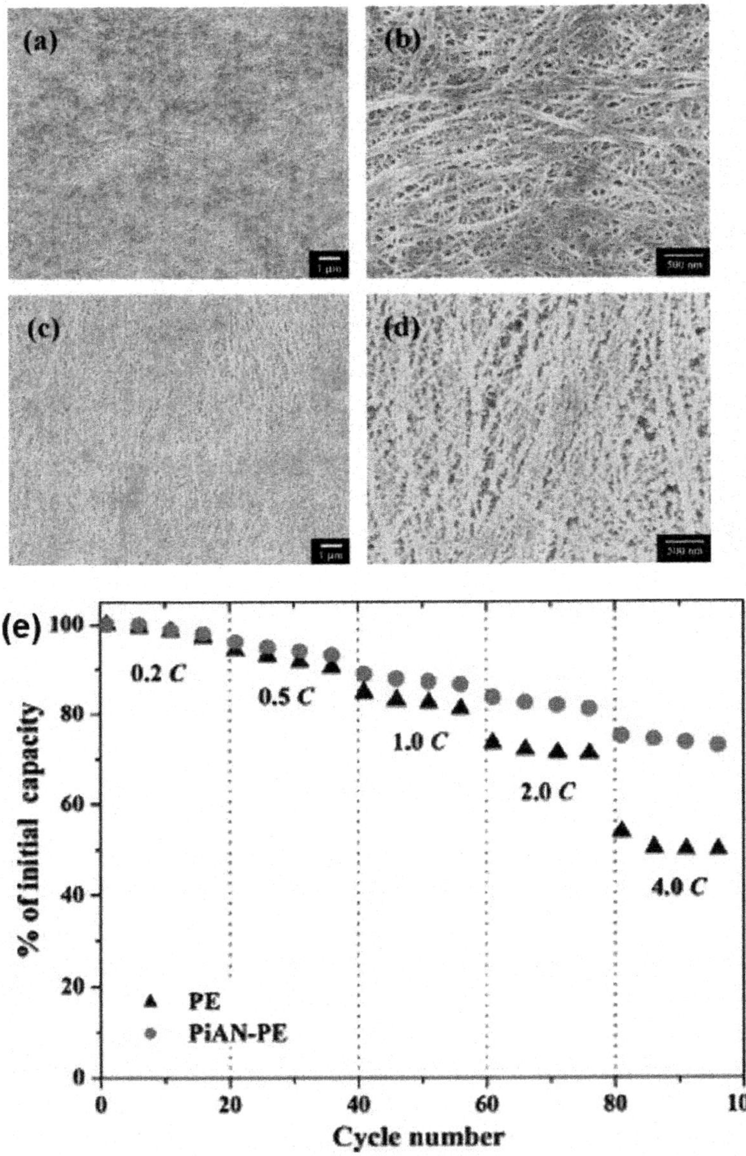

Figure 11.13 SEM images of the surfaces for (a and b) PE and (c and d) PiAN-PE separators, respectively. (e) Rate capability of the lithium-ion polymer cells with the PE or the PiAN-PE separator. Reproduced with permission from Ref. [8]. Copyright 2009, Elsevier.

Various techniques can be used to graft hydrophilic groups on the separator's surface, including plasma, ultraviolet irradiation, and electron beam irradiation. For example, Kim et al. [8] used plasma technology to graft acrylonitrile (AN) monomer to prepare modified PE separator. The plasma-induced acrylonitrile coated PE (PiAN-PE) results in a hydrophilic surface, which improves the wettability and retention of the separator to electrolytes. The SEM images of PE and PiAN-PE separators and their battery performances are shown in Fig. 11.13. In short, grafting hydrophilic monomers to separator surface is a simple but feasible method to change the structure and property of polyolefin separators.

The coating method uses a specific machine or appliance to coat a uniformly mixed slurry on the surface of the base separator. This method is simple and easy to implement, relatively inexpensive, and is suitable for mass preparation of composite modified separators in practical applications. Guo et al. introduced $LiPF_6$, a commonly used lithium salt in the electrolyte, onto the surface of commercial Al_2O_3/PE separators by coating and used the modified composite separators for quasi-solid lithium–sulfur (selenium) batteries [9]. The results display that this optimized composite electrolyte can achieve longer cycle life in lithium–sulfur (selenium) batteries, showing a reversible specific capacity close to 1000 mAh/g at a current density of 0.5 C, and 100% DOD (depth of discharge) cycle life exceeded 500 cycles. The lithium–sulfur soft pack battery based on this composite electrolyte can cycle stably under a high areal load of 8 mg/cm, presenting good industrial application potential.

11.2.7 Wet Papermaking

The wet papermaking process is a commonly used method for manufacturing paper-based separator materials. The preparation method has the advantages of environmental protection, simplicity, and low cost. Paper-based separators use cellulose and its derivatives as raw materials and have excellent wettability, strong mechanical strength, and high thermal/chemical stability.

The wet papermaking process is based on the traditional papermaking process, where papermaking pulp fibers, nanofibers, or synthetic fibers mix, disperse with short and thin fibers and binders in a certain proportion to form the slurry, and then spread the slurry with transfer coating on the carrier. After dehydration/solvent, drying, and winding, a separator with high porosity, thermal stability, and good wettability is formed. For example, Zhang et al. [10] used a wet papermaking process to prepare a flame-retardant cellulose composite separator with excellent wettability and liquid absorption, which significantly reduced the cost of preparing the separator (Fig. 11.14).

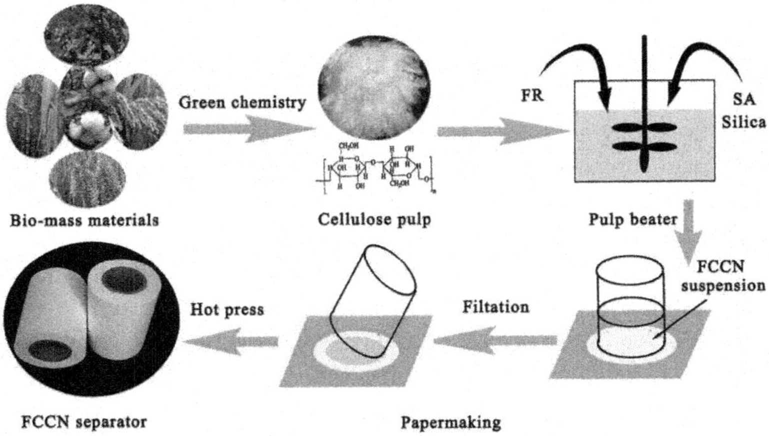

Figure 11.14 Schematic illustration for the preparation process of the flame-retardant cellulose composite separator. Reproduced with permission from Ref. [10]. Copyright 2014.

11.2.8 Magnetron Sputtering Method

The magnetron sputtering method involves filling of a certain amount of argon gas into a certain equipment in a high vacuum condition and applying a higher DC voltage between cathode and anode, so that a magnetron-type glow is generated in the coating chamber. The discharge causes the argon gas to be ionized and generate argon ions. The argon ions in the coating chamber are accelerated, and they bombard the surface of the cathode target, sputtering the surface atoms of the target material and depositing on the substrate surface to form a thin separator. By regulating experimental parameters, such as changing target materials and sputtering times, the separator with various materials and varied thickness can be obtained.

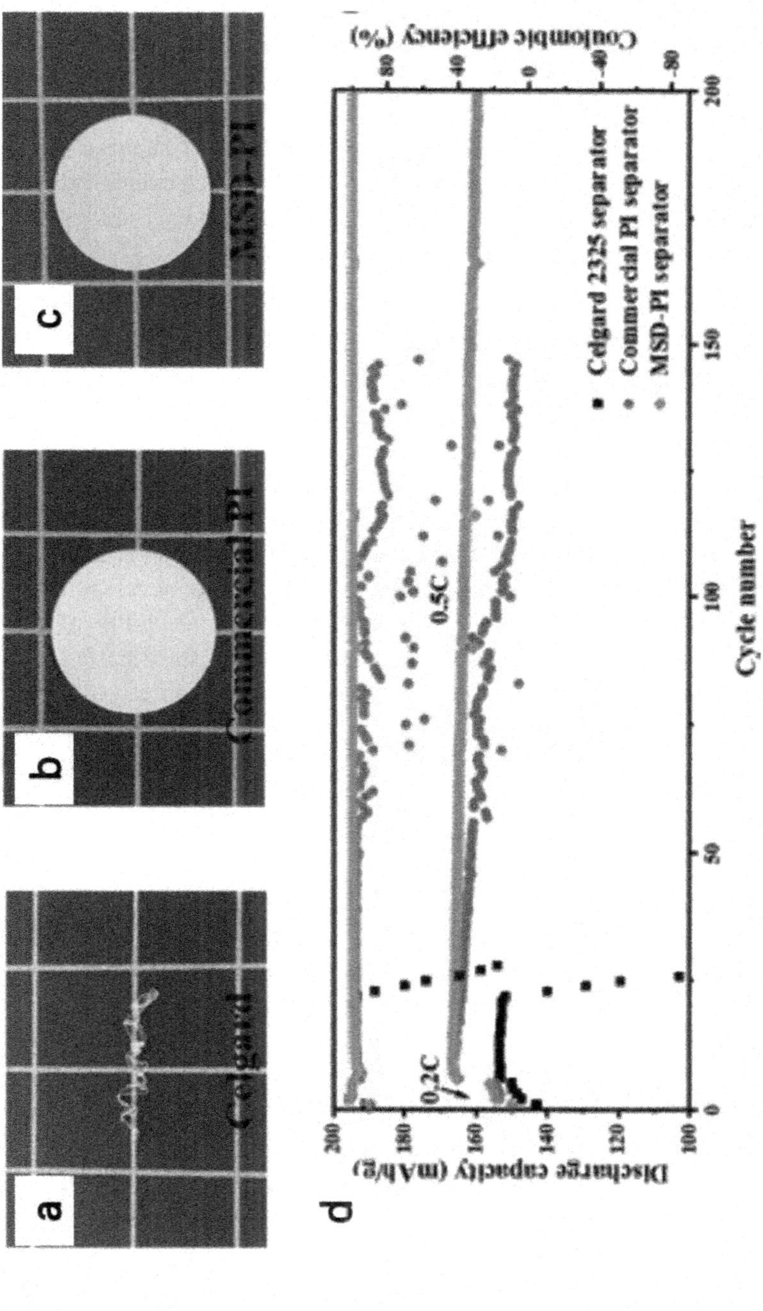

Figure 11.15 Digital pictures of various separators after heat treatment for 30 min under 200°C: (a) Celgard 2325 separator, (b) commercial PI separator, (c) MSD-PI separator. (d) High-temperature electrochemical performances: long-term cycle performance of cells with Celgard 2325 separator, commercial PI separator, and MSD-PI separator at 110°C. Reproduced with permission from Ref. [11]. Copyright 2021, Elsevier.

Hu et al. [11] used magnetron sputtering deposition technology to modify both sides of a commercial polyimide (PI) separator with silicon nitride (Si_3N_4). Consequently, the modified PI composite separator (MSD-PI) shows higher thermal stability and ionic conductivity (Fig. 11.15a–c). More importantly, compared with the battery using Celgard separator, the MSD-PI separator displays satisfying thermal stability and excellent electrochemical performance, especially the long-term cycling performance under high-temperature environment (Fig. 11.15d).

11.3 Performance Indexes and Test Techniques of Separators

Advanced characterization technology can clearly reveal the mechanism of sodium-ion transport near the separator, describe the structure–function relationship between separator parameters and the battery performance, and more effectively guide the design and development of separators for SIBs. For instance, newly developed optical sensor technology in separator verified the feasibility of *operando* decoding of chemical and thermodynamic information in SIBs, strongly promoting the progress of battery safety and reliability. Nowadays, various characterization technologies and equipment have been able to evaluate the overall performance of the separator, which provides a reference for more targeted performance modification strategies to separators in SIBs.

11.3.1 Thickness

The thickness is one of the most basic parameters for a separator, which is usually inversely proportional to the ion permeability but proportional to the mechanical properties. Therefore, the separator thickness should be reduced as much as possible to improve the battery performance under the condition of satisfying the mechanical strength.

Nowadays, separators with a thickness ranging from 16 to 30 μm are commonly used in commercialized batteries for different applications. Generally, the battery with higher current density and higher power requires a thick separator. Besides, the thickness uniformity is also a crucial parameter for the separator to achieve high performance, and an error less than 1 μm is usually required.

11.3.1.1 Micrometer

A micrometer is a precise length-measuring instrument, which is divided into mechanical and electronic types. The screw micrometer is made based on the principle of screw amplification, where the screw rotates in the nut for one round, and the screw advances or recedes a pitch along the axis of rotation. Therefore, the tiny distance moved along the axis can be represented by readings on the circumference. The typical structure of a micrometer is shown in Fig. 11.16.

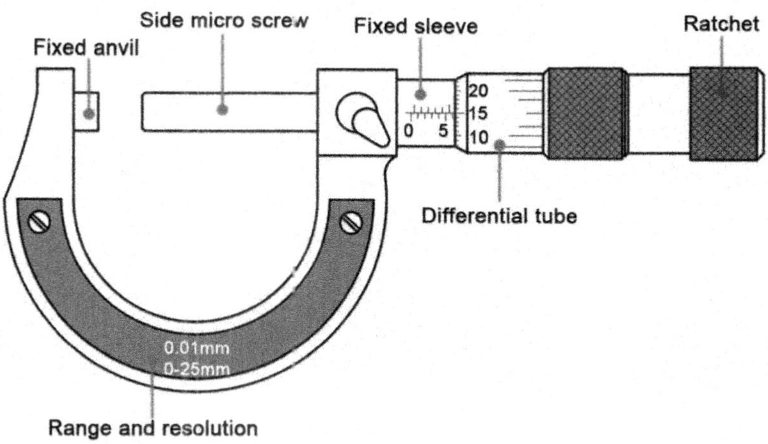

Figure 11.16 The schematic structures of a micrometer.

11.3.1.2 Scanning electron microscope

An electron beam emitted by an electron gun accelerates under the action of an electric field and is focused into a finer electron beam with 5 nm or less. The electron beam scans the sample sequentially and excites it to produce various physical signals (secondary electrons, back-scattered electrons, absorbed electrons, Auger electrons, etc.). The signal detector collects these physical signals and converts them sequentially and proportionally into video signals. Through the detection of some signals, video amplification, and signal processing, the scanning image can reflect the surface characteristics of the sample on the display screen.

According to the preceding mechanism, the separator's thickness can be measured by an image of the cross-sectional scanning electron microscope (SEM), as shown in Fig. 11.17.

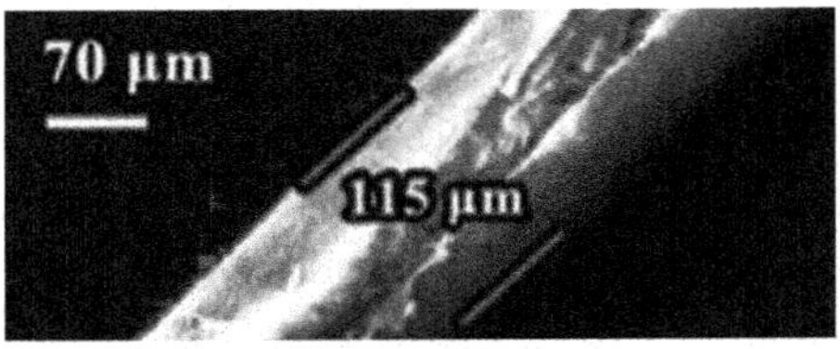

Figure 11.17 Cross-sectional SEM images of modified polyacrylonitrile/silica-aerogel (M-PSA) separator. Reproduced with permission from Ref. [12]. Copyright 2019, Elsevier.

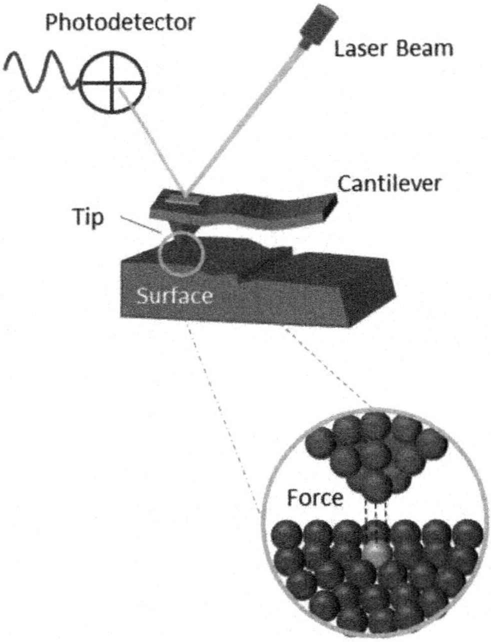

Figure 11.18 The schematic diagram of AFM test.

11.3.1.3 Atomic force microscope

An atomic force microscope (AFM) is an analytical instrument that can be used to study the surface structure of solid materials, including insulators. Although there is a term "microscope" in AFM, it does not "see" any microscopic object just like the optical and electron microscopes. Instead, the surface structure and properties of matter are studied by the extremely weak interatomic interaction force between the sample surface and a micro force sensitive element. A nanoscale probe is fixed to an elastic cantilever beam that is force sensitive, and the other end of the cantilever beam is fixed. When the sample is close to the tip, the interaction force between the tip atom and the surface atom of the sample deforms the cantilever beam. The deformation is detected by a laser detector and transmitted to the detection equipment. The interaction force between the tip and the surface atom can be obtained by converting the deformation signal into a photoelectric signal [13, 14]. The principle of AFM test is shown in Fig. 11.18.

11.3.1.4 Other methods

In addition to the common micrometer, SEM and AFM detection methods, the hand-held magnetic induction coating thickness meter, the eddy current coating thickness meter, and the fluorescence X-ray meter coating thickness meter can also be used for the separator thickness measurement.

11.3.2 Porosity

The porosity of a separator is defined as the ratio of the volume of micropores to the total volume of the separator, which is another important parameter of the separator. Theoretically, some parameters, such as permeability, liquid absorption rate, electrochemical impedance, and so on, are also related to porosity. At present, the porosity-measuring methods include suction method, direct calculation method, and instrument test method.

11.3.2.1 Suction method

The suction method is commonly used in the laboratory because it is easy to operate. A solvent with excellent wettability, such as anhydrous ethanol, hexadecane, n-butanol, and so on, should be selected to test separator porosity since the results are related to the wettability of the separator. First, for anhydrous ethanol, weigh the dry separator's mass (expressed as μ_0) before the test. Immerse the separator completely in anhydrous ethanol for a certain period, and then take out the immersed separator. After gently wiping the anhydrous ethanol on the separator's surface with a filter paper, measure the separator's quality (expressed as μ). In this way, the porosity of the separator can be calculated by Eq. (11.1).

$$\varepsilon = \frac{(\mu - \mu_0)\rho}{\rho\mu + (\rho - \rho_0)\mu_0} \tag{11.1}$$

where ε is the separator's porosity (%), μ_0 and μ are the weight of separator before and after immersing in solvents (g), respectively. ρ and ρ_0 represent the density of separator and anhydrous ethanol (g/cm^3), respectively.

11.3.2.2 Direct calculation method

To date, the following calculation method is used by most separator manufacturers. Only parameters of matrix quality and material size need to be known; the porosity of the separator can be given by Eq. (11.2).

$$P = \left(1 - \frac{M/V}{\rho}\right) \times 100\% \tag{11.2}$$

where P is the separator porosity (%), M is the separator mass (g), V is the separator volume (cm^3), and ρ is the separator density (g/cm^3).

11.3.2.3 Instrument test method

Though some instruments can achieve good precision, they have to be operated at a high cost by highly trained technicians, which are often found in some great separator manufacturers and some qualified scientific research teams. The commonly used equipment mainly includes capillary flow analyzer, mercury hydrometer, and water hydrometer. The measurement results are closely related to the measurement principle and experimental conditions. These facilities can effectively measure not only the pore size and distribution, but also the gas permeability, liquid permeability, surface area, integrity and other parameters, which are of great benefit for the characterization of separator microstructure.

11.3.3 Average Pore Size and Size Distribution

In order to keep the battery running continuously and stably, the current density in batteries should be uniform and stable, so the pore sizes of the separators are a crucial parameter. If the pore size is too small, the ion permeability will be limited, which will increase the internal resistance and reduce the overall performance of batteries. On the contrary, it will increase the ion permeability and cause the risk of electrode particles penetration into separators. When the pore size distribution of the separator is not uniform, it is easy to cause uneven metal deposition and generates dendrites, which can puncture the separator and cause safety problems such as short circuits or even explosion. The pore size and distribution are mainly measured by SEM, capillary flow porosimeter, or mercury porosimeter.

11.3.3.1 Mercury porosimeter

Mercury is non-wettable for most solid materials and requires additional pressure to enter solid pores. The pore radius into which mercury can insert is inversely proportional to the external pressure. The larger the external pressure, the smaller the pore radius that mercury can enter. Mercury is filled into large pores, mesopores, and small holes. Measuring the amount of mercury entering the hole under different external pressure can detect the size of the corresponding pore. However, because of the toxicity of mercury, this method should no longer be developed or used.

11.3.3.2 Capillary flow porometer

a. The separator pore is completely wetted and filled with liquid, and positive pressure is formed in the pore due to the capillary phenomenon.
b. Place the separator in a sealed tank and pressurize the liquid with gas pressure to be extruded from the capillary.
c. The pore diameter of separators can be obtained according to the Laplace equation, where describes the relationship between the pressure applied and the pore's diameter when the liquid in a single pore is completely extruded by the capillary pores. The Laplace equation is shown in Eq. (11.3).

$$d = \frac{4\gamma\cos\theta}{\Delta P} \tag{11.3}$$

where d is the pore diameter (m), ΔP is the pressure difference (Pa), γ is the surface tension of the liquid (N/m), and θ is the contact angle between separator and the liquid (°).

At different pressures, the liquid in separators will be extruded one after another and a certain gas penetration flow rate is produced. The pore size can be calculated according to the relationship between pressure and flow rate.

11.3.3.3 N_2 isothermal adsorption and desorption curves

The principle of gas adsorption method to measure the specific surface area is based on the adsorption characteristics of gas on solid surface. Under certain pressure, the separator surface (adsorbents) has reversible physical adsorption on gas molecules (adsorbents) at ultra-low temperature. Moreover, there is a certain equilibrium adsorption capacity corresponding to a certain pressure. By measuring the equilibrium adsorption capacity, the specific surface area of the separator is equivalent to that of the theoretical model.

Mesoporous analysis usually adopts the Barrett–Joyner–Halenda (BJH) model, which is the application of Kelvin equation in the cylinder model. It is mainly based on capillary condensation phenomenon and combined with the Kelvin equation. As long as the gas isothermal adsorption curve is measured, the pore volume–aperture distribution, total pore volume, and average pore size can be calculated in turn.

11.3.4 Mechanical Properties

Considering the winding and packing of separators during the production, assembly, and disassembly processes of batteries, and the repeated charge/discharge in the practical applications, the separator must have enough mechanical strength to prevent the damage caused by the physical impact, puncture, wear, and compression. The mechanical properties of separators mainly include the puncture strength and tensile strength. At a certain speed (3–5 m/min), a needle with a diameter of 1 mm is let to the annular fixed separator; the maximum force applied to the needle through the separator is called the puncture strength. Tensile strength is a parameter that reflects the dimensional stability of a separator under external force. If the tensile strength is not enough, it is not easy to restore the original size after the separator is deformed, which will lead to the short circuit of batteries.

11.3.4.1 Universal electronic tension machine

A universal electronic tension machine is suitable for measuring the tensile, compression, stripping, and puncture strengths of metal and nonmetallic materials. The universal electronic tension machine works in an experimental process of pulling, pressing, bending, and shearing with the moving beam. The force data of sensor, the displacement, deformation data of encoder, grating ruler or extension meter are collected. Finally, the experiment results are calculated by manual or computer calculation to obtain the puncture strength.

11.3.4.2 Tension tester

The separator is clamped between two clamps of the tension tester, and then the two clamps separate and stretch the separator at a certain speed to measure stress change until it is destroyed. Then the force value change and displacement change during the test are collected by the force value sensor located on the moving chuck and the displacement sensor to calculate the tensile, tear, deformation rate, and other performance indexes of the separator.

11.3.5 Wettability

Wettability refers to the affinity of electrolytes with separator, that is, the ability of electrolytes to spread on the separator. Good wettability is also the premise of electrochemical performance because ion transport relies on electrolytes. If the separator cannot be well wetted and covered with electrolytes, ions cannot transfer freely and the ionic conductivity of batteries will be extremely low, causing the huge internal resistance. The electrolyte uptake, electrolyte immersion height, and contact angle of separator are several indexes to judge its wettability.

11.3.5.1 Weighing method (electrolyte uptake)

The electrolyte uptake of the separator can be calculated by the weight change before and after the separator is immersed with the electrolyte. The higher the electrolyte uptake, the stronger the ability of the separator to absorb the electrolyte. The calculation is based on Eq. (11.4):

$$\eta = \frac{W - W_0}{W_0} \times 100\% \tag{11.4}$$

where η represents the electrolyte uptake of the separator (%), and W and W_0 are the weight of the separator before and after impregnation with the electrolytes (g), respectively.

11.3.5.2 Time method (electrolyte immersed height)

The electrolyte immersed height can be determined by immersing the long-strip separator into the electrolyte vertically to observe the separator height wetted by the electrolytes. The higher the electrolyte immersion height, the stronger the ability of the separator to absorb the electrolytes.

11.3.5.3 Instrument test method: contact angle

The contact angle refers to the drop of a liquid on a solid horizontal plane, and the three solid, liquid, and gas phases form a junction point on the solid plane, in which the angle between the tangents of the gas–liquid interface and the solid–liquid interface is named the contact angle. The contact angle of liquid to solid is mainly measured by a contact angle tester, that is, the wettability of liquid to solid [15, 16]. The schematic contact angle of a solid–liquid interface is shown in Fig. 11.19.

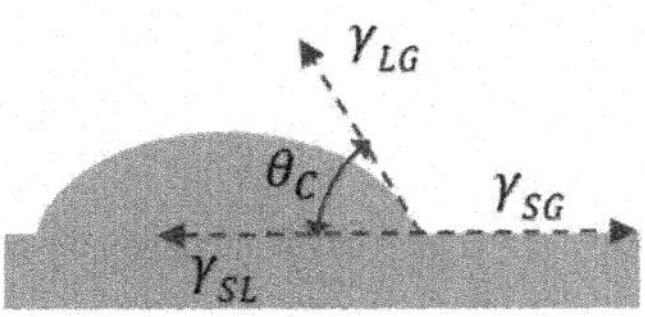

Figure 11.19 The schematic contact angle of a solid–liquid interface.

A syringe needle is attached to the contact angle tester, and the liquid is added drop by drop onto the separator. The droplets will attach to the separator surface and cast a shadow. The projection screen micrometer uses optical amplification to project images to the screen for measurement. This projection screen micrometer has an adjustable specimen clip that can be aligned vertically or axially to the image. The image can be adjusted horizontally by sliding the screen. The locking knob can fix the projection droplet in position. For the droplet angle reading, it is necessary to find the tangent between the contact point at the corner of the image and the highest point of the image, and to measure the angle with a specially calibrated scale. The wettability of separators can be directly compared by the contact angle.

11.3.6 Thermal Stability

The battery will produce much heat when heated, or in the external/internal short circuits. If the separator shrinks and melts at this time, the cathode and anode will contact each other and cause a short circuit with possible fire and explosion. Therefore, thermal stability is an important parameter for separators to determine the safety performance of batteries.

11.3.6.1 Thermal shrinkage

To test the heat tolerance of the separator, it needs to put a certain size of separator in a high-temperature environment and slowly increase the temperature from a certain initial temperature to the set temperature. The thermal stability can be expressed by the area shrinkage of separators based on Eq. 11.5:

$$Q = \frac{S_1 - S_2}{S_1} \times 100\% \tag{11.5}$$

where Q is the area retention rate of separator (%), S_2 is the area of the separator after heating at a certain temperature (m^2), and S_1 represents the initial area of the separator before heating (m^2).

11.3.6.2 Differential scanning calorimetry

Differential scanning calorimetry (DSC) can be used to measure the temperature and heat flux related to the internal thermal transition of the material, and the closed pore and melting temperature of the separator can be measured by it.

The relationship between heat flow or power difference and temperature or time under program-controlled temperature can be measured by this method. The heat effect and heat change occur in the heating process, and the heat change can be compensated by the input electric energy in the process, whereas the temperature of the sample and the reference object can be restored to balance. Therefore, as long as the amount of electricity compensated is recorded, the amount of heat generated by the thermal effect (absorption or release) of the sample can be known.

11.3.6.3 Thermogravimetric analysis

Thermogravimetric (TG) analysis is a method of controlling the temperature of the instrument by programing to measure the relationship between the mass of matter and temperature or time. By analyzing the TG curves, we can know the information of compositions, thermal stability, and the thermal decomposition temperature. The resulting products are related to the quality of the sample and its possible intermediate products. Its main characteristics, i.e., strong quantitative, can accurately measure the mass change and the change rate. According to this characteristic, as long as the mass of matter changes with the heating process, it can be studied by TG analysis. Since there are many physical and chemical changes, such as sublimation, vaporization, adsorption, desorption, and gas–solid reactions, we can use the TG analysis to evaluate the thermal stability of the separator.

11.3.7 Electrochemical Performance

The pore structure of the separator provides ion transport channels and stores the electrolyte, ensuring rapid sodium-ion transport inside the battery. A good separator is a prerequisite for high-performance batteries, though it is not involved in the electrochemical reaction of sodium ion. Thus, it is significant to realize the performance of the separator in the electrochemical condition.

11.3.7.1 Linear sweep voltammetry

Linear sweep voltammetry (LSV) is a commonly used electrochemical research method, in which the electrode potential is controlled by performing a one-time scan in a triangular waveform at different rates and time. The set potential range enables electrochemical reactions on the electrode and records the current-potential change curve during the scanning.

LSV is usually adopted to determine the possible electrochemical reactions of an electrode system. The anodic current peak on the LSV curve indicates that the electrode has an oxidation reaction, while the cathodic current peak indicates the occurrence of a reduction reaction. The potential range corresponding to the current peak can help determine whether electrochemical reaction occurs, and the difference between the equilibrium potential of reaction and the reaction indicates the difficulty of the reaction. Therefore, the electrochemical stability of separator in electrolytes can be determined by the LSV technique.

11.3.7.2 Electrochemical impedance spectroscopy

Electrochemical impedance spectroscopy (EIS) is an original method to study the frequency response characteristics of linear circuit networks in electric power. Because impedance also exists in electrochemistry, it is introduced in the field of battery.

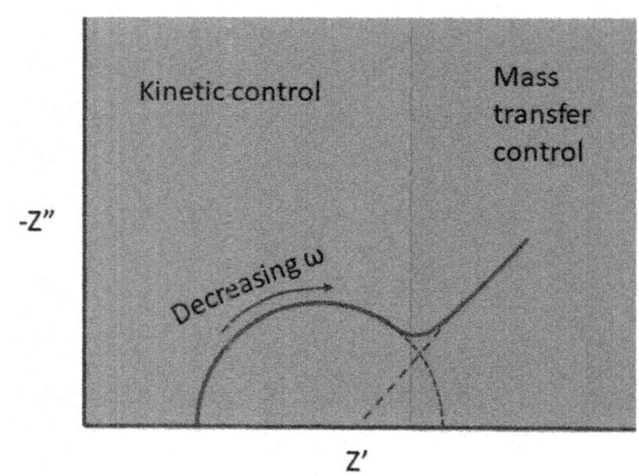

Figure 11.20 AC impedance diagram.

When the electrode system is disturbed by a sinusoidal voltage (current) AC signal, a corresponding current (voltage) response signal will be generated, and the impedance of the electrode can be obtained through the signal. The impedance spectrum generated by a series of frequency sine wave signals is called EIS. The semicircular area (high-frequency area) is controlled by the electrode reaction kinetic (charge transfer process), and the oblique line (low-frequency area) is controlled by the electrode reactant or product diffusion process. Through the Nyquist diagram (Fig. 11.20), the charge transfer resistance, ohmic resistance, and ion conductivity σ of the entire system can be obtained [17].

11.3.7.3 Ionic conductivity

Ionic conductivity is the ability of ions to pass through the electrolyte and separator between cathode and anode. It is one of the basic electrochemical properties of the separator–electrolyte system, and another key factor that should be considered whether the separator can be practically applied.

The ionic conductivity (σ) and the ionic resistivity (ρ) are reciprocal to each other. The ionic resistance measured by experiments is usually the sum of separator resistance and the electrolyte resistance, as shown in Eq. (11.6):

$$R_{\text{total}} = R_{\text{separator}} + R_{\text{electrolyte}} \tag{11.6}$$

In order to facilitate the calculation, the influence of $R_{\text{electrolyte}}$ can be ignored, that is, $R_{\text{total}} = R_{\text{separator}}$. According to the reciprocal of ion conductivity and ion resistivity, that is $\sigma = 1/\rho$, the calculation of ion conductivity can be obtained like Eqs. (11.7) and (11.8):

$$\rho = \frac{R \times S}{d} \tag{11.7}$$

$$\sigma = \frac{d}{R \times S} \tag{11.8}$$

where d is the average thickness of the separator (m), S is the effective area of the separator (m^2, the area where separator is in contact with the electrode slice), and R is the resistance obtained from the AC impedance diagram (Ω).

11.3.7.4 Ion diffusion coefficient

Diffusion is an important form of mass transfer. The insertion and extraction of sodium ions in the electrode material is a typical diffusion movement. Meanwhile, the chemical diffusion coefficient D_{Na} largely determines the reaction rate and affects the overall performance of batteries. Therefore, determining the chemical diffusion coefficient is of great significance to the study of the electrochemical performance of batteries.

Cyclic voltammetry

The test principle of cyclic voltammetry is similar to that of previous sweep voltammetry. The peak current in the cyclic voltammetry diagram conforms to the Randles–Sevcik equation, which is shown in Eq. (11.9).

$$I_P = 2.69 \times 10^5 \times n^{(3/2)} \times A \times D^{(1/2)} \times C \times V^{(1/2)} \quad (11.9)$$

where I_p is the peak current (A), n is the number of transferred electrons, A is the surface area of the positive electrode (cm^2), C is the concentration of sodium ions (mol cm^3), V is the scan rate of cyclic voltammetry (V/s), and D is the ion diffusion coefficient (cm^2/s). The ion diffusion coefficient D can be calculated based on the parameter information obtained by the cyclic voltammetry measurement [18].

Galvanostatic intermittent titration technique

The galvanostatic intermittent titration technique (GITT) is composed of a series of "pulse + constant current + relaxation." The relaxation process means that no current flows through the battery during this period. Therefore, there are two main parameters set by GITT: current intensity (i) and relaxation time (τ). The typical diagram of GITT is shown in Fig. 11.21.

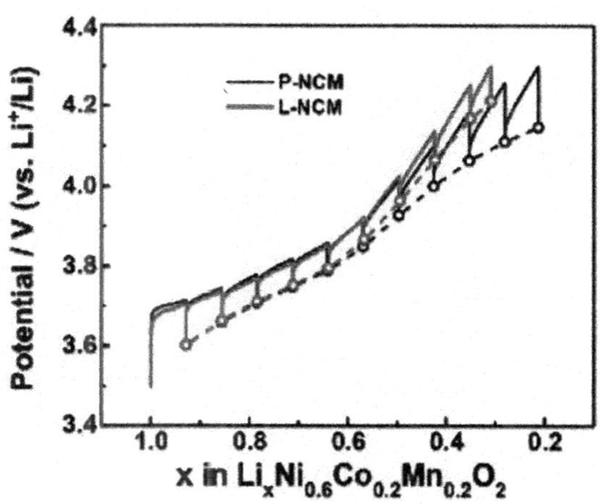

Figure 11.21 The diagram of GITT. Reproduced with permission from Ref. [19]. Copyright 2018, American Chemical Society.

First, a positive current pulse is applied to the battery system, and the battery potential rises rapidly. Subsequently, the charging current is maintained constant, and the potential rises slowly. This is also the source of the "constant current" in the name of the constant current intermittent titration technique. Then, the charging current is interrupted, and the potential drops rapidly, and the value of the drop is proportional to the IR drop. Finally, it enters the relaxation process. During this period, the composition in the electrode tends to be uniform through ion diffusion, and the potential slowly drops until it is balanced again. The preceding process is repeated until the battery is fully charged. The discharge process is opposite to the charging process [20].

Calculation formula:

$$D = \frac{4}{\pi}\left(\frac{iV_m}{Z_A FS}\right)^2 \left[\frac{dE / d\delta}{dE / d\sqrt{t}}\right]^2 \tag{11.10}$$

where i is the current value (A), F is the Faraday constant (96,485 C/mol), Z_A is the charge of the ion (sodium ion is 1), S is the electrode/electrolyte contact area (m^2), $dE/d\delta$ is the slope of the Coulomb titration curve, and $dE / d\sqrt{t}$ is the relationship between electric potential and time.

In order to simplify the solution, it is considered that the applied current is tiny, and the relaxation time τ is very short, $dE / d\sqrt{t}$ has a linear relationship, and as such, the preceding formula can be simplified to

$$D = \frac{4}{\pi\tau}\left(\frac{n_m V_m}{S}\right)^2 \left[\frac{\Delta E_s}{\Delta E_t}\right]^2 \tag{11.11}$$

Among them, τ is the relaxation time (s), n_m is the number of moles (mol), V_m is the molar volume of the electrode material (m^3/mol), S is the electrode/electrolyte contact area (m^2), ΔE_s is the voltage change caused by the pulse (V), and ΔE_t is the voltage of constant current charging (discharging) variety (V). The values of ΔE_s and ΔE_t are obtained from the GITT diagram.

11.3.7.5 Ion transference number

Among the many ions in the electrolyte system, the transference of Na^+ is the most meaningful. The ion transference number is defined as

$$t = \frac{D_+}{D_+ + D_-} \tag{11.12}$$

where D_+ and D_- are the ion diffusion coefficients of cations and anions (cm^2/s), respectively. It is known from the definition that as soon as the Na^+ transference number is closed to 1, the battery charge and discharge are much faster. The ion transference number is an important indicator for judging whether the battery can achieve effectively rapid charging and discharging.

It is difficult to measure the transference number of Na^+ in the electrolyte, especially when both anions and cations are diffused in the electrolyte. This makes the measurement of the transference number of ions get more difficult, and the accuracy rate is lower. This is also the important reason why the current majority of battery enterprises does not use ion transference numbers to characterize for the electrolytes.

At present, the ion transference number (t_{Na+}) in a battery is mainly measured by a combination of AC impedance technology and DC polarization technology. In this measurement method, the charge transfer impedance is first measured by the AC impedance technique, and then a certain voltage (ΔV = 10 mV or other) is applied to the battery at a constant potential to obtain the function image between the polarization current and time. Then the load transfer impedance of battery is tested, and the ion transference number is calculated from the parameter information obtained from the preceding test. The calculation formula is displayed as follows:

$$t_{Na^-} = \frac{I_S(\Delta V - I_0 R_0)}{I_0(\Delta V - I_s R_s)} \tag{11.13}$$

where ΔV is the potential difference (V), I_0 is the initial current obtained (A), I_S the steady-state current (A), and R_0 and R_S are the initial and steady-state load transfer resistance before and after polarization (Ω), respectively [21, 22].

11.4 Components and Development of Separators

In the past few years, there have been significant advances in the research and commercial production of separators in SIBs. In order to achieve the application goal of SIBs, it is necessary to realize the scientific problems in the development of separators and optimize the components and structure of the separator.

11.4.1 Analysis of Separator Components

At present, the widely used separators in SIBs mainly include polyolefin separators, such as PP/PE separator, tri-layer PP/PE/PP separator, and inorganic GF separator. In addition, some domestic and foreign enterprises have also developed other battery separators, including polyvinylidene fluoride (PVDF) and PI. The related molecular structures of separators are shown in Fig. 11.22.

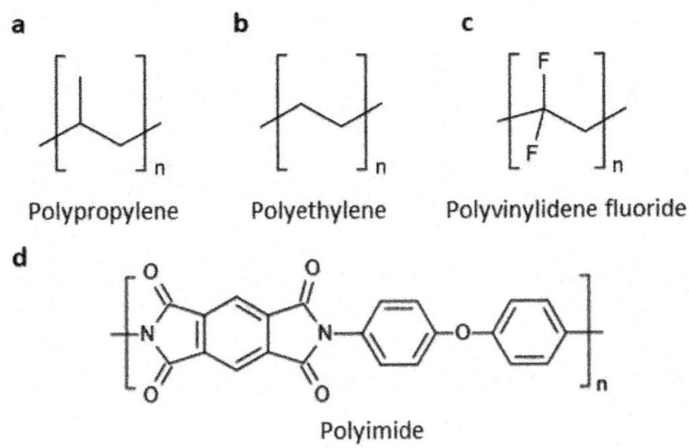

Figure 11.22 The molecular structures of (a) PP, (b) PE, (c) PVDF, and (d) PI.

11.4.1.1 Polyolefin separator

PP is a kind of polymer formed by the copolymerization of propylene monomer. PP material has high crystallinity, regular structure, excellent mechanical properties, and chemical stability. The melting point of PP is about 165°C, and the dielectric constant is 2.15. The excellent insulation performance is not affected by environmental humidity.

PE is a kind of polymeric compound formed by the addition polymerization of vinyl monomer. Its molecular structure is linear and composed of several methylene groups. The melting point of PE is about 130°C, and the dielectric constant is 2.3. When the temperature of batteries rises up, a PE separator can easily melt, resulting in short circuits of batteries.

Tri-layer PP/PE/PP possesses multilayer structure and high safety. When the temperature is close to the melting temperature of PE, the PE layer first melts and blocks the ion transport pathway. Whereas the PP layer can maintain its dimensional stability, which prevents short circuit caused by contact between cathode and anode.

A polyolefin separator usually has a relatively thin thickness (16–32 μm) and moderate porosity (about 40%). Under the condition of ensuring the smooth transport of sodium ion inside the batteries, its small pore structure can prevent the growth and penetration of sodium dendrite more effectively. Furthermore, good chemical/electrochemical stability of polyolefin material also allows its separator to work stably under strong oxidation or reduction conditions without the occurrence of side reactions with electrolytes and electrodes. The detailed physical parameters of the polyolefin separators are presented in Table 11.2.

11.4.1.2 Glass fiber

Glass fiber (GF) is an inorganic nonmetallic material, containing SiO_2, Na_2O, Al_2O, B_2O_3, and nonmetallic oxides. Therefore, a GF separator has excellent heat-resistance (melting point more than 500°C) and flame-retardant properties compared with a polyolefin separator. Importantly, good wettability to electrolytes and high electrolyte uptake ensure efficient sodium-ion transport between cathode and anode. However, high cost and poor mechanical strength limit the development and large-scale application of GF separators. More seriously, large thickness (approximately 200–800 μm) usually leads to large resistance and low energy density of batteries, and the excessive porosity with large pore size not only requires a mass of electrolytes, but provides enough room for sodium dendrite growth, damaging the cycle life of batteries. The physical parameters of different types of GF separators are presented in Table 11.3.

11.4.1.3 Polyvinylidene fluoride and its copolymers

Polyvinylidene fluoride (PVDF) and its copolymers are partially fluorinated semicrystalline polymers in which the amorphous

Table 11.2 Physical parameters of commercialized polyolefin separator

Manufac-turers	Type	Materials	Number of Layers	Thick-ness/ (μm)	Air Perme-ability/ (sec/ 100 cc)	Porosity/ (%)	Pore Size /(μm)	Tensile Strength /(kg/cm^2)	Puncture Strength/ (N)	Contrac-tility	Other
Celgard	2400	PP	Monolayer	25	620	41	0.043	TD 140 MD 1420	4.5	90°C, 1 h TD 0% MD 5%	
	2500	PP	Monolayer	25	200	55	0.064	TD 135 MD 1055	3.35	90°C, 1 h TD 0% MD 5%	
	2325	PP/PE/PP	Tri-layers	25	620	39	0.028	TD 150 MD 1700	3.8	90°C, 1 h TD 0% MD 5%	
	2320	PP/PE/PP	Tri-layers	20	530	39	0.027	TD 165 MD 2050		0%	

Manufacturers	Type	Materials	Number of Layers	Thickness/ (μm)	Air Permeability/ (sec/ 100 cc)	Porosity/ (%)	Pore Size /(μm)	Tensile Strength /(kg/cm²)	Puncture Strength/ (N)	Contractility	Other
AsahiKASEI	Jas20600	PE	Monolayer	20	502	42.31	0.118	TD 144 MD 1530			
		PE	Monolayer	16±2.0	230±70	45±5		TD 1200 MD 1200	>4	90°C, 1 h TD ≤ 1% MD ≤ 3%	
		PE	Monolayer	25±1.0	200±30	45±5		TD≥1200 MD≥1200	>4	90°C, 1 h TD ≤ 1% MD ≤ 3%	
Green	GRE-16TP1	Ceramic coating		16	380			TD 230 MD 1050	2.7	90°C, 1 h TD 0.2% MD 0.4%	Coating thickness: 4 μm
	GRE-20TP1	Ceramic coating		20	380			TD 220 MD 950	3.2	90°C, 1 h TD 0.2% MD 0.4%	Coating thickness: 4 μm
	GRE-25TP1/ TP2	Ceramic coating		25	380c			TD 220 MD 950	3.9	90°C, 1 h TD 0.2% MD 0.5%	Coating thickness: 5 μm

(Continued)

Table 11.2 *(Continued)*

Manufacturers	Type	Materials	Number of Layers	Thickness/ (μm)	Air Permeability/ (sec/ 100 cc)	Porosity/ (%)	Pore Size /(μm)	Tensile Strength /(kg/cm^2)	Puncture Strength/ (N)	Contractility	Other
Green	GRE-30TP1/ TP2	Ceramic coating		30	380			TD 230 MD 950	3.8	90°C, 1 h TD 0.2% MD 0.3%	Coating thickness: 5 μm
	GRE-32TP1/ TP2	Ceramic coating		32	400			TD 230 MD 950	3.8	90°C, 1 h TD 0.2% MD 0.3%	Coating thickness: 7 μm
SK Innovation		PE	Monolayer	16	200	44		TD1200 MD 1200		90°C, 1 h TD 1% MD 3%	
		Ceramic coating		16	215	43		TD 1855 MD 1960		90°C, 1 h TD 0.3% MD 1.2%	Coating thickness: 4 μm

Table 11.3 Physical parameters of glass fiber separator

	Type	Thick-ness/ (μm)	Pore Size/ (μm)	Air Velocity/ Gurelry(s)	Applicable Tempera-ture/(°C)	Mass/ (g/m²)
	GF/A	260	1.6	4.3	<500	55
	GF/B	675	1.0	12	<500	143
	GF/C	260	1.2	6.7	<500	55
Whatman	GF/D	675	2.7	2.2	<500	121
	GF/F	420	0.7	3.7	<500	75
	GF/M	800	/	14.2	<500	178
	934-AH	435	1.5	3.7	<500	64

phase lies between the spherulite-aligned wafers. It can crystallize in different crystalline phases with different temperature and processing conditions. The crystalline β phase of PVDF and its copolymers are ferroelectric, piezoelectric, and pyroelectric. The α phase is thermodynamically the most stable when the materials are obtained directly from the melt. For example, Young's modulus of PVDF is 1500–3000 MPa, and high dielectric constant (6–8) can significantly assist the dissociation of electrolyte salts. The addition of monomers such as hexafluoropropylene and tetrafluoropropylene can further increase the fluorine content, reduce the crystallinity, leading to the easy absorption of the electrolytes in the amorphous region during the swelling process, and the increase in the ionic conductivity.

Among PVDF and its copolymers, polyvinylidene fluoride-hexafluoropropylene (PVDF-HFP) as separator material has attracted much attention. Because of the high affinity to liquid electrolytes, excellent electrochemical stability, and adhesion to electrodes, the amorphous hexafluoropropylene phase can help to capture a large amount of electrolytes, while the PVDF crystal phase can act as a mechanical support for PVDF-HFP separator. High dielectric constant and electron absorption of fluorine atoms in polymer skeleton are beneficial to the dissociation of electrolyte salts [15].

11.4.1.4 Polyimide

Polyimide (PI) is a class of polymers containing imide rings (-CO-N-CO-) on its main chain. The polar functional groups in PI enable good wettability with electrolytes. Meanwhile, PI has excellent heat resistance in the range of 200–400°C and fire resistance. Moreover, its volume resistance is as high as 10^{17} Ω cm, presenting good electrical insulation. However, PI molecules contain more polar groups, which are soluble in acid and alkaline. In addition, the cost of PI materials is relatively high.

11.4.2 Requirements and Development Status of Separators

11.4.2.1 Requirements of separators

The physical parameters and chemical properties of the separator strongly affect the performance of batteries. For high-performance SIBs, the separator should meet the following conditions:

a. **Electronic insulation and chemical/electrochemical stability:** First, a separator must be electrically insulated to effectively isolate the electronic conduction between anode and cathode. Moreover, the separator must have high chemical and electrochemical stability to ensure that it will not change during the immersion of electrolyte and the charge–discharge processes of batteries.
b. **Good electrolyte wettability:** The electrolyte wettability of a separator is important for sodium-ion transmission because the electrolyte is the medium that conducts sodium ions in batteries. The wettability performance of separators mainly includes contact angle and electrolyte uptake. The contact angle of separators to the electrolytes determines the injection time during the assembly process of batteries. And high electrolyte uptake can ensure adequate supply of sodium ion from electrolytes and improve the ionic conductivity of separators.

c. **Suitable porosity:** The separator should have appropriate porosity to provide ion transmission channels and storage electrolytes. Low porosity will hinder sodium-ion transmission. However, excessive porosity is accompanied by poorer mechanical strength, which will easily lead to separator damage and dendrite penetration.

d. **Appropriate pore structure:** The pore size of a separator is preferably less than 1 μm. Larger pore size can cause self-discharge of the batteries and may also cause sodium dendrite growth through the separator. The pore size distribution of separators also affects the transport rate of sodium ions and the deposition site of sodium metal. The uniform pore size distribution can ensure the stable current density and uniform deposition of sodium, inhibiting sodium dendrite growth.

e. **Excellent thermal stability:** The thermal stability of the separators is an important property affecting the battery safety. The battery will produce heat during charge and discharge process, especially under the condition of collision or abuse (e.g., overcharge, overdischarge), arising the temperature inside the battery rapidly. When the temperature reaches the melting point of the separator materials, the separator will shrink and even melt, which will cause a large range of short circuits inside the battery, and further worsen the battery safety.

11.4.2.2 Development status of separators

Now, researchers are adopting various strategies to obtain a separator that meets the preceding requirements. In fact, the battery performance will be greatly limited if the synergistic development of separators for SIBs is ignored. Fortunately, many methods have been proposed to improve the performance of separators in SIBs.

a. Modification of commercial polyolefin separator

PVDF has high thermal, electrochemical stability, mechanical strength, and good wettability. When PVDF is coated on both sides of PP separator, a sandwich structure can be formed by the electrospinning process. Figure 11.23 shows that compared with PP separator, the modified separator shows improved thermal stability (50% weight loss at 400°C, higher than 350°C when uncoated) and dimensional stability. The lyophilic nature of PVDF can also effectively improve the surface wettability of the separator [23].

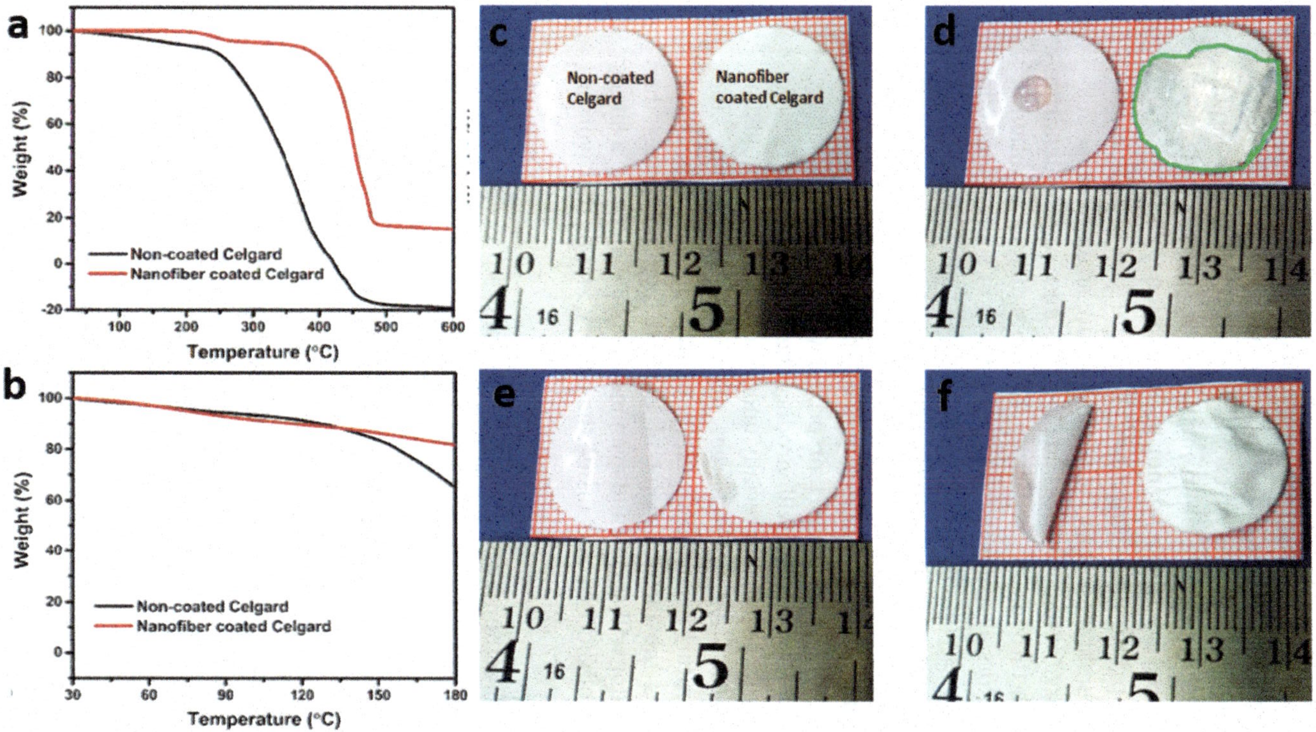

Figure 11.23 (a) TG curves of non-coated and nanofiber-coated PP Celgard membrane under an air atmosphere at a heating rate of 10°C/min. (b) TG curves of gel polymer electrolytes (GPEs) under nitrogen atmosphere at a heating rate of 10°C/min. (c, d) Photographs of non-coated and nanofiber-coated PP Celgard separator before and after dropping liquid electrolyte. (e, f) Thermal shrinkage of non-coated and nanofiber-coated PP Celgard separator membrane at 100°C and 160°C exposed for 4 h in a vacuum chamber. Reproduced with permission from Ref. [23]. Copyright 2020, Elsevier.

In addition to PP separators, PE separators can also be modified. For example, SiO_2 is coated on the surface of PE separators by a simple chemical vapor deposition (CVD) method to form a thin SiO_2 layer (tens of nanometers). SiO_2 keeps the separator stable at 140°C; at the same time, the SiO_2 will also enter the pores of the PE separator to improve its wettability to electrolytes, promoting the transmission of sodium ions [24].

Surface grafting on PP separators is another effective modification strategy. The polymer chain can be fixed on the surface of PP separators by the "graft-filter" and ultraviolet light, which can ensure the porous structure of polymer, and effectively adsorb the electrolyte. When such a modified separator is used in a room-temperature Na–S battery, the polymer on the surface of the modified separator repels polysulfide by electrostatic interaction and inhibits sodium dendrites growth [25].

b. Development of other polymer substrates

Natural cellulose is abundant and conforms to the concept of green chemistry. Therefore, cellulose-derivatives-based separators have been studied widely. A porous separator with high dimensional stability (the size remains unchanged at 180°C) was prepared by Lizundia et al. [26] via cross-linking carboxymethyl cellulose and hydroxyethyl cellulose. The abundant oxygen-containing groups on the separator can fix the electrolytes. Meanwhile, a large number of carboxyl groups promote the transfer of sodium ions and enhance the electrochemical performance of SIBs.

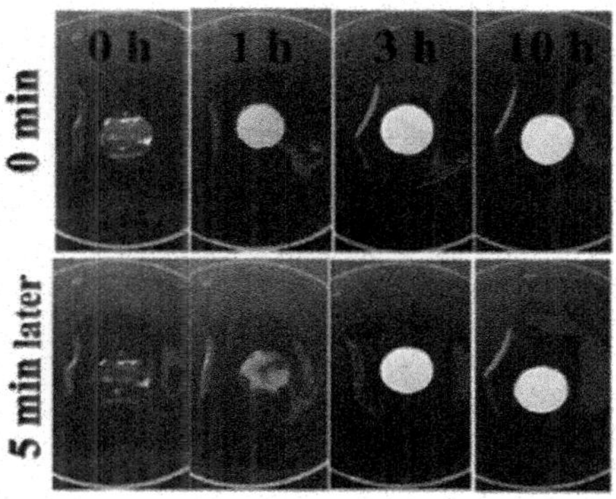

Figure 11.24 Stabilities of the modified flexible cellulose acetate separators (immersed in sodium hydroxide/water solution for 0, 1, 3, and 10 h) in electrolyte (EC/PC (volume ratio 1:1)). Reproduced with permission from Ref. [27]. Copyright 2018, American Chemical Society.

As shown in Fig. 11.24, substituting a certain number of acetate groups on the surface of the cellulose acetate separator with hydroxyl groups can also enhance its chemical stability. Due to the moderate content of hydrophobic acetate groups, the wettability of separators to electrolytes is improved obviously [27].

Chitin is another natural polymer, which possess excellent thermal stability and mechanical strength. The pore size on chitin can be controlled by adjusting the content of the pore-forming agent (sodium dihydrogen citrate, SDCA). Moreover, the chitin separator is composed of tightly packed sugar chains and a large number of hydrogen bonds, which can effectively improve the thermal stability and size retention of the material (Fig. 11.25) [28]. Similar to cellulose, a large number of oxygen-containing groups can fix the electrolytes and promote the transfer of sodium ions, which is superior to the PP separator.

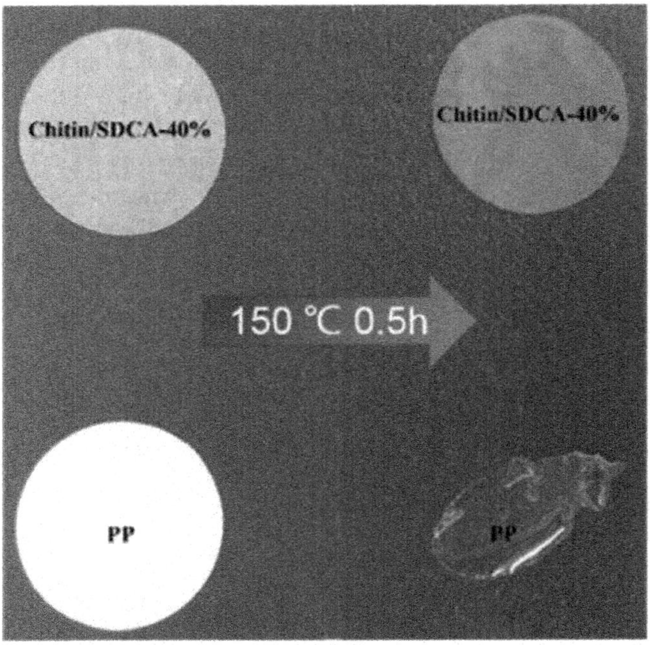

Figure 11.25 Photos of the PP separator and the chitin/SDCA-40% separator before (left) and after (right) thermal treatment at 150°C for 0.5 h. Reproduced with permission from Ref. [28]. Copyright 2017, American Chemical Society.

Commercial polymers can prepare a separator that can also solve the source problem of separator materials. For example, commercial polyacrylonitrile (PAN) has a high melting point (over 290°C), low cost, good affinity to electrolytes, and high electronic insulation. However, PAN has poor chemical stability in organic electrolytes, which is also an important reason for hindering development. A method was proposed by Chen et al. to combine PAN with silica aerogel (Fig. 11.26). Due to the hydrolysis of nitrile groups and the *in situ* growth of silica, this composite separator can be maintained in the conventional electrolyte [12]. In addition, Al_2O_3 can also be coated on cellulose–PAN fiber to further enhance its thermal stability and mechanical strength (Fig. 11.27) [29].

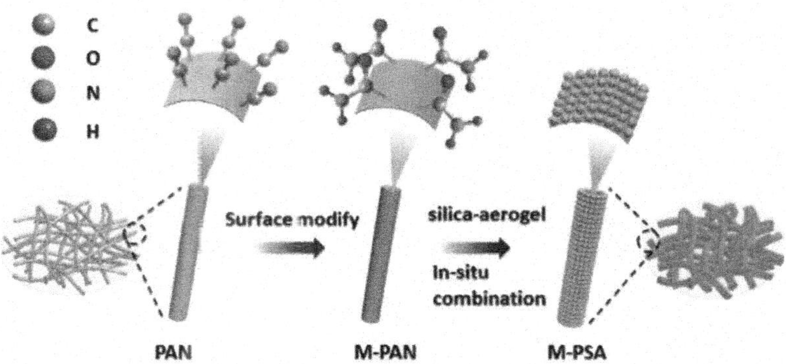

Figure 11.26 Schematic diagram of the synthesis process of a high-safety modified polyacrylonitrile/silica-aerogel (M-PSA) separator. Reproduced with permission from Ref. [12]. Copyright 2019, Elsevier.

c. Preparation of inorganic composite separators

The porous ceramics membrane (PCM) has a large number of cavities to absorb electrolytes and improve the ionic conductivity of separators. For example, porous ceramic PVDF-HFP/polybutyl methacrylate separator blended with barium titanate ($BaTiO_3$) is prepared by solution casting method. Compared with the GF separator, the sodium-ion transfer number, electrochemical stability, and mechanical stability (Fig. 11.28) have been significantly improved. Therefore, it can be used as a promising substitute for commercial separators in SIBs [30].

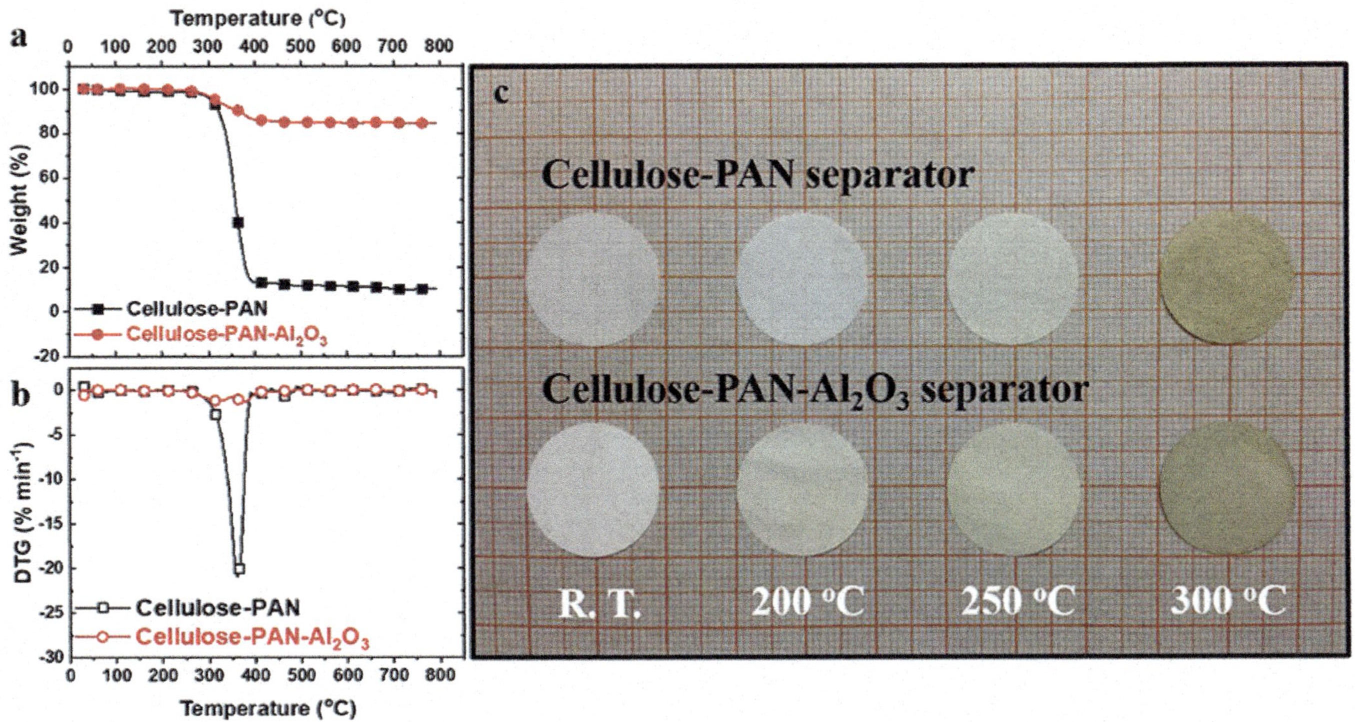

Figure 11.27 (a) TG and (b) derivative thermogravimetry curves of bare and cellulose–PAN-Al_2O_3 composite separator. (c) Thermal shrinkage rate of bare and cellulose–PAN-Al_2O_3 composite separator after heat treatment in the temperature range of 25–300°C. Reproduced with permission from Ref. [29]. Copyright 2020.

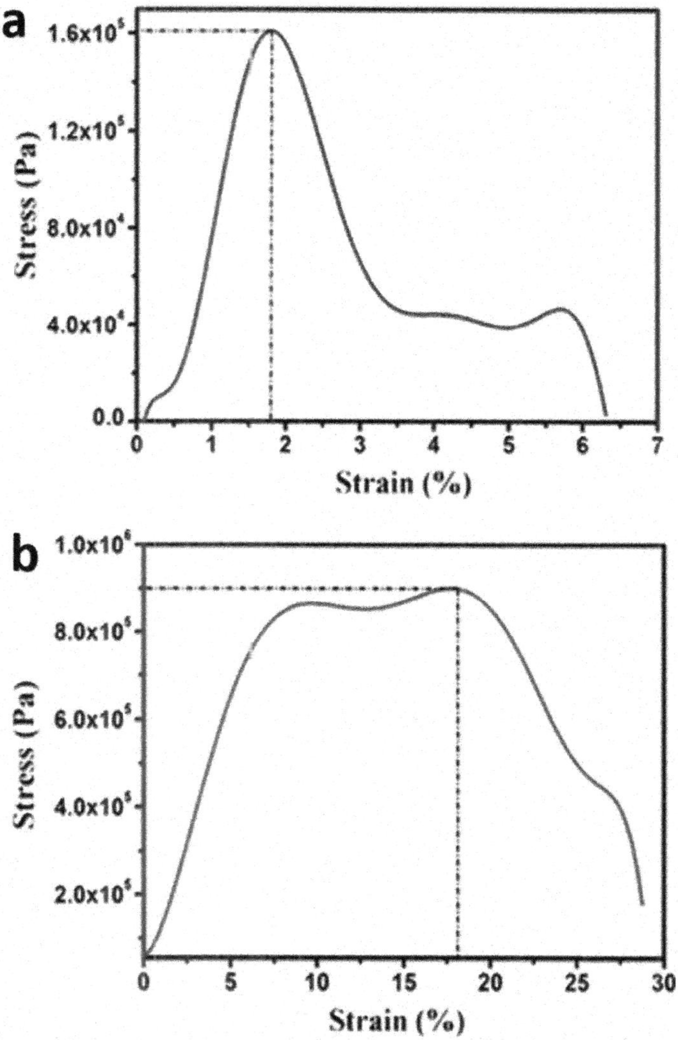

Figure 11.28 Stress–strain behavior of the membranes: (a) GF and (b) PCM. Reproduced with permission from Ref. [30]. Copyright 2019, American Chemical Society.

The preceding sections discuss the three areas that have been extensively studied. Among them, polymers and materials with different characteristics are used to prepare composite separators. This optimized method essentially improves the performance of separators and retains some beneficial characteristics from the original separator. The preparation of composite separator can improve the thermal stability and dimensional stability of separators at high temperatures and increase the separator's ability to absorb electrolyte. Indeed, these strategies can not only optimize the separator performance in SIBs, but also guide the development of separator in other battery systems and even solid polymer electrolytes.

11.5 Influence of Separators on Battery Performances

11.5.1 Electrochemical Performance

11.5.1.1 Influence on voltage

Separator thickness should be thin to ensure short ion-migration channels; otherwise, the electrochemical polarization in batteries will be serious, and the low-temperature voltage platform will be relatively low. An oversized pore structure will accelerate the self-discharge process, reducing the consistency of voltage before and after. In addition, the fibrous organic components in separators, due to their interaction with electrolytes, may have a unique effect on the electrolytes, thus changing the normal charge/discharge voltage of batteries. However, reliably quantifying the separator–electrolyte interaction with further experimental derivation or theoretical simulation remains extremely challenging.

11.5.1.2 Influence on internal resistance

The separator thickness is closely related to the internal resistance of batteries. The thinner the separator, the smaller the internal resistance of batteries, which is more conducive to the realization of high-power charging and discharging of batteries. The separator thickness should be as small as possible under certain mechanical strength, but the effective thickness and extrusion resistance should also be considered simultaneously for a separator. For commercial LIBs, 25 μm is generally adopted as the standard polyolefin separator thickness. With the increasing demand for portable products, the separators with the thickness of 16 μm or thinner have been widely used. For power batteries, the mechanical requirements and safety of the assembly process require thicker separators. In addition, the air permeability of separators is proportional to the internal resistance of batteries. The higher the air permeability of separators, the larger the internal resistance of batteries. At the same time, the separator materials have a certain micropore structure, and excessively small micropores can also increase the resistance of batteries. If the pore is too large, short circuits caused by accidental contact between cathode and anode or dendrite growth can easily occur. In general, a submicron aperture separator is sufficient to prevent the direct passage of electrode particles.

When assembling a battery, the electrolyte should also be fully infiltrated. Otherwise, poor electrolyte wettability to separators can also increase the internal resistance of the assembled battery.

11.5.2 Safety Performance

The safety of batteries has always been a concern for the public, especially in the recent years, after incidents of mobile phones exploding while charging, electric car overheating/burning, and other incidents. Battery safety has much to do with the design of battery packs and abuse conditions. For a single battery, in addition to the cathode and anode materials, the separator and electrolyte have a great relationship with battery safety.

11.5.2.1 Influence on heat resistance

When an exothermic reaction occurs in batteries, such as self-heating, overcharging, or external short circuit, a large amount of heat will be generated, which results in the closure of the separator micro-hole, and form a circuit break to block the continuous transmission of ions. The temperature at which the pores of the separator are closed is called the closed pore temperature. Since the internal temperature of batteries will rise during actual continuous use, the closed pore temperature of separators should not be too high.

Another safety index is the broken film temperature. When temperature rises above the closed pore temperature caused by the internal self-heating and external short circuits, the separator ruptures and an internal short circuit occurs in the battery. The temperature at the rupture of the separator is the broken film temperature. Generally, it is better to break a separator at a higher temperature.

In the process of thermal runaway, the battery will undergo successively capacity decay, solid-electrolyte interphase layer decomposition, anode and electrolyte reaction, separator melting, cathode and electrolyte decomposition reaction, electrolyte combustion, and other processes. In the process of temperature rise, the micropores on separator will close first to block the current, and the separator will shrink when the temperature continues to rise to reach the melting point of the separator. The temperature rise of the cell slows down as the separator melts, because the separator absorbs heat. Therefore, the closed pore temperature and the broken film temperature of separators are vital to ensure the safety of batteries. After repeated experiments, it has been widely accepted that the internal resistance of the battery first increases and then decreases during the separator melting process [1]. As shown in Fig. 11.29, the internal resistance increases significantly at 135°C, and the PE-based separator has contracted to form the obturator effect. In the process of heating up, the separator closes the pores, blocking the current to some extent to achieve the blocking effect [31].

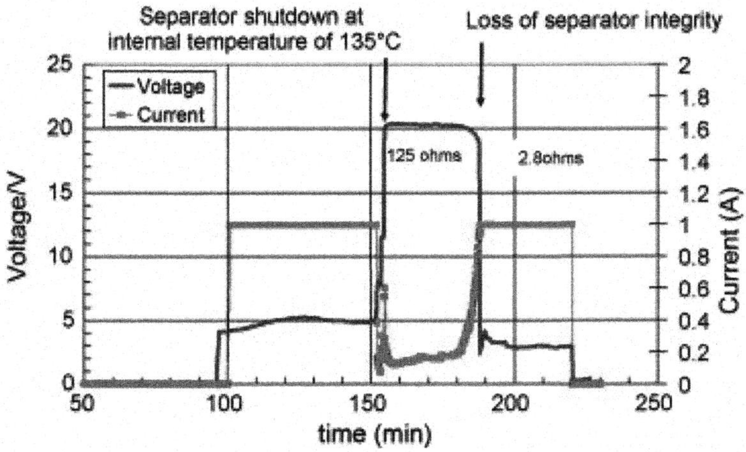

Figure 11.29 Characteristics of obturator foramen process of separator. Reproduced with permission from Ref. [31]. Copyright 2007, Elsevier.

11.5.2.2 Influence on mechanical safety performance

The mechanical properties of separators also affect the safety of batteries. If a separator ruptures in the assembling and working processes of batteries, internal short circuits may occur in batteries, so a certain mechanical strength is required for a separator. Separators with robust mechanical strength are better at separating the anode from the cathode when the battery is hit by an external object, so thicker separators also mean better safety. It is usually guaranteed that the battery pack will only puff up when it is hit by a foreign object, rather than explode directly. The mechanical strength of separators can be measured by puncture and tensile strength.

Puncture strength is an important physical index of separators, which is associated to thickness, tensile force, thermal bonding, and other factors. It reflects maximum resistance when the separator is pierced by a sharp object. Adequate puncture strength can prevent dendrite growth and piece burr from penetrating the separator and causing short circuits.

The tensile strength of separators is related to the preparation process of separators. With uniaxial stretching, the separator has different strengths in stretching direction and vertical direction. When biaxial stretching is used, the separator will be similar in both directions. This contraction will increase the probability of cathode and anode contact during the process of manufacturing the batteries.

11.5.2.3 Influence on dendritic inhibition

Metal dendrites form along the pores in separators, so reducing the separator pores can prevent their formation and growth. However, this strategy would greatly reduce ion permeability and affect the charge/discharge rate. Therefore, the separator design should pay attention to not only inhibiting the dendrite growth, but also maintaining the ionic conductivity. The dendrite growth can be inhibited by two strategies: First, the pore size distribution of separators can be controlled to ensure the uniform distribution of pores, and then the uniform deposition of metal can be obtained to avoid the dendrite growth. Second, "killing mental" reagent is added in the separator to consume the generated mental dendrite to avoid the separator being punctured.

11.5.2.4 Improved strategy

Adding special inorganic ceramic materials or coating layer on separators can provide the separator with characteristics that the original separator does not possess. Compared with the original separator, some composite separators with inorganic ceramic materials usually lead to the improvement of heat resistance, wettability to electrolytes, and ion transport performance. For example, Kim et al. used simple CVD-based nano-SiO_2 particles uniformly coated on the surface of polyethylene separator [24], resulting in a modified separator that showed better breakdown resistance and thermal insulation performance, as well as improved wetting characteristics and safety. Of course, this approach also has obvious disadvantages: (a) The separator thickness will gradually thicken with the increase in the types of materials. The use of a variety of materials will greatly increase the cost of separators, including the raw material cost, and the technical cost caused by the increase in operation difficulty. (b) The difference in surface properties between inorganic coating and organic separator will lead to poor compatibility. This approach is still difficult to apply widely in terms of cost and operational difficulties.

Developing voltage-sensitive separators is also effective in improving battery safety. For instance, Ai et al. proposed the preparation of a voltage-sensitive separator [32]. When the battery is overcharged, the separator is electropolymerized to become an electronic conductor by using the conductive ability of the polymer at a certain potential, which promotes the internal short circuits of batteries and prevents a further voltage increase, thus, improving battery safety. Under normal charge and discharge conditions, the composite separator has no negative effect on battery performance. However, under the condition of long-time overcharge, the voltage-sensitive separator can effectively clamp the charging voltage of batteries, providing reversible overcharge protection for the battery.

Starting from the inherent dendrite phenomenon of sodium anode, Liang et al. [33] designed a new tri-layer separator to effectively inhibit the growth of sodium dendrites, thus improving the internal safety of batteries. As shown in Fig. 11.30, the middle layer containing TiO_2 nanoparticles can react with sodium dendrites, avoiding the penetration of sodium dendrite to the separator.

Usually, the multilayer separator is used to extend the operating temperature range of separators, while the sandwich-type separator is more suitable for reducing the damage caused by dendrite.

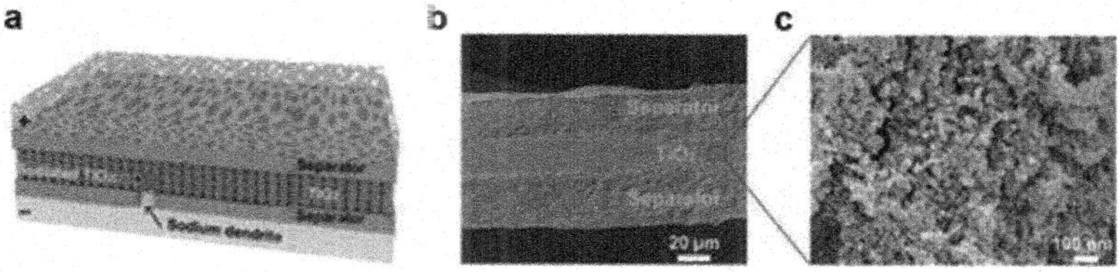

Figure 11.30 (a) Schematics of suppressed dendrites by the TiO_2 middle layer of sandwich-type separator. SEM images of (b) the sandwich-type separator and (c) the middle layer containing TiO_2 nanoparticles. Reproduced with permission from Ref. [33]. Copyright 2018 Royal Society of Chemistry.

11.6 Future Research Direction of Separators

Separators have been somewhat neglected for a long time, but in fact they play a crucial role in high-performance and high-safety sodium-ion battery systems. The high performance of separators is expected to establish a viable path to further promote the commercial application of SIBs. With the progress of manufacturing and microelectronic technology, the development of next-generation separator for SIBs has attracted widespread attention. Among them, typical strategies include electrode-separator integration production and the development of intelligent separators.

11.6.1 Electrode-Separator Integration Production

In the assembly process of batteries, the electrode material and separator are assembled in sequence, which is undoubtedly cumbersome and increases the labor and mechanical costs of the process [34]. The integration of electrode-separator materials refers to the preparation of separator's precursor slurry coated on the cathode, which simplifies the step process as well as helps to obtain tight contact between separators and electrodes. When it is applied to anode, some functional separator material can inhibit the dendrite growth. Whether from the perspective of laboratory or factory, this novel battery shows very important advanced significance.

In addition, some electrode materials have higher requirements for the assembly environment due to their high sensitiveness. Therefore, the production process of electrode-separator integration in the same environment first shortens the material transfer time and reduces the unpredictable changes. Second, the process of assembling battery will be significantly simplified. However, the real realization of electrode-separator integration production still needs continuous efforts, and the related instruments will undoubtedly pose higher operational difficulty.

11.6.2 Intelligent Response of Separators

Real-time monitoring of dynamic chemical and thermal state changes of batteries during operation is the key to ensuring their safety, especially for large-scale SIB energy storage system, which also highlights the significance of developing an intelligent separator. An effective noninvasive technique was used by Tarascon et al. to inject intelligent microdevices with sensing function into the separator and measure several important parameters of the battery with temporal and spatial resolution [35]. In addition, designing novel separator with self-response characteristic to various adverse environmental changes is also the key to the gradual realization of intelligent batteries in the future. The external intelligent detection components cannot make corresponding response strategies for internal changes in time, which to a large extent cannot guarantee the safety of the whole battery. In the case of defects caused by environmental changes in the separator, the intelligent response makes up for the defects.

In order to further improve the separator's contribution to the safety of the whole battery, the development of separators with repeated response mechanism will gain more and more attention in the future. When the battery works improperly, the separator with unique expansion and separation feature is expected to cut off the circuit (or separate the conductive material), ensuring the safety performance of the whole battery.

11.6.2.1 Voltage-response separator

A voltage-responsive separator can adjust voltage when the battery overcharges, protecting the battery from the danger of voltage runaway. During this process, the voltage-responsive polymer changes from an insulating phase to a conductive phase when the voltage overcharges and reduces the voltage by forming an internal micro-short circuit to prevent the battery voltage from running out of control. When the voltage drops to the normal voltage, the polymer restores its insulation. The battery can continue to be used and have strong reversibility. However, it is a problem that needs to be solved to achieve good polymer dispersion on the separator with redox activity.

11.6.2.2 Self-extinguishing separator

Preparing a self-extinguishing separator by embedding microspheres with flame-retardant materials or heat-sensitive materials in the separator may be a new attempt to modify the separator. It is well known that the liquid electrolytes used in SIBs generally contain organic solvents. If the battery is thermally runaway, it may ignite the electrolyte, causing possible fire and explosion [36]. Flame retardants (such as trimethyl phosphate (TMP), phosphazene, etc.) can be added to obtain the modified electrolyte with good non-combustibility. However, they are usually electrochemically decomposed on the carbon-based anode, which would decrease the battery interface compatibility. In self-extinguishing separators, the microspheres will gradually collapse and release the flame-retardant materials to extinguish the fire when the temperature rises due to the thermal runaway of the battery. This method may effectively overcome the negative effect of flame-retardant additives in electrolytes on battery performance.

References

1. Arora, P. and Zhang, Z. (2004). Battery separators, *Chem. Rev.*, **104**, 10, 4419–4462.
2. Yanilmaz, M., Lu, Y., Dirican, M., Fu, K., and Zhang, X. (2014). Nanoparticle-on-nanofiber hybrid membrane separators for lithium-ion batteries via combining electrospraying and electrospinning techniques, *J. Membr. Sci.*, **456**, 57–65.
3. Jeong, H., Kim, D., Jeong Y., and Lee, S. (2010). Effect of phase inversion on microporous structure development of Al_2O_3/poly(vinylidene

fluoride-hexafluoropropylene)-based ceramic composite separators for lithium-ion batteries, *J. Power Sources*, **195**, 18, 6116–6121.

4. Park, M. and Kim, C. (2014). Fabrication of polyethylene microporous membranes using triethylolpropane tris(2-ethylhexanoate) as a novel diluent by a thermally induced phase separation process, *J. Membr. Sci.*, **449**, 127–135.
5. Xiang, H., Chen, J., Li, Z., and Wang, H. (2011). An inorganic membrane as a separator for lithium-ion battery, *J. Power Sources*, **196**, 20, 8651–8655.
6. Jung, Y., Cavanagh, A., Gedvilas, L., Widjonarko, N., Scott, I., Lee, S., Kim, G., George, S., and Dillon, A. (2012). Improved functionality of lithium-ion batteries enabled by atomic layer deposition on the porous microstructure of polymer separators and coating electrodes, *Adv. Energy Mater.*, **2**, 8, 1022–1027.
7. Cho, T., Tanaka, M., Ohnishi, H., Kondo, Y., Yoshikazu, M., Nakamura, T., and Sakai, T. (2010). Composite nonwoven separator for lithium-ion battery: Development and characterization, *J. Power Sources*, **195**, 13, 4272–4277.
8. Kim, J., Lee, Y., and Lim, D. (2009). Plasma-modified polyethylene membrane as a separator for lithium-ion polymer battery, *Electrochim. Acta.*, **54**, 14, 3714–3719.
9. Wang, W., Zhang, J., Yin, Y., Duan, H., Chou, J., Li, S., Yan, M., Xin, S., and Guo, Y. (2020). A rational reconfiguration of electrolyte for high-energy and long-life lithium-chalcogen batteries, *Adv. Mater.*, **32**, 23, 2000302.
10. Zhang, J., Yue, L., Kong, Q., Liu, Z., Zhou, X., Zhang, C., Xu, Q., Zhang, B., Ding, G., Qin, B., Duan, Y., Wang, Q., Yao, J., Cui, G., and Chen, L. (2014). Sustainable, heat-resistant and flame-retardant cellulose-based composite separator for high-performance lithium ion battery, *Sci. Rep.*, **4**, 3935.
11. Liao, C., Wang, W., Wang, J., Han, L., Qiu, S., Song, L., Gui, Z., Kan, Y., and Hu, Y. (2021). Magnetron sputtering deposition of silicon nitride on polyimide separator for high-temperature lithium-ion batteries, *J. Energy Chem.*, **56**, 1–10.
12. Zhang, L., Feng, G., Li, X., Cui, S., Ying, S., Feng, X., Mi, L., and Chen, W. (2019). Synergism of surface group transfer and in-situ growth of silica-aerogel induced high-performance modified polyacrylonitrile separator for lithium/sodium-ion batteries, *J. Membr. Sci.*, **577**, 137–144.

13. Binnig, G., Quate, C., and Gerber, C. (1986). Atomic force microscope, *Phys. Rev. Lett.,* **56**, 9, 930–933.

14. Jia, Z., Dai, C., and Chen L. (2006) *Electrochemical Measurement Method,* 1st Ed. (Chemical Industry Press, China).

15. Lee, H., Yanilmaz, M., Toprakci, O., Fu, K., and Zhang, X. (2014). A review of recent developments in membrane separators for rechargeable lithium-ion batteries, *Energy Environ. Sci.,* **7**, 12, 3857–3886.

16. Zhang, L. (2019). Preparation and performance research of new type sodium ion battery separator. *Master's Thesis,* China.

17. Wang, P. and Xu, Q. (2005) *Material Research Methods,* 1st Ed. (Science Press, China).

18. Wang, M., Chen, X., Wang, H., Wu, H., Jin, X., and Huang, C. (2017). Improved performances of lithium-ion batteries with a separator based on inorganic fibers, *J. Mater. Chem. A,* **5**, 1, 311–318.

19. Liang, J., Zeng, X., Zhang, X., Wang, P., Ma, J., Yin, Y., Wu, X., Guo, Y., and Wan, L. (2018). Mitigating interfacial potential drop of cathode-solid electrolyte via ionic conductor layer to enhance interface dynamics for solid batteries, *J. Am. Chem. Soc.,* **140**, 22, 6767–6770.

20. Shen, Z., Cao, L., Rahn, C., and Wang, C. (2013). Least squares galvanostatic intermittent titration technique (LS-GITT) for accurate solid phase diffusivity measurement, *J. Electrochem. Soc.,* **160**, 10, A1842–A1846.

21. Perera, K. and Vidanapathirana, K. (2017). Impedance spectroscopy, DC polarization, XRD and SEM studies on an ionic liquid based gel polymer electrolyte to be used for dye sensitized solar cells, *Materials Discovery,* **7**, 30–33.

22. Zhang, J., Wen, H., Yue, L., Chai, J., Ma, J., Hu, P., Ding, G., Wang, Q., Liu, Z., Cui, G., and Chen, L. (2017). In situ formation of polysulfonamide supported poly(ethylene glycol) divinyl ether based polymer electrolyte toward monolithic sodium ion batteries. *Small,* **13**, 2, 1601530.

23. Janakiraman, S., Khalifa, M., Biswal, R., Ghosh, S., Anandhan, S., and Venimadhav, A. (2020). High performance electrospun nanofiber coated polypropylene membrane as a separator for sodium ion batteries. *J. Power Sources,* **460**, 228060.

24. Kim, J., Heo, J., and Park, J. (2017). Tailored metal oxide thin film on polyethylene separators for sodium-ion batteries, *J. Electrochem. Soc.,* **164**, 9, A1965–A1969.

25. Zhou, D., Tang, X., Guo, X., Li, P., Shanmukaraj, D., Liu, H., Gao, X., Wang, Y., Rojo, T., Armand, M., and Wang, G. (2020). Polyolefin-based janus separator for rechargeable sodium batteries. *Angew. Chem. Int. Ed. Engl.*, **59**, 38, 16725–16734.

26. Casas, X., Niederberger, M., and Lizundia, E. (2020). A sodium-ion battery separator with reversible voltage response based on water-soluble cellulose derivatives. *ACS Appl. Mater. Interfaces,* **12**, 26, 29264–29274.

27. Chen, W., Zhang, L., Liu, C., Feng, X., Zhang, J., Guan, L., Mi, L., and Cui, S. (2018). Electrospun flexible cellulose acetate-based separators for sodium-ion batteries with ultralong cycle stability and excellent wettability: the role of interface chemical groups. *ACS Appl. Mater. Interfaces,* **10**, 28, 23883–23890.

28. Zhang, T., Shen, B., Yao, H., Ma, T., Lu, L., Zhou, F., and Yu, S. (2017). Prawn shell derived chitin nanofiber membranes as advanced sustainable separators for Li/Na-ion batteries. *Nano Lett.*, **17**, 8, 4894–4901.

29. Jo, J., Jo, C., Qiu, Z., Yashiro, H., Shi, L., Wang, Z., Yuan, S., and Myung, S. (2020). Nature-derived cellulose-based composite separator for sodium-ion batteries. *Front. Chem.* **8**, 153.

30. R, A., Vijaya Kumar Saroja, A. P., and Sundara, R. (2019). Barium titanate-based porous ceramic flexible membrane as a separator for room-temperature sodium-ion battery. *ACS Appl. Mater. Interfaces,* **11**, 4, 3889–3896.

31. Roth, E., Doughty, D., and Pile, D. (2007). Effects of separator breakdown on abuse response of 18650 Li-ion cells, *J. Power Sources,* **174**, 2, 579–583.

32. Ai, X., Lin, C., Liao, Q., Liu, B., and Yang, H. (2006). A potential-sensitive separator capable of providing a reversible overcharge protection for lithium ion batteries. *J. Wuhan University*, **52**, 2, 149–153.

33. Yang, H., Sun, J., Wang, H., Liang, J., and Li, H. (2018). A titanium dioxide nanoparticle sandwiched separator for Na-O_2 batteries with suppressed dendrites and extended cycle life, *Chem. Commun.*, **54**, 32, 4057–4060.

34. Song, K., Liu, J., Dai H., Zhao, Y., Sun, S., Zhang, J., Qin, C., Yan, P., Guo, F., Wang, C., Cao, Y., Li, S. and Chen, W. (2021). Atomically dispersed Ni induced by ultrahigh N-doped carbon enables stable sodium storage, *Chem*, **7**, 1–11

35. Huang, J., Albero Blanquer, L., Bonefacino, J., Logan, E. R., Alves Dalla Corte, D., Delacourt, C., Gallant, B. M., Boles, S. T., Dahn, J. R., Tam, H. Y., and Tarascon, J. M. (2020). Operando decoding of chemical and thermal events in commercial Na(Li)-ion cells via optical sensors, *Nat. Energy*, **5**, 9, 674–683.

36. Zhang, L., Li, X., Yang, M. and Chen, W. (2021). High-safety separators for lithium-ion batteries and sodium-ion batteries: advances and perspective, *Energy Storage Mater.*, **41**, 522–545.

Chapter 12

Advanced Electron Microscopy Characterization of Sodium-Ion Battery Materials

Ruth Knibbe, Ming Li, and Lingbing Ran
School of Mechanical and Mining Engineering, University of Queensland, Australia
ruth.knibbe@uq.edu.au

12.1 Introduction

Electron microscopy is a key characterization tool in the development of battery materials. Standard scanning electron microscopy (SEM) can be used to reveal complex microstructures, and transmission electron microscopy (TEM) can provide a higher-resolution insight into the microstructures, coupled with information on crystal structure, disorder, and valence states. More recently, the development of *in operando* TEM characterization has opened the possibility to observe battery materials under dynamic conditions.

Handbook of Sodium-Ion Batteries: Materials and Characterization
Edited by Rohit R. Gaddam and X. S. (George) Zhao

ISBN 978-981-4968-15-7 (Hardcover), 978-1-003-30874-4 (eBook)
www.jennystanford.com

This chapter will focus on advanced TEM characterization of sodium-ion battery (NIB) materials, including high-resolution scanning TEM, electron energy loss spectroscopy, and *in operando* TEM. One section will focus on developing techniques that have not yet been used in NIBs; this includes cryo-TEM, 4D-scanning TEM (4D-STEM), electron holography, and electron tomography.

One major benefit of TEM is the relatively high spatial resolution that can be obtained. This allows the research to focus on nanoscale features that cannot be collected through other techniques. For example, the observation of surface defects will have a direct impact on the electrokinetics of the NIB material. However, the high resolution also carries a limitation as the TEM operator needs to be judicious in selecting images that are representative of the entire sample. Using TEM in conjunction with techniques such as X-ray photoelectron spectroscopy (XPS) and X-ray diffraction (XRD) is key in providing a detailed insight into both bulk crystal structure and crystal defects.

Despite the similarities between Na^+ and Li^+ battery systems, there are also notable differences, namely, the ionic radius, electronegativity, and preferred coordination [1]. As such, the same material will behave distinctly different if used for a lithium-ion battery (LIB) in comparison with NIBs. For example, sodiation of Se is a multistep alloying mechanism, whereas the lithiation mechanism is a single-step alloying process [2].

The cycling stability of sodiated materials is also problematic. The large ionic radius encourages large stresses during sodiation/desodiation processes; this in turn can introduce undesirable fractures at the interface between sodiated/desodiated regions. For example, 2D MoS_2 offers an intercalation storage mechanism for both Na^+ and Li^+. However, the large volume expansion during MoS_2 sodiation encourages fracturing of the protective solid electrolyte interphase (SEI). This forces the SEI to renew, which reduces the coulombic efficiency [3, 4].

Furthermore, there are still some anomalies in regard to kinetics. Typically, lithiation processes proceed faster than sodiation. However, this is not the case for Co_9S_8, where it has been found that the sodiation process is faster than the lithiation process. This has been ascribed to the improved conductivity of the new phases that form [5]. This hypothesis is important in materials design. By controlling the conductivity of intermediate phase, kinetics can be improved.

12.2 High-Resolution Scanning TEM and Electron Energy Loss Spectroscopy

New electron guns and detectors in TEMs are allowing researchers to observe materials at a higher resolution with a lower electron dose. Aberration-corrected TEMs are becoming more commonplace, which further pushes the imaging resolution through correcting for spherical and/or chromatic aberrations in the electron beam. Such powerful tools can be used with standard TEM techniques such as electron diffraction, conventional imaging, energy dispersive X-ray spectroscopy (EDS), and electron energy loss spectroscopy (EELS).

With aberration-corrected TEMs, a larger emphasis has been placed on the improved quality and signal when operating in scanning TEM (STEM) mode. In high-resolution STEM (HRSTEM), researchers can gain an atomic-level insight into newly designed materials. In comparison to conventional high-resolution TEM images, the interpretation of HRSTEM images can be relatively easy. For high-angle annular dark-field (HAADF) HRSTEM, the intensity of atom columns is reflective of the average atomic number; the brighter columns are associated with heavier atoms. This still makes imaging of low atomic number atom columns, such as O, Li, and Na, difficult or impossible.

Another powerful tool in the TEM for characterization of battery systems is EELS. In the simplest format, sample composition can be obtained using EELS. However, EELS can also be used to assess valence states and coordination of various atoms/atom columns. In battery materials, this is used to look at transition metals, which may exist in a combination of valence states dependent on the state of charge.

These techniques have been used for *ex situ* characterization to understand the structure of the pristine material and also the changes occurring in the material during sodiation and desodiation. HRTEM combined with EELS can elucidate sodiation staging processes, charge ordering, ion redistribution, valence and coordination changes, and phase changes.

12.2.1 HRSTEM and EELS for NIB Characterization

Aberration-corrected TEM was used to observe the structure of Mn- and Fe-doped NAISCON-type $Na_3V_2(PO_4)_3$—an NIB cathode material. Mn doping and Fe doping are used to replace the expensive V and also to increase the voltage stability. However, due to the Jahn–Teller distortion associated with Mn, it is not desirable to completely replace the V with Mn [6].

In the Mn-doped structure, the Na ions are identified in two locations—with a six-fold and a 10-fold coordination. As the 10-fold coordinated Na has weaker bonding, it is easier to sodiate/desodiate this site in comparison to the six-fold coordinated Na. This improves the Na diffusion kinetics. The Fe-doped structure is a little different. The FeO_6 octahedron distorts the crystal structure to monoclinic. It is interesting that despite the similar atomic radius between Fe^{3+} and V^{3+}, the Fe-doped structure adopts a monoclinic structure, similar to the $Na_3Fe_2(PO_4)_3$, instead of $Na_3V_2(PO_4)_3$.

These differences are also highlighted through HRSTEM work, which highlights the location of the Na, P, and Mn/V in the $Na_3V_2(PO_4)_3$ structure. In the HRSTEM images, the atom column intensity makes it possible to see the Mn/V atom columns. However, due to the low atomic number, it is difficult to see the Na and O columns in most projections. Despite this, the space available for Na^+ transportation can be easily observed.

Huang et al [8] used aberration-corrected STEM to understand the phase transitions in layered transition-metal (TM) oxides (Na_xTMO_2 (TM = Fe, Mn, Co, Ni, Cu, Ti, etc.))—an NIB cathode material. These structures have a 2D Na^+ pathway. In the study, a stoichiometry of $Na_{0.62}Li_xMn_{0.66}Ni_{0.17}Co_{0.17}O_2$ (x = 0 and 0.18) was used. The Li^+ incorporated into the Na^+ site is not redox active but is added to improve phase stability.

Na_xTMO_2 has three possible phases: P3, P2, and O'3 [9]. The O and the P refer to octahedral and prismatic cation coordination, respectively. The 3 and 2 refer to the number of TM layers per unit cell. In the P3 and P2 phases, the TMO_2 and the NaO_2 layers stack in a sandwich structure. The P2 structure has two sodium occupancy sites, whereas the P3 structure has only one. In the P2 and P3 structures, the sodium ions can diffuse along the face-shared, neighboring triangular prism sites. The O′3 phase consists of an alternate-mixed Mn^{4+}/Li^+ and Li^+ layer where all the metal ions have an octahedral coordination.

By imaging down the [110] and [310] zone axis, it is possible to identify the P3, P2, and O′3 phases. The P3/P2 transition can be observed by imaging down the [110] zone axis and monitoring a change in the rectangular configuration. The intergrowth orientation between the P3/P2/O′3 tri-phase was identified as <110>P3//<110>P2//<310>O′3. The tri-phase cathode material (P3/P2/O′3) has a higher high first cycle and charge capacity than the dual-phase P3/P2 cathode material. This is attributed to the ability to activate more sodium sites and for the partial Li^+ extraction from the O′3 phase at high voltage (Fig. 12.1).

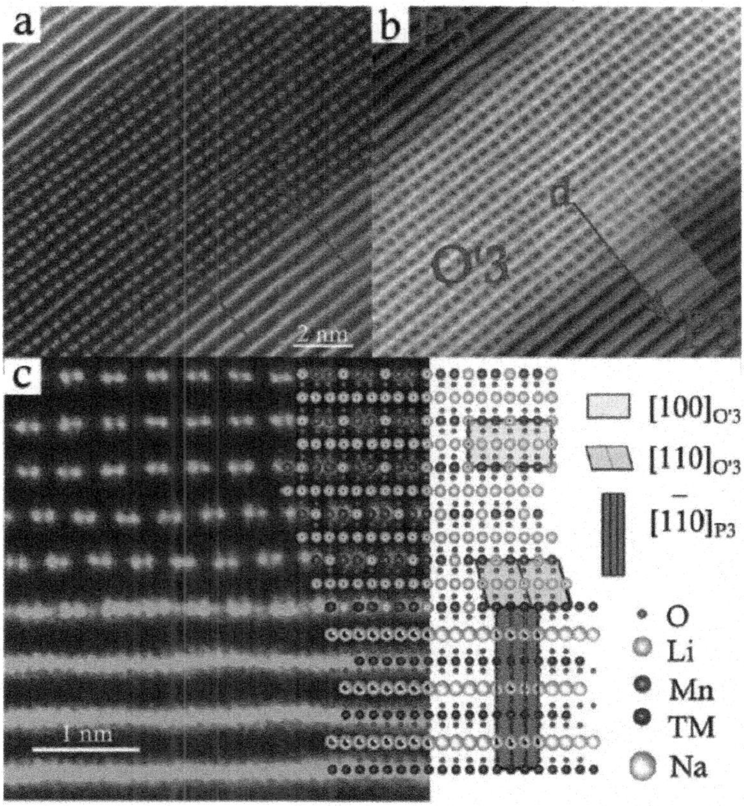

Figure 12.1 Atomic-resolution images of the as-prepared Na-P3/Li-O′3 composite. (a) HAADF-STEM and (b) ABF-STEM images. (c) Enlarged HAADF-STEM image from the rectangle in (a) and crystal structure model showing that the Na-P3 phase along the [1 1 0] zone axis is coherent with the Li-O′3 phase along different zone axes [7].

This work was extended by the group with nano beam diffraction (NBD), which identified that the P3 and the O′3 phases are due to the oxygen sublattice. Through HRTEM, the O′3 was found to play a key role in suppressing the deleterious P3 to P3″ phase transition. The P3" phase has the same space group as the P3 phase, but a larger interplanar spacing of 7.1 versus 5.6 Å. The larger spacing of the P3″ can accommodate the electrolyte solvent molecule, which leads to irreversibility. However, if the O′3 phase is present, the lattice strain can suppress the P3–P3″ phase transition.

Complementary work was done by Jung et al. [10]. The group produced a nanostructure P2-$Na_{2/3}MnO_2$ (NMO) cathode material with good capacity and rate capability. In this work, EELS was used to calculate the Mn L_3/L_2 white line ratios, which provides information on the Mn oxidation state. The average Mn oxidation state changed from 3.74 to 3.29 across the grain boundary, suggesting that two different NMO polymorphs exist: an α and a β NMO phase. They argue that the grain boundaries between the nanodomains facilitate the Na^+ diffusion.

Li et al. used HRSTEM to present evidence of Jahn–Teller distortions in the NIB cathode material, $Na_{5/8}MnO_2$ [11] (Fig. 12.2). The Jahn–Teller distortions induce long-range ordering between the occupied and unoccupied sodium sites. In the HAADF-STEM images taken, the bright intensity reflects the Mn atom columns, and the weak intensity is from the Na atom columns as observed in the [010] direction. For the Na atom columns, the authors note that the intensity shows one slightly brighter intensity, followed by three reduced intensity atom columns. This reflects the Na concentration in these atom columns.

The Mn valence state was determined using L_3/L_2 white line ratios from STEM-EELS measurements. A high ratio indicates a valence state close to Mn^{3+}, whereas a low intensity suggests an Mn^{4+} charge. From these ratios, three different columns are observed: high intensity (reflecting Mn^{3+}), low intensity (reflecting Mn^{4+}), and intermediate intensity (reflecting a mixed $Mn^{3+/4+}$ valence state). The $Mn^{3+}O_6$ octahedron can distort due to the Jahn–Teller effect, whereas the $Mn4^{r+}O_6$ octahedron does not. The periodic valence state change of the Mn leads to a periodic distortion due to the Jahn–Teller effect, which can be seen by the slight undulation in the Mn layers. This, in turn, impacts the Na and Na vacancy ordering. These observations are distinctly different to other alkali-ion systems.

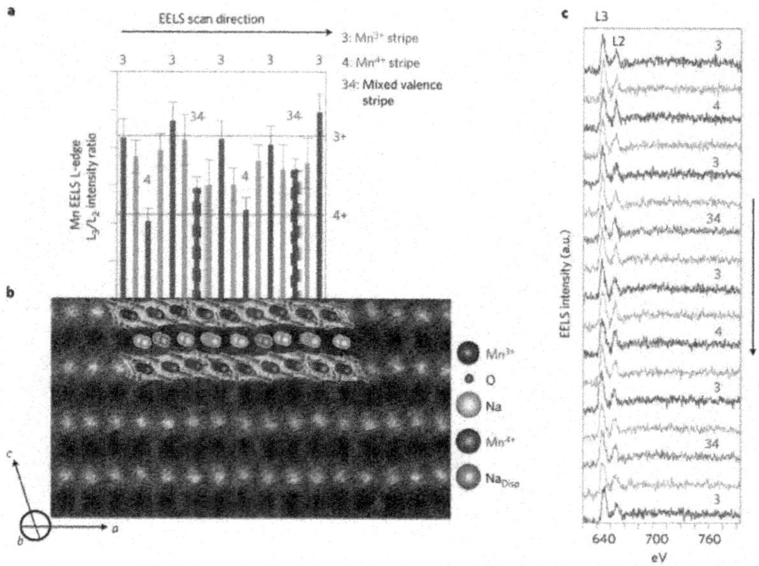

Figure 12.2 STEM image of cooperative Jahn–Teller effect. (a) Mn EELS L3/L2 peak ratio at each Mn [010] atomic column site along the STEM-EELS scanning direction of [100]; the dashed lines show the ratios that correspond to the standard MnO_2 and Mn_2O_3 samples. (b) STEM image along the [010] zone axis shows the periodic distortion of the Mn ab plane and the intensity modulation of the Na plane, a result quantitatively consistent with the superstructure model in the inset. (c) Mn L2,3-edge EEL spectra after background subtraction that provide the L3/L2 ratios shown in (b) [11].

Two-dimensional (2D) metal dichalcogenides, MX_2, compounds (M is a transition metal and X is the chalcogen) can exist in three structural polytypes 1T, 2H, and 3R, where 1, 2, and 3 refer to the number of layers and T, H, and R refer to the trigonal, hexagonal, and rhombohedral unit cells, respectively [12, 13]. Many metal chalcogenides have a layered crystal structure with adjacent layers bonded by weak van der Waals force. It is this layered structure that provides highways for the alkali ions.

Panda et al. [14] presented a detailed aberration-corrected STEM investigation of the NIB anode material $MoTe_2$ (Fig. 12.3). The interlayer spacing of $MoTe_2$ is around 0.7 nm, which is considerably larger than graphite. In this work, position-averaged convergent beam electron diffraction (PACBED) was used to determine the sample thickness. This information was, in turn, used to simulate the HAADF and ABF images using a multislice method. For imaging, a combination of the HAADF, ABF, and BF detectors was used.

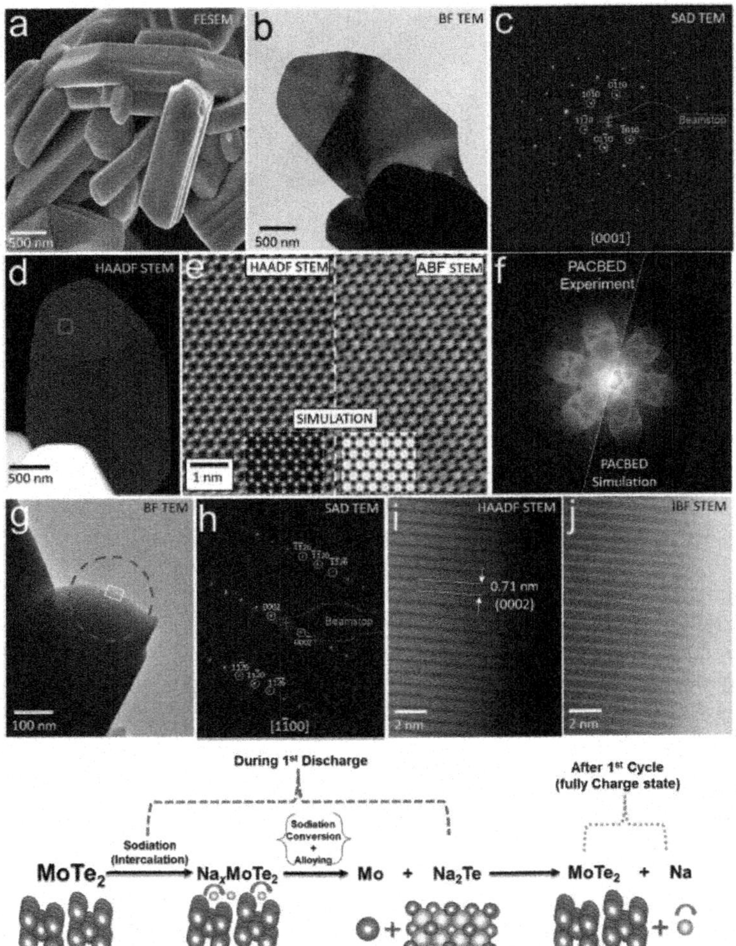

Figure 12.3 Electron microscopy studies of the as-prepared $MoTe_2$ sample. (a) FE-SEM image of the as-prepared $MoTe_2$ sample. (b) Bright-field TEM image of a thin area of a flat crystal. (c) SAD pattern from the indicated area in (b). (d–e) Low magnification HAADF-STEM, atomic-resolution HAADF, and ABF-STEM images at the [0001] zone axis; the inset shows the simulation images for the same thickness and orientation. (f) PACBED experimental patterns for the area in (e) and the simulated pattern for 24 nm. (g) Bright-field TEM image of the end of an edge of a crystal. (h) SAD pattern from the area that is indicated in (g). (i–j) HAADF and IBF STEM images, taken from a marked box in image (g), along the [0002] direction [14].

These methods were used to identify that $MoTe_2$ forms a 2H structure. The structure was imaged down the [0001] and $[1\bar{1}00]$ directions. The lack of additional reflections in the selected area diffraction (SAD) in combination with the HRTEM images confirmed the 2H structure.

The group also used HRTEM to image the $MoTe_2$ after multiple charge/discharge cycles. The SAD patterns show an increase in the amorphous nature, which the group attributed to the formation of $MoTe_2$ nanocrystals after repeated sodiation/desodiation. This HRTEM work was coupled with S-XRD, XPS, and X-ray absorption near edge structure (XANES) work to propose the charge/discharge mechanism.

Wang et al. [4] used *in operando* and *ex situ* methods to map the transition of semiconducting 2H-MoS_2 to 2H-$Na_{0.5}MoS_2$ and subsequent transition to the electronically conducting 1T-$NaMoS_2$. In addition, the Na^+ occupancy was also clarified at the atomic scale. The group used a combination of HRTEM with SAD to characterize the transition of the 2H to the 1T phase. They proposed that the Na intercalation introduces in-plane strain. This strain decreases the Mo–Mo distance and encourages the 1T-MoS_2 nucleation at the boundary. They also observed that the 1T-MoS_2 is more likely to grow along specific lattice planes, such as the (100).

The Na occupancy was also observed using HRTEM. Although the authors proposed a clean hypothesis in regard to the Na intercalation mechanism, due to the disorder in the sodiated samples, it is difficult to conclude the precise mechanism from the HRTEM images. The authors suggested that sodiation starts in every other MoS_2 interlayer. Once these layers are fully sodiated, the empty layers are sodiated. This is similar to the staging feature observed for graphite intercalation. In conjunction with this, the sulfur planes glide along an intralayer atomic plane to form 1T-MoS_2, resulting in a straight chain of S-Mo-S. The intercalation is reversible up to $Na_{1.5}MoS_2$. After this point, the structure decomposes to Na_xS and metallic Mo.

Chen et al. [15] also used a combination of *in operando* TEM and aberration-corrected STEM to further clarify the MoS_2 transition. They observed that during sodiation, the 2H MoS_2 shows a clear phase transition to 1T Na_xMoS_2. However, during desodiation, the system shows a solid–solution transition. It is this irreversibility that is considered central to the low energy efficiency of 2D metal dichalcogenide materials in NIBs. This sodiation process is distinctly different to what occurs during MoS_2 lithiation. During lithiation, MoS_2 transitions from the pristine 2H to a 1T phase and finally to an expanded 1T' phase.

A three-phase separation mechanism has been proposed by Sun et al. [16, 17] during the sodiation of a spinel $Li_4Ti_5O_{12}$ NIB anode material. The group used HRSTEM annular bright-field (ABF) images to identify either the $Li_4Ti_5O_{12}$ phase or the $Li_7Ti_5O_{12}$ phase. The ABF imaging allows for the low-contrast Li columns to be imaged, which is not possible using HRSTEM-HAADF imaging. More specifically, the 8a Li sites and the 16c Li sites can be observed in $Li_4Ti_5O_{12}$ and $Li_7Ti_5O_{12}$, respectively, which identifies the different phases.

The sodiated structure is also imaged using HRSTEM-HAADF and ABF images. The increased intensity of the 16c site in the ABF HRSTEM image indicates that the site is now sodiated (Na_6Li). As such, when looking at a larger area of the sample, the different regions can be identified—the sodiated phase and the parent $Li_4Ti_5O_{12}$ and $Li_7Ti_5O_{12}$ phases. The identification of the sodiated/unsodiated interfacial region through HRSTEM allows battery developers to see these clean and defect-free regions, which can easily accommodate repeated cycling.

Although TEM is a routine tool for many NIB research groups, advanced HRSTEM and EELS are only used by a few groups. To obtain these high-end results, advanced instrumentation coupled with a skilled operator and excellent TEM sample preparation is of paramount importance. This detailed characterization will continue to be important in characterizing the defect chemistry in pristine, sodiated, and cycled NIB materials. As sodiated and cycled samples have a large amount of disorder, this presents challenges for HRSTEM and EELS characterization. However, extending HRSTEM and EELS characterization with diffraction techniques (including 4D-STEM techniques) greatly assists in tracking these complex changes. Finally, the combination of more stable electron guns and more sensitive electron detectors allows for samples to be imaged under low electron doses, which can avoid artefacts forming through beam-induced damage.

12.2.2 *In operando* TEM

In traditional TEM, the sample is in a vacuum. With the *in operando* TEM, two methods have been developed for battery work: an open-cell and a closed-cell configuration (Fig. 12.4). The open cell uses a solid electrolyte or a low-vapor-pressure ionic liquid. In the open-cell configuration, the material is still subjected to the TEM vacuum. The closed-cell configuration uses a liquid electrolyte. To allow for the liquid electrolyte, the material is sandwiched between two electron transparent layers—most commonly silicon nitride but in some cases graphene or silicon carbide layers have been used. In both configurations, a potentiostat can be connected to the material, which can be used to expose the material to potentials or currents typically encountered in a battery.

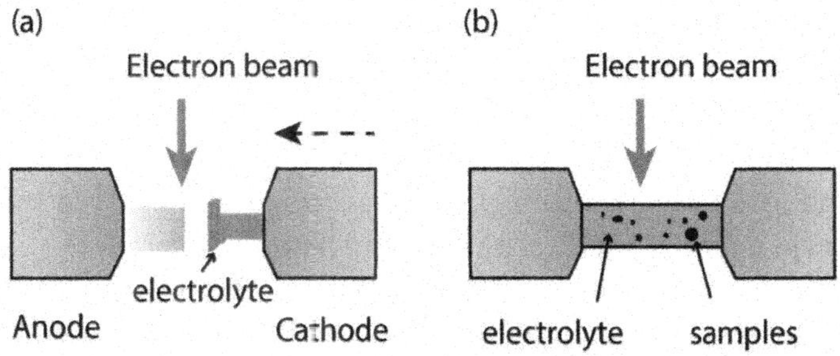

Figure 12.4 Schematic of (a) open-cell and (b) closed-cell configuration.

Authors working in the *in situ* and *in operando* TEM use slightly different terminology to distinguish their work. *In situ*, often, will refer to materials that are just in a native environment; for battery materials, this would be looking at the material in the electrolyte. However, the term *in operando* is used to distinguish *in situ* methods, in that the material has both the electrolyte environment and an applied bias/current. To maintain consistency, we will refer to work where both a bias and electrolyte is used as *in operando* TEM.

With the open-cell configuration, high-resolution images can be obtained. However, the quality of these images will be impacted by the limited ability to tilt down the zone axis (if necessary) and the sample drift (which reduces imaging resolution). For the closed-cell configuration, the image resolution is dramatically impacted by the liquid layer and the electron transparent windows [18]. Work has been done to improve this imaging resolution by limited liquid layer thickness, but this introduces other limitations such as electrolyte supply.

It is also important to note that the open-cell configuration is simpler than what the material experiences during battery testing. During typical battery testing, the material surface is saturated with a liquid electrolyte with an associated secondary electrolyte interphase (SEI). In comparison, in the open-cell configuration, the sample is exposed to the TEM vacuum. By changing from a full surface coverage to a single point of contact in the open-cell configuration, it is possible that the reaction kinetics are impacted. This, in turn, may alter the observed lithiation/sodiation behavior. Researchers have investigated this issue and recommended the use of open-cell and closed-cell configurations in parallel [19]. However, due to the higher resolution and comparative ease of use, all *in operando* NIB studies to date have used an open-cell configuration.

These *in operando* techniques are designed to collect dynamic processes, which would be missed using traditional *ex situ* methods. Phase transitions and separation, volume change, and defect formation are among several processes that may be observed during the *in operando* TEM characterization of battery materials.

For NIBs, the *in operando* TEM work has only been used for the characterization of anode materials in an open-cell configuration. For the Na-O_2 system, a closed-cell *in operando* system has been used [20]. This work is coupled with HRSTEM. This allows for both kinetic and complex defect structure development to be explored. The combination of *in operando* and *ex situ* methods, at present, is a judicious approach to getting the best understanding of the underlying mechanisms.

12.2.3 *In operando* TEM for NIB Characterization

12.2.3.1 Carbonaceous materials

Due to the limited spacing and instability of intercalation compounds, graphite cannot be reversibly sodiated [1]. As such, disordered or expanded carbonaceous materials are of greater interest for NIBs [21]. Through *in operando* TEM, the sodiation process of various carbonaceous materials, including CNF [22, 23], r-rGO [24], doped graphene [25, 26], and EG [27], has been investigated. These studies have been observed to reveal the rate-limiting pathways, expansion, and potential sodium plating processes. This, in turn, can be used to design new and improved materials that accentuate the beneficial defects in carbon while ameliorating the negative properties.

Liu et al. [22] used *in operando* TEM to investigate the sodiation of carbon nanofibers (CNFs) with both crystalline (c-CNF) and disordered (d-CNF) regions. During sodiation, the c-CNF and d-CNF layers expand 13.7% and 18.7%, respectively, indicating an increased sodium uptake in the disordered region. It is argued that the uptake in the d-CNF region is three times that of the c-CNF region due to the large number of defects, which can accommodate the sodium ion. Although degradation was not directly observed in the d-CNF region of the CNF, long-term capacity fading was ascribed to the formation of cracks at the c-CNF/d-CNF interface. Li et al. [23] expanded on the CNF work to reveal the dynamic nature of the sodiation/desodiation of individual CNF structures. It was revealed that the mesopores in the CNF present preferential sites for metallic Na deposition and growth. Unfortunately, this Na plating process is faster than the relatively slow CNF sodiation process.

Na metal clusters have also been observed on rapidly reduced graphene oxide (r-rGO) [24]. However, these Na metal clusters were relatively small (ca. 10) in comparison to the large 500 nm deposits observed by Li et al. [23]. For r-rGO, three main steps were observed during sodiation. First, Na ions were stored on defective sites from 2.0 to 0.3 V. Subsequently, the r-rGO structure was intercalated below 0.3 V and Na metal clusters grew at very low voltages. In addition to these reversible processes, an irreversible, thin, Na_2O layer formed on the r-rGO surface. This oxide layer is undesirable as it not only consumes Na, but also encourages the r-rGO nanosheets to collapse.

Although defects are beneficial in increasing the Na^+ storage capacity in carbonaceous materials [22, 27], highly disordered defects can provide an excessively strong Na^+ binding energy. This traps the Na^+, causing irreversibility [28]. The Na^+ storage capacity can also be altered by introducing dopants. Yang et al. [25] used *in operando* TEM and density functional theory (DFT) to demonstrate that the phosphorous doping of graphene increases the interplanar spacing, which allows for reversible sodiation. The P-doped graphene also facilitates the formation of a thin and stable SEI, which mitigates ongoing side reactions between the graphene and the electrolyte.

Nitrogen doping also benefits graphene for NIB applications through improving the Na^+ binding energy, introducing metallic conduction and facilitating Na^+ diffusion [26]. Zhang et al. [26] observed a reversible expansion and contraction of the N-doped graphene during cycling using *in operando* TEM. The ability of the structure to reduce stresses during sodiation makes N-doped graphene a suitable NIB anode candidate.

12.2.3.2 Phosphorus and Phosphorene

Phosphorus is an attractive anode material for NIBs with a high theoretical specific capacity of 2596 mAh/g. Phosphorus exists in three main allotropes: white, red, and black [29]. Phosphorene refers to a single to a few layers of black phosphorus materials—a relationship similar to that of graphene and graphite. Several groups have used *in operando* TEM to further understand the sodiation behavior of phosphorous and phosphorene. In 2015, Sun et al. [30] used *in operando* TEM to observe the sodiation process of black phosphorous. After 170 seconds, the black phosphorous particle had expanded 92% in the *y*-axis, while the *x*-axis appeared unchanged. These large volume expansions created large internal stresses and can lead to pulverization of the anode material during repeated cycling. As such, from the *in operando* TEM work, it was proposed that by sandwiching several layers of black phosphorous (phosphorene) between several graphene layers, the internal stresses in bulk black phosphorous could be minimized. The composite was made of several layers of phosphorene sandwiched between graphene layers. The graphene provides both a mechanical support and an electrical pathway to the phosphorene.

This work was expanded on by Nie et al. who observed few-layer thick phosphorene flakes [31]. Distinctive gray-contrast stripes were observed after the initiation of the sodiation process. These stripes are correlated with a high sodium concentration using EELS. Sodium ions prefer to diffuse along the [100] direction rather than inside few-layer phosphorene. As such it is proposed that the sodium ions diffuse along this direction before intercalating. The sodiation process of monolayer phosphorene was also observed using *in operando* TEM. Similar to the sodiation of few-layer phosphorene, the sodiation stripe in monolayer phosphorene is also along the [100] direction. As the sodium diffusion energy barrier for monolayer phosphorene is lower than that of few-layer phosphorene, it is demonstrated that monolayer phosphorene-based NIBs would exhibit a higher rate capability.

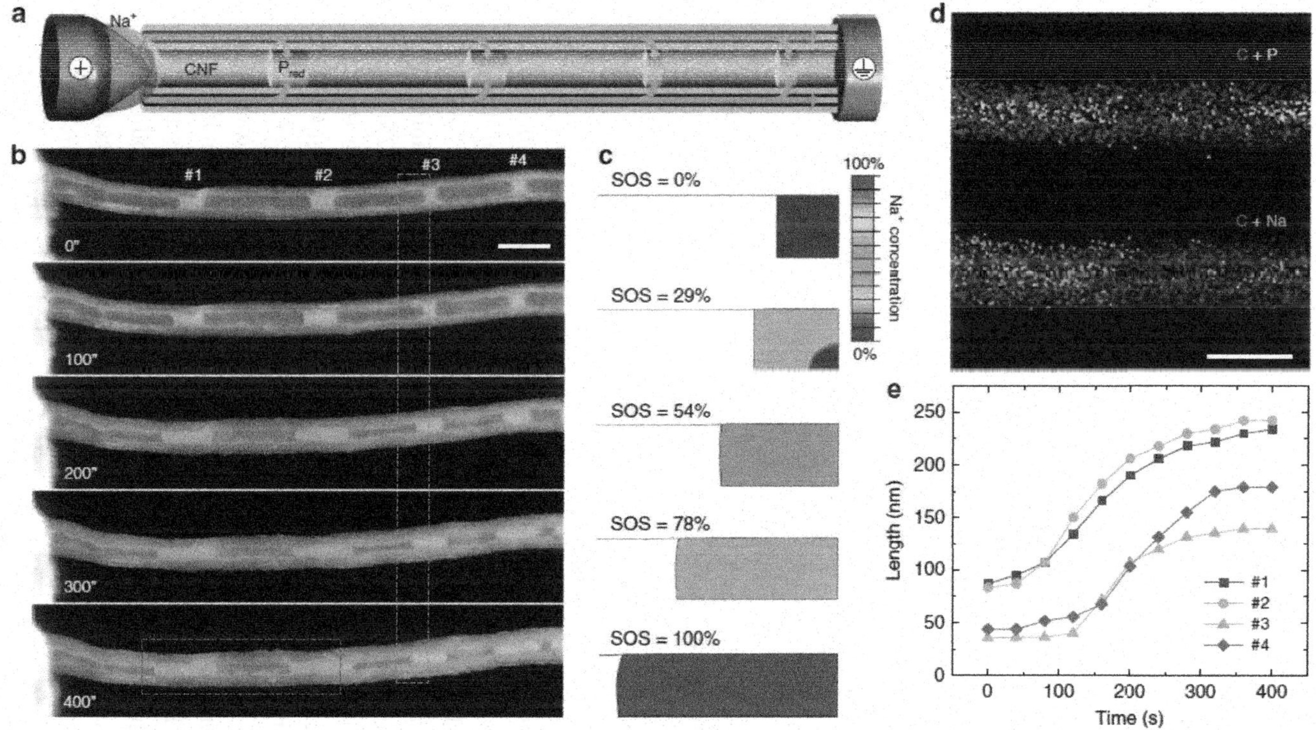

Figure 12.5 *In operando* TEM experiment of red phosphorous embedded in a CNT. Schematic showing the *in operando* TEM experimental setup of an open-cell configuration. (b) STEM image series captured in real time showing red phosphorous volume expansion during sodiation. Scale bar is 200 nm. (c) Simulated SOS according to the red phosphorous morphological evolution of the region labeled with the blue rectangle in (b). (d) EDS maps of phosphorus (green), sodium (yellow), and carbon (red) elements of the region labeled with the red rectangle in (b). Scale bar is 200 nm. (e) Length change of four red phosphorous segments marked in (b) as the function of time during the sodiation process [32].

More recently, Liu et al. (2020) looked at the sodiation/desodiation of red phosphorous embedded in a CNT [32]. *In operando* TEM captured the liquid-like characteristics of the phosphorous during sodiation (Fig. 12.5). As such, embedding red phosphorous has a two-fold impact. It contains the liquid-like sodiated phorphorous and also improves the poor electrical conductivity. This liquid-like material is effective in decreasing detrimental durability effects of fracture initiation and propagation during cycling. Despite this, further work is required to improve the electrode/electrolyte interface to avoid deleterious side reactions.

12.2.3.3 Metal oxides

Although metal oxides have a high theoretical capacity, many metal oxides show a low capacity and limited cycle life when used in NIBs. This is due to limited kinetics, development of complex defect structures, volume expansion, and associated crack formation. To date many oxide materials have been observed under open-cell *in operando* TEM configuration exploring the nanoscale effects that occur during sodiation/desodiation.

In 2010, Huang et al. [33] demonstrated the first use of *in operando* TEM for LIBs by monitoring the lithiation process in an SnO_2 nanowire electrode in an open-cell configuration. During lithiation, a reaction front migrates along the SnO_2 nanowire. This reaction front is characterized by a high density of dislocations, which separates the non-lithiated and lithiated regions of the nanowire. The term coined for this front is the "Medusa zone." It is this medusa zone that is found to be integral in the cycling stability of this otherwise brittle material and accommodates the ca. 250% volume expansion during lithiation. It is interesting to note that the reaction mechanism during sodiation is notably different in comparison to the lithiation process.

During the *in operando* TEM observation of SnO_2 nanowire sodiation, initially Sn nanoparticles form in a dispersed Na_2O matrix [34] (Fig. 12.6a–e). Afterward, the sodiation process follows two stages. In the first stage, the Sn nanoparticles are alloyed with sodium to form an amorphous Na_xSn alloy phase. As sodiation continues, the amorphous Na_xSn alloy phase is converted to crystalline $Na_{15}Sn_4$, which is dispersed in the Na_2O matrix. During the desodiation process, the $Na_{15}Sn_4$ is dealloyed back to Sn nanoparticles. This desodiation process leaves behind Sn nanoparticles and cavities in an Na_2O matrix. The discontinuous Sn nanoparticle network combined with the poor mechanical properties of the desodiated structure gives the material a poor coulombic efficiency and cycling stability [35]. It is interesting, but unsurprising, to note that the sodiation process is more than an order of magnitude less than the lithiation process of SnO_2.

Unlike for SnO_2, the sodiation and lithiation processes for Co_3O_4 are relatively similar, following an intercalation–conversion process. Using *in operando* TEM, it is shown that Co + Na_2O is formed after sodiation [36]. Unfortunately, the kinetics of sodiation is relatively sluggish in comparison to lithiation.

In operando TEM provided clear insights into the origin of the poor performance of NiO during sodiation [37]. During NiO sodiation, an Na_2O layer forms on the NiO nanosheets, while Ni nanoparticles form at the Na_2O/NiO interface. During sodiation, the Na_2O growth continues, but concomitantly this impedes the sodiation of the underlying NiO. This gives NiO its poor performance during sodiation/desodiation. This is dissimilar to the observation of finger-like reaction front during the lithiation of NiO.

Zhu et al. [38] also found that Na_2O played an important role in the irreversible sodiation of the spinel, $NiCo_2O_4$ nanorods. During the initial sodiation, the spinel is decomposed into CoO and NiO nanoparticles (both with a rocksalt phase structure) with an Na_2O layer. During desodiation, the original $NiCo_2O_4$ phase did not recover. Due to the incomplete reduction and associated Na_2O formation, the spinel nanorods displayed a poor discharge capacity and initial coulombic efficiency. Furthermore, metallic Na deposition was also observed. It is interesting to note that despite the ca. 100% volume expansion during sodiation, no deleterious fractures were observed.

Another metal oxide with the promise for NIBs is CuO, which has a theoretical capacity of 4600 mAh/g. In 2015, Lui et al. [39] investigated the two-step and incomplete sodiation of CuO nanowires. The authors found that in the initial step, Cu_2O and Na_2O are formed. On further sodiation, $Na_6Cu_2O_6$ is formed and with Cu nanoparticles embedded in an Na_2O matrix. This is different from the findings of Zhang et al. [40] who studied the transformations of single-crystal CuO nanowires during sodiation. They found that initially the CuO transforms to a nanocrystalline Cu_2O with an Na_2O

layer coating. On further sodiation, Cu_2O is reduced to Cu to form a polycrystalline Cu and Na_2O material. This reduction is associated with a relatively large volume expansion of ca. 230%. During desodiation, a polycrystalline CuO phase is formed, but interestingly the shape remained unchanged. Although a mixed-valence copper oxide phase was observed at 1.3 V in cyclic voltammetry, this phase was not observed during the *in operando* TEM studies.

Asayesh-Ardakani et al. [41] compared the sodiation and lithiation behavior of ZnO nanowire. During lithiation, cracks were formed at the reaction front, which encouraged lithiation in particular directions (Fig. 12.6). In addition to the cracks, Zn particles were also formed. This unfavorable combination led to a reduced capacity during cycling. The sodiation behavior of the ZnO was, however, quite different. The ZnO nanowire sodiation occurred in two steps: an incubation step and an inward sodiation step. During incubation, the sodium diffuses along the nanowire surface to form an Na_2O layer. The sodium then diffuses into the nanowire, which encourages dislocations to form. These dislocations were clearly observed using HAADF imaging. During the second "inward sodiation" step, the sodiation is facilitated by the dislocations along the nanowire axis, resulting in a fast transverse amorphization. Although the authors describe an amorphization process, there are rings in the SAD, which would suggest that it is a polycrystalline formation rather than an amorphization formation. The authors concluded that the formation of the dislocations, rather than the cracks at the interface, improves the cycling stability during sodiation in comparison to lithiation. The observation of the dislocation front at the advancing interface is similar to the "Medusa zone" observed by Huang et al. (2010) during SnO_2 lithiation [33].

In a similar study, Xu et al. [42] compared the sodiation of single- and polycrystalline ZnO nanowire (sc-ZnO and pc-ZnO, respectively). The grain boundaries in the pc-ZnO improved the reaction kinetics. In comparison, the sodiation process in sc-ZnO was found to be diffusion limited. For the sc-ZnO, during sodiation Na initially coated the surface before converting to Na_2O. At the end of sodiation, the $NaZn_{13}$ nanocrystals are found dispersed through the Na_2O matrix. This conversion reaction is associated with a ca. 150% volume expansion and forms problematic cracks at the interfaces. These cracks were not observed in the pc-ZnO nanowire as the grain

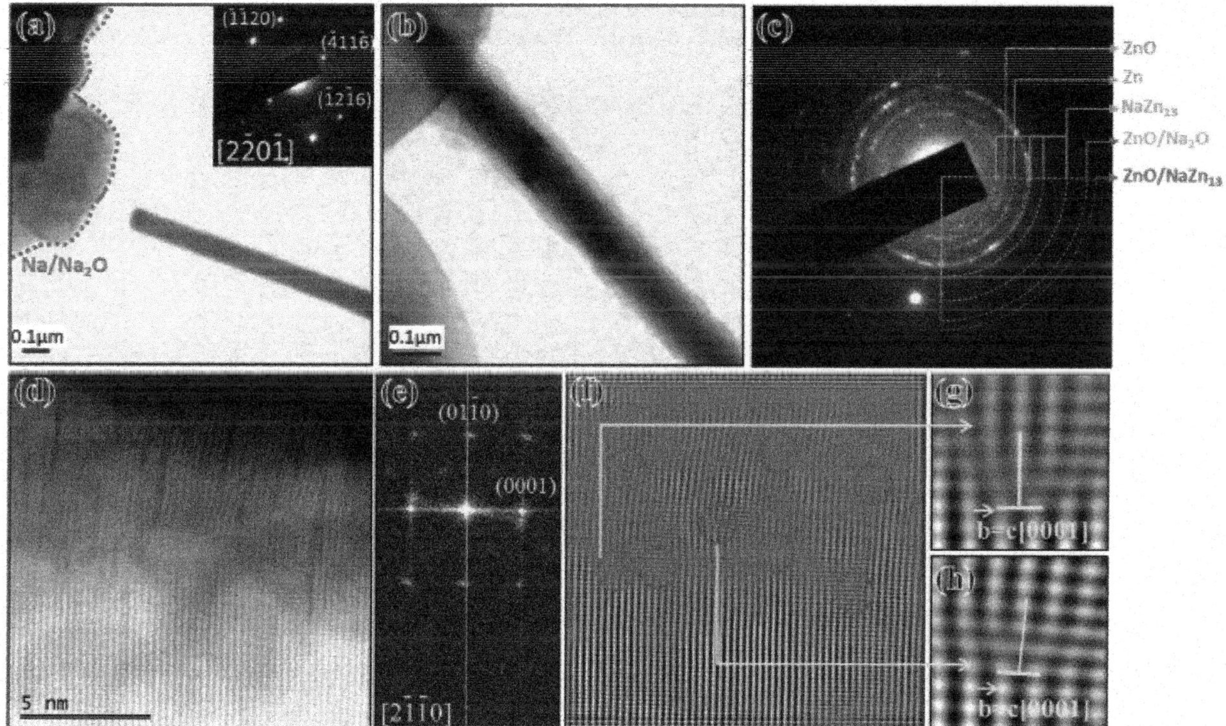

Figure 12.6 *In operando* TEM sodiation of individual ZnO nanowire. (a) Pristine nanowire and SAD pattern (inset) of the nanowire showing single-crystal structure. (b) ZnO nanowire after the sodiation process with radial expansion. (c) SAD pattern of sodiated nanowire with the formation of Zn and $NaZn_{13}$ alloy. (d, e) HAADF image of sodiated nanowire (d) and FFT of the image with the generation of high-density dislocations corresponding to distortion of (0001) lattice plane (indicated by the red circles) (e). (f) Inverse FFT image from (0110) and (0002) spots in part (e), showing clearer dislocations on the nanowire surface. (g, h) High-resolution images of the dislocations with a Burger's vector indicated [41].

boundaries facilitate and even Na^+ diffusion. The lack of cracks and improved reaction kinetics improved the reversibility of the sodiation/desodiation of the pc-ZnO in comparison to the sc-ZnO. It is interesting to note that Xu et al. did not observe the dislocation formation that Asayesh-Ardakani et al. observed [41].

Molybdenum trioxide (α-MoO_3) nanobelts were found to experience an irreversible phase transformation during the initial sodiation/desodiation process [43]. During sodiation, the orthorhombic α-MoO_3 phase is replaced by an amorphous Na_xMoO_2. This amorphous phase is subsequently converted to crystalline $NaMoO_2$ and Na_2O. In the final sodiation stage, polycrystalline Mo is formed with an Na_2O layer. Although no fractures are observed upon desodiation, neither the original nanobelt shape nor phase reform. As such, this material shows a large capacity loss coupled with a poor Coulombic efficiency.

Although metal oxides can have a high theoretical capacity, *in operando* TEM characterization has demonstrated that poor kinetics and the inability of the brittle metal oxide structures to accommodate stresses during cycling limit the practical use.

12.2.3.4 Metal chalcogenides

Metal chalcogenides in both layered and nonlayered forms are of interest due to their promising electrocatalytic properties and unique crystal structures. Molybdenum disulfide (MoS_2) is a layered metal chalcogenide, which has been investigated by several groups using *in operando* TEM [3, 44, 45]. In one study, Gao et al. [3] showed that MoS_2 nanosheets were transformed from 2H-MoS_2 to 1T-$NaMoS_2$, through a two-phase reaction mechanism. In the initial step, the new $NaMoS_2$ phase nucleates. This phase then propagates within the MoS_2 interlayers, which disassembles the lamellar structure into fine nanodomains. These observed nanodomains suppress the volume expansion during sodiation, which improves the long-term stability.

TiS_2 is a light-weight layered metal chalcogenide. Han et al. [46] showed that TiS_2 nanosheets sodiate via an intercalation–conversion reaction process. In the first step, the TiS_2 is intercalated to form Na_xTiS_2. It was found that this phase transformation occurs within the TiS_2 interlayers. To accommodate stresses, Na ions occupy the sulfide octahedral sites in the alternate (0001) planes.

For $0.2 < x < 0.25$, this is characterized as a O1–P3 structure with a 2 × 2 superstructure. As sodiation continues to lower voltages, a conversion reaction to Na_2S + Ti occurs. In this stage, the lamellar structure is lost and nanodomains are formed. This is only partially recoverable and as such, an undesirable conversion reaction. However, this is somewhat different to the observation by Wang et al., who found that intercalated Na_xTiS_2 phases went through a phase transformation from a O1–P3 structure ($0.17 < x < 0.33$) to a P3 structure ($0.38 < x < 0.72$) and finally to a O3 structure ($0.79 < x < 1$) [47].

Due to the reactivity of Sn with Na, SnS_2 undergoes an intercalation–conversion–alloying reaction pathway. During the sodiation of a single-crystal tin sulfide (sc-SnS_2), a high density of defects is formed between the unsodiated and sodiated phases [48]. Unlike in the case of the lithiation of SnO_2, these defects were associated with nanosheet distortion and crack formation. Crack formation is an irreversible and as such undesirable defect.

The antimony chalcogenides (Sb_2X_3, where X is the chalcogenide) also form layered structures and typically follow an intercalation–conversion–alloying reaction pathway [49]. Using *in operando* TEM, Yang et al. [50] found fractures formed at the sodiation reaction front in Sb_2Te_3 NPs. These fractures interrupt the important electronic pathway. A carbon coating on the Sb_2Te_3 NPs increased the electronic conductivity and suppressed fracture growth, improving the capacity and cycling stability of the carbon-coated Sb_2Te_3 NPs.

Carbon coating is a commonly used method to improve the electronic conductivity and consequently the kinetics of active materials. It was also used for Sb_2S_3 nanorods and monitored using *in operando* TEM methods [51]. During sodiation, a clear reaction front is observed between the sodiated and unsodiated regions of the nanorod. The TEM images are supported with radially integrated selected area diffraction patterns to understand the Sb_2S_3 phase evolution during the first sodiation and desodiation processes. Two distinctive stages were observed: first an intercalation process followed by an alloying/conversion process. The initial intercalation process is fast and transforms the crystalline Sb_2S_3 to an amorphous $Na_xSb_2S_3$. The fast intercalation process is attributed to the weak van der Waals forces between the Sb_2S_3 layers. From DFT work, it is predicted that this amorphous phase has good electronic and ionic conductivity. The following alloying/conversion stage is slower. During this step, Na_2S, Na_3Sb, and a small amount of Sb are formed. During desodiation, the parent crystalline Sb_2S_3 is not recovered. Instead, an amorphous Sb_2S_3 phase is left behind.

Metal chalcogenides may also have a nonlayered structure. These materials cannot be easily intercalated, as such they are most commonly categorized as conversion-type electrodes.

One of the earliest *in operando* TEM studies of sodiation of conversion-type metal chalcogenides were carried out by Su et al. in 2014 [52]. The group investigated two types of Co_9S_8-filled CNTs: CNTs with an open end and CNTs with a closed end. Similar to the observations made with other transition metal sulfides, metallic Co NPs are formed in an Na_2S matrix during sodiation. During the sodiation of the open-ended CNTs, the Co_8S_9 particles expand out the end of the CNT. However, during the sodiation of the closed-ended CNTs, the Co_8S_9 particles expand and fracture, leading to cycling instability. As previously mentioned, it is interesting to note that the sodiation of Co_8S_9 is faster than lithiation. The improved kinetics is attributed to a higher conductivity of the reaction phases, but also presents an opportunity for materials design [5]. A similar irreversible conversion was observed for nickel sulfide (Ni_3S_2) nanorods [53]. During sodiation, Ni and Na_2S are formed, which is associated with a large volume expansion and consequently pulverization. This pulverization can be ameliorated by carbon coating the Ni_3S_2 nanorods.

Boebinger et al. [55] observed that during Cu_2S sodiation, Cu nanoparticles are formed within an Na_2S matrix. During this initial sodiation process, the Na insertion is facilitated by a fast diffusion of the Cu ions. However, only a limited amount of Na is intercalated into the Cu_2S structure, before the Cu_2S NPs are converted to Cu NPs. These Cu NPs are dispersed either within the Na_2S matrix or on the Na_2S surface. Due to the small size and uniform distribution of these phases, the structure demonstrates good stability during subsequent sodiation and desodiation processes. Boebinger et al. expanded on their initial work through an extensive comparative *in operando* TEM study of lithiation, sodiation, and potassiation of FeS_2 nanocubes (Fig. 12.7) [54]. A distinguishing feature is that the sodiation of the FeS_2 nanocubes sodiation product encapsulates the FeS_2 in an oval structure with rounded edges. The lack of sharp edges avoids stress concentration points (which are observed during lithiation). This, in turn, limits fracture propagation through the sodiated conversion product, improving long-term stability.

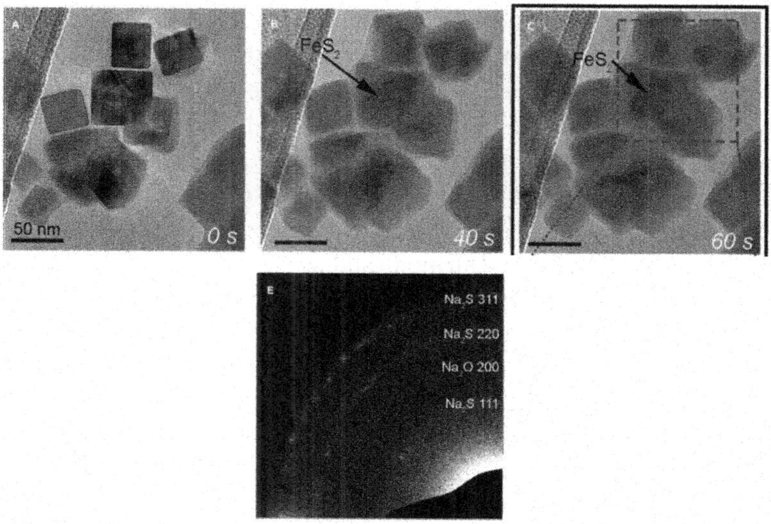

Figure 12.7 Sodiation of FeS_2 particles. (a) Four unreacted FeS_2 particles in the upper half of the frame, and the sodiation front is approaching from the lower-right corner of the frame. (b) The particles are shown in the midst of sodiation, and a sharp reaction front is visible between the unreacted crystals and the reacted phase. (c) Continuation of the sodiation reaction. (d) SAD pattern of the sodiated material showing diffused rings that indicate the presence of Na_2S and Na_2O [54].

12.2.3.5 Others

Limited work has been done using *in operando* TEM of fluoride and phosphide systems [56, 57]. *In operando* TEM investigation of FeF_2 showed that the sodiation of FeF_2 does not follow the same pathway as the lithiation [56]. During lithiation, FeF_2 converts to Fe and LiF, whereas during sodiation, the conversion reaction to Fe and NaF is limited to the particle surface. However, internally, the NaF phase is also associated with some iron (Na_3FeF_6).

12.2.3.6 Metalloids

Metals or metalloids can alloy with Na to provide a high capacity. Group 14 and 15 metals are of particular interest. These materials have good electronic conductivity and ductility, but large volume changes and limited internal Na^+ diffusion can limit the capacity and long-term stability.

Lu et al. [58] demonstrated the significance of the electrode microstructure through comparing the *in operando* sodiation pathways of single-crystal and polycrystalline Ge nanowires. For the crystalline Ge nanowires, the sodiation processes were controlled by limited Na^+ surface diffusion. However, by introducing grain boundaries, more diffusion pathways were provided for the Na ions in the polycrystalline Ge. As such, the sodiation process was substantially faster. The structure also accommodated a relatively large volume expansion. During desodiation, the heavily expanded polycrystalline Ge nanowire becomes riddled with pores.

Large volume expansions and phase changes can introduce internal stresses but can also lead to large particle morphology changes during cycling. Wang et al. [59] observed that Sn nanoparticles went through four phase changes during sodiation. Initially $NaSn_2$ is formed, followed by Na_9Sn_4, then Na_3Sn and finally $Na_{15}Sn_4$. Despite a huge volume expansion of 445% associated with this final phase formation, no cracks or fractures were observed. Naturally, due to the large volume expansion, the initial shape of $NaSn_2$ was not recovered. Instead during cycling, $NaSn_2$ transitioned from a spherical to an irregular shape.

Li et al. [60] revealed the two-stage sodiation process for Sb thin film on a TiN-coated Ge nanowire. The TiN and Ge nanowire were not involved in the sodiation process. At high voltages, $Na_{0.9}Sb$ is formed, and at lower voltages, this phase was sodiated to Na_3Sb. The Sb film was associated with a large volume expansion (ca. 260%), but as this was not fully recovered during desodiation, it indicates issues with long-term stability.

In an extension of this work, Xie et al. [61] studied various coatings on the TiN-coated Ge nanowire. The nanowires were coated with Sb/Sn, Sn/Sb, or a β-SnSb layer. Due to differences in the insertion potential, the Sb layer is first sodiated, followed by the Sn layer. The Sb layer converts to Na_3Sb, whereas the Sn is converted to $Na_{15}Sn_4$. During desodiation, a polycrystalline β-SnSb phase is formed. During subsequent cycles, Sn nanocrystals exsolve leaving an Sb-rich β-SnSb. Despite this, the material shows good stability. Naturally, as the alloyed layers are converted to β-SnSb during the first sodiation process, the various nanowires show a similar electrochemical performance.

Nie et al. [62] found a remarkably different behavior when observing the β-Zn_4Sb_3 nanowire sodiation. The nanowire showed both an irreversible shape and phase transformation. During sodiation, a mixture of $NaZn_{13}$ and Na_3Sb is formed, with a modest volume change (ca. 160%). However, on desodiation, a complex desodiated structure of NaZnSb and Zn is formed. This newly formed lamellar phase can provide a rapid diffusion pathway and Na^+ storage location between the (002) planes. This structure is key to the excellent long-term performance and stability of these NWs.

Gutierrez-Kolar et al. [63] presented a unique method to use *in operando* TEM data to quantify sodiation kinetics in an Sn–Sb alloy. They clearly demonstrated that the microstructural changes have a direct impact on the Na^+ mobility in the investigated Sn–Sb alloy. The application of a potential increases the speed of the Na^+ diffusion, by one order of magnitude. The combination of simulation and modeling of this work demonstrates the power of *in operando* TEM methods in obtaining quantitative results, which can be directly used in NIB modeling.

12.2.3.7 Conclusion

The use of *in operando* TEM has been central in understanding the nanoscale transformations that occur during sodiation and desodiation of NIB active materials. The ability to track rate-limited kinetics, phase nucleation and transformation, volume expansion, dislocation, and fracture formation has allowed researchers to design and develop improved active materials.

There are still limitations to these methods. The closed-cell configuration is limited by resolution due to the presence of liquid. The open-cell configuration does not suffer as much from limited resolution, but the lack of the SEI layer may have a dramatic impact on the observed mechanism. Beyond this, as with all TEM characterization methods, beam-induced effects need to be systematically studied and considered. However, with continued developments in low-dose detectors and improved MEMS chip design, it is possible that some of these issues will be alleviated in the coming years, making these *in operando* TEM methods for NIB active material characterization more robust.

12.3 Emerging Tools for NIB Characterization

High-end TEM techniques, including HRSTEM combined with EELS and *in operando* TEM, have been successfully used to characterize NIB materials. However, new and emerging techniques, including cryo-TEM, 4D STEM, electron holography, and electron tomography, have been used in the field of LIBs, but not yet used for NIBs. In the following section, these techniques will be introduced in conjunction with some LIB applications. This will be used to highlight how these techniques can be used to probe NIB materials in the future.

12.3.1 Cryo-TEM

Cryo-TEM has opened opportunities for imaging air-sensitive and beam-sensitive materials. In the field of batteries, this work has benefitted from major successes in biological cryo-TEM. Cryo-TEM is complementary to *in operando* tools, which have been either limited by the lack of liquid (in the open-cell configuration) or limited by resolution (in the closed-cell configuration). The cryo-TEM instrument coupled with the development of cryo-TEM sample preparation workflows allows for intermediate, metastable phases to be captured in batteries.

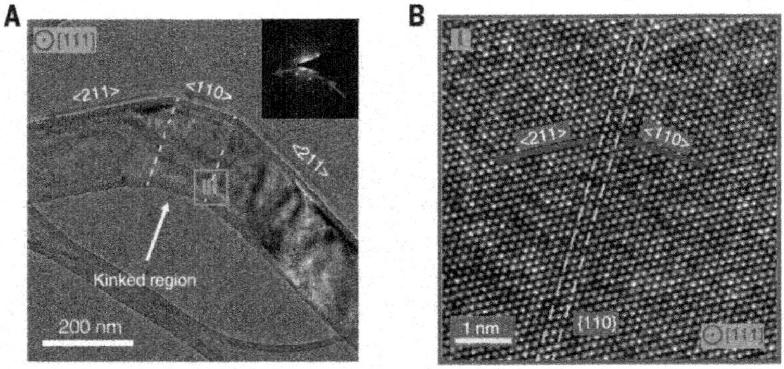

Figure 12.8 Cryo-TEM atomic-resolution images of kinked Li metal dendrite and SEI interface. (a) TEM image of kinked Li metal dendrite that changes from a <211> to <110> and back to <211> growth direction. Inset shows corresponding SAD pattern. The dashed yellow lines indicate change in growth direction.. (b) Magnified images of red region I [64].

At room temperature, Li metal is very reactive and corrodes when exposed to air [65]. Furthermore, Li has a low melting point and weak atomic bonding, making Li unstable when exposed to an electron beam [66]. Li et al. used cryo-TEM to observe Li dendrites formed on a Cu TEM grid [64]. The observed dendrites showed clear faceted growth and due to the cryo-TEM operation, it was possible to clearly observe the Li columns using high-resolution imaging (Fig. 12.8). This improves on previous work completed under either non-cryo transfer conditions or imaged under standard TEM conditions.

The use of cryo-TEM in alkali-ion batteries is limited to only a few studies, including Li dendrite morphology characterization [64, 67], imaging of Li atomic columns [64], and characterization of SEI phases [67–70]. Despite this interesting work being completed, no work has been done to date for NIB systems.

12.3.2 4D STEM and Electron Holography

4D STEM is a developing field in electron microscopy and is an extension of the traditional STEM and diffraction techniques. The electron beam is rastered over the sample. At each position, a diffraction pattern is collected using a pixelated detector (often referred to as a direct electron detector (DED)). The DEDs have an improved quantum efficiency and signal-to-noise ratio in comparison to their traditional charge couple device (CCD) predecessors. Furthermore, they can push imaging frame rates to 400 frames per second (fps). From these diffraction patterns, detailed sample information, including electric and magnetic field, strain, phase, and orientation [71, 72], can be obtained. Furthermore, the use of electron ptychography has pushed the imaging resolution [73] with the possibility of observing light elements in LIB cathode materials [74]. Due to the nature of these detectors, the beam intensity can be reduced, which reduces beam-induced sample damage.

Complementary to 4D STEM is electron holography. Using electron holography, the beam phase can be extracted (instead of just the beam intensity). With this additional information, imaging resolution can be pushed and information about magnetic and electric domains can be extracted. This has been used in LIBs to understand the charge distribution during lithiation [75] or to investigate the charge distribution at the solid electrode–electrolyte interface [76–81].

12.3.3 Electron Tomography

Electron tomography (ET) allows for a 3D reconstruction to reveal features such as grain topography, sample porosity, and phase distribution. Tomography can be constructed from data generated from the focused ion beam-SEM (FIB-SEM) system, or from using a tomography holder in a TEM. A large sample can be investigated using the FIB-SEM, but naturally this limits the spatial resolution. Shim et al. used an FIB-SEM technique to provide information about three different Li-ion battery microstructures: commercial, core–shell, and hollow [82]. These grains were around 10 micron in size but contained complex microstructures with different porosity and crystal sizes and shapes. The FIB-SEM revealed that the commercial powder contained large primary particles around 500 nm in size. The core–shell structure contained lenticular crystals in the shell and fine (100 nm) crystals in the core, whereas the hollow showed lenticular structures in the shell and a relatively empty core. This work was conducted in conjunction with aberration-corrected HRSTEM and TEM-EELS, which provided information on the oxidation states of the transition metals used in the core–shell structure. The combination of these complementary techniques provides a deep insight into the complex structure of these materials. More recently, the development of cryo-FIB techniques has opened the possibility of milling and reconstructing materials that previously were not possible to mill with a conventional FIB [67]. This will open many opportunities in both LIB and NIB systems.

Electron tomography has also been used to monitor and map the redistribution of elements in LIB phases. One example by Lin et al. looks at the incorporation of 4d-containing cathodes and the impact on stability (Fig. 12.9) [83]. The 4d and 5d transition metals are used to stabilize the crystal structure from oxygen release. However, the authors highlighted that the work to date was limited to pristine work. They used a combination of STEM–EDS tomography to show that the pristine phase shows an Mn-rich surface and an Ru-rich core. The pristine phase is also relatively homogenous with rounded surfaces. However, after cycling the cathode grains look distinctly different. The grain morphology is jagged and riddled with porosity. There is obvious phase segregation. The surface has become enriched with Ru, and the elements appear to be less homogeneously distributed. This segregation was attributed to the loss of oxygen, which can lead to surface reduction. When the surface is reduced, the Ru is no longer soluble in the parent phase and exsolves. The electron tomography work was key in revealing the distribution and size of these secondary phases both before and after cycling. Despite the work in the field of LIBs, electron tomography has not yet been used for NIBs.

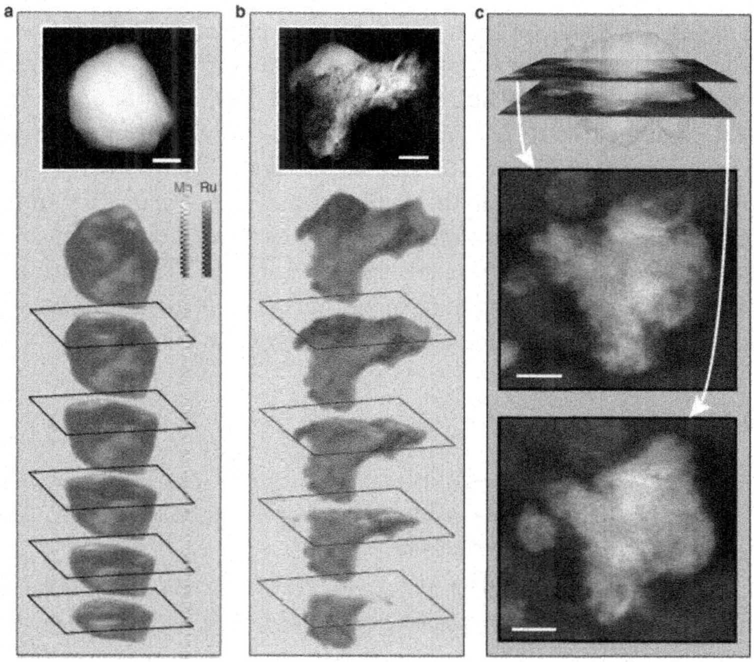

Figure 12.9 3D visualization of the chemical and structural changes of $Li_2Ru_{0.5}Mn_{0.5}O_3$ (LRMO) before and after extended charge/discharge cycling. 3D distribution of Ru and Mn of (a) pristine LRMO and (b) LRMO after 97 charge/discharge cycles reconstructed by STEM-EDX tomography. (c) Annular dark-field STEM tomographic (ADF-STEM) reconstruction of an LRMO primary particle after 97 cycles. Two cross-sectional images of the three-dimensional reconstruction are presented to demonstrate that a porous structure had developed throughout the particle. Scale bar = 100 nm [83].

12.4 Future Outlook

Advanced TEM characterization, including HRSTEM, EELS, and *in operando* TEM, has played a central role in revealing complex structures of NIB materials, including interfacial defects between sodiated and desodiated regions, phase changes, defect structures, charge ordering, ion redistribution, and crack formation. The development of 4D-STEM and cryo-TEM techniques provides growing opportunities to characterize beam-sensitive materials.

However, strong work to date in the field uses a combination of complementary tools to mechanistic insight. For example, Lutz et al. used a range of both *ex situ* and *in operando* TEM tools to provide a detailed understanding of mechanisms in Na–O_2 battery systems. Furthermore, Gutierez-Kolar et al. [63] used *in operando* TEM combined with DFT and simulation to provide a quantitative assessment of kinetics in NIB SbSn electrodes.

It is anticipated that in the coming years, *in operando* TEM tools and methods will be developed to further enhance spatial and temporal resolution. In parallel, cryo-TEM and 4D-STEM workflows will be further developed to become routine tools for the advanced understanding of the complex mechanisms of NIB materials.

References

1. Wang, Z., S. M. Selbach, and T. Grande, Van der Waals density functional study of the energetics of alkali metal intercalation in graphite. *RSC Advances*, 2014. **4**(8): 4069–4079.
2. Li, Q., et al., Electrochemistry of selenium with sodium and lithium: Kinetics and reaction mechanism. *ACS Nano*, 2016. **10**(9): 8788–8795.
3. Gao, P., et al., Atomic-scale probing of the dynamics of sodium transport and intercalation-induced phase transformations in MoS_2. *ACS Nano*, 2015. **9**(11): 11296–11301.
4. Wang, L., et al., Atomic mechanism of dynamic electrochemical lithiation processes of MoS_2 nanosheets. *Journal of the American Chemical Society*, 2014. **136**(18): 6693–6697.
5. Yuan, Y., et al., Understanding materials challenges for rechargeable ion batteries with in situ transmission electron microscopy. *Nature Communications*, 2017. **8**.

6. Zhou, W., et al., $Na_xMV(PO_4)_3$ (M = Mn, Fe, Ni) structure and properties for sodium extraction. *Nano Letters*, 2016. **16**(12): 7836–7841.
7. Huang, Q., et al., Tailoring alternating heteroepitaxial nanostructures in Na-ion layered oxide cathodes via an in-situ composition modulation route. *Nano Energy*, 2018. **44**: 336–344.
8. Huang, Q., et al., Improving the electrochemical properties of the manganese-based P3 phase by multiphasic intergrowth. *Inorganic Chemistry*, 2018. **57**(24): 15584–15591.
9. Delmas, C., C. Fouassier, and P. Hagenmuller, Structural classification and properties of the layered oxides. *Physica B+C*, 1980. **99**(1–4): 81–85.
10. Jung, E., et al., Synthesis of nanostructured P2-$Na_{2/3}MnO_2$ for high performance sodium-ion batteries. *Chemical Communications*, 2019. **55**(33): 4757–4760.
11. Li, X., et al., Direct visualization of the Jahn–Teller effect coupled to Na ordering in $Na_{5/8}MnO_2$. *Nature Materials*, 2014. **13**(6): 586–592.
12. Cho, J. S., et al., Carbon/two-dimensional $MoTe_2$ core/shell-structured microspheres as an anode material for Na-ion batteries. *Nanoscale*, 2017. **9**(5): 1942–1950.
13. Zhou, T., et al., Enhanced sodium-ion battery performance by structural phase transition from two-dimensional hexagonal-SnS_2 to orthorhombic-SnS. *ACS Nano*, 2014. **8**(8): 8323–8333.
14. Panda, M. R., et al., Blocks of molybdenum ditelluride: A high rate anode for sodium-ion battery and full cell prototype study. *Nano Energy*, 2019. **64**.
15. Chen, S., et al., Atomic structure and migration dynamics of MoS_2/Li_xMoS_2 interface. *Nano Energy*, 2018. **48**: 560–568.
16. Sun, Y., et al., Direct atomic-scale confirmation of three-phase storage mechanism in $Li_4Ti_5O_{12}$ anodes for room-temperature sodium-ion batteries. *Nature Communications*, 2013. **4**.
17. Wen, Y., et al., Microstructure dynamics of rechargeable battery materials studied by advanced transmission electron microscopy. *NPG Asia Materials*, 2017. **9**(3).
18. Li, M. and R. Knibbe, A study of membrane impact on spatial resolution of liquid in situ transmission electron microscope. *Microscopy and Microanalysis*, 2019. **26**(1): 126–133.
19. Gu, M., et al., Demonstration of an electrochemical liquid cell for operando transmission electron microscopy observation of the lithiation/delithiation behavior of Si nanowire battery anodes. *Nano Letters*, 2013. **13**(12): 6106–6112.

20. Lutz, L., et al., Operando monitoring of the solution-mediated discharge and charge processes in a Na-O_2 battery using liquid-electrochemical transmission electron microscopy. *Nano Letters*, 2018. **18**(2): 1280–1289.

21. Doeff, M. M., et al., Electrochemical insertion of sodium into carbon. *Journal of the Electrochemical Society*, 1993. **140**(12): L169–L170.

22. Liu, Y., et al., In situ transmission electron microscopy study of electrochemical sodiation and potassiation of carbon nanofibers. *Nano Letters*, 2014. **14**(6): 3445–3452.

23. Li, X., et al., In-situ electron microscopy observation of electrochemical sodium plating and stripping dynamics on carbon nanofiber current collectors. *Nano Energy*, 2017. **42**: 122–128.

24. Wan, J., et al., In situ transmission electron microscopy observation of sodiation-desodiation in a long cycle, high-capacity reduced graphene oxide sodium-ion battery anode. *Chemistry of Materials*, 2016. **28**(18): 6528–6535.

25. Yang, Y., et al., "Protrusions" or "holes" in graphene: Which is the better choice for sodium ion storage? *Energy and Environmental Science*, 2017. **10**(4): 979–986.

26. Zhang, C., et al., Amorphous phosphorus/nitrogen-doped graphene paper for ultrastable sodium-ion batteries. *Nano Letters*, 2016. **16**(3): 2054–2060.

27. Wen, Y., et al., Expanded graphite as superior anode for sodium-ion batteries. *Nature Communications*, 2014. **5**.

28. Jian, Z., et al., Insights on the mechanism of Na-ion storage in soft carbon anode. *Chemistry of Materials*, 2017. **29**(5): 2314–2320.

29. Corbridge, D. E. C., *Phosphorus: Chemistry, Biochemistry and Technology*. CRC Press, 2013.

30. Sun, J., et al., A phosphorene-graphene hybrid material as a high-capacity anode for sodium-ion batteries. *Nature Nanotechnology*, 2015. **10**(11): 980–985.

31. Nie, A., et al., Selective ionic transport pathways in phosphorene. *Nano Letters*, 2016. **16**(4): 2240–2247.

32. Liu, Y., et al., Red-phosphorus-impregnated carbon nanofibers for sodium-ion batteries and liquefaction of red phosphorus. *Nature Communications*, 2020. **11**(1).

33. Huang, J. Y., et al., In situ observation of the electrochemical lithiation of a single SnO_2 nanowire electrode. *Science*, 2010. **330**(6010): 1515–1520.

34. Gu, M., et al., Probing the failure mechanism of SnO_2 nanowires for sodium-ion batteries. *Nano Letters*, 2013. **13**(11): 5203–5211.

35. Park, J., et al., Charge-discharge properties of tin dioxide for sodium-ion battery. *Materials Research Bulletin*, 2014. **58**: 186–189.

36. Chen, D., et al., Two-dimensional holey Co_3O_4 nanosheets for high-rate alkali-ion batteries: From rational synthesis to in situ probing. *Nano Letters*, 2017. **17**(6): 3907–3913.

37. He, K., et al., Sodiation kinetics of metal oxide conversion electrodes: A comparative study with lithiation. *Nano Letters*, 2015. **15**(9): 5755–5763.

38. Zhu, C., et al., Identifying the conversion mechanism of $NiCo_2O_4$ during sodiation–desodiation cycling by in situ TEM. *Advanced Functional Materials*, 2017. **27**(17).

39. Liu, H., et al., In situ observation of the sodiation process in CuO nanowires. *Chemical Communications*, 2015. **51**(52): 10443–10446.

40. Zhang, L., et al., In situ transmission electron microscopy study of the electrochemical sodiation process for a single CuO nanowire electrode. *RSC Advances*, 2016. **6**(14): 11441–11445.

41. Asayesh-Ardakani, H., et al., In situ TEM investigation of ZnO nanowires during sodiation and lithiation cycling. *Small Methods*, 2017. **1**: 8.

42. Xu, F., et al., In situ TEM probing of crystallization form-dependent sodiation behavior in ZnO nanowires for sodium-ion batteries. *Nano Energy*, 2016. **30**: 771–779.

43. Xia, W., et al., Probing microstructure and phase evolution of α-MoO_3 nanobelts for sodium-ion batteries by in situ transmission electron microscopy. *Nano Energy*, 2016. **27**: 447–456.

44. Hu, Z., et al., MoS_2 nanoflowers with expanded interlayers as high-performance anodes for sodium-ion batteries. *Angewandte Chemie - International Edition*, 2014. **53**(47): 12794–12798.

45. Zhang, L., et al., In situ TEM observing structural transitions of MoS_2 upon sodium insertion and extraction. *RSC Advances*, 2016. **6**(98): 96035–96038.

46. Han, B., et al., Tracking sodium migration in TiS_2 using in situ TEM. *Nanoscale*, 2019. **11**(15): 7474–7480.

47. Wang, X., et al., In situ electron microscopy investigation of sodiation of titanium disulfide nanoflakes. *ACS Nano*, 2019. **13**(8): 9421–9430.

48. Gao, P., et al., In situ atomic-scale observation of reversible sodium ions migration in layered metal dichalcogenide SnS_2 nanostructures. *Nano Energy*, 2017. **32**: 302–309.

49. Cui, J., H. Zheng, and K. He, In situ TEM study on conversion-type electrodes for rechargeable ion batteries. *Advanced Materials*, 2020. **33**(6).

50. Yang, Z., et al., Facile synthesis and in situ transmission electron microscopy investigation of a highly stable Sb_2Te_3/C nanocomposite for sodium-ion batteries. *Energy Storage Materials*, 2017. **9**: 214–220.

51. Yao, S., et al., Unveiling the unique phase transformation behavior and sodiation kinetics of 1D van der Waals Sb_2S_3 anodes for sodium ion batteries. *Advanced Energy Materials*, 2017. **7**(8).

52. Su, Q., et al., In situ transmission electron microscopy observation of electrochemical sodiation of individual Co_9S_8-filled carbon nanotubes. *ACS Nano*, 2014. **8**(4): 3620–3627.

53. Xia, X., et al., Improved Na-storage cycling of amorphous-carbon-sheathed Ni_3S_2 arrays and investigation by in situ TEM characterization. *Materials Today Energy*, 2017. **5**: 99–106.

54. Boebinger, M. G., et al., Avoiding fracture in a conversion battery material through reaction with larger ions. *Joule*, 2018. **2**(9): 1783–1799.

55. Boebinger, M. G., et al., Distinct nanoscale reaction pathways in a sulfide material for sodium and lithium batteries. *Journal of Materials Chemistry A*, 2017. **5**(23): 11701–11709.

56. He, K., et al., Sodiation via heterogeneous disproportionation in FeF2 electrodes for sodium-ion batteries. *ACS Nano*, 2014. **8**(7): 7251–7259.

57. Shi, S., et al., A covalent heterostructure of monodisperse Ni_2P immobilized on N, P-co-doped carbon nanosheets for high performance sodium/lithium storage. *Nano Energy*, 2018. **48**: 510–517.

58. Lu, X., et al., Germanium as a sodium ion battery material: In situ TEM reveals fast sodiation kinetics with high capacity. *Chemistry of Materials*, 2016. **28**(4): 1236–1242.

59. Wang, J. W., et al., Microstructural evolution of tin nanoparticles during in situ sodium insertion and extraction. *Nano Letters*, 2012. **12**(11): 5897–5902.

60. Li, Z., et al., Coupling in situ TEM and ex situ analysis to understand heterogeneous sodiation of antimony. *Nano Letters*, 2015. **15**(10): 6339–6347.

61. Xie, H., et al., β-SnSb for sodium ion battery anodes: Phase transformations responsible for enhanced cycling stability revealed by in situ TEM. *ACS Energy Letters*, 2018. **3**(7): 1670–1676.

62. Nie, A., et al., Ultrafast and highly reversible sodium storage in zinc-antimony intermetallic nanomaterials. *Advanced Functional Materials*, 2016. **26**(4): 543–552.
63. Gutiérrez-Kolar, J. S., et al., Interpreting electrochemical and chemical sodiation mechanisms and kinetics in tin antimony battery anodes using in situ transmission electron microscopy and computational methods. *ACS Applied Energy Materials*, 2019. **2**(5): 3578–3586.
64. Li, Y., et al., Atomic structure of sensitive battery materials and interfaces revealed by cryo-electron microscopy. *Science*, 2017. **358**(6362): 506–510.
65. Markowitz, M. M. and D. A. Boryta, Lithium metal-gas reactions. Interaction of lithium metal with air and its component gases. *Journal of Chemical and Engineering Data*, 1962. **7**(4): 586–591.
66. Egerton, R. F., P. Li, and M. Malac, Radiation damage in the TEM and SEM. *Micron*, 2004. **35**(6): 399–409.
67. Zachman, M. J., et al., Cryo-STEM mapping of solid–liquid interfaces and dendrites in lithium-metal batteries. *Nature*, 2018. **560**(7718): 345–349.
68. Huang, W., et al., Evolution of the solid-electrolyte interphase on carbonaceous anodes visualized by atomic-resolution cryogenic electron microscopy. *Nano Letters*, 2019. **19**(8): 5140–5148.
69. Wang, X., et al., New insights on the structure of electrochemically deposited lithium metal and its solid electrolyte interphases via cryogenic TEM. *Nano Letters*, 2017. **17**(12): 7606–7612.
70. Xu, Y., et al., Atomic to nanoscale origin of vinylene carbonate enhanced cycling stability of lithium metal anode revealed by cryo-transmission electron microscopy. *Nano Letters*, 2020. **20**(1): 418–425.
71. Brunetti, G., et al., Confirmation of the domino-cascade model by $LiFePo_4/FePo_4$ precession electron diffraction. *Chemistry of Materials*, 2011. **23**(20): 4515–4524.
72. Ophus, C., Four-dimensional scanning transmission electron microscopy (4D-STEM): From scanning nanodiffraction to ptychography and beyond. *Microscopy & Microanalysis*, 2019. **25**: 563–583.
73. Gao, W., et al., Real-space charge-density imaging with sub-Angstrom resolution by four-dimensional electron microscopy. *Nature*, 2019. **575**(7783): 480–484.
74. Lozano, J. G., et al., Low-dose aberration-free imaging of Li-rich cathode materials at various states of charge using electron ptychography. *Nano Letters*, 2018. **18**(11): 6850–6855.

75. Gan, Z., et al., Direct mapping of charge distribution during lithiation of Ge nanowires using off-axis electron holography. *Nano Letters*, 2016. **16**(6): 3748–3753.
76. Yamamoto, K., et al., Dynamic visualization of the electric potential in an all-solid-state rechargeable lithium battery. *Angewandte Chemie International Edition*, 2010. **49**(26): 4414–4417.
77. Yamamoto, K., et al., Direct observation of lithium-ion movement around an in-situ-formed negative-electrode/solid-state-electrolyte interface during initial charge-discharge reaction. *Electrochemistry Communications*, 2012. **20**(1): 113–116.
78. Aizawa, Y., et al., In situ electron holography of electric potentials inside a solid-state electrolyte: Effect of electric-field leakage. *Ultramicroscopy*, 2017. **178**: 20–26.
79. Hirayama, T., et al., Advanced electron holography techniques for in situ observation of solid-state lithium ion conductors. *Ultramicroscopy*, 2017. **173**: 64–70.
80. Wen, Y., et al., Interface charges boosted ultrafast lithiation in $Li_4Ti_5O_{12}$ revealed by in-situ electron holography. *Journal of Energy Chemistry*, 2018. **27**(5): 1397–1401.
81. Nomura, Y., et al., Direct observation of a Li-ionic space-charge layer formed at an electrode/solid-electrolyte interface. *Angewandte Chemie International Edition*, 2019. **58**(16): 5292–5296.
82. Shim, J.-H., et al., Hierarchically structured core–shell design of a lithium transition-metal oxide cathode material for excellent electrochemical performance. *Applied Materials & Interfaces*, 2019. **2019**: 4017–4027.
83. Lin, R., et al., Anomalous metal segregation in lithium-rich material provides design rules for stable cathode in lithium-ion battery. *Nature Communications*, 2019. **10**(1).

Chapter 13

Synchrotron Radiation-Based X-Ray Characterizations of Sodium-Ion Battery

Yu Zhou[b,a] **and Li Song**[a]
[a]*National Synchrotron Radiation Laboratory, CAS Center for Excellence in Nanoscience, University of Science and Technology of China, Hefei 230029, China*
[b]*The First Affiliated Hospital of Anhui Medical University, Hefei 230022, China*
song2012@ustc.edu.cn

13.1 Introduction

Due to the ever-increasing demand for energy supply in transportation, electronics, and grid energy storage, the rechargeable lithium-ion batteries (LIBs), which are ubiquitous in modern society, are expected to grow continuously [1–3]. However, challenges such as safety, higher power, and capacity are still needed to be resolved urgently. Besides the expansion development of LIBs, lithium reserves are becoming scarce with a rising price trend [4]. As one of its promising complements, sodium-ion batteries (SIBs)

Handbook of Sodium-Ion Batteries: Materials and Characterization
Edited by Rohit R. Gaddam and X. S. (George) Zhao

ISBN 978-981-4968-15-7 (Hardcover), 978-1-003-30874-4 (eBook)
www.jennystanford.com

with Na^+ charge carriers have attracted world's attention in recent years. In comparison, both sodium (Na) and lithium (Li) are alkali metals and thus possess similar physical and chemical properties (Table 13.1) [5–7]. Unlike Li mines, sodium sources are widely distributed on earth with abundant reserves and comparatively low prices. Moreover, the less costly aluminum can be employed as both anode and cathode current collectors for SIBs, which cannot work for LIBs because of alloying effect [8]. Additionally, sodium was treated as one of the most appealing rechargeable battery system candidates for energy storage at the end of the 1970s and the beginning of the 1980s[9]. Exploratory works have mainly focused on the interaction behaviors of sodium in disulfide and diselenide compounds [10, 11].

Table 13.1 Comparison of physical properties between sodium and lithium [5–7]

Category	Sodium	Lithium
Ionic radius/nm	0.102	0.076
Atomic weight/(g/mol)	23	6.94
Theoretical capacity/(mAh/g)	1166	3861
First ionization energy/(kJ/mol)	495.8	520.2
E° (vs.SHE)/V	-2.71	-3.04
Abundance/%	2.64	0.01
Distribution	Everywhere	70% in South America
Cost/(RMB/kg)	~2	~40

As shown in Fig. 13.1, SIBs generally consist of four syntrophic components: cathode materials, anode materials, porous films between the two electrodes as separators and sodium salt dissolved in an organic liquid as electrolytes [12–15]. The principle of SIBs is the Na^+ shuttle between the cathode and anode electrodes during operation. During charging process, Na-ions flow from the cathode to the anode through the electrolyte, accompanying the electron migration from anode to cathode through external circuits. Discharging process is the reverse of the charging process [5, 16, 17].

A large amount of electrode materials for SIBs have been developed, including layer-structured transition metal (TM) oxides [18–21], polyanionic compounds [22–26], Prussian blue analogs [27–29], and organic-based materials [30–32] for cathode materials and carbonaceous materials [33–36], alloy metals [37–39], phosphorus [59–61], phosphate [43–45], metal oxides [46–50], and sulfides [51–55] for anode materials, which boosts their future commercialization.

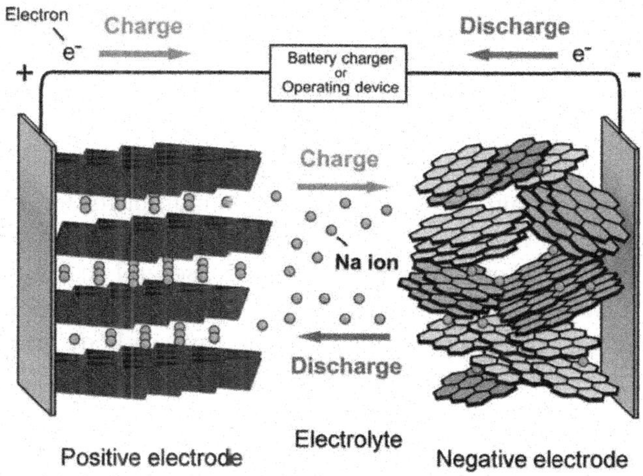

Figure 13.1 Schematic illustration of SIBs. Reprinted with permission from Ref. [5]. Copyright 2014, American Chemical Society.

13.1.1 Challenges in SIBs

In comparison with LIBs, energy density drawbacks in commercializing SIBs can be ascribed to several key facts [14, 15, 56, 67], such as (1) larger ionic radius of Na^+(1.02 Å) as compared to 0.76 Å of Li^+ (Consequently this causes big volume change and phase instability during sodiation/disodiation); (2) larger molar mass of Na (22.990 g/mol) as compared to 6.941 g/mol of Li and (3) higher reduction potential (−2.71 V versus SHE) of Na^+/Na as compared to (E = −3.04 V versus SHE) of Li . The cost of a typical LIB having $LiFPO_4$ as cathode and commercial carbon as anode is about 0.54 RMB/Wh with an energy density of 165 Wh/kg. In order to guarantee SIBs with a price comparable to commercial LIBs, a proposed SIB with a hard-carbon anode should have an estimated energy density not exceeding 120Wh/kg [69]. On the other hand, the available cathode materials (including O_3-$NaMO_2$, $Na_2FeFe(CN)_6$, $Na_2MnFe(CN)_6$, $Na_3V_2(PO_4)_3$, $Na_3V_2(PO_4)F_3$, and so on) have shortcomings such as poor cycling, difficulty in preparation, and toxicity of constituent materials. Thus, it is highly desirable to develop electrode materials for SIBs with high working voltage, excellent cycling and rate performance, long shelf life, and high safety.

As known, an in-depth understanding of electrode properties and electrochemical evolution in the working condition of a battery is fundamental for rationally designing and optimizing electrode materials [15, 70]. Various physical and chemical changes simultaneously take place in an operational battery. In this regard, changes in a certain property may result in multiple changes in the performance of batteries. Similarly, the same performance of SIBs or other batteries may also be influenced by several structures and dynamic factors [56, 68].

For instance, complex relationships between cathode properties and the material's performance are shown in Fig. 13.2 [56, 71]. The properties of cathode materials mainly include crystal/interior structure, dynamic, and material characteristics. Taking structure evolution as an example, since most of the cathode materials for SIBs, such as layered metal oxide and polyanionic compounds, contain TM. The insertion/extraction of Na^+ upon cycling might lead to skeleton cation and anion migration and further cause cation mixing, stacking fault, and phase separation. The structure evolution will severely make an impact on mechanical properties, cycle stability, and rate performance.

Conversely, the test parameter setting also influences the structure evolution. As for Mn-based cathode materials, an inevitable topic is the existence of the Jahn–Teller effect, which can affect charge distribution and electrostatic repulsion. The element doping and cut-off voltage setting have been proposed to partially solve the problem based on a better understanding of electronic/phase structure evolution with the requirement of different characterization techniques [72].

In other cases, the performance of SIBs is dependent on morphology and particle size. Design of unique morphology such as nanotube [73], nanowire [74], nanosphere and core–shell [75, 76] electrode materials, which may benefit from electron/ionic transfer, volume change buffer, and active sites exposure at one of several aspects. As for particle size, nanometric materials may exhibit considerably different phase behavior compared to conventional electrode materials. Simultaneously, nanomaterials also mean high specific area contact with an electrolyte, which brings a series of reactions during the battery operation [77, 78]. An analysis of the preceding issues callfor different characterization techniques.

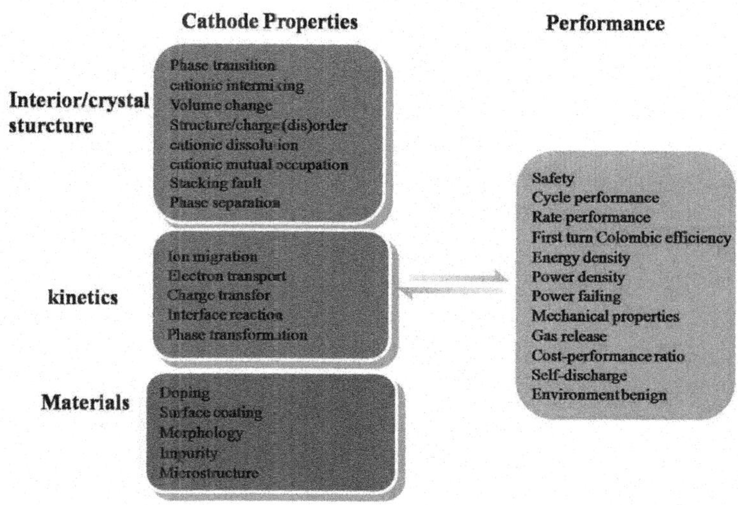

Figure 13.2 Complex relationships between cathode properties and performance [56, 71].

The interfacial and surface characteristics of components play a crucial role in determining battery performance for LIBs [78] and SIBs [79]. Such characteristics mainly correlate with electrode/electrolyte interphases, including solid electrolyte interphase (SEI) on the anode surface or cathode electrolyte interphase (CEI) on the cathode surface. The SEI/CEI originates from a chemical reaction between bulk electrode materials and electrolyte and possesses the cation penetration while electronic insulation characteristics. A stable SEI can break off the continuous decomposition of electrolyte, preventing further surface passivation, which severely consumes the cation in both electrolyte and cathode, resulting in capacity degradation. Forming a stable SEI is also vital for volume to suppress the alloy-based anode materials upon cycling [80].

In addition to SEI, CEI on the cathode surface also severely impacts the functioning of batteries. However, little attention has been given to the CEI , especially for SIBs. The CEI in SIBs contributes to the Na^+ transport kinetics, cell impedance, and capacity retention. Besides, dissolved ions from the cathode may migrate to the anode surface, hence influencing the properties of SEI [15]. Hence, a well understanding of CEI's formation mechanism is also necessary, except for conventional research mainly focusing on methods to avoid electrolyte decomposition and TM dissolution using different methods such as electrolyte addictive, doping, and surface coating.

However, SEI and CEI are usually very thin (ranging from several nanometers to a hundred nanometers) [81]. Meanwhile, the composition, thickness, morphology, and solubility of SEI/CEI often vary with the operation of the batteries [56]. Thus, probing SEI/CEI requires surface-sensitive techniques, preferably those with depth-profiling capabilities.

In a word, composition–structure–performance relationships and redox reaction mechanisms in batteries are very complex and changeable [15, 56, 81, 82]. To draw a complete picture, different sophisticated characterization techniques must be employed to detect the batteries in equilibrium and nonequilibrium states. Synchrotron-radiation-based X-ray techniques provide unique, nondestructive techniques to solve questions in batteries across a wide range of length scales (from atoms to microns to mm and cm) [81, 82]. Besides, *ex situ*/*in situ* experiments are compatible with a broad set of sample environments such as temperature, pressure, and reactive environment to mention a few, providing another advantage for battery study [83], and details will be demonstrated in the following sections.

13.2 Synchrotron-Radiation-Based X-Ray Characterizations in SIBs

With nearly 80 years of development, synchrotron radiation (SR) light source has been developed mainly with high brilliance and high energy; more details can be found in previous reviews and books [84–86]. The synchrotron radiation spectrum across a wide range of wavelengths with pertained high brightness, high flux, and high collimation, from infrared (IR) through the visible and ultraviolet range, to the soft and hard X-ray part of the electromagnetic spectrum (Fig. 13.3) [81]. In the following parts, we will briefly summarize the focus on synchrotron-based X-ray techniques that have been employed for batteries.

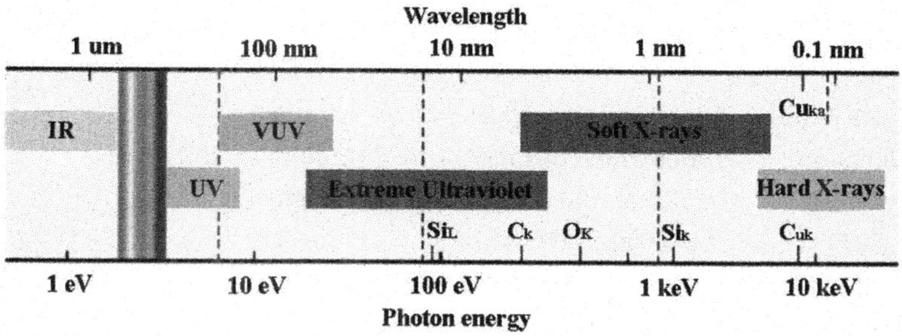

Figure 13.3 Range schematic of the synchrotron radiation spectrum. Reproduced with permission from Ref. [81]. Copyright 2017, American Chemical Society.

Many state-of-the-art X-ray techniques have been developed for various types of measurements, such as scattering [87–90], spectroscopy [91–95], and imaging techniques [96–98], as concluded in Table 13.2. With the advantage of synchrotron radiation light source, an in-depth understanding of the complicated issues is involved for studying and exploring the working mechanism of SIBs using these techniques. For example, it may take several days or even more to collect the delicate phase structure such as atomic occupation information using laboratory X-ray diffractometer, while precise information can be achieved, and time will be shortened to a few hours or even several minutes in case of using synchrotron X-ray beamlines with high brilliance and high flux [99].

Compared to conventional characterization, another superiority of synchrotron-based techniques is their capability to detect the local structure of materials without long-range ordering structure or with amorphous phases [100–102]. The metastable intermediate phase, which usually occurs in the phase transformation in batteries upon the cycling process, can be conveniently captured by *in situ* synchrotron-based techniques [103, 104].

Table 13.2 Examples of synchrotron-based X-ray scattering and spectroscopy techniques with capabilities/limitations

Technique	Capability	Limitation	Reference
X-ray diffraction (XRD)	Average structure information; phase identification; lattice parameter	Limited for amorphous or short-range structure	118–124, 126, 127, 136, 137
Pair distribution function analysis (PDF)	Structure information for long and short range; atom bond distance and coordination local ordering/disordering; useful for amorphous materials	Limited resources; require careful design of *in situ* test	129–134
Hard X-ray absorption spectroscopy (hXAS)	Oxidation states and local environment of specific elements (bond length and coordination)	Limited for light elements	152, 154–159, 161
Soft X-ray absorption spectroscopy (sXAS)	Probe K-edge of light elements and L-edge of heavy elements; surface sensitive, detect different depth with different test mode	Limited for hard elements; difficult to operate *in situ* experiment	167–171, 192, 175, 176
Soft X-ray emission spectroscopy (XES)	Electronic configurations		164, 165, 172
Resonant inelastic X-ray scattering (RIXS)	Electronic structure, element-specific bulk sensitive	Limited sources	166, 177–182
Hard X-ray photoelectron spectroscopy (HAXPES)	Elemental composition and identification of species at surfaces and interfaces		185–187, 190–195, 202

Technique	Capability	Limitation	Reference
Soft X-ray photoelectron spectroscopy (SOXPES)	Elemental composition and identification of species at surface	Risk of sample damage; difficult *in situ* cells design	234, 212
Transmission X-ray microscopy (TXM)	Morphology and structure evolution monitoring through 2D/3D imaging; element/chemistry distribution; able to apply for real batteries test.	Weak image contrast for light elements; complex *in situ* cells design	105, 215, 218, 221–234

Unlike scattering and spectroscopy techniques, the imaging method provides a visual style to study materials [15, 81]. Synchrotron-based X-ray can also be combined with imaging techniques, which can be used *in situ* while providing an intuitive way to understand the structure, morphology change, and element distribution of battery materials. Another feature of the synchrotron-based X-ray imaging technique is the possibility of 3D investigation for battery material, which is essential for analyzing inhomogeneous reaction in the electrode materials and can monitor changes such as volume change, strain, and crack formation during sodiation/disodiation [83]. When combined with spectroscopy methods, chemical information such as oxidation states variation can also be obtained. Except for the advantage of high spatial and time resolution, compared to conventional imaging gain style, synchrotron-based X-ray imaging techniques might provide a nearly nondestructive way to study battery materials [105].

As described earlier, *in situ/in operando* tests are also available using synchrotron-based techniques that provide vital information involving redox mechanism, phase transition ion diffusion, formation, breakage of chemical bonding, and surface reconstruction [106–109]. Thus, the design of suitable cells is necessary for *in situ/in operando* test to fit specific experimental techniques. The design of test cells should comply with the following rules and beyond [81, 82, 110]: (1) The cell should be easy to assemble/disassemble at synchrotron facilities and highly reproducible for electrochemical testing.

(2) Allow X-rays to reach the sample region and the detector. (3) Interference from inactive cell components to outgoing X-ray signals should be avoided, especially for X-ray detection in transmission mode. (4) Special requirements such as rotation of samples in the TXM test should also be considered during the experiment. Success cases of cells designed for synchrotron-based X-ray techniques will also be present in various parts.

13.2.1 Synchrotron-Radiation-Based X-Ray Diffraction

As a conventional characterization method, X-ray diffraction has been widely used for structure analysis of crystal or partially crystalline materials based on X-ray scattering [14]. Structural information of materials obtained from XRD is mainly extracted from a series of peaks, which are characterized by their position, intensity, and profile. We can generally extract information such as lattice parameters, space group, crystal/phase structure, and atomic occupancy correspond to XRD position, intensity, and profile, respectively [111, 113]. Except for conduct in the in-house laboratory, XRD can also be achieved using synchrotron-radiation-based X-ray scattering. Compared to laboratory XRD, synchrotron-based X-ray diffraction (SXRD) has the following advantages: (1) High brilliance, which simultaneously means large penetration depth and a large wave vector allowing data to be collected in a small angular range, such as the small-angle X-ray scattering technique (SAXS). (2) High collimation can naturally provide high angular resolution desirable for XRD measurements. In contrast, laboratory-based high angular resolution XRD generally obtained through compromising the flux severely damages data authenticity. (3) The widely tunable energy range of synchrotron light source makes XRD possible to be measured at the expense of angular resolution. (4) The high spatial and time resolution features of SR allows *in situ* experiment conduct on synchrotron facilities feasible and efficiently [81, 82, 112].

It is well known that the design of a suitable measuring instrument is fundamental for collecting precise XRD signal free from interference. Modified coin cells are a common and convenient approach through opening small holes on the shell and sealing the holes with beryllium foil, mylar sheets, and Kapton tape, allowing X-rays to pass through without evident absorption [110, 114–117]. Another design, the Swagelok-type *in situ* cell, is shown in Fig. 13.4d

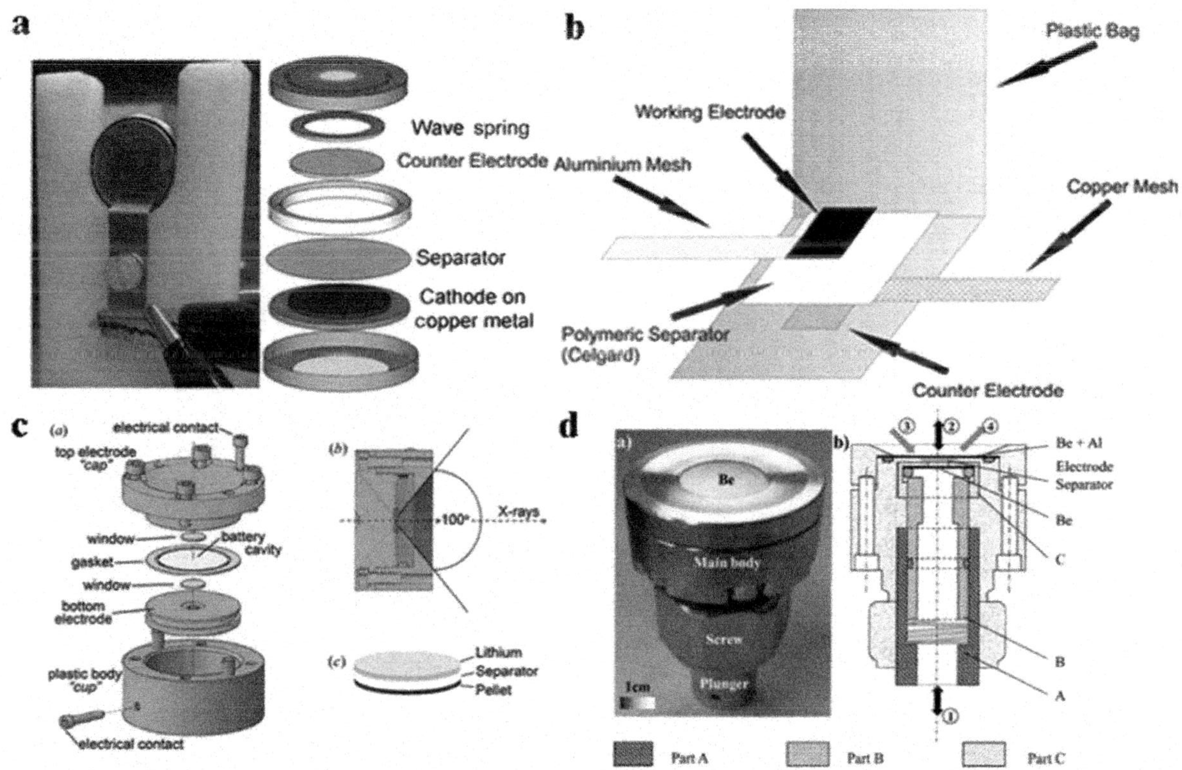

Figure 13.4 Schematic images of designing cells for *in situ* X-ray measurement. (a) Modified coin cell. Reproduced with permission from Ref. [115]. Copyright 2016, Elsevier. (b) "Coffee bag" type cell. Reproduced with permission from Ref. [117]. Copyright 2007, Wiley-Blackwell. (c) AMPIX cell. Reproduced with permission from Ref. [114]. Copyright 2012, International Union of Crystallography. (d) Reproduced with permission from Ref. [116]. Copyright 2010, Electrochemical Society.

[116]. Borkiewicz et al., at the Argonne National Laboratory, designed a multipurpose *in situ* X-ray (AMPIX) cell (Fig. 13.4b) [114], which consists of a cup-shaped body, bottom, and top two electrodes, X-ray transmission windows, and a flat annular gasket sandwiched between the electrodes. The as-designed AMPIX cell is suitably applied for SXRD, synchrotron-based PDF, and XAS tests.

13.2.1.1 Phase and structure evolution monitoring

SXRD is a well-established technique for studying the crystal structure, cation mixing, and phase transitions of electrode materials. It has been applied in phase and structure evolution monitoring in different kinds of cathode materials such as layered TM oxide, polyanion compounds, and Prussian blue analogs [118–123], via *in situ* or *ex situ* manner. Among numerous cathode materials, polyanion framework compounds have attracted extensive attention due to their high capacity and long-term cycling stability [120]. Du et al. reported a NASICON-structured $Na_2VTi(PO_4)_3$ with a high rate capacity and long cycle stability [121]. SXRD and *in situ* SX-RD were applied to study the crystal structure and phase transformation upon cycling, respectively.

The Rietveld refinement SXRD in Fig. 13.5b demonstrates the crystal structure of $Na_2VTi(PO_4)_3$ in the space group R3c. The schematic crystal structure in Fig. 13.5a represents corner-sharing $[TiO_6]/[VO_6]$ octahedral; $[PO_4]$ tetrahedral consists of the crystal structure; V^{3+} and Ti^{4+} ions randomly occupy the octahedral 12c site, while Na^+ occupy the 6b and 18e sites. Figure 13.5c represents the counterplots of *in situ* SXRD of $Na_2VTi(PO_4)_3$ at different cycle states, which illustrate the initial charge process in the region I, Na^+ extraction and formation of a new $Na_{2-x}VTi(PO_4)_3$. The analysis suggests that the extracted Na^+ mainly originates from the 18e sites, which have weak bonding energy with their neighboring atoms or polyanion. With the discharge proceeding, features suggest that a reversible phase transition occurs between 4.5 and 2.4 V and another phase transformation in region II. It is worth noting that the Na^+ (which occupied the 6b sites) disappeared and transitioned to the 18e sites, which makes $Na_2VTi(PO_4)_3$ perform a high reversibility phase transition.

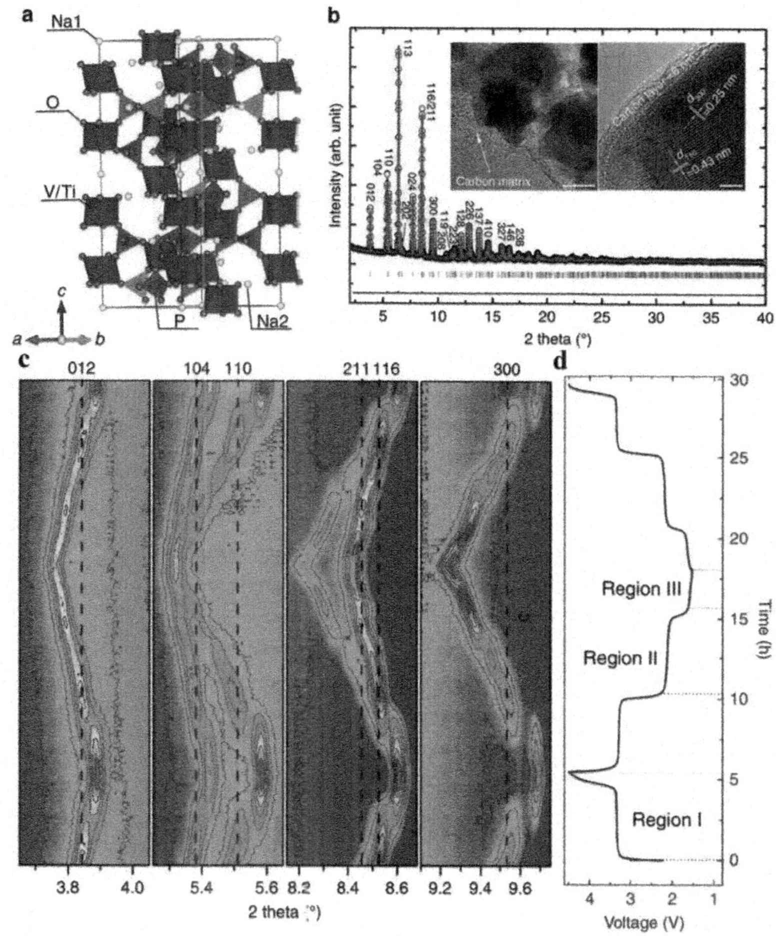

Figure 13.5 (a) Schematic crystal structure and (b) Rietveld refinement SXRD of $Na_2VTi(PO_4)_3$. (c) Contour plots of *in situ* SXRD of $Na_2VTi(PO_4)_3$ at different cycle states of the initial charge/discharge process and the charging process of the subsequence cycle. (d) Corresponding charge–discharge profiles. Reprinted with permission from Ref. [121]. Copyright 2017, Nature Publishing Group.

Cui et al. conducted *in situ* SXRD to understand and improve the poor battery performance in a past test of an organic $Na_2C_6O_6$ SIB electrode, which possesses a theoretical capacity of 501 mAh/g [124]. SXRD presented in Fig. 13.6 was collected using diethylene glycol dimethyl ether (DEGDME) as an electrolyte. In the initial discharge process, α-$Na_2C_6O_6$ transformed to γ-$Na_2C_6O_6$ at the 2.0 V plateau.

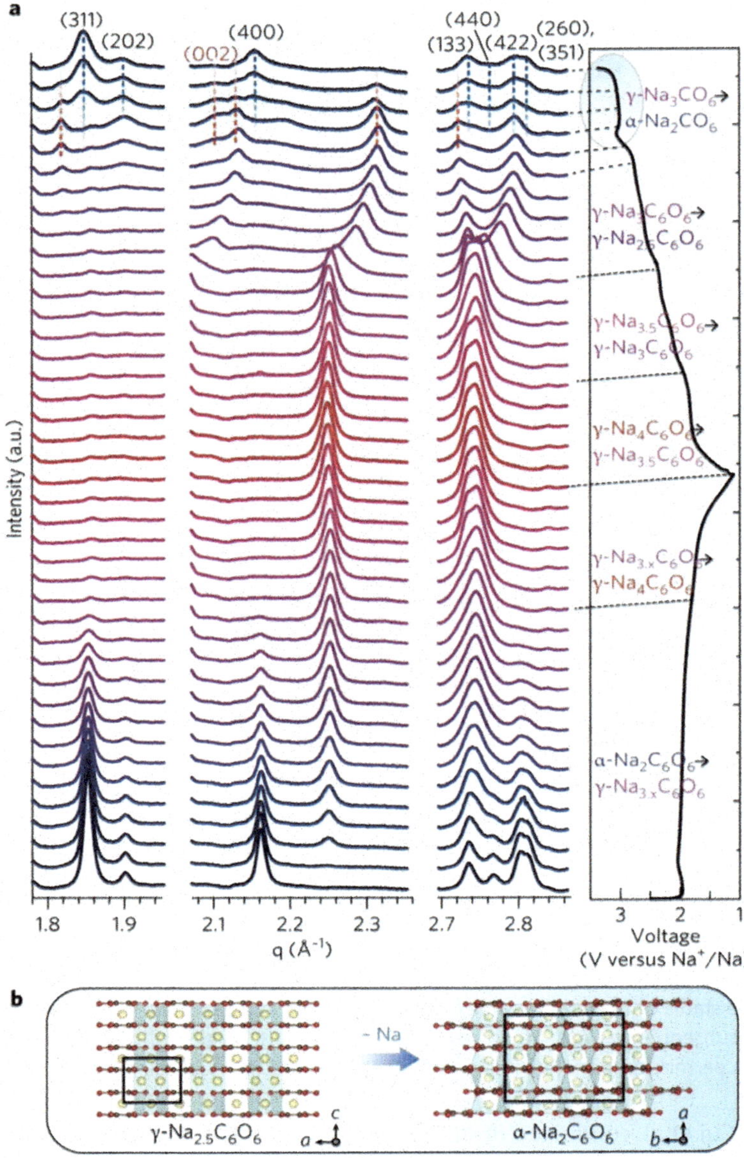

Figure 13.6 (a) *In situ* SXRD at different cycle states corresponds to the voltage profiles at the initial discharge/charge process. Crystal structure changes from γ-$Na_{2.5}C_6O_6$ to α-$Na_2C_6O_6$. Reprinted with permission from Ref. [124]. Copyright 2017, Nature Publishing Group.

The intensity of α-$Na_2C_6O_6$ decreased; meanwhile, the peak intensity of γ-$Na_2C_6O_6$ increased, and the α phase completely disappeared after full discharge. With the extraction of Na^+, peaks appeared in the q range of 2.7 to 2.8, suggesting that multiple phase transition occurred, and peaks of α phase evolved again at 2.9 V at the initial charge process.

The *in situ* SXRD evidence demonstrates the irreversibility of nano-$Na_2C_6O_6$ in DEGDME. Although the XRD results show that even if the voltage reached 3.2 V, peaks of nano-$Na_2C_6O_6$ cannot be observed in the initial charge process in propylene carbonate (PC) electrolyte. The electrochemical test and *ex situ* SEM analysis claimed a $Na_2C_6O_6$ electrode with four sodium transitions, and a reversible capacity of 484 mAh/g was produced. They suggest two ways to achieve excellent $Na_2C_6O_6$ SIB electrode namely: reducing particle size and using suitable electrolytes such as DEGDME.

13.2.1.2 Thermal stability analysis

As one of the essential energy storage system members, secondary batteries provide energy through store and release electrical power. However, the storage and release of electric power in SIBs is inevitably accompanied by heat generation, which seriously affects the safety of batteries [125]. Hence, it is vital to monitor phase transformation evolution during heating to provide valuable insights into evaluating the safety characteristics of materials from a structural point of view. Thermal stability and phase transition processes of $NaCrO_2$ and $Na_{0.5}CrO_2$ have been monitored by the means of high-temperature synchrotron X-ray diffraction (HT-SXRD) method, as shown in Fig. 13.7.

The HT-SXRD patterns of $NaCrO_2$ conducted in a temperature range of 27°C to 527°C are represented in Fig. 13.9b,c, demonstrating no evidence of big change in the main crystal phase of $NaCrO_2$, which can be concluded from the XRD peak shift to the lower angular range. The HT-SXRD patterns of $Na_{0.5}CrO_2$ in the temperature range of 27–573°C are shown in Fig. 13.7d, implying obvious phase changes of $Na_{0.5}CrO_2$ that can be described as follows: (1) Monoclinic turned into rhombohedral at 207°C due to the loss of Na/vacancy order; (2) phase segregation to two different P3-Na-rich and Na-free P3-layered phases; (3) oxygen loss resulted in the formation of corundum-type Cr_2O_3 and O3-type $NaCrO_2$ phases above 500°C. They demonstrated that the thermal stability of $NaCrO_2$ originates from the superior chemical stability of Cr^{4+} in CrO_2 slabs at octahedral sites, and the O3-type $NaCrO_2$ is treated as safety electrode materials for SIBs.

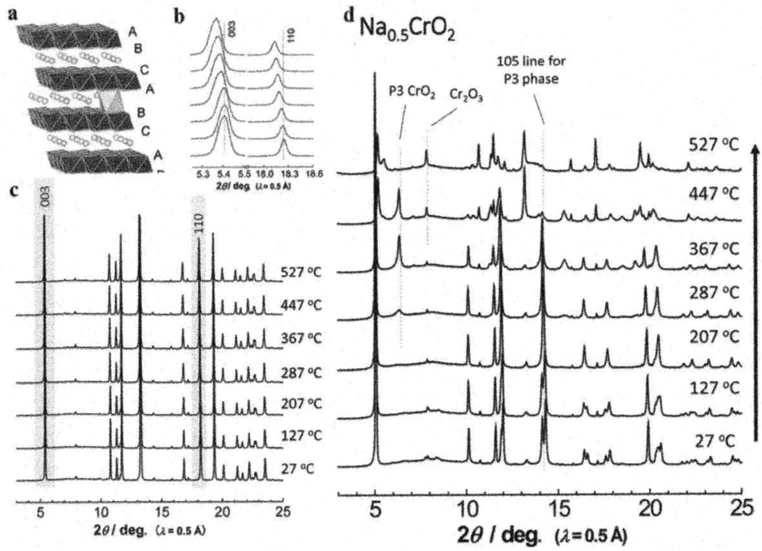

Figure 13.7 (a) Schematic illustration of O3-type $NaCrO_2$. (b, c) HT-SXRD and magnified patterns of $NaCrO_2$ conducted in a temperature range of 27°C to 527°C. (d) HT-SXRD patterns of $Na_{0.5}CrO_2$ were conducted in a temperature range of 27°C to 573°C. Reprinted with permission from Ref. [126]. Copyright 2016, American Chemical Society.

Delmas et al. have also used HT-SXRD to unveil the structural evolution of the O′3-$Na_{1/2}VO_2$ phase as a function of temperature between 25 and 270°C [127]. The HT-SXRD patterns in Fig. 13.8 at different temperatures clearly show the irreversible phase transition of O′3 to P′3 during the first heating to 145°C. On subsequent additional *in situ* heating of the same powder sample to 230°C, a single homogeneous phase diffraction pattern P′3 was observed. With the continued heating of the sample up to 270°C, many spilt peaks merged, indicating that at higher temperatures, the increased mobility of Na^+ enables them to impede the Na ordering between VO_2 layers resulting in the structure of $Na_{\sim 1/2}VO_2$ transformed to a more symmetric structure. This contributes to the structural rearrangements and a transformation into a more thermodynamically stable compound at a higher temperature to induce defects due to the produced process by chemical or electrochemical alkali extraction.

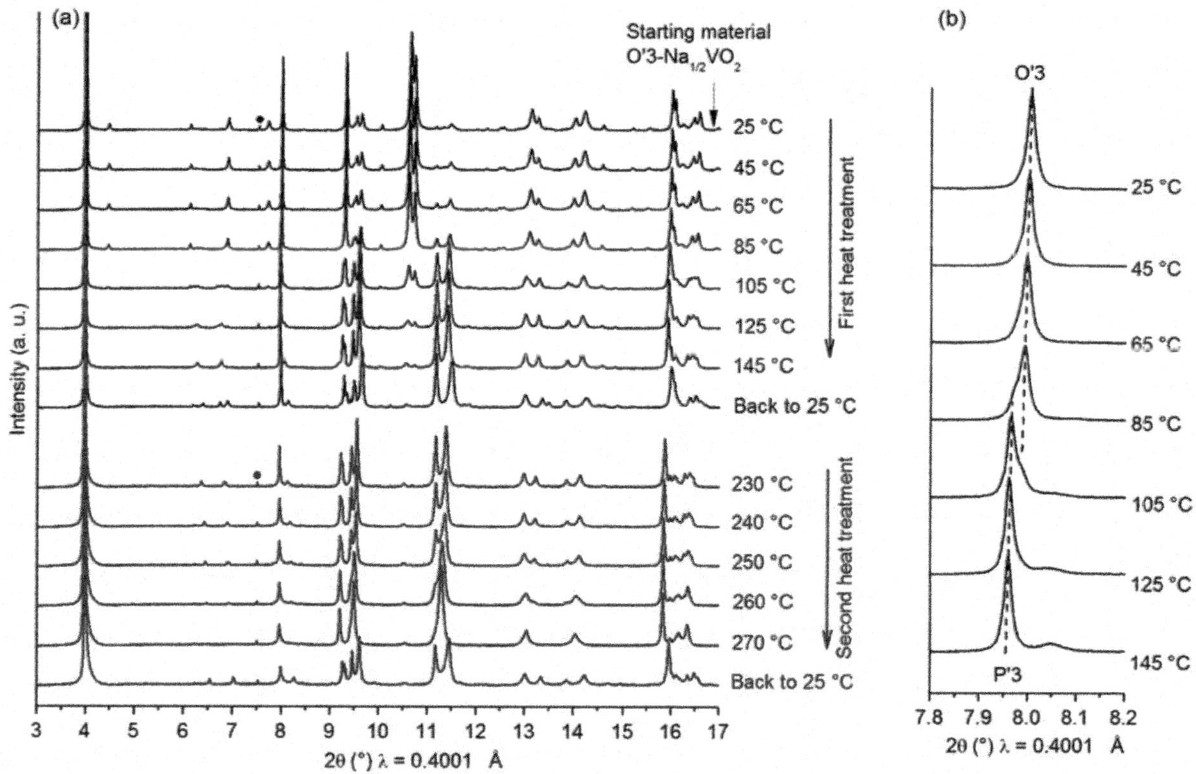

Figure 13.8 (a) *In situ* HT-SXRD patterns start from O'3-$Na_{1/2}VO_2$ during the first heat treatment (in black) and the second heat treatment (in blue). (b) Magnified diffraction patterns recorded during the first heat treatment. Reproduced with permission from Ref. [127]. Copyright 2015, American Chemical Society.

13.2.1.3 Pair distribution function

The pair distribution function (PDF) is also a scattering technique gaining extensive application in recent years, which takes into account both the Bragg peaks and the diffuse scattering components of neutron or X-ray data [15, 132]. The ability to detect sample's local structure is quite similar to the EXAFS analysis technique. It is worth noting that the distance range detectable by PDF can extend to several nanometers. In general, PDF data are collected by detecting the scattering intensity as a function of momentum transfer (Q). In comparison to laboratory X-ray source, synchrotron-based X-ray with short wavelength allows PDF data to be collected in a much smaller angular range, with a large Q_{max}, which leads to a higher spatial resolution [128]. Unlike XRD, which only provides average crystal structure by the Rietveld method, PDF provides local structure information on the distance between atoms, along with the possibility of finding two atoms at a certain distance. That also means that samples for PDF tests do not need to have a long-range order, specifically crucial for the study of nanocrystalline or amorphous battery materials [81, 130, 131, 134]. With the availability of synchrotron-based high-energy X-ray sources, the synchrotron-based PDF method coupled with computational modeling and simulation has been approved as a powerful tool to understand in depth the atomic structures of battery materials. For example, Delmas et al. probed the vanadium clustering/de-clustering phenomenon in P2-$Na_{1/2}VO_2$ combining the use of SXRD and synchrotron-based PDF at 300 and 350 K [129]. The reduced PDFs, G(r), for P2-$Na_{1/2}VO_2$ at 300 and 350 K are shown in Fig. 13.9b. It demonstrates that the main structural changes between 300 and 350 K are related to the vanadium–vanadium correlations. With further analysis they suggested that the triangular lattice formed by vanadium ions could be distorted by the formation of clusters of pseudotrimers, which resulted in an enhanced charge transport and electronic conductivity of P2-$Na_{1/2}VO_2$.

Olivine-$NaFePO_4$ has also awaken the interest of researchers due to the versatile application of its analog $LiFePO_4$. Xiang et al. found that the phase crystalline fraction of $NaFePO_4$ dropped upon cycling with the refinement of the SXRD. Synchrotron-based PDF was employed to elucidate this phenomenon [133]. The PDF results in Fig. 13.9e, f clearly show an amorphous phase between Na_xFePO_4 and $FePO_4$. They also suggest that the intermediate amorphous phase formation can efficiently buffer the large lattice mismatch between primary phases in olivine $NaFePO_4$ as a strain-accommodation mechanism.

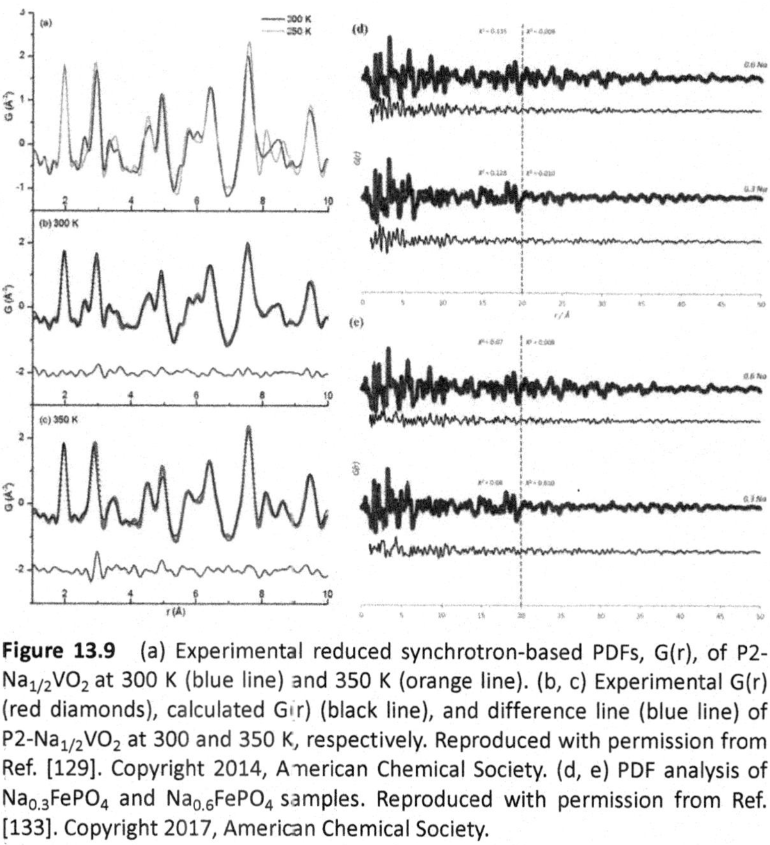

Figure 13.9 (a) Experimental reduced synchrotron-based PDFs, G(r), of P2-$Na_{1/2}VO_2$ at 300 K (blue line) and 350 K (orange line). (b, c) Experimental G(r) (red diamonds), calculated G(r) (black line), and difference line (blue line) of P2-$Na_{1/2}VO_2$ at 300 and 350 K, respectively. Reproduced with permission from Ref. [129]. Copyright 2014, American Chemical Society. (d, e) PDF analysis of $Na_{0.3}FePO_4$ and $Na_{0.6}FePO_4$ samples. Reproduced with permission from Ref. [133]. Copyright 2017, American Chemical Society.

13.2.1.4 Joint use of SXRD and neutron diffraction

It is well known that X-rays interact with electrons in materials and the intensity of scattered X-rays from a material is proportional to the atomic number of the material. Neutron scattering activity is derived from nuclear interaction between neutron and nucleus [81, 135]. The scattering power of X-rays and neutrons of an atom is generally described by $\sigma=4\pi b^2$, where b represents the scattering length. X-ray and thermal neutron scattering lengths of some selected elements are shown in Fig. 13.10, from which we can conclude that neutron scattering is very powerful in studying low-Z elements and can distinguish nearby elements [112]. Therefore, Neutron diffraction provides complementary information of XRD.

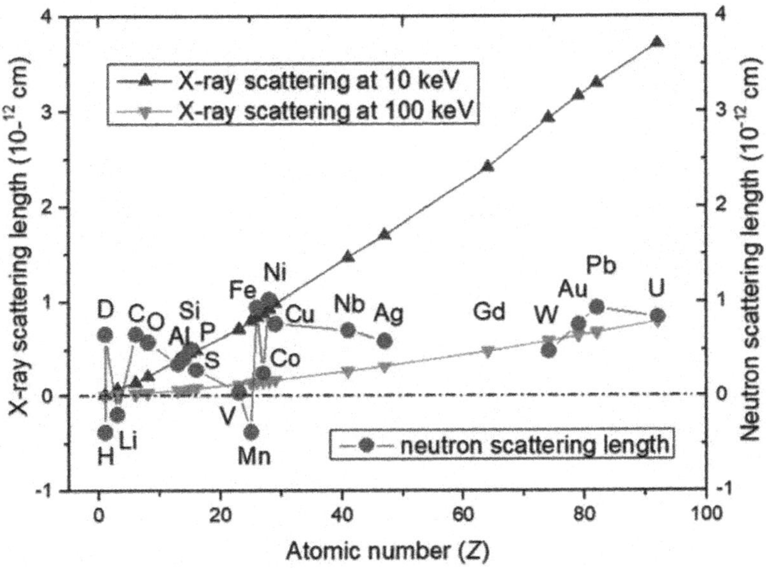

Figure 13.10 X-ray and neutron scattering length of selected elements. Reproduced with permission from Ref. [112]. Copyright 2018, WILEY-VCH.

To understand the relationship between structure and electrochemical activity of a promising SIB cathode $Na_{2/3}(Fe_{1/2}Mn_{1/2})O_2$, Guo et al. employed SXRD and neutron powder diffraction (NPD) to characterize the as-prepared samples [137]. Both SXRD and NPD in Fig. 13.11 demonstrate that the NFMO has a hexagonal $P6_3/mmc$ space group. Interestingly, crystallographic details of the NFMO obtained from SXRD and NPD represent different lattice parameters, with NPD (yielding a larger *a* and a smaller *c* parameter) and dissimilar sodium occupancies with more extensive Na content determined from the NPD data approaching the expected value. They attributed the result difference to the existence of moisture, which leads to the decomposition of $Na_{2/3}(Fe_{1/2}Mn_{1/2})O_2$ when samples exposed in the air during the SXRD measurement. They suggest that the NPD data are more advantageous in understanding the structure than the XRPD data.

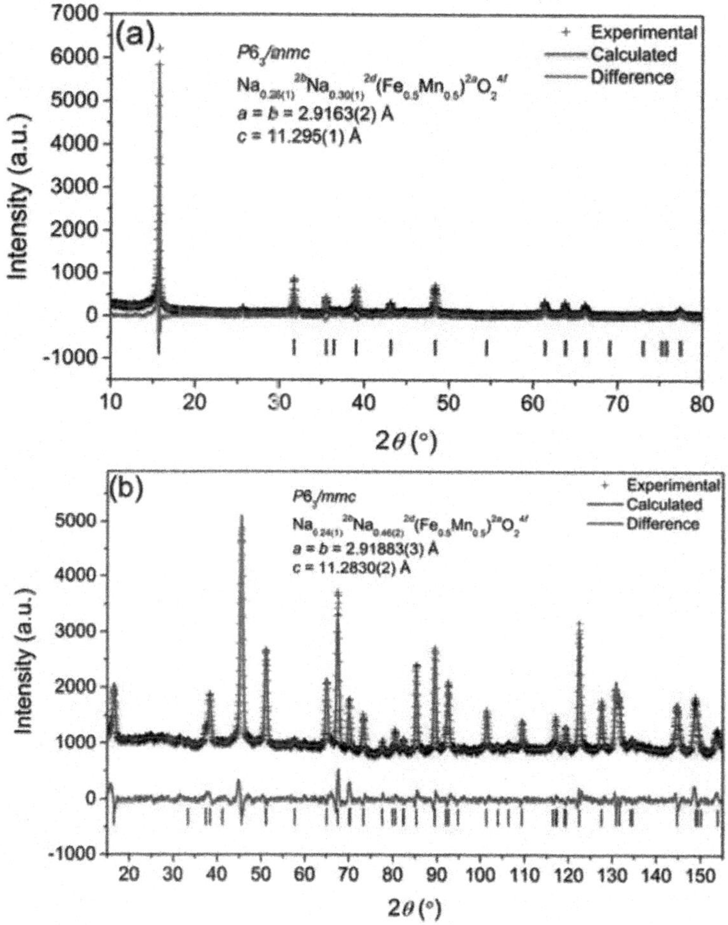

Figure 13.11 Rietveld refinement profiles for the as-prepared $Na_{2/3}(Fe_{1/2}Mn_{1/2})O_2$ (NFMO) powder using (a) SXRD and (b) NPD data. Reproduced with permission from Ref. [137]. Copyright 2015, American Chemical Society.

Grey et al. have synthesized P2-$Na_{2/3}Mn_{1-y}Mg_yO_2$ SIB cathode materials with high reversible capacity and rate capability compared with P2-$Na_{2/3}MnO_2$ [136]. *Ex situ* synchrotron XRD and neutron diffraction were applied to investigate in depth the effect of Mg doping on the electrochemical performance and structural stability of SIB cathodes. As presented in Fig. 13.12, after fully charged,

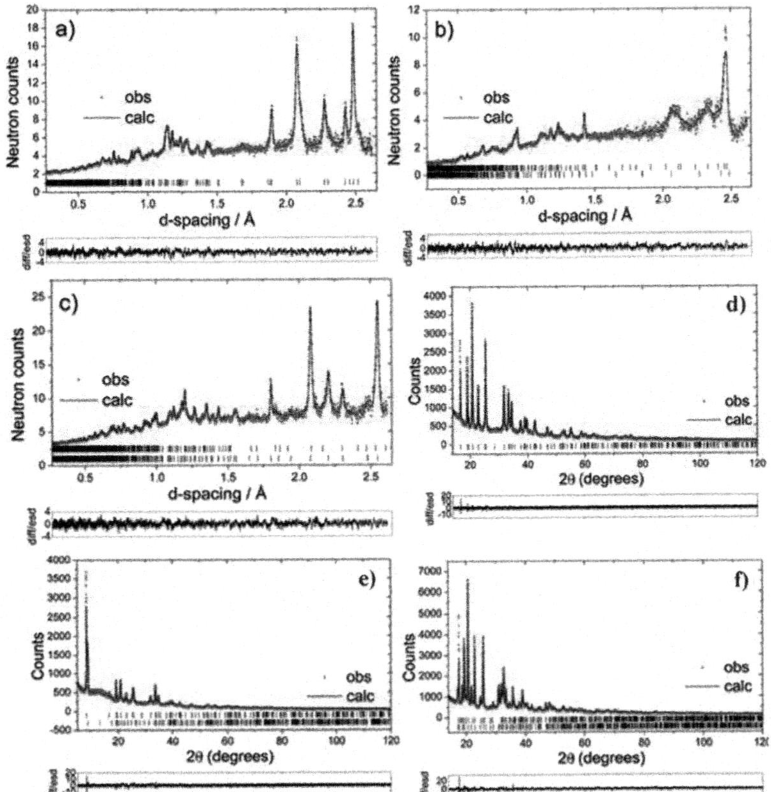

Figure 13.12 Fitted powder neutron diffraction (NPD) patterns for $Na_{0.67}Mn_{0.95}Mg_{0.05}O_2$ (a) before the end-of-charge voltage plateau, (b) after the end-of-charge voltage plateau, and (c) after fully discharged. Fitted SXRD for $Na_{0.67}Mn_{0.9}Mg_{0.1}O_2$. (d) Before the end-of-charge voltage plateau. (e) After the end-of-charge voltage plateau. (f) After fully discharged. Reproduced with permission from Ref. [136]. Copyright 2016, The Royal Society of Chemistry.

the Mg-doped materials do not fully transform to the OP4 phase, which comprises a primary P2 phase and a minor disordered OP4 phase with the ratio of 60:40 and 65:35 for $Na_{0.67}Mn_{0.95}Mg_{0.05}O_2$ and $Na_{0.67}Mn_{0.9}Mg_{0.1}O_2$, respectively. In the initial discharge process, the OP4 phase disappeared, an ideal P2 structure and a P2'orthorhombic phase were formed during Na reinsertion for $Na_{0.67}Mn_{0.95}Mg_{0.05}O_2$. However, for the $Na_{0.67}Mn_{0.9}Mg_{0.1}O_2$, the phase remains orthorhombic until at the final stage of discharge. After full discharge, the Mg-doped materials exhibit a two-phase region at high Na content (x>0.9); two phases exist in $Na_{0.67}Mn_{0.95}Mg_{0.05}O_2$ and $Na_{0.67}Mn_{0.9}Mg_{0.1}O_2$ with Na stoichiometry of 1, and a minor phase with a lower Na having varying content of 78% to 22% and 88% to 22%, respectively. In summary, the doping of Mg suppresses the Jahn–Teller effect of Mn^{3+}. Besides, Mg substitution also increases the voltage range and stability due to the P2 phase. The occurrence of oxygen layer glides was delayed, leading to the formation of an OP4 phase. The structure, especially the compounds doped with 5% Mg, obtained an excellent cycle and rate performance.

13.2.2 Synchrotron-Radiation-Based X-Ray Absorption Spectroscopy

It is well known that when a beam of X-ray travels through a sample, the intensity of the X-ray attenuates due to the interaction between the specimen and the X-ray. The relationship between the incoming X-ray intensity (I_0), outgoing X-ray intensity (I), and absorption coefficient can be described through the Bouguer–Lambert–Beer law [81, 140]:

$$I = I_0 \exp(-\mu x)$$

When the incident X-ray photon energy is high enough to excite core electrons from the probing element, the linear relationship between scanning energy and absorption will be interrupted. A sharp increase in absorption can be identified as the absorption edge. Corresponding to the absorption edge, there exists a fine structure of the absorption coefficient, namely, fine X-ray absorption. X-ray absorption spectroscopy (XAS) is a powerful technique through which we can measure the X-ray absorption coefficient of a certain element in samples as a function of the X-ray photon energy for studying the structure of materials and the oxidation state of certain elements in given materials [82, 138, 139, 141]. Based on the energy level of probing X-ray sources, XAS can roughly be classified as soft XAS corresponding to energy below 2 KeV and hard XAS starting from about 5 KeV, respectively [81, 142].

According to the multiple and single scattering effects between coordination atoms, absorption atom, and excited photoelectrons, a full XAS spectrum can be divided into two regions: The X-ray absorption near-edge structure (XANES) refers to energy ranges from the absorption edge to 50 eV above the edge, and the extended X-ray absorption fine structure (EXAFS) typically lies in the range of 50 eV above the edge, to as high as 1000 eV [143]. The XANES section consists of pre-edge, edge, and XANES, which can provide information, including oxidation state, local geometric and electronic structure of the probed atoms. In the EXAFS parts, the kinetic energy of excited photoelectrons is determined by single back-scattering due to the low-amplitude photoelectron scattering. Therefore, EXAFS typically offers plane local geometries such as bond length, coordination number around the absorption atom. Compared with XRD, which is only effective for materials with long-range order, the XAS technique, however, does not need to consider the condition of the test samples. That is to say, the test sample can be liquid, solid, gas, or even amorphous materials. As an essential characterization technique, XAS possesses a great advantage since its elemental selectivity is particularly important when probing materials containing different elements with a similar X-ray scattering cross section [144]. The incredible power of XAS generally plays a crucial role in understanding the charge compensation process in electrode materials, which may possess multiple active elements [81]. Besides, XAS can also be collected upon battery cycling and provide information about the structural changes and kinetic information of electrode materials [145, 146]. Overall, XAS plays a key role in discovering and designing novel electrode materials.

To avoid the interference of false signals during XAS measurement, it is vital to setup the XAS detect mode [81, 82, 147]. To date, the existing detection method includes the transmission mode, fluorescence detection mode, and total electronic yield model (TEY) [148]. The transmission mode is the most widely used test method, which acquires the absorbing coefficient by testing the incident's intensity and outgoing X-ray. In the fluorescence detection mode, the absorption coefficient is obtained via probing the intensity of the as-produced fluorescence and ingoing the X-ray. While in the TEY mode, the XAFS signal is detected through the Auger electron and secondary electron. Generally, the choice of the test mode depends on the sample thickness and the probed element's concentration in the sample. Generally, if the test sample is thin enough and has sufficient elements to be probed, the transmission mode is preferred. Otherwise, the fluorescence detection mode is chosen. The TEY mode is usually applied for the XAS signal collection of thin-film materials or surface structure information acquisition.

13.2.2.1 Hard X-ray absorption spectroscopy (hXAS)

13.2.2.1.1 Local and electronic structure analysis

In light of XAS with high sensitivity for probing electronic structure and local geometries of materials, these involve two parts of XANES and EXAFS. Thus, it has been widely applied to probe the relationships between electronic structure, local environment, and electrode materials [149–151, 173, 174]. Huang et al. have synthesized an Fe/Mn mixed alluaudite $Na_2Fe_{3-x}Mn_x(PO_4)_3$ microcompounds and used them as cathode materials for SIBs [152]. To illustrate the influence of the local structure and electronic structure on electrochemical performance, they employed sXAS and hXAS combined with Mössbauer spectroscopy The EXAFS spectra at the K-edges of Fe and Mn show the local geometry around the absorbing atoms (Fig. 13.13). The difference in peak position and intensity in Fig. 13.13a,b illustrates that the Mn concentration directly influences the bond length and degree of disorder. However, when the Mn/Fe ratio increases, the local structures turn to be more stable. The Mn/Feratio's variation and influence on Fe and Mn electronic state were also analyzed by carrying out XANES and soft XAS measurements. The XANES in Fig. 13.13 showing the exact pre-edge feature located before the white line for both Fe and Mn, which originates from the electronic excitation of the 1s orbital to the 3d orbital, are sensitive to symmetry variations. An increase in both the Mn and Fe pre-edge intensities indicates the distortion of MO_6 (M=Fe or Mn) octahedral. On the other hand, the Fe K-edge moves to high energy demonstrating an increase in the Fe oxidation state. While the Mn K-edge shows no visible shifts, the Fe and Mn oxidation states are in accord with the Mössbauer spectra results. The analysis of Mn L-edges XAS spectra, shown in Fig. 13 13f, suggests that the average valence state of Fe in the MO_6 octahedral can be regarded as between Fe^{2+} and Fe^{3+}, and the Mn^{2+} is the main component of the as-prepared samples. To further demonstrate the relationship between electronic and local structure and electrochemical performance, the samples prepared earlier have been tested as cathode materials for SIBs. The results in Fig. 13.13 represent the difference in the cycle performance of samples as a function of Mn concentration. This work has proved that local and long-range structures have a significant impact on electrochemical performance.

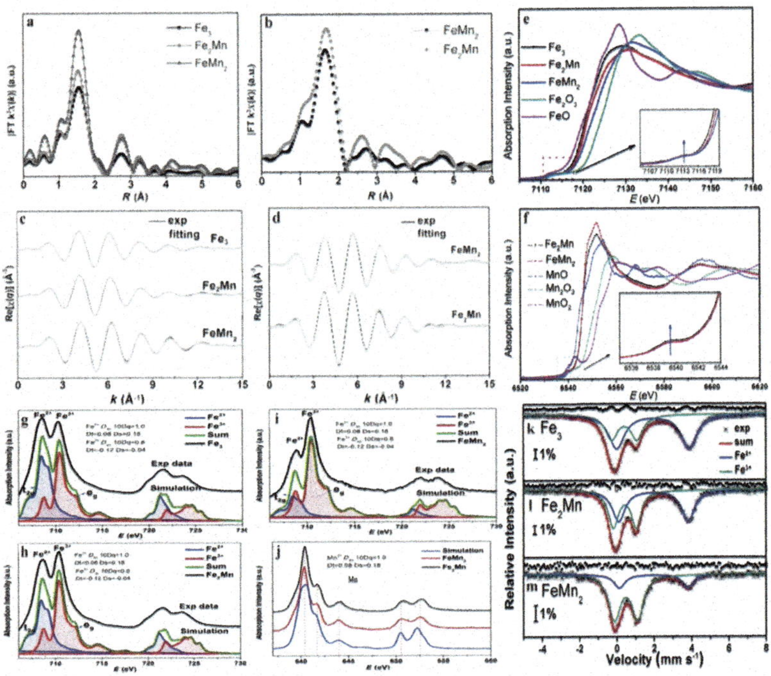

Figure 13.13 (a–d) Fourier-transformed, k^3-weighted EXAFS spectra of alluaudite compounds with different Fe/Mn ratios: (a) for Fe and (b) for Mn K-edges. Comparison between experimental and theoretical EXAFS signals of $Na_2Fe_{3-x}Mn_x(PO_4)_3$ samples at (c) Fe and (d) Mn K-edges. (e, f) XANES spectra of three isostructural alluaudite compounds at (e) Fe K-edge and (f) Mn K-edge. Insets: magnification of the pre-edge features. (f) Mn L-edge sXAS spectra of alluaudite$Na_2Fe_{3-x}Mn_x(PO_4)_3$ samples with different Mn dopants; Fe L-edges XAS software on (g) $Na_2Fe_3(PO_4)_3$, (h) $Na_2Fe_2Mn(PO_4)_3$, and (i) $Na_2FeMn_2(PO_4)_3$. Mössbauer spectra of alluaudite $Na_2Fe_{3-x}Mn_x(PO_4)_3$ samples with different Mn contents (x) at room temperature for x=0 (j), 1 (k), and 2(l). Observed and calculated patterns are presented as black circles and lines, respectively. (m) Cycle performance at 5 mA/g of different alluaudite samples with Fe/Mn ratios. Reproduced with permission from Ref. [152]. Copyright 2015, Wiley-VCH.

Manganese-based layered oxides have received extensive attention as cathode materials due to their high theoretical specific capacity. However, the Jahn–Teller distortion arising from the Mn^{3+}

centers destabilizes the host structure and severely deteriorates the cycling life, usually occurring in the Mn-based layer cathode materials. To suppress the Jahn–Teller effect, Kang et al. have reported a Zn-doped strategy for P2-$Na_{0.833}[Li_{0.25}Mn_{0.75}]O_2$(NLMO)-layered oxide material, which shows superior cycling performance [153]. *Ex situ* EXAFS and XANES were applied to verify the inhibitory effect of the Jahn–Teller distortion with the contrast of the Mn K-edge of NLMO and NLMO-Zn samples at different cycle states (Fig. 13.14). The magnified pre-edge region presented in Fig. 13.14a clearly shows two peaks that look similar for both pristine NLMO and NLMO-Zn. After full discharge to 1.5 V, these two peaks of the two samples shift to low energies. A notable contrast is that both the pre-edges of NLMO shift to lower energies than the pre-edges of NLMO-Zn, which implies that the Mn in NLMO has a lower average oxidation state closer to the Jahn–Teller active Mn^{3+}. This demonstrates that the NLMO may suffer from severe Jahn–Teller distortion. Fourier-transformed (FT) magnitude plots of the Mn K-edge EXAFS spectra shown in Fig. 13.14b offer further certification. For NLMO in the fully discharged state, an additional peak at 0.98 Å (circled in dashed lines) in the fully discharged state for NLMO resulting from Jahn–Teller distortion can be resolved, whereas the peak was not observed in NLMO-Zn in the fully charged and discharged states. Evidence illustrates the suppression of Jahn–Teller distortion by the doped Zn. Thus, they suggested that Zn doping may be an effective and broadly applicable strategy for Mn-based cathode materials.

Song et al. conducted another case of using hXAS to analyze the local structure. As we all know, conversion-reaction-based anode material possesses high specific capacities and has been thoroughly studied. However, those materials generally suffer from volume expansion, which results in low columbic efficiency and poor cycle performance. To solve this problem, Song et al. employed a simple strategy to produce an intercalation pseudocapacitive-type layered vanadium nitride (VN) by one-step ammonification of V_2C MXene [154]. The prepared sample delivered remarkably long cycling stability over 7500 cycles without capacity attenuation at 500 mA/g. XANES spectra were applied to verify the local structure of layered VN, as shown in Fig. 13.14c,d. The Fourier-transformed V K-edge EXAFS spectra of R-space revealed the increase in V–N and V–V bond length of VN compared with V_2C MXene. The wavelet transforms

(WTs) of V_2C and VN were applied to present the V K-edge EXAFS oscillations. The results demonstrated fewer Al atoms bonding with V of VN than V_2C MXene, which plays a key role for the structural stability during charge and discharge processes.

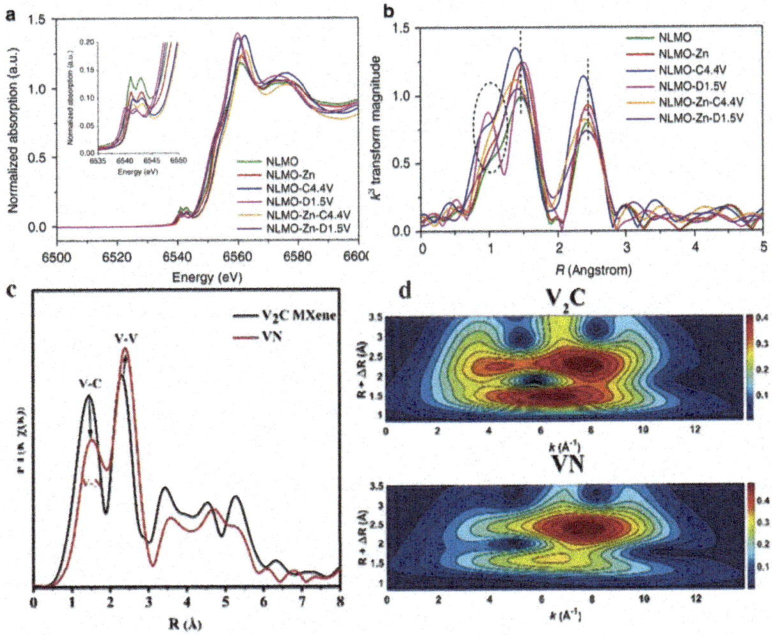

Figure 13.14 (a, b) Mn K-edge XANES spectra for $Na_{0.833}[Li_{0.25}Mn_{0.75}]O_2$(NLMO) and Zn-doped NLMO (NLMO-Zn) before cycling as well as in the fully charged and discharged states. (b) Fourier-transformed (FT) magnitude of Mn K-edge EXAFS spectra for NLMO and NLMO-Zn before cycling as well as in the fully charged and discharged states. Reproduced with permission from Ref. [153]. Copyright 2019, Nature Publishing Group. (c) Fourier-transformed V K-edge EXAFS spectra for V_2C and VN. (d) Wavelet transforms (WTs) of V_2C and VN, respectively. Reproduced with permission from Ref. [154]. Copyright 2020, Wiley-VCH.

13.2.2.1.2 Monitoring charge compensation

Element selectivity feature is one of the unique advantages of XAS, which is especially crucial for electrode materials containing multiple elements to distinguish which element acts as the redox center [155, 156]. Komaba et al. employed hXAS to analyze the charge compensation of the P2-$Na_x[Fe_{1/2}Mn_{1/2}]O_2$ system [157]. Figure 13.15a clearly shows that the Mn K-edge shifts to high energy

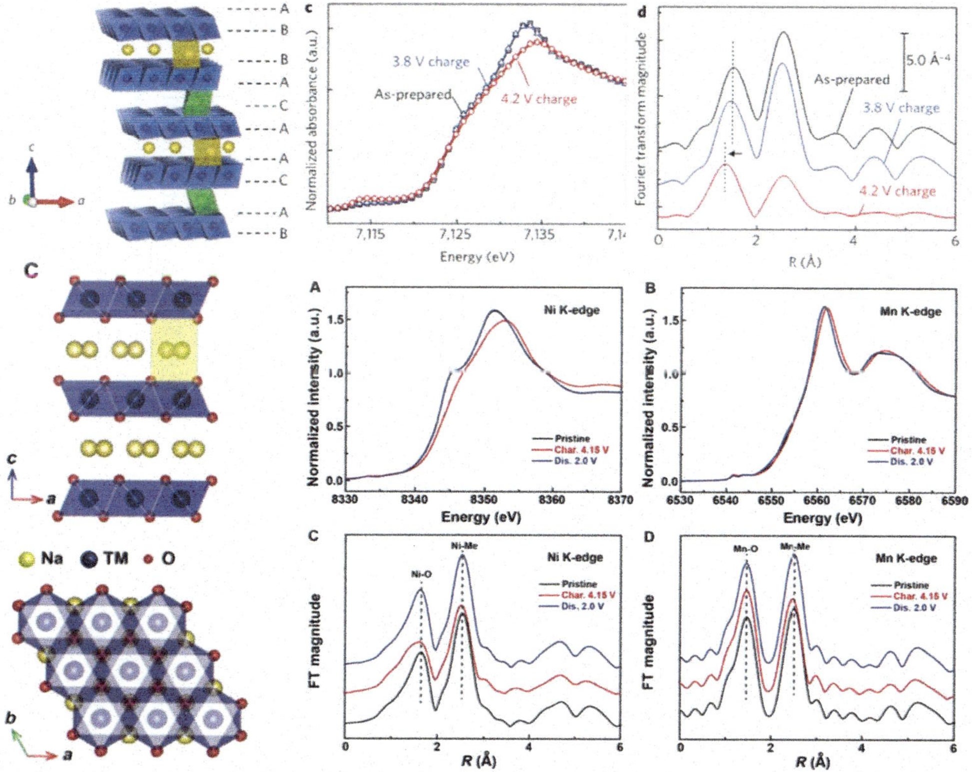

Figure 13.15 (a) Schematic illustration of P2-$Na_x[Fe_{1/2}Mn_{1/2}]O_2$. (b, c) Mn and Ni K-edge XANES and Fourier-transformed EXAFS spectra of the P2-$Na_x[Fe_{1/2}Mn_{1/2}]O_2$ samples at cut-off voltages of 3.8 V and 4.2 V, respectively. Reprinted with permission from Ref. [157]. Copyright 2012, Springer Nature. (d) Schematic illustration of *ex situ* Ni K-edge (e) and Mn K-edge (f) XANES spectra of P2-NaNMT electrodes collected at different charge/discharge states. Fourier-transformed EXAFS spectra at the (g) Ni K-edge and (h) Mn K-edge of Na P2-NaNMT electrodes collected at different charge/discharge states. Reprinted with permission from Ref. [158]. Copyright 2019, American Association for the Advancement of Science.

after charging to 3.8 V, demonstrating that the charge compensation contributes to the oxidation of Mn. However, the Fe K-edge shift can be found when the electrode was charged over 4.2 V. To illustrate this phenomenon, the Fourier-transformed (FT) magnitude of Fe K-edge EXAFS was further analyzed, as presented in Fig. 13.16. The Fe–O local environment is markedly changed after oxidation to 4.2 V, where the intensity of the peaks for both the first and second coordination shells, Fe–O and Fe–Fe (Mn), is reduced, indicating the distortion around Fe and also that the bond length of Fe–O is reduced after oxidation to 4.2 V. This change will incur noncooperative Jahn–Teller effect on the oxidized high-spin $Fe^{4+}(t^3_{2g}e^1_g)$.

Another application of hXAS to investigate the redox center was conducted by Guo et al. [158]. They designed an SIB cathode material by Ti partial substituting Mn in P2-$Na_{2/3}Ni_{1/3}Mn_{2/3}O_2$ (NaNM). The as-prepared P2-$Na_{2/3}Ni_{1/3}Mn_{1/3}Ti_{1/3}O_2$ (NaNMT) with an Na^+/vacancy disordering structure displayed satisfactory battery performance. *Ex situ* hXAS (Fig. 13.15) was performed to elucidate the charge compensation mechanism of NaNMT upon sodiation/desodiation. The Ni K-edge spectra shifted toward the higher-energy region after charging to 4.15 V. In contrast, the Mn K-edge in Fig. 13.15 shows no evident changes, indicating that the tetravalent manganese is electrochemically inactive and Ni^{2+}/Ni^{3+} acts as the redox center. An analysis of the Fourier-transformed Ni and Mn K-edges spectra further demonstrated the conclusion, which is consistent with the *ex situ* XANES results.

Lin et al. have applied the hXAS technique to ascertain the charger compensation of the layered oxide $Na_{0.75}Co_{0.125}Cu_{0.125}Fe_{0.125}Ni_{0.125}Mn_{0.5}O_2$ (CCFNM) with complex chemical and intergrowth of ordered P2 and P3 phases [159]. The EXAFS spectra at different cycle states in Fig. 13.16 show that the Ni K-edge shifts to a higher energy upon charging, which reversibly shifts back to its initial energy position upon discharging, demonstrating that Ni showed reversible redox behavior during charge and discharge. The Mn K-edge shows an unapparent shift during the first charge, while the edge energy shifts to the low energy upon discharging. Co also seemed to be redox active through the analysis of the pre-edge. As for Cu and Fe, their K-edge showed minimal energy shift. However, it should be noted that the Cu K-edge has more intensity of pre-edge peak, which is associated with the local structural distortion and with the reduction in symmetry surrounding a TM ion from centrosymmetric to non-centrosymmetric. Fourier-transformed EXAFS was also employed to study the local structural evolution of CCFNM at different charge/discharge states (Fig. 13.16). The Ni–O bond length showed a reversible evolution, whereas the Co–O, Mn–O, and Fe–O bond lengths remained mostly unchanged upon cycling. The hXAS results suggested that Ni, Co, and Mn are redox active and contributed to the charge compensation. This work also highlighted the importance of decoupling the contribution of each TM to performance degradation.

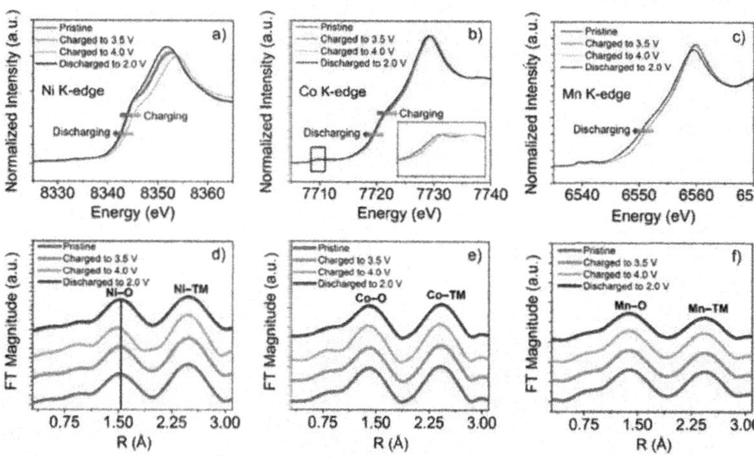

Figure 13.16 (a–c) Ni, Co, and Mn K-edge XANES spectra at different charge/discharge states upon cycling. (d–f) Fourier-transformed EXAFS of Ni (d), Co (e), and Mn (f) at different states during cycling. Reprinted with permission from Ref. [159]. Copyright 2019, American Chemical Society.

13.2.2.1.3 *Dynamic study by using* in situ/in operando *hXAS*

The time-resolved XAS instrument reduces the experimental time scale to as short as milliseconds, making it possible to probe the dynamic nature of battery materials in a fast-transient process [144, 145]. Cells used for *in situ* hXAS measurements are similar to *in situ* XRD. Since hXAS is generally obtained through transmission mode, it requires two X-ray transparent windows at symmetric positions at the top and bottom of the cell. For example, two types of *in situ* cells with Kapton windows on both sides of the modified coin cell and pouch cell, which allows X-rays to penetrate through the whole sample without interrupting the signal noises, are illustrated in Fig. 13.17 [83].

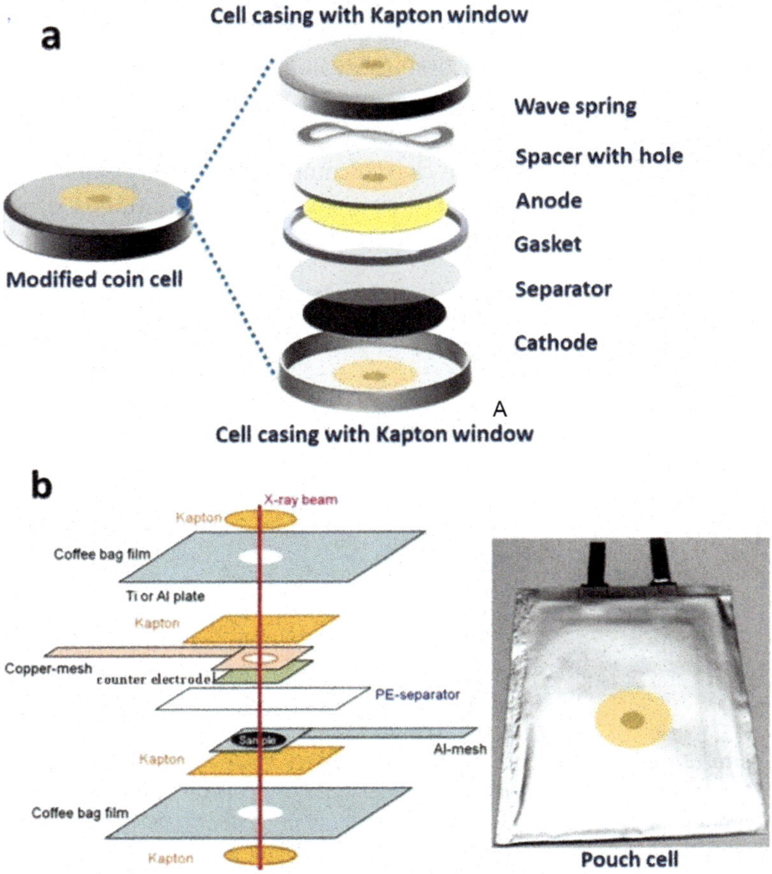

Figure 13.17 (a) Schematic illustration of a modified coin cell and (b) pouch cell for *in situ* hXAS test. Reprinted with permission from Ref. [83]. Copyright 2018, Nature Publishing Group.

In recent years, vanadium-based electrode materials have also awakened researchers' interest and have been intensively studied [19, 23, 26, 75]. Du et al. took advantage of *in situ* hXAS to unravel the underlying charge compensation behavior of a NASCION-type cathode material, $Na_2VTi(PO_4)_3$, prepared by a simple sol-gel method and having high-performance specific capacity, ultralong life, and a superior rate capability [121]. As shown in Fig. 13.18,

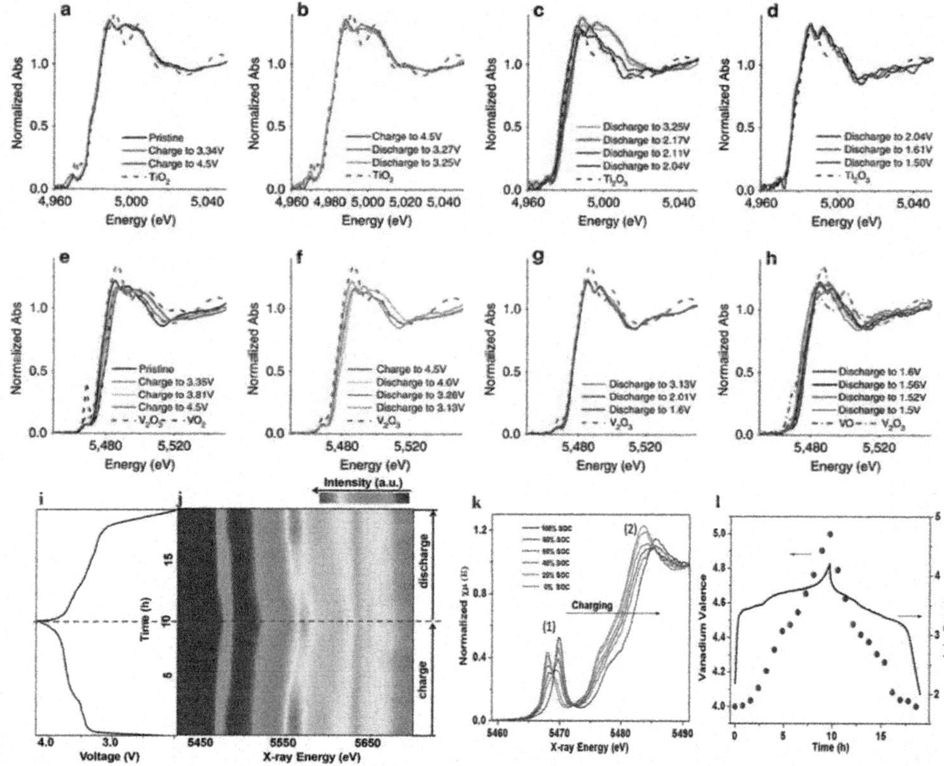

Figure 13.18 Schematic illustration of *in situ* Ti (a–d) and V (e–h) K-edge XANES spectra of $Na_2VTi(PO_4)_3$ at different states in the initial cycle. Reprinted with permission from Ref. [121]. Copyright 2017, Nature Publishing Group. (i–g) The voltage profile and corresponding normalized XANES spectra (2D contour plot) of the coin cell during the charge–discharge process of the first cycle. (k) *In situ* V K-edge XANES spectra during the initial charge process. (l) The valence change in vanadium and the voltage profiles of $Na/NaVOPO_4$ coin cells as a function of time. Reprinted with permission from Ref. [120]. Copyright 2018, Elsevier.

in situ Ti and V K-edge XANES spectra were collected at different cycle states between 1.5 V and 4.5 V. From the XANES spectra, the Ti K-edge shifts to higher energy, and after full charge, V^{3+} progressively oxidized and converted to V^{4+}. In contrast, the valence state of the Ti is robust since no significant changes in the edge energy position and shape are observed in the Ti K-edge XANES spectra. During the initial discharge, the V K-edge shifts to lower energy till 3.1 V and overlapped with the reference V_2O_3, indicating that V^{3+}/V^{4+} was the dominant redox couple accounting for the charge compensation. The Ti K-edge still shows no evidence of change until the discharge voltage processed below 3.1 V. The Ti^{3+}/Ti^{4+} redox couple plays a dominating role in electron transfer in the voltage window of 2.0–3.1 V.

Cao et al. have fully employed *in situ* XANES to monitor the V valence and structure evolution of a layered structure $NaVOPO_4$ showing high voltage and long lifespan as cathode for SIBs [120]. The 2D contour plots of the normalized XANES spectra presented in Fig. 13.18 proved its high reversibility process during charge and discharge. The similar reversible behavior for the V pre-edge, the main K-edge, and the edge resonance indicates that V^{4+} oxidized to V^{5+} during the initial charge, and V^{5+} reduced back to the V^{4+} original state. Change in valence during the charge–discharge process was also analyzed by linear combination fitting, showing Fig. 13.18, which clearly shows variation in the valence state of vanadium, indicating the high reversibility of valence states. Above *in situ* synchrotrons, XANES characterization confirmed the electronic and local structural reversibility of the $NaVOPO_4$ materials upon cycling.

In situ hXAS has also been applied to investigate the Na storage mechanism. Recently, Chen et al. have prepared a polyoxometalate (POM) V-based cluster ($Na_7[H_2PV_{14}O_{42}]\cdot nH_2O$) with unique structures (Fig. 13.19a) as anode materials for SIBs and carried out *in situ* XANES studies to investigate the valence change in $Na_7[H_2PV_{14}O_{42}]\cdot nH_2O$ upon cycling [160]. The energy shift of V K-edge XANES spectra (shown in Fig. 13.19b) indicates that the electrode was reduced during the discharge process. Further analysis demonstrates that V's valence changed from the initial V^{5+} to a valence between V^{3+} and V^{4+}. The average V valence of $Na_7[H_2PV_{14}O_{42}]$ anodes as a function of time (shown in Fig. 13.19) indicates that V^{5+} ions in $Na_7[H_2PV_{14}O_{42}]$ are reduced to $V^{3.7+}$. Therefore, a high theoretical capacity of

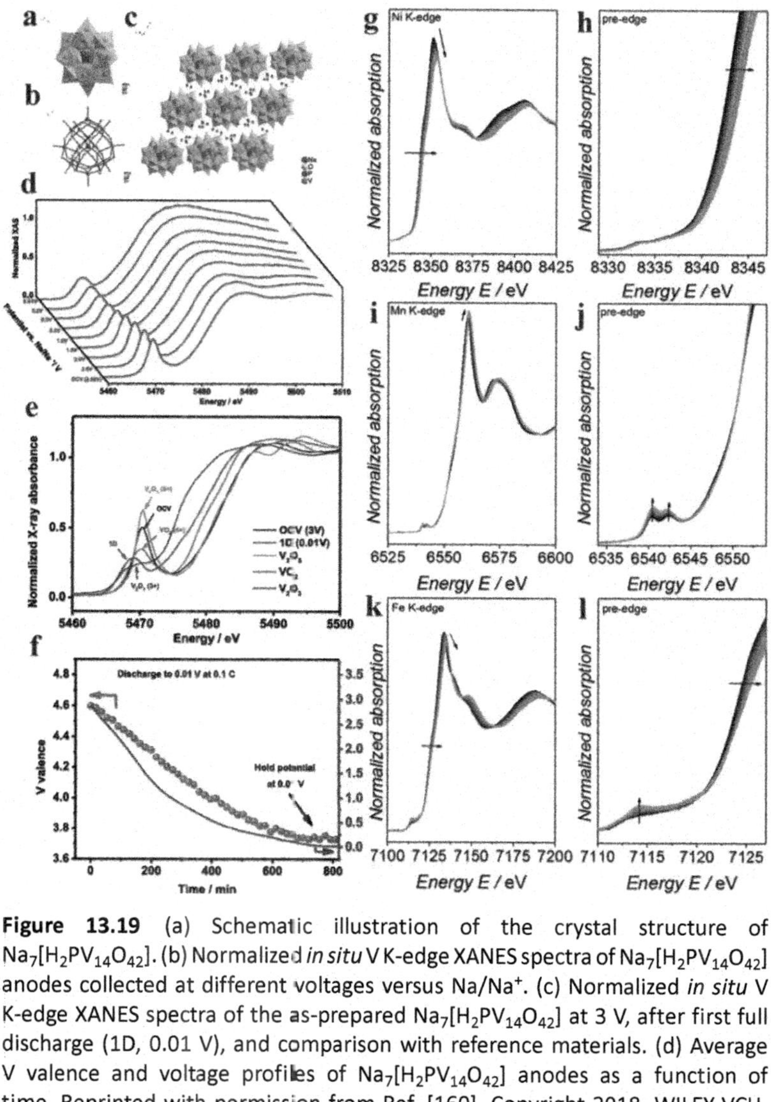

Figure 13.19 (a) Schematic illustration of the crystal structure of $Na_7[H_2PV_{14}O_{42}]$. (b) Normalized *in situ* V K-edge XANES spectra of $Na_7[H_2PV_{14}O_{42}]$ anodes collected at different voltages versus Na/Na^+. (c) Normalized *in situ* V K-edge XANES spectra of the as-prepared $Na_7[H_2PV_{14}O_{42}]$ at 3 V, after first full discharge (1D, 0.01 V), and comparison with reference materials. (d) Average V valence and voltage profiles of $Na_7[H_2PV_{14}O_{42}]$ anodes as a function of time. Reprinted with permission from Ref. [160]. Copyright 2018, WILEY-VCH. *In situ* XANES spectra for the Ni K-edge (e) and (f) pre-edge, Mn (g) K-edge, and (h) pre-edge, as well as the Fe (i) K-edge and (j) pre-edge of P2/P3/O3-$Na_xMn_{0.5}Ni_{0.3}Fe_{0.1}Mg_{0.1}O_2$. Reprinted with permission from Ref. [161]. Copyright 2019, WILEY-VCH.

308 mAh/g can be achieved through 18 electrons stored in the reduction process of $Na_7[H_2PV_{14}O_{42}]$ during the sodiation process. It

is well known that single-phase materials generally allow for a direct correlation of the findings obtained and the compound studied, which is different from materials with mixtures of different phases and possesses complex structure. However, from the perspective of material design, it is necessary to elucidate the origin of capacities. Recently, Passerini et al. have employed *in situ* XAS to elucidate the redox center of an SIB cathode material with mixtures of different phases, P2/P3/O3-$Na_xMn_{0.5}Ni_{0.3}Fe_{0.1}Mg_{0.1}O_2$ [161]. The Ni and Fe K-edge spectra clearly show shifts in the edge position. In contrast, the Mn K-edge position remains virtually unaffected. They suggested that Ni and Fe act as the redox center, and the persistence of the tetravalent oxidation state for manganese is anticipated to prevent the detrimental disproportionation and dissolution mechanism of trivalent manganese. This also demonstrates that the existence of the P phase in the P2/P3/O3-type material appears essential to suppress manganese redox activity, resulting in superior cycle performance.

As a powerful technique for detecting the electronic structure and local geometries of materials, as well as tenable *in situ* tests, hXAS will play a critical role in the design of electrode materials with the development of synchrotron light sources and new test methods. When combined with X-ray imaging techniques, a directly visible effect of chemical phase distribution or evolution will be presented, which will be discussed in the following part.

13.2.2.2 Soft X-ray spectroscopy

The electronic states of electrode materials determine the intrinsic voltage limit and the phase transition process during the battery operation from solid-state physics. Therefore, the battery performance can be efficiently optimized by tailoring the electrode materials' relevant electronic states. Soft X-Ray Spectroscopy (SXS) techniques can detect the vital electronic states of battery electrode materials in the Fermi level's vicinity and their properties and evolution.

Core-level SXS takes advantage of soft X-ray to excite core-level electronics to a high level and thus produce core–hole function as a detection probe to survey electronic state in the vicinity of the Fermi level. SXS techniques typically include Soft X-Ray Absorption Spectroscopy (sXAS), Nonresonant Soft X-Ray Emission

Spectroscopy (XES), and Resonant Inelastic X-Ray Scattering (RIXS). The advantages of SXS are its natural elemental and orbital sensitivity since the SXS techniques mentioned earlier are all core-level spectroscopy. Another benefit of SXS is its ability to study 3d TMs, the main redox center of battery materials. Note that modern SXS techniques, with their powerful features and capabilities, make them suitable for the analysis of battery materials.

Figure 13.20 shows the atomic schemes of the core-level XAS, XES, and RIXS. During XAS measurement, with the incident X-ray's absorption, the core electron is excited to an unoccupied state upon the Fermi level. sXAS generally collects in two modes: the total yield of fluorescence photons (TFY) and total electron yield (TEY). TEY is a photon-in-electron-out (PIEO) detection mode, which is very surface sensitive with a probe depth of several nanometers. In contrast, TFY is a photon-in-photon-out (PIPO) detection mode with a probe depth of about 150 nm by detecting the exciting photon. The detection depth contrast of TEY and TFY modes makes it suitable for probing surface and bulk states of the specimen, respectively. With the progress in the *in situ* instruments, the *in situ* sXAS test, restricted by the limited probe depths, leads to practical sXAS, which must be conducted in an ultrahigh vacuum chamber. Due to the *in situ* instruments' progress, *in situ* sXAS also has practical applications in electrode material analysis.

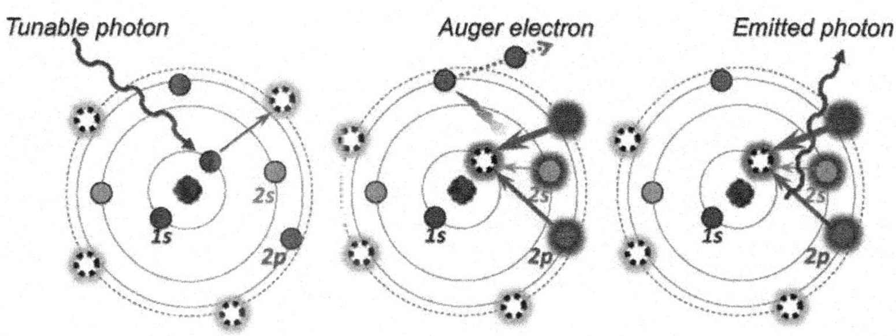

Figure 13.20 Atomic schemes of the core-level XAS, XES, and RIXS. (a) Core-electron excitation, (b) photon-in-electron-out XAS, and (c) photon-in-photon-out XAS, XES, and RIXS. Reprinted with permission from Ref. [165]. Copyright 2013, Elsevier.

XES is also a PIPO technique, which provides information about occupied valence electrons by detecting the decay process of the occupied electron states. On the other hand, tuning the excitation energy near the region of the element absorption threshold makes the core electron resonantly excited to a specific energy level in the conduction band rather than escape from the material system. Hence, the resulting emission spectra feature strong excitation energy dependence and provide the needed information of both unoccupied and occupied valence states. Unlike XAS, which detects the first-level optical processes, XES and RIXS are achieved by probing second-level optical processes with low luminous efficiency. Thanks to the development of the third-generation light source with high brightness and photonic detector efficiency, XES and RIXS have become feasible for material analysis. RIXS is elemental and orbital sensitive, making it possible to distinguish between the excited states of elements and atoms in a specific chemical environment. As a state-of-art technique, RIXS has shown promising applicability in the *in situ* characterization, spatial resolution, and time dynamic response of battery materials in the future.

13.2.2.2.1 Probing electron/valence state of electrode materials

Soft XAS is a powerful tool for investigating the electronic states of battery materials [162]. Soft XAS provides element-specific information sensitive to the valence states, spin state, and chemical bond configuration of 3d TMs (Ms), and also the low-Z element K-edges through dipole-allowed 2p to 3d (L-edge) electron excitations and1s to 2p transitions, respectively.

As shown in Fig. 13.21a, a quantitative analysis of Mn L-edges XAS on a series of $Na_{0.44}MnO_2$ electrodes at different SOCs of the first discharge process was performed. The sXAS spectra demonstrate the formation of Mn^{3+} and Mn^{2+} during the discharge process, and the calculated concentrations of Mn^{4+} decrease monotonously with the sodium insertion till the voltage reaches 2.6 V (Fig. 13.21b, c). However, when the voltage drops below 2.6 V, Mn^{2+} concentration increases rapidly. They suggested that Mn^{2+} formation occurs only on the surface with the combined analysis of XAS and hard X-ray Raman spectroscopy. They also proved that the Mn^{2+} compounds became electrochemically inactive after extended cycles, resulting in capacity fading. It is gratifying that cycling $Na_{0.44}MnO_2$ above 3 V could efficiently suppress the Mn^{2+} formation [167].

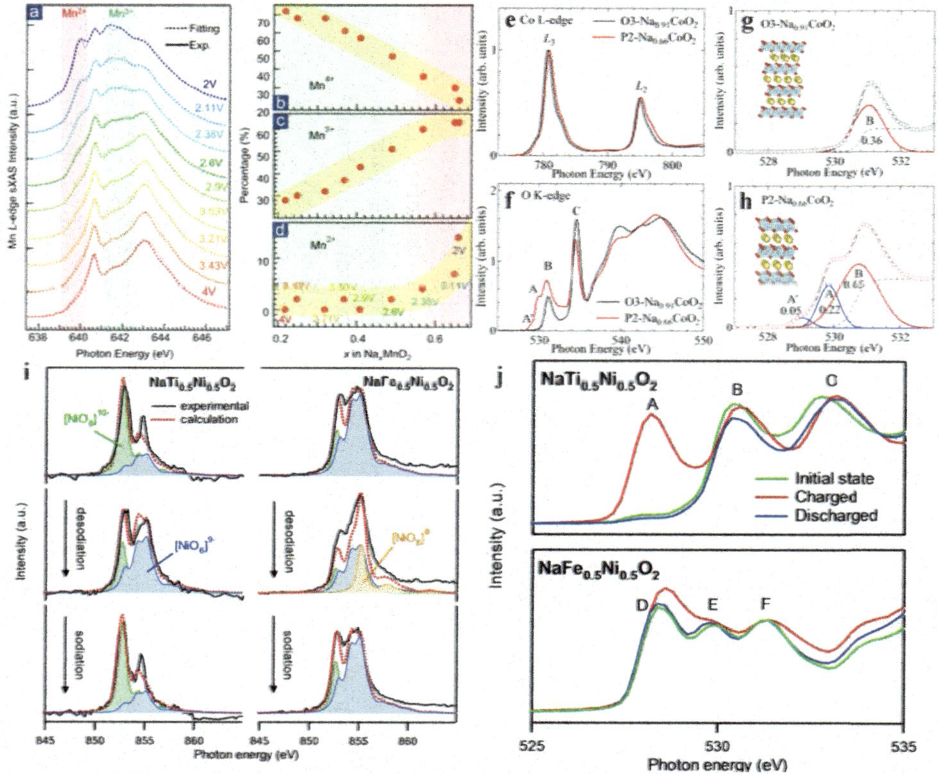

Figure 13.21 Quantitative analysis of Mn L-edge sXAS collected on a series of $Na_{0.44}MnO_2$ electrode cycled to different electrochemical states. (b–d) Evolution of Mn^{4+}, Mn^{3+}, and Mn^{2+} concentration in Na_xMnO_2 electrodes plotted as a function of sodium concentration. Reprinted with permission from Ref. [167]. Copyright 2015, Elsevier. (e, f) Co L-edge and (b) O K-edge spectra recorded in the TEY mode of O3-$Na_{0.91}CoO_2$ and P2-$Na_{0.66}CoO_2$. (g, h) O K-edge spectra recorded in the TEY mode of (g) O3-$Na_{0.91}CoO_2$and (h) P2-$Na_{0.66}CoO_2$ in the pre-edge region. Reprinted with permission from Ref. [168]. Copyright 2018, Elsevier. (i) Ni L_3-edge X-ray absorption spectra obtained in the PFY mode for the pristine, desodiated (x = 0.5), and sodiated (x=1.0) $Na_xTi_{0.5}Ni_{0.5}O_2$ and $Na_xFe_{0.5}Ni_{0.5}O_2$. (j) Oxygen K-edge X-ray absorption spectra for the pristine (green lines), desodiated (red lines, x=0.5), and sodiated (blue lines, x=1.0). Reproduced with permission from Ref. [169]. Copyright 2016, American Chemical Society.

Related studies suggest that the charge dynamics in battery electrode materials is determined by the electronic states, which plays a key role in optimizing battery performance. Therefore, it is important to understanding the electronic states of battery materials. Recently, electronic states in oxidized Na_xCoO_2 (O3-$Na_{0.91}CoO_2$ ($CoO_2^{-0.91}$) and P2-$Na_{0.66}CoO_2$ ($CoO_2^{-0.66}$)) have been revealed by X-ray absorption spectroscopy coupled with *ab initio* calculation [168]. Figure 13.21e shows the L-edge, including the L_2 and L_3 edges, ascribed to the dipole-allowed transition from the Co $2p_{1/2}$ (Co $2p_{3/2}$) to Co 3d orbitals. The Co L-edge spectra of O3-$Na_{0.91}CoO_2$ and P2-$Na_{0.66}CoO_2$ possess similar features. The O K-edge spectra in Fig. 13.21f were divided into the pre-edge (529–536 eV) and main (>536 eV) regions. The peak located at around 531 eV for both compounds is labeled as peak B. The peaks observed at around 530 eV and 529 eV in P2-$Na_{0.66}CoO_2$ were labeled as peak A and peak A'. Spectral weights of the peaks A and B were analyzed, with the aid of the magnified O K-edge spectra recorded in TEY mode. The spectral variation suggests that the unoccupied O 2p state increases with the oxidization of the Na_xCoO_2.

As mentioned earlier, the electronic states of electrode material determine the intrinsic voltage limit, and raising the operating potential of the cathode materials in sodium-ion batteries is a crucial challenge. Since sXAS can detect electronic states, Yamada et al. applied sXAS to assess the origin of the unusually lower $E(Ni^{4+}/Ni^{3+})$ compared with $E(Ni^{3+}/Ni^{2+})$ in $NaTi^{4+}{}_{0.5}Ni^{2+}{}_{0.5}O_2$ and $NaFe^{3+}{}_{0.5}Ni^{3+}{}_{0.5}O_2$ [169]. Both the Ni L_3-edge XAS spectrum in $NaTi_{0.5}Ni_{0.5}O_2$ and $NaFe_{0.5}Ni_{0.5}O_2$Fe L-edge XAS spectra show little changes during a reversible change in the electronic state of the $[NiO_6]^{n-}$ cluster. However, the electronic states of Ti and Fe showed no evident changes during desodiation/sodiation. They further calculated the electronic states of $[NiO_6]^{n-}$ clusters (n = 8, 9, and 10) by using the Configuration-Interaction Full-Multiplet (CIFM) calculation. Based on the CIFM calculation of Ni L_3-edge, Ni is not oxidized to the extent expected or barely oxidized to Ni^{4+} in $NaFe_{0.5}Ni_{0.5}O_2$ and $NaTi_{0.5}Ni_{0.5}O_2$, demonstrated that the hybridization between the oxygen 2p and transition metal 3d orbitals plays a critical role in determining the redox potential of $NaMO_2$.

13.2.2.2.2 Investigate surface chemical by sXAS and XES

As described earlier, interfacial chemistry is vital for alkali-metal-ion batteries, which seriously impacts battery performance through different reactions, including the generation of SEI and CEI, TM dissolution, segregation, electrolyte decomposition, and surface reconstruction of the electrode, along with other factors such as crack formation and the dissolution of TM cations. Many of these phenomena, such as surface degradation, are related to the orbital hybridization between TM cations and oxygen (O) anions. While XAS and XES are potent tools for detecting the TM 3d electrons, O K-edge analysis can provide information on the 2p electron. Especially, the surface-sensitive XAS collected at TEY mode can detect a depth below 10 nm, which could be applied to distinguish the different reaction processes between the bulk and surface of electrode materials.

Recently, Lin et al. have reported a sodium layered oxide material $Na_{0.9}Cu_{0.2}Fe_{0.28}Mn_{0.52}O_2$ (CFM-Cu) with a high degree of compositional heterogeneity, which could also deliver stable electrochemical performance [170]. They employed sXAS in the TEY mode to obtain information on the evolution of the surface TM oxidation states at different cycle voltages. As shown in Fig. 13.22a, b, the valence states of Cu and Fe remained relatively unchanged throughout the charging processes. The challenge of the Fe L-edge spectrum is amplified by the fact that it significantly overlaps with that of the FK-edge, which originates from the electrolyte decomposition, poly(vinylidene difluoride) (PVDF) binder, or $NaPF_6$ residual. The Mn in the as-prepared compound shows high activity, which was gradually oxidized as the charging proceeded. The TM3d-O2p hybridization present as pre-edge (shown in Fig. 13.22d) followed a similar trend as the Mn-redox process. The concurrent evolution of the TM3d-O2p hybridization and the Mn oxidation state implies that the surface reaction of CFM-Cu mainly occurs in Mn and O.

They further conducted sXAS combined with XRS to understand Li substitution's role in altering the TM 3d-O 2p hybridization and the surface chemistry of compositionally heterogeneous $NaLi_{0.045}Cu_{0.185}Fe_{0.265}Mn_{0.505}O_2$ cathode [171]. XRS detects the lattice oxygen and TMs' electronic environment throughout the entire bulk of the electrode material, which indicates that more unoccupied states were created in the hybridized orbitals. In contrast to XRS,

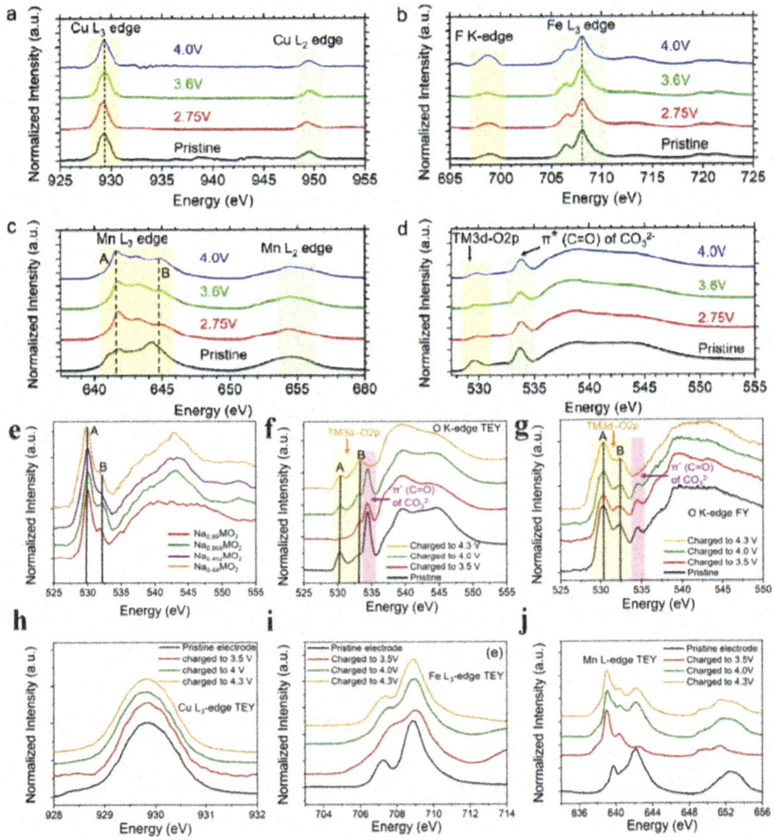

Figure 13.22 (a) Cu L-edge, (b) Fe L-edge, (c) Mn L-edge XAS of CFM-Cu collected at TEY mode as a function of charge voltage. (d) O K-edge at different states of charge in the first charge cycle at the C/10 rate. Reprinted with permission from Ref. [170]. Copyright 2018, The Royal Society of Chemistry. (e) OK-edge XRS of Na-$Li_{0.045}Cu_{0.185}Fe_{0.265}Mn_{0.505}O_2$ at different charge states. (f, g) O K-edge sXAS detected at TEY and TFY modes during charge. The peaks in the yellow shaded region originated from the TM 3d-O 2p hybridization, while the peak in the pink shaded region originated from the oxygen of the surface carbonate species. (h) Cu L_3-edge, (i) Fe_3-edge, and (g) Mn L-edge sXAS spectra in the TEY mode of Na-$Li_{0.045}Cu_{0.185}Fe_{0.265}Mn_{0.505}O_2$ at different charge states. Reproduced with permission from Ref. [171]. Copyright 2019, American Chemical Society.

sXAS in TEY and TFY modes probes information about the surface and subsurface. The sXAS spectra in Fig. 13.22 show the variation in TM 3d-O 2p, and the pre-edge peak area decreases when the electrode was charged to 3.5 V. While above 3.5 V, the area shows an increasing tendency, which can be because the electrolyte has a natural tendency to get oxidized and, in turn, reduce the surface TMs. The sXAS spectra of Cu and Fe show no evidence of change during the charging process, whereas Mn experienced dramatic changes in the oxidation states (Fig. 13.22g). The variation in the valence states of TM is similar to that of $Na_{0.9}Cu_{0.2}Fe_{0.28}Mn_{0.52}O_2$, and Mn plays the most predominant role in the surface chemistry of both the cathode compounds. They also concluded that surface Mn reduction is a significant barrier to Mn-containing sodium cathodes' practical applications.

The theoretical capacity of red phosphorus is as high as 2596 mAh/g within a desirable operating voltage range. Hence, red phosphorus has been treated as a promising anode candidate for SIBs. However, red phosphorus with intrinsic low electronic conductivity usually suffers significant volume change during sodiation/desodiation, resulting in poor cycle life and rate capability [41, 42]. Additionally the electronic structure of phosphorus is $3s^2 3p^3$ and it is easily oxidized and generally exists as an inorganic phosphate. Sun et al. have employed sXAS and XES to understand the chemical evolution of passivated surface in air/water conditions [172]. In P L-edge sXAS (Fig. 13.23e), the increasing intensity ratios of the peak at 137.7 eV to the peak at 136.9 eV (oriented from $2p_{3/2}$ and $2p_{1/2}$ levels after the spin-orbit splits 2p into the first unoccupied 3s-like antibonding state, respectively, for electrodes aged in air) verify the structural distortion of phosphate tetrahedron. XES collected with different excitation energies, as presented in Fig. 13.23a,d, was applied to understand the surface phase evolution of phosphorus electrodes in air. With the increasing aging time in air, the peak energy shift in Fig. 13.23a and the peak intensity variation in Fig. 13.23b demonstrate structural distortion of phosphate after the continuous reactions between phosphorus and oxygen. Simultaneously, the decreased peak intensities for the lower-energy P–P bonding and increased peak intensities for the high-energy P–O bonding can be observed with a prolonged exposure time of red phosphorus in ambient air, indicating a gradual increase in phosphate thickness on the surface of the initial phosphorus. P L-edge sXAS was also applied to study the structural evolution and phase transformation of phosphorus/carbon composite in water and oxygen after 24 h. The results indicated partial degradation in water and severe oxidation in oxygen for phosphorus.

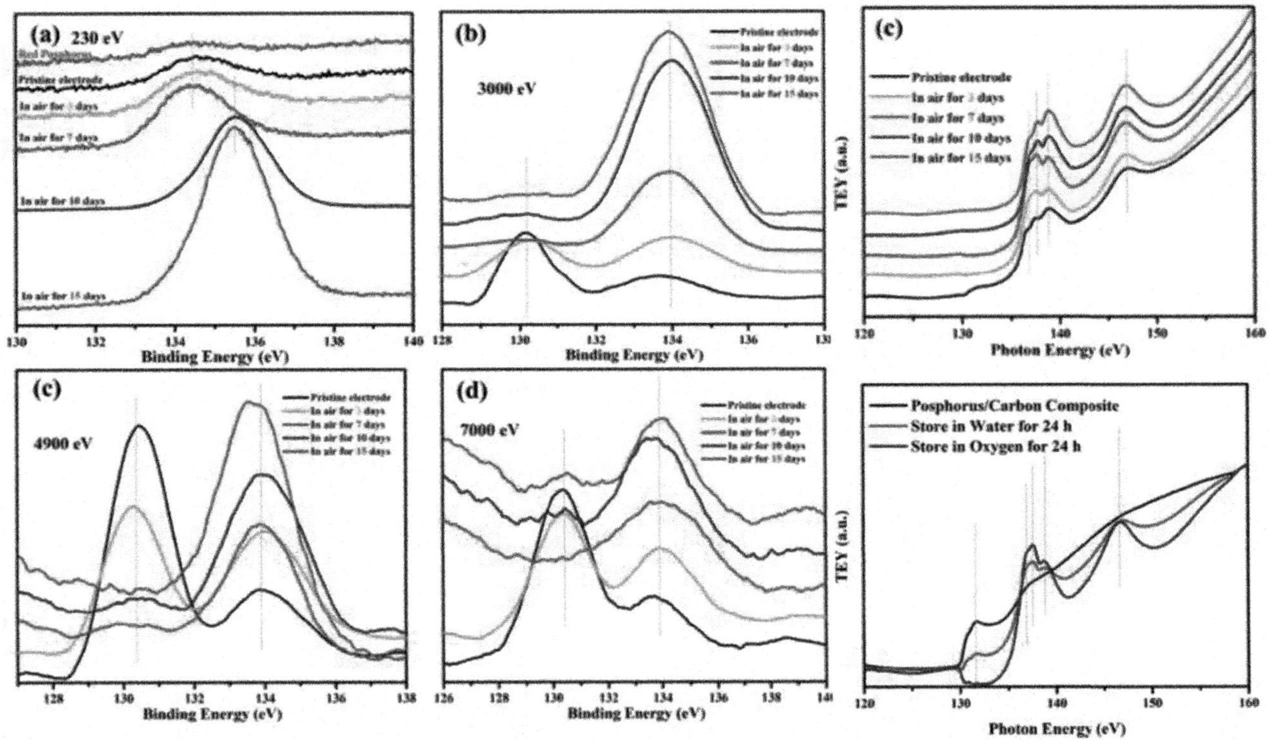

Figure 13.23 (a–d) P 2p X-ray photoemission spectroscopies detected at different photon energies; (e) P L-edge sXAS for the red phosphorus of the pristine electrode and the electrodes in the air for different days. (f) P L-edge sXAS for the pristine phosphorus@ carbon composite and phosphorus/carbon composite stored in water and oxygen for 24 h. Reproduced with permission from Ref. [172]. Copyright 2019, American Chemical Society.

Moreover, the oxidation of phosphorus in different conditions was also verified by the emerging peaks at different energies. Based on the spectral analysis and electrochemical tests, they suggest that the formed insulation phosphate layer severely deteriorates the electrochemical reversibility and stability of the phosphorus/carbon electrodes. The conclusion provides an essential guiding role in storing conditions and interfacial designs of phosphorus-based anode materials for high-performance SIBs.

13.2.2.2.3 Redox mechanism monitoring

Soft XAS is a powerful technique to probe the TM 3d states through the strong dipole transition directly. Due to the element-sensitive and well-defined multiplex features of TM L-edge sXAS to the TM states, sXAS has been extensively applied to the TM redox analysis in electrodes [177, 178]. Yang et al. have studied the influence of interstitial water molecules on the influence of the spin-state-related TM redox in $Na_xMnFe(CN)_6{\cdot}zH_2O$ (NaMnHFC) systems through sXAS and theoretical calculations [175]. First, they set a benchmark of $Fe^{2+/3+}$ and $Mn^{2+/3+}$with distinct spin states (Fig. 13.24a). The spectral features are interpreted by LS Fe and HS Mn calculations, which demonstrates that the Fe and Mn in $Na_{2-x}Mn[Fe(CN)_6]$ are at the C- and N-coordinated sites and takes the LS and HS states, respectively. The electrochemical property of $Na_{2-x}Mn[Fe(CN)_6]$, as revealed by sXAS results, illustrated that the low-spin configuration of the Fe fundamentally increases its relative electrochemical potential, while Mn is reduced on the electrode surface, which may result in capacity loss.

The Fe redox behavior in the hydrated and non-hydrated systems was obtained by analyzing the evolution of the intermediate states' sXAS line shape. Fe L_3-edge sXAS spectra presented in Fig. 13.24 show that the $Fe^{2+/3+}$ redox reaction has been completed at the low voltage plateau from H1 to H3. Simultaneously, the line shape in the non-hydrated sample presents a gradual change in Fe state Fe^{2+}. As for Mn, it can be observed that the $Mn^{2+/3+}$ redox reaction in the hydrated sample mainly occurs at the high-voltage regime. Like the observed results in Fe L_3-edges XAS spectra of the non-hydrated sample, the Mn L_3-edge line shape also shows a continuous change. The analysis was verified by calculating the concentration of the oxidation states of Fe^{3+} and Mn^{3+} at different charge states.

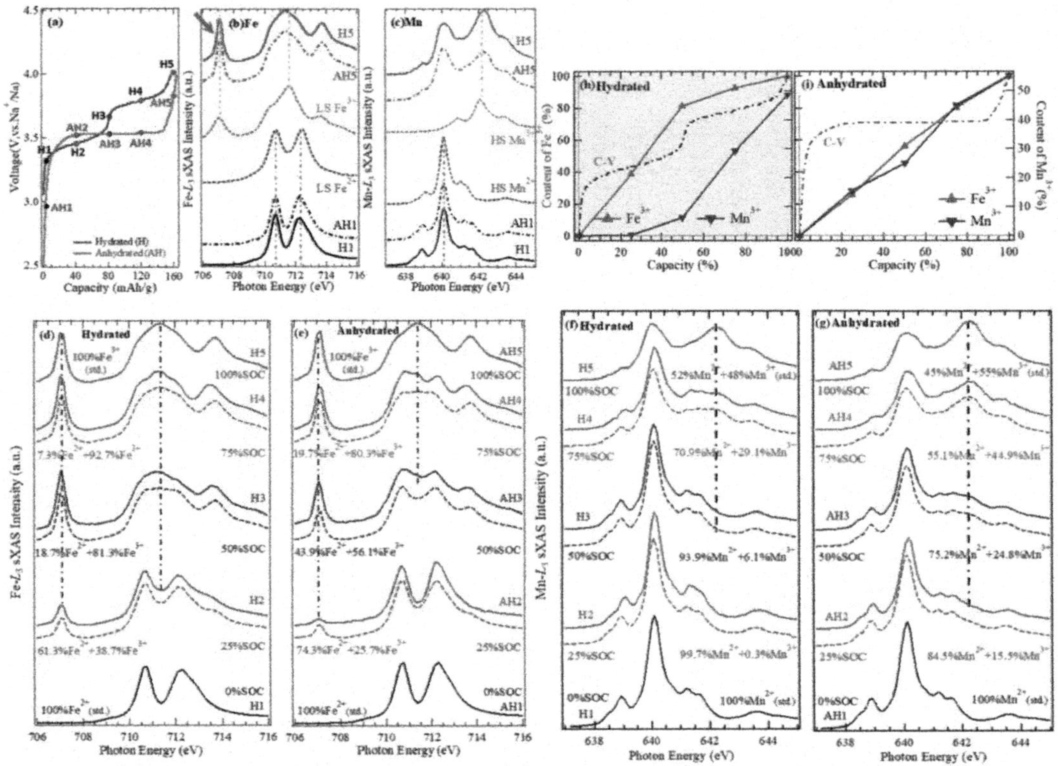

Figure 13.24 (a) Two sets of samples with five different SOCs are defined here for hydrated (H1 to H5) and non-hydrated (AH1 to AH5) $Na_{2-x}Mn[Fe(CN)_6]$; H1 and AH1 refer to Fe^{2+} and Mn^{2+} in a fully discharged state, and H5 and AH5 refer to Fe^{3+} and Mn^{3+} in a fully charged state. (b, c) Fe and Mn L_3-edge sXAS spectra are compared with the calculated spectra of (low spin) LS Fe^{2+}/Fe^{3+} and (high spin) HS Mn^{2+}/Mn^{3+}, respectively. (d–g) Fe and Mn L_3-edge sXAS spectra collected at different SOCs: (d, f) hydrated samples; (e, g) non-hydrated samples. (h) Percentage of the oxidized TM^{3+} states in (a) hydrated samples and (b) non-hydrated samples. Reproduced with permission from Ref. [175]. Copyright 2017, American Chemical Society.

Hence, they suggested that for the hydrated system, $Fe^{2+/3+}$ and $Mn^{2+/3+}$ redox reactions occur separately at the low potential and high potential of the charging process, leading to two separated voltage plateaus. While for the non-hydrated system, a continual oxidation reaction takes place for both Fe and Mn as a function of the SOC changes. The determining factors on the modification of the TM redox mainly contribute to the spin states, the crystal field, and the oxidation energy of TMs.

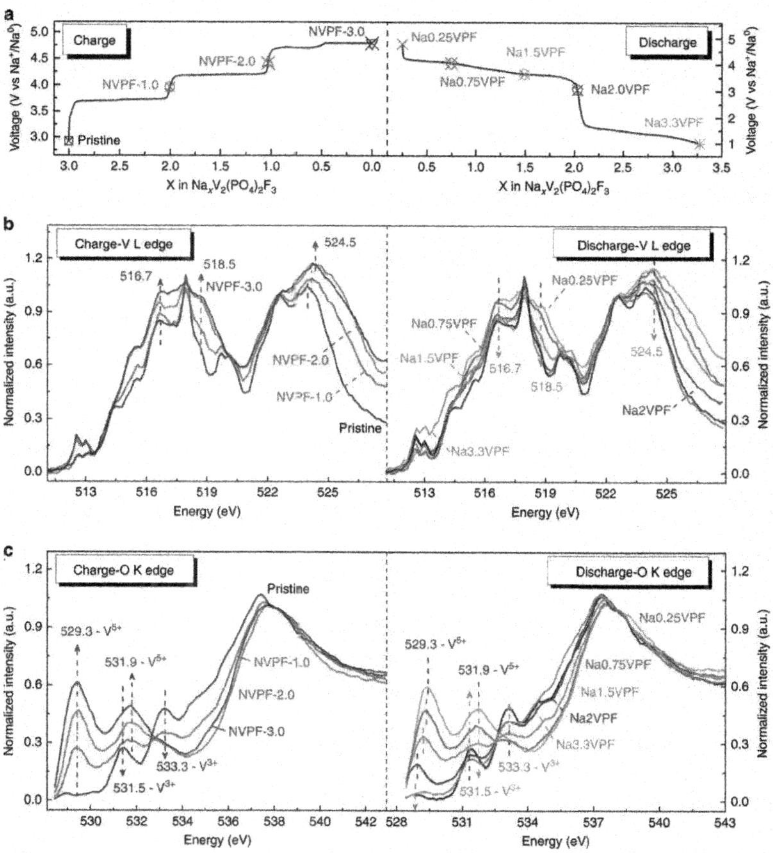

Figure 13.25 (a) The voltage-composition curve of the NVPF-3.0 sample cycled at C/10. Plots at different cycling states on the curves were chosen for the sXAS test. The normalized (b) V L-edge and (c) O K-edge sXAS of the corresponding charge (left) and discharge (right) samples obtained in the TFY mode. Reprinted with permission from Ref. [176]. Copyright 2019, Nature Publishing Group.

Recently, Tarascon et al. used sXAS to study the charge compensation of $Na_3V(PO_4)_2F_3$ (NVPF). Since the as-prepared NVPF can reversibly extract/uptake three Na^+ per formula, a new disordered tetragonal "NVPF" phase is formed upon cycling [176]. As shown in Fig. 13.25, both the V L-edge and O K-edge sXAS spectra demonstrate the V function as the redox center. The spectral variation of intensity and peak position improved the reversible behavior of V, through the whole cycling process between 4.8 V and 1 V. Noting the appearance of a well-defined peak at the O K-edge consistent with the emergence of a single valence and environment of V, they fitted the O K-edge sXAS spectra and observed a continuous increase upon extraction. This demonstrates the V^{4+} disproportionation to V^{3+} and V^{5+} during the desodiation process. Opposite variation occurred during the discharge, not from the V^{5+} t_{2g} state. Hence, a new irreversible $Na_xV(PO_4)_2F_3$ phase of tetragonal symmetry (I4/mmm space group) was formed after full charge. The unique sodium storage mechanism of the as-prepared $Na_3V(PO_4)_2F_3$ displayed excellent cycle and suitable rate capabilities.

13.2.2.2.4 sXAS and RIXS applied for oxygen redox reaction probing

It is well known that in TM oxide-based cathode, the redox center is typically associated with only the 3d valence states of the 3d TMs in oxide cathodes. In conventional wisdom, an oxygen redox reaction is generally associated with the degradation of the cathode materials due to the dimerization of anions, forming, e.g., O_2^{2-}, which is usually detrimental to the stability and kinetics of batteries, leading to irreversible capacity fading [163, 166]. However, recent research on oxygen activities suggests that it could be possible to avoid irreversible oxygen redox reactions to achieve high-energy-density Li-ion and Na-ion batteries [177–180]. Thus, emerging sXAS and RIXS techniques, which can directly detect information about the O-2p and TM-3d valence states in different modes, have been extensively applied to understand the depth mechanism of the oxygen redox reaction.

Yamada et al. have synthesized an ordered Na_2RuO_3 with honeycomb-ordered $[Na_{1/3}Ru_{2/3}]O_2$ slabs, which delivered higher capacity than the "disordered" Na_2RuO_3 with randomly distributed $[Na_{1/3}Ru_{2/3}]O_2$ slabs (Fig. 13.28a) [181]. sXAS was employed to interpret the origin of the excess capacity. Ru L-edge sXAS in Fig. 13.26 demonstrates that the oxidation from Ru^{4+} to Ru^{5+}occurs at the initial potential plateau. Whereas the peak position of the 2p-e_g did not change at the second plateau, suggesting no substantial change in the valence state of Ru. As for the O K-edge shown in Fig. 13.26c, three peaks were observed at 534 eV, 529 eV, and 532 eV.

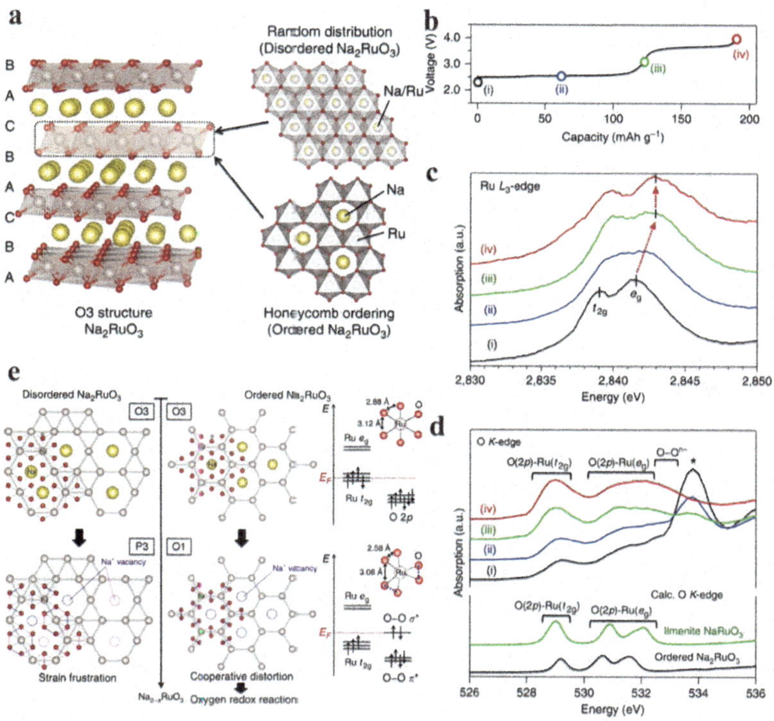

Figure 13.26 (a) Schematic illustration of disordered Na_2RuO_3 and ordered Na_2RuO_3. (b) States of charge of the samples for sXAS. (c) Ruthenium L_3-edge and (d) oxygen K-edge X-ray absorption spectra for various O-Na_xRuO_3 compositions: (i) $x = 2.0$ (black), (ii) $x = 1.5$ (blue), (iii) $x = 1.0$ (green), and (iv) $x = 0.62$ (red). (e) Schematic representation of the structural changes during charge–discharge for disordered Na_2RuO_3 and ordered Na_2RuO_3. Reprinted with permission from Ref. [181]. Copyright 2016, Nature Publishing Group.

These peaks can be ascribed to the unoccupied hybridized orbitals of O 2p-Na 3p, O 2p-Ru t_{2g}, and O 2p-Ru e_g, respectively. The O 2p-Na 3p signal was diminished at the first voltage plateau due to the extraction of Na^+, whereas the peak intensity of O 2p-Ru t_{2g} increased, indicating the generation of a hole on the O 2p-Ru t_{2g} orbital. With the desodiation process proceeding to the second plateau, the peak intensity of the O 2p-Ru t_{2g} shows no evidence of change, whereas the peak around 532 eV broadens as a result of the new absorption band emerging around 533 eV, which can be explained by the formation of a new chemical bond involving oxygen. Schematic representation of the structural changes during charge–discharge for disordered Na_2RuO_3 and ordered Na_2RuO_3 suggests that the short O–O bonding plays a crucial role in triggering the oxygen redox reaction in ordered Na_2RuO_3 and leading to a 30% extra capacity.

Recently, Yang et al. have developed an ultra-high-efficiency soft X-ray with full-range mapping of RIXS (mRIXS), which was proved to be a sensitive and reliable probe methods of the TM and O-redox in bulk battery electrodes by resolving energy of the fluorescence photon [182]. Since the O K-edge sXAS suffers from overwhelming effect of TM-O hybridization, it severely limits the reliability of the O K-edge detection. In addition, to avoid the problem of line shape distortion in sXAS due to the self-absorption effect in detecting TM-3d valence states, they applied soft X-ray inverse partial fluorescence yield (iPFY) as a non-distorted bulk probe of TM-3d states. Subsequently, mRIXS-iPFY and mRIXS-sPFY, which can selectively integrate the mRIXS features in a specific emission-energy region, were conducted to quantify and analyze the evolution of both the oxygen and TM redox states of $Na_{2/3}Mg_{1/3}Mn_{2/3}O_2$ SIB cathode material.

The Mn-L_3 iPFY and TEY spectra of the initial cycle (Fig. 13.27a) demonstrate that Mn^{2+} has not been observed in the bulk electrode, and $Mn^{4+/3+}$ is the main redox couple in bulk $Na_{2/3}Mg_{1/3}Mn_{2/3}O_2$. In contrast, the surface Mn states show a different trace by analyzing the variation of Mn valence states in different cycles (Fig. 13.27c, d), which suggests that surface Mn is partially electrochemically active. The Mn-L_3 iPFY and TEY spectra suggest that the majority of the capacity of $Na_{2/3}Mg_{1/3}Mn_{2/3}O_2$ electrodes is not from Mn redox.

The O K-edge mRIXS in sPFY mode in Fig. 13.27e,f shows that the oxygen oxidation starts at 4.2 V during the initial charge and stops at 2.5 V. During the process, Mn valence states show no significant change. Quantifications of Mn-redox and changing O-redox illustrate that the corresponding charge compensations of Mn (blue in Fig. 13.27g) and O (red in Fig. 13.27h) redox are 0.03 mol and

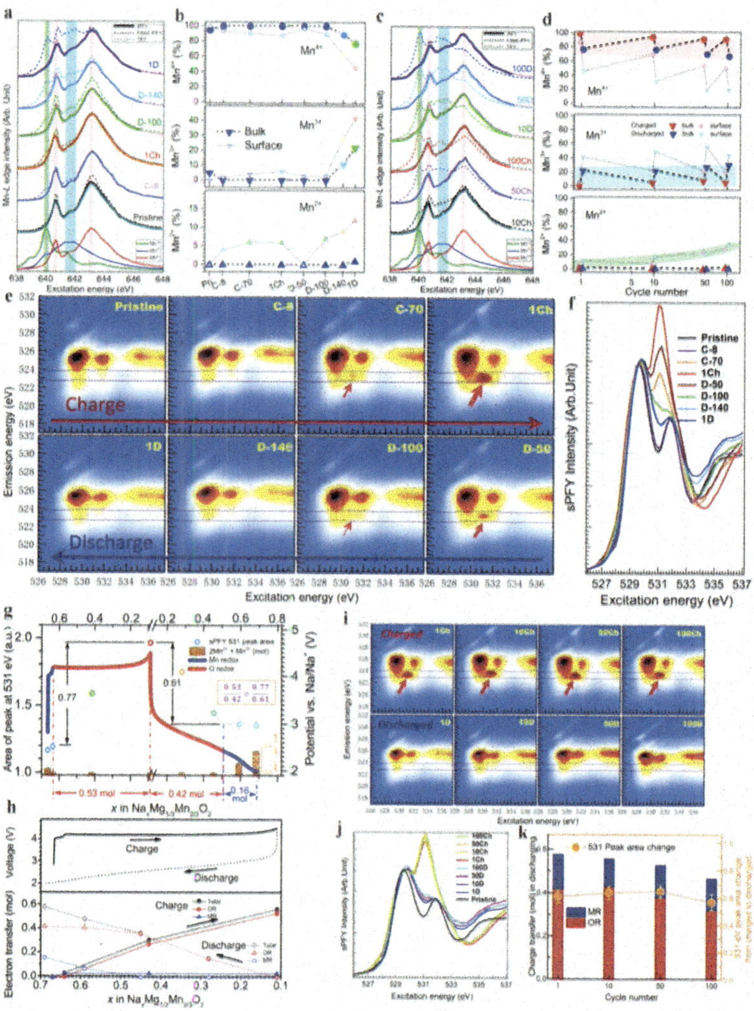

Figure 13.27 (a, c) Mn-L_3 iPFY and TEY spectra and the quantified bulk and surface Mn valence states of$Na_{2/3}Mg_{1/3}Mn_{2/3}O_2$ at different SOC during the initial charge/discharge and after different cycles. (b, d) Fitting results of the Mn valence distributions in bulk and on the surface at different cycle states. (e) O-K mRIXS in sPFY modes of $Na_{2/3}Mg_{1/3}Mn_{2/3}O_2$ at different SOC during the initial cycle. (f) sPFY spectra extracted from mRIXS. (g) Deciphering the total electrochemical capacity by independent quantifications of Mn-redox and changing O-redox states (h) Accumulated electron transfer of Mn-redox and O-redox during charging/discharging. (i) O-K mRIXS images of a series of charged (top) and discharged (bottom) electrode after the 1st, 10th, 50th, and 100th cycles. (j) sPFY spectra extracted from mRIXS by integrating the

(Continued)

characteristic 523.7 eV emission-energy range. (k)The quantified electron transfer of Mn-redox (blue bar), the expected value of O-redox contribution (red bar) calculated by subtracting Mn-redox contribution from total capacity. Reprinted with permission from Ref. [182]. Copyright 2018, Elsevier.

0.53 mol during charging, and 0.16 mol and 0.42 mol during discharge, from electrochemical experiments (*x*-axis of Fig. 13.27). Due to the O-redox feature of $Na_{2/3}Mg_{1/3}Mn_{2/3}O_2$, the cycle ability of oxygen redox upon extended cycles was also calculated through peak area analysis of the sPF spectra. Figure 13.27k clearly shows that the O-redox activities are very stable up to 50 cycles and then start to decay with about 87% of the O-redox sustained at the 100th cycle. This work has provided a new characterization method for an in-depth and precise understanding of oxygen redox reaction.

As elaborated earlier, SXS with nature surface and element sensitivity has been extensively applied in material optimization and design, with the rapid development of synchrotron-based light sources and new SXS techniques. Regrettably, *in situ* SXS tests reported in LIB research are rarely employed for studying SIB electrode materials, mainly due to the necessity of a high vacuum chamber and the limitation of measurement of test instrument design.

13.2.3 Synchrotron Radiation-Based X-Ray Photoelectron Spectroscopy

As one of the standards and widely applied techniques in surface analysis, X-ray photoelectron spectroscopy (XPS) has come a long way from its origins. Einstein's explanation of the photoelectric effect is the development cornerstone of XPS [183]. The XPS data acquisition process can be briefly depicted as follows: Excited photoelectrons liberated from the material surface are captured by energy analyzer detection. Thus, the XPS spectra can be achieved by taking the kinetic energy (E_{Kinetic}) and intensity of the photoelectrons expelled from the *X*-axis and *Y*-axis of the material, respectively. However, the binding energy (E_{Binding}) of atoms to electrons rather than E_{Kinetic} reflects the inherent properties of matter. Based on the principle of conservation of energy, people can obtain the parameters that reflect the actual physical meaning by converting measured physical quantities through the following formula, in which W_f is the work function of the sample [188]:

$$E_{Binding} = h\nu - E_{Kinetic} - W_f$$

According to the preceding formula, by changing the energy of incident light ($h\nu$), the detected energy level will also change. The intrinsic surface sensitivity of PES is due to the short mean free path of the electrons in the material, as shown in Fig. 13.28. It presents the universal curve showing the inelastic mean free path (IMFP) versus the kinetic energy of the excited photoelectron [188]. It is clearly shown that the probe depth and the outgoing electron kinetic energies are closely related to the incident energy.

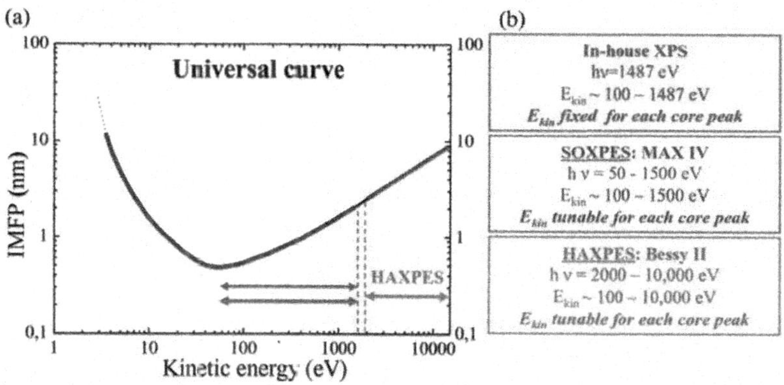

Figure 13.28 (a) The universal curve represents the inelastic mean free path (IMFP) versus the kinetic energy of the excited photoelectron. (b) Illustration of the energy range ($h\nu$ and $E_{Kinetic}$) reached by using monochromatic (Al Kα) or synchrotron (soft X-ray and hard X-ray) sources. For a particular beamline, constraints in the monochromator configuration limit the accessible energy range. Soft X-ray beamlines are equipped with grazing incidence grating monochromators, while photon energies above 2000 eV crystal monochromators are required. Reproduced with permission from Ref. [188]. Copyright 2016, Electrochemical Society.

Traditional in-house laboratory XPS generally uses fixed photon source (typically Mg or Al Kα X-rays at 1253.6 and 1486.6 eV, respectively) with a resolution of about 1 eV [81, 184]. Thanks to the tunable feature of synchrotron sources, it provides the opportunity to detect the depths of different test samples. Besides, the high flux and high brilliance also enable routine high-resolution XPS [185, 186, 190]. Thus, synchrotron-based XPS is suitable to detect the chemical composition and valence state of battery materials or surface and bulk close to the material surface. Moreover, synchrotron sources with an excellent temporal resolution allow for interface studies approaching *in situ* conditions [187]. Based on the incident X-ray energy, researchers usually define measurements using photon energies below 2000 eV as soft X-ray photoelectron spectroscopy (abbreviated as SOXPES). Simultaneously, hard X-ray photoelectron spectroscopy (abbreviated as HAXPES), commonly referred to as XPS, uses photon energies up to 10 keV [81].

13.2.3.1 Hard X-ray photoelectron spectroscopy

In comparison with SOXPES, HAXPES implies a high-energy X-ray, with detect depth of about several tens of nanometers in a nondestructive manner [190]. The application of HAXPES in the research of battery materials started in the mid-2000s. After that, more and more studies utilized HAXPES in the research of SIBs electrode materials. Such as SEI composition analysis, Na storage mechanism into hard-carbon and binder effect evaluation et.al. In recent years, with the extensive study of SIBs, HAXPES has also been used to extract surface and interfacial information [185–187].

13.2.3.1.1 SEI/CEI composition analysis and evolution probing

The electrode/electrolyte interphase in batteries generally refers to solid electrolyte interphase (SEI) or cathode electrolyte interphase (CEI). Variation in SEI/CEI thickness, composition, structure and morphology, and SEI functions (stability, solubility, ion transfer properties, etc.) severely impact the stability and irreversibility of electrode materials, subsequently influencing the electrochemical performance [189, 191, 192]. Hence, it is vital to characterize and understand the surface reactions to achieve a stable SEI/CEI composition.

HAXPES can simultaneously detect information about the interface and the bulk material beneath the interface, as well as the surface-sensitive features, indicating it as one of the most suitable methods to study the interface of batteries. Younesi et al. have investigated Sn_4P_3, a promising candidate with high-performance capacity, by using the HAXPES method to illuminate the SEI properties during the first 1.5 cycles [193]. The used half-cells

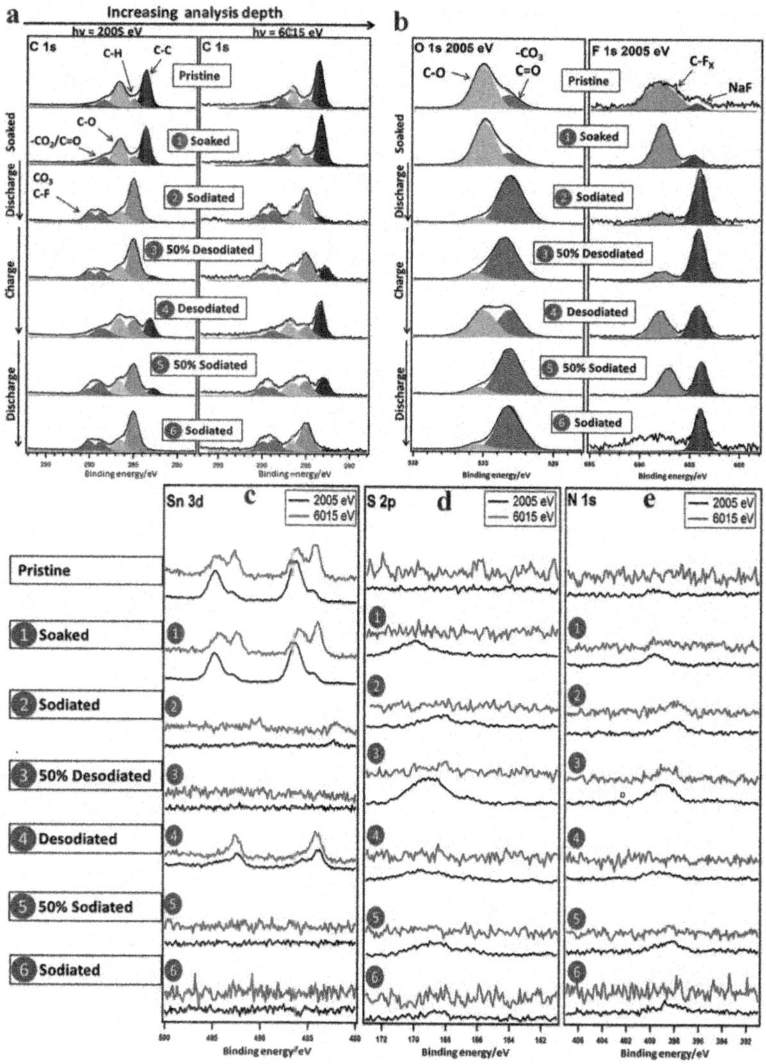

Figure 13.29 HAXPES curve fitted C 1s spectra of Sn_4P_3 electrodes cycled in 1 M NaFSIEC/DEC (1:1) with 5vol% FEC with excitation energies of 2005 eV (a) and 6016 eV (b). (c) O 1s and F 1s (d) spectra of Sn_4P_3 electrodes cycled in 1 M NaFSI in EC/DEC (1:1) with 5vol% FEC at different SOC, measured by photon energy of 2005 eV. Sn 3d (e), S 2p (f), and N 1s spectra (g) of pristine and soaked Sn_4P_3 electrodes, as well as Sn_4P_3 electrodes, stopped at different SOC in sodium half-cells. Black and redlines represent 2005 and 6015 eV excitation energies, respectively. Reproduced with permission from Ref. [193]. Copyright 2017, Elsevier.

versus metallic sodium were cycled in a 1 M NaFSI in EC:DEC (with a mole ratio of 1:1) electrolyte, including 5 vol% FEC as SEI-forming additive. The HAXPES spectra of C 1s, O1s, F1s, Sn 3d, S 2p, and N1s were thoroughly studied at different cycle states, as presented in Fig. 13.29. Subsequently, SEI thickness and composition information were extracted, and it was concluded that with the cycle proceeding, the SEI composition and thickness of the Sn_4P_3 electrodes change. Based on the HAXPES C 1s and Sn 3d spectra and the probe depth using different excitation energies, they inferred that the SEI thickness was enough to obscure both the carbon black and Sn_4P_3 signals. In contrast, the signals reappeared during the desodiation process.

The SEI composition can be analyzed from the F 1s and O 1s spectra. The generation of carbonates or oxides while sodiation and displaying organic C–O and C=O bonds while desodiationin combination with reappearing substrate signals during charge illustrates the loss of SEI compounds instead of one layer covering another recognized layer. Thus, they concluded that the SEI is not retained throughout the full sodiation/desodiation cycle in the electrolyte used in the research. They have also investigated the capacity-fading mechanism of tin phosphide anodes in SIBs in the subsequent work. They assessed the influence on the performance of Sn_4P_3anode of two different electrolyte salts of sodium-bis(fluorosulfonyl)imide (NaFSI) and sodium hexafluorophosphate ($NaPF_6$) mixed with ethylene carbonate and diethylene carbonate with 10 vol% fluoroethylenecarbonate as the electrolyte. The HAXPES results also showed that the thickness of the SEI changed during cycling. Besides, NaFSI is more effective than $NaPF_6$ in passivating the tin surface when the electrodes are desodiated at 2.5 V potential. When the desodiation potential was too high, the SEI on tin phosphide electrodes in $NaPF_6$-based electrolyte cells was stripped from the tin particles. The authors suggested that carbonate-based solvents in combination with either salt could be fully functional for a tested system.

Like the SEI on the anode part, a chemical reaction(s) will also occur at the cathode/electrolyte interface at a high voltage. However, little attention has been given to CEI in LIBs since the cathode materials usually operate within the stable voltage range of the organic electrolyte (<4.3 V). Different from its counterpart,

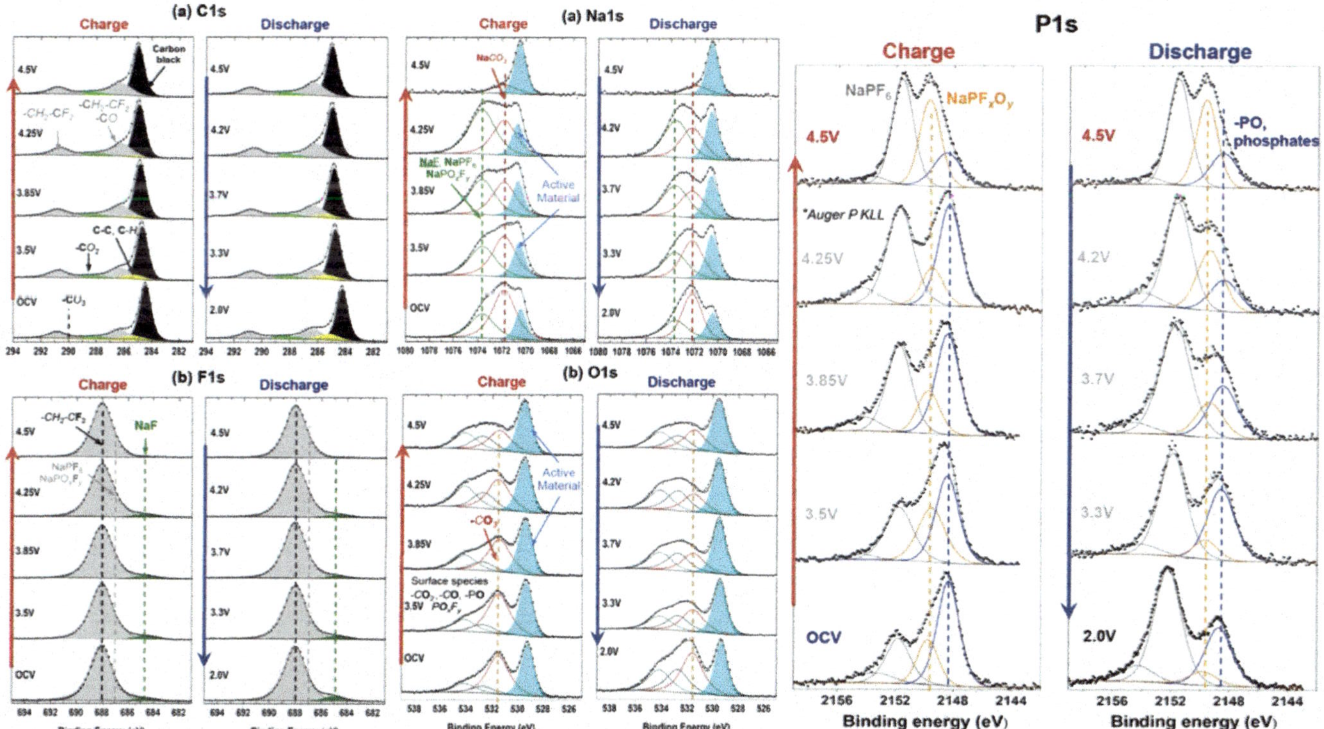

Figure 13.30 The C1s (a), F1s (b), Na 1s (c), O1s (d), and P 1s (e) HAXPES spectra for $Na_xCo_{2/3}Mn_{2/9}Ni_{1/9}O_2$ at different sodiation/desodiation states. The cut-off voltage ranges from 2.0 to 4.5 V. These spectra were recorded with a photon energy of 4000 eV. Reproduced with permission from Ref. [194]. Copyright 2016, Wiley-VCH.

the cathode Coulombic efficiency in SIBs is usually lower than LIB cathodes when charged to 4 V or above. This mainly attributes to the Na/Na's higher standard potential (≈0.3 V) than the Li/Li [15, 57, 58]. Thus, CEI is crucial to investigate during the cycle process, which primary influence the Na-transport kinetics, cell impedance, and capacity retention. Doubaji et al. used two different excitation energies, 1486.6 eV and 4000 eV, to probe the interface of the pristine $Na_xCo_{2/3}Mn_{2/9}Ni_{1/9}O_2$ sample without contact with an electrolyte (Fig. 13.30a) [194]. The C1s and F1s core levels show that Na_2CO_3 and NaF are present on the electrode's surface, which can be contributed to the synthesis process and the decomposition of PVDF. Performance degradation was also carried out by extracting information from the C1s, F1s, Na1s, and O1s core-level spectra obtained upon cycling. The results demonstrated inorganic compositions, including Na_2CO_3 and NaF, of the CEI variation during charge and discharge.

In contrast to Na_2CO_3 and NaF, the other oxygenated and carbonaceous species (–CO and –CO_2) were barely affected by the cycling. Other CEI ingredients such as $NaPF_6$ and its degradation products, including fluorophosphates and phosphates, were also investigated by analyzing the P1s core-level HAXPES spectra in Fig. 13.30e. It was found that phosphates were the main degradation products at low potential, while fluorophosphates ($NaPO_xF_y$) were found to be the majority at high potential. A detailed study on the electrode/electrolyte interface on cathode materials emphasized the importance of CEI in SIBs.

13.2.3.1.2 Redox process monitoring

As one of the promising candidates of SIB cathodes, layered oxides with the general formula Na_xMO_2 (M=Co, Ni, Mn, Al, Fe, etc.) have been proposed. An understanding of the redox reaction and charge compensation mechanisms is highly relevant when designing SIB cathodes. However, the redox reaction is usually a complicated process, especially with the incorporation of various TMs. Hence, the oxidation state of the TM ions and their evolution with electrochemical cycling is difficult to detect [60, 62]. Recently, Younesi et al. have studied the charge compensation mechanism in P2-$Na_{0.56}Mg_{0.04}Ni_{0.19}Mn_{0.70}O_2$ with the combined use of XAS, HAXPES, and RIXS [195]. The HAXPES spectra of Mn 3s use an excitation energy of 2350 eV, and Ni 2p and O1s with surface-sensitive photon energies (1090 eV, 2350 eV, and 7050 eV) were obtained at different cycle states. Since no strong conclusion could be drawn from the Ni 2p HAXPES, only Mn 3s and O1s HAXPES spectra have been cited in this part (Fig. 13.31).They predicted Mn's oxidation with the multiplet splitting (MS) gap energy since the relationship between the MS and the oxidation is linear. As shown in Fig. 13.31, Mn^{4+} is mainly present in the desodiated samples. After the full cycle, Mn goes through a reduction process and Mn^{4+} is reduced to Mn^{2+}. In the second cycle, Mn^{2+} is partially oxidized to Mn^{4+} during desodiation.

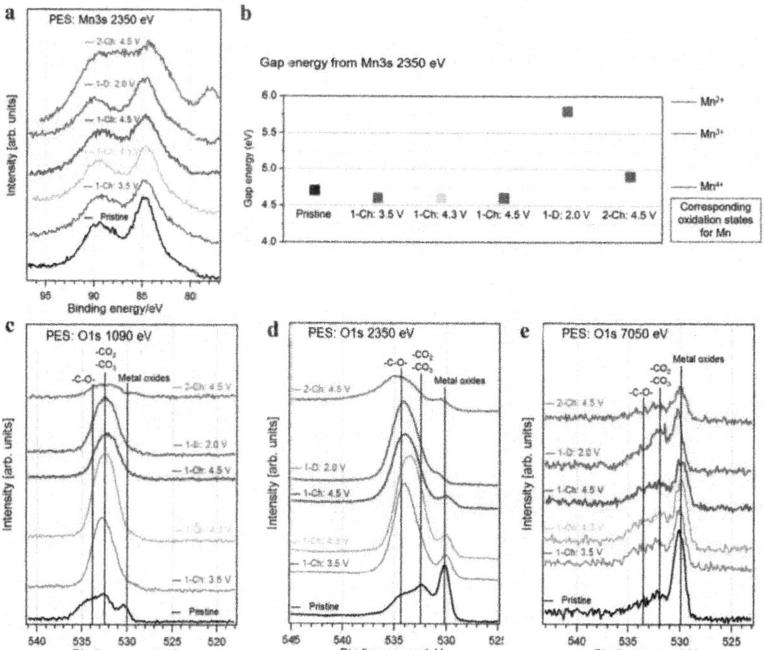

Figure 13.31 HAXPES results of Mn 3s and O1s. (a) HAXPES spectra at different cycle states of Mn 3s at 2350 eV and (b) with corresponding gap energies to identify the oxidation state. (d, e) The O1s spectra were measured at excitation energies of 1090 eV, 2350 eV, and 7050 eV, respectively. Reproduced with permission from Ref. [195]. Copyright 2019, Nature Publishing Group.

The HAXPES results are in good agreement with the XAS measurements. The HAXPES O1s spectra (Fig. 13.31e) show intense peaks associated with carbonate species during desodiation and sodiation processes, which correspond to electrolyte decomposition products. Consolidating results from both XAS and PES measurements for Mn, Ni, and O demonstrate the Mn and Ni's charge compensation mechanism can be described as follows: During the first desodiation, bulk Mn^{2+}isoxidized to Mn^{4+} and surface Mn^{4+} remains mostly unchanged. The surface Ni does not appear to change throughout the first charge process. During the first sodiation, both surface and bulk Mn become redox activated. On sodiation (charge) in the second cycle, surface Ni^{2+} remains unchanged but bulk Ni^{2+} is oxidized to Ni^{3+}due to the formation of electrolyte decomposition products. The redox mechanism can be briefly summarized as follows: Bulk Mn and Ni have been oxidized, whereas surface TMs remain unchanged during the first desodiation cycle due to the generated electrolyte decomposition products. However, during the second charge process, both surface and bulk Mn and N become partially activated to participate in redox processes. The research has enhanced the understanding of redox reactions and charge compensation mechanisms for this P2-$Na_{0.56}Mg_{0.04}Ni_{0.19}Mn_{0.70}O_2$ material.

13.2.3.1.3 Evaluation of electrolyte additive effect

As mentioned earlier, SEI plays a critical role in the electrochemical performance of SIBs. The quality of an ideal ionically conducting while electronically insulating, SEI mainly influenced by traits including: complete, dense, stable and mechanical flexible. one standard method to modify SEI layer is achieved by addition of film-formation additives [196–199], e.g., fluoroethylene carbonate (FEC,) and vinylene carbonate (VC) [200, 201]. The effect on the composition and formation of the interphase of FEC and VC additives applied to black phosphorus electrode in conventional 1.0 M $NaPF_6$/EC-DEC with a volume ratio of 1:1 was investigated. Figure 13.32 shows the HAXPES spectra of C 1s, O 1s, and P 1s for the black P/PANa (poly(sodium acrylate)) electrodes after the first cycle and 20th cycle in the SIBs with an excitation energy of 7938.9 eV.

Analysis results of the HAXPES spectra showed the disappeared P^0 and Na_xP peaks in the additive-free sample after 20 cycles. The SEI generated in the additive-free electrolyte was thicker and inhomogeneous compared to that formed in the electrolyte with FEC or VC additives, which protects the electrode against electrolyte

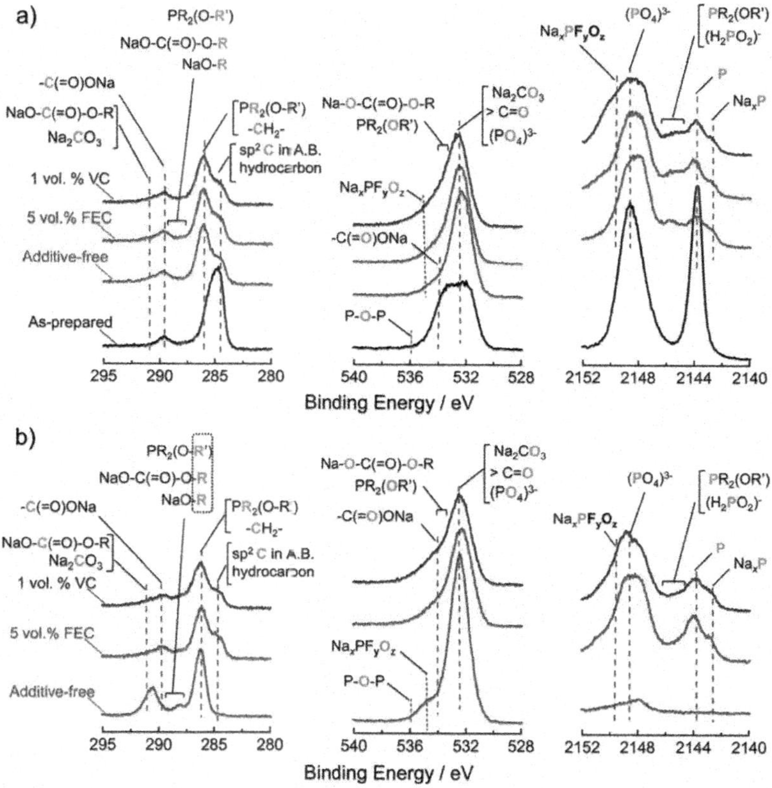

Figure 13.32 HAXPES spectra of C 1s, O 1s, and P 1s regions for the black P/PANa electrodes after the first cycle (a–c) and after the 20th cycle (d–f) in the Na cell. The excitation energy is 7938.9 eV. Electrode as-prepared (black), additive-free (green), with 5% FEC (red) and 1% VC (blue). Reproduced with permission from Ref. [202]. Copyright 2016, American Chemical Society.

decomposition. Furthermore, using time-of-flight secondary ion mass spectrometry (TOF-SIMS), they suggested that the SEI layer in the electrolyte with VC additive consisted of both organic and inorganic compounds. In contrast, signals of mainly inorganic compounds were detected on the SEI layer formed in FEC with electrolyte additive and additive-free electrolyte. Besides, the cycle performance indicated that compared with FEC-added SIB cells, the VC-added electrolyte significantly improves the reversible capacity and achieves longer cycle life for black P electrodes with $NaPF_6$ in EC/DEC. Based on the test result, the authors proposed that VC can be a promising candidate as an electrolyte additive to realize rechargeable SIBs.

13.2.3.2 Soft X-ray photoelectron spectroscopy

SOXPES is referred to as measurements utilizing photon energies below 2000 eV. The soft X-ray region provides high resolution and high cross section, giving SOXPES the ability to detect most of the light elements apart from H and He as well as detect depth range from a few nanometers to below a nanometer [142, 164].

13.2.3.2.1 Surface layer evolution monitoring

The SOXPES method with high surface sensitivity is usually applied to distinguish the atomic, chemical environments, and oxidation states during the formation of the surface layer [203–207]. A recent study on the evolution of surface layer on graphite during electrochemical sodium–tetraglyme co-intercalation was conducted by Valvo et al. using the SOXPES technique (Fig. 13.35) [234]. The SOXPES spectra were tested with the use of sodium-bis(fluorosulphonyl)imide (NaFSI) salt dissolved in tetraethylene glycol dimethyl ether (TEG-DME) as an electrolyte, which confirmed the formation of SEI upon cycling. The existence of hydrocarbon in the as-formed SEI was confirmed based on the analysis of C1s and O1s PES spectra. They suggested that the hydrocarbon derived from either the binder or the solvent decomposition induced by the resolution of NaFSI at potentials below 0.5 V versus Na^+/Na and seems to be a critical factor in the formation of the surface layer. The existence of bulk graphite PES peak even after electrochemical cycling demonstrates that the formed SEI seems to be fairly thin, which can be calculated in the range of 3–8 nm. From the analysis of the changes in the relative intensity of the C 1s and O 1s spectra, it can be concluded that the thin SEI facilitates the co-insertion of the Na^+ and the solvent molecules (glymes) into the bulk graphite and thus leads to the formation of the so-called "ternary graphite intercalation compounds" (tGICs). This work represents a valuable guideline for future work to focus on electrolyte salt contributions to the SEI formation.

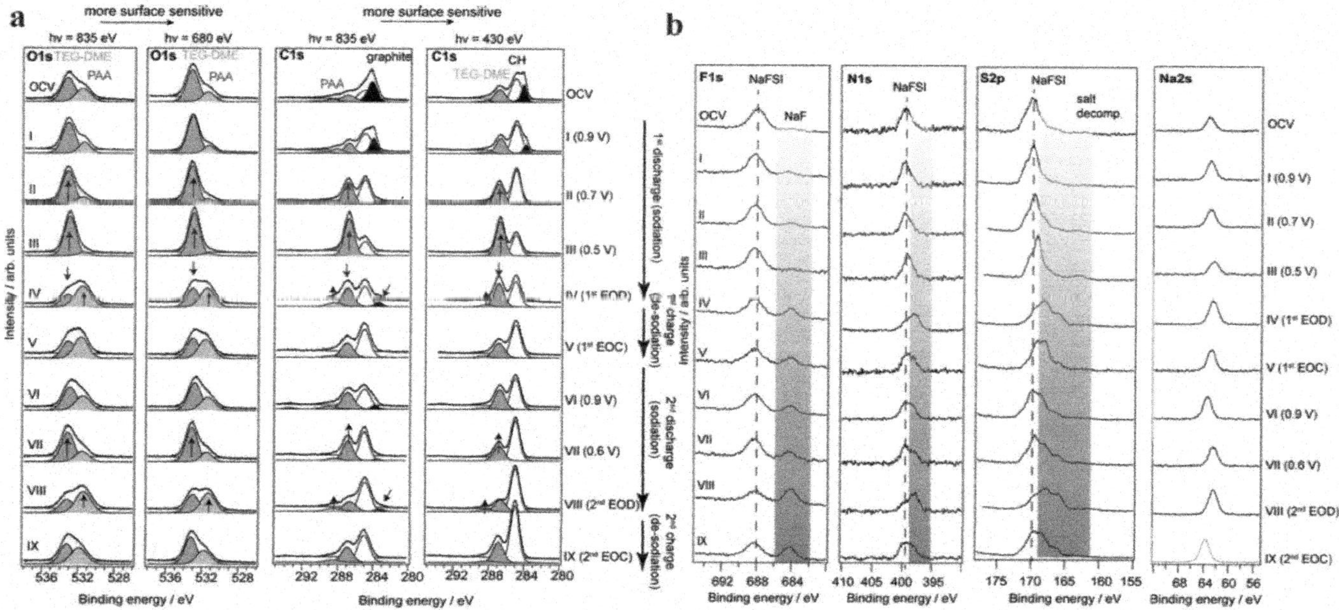

Figure 13.33 O 1s, C 1s, F 1s, N 1s, S 2p, and Na 2s SOXPES spectra of the electrode samples recorded at different cycle states. (a, b) PES spectra for O 1s, C 1s obtained with two different photon energies. (c) F 1s, N 1s, S 2p, and Na 2s SOXPES spectra of the various electrode samples recorded at 835 eV excitation energy. Reproduced with permission from Ref. [234]. Copyright 2017, American Chemical Society.

13.2.3.2.2 Evaluation of binder effect

Binders (usually a polymer compound) connect cathode/anode active materials, conductive agent, and current collector to prepare sheet electrodes, as one of the essential value-added materials in batteries, which play an important role in battery performance [208–211]. However, a few studies have focused on how binders affect the electrochemical and interfacial properties of batteries. Komaba et al. have studied the influence of different binders such as PVDF, sodium carboxymethyl cellulose (CMC), and monofluoroethylene carbonate (FEC) additive. The reversibility, surface layer, and initial irreversible capacity for hard carbon with the combined use of SOXPES, HAXPES, and electrochemical tests are shown in Fig. 13.34 [212]. In Fig. 13.34, the HAXPES spectra for PVDF are shown. The intensity of Na_xPF_y and $Na_xPO_yF_z$ become slower due to the FEC additive and is further decreased for CMC, illustrating that electrolyte decomposition is suppressed mostly for CMC. The SOXPES measurements have

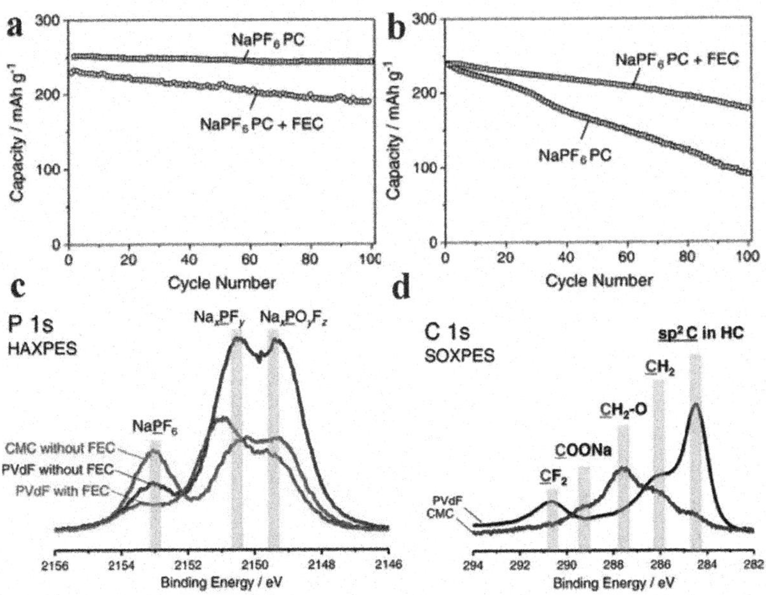

Figure 13.34 (a, b) Cycle performance of hard-carbon electrodes in $NaPF_6$ PC solutions with or without FEC addition: (a) CMC and (b) PVDF binders. (c) P 1s for the hard carbon tested in $NaPF_6$ PC after 1st cycle and (d) SOXPES spectra of C 1 s for the as-prepared hard-carbon electrodes with PVDF and CMC binder. Reproduced with permission from Ref. [212]. Copyright 2014, Elsevier.

uncovered the difference in surface layers based on CMC and PVDF for the as-prepared electrodes, shown in Fig. 13.34. The nearly disappeared hard-carbon peak demonstrates that the hard carbon was uniformly covered by a CMC binder film, resulting in the cycle stability shown in Fig. 13 34a.

13.2.4 Synchrotron Radiation-Based X-Ray Imaging Techniques

In-depth understanding and exploration of battery materials with high safety and energy density require systematic exploration of the change in phase transition morphology and chemical composition during the electrochemical process [15, 213, 214]. Real-time observation of the chemical composition, phase, structure/morphology evolution during electrochemical reaction demands high spatial and time resolution. Imaging techniques have the advantage of giving information intuitively and offer "a seeing is believing" manner of understanding the variety during battery operation [81, 105]. Traditional characterization techniques like TEM have partially realized the functions mentioned earlier. However, the limitation is that nonvolatile ionic or Na_2O results in an electrochemical process different from real batteries based on organic electrolytes; this is due to the necessity to maintain high vacuum in the TEM chamber [14, 218]. X-ray imaging has been developed since the late 19th century [120], especially in recent years, by virtue of the advent of third-generation synchrotron light sources and beyond, which can generate high brilliance, wide energy bandwidth, short photon pulse, and coherence X-ray, which has largely propelled the advancement of synchrotron microscopy [219, 220]. Additionally, X-ray imaging with highly penetrative power and elemental sensitivity is significantly nondestructive due to quite large spot size. Also, during the electrochemical process, qualitative and quantitative changes can be monitored by designing an *operando* experiment. Thus, synchrotron-based X-ray imaging techniques have been extensively applied in the research and exploration of battery materials [221–226].

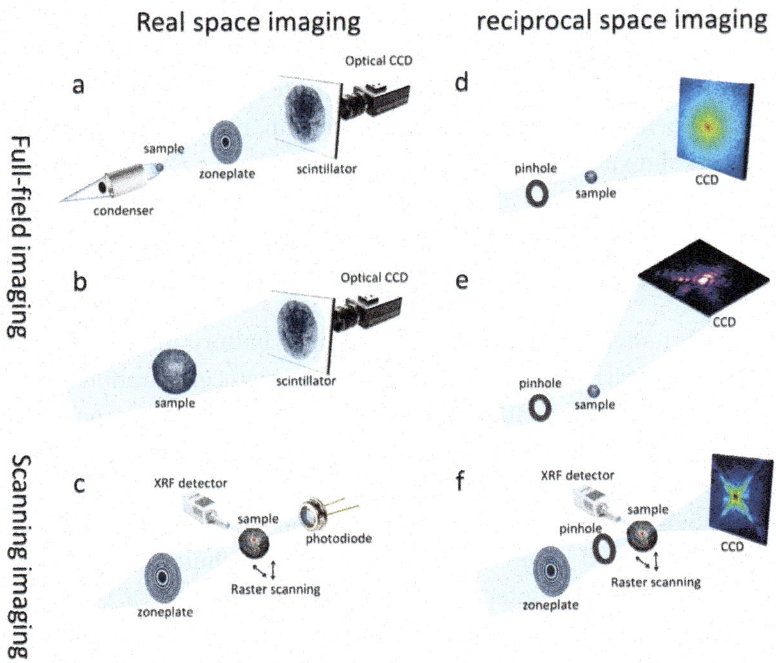

Figure 13.35 Schematic of different X-ray imaging methods. (a, b) The nanoscale and microscale full-field transmission X-ray microscopy, respectively. (c) Scanning micro/nano probe, in which the size of the focal spot determines the resolving power. (d) Planewave coherent diffraction imaging; (e) Bragg coherent diffraction imaging, in which the diffraction pattern surrounding the Bragg peak is recorded. (f) Scanning coherent diffraction imaging method, also known as ptychography. Reproduced with permission from Ref. [81]. Copyright 2017, American Chemical Society.

Emerging synchrotron-based X-ray imaging techniques mainly include X-ray projection imaging, transmission X-ray microscopy (TXM), scanning transmission X-ray microscopy (STXM), micro-X-ray fluorescence (XFM), tender and soft X-ray imaging, and coherent diffraction imaging (CDI). Schematic drawings of different X-ray imaging methods are given in Fig. 13.37. Features and limitations of these techniques are listed in Table 13.3 [81, 105, 218].

Table 13.3 Synchrotron-based X-ray imaging techniques for battery research [81, 105, 218]

Technique	Measures	Spatial Resolution	Time Resolution	Energy Range	Sample Thickness	Advantages	Limitations
TXM	Element; chemistry; 2D/3D imaging	20–30 nm	A few seconds to a few minutes	5–11 keV	Tens of nanometers to tens of micrometers	Chemical analysis	Weak image contrast for light elements
STXM	Element (including light elements)	12–40 nm	Hundreds of seconds to a few hours	<2 keV	Tens of nanometers to hundreds of nanometers	Chemical analysis	Thin sample, vacuum chamber, light-element mapping
XFM	Chemistry; 2D/3D imaging, element chemistry	Sub-micrometer	A few minutes to a few hours atomic	20–60 keV	Tens of micrometers	Trace element detection (down to ppm concentration)	Low spatial resolution
CDI	2D/3D imaging; 2D/3D electron density; strain mapping	Several nanometers	A few seconds	5–15 keV	Hundreds of nanometers to tens of micrometer	Strain and structural analysis	Limits of partial coherence; does not work for non-crystalline samples; beam damage

Depending on the setup of the imaging systems and information extraction principles, synchrotron-based X-ray imaging techniques can generally be classified into real-space imaging methods and reciprocal-space imaging methods [215, 216]. The images acquired from real-space imaging methods are visualized directly, while the images obtained from the real-space imaging methods need a reconstruction process. In terms of resolution, the ultimate spatial resolution can be on the order of 10 nm for soft X-rays, which is limited by the quality of the optics place between samples and recorder. Since the reciprocal-space imaging methods get images using the coherence property of X-ray, the limitation of optics is eliminated, bringing a higher spatial resolution that can be as high as 0.1 nm. X-ray imaging techniques can also be categorized as full-field and scanning X-ray imaging methods. Full-field X-ray imaging methods are images of record methods without mechanical movements during the test.

On the other hand, the scanning X-ray imaging methods are acquired with a raster scan of the samples [81, 83, 218]. This part will mainly focus on real-space imaging methods, including TXM and X-ray tomography, which are mainly mentioned in SIB research papers. Detailed information and application of other real-space imaging techniques and reciprocal-space imaging methods can be found in previous reviews and research papers [81, 105, 215-217, 225].

13.2.4.1 Applications of transmission X-ray microscopy in SIBs

TXM has been applied to acquire 2D and 3D morphological and chemical information about SIBs [218]. The typical work mechanism of TXM is shown in Fig. 13.37. In TXM, the synchrotron hard X-ray beam is first focused down to micrometers by a capillary condenser. The focusing X-ray beam then illuminates the sample and a Fresnel zone plate magnifies the X-ray beam. Thereafter, the incoming X-ray is converted into visible photons by the scintillator, and a charge-coupled device (CCD) is used to record the X-ray signals [81]. As a full-field technique, the resolution of TXM is determined by the width of the outmost zone of the zone plate, and the field of view set by the focused X-ray beam size. The typical resolution of TXM is 20–30 nm. Since the penetrative characteristics of hard X-ray and

TXM possess nearly nondestructive features combined with fast data acquisition speed, TXM is a suitable method for the *in operando* electrochemical test [215].

Absorption contrast and phase contrast are the most commonly used mechanisms for full-field TXM. Absorption contrast is based on the attenuation of X-ray intensity to form a projection image on a detector. When the photon energies cross the element absorption edge, X-ray absorption occurs. After detection by a CCD coupled with a scintillator, microscope images are achieved [218]. Chemical phase information can also be extracted by post-processing the XANES data. Although absorption contrast is a powerful mode to detect the chemical phase, oxidation state, and morphology changes in battery materials, it is not adequate to probe light elements due to the low absorption in the hard X-ray range. Taking advantage of the phase shift in a propagated X-ray beam, the phase contrast mode has been proved to be a sensitive way to probe information about low-atomic-number materials such as Li and Na [81, 83].

In addition to the 2D morphological and chemical mapping by TXM, another unique capability of the TXM technique is 3D tomography imaging. In 3D X-ray tomography, a series of images can be captured by viewing the samples from different angles. The process is realized by rotating the specimen along an axis perpendicular to the incoming X-ray beam. Then those 2D images are reconstructed into 3D images, which provide 3D morphology and structure information. Combined with high penetration, high flux, and nearly nondestructive synchrotron-based X-ray, the 3D X-ray tomography can also be operated in an *in operando* way [105, 218].

13.2.4.1.1 Research on morphological changes

As we all know, alloy and conversion anode materials such as Sn, Ge, Sb, Si, and P and conversion anode materials such as TM oxide and TM always suffer from volume change during Na/Li insertion and extraction [37–41, 59, 62–66]. This results in severe attenuation of battery capacity and shelf life, limiting their application as commercial battery electrode materials. Hence, an in-depth understanding of the morphology and volume changes during Na/Li insertion and extraction of such materials in lithium/sodium ions is essential. The TXM method has unique advantages since it can observe a thicker battery sample and detect a broader observation range than atomic force microscopy (AFM) and TEM. The first application of TXM imaging in battery is to observe the morphology evolution of electrode materials.

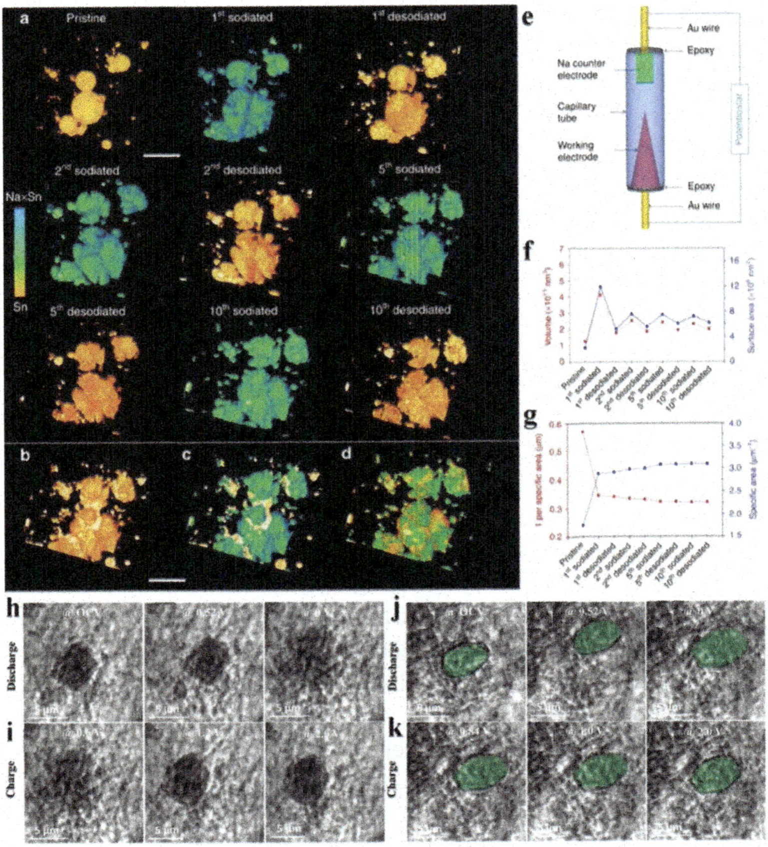

Figure 13.36 (a) Microstructural reversibility of Sn anode during 10 sodiation–desodiation cycles. (b–d) The overlapped images of (b) pristine/second desodiated, (c) pristine/tenth sodiated, and (d) second sodiated/tenth desodiated. (e) Schematic of electrochemical cells for Na-ion batteries. (f) Volume and surface area change during the electrochemical cycles. (g) Quantitative specific area changes during the electrochemical cycles. Scale bar, 10 mm. Reproduced with permission from Ref. [227]. Copyright 2015, Nature Publishing Group. (h–i) TXM images of bulk Sn during (h) sodiation and (i) desodiation. (j, k) TXM images of micrometer-sized nanoporous Sb/C anode during (j) sodiation and (k) desodiation. The colors in (j) and (k) have been adjusted for better visualization. Reproduced with permission from Ref. [228]. Copyright 2018, American Chemical Society.

Utilizing hard X-ray tomography, Wang et al. from the Brookhaven National Laboratory tracked the *in situ* Na (Li) interaction/extraction process of tin (Sn) anodes [223, 227]. To overcome the problem of block X-ray during rotation of traditional test cells, they designed an improved half-cell by sealing Na and counter-electrode materials in a quartz capillary (1 mm diameter) with two gold rods connected with the two electrodes (Fig. 13.36) [223]. The small quartz tube battery allows a 180° rotation without blocking the X-ray beam, hence becoming convenient to acquire 3D reconstruction images. Taking the advantages of this technique, 3D morphology evolution probing has been accomplished. They concluded that the microstructural damage of Sn anode mainly occurred in the sodiation process with 3D surface view, along with the 2D experiment. While in LIBs, the structural failure of Sn mainly occurs in the de-lithiation process. They further analyzed Sn's sodium intercalation/deintercalation process in the first 10 cycles of charge and discharge correlating with the X-ray tomography analysis and charge–discharge performance. As shown in Fig. 13.36, with quantitative analysis of volume and specific surface area change, they concluded that Sn anode's morphological degradation mainly occurs in the first cycle. In the second cycle, the microstructure reaches an equilibrium state. Compared with structural evolution of Sn anode in LIBs, contrary to what people generally think, the microstructures showed a relatively uniform sodiation and desodiation mechanism and maintained microstructural integrity upon cycling in SIBs. However, the microstructure cannot reach equilibrium after the first cycle for LIB, and further damage may occur during the following cycles and deteriorates the performance.

Wu et al. synthesized a micrometer-sized nanoporous Sb/C anode composite [228] containing 10 wt% C; it possesses micrometer-sized secondary particles. The electrode material shows outstanding cycling stability and exhibits a gravimetric capacity when tested as an anode for SIBs. The *in situ* TXM analysis at different potentials during the first sodiation–desodiation cycle revealed the fracture-free reversible feature of Sb, totally different from Sn and Si that fractured during Li/Na-alloying processes. Benefitting from the construct of unique microstructure and carbon composite, nanoporous Sb/C anode showed an even smaller volume and dimensional variation upon cycling, thus considerably reducing the deformation of the composite particles during the sodiation/desodiation cycle.

Conversion reaction mechanism-based anode materials generally have a high-performance specific capacity and are potential candidates for practical applications in SIBs. However, these materials usually suffer from volume changes and result in capacity fading during cycling. Hence, it is valuable to have an in-depth understanding of the sodiation/disodiation mechanism of these electrode materials. Recently, Wang et al. have synthesized Fe_3S_4 flower-like hierarchitectures via a one-step hydrothermal approach. It shows an ultrahigh reversible capacity of 762 mA/hg close to the theoretical value with an impressive first-coulomb efficiency of up to 96.3% [96]. It also exhibits excellent cycle stability with no obvious capacity fading after 500 cycles at a current density of 4 A/g. Except for analysis of the ether-based electrolyte function by HRTEM, they emphasized the important role of microstructure control using 3D X-ray tomography. As a comparison, they also synthesized Fe_3S_4 microspheres by simply changing the reaction condition (Fig. 13.37a). Figure 13.37e shows an apparent disparity in the cycle stability between microspheres and flower-like Fe_3S_4, when tested as an anode of SIBs. By utilizing the powerful 3D X-ray tomography technique, they have further characterized the Fe_3S_4 microspheres and 3D flower-like Fe_3S_4 nano-structured particle after cycling. The results noted that the hierarchitectures formed by the nanosheets of 3D flower-like Fe_3S_4 nanostructure are still clearly visible after cycling. Simultaneously, the microsphere Fe_3S_4 cracked during the cycling process, as shown in Fig. 13.37a, indicating that the 3D flower-like Fe_3S_4 anode overcomes the volume expansion during the cycling process. Taking advantage of the inherent high electronic conductivity in Fe_3S_4, they demonstrated the paramount importance of the synergistic strategy to improve the stability of SIBs through X-ray nano-tomography.

Wang et al. have unraveled the origins of an unusual "unreactive core" behavior that renders CuO inactive in the process of sodium-ion insertion with a synergistic combination of various synchrotron-based XRD, XAS, and 3D X-ray tomography [229]. The close-up views of the X-ray tomography of battery particle with different orientations in Fig. 13.37 show that the discharged particle preserves the microstructural integrity. They revealed that the formation is associated with the microstructural integrity. The cut-view in Fig. 13.37 displays the details of internal and external phases, which indicates that negligible structural degradation occurs. The sliced view in Fig. 13.37 also shows only the top and bottom parts in which the components were fully desodiated to a single phase. The existing "core–shell" phenomenon seriously hinders further sodiation in the subsequent cycle and results in poor cycle performance. However, they suggested that undesirable process can be eliminated by tuning the nanostructure of battery particles.

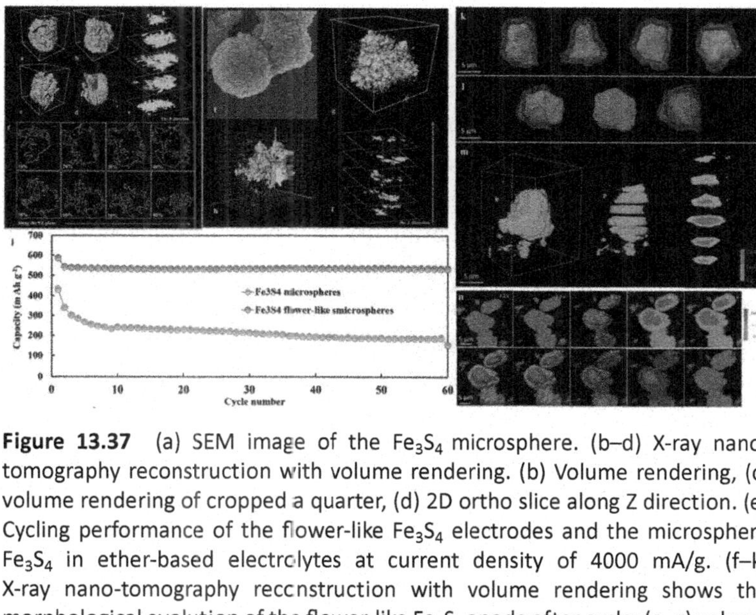

Figure 13.37 (a) SEM image of the Fe_3S_4 microsphere. (b–d) X-ray nano-tomography reconstruction with volume rendering. (b) Volume rendering, (c) volume rendering of cropped a quarter, (d) 2D ortho slice along Z direction. (e) Cycling performance of the flower-like Fe_3S_4 electrodes and the microsphere Fe_3S_4 in ether-based electrolytes at current density of 4000 mA/g. (f–k) X-ray nano-tomography reconstruction with volume rendering shows the morphological evolution of the flower-like Fe_3S_4 anode after cycle: (a–c) volume rendering, (d) cropped of one-eighth volume rendering, (e) 2D ortho slice along Z direction, (f) clipping plane of iso-contour slice along YZ plane. Reproduced with permission from Ref. [96] Copyright 2019, Elsevier. X-ray nano-tomography of unreactive core CuO particle in SIBs. (k–l) 3D views of unreactive core CuO particle from different angles. (m) Cut-away view of the unreactive core CuO. (n) Sliced view of the unreactive core CuO. Reproduced with permission from Ref. [229]. Copyright 2019, American Chemical Society.

13.2.4.1.2 *Phase transformation monitoring and elemental distribution analysis*

As mentioned earlier, phase transmission plays an essential role in applying electrode material design and test parameter settings.

In operando synchrotron-based TXM techniques provide a visual approach to monitor the phase transformation and elemental distribution during SIBs operation. Wang et al. interpreted that the origin of irreversible capacity degrades with the use of *in operando* TXM methods [230]. A sodium half-cell repacked from a typical 2032-type coin cell with 3 mm diameter Kapton windows on both sides was applied in the *in operando* test. The TXM-XANES images were *in operando* collected at different charge and discharge states. The chemical phase mapping shown in Fig. 13.38 was obtained, thoroughly processed, and fitted with the selected standard reference spectra. Three consecutive phases, including $NaNiO_2$, partly charged Na_xNiO_2, and fully charged state $Na_{0.19}NiO_2$, were investigated during the charging process. The chemical phase distribution showed that the phase transformations occur at all directions, and a "core–shell" reaction mechanism is suggested. Mixed color pixels in the middle of the particles were observed after the first cycle's discharge, which shows an irreversible phase conversion. To illustrate more on the irreversible mechanism, quantitative analysis of the chemical phase was conducted. As shown in Fig. 13.38, at high-voltage zones of above 4.0 V, the XANES spectra just exhibit a slight shape change with little absorption edge shift. The phenomenon is in agreement with the Ni K-edge spectra in $Na_x[Ni_{1/3}Mn_{2/3}]O_2$ ($0 < x < 2/3$) and $Na_{2/3}[Ni_{1/3}Mn_{2/3-x}Ti_x]O_2$ ($0 < x < 2/3$) reported by Meng et al. [231, 232]. They suggested partial Ni-ion migration from the octahedral sites to the inter-slab space, and the structural distortions may be responsible for their reversible capacity loss when $NaNiO_2$ is charged above 4.0 V.

Ma et al. used TXM-XANES techniques to verify further the presence of a nonequilibrium solid solution of $Na_{1-\delta}Ni_{1/3}Fe_{1/3}Mn_{1/3}O_2$ to find the composition of the solid solution (Fig. 13.38h–l) [233]. The 3D Ni XANES chemical map of the $Na_{1-\delta}Ni_{1/3}Fe_{1/3}Mn_{1/3}O_2$ was obtained over the potential range from 2.8 to 4.6 V. The collected TXM mapping clearly shows the existence of a substitutional solid solution of Na-poor $Na_{1-\delta}Ni_{1/3}Fe_{1/3}Mn_{1/3}O_2$. The compositions span between the two thermodynamic phases, $NaNi_{1/3}Fe_{1/3}Mn_{1/3}O_2$ and $Na_{1-\delta}Ni_{1/3}Fe_{1/3}Mn_{1/3}O_2$.

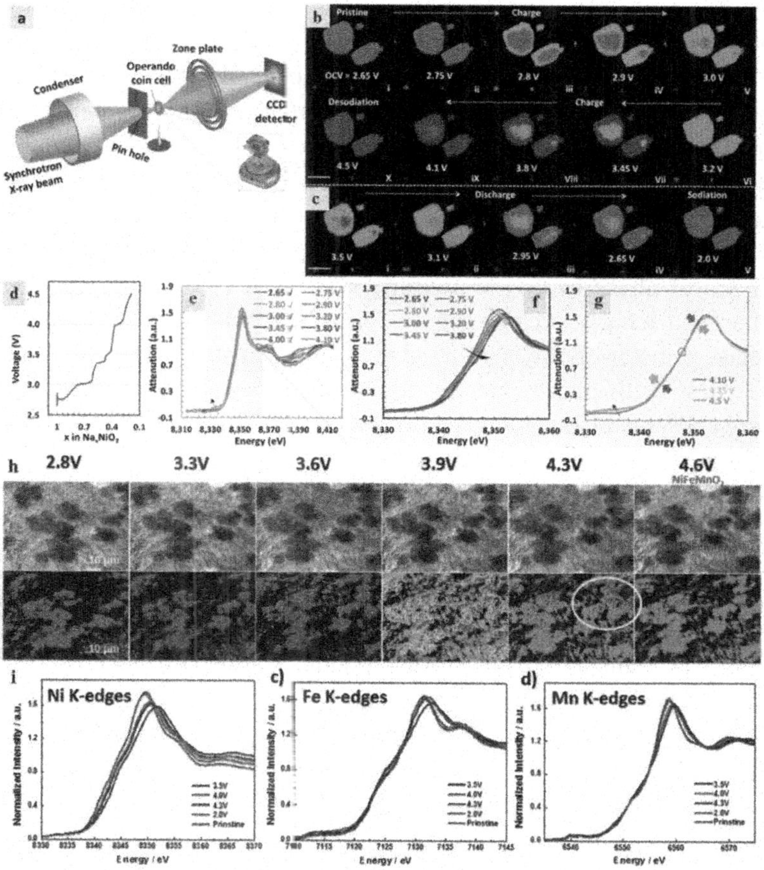

Figure 13.38 (a) Schematic illustration of the TXM experimental setup. (b, c) *In operando* hard X-ray spectra imaging of two-dimensional microstructural evolution of $NaNiO_2$ cathode (b) the charge process and (c) the discharge process. Scale bar: 10 μm. (d–g) *In operando* XANES spectra during the first charge process: (d) the charge profile of *in operando* cell and (e) the *in operando* XANES spectra collected at various states of charge. (f, g) The enlarged profiles from the selected area (the shadow section in b). Reproduced with permission from Ref. [230]. Copyright 2017, Elsevier. (h, i) Morphology evolution and chemical phase mappings of $Na_{1-\delta}Ni_{1/3}Fe_{1/3}Mn_{1/3}O_2$ using *in operando* TXM during charge. *Ex situ* XANES spectra evolution for (j) Ni, (k) Fe, and (l) Mn K-edges of the pristine $NaNi_{1/3}Fe_{1/3}Mn_{1/3}O_2$ and the $Na_{1-\delta}Ni_{1/3}Fe_{1/3}Mn_{1/3}O_2$ samples obtained at different charge states. Reproduced with permission from Ref. [233]. Copyright 2017, Wiley-VCH.

As aforementioned, the irreversible electrochemical behavior and large voltage hysteresis resulting in undesirable round-trip energy loss and low columbic efficiency commonly happen in conversion-reaction-based electrode materials. To seek solution to these challenges, Wang et al. selected FeS as a model battery material. They used TXM-XANES, 3D X-ray tomography, and other techniques to uncover the conversion reaction in sodium metal sulfide batteries. Direct visualization in Fig. 13.39a–o shows the chemical phase evolution and distribution of FeS during the first charge–discharge cycle. It can be clearly seen that at the early stage, phase transformation is initiated at the local surface of the particles. A core–shell phase model demonstrated with the discharge proceeds and phase boundaries propagate toward the particle core, accompanied by volume expansion.

Finally, the nearly pure iron phase was achieved. Subsequent charge process TXM images demonstrated that most of the particles transform back to the iron sulfide phase accompanied by sodium-ion extraction and volume shrinkage. However, a few different phase regions existing in the core contribute to the sluggish diffusion of Na ions in the bulk and result in a few Na ions trapped inside. Based on the standard spectra of the pure FeS and Fe samples, the chemical composition information was obtained with a linear combination fitting method (Fig. 13.39). The fitting results indicate the two-phase process conversion reaction of FeS during the sodiation process of the first cycle. During the sodiation process, microstructural fracture appears due to volume expansion. They emphasized that friction and cracks destroyed the structural integrity of FeS. However, it is favorable for the electrolyte to penetrate the electrode materials' inner part, hence stimulating the inner conversion reaction between FeS and Fe. The understanding of the reaction mechanism demonstrates the critical role of the microstructure design of electrode materials.

In recent years, synchrotron-based image methods have been widely used in the research of battery materials, due to the indirect visualization function. However, there are limitations such as temporal and spatial resolution. Since X-ray imaging capabilities with around 10 nm spatial resolution and sub-second time resolution are achievable, sustainable development is required to cater to higher requirements put forward by high energy storage

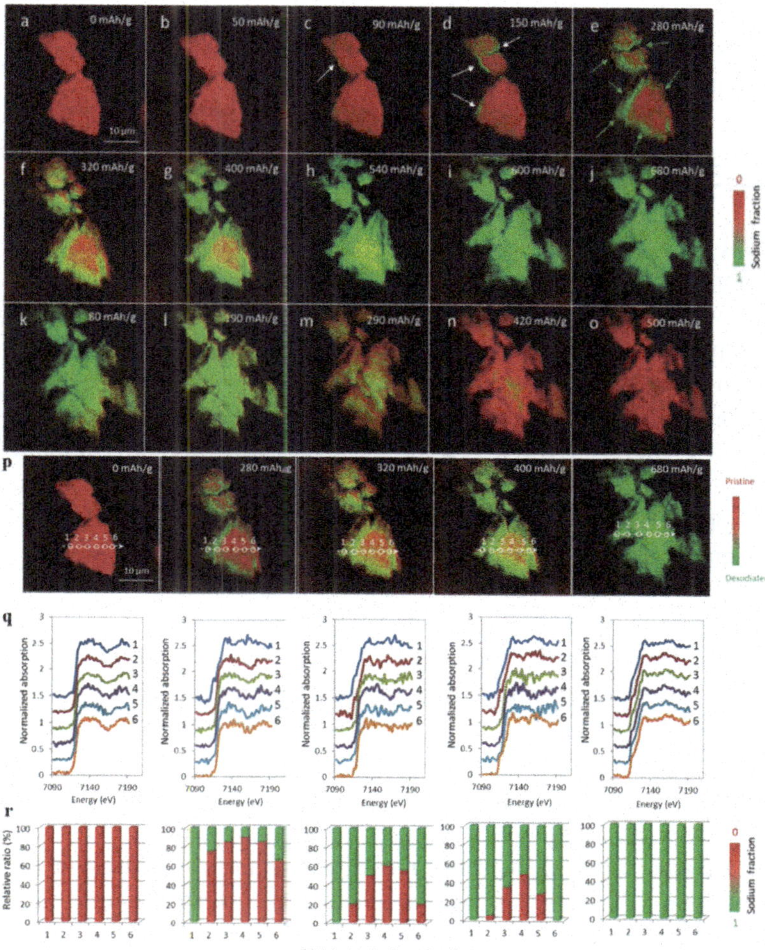

Figure 13.39 *In operando* 2D maps of FeS phase transformation at the first discharge/charge in Na-FeS batteries. The sodiation process (a–j) shows both morphological expansion/cracks and chemical phase transformation from FeS (red) to Fe (green). The descdiation process indicates a partial irreversibility in the core of the sample. (p–r) Phase composition analysis of selected states during the first sodiation cycle in Na-FeS batteries. Reproduced with permission from Ref. [234]. Copyright 2017, Wiley-VCH.

requirements. Another suggestion is related to possible beam damage in spite of the nearly nondestructive nature of X-ray beam. The high-flux synchrotron X-ray source provides the opportunity for time-resolved dynamic studies. However, it is not suitable for use in organic liquid electrolytes. Hence, it needs careful design of the *in operando* experiment by limiting the exposure dose and using more stable electrolytes.

13.3 Conclusion and Outlook

As a promising candidate for energy storage systems, SIBs aim to compete with commercial LIBs because of low assembly cost and the vast availability of Na source. Despite many achievements in the past few years, there is still a big gap between SIBs and state-of-the-art LIBs on energy density. One of the essential requirements is improving the energy density by developing advanced cathodes, anodes, and electrolytes. Another goal is to gain an in-depth understanding of the redox mechanism and chemical/physical properties of battery materials with the assistance of advanced diagnostic tools.

The past decades have witnessed significant progress and breakthroughs of diverse characterization techniques based on synchrotron radiation techniques. Those sophisticated techniques have played an essential role in the in-depth understanding of the redox mechanism and composition–structure–performance relationships of batteries with *ex situ*/*in situ* or *in operando* styles. In this chapter, widely used techniques such as SXRD, PDF, hXAS, SXS, PES, and synchrotron-based imaging techniques have been briefly introduced from basic concepts to practical applications in SIB electrode materials. The synchrotron-based techniques and their detailed application in SIBs have been roughly highlighted in this chapter. The purpose is to provide specific references to the research on the application of these synchrotron-based techniques in SIBs and other secondary batteries.

In fact, synchrotron-based techniques in battery research have achieved tremendous progress. However, there are still limitations and big room for improvement in these sophisticated characterization techniques. More attention should be given during synchrotron-related characterizations. As earlier mentioned , synchrotron X-ray based techniques are generally treated as nondestructive. However, during the application of synchrotron X-rays, especially for soft

X-ray tests, X-ray concentrated onto a small thin surface region *in situ*, will also results in X-ray beam damage, extremely serious for *in situ* experiments, which employ high-flux light sources. On the other hand, chemical and physical properties of materials in operating battery systems are quite complex and dynamic. Hence, one needs to pay much attention in selecting the proper method and experimental setup to ideally match the materials to achieve reliable and precise information. For example, in XAS tests, when the transmission mode is used, the thickness of test samples should be adjusted to make sure enough photon counting in the ionization chamber. High concentration and thickness should be taken into consideration due to the self-absorption effect for the fluorescence test mode.

The requirement of higher spatial and temporal resolution is inspiring the next-generation synchrotron facilities. Especially for battery systems, intermediate phase transition plays a key role in cycle stability and failure mechanism. A precise probe will be possible with the development of fourth-generation light source with high brightness and coherence, as well as small size of light sport. It is well known that *in situ/in operando* tests provide a nonequilibrium style to monitor a variety of variations in operational batteries and provide important information for battery material design and optimization. Thus, it is highly desirable to efficiently combine various synchrotron-based techniques to gain information as rich as possible from the tested samples. Moreover, it is also vitally important to develop *in situ* techniques beyond the methods mentioned earlier, such as *in situ* infrared spectroscopy, Raman spectroscopy, and sum-frequency generation spectroscopy. Besides, proper design of test cells for specific characterization techniques is also necessary and important. In addition, combine use of different techniques (such as synchrotron-based techniques combine use with neutron scientific facilities, TEM, AFM, mass spectrometry, et. al) to mention a few draw a complete map and more comprehensive information for SIBs materials.

In summary, there is still a long way to go in the practical application of SIBs with potentially higher performance to meet the requirements of energy storage. The rapid development of synchrotron light sources and the corresponding advanced techniques could significantly accelerate the practical use of SIBs.

Acknowledgment

This chapter is based on many valuable articles and related books cited here. We would like to express our sincere thanks and respect to the authors. Due to the limitation of the authors' knowledge, not all important and representative works have been included here. We thank the support from the National Key R&D Program of China (2020YFA0405800 and 2017YFA0303500), NSFC (U1932201), CAS International Partnership Program (211134KYSB20190063) and Collaborative Innovation Program of Hefei Science Center (2019HSC-CIP002), The University Synergy Innovation Program of Anhui Province (GXXT-2020-002) and USTC. L.S. acknowledges the support from Key Laboratory of Advanced Energy Materials Chemistry (Ministry of Education), Nankai University (111 project, B12015). We thank Dr. Changda Wang, Dr. Mukhtar Lawan Adam, Peter J. Chimtali and Shiqiang Wei for kind helps.

References

1. Goodenough, J. B. and K.-S. Park (2013) The li-ion rechargeable battery: A perspective. *J. Am.Chem. Soc.,* **135,** pp. 1167–1176.
2. Li, M., J. Lu, Z. Chen, and K. Amine (2018) 30 years of lithium-ion batteries. *Adv. Mater.*, **30**, pp. 1800561–1800585.
3. Ong, S. P., V. L. Chevrier, G. Hautier, A. Jain, C. Moore, S. Kim, X. Ma, and G. Ceder (2011) Voltage, stability and diffusion barrier differences between sodium-ion and lithium-ion intercalation materials. *Energy Environ. Sci.,* **4,** pp. 3680–3689.
4. Battistel, A., M. S. Palagonia, D. Brogioli, F. La Mantia, and R. Trócoli (2020) Electrochemical methods for lithium recovery: A comprehensive and critical review. *Adv. Mater.,* **32**, pp. 1905440–1905463.
5. Yabuuchi, N., K. Kubota, M. Dahbi, and S. Komaba (2014) Research development on sodium-ion batteries. *Chem. Rev.,* **114,** pp. 11636–11682.
6. Wang, L. P., L. Yu, X. Wang, M. Srinivasan, and Z. J. Xu (2015) Recent developments in electrodematerials for sodium-ion batteries. *J. Mater. Chem. A,* **3**, pp. 9353–9378.
7. Chayambuka, K., G. Mulder, D. L. Danilov, and P. H. L. Notten (2018) Sodium-ion battery materials and electrochemical properties reviewed. *Adv. Energy Mater.,* **8**, pp. 1800079–1800128.

8. Vaalma, C., D. Buchholz, M. Weil, and S. Passerini (2018). A cost and resource analysis of sodium-ion batteries. *Nat. Rev. Mater.,* **3**, pp. 18013–18024.
9. Whittingham, M. S. (1978) Chemistry of intercalation compounds: Metal guests in chalcogenide hosts. *Prog. Solid State Chem.*, **12**, pp. 41–99.
10. Abraham, K. M. (1982) Intercalation positive electrodes for rechargeable sodium cells. *Solid State Ionics*, **7**, pp. 199–212.
11. Delmas, C. (2018) Sodium and sodium-ion batteries: 50 years of research. *Adv. Energy Mater.,* **8**, pp. 1703137–1703146.
12. Hwang, J. Y., S. T. Myung, and Y. K. Sun (2017) Sodium-ion batteries: Present and future. *Chem. Soc. Rev.*, **46,** pp. 3529–3614.
13. Deng, J., W.-B. Luo, S.-L. Chou, H.-K. Liu, and S.-X. Dou (2018) Sodium-ion batteries: From academic research to practical commercialization. *Adv. Energy Mater.,* **8**, pp. 1701428–1701445.
14. Shadike, Z., E. Zhao, Y.-N. Zhou, X. Yu, Y. Yang, E. Hu, S. Bak, L. Gu, and X.-Q. Yang (2018) Advanced characterization techniques for sodium-ion battery studies. *Adv. Energy Mater.,* **8**, pp. 1702588–1702617.
15. Xu, G.-L., R. Amine, A. Abouimrane, H. Che, M. Dahbi, Z.-F. Ma, I. Saadoune, J. Alami, W. L. Mattis, F. Pan, Z. Chen, and K. Amine (2018) Challenges in developing electrodes, electrolytes, and diagnostics tools to understand and advance sodium-ion batteries. *Adv. Energy Mater.,* **8**, pp. 1702403–1702466.
16. Nayak, P. K., L. Yang, W. Brehm, and P. Adelhelm (2018) From lithium-ion to sodium-ion batteries: Advantages, challenges, and surprises. *Angew Chem. Int. Edit.,* **57**, pp.102–120.
17. Sun, Y., S. Guo, and H. Zhou (2018) Exploration of advanced electrode materials for rechargeable sodium-ion batteries. *Adv. Energy Mater.,* **9**, pp. 1800212–1800230.
18. Liang, L., X. Sun, D. K. Denis, J. Zhang, L. Hou, Y. Liu, and C. Yuan (2019) Ultralong layered $NaCrO_2$ nanowires: A competitive wide-temperature-operating cathode for extraordinary high-rate sodium-ion batteries. *ACS Appl. Mater. Inter.,* **11**, pp. 4037–4046.
19. Guignard, M., C. Didier, J. Darriet, P. Bordet, E. Elkaim, and C. Delmas (2013) P2-$Na_{(x)}VO_2$ system as electrodes for batteries and electron-correlated materials. *Nat. Mater.,* **12**, pp. 74–80.
20. Mortemard de Boisse, B., J. H. Cheng, D. Carlier, M. Guignard, C. J. Pan, S. Bordère, D. Filimonov, C. Drathen, E. Suard, B. J. Hwang, A. Wattiaux, and C. Delmas (2015) O3–$Na_xMn_{1/3}Fe_{2/3}O_2$ as a positive electrode material

for Na-ion batteries: Structural evolutions and redox mechanisms upon Na^+ (de)intercalation. *J. Mater. Chem. A,* **3**, pp. 10976–10989.

21. Xiao, Y., Y. F. Zhu, H. R. Yao, P. F. Wang, X. D. Zhang, H. Li, X. Yang, L. Gu, Y. C. Li, T. Wang, Y. X. Yin, X. D. Guo, B. H. Zhong, and Y. G. Guo (2019) A stable layered oxide cathode material for high-performance sodium-ion battery. *Adv. Energy Mater.,* **9**, pp. 1803978–1803976.
22. Barpanda, P., G. Oyama, S. Nishimura, S. C. Chung, and A. Yamada (2014) A 3.8-V earth-abundant sodium battery electrode. *Nat. Commun.,* **5**, pp. 4358–4366.
23. Li, F., Y. Zhao, L. Xia, Z. Yang, J. Wei, and Z. Zhou (2020) Well-dispersed $Na_3V_2(PO_4)_2F_3$@rGO with improved kinetics for high-power sodium-ion batteries. *J. Mater. Chem. A,* **8**, pp. 12391–12397.
24. Liu, X., W. Zuo, B. Zheng, Y. Xiang, K. Zhou, Z. Xiao, P. Shan, J. Shi, Q. Li, G. Zhong, R. Fu, and Y. Yang (2019) P2-$Na_{0.67}Al_xMn_{1-x}O_2$: Cost-effective, stable and high-rate sodium electrodes by suppressing phase transitions and enhancing sodium cation mobility. *Angew Chem. Int. Edit.,* **58**, pp. 18086–18095.
25. Xu, Y., Q. Wei, C. Xu, Q. Li, Q. An, P. Zhang, J. Sheng, L. Zhou,and L. Mai (2016) Layer-by-layer $Na_3V_2(PO_4)_3$ embedded in reduced graphene oxide as superior rate and ultralong-life sodium-ion battery cathode. *Adv. Energy Mater.,* **6**, pp. 1600389–1600398.
26. Zhu, C., C. Wu, C.-C. Chen, P. Kopold, P. A. van Aken, J. Maier, and Y. Yu (2017) A high power–high energy $Na_3V_2(PO_4)_2F_3$ sodium cathode: Investigation of transport parameters, rational design and realization. *Chem. Mater.,* **29**, pp. 5207–5215.
27. Wang, H., L. Wang, S. Chen, G. Li, J. Quan, E. Xu, L. Song, and Y. Jiang (2017) Crystallographic-plane tuned Prussian-blue wrapped with RGO: A high-capacity, long-life cathode for sodium-ion batteries. *J. Mater. Chem. A,* pp. 3569–3577.
28. Xie, B., P. Zuo, L. Wang, J. Wang, H. Huo, M. He, J. Shu, H. Li, S. Lou, and G. Yin (2019) Achieving long-life Prussian blue analogue cathode for Na-ion batteries via triple-cation lattice substitution and coordinated water capture. *Nano Energy,* **61**, pp. 201–210.
29. You, Y., X. Yu, Y. Yin, K.-W. Nam, and Y.-G. Guo (2014) Sodium iron hexacyanoferrate with high Na content as a Na-rich cathode material for Na-ion batteries. *Nano Res.,* **8**, pp. 117–128.
30. Chen, Y. and S. Manzhos (2016) A comparative computational study of lithium and sodium insertion into van der Waals and covalent tetracyanoethylene (TCNE)-based crystals as promising materials for

organic lithium and sodium ion batteries. *Phys. Chem. Chem. Phys.,* **18,** pp. 8874–8880.

31. Xiong, W., W. Huang, M. Zhang, P. Hu, H. Cui, and Q. Zhang (2019) Pillar[5]quinone-carbon nanocomposites as high-capacity cathodes for sodium-ion batteries. *Chem. Mater.,* **31,** pp. 8069–8075.

32. Zhu, L., Y. Niu, Y. Cao, A. Lei, X. Ai, and H. Yang (2012) n-type redox behaviors of polybithiophene and its implications for anodic Li and Na storage materials. *Electrochimica Acta,* **78,** pp. 27–31.

33. Jache, B. and P. Adelhelm (2014) Use of graphite as a highly reversible electrode with superior cycle life for sodium-ion batteries by making use of co-intercalation phenomena. *Angew Chem. Int. Edit.,* **53,** pp. 10169–10173.

34. Jian, Z., C. Bommier, L. Luo, Z. Li, W. Wang, C. Wang, P. A. Greaney, and X. Ji (2017) Insights on the mechanism of Na-ion storage in soft carbon anode. *Chem. Mater.,* **29,** pp. 2314–2320.

35. Novoselov, K. S., V. I. Fal'ko, L. Colombo, P. R. Gellert, M. G. Schwab, and K. Kim (2012) A roadmap for graphene. *Nature,* **490,** pp. 192–200.

36. Stevens, D. A. and J. R. Dahn (2001) The mechanisms of lithium and sodium insertion in carbon materials. *J. Electrochem. Soc.,* **148,** pp. A803–813.

37. Allan, P. K., J. M. Griffin, A. Darwiche, O. J. Borkiewicz, K. M. Wiaderek, K. W. Chapman, A. J. Morris, P. J. Chupas, L. Monconduit, and C. P. Grey (2016) Tracking sodium-antimonide phase transformations in sodium-ion anodes: Insights from operando pair distribution function analysis and solid-state NMR spectroscopy. *J. Am. Chem. Soc.,* **138,** pp. 2352–2365.

38. Liu, Y., N. Zhang, L. Jiao, and J. Chen (2015) Tin nanodots encapsulated in porous nitrogen-doped carbon nanofibers as a free-standing anode for advanced sodium-ion batteries. *Adv Mater,* **27,** pp. 6702–6707.

39. Yang, H., L. W. Chen, F. He, J. Zhang, Y. Feng, L. Zhao, B. Wang, L. He, Q. Zhang, and Y. Yu (2020) Optimizing the void size of yolk-shell Bi@Void@C nanospheres for high-power-density sodium-ion batteries. *Nano Lett.,* **20,** pp. 758–767.

40. Dahbi, M., N. Yabuuchi, M. Fukunishi, K. Kubota, K. Chihara, K. Tokiwa, X.-f. Yu, H. Ushiyama, K. Yamashita, J.-Y. Son, Y.-T. Cui, H. Oji, and S. Komaba (2016) Black phosphorus as a high-capacity, high-capability negative electrode for sodium-ion batteries: Investigation of the electrode/electrolyte interface. *Chem. Mater.,* **28,** pp. 1625–1635.

41. Li, W., Z. Yang, M. Li, Y. Jiang, X. Wei, X. Zhong, L. Gu, and Y. Yu (2016) Amorphous red phosphorus embedded in highly ordered mesoporous carbon with superior lithium and sodium storage capacity. *Nano Lett.*, **16**, pp. 1546–1553.

42. Wu, Y., Z. Liu, X. Zhong, X. Cheng, Z. Fan, and Y. Yu (2018) Amorphous red phosphorus embedded in sandwiched porous carbon enabling superior sodium storage performances. *Small,* **14**, pp. 1703472–1703440.

43. Mogensen, R., J. Maibach, W. R. Brant, D. Brandell, and R. Younesi (2017) Evolution of the solid electrolyte interphase on tin phosphide anodes in sodium ion batteries probed by hard X-ray photoelectron spectroscopy. *Electrochimica Acta,* **245**, pp. 696–704.

44. Sun, X., W. Li, X. Zhong, and Y. Yu (2017) Superior sodium storage in phosphorus@porous multichannel flexible freestanding carbon nanofibers. *Energy Storage Mater.,* **9**, pp. 112–118.

45. Wu, C., P. Kopold, P. A. van Aken, J. Maier, and Y. Yu (2017) High performance graphene/Ni_2P hybrid anodes for lithium and sodium storage through 3D yolk-shell-like nanostructural design. *Adv. Mater.,* **29**, pp. 1604015–1604021.

46. Patra, J., H.-C. Chen, C.-H. Yang, C.-T. Hsieh, C.-Y. Su, and J.-K. Chang (2016) High dispersion of 1 nm SnO_2 particles between graphene nanosheets constructed using supercritical CO_2 fluid for sodium-ion battery anodes. *Nano Energy,* **28**, pp. 124–134.

47. Rahman, M. M., A. M. Glushenkov, T. Ramireddy, and Y. Chen (2014) Electrochemical investigation of sodium reactivity with nanostructured Co_3O_4 for sodium-ion batteries. *Chem. Commun.,* **50**, pp. 5057–5060.

48. Xiong, H., M. D. Slater, M. Balasubramanian, C. S. Johnson, and T. Rajh (2011) Amorphous TiO_2 nanotube anode for rechargeable sodium ion batteries. *J. Phy. Chem. Lett.,* **2**, pp. 2560–2565.

49. Yu, L., J. Liu, X. Xu, L. Zhang, R. Hu, J. Liu, L. Ouyang, L. Yang, and M. Zhu (2017) Ilmenite nanotubes for high stability and high rate sodium-ion battery anodes. *ACS Nano,* **11**, pp. 5120–5129.

50. Zhang, N., X. Han, Y. Liu, X. Hu, Q. Zhao, and J. Chen (2015) 3D porous γ-Fe_2O_3@C nanocomposite as high-performance anode material of Na-ion batteries. *Adv. Energy Mater.,* **5**, pp.1401123–1401130.

51. Li, Y., Y. Liang, F. C. Robles Hernandez, H. Deog Yoo, Q. An, and Y. Yao (2015) Enhancing sodium-ion battery performance with interlayer-expanded MoS_2-PEO nanocomposites. *Nano Energy,* **15**, pp. 453–461.

52. Liu, Z., A. Daali, G. L. Xu, M. Zhuang, X. Zuo, C. J. Sun, Y. Liu, Y. Cai, M. D. Hossain, H. Liu, K. Amine, and Z. Luo (2020) Highly reversible sodiation/desodiation from a carbon-sandwiched SnS_2 nanosheet anode for sodium ion batteries. *Nano Lett.,* **20**, pp. 3844–3851.

53. Tabassum, H., C. Zhi, T. Hussain, T. Qiu, W. Aftab, and R. Zou (2019) Encapsulating trogtalite $CoSe_2$ nanobuds into BCN nanotubes as high storage capacity sodium ion battery anodes. *Adv. Energy Mater.,* **9**, pp. 1901778–1901788.

54. Tao, S., D. Wu, S. Chen, B. Qian, W. Chu, and L. Song (2018) A versatile strategy for ultrathin SnS_2 nanosheets confined in a N-doped graphene sheet composite for high performance lithium and sodium-ion batteries. *Chem. Commun.,* **54**, pp. 8379–8382.

55. Wu, L., H. Lu, L. Xiao, X. Ai, H. Yang, and Y. Cao (2015) Improved sodium-storage performance of stannous sulfide@reduced graphene oxide composite as high-capacity anodes for sodium-ion batteries. *J. Power Sources,* **293**, pp. 784–789.

56. Zhao, C., Y. Lu, Y. Li, L. Jiang, X. Rong, Y.-S. Hu, H. Li, and L. Chen (2017) Novel methods for sodium-ion battery materials. *Small Methods,* **1**, pp. 1600063–1600090.

57. Xiao, Y., N. M. Abbasi, Y. F. Zhu, S. Li, S. J. Tan, W. Ling, L. Peng, T. Yang, L. Wang, X. D. Guo, Y. X. Yin, H. Zhang, and Y. G. Guo (2020) Layered oxide cathodes promoted by structure modulation technology for sodium-ion batteries. *Adv. Func. Mater.,* **30**, pp. 2001334–2001361.

58. Xiang, X., K. Zhang, and J. Chen (2015) Recent advances and prospects of cathode materials for sodium-ion batteries. *Adv Mater,* **27**, pp. 5343–5364.

59. Bekaert, E., L. Buannic, U. Lassi, A. Llordés, and J. Salminen. (2017) *Emerging Nanotechnologies in Rechargeable Energy Storage Systems.* "Electrolytes for Li- and Na-ion batteries: Concepts, candidates, and the role of nanotechnology," (Elsevier), pp. 1–35.

60. Li, W., Z. Yang, M. Li, Y. Jiang, X. Wei, X. Zhong, L. Gu, and Y. Yu (2016) Amorphous red phosphorus embedded in highly ordered mesoporous carbon with superior lithium and sodium storage capacity. *Nano Lett.,* **16**, pp. 1546–1553.

61. Wu, Y., Z. Liu, X. Zhong, X. Cheng, Z. Fan, and Y. Yu (2018) Amorphous red phosphorus embedded in sandwiched porous carbon enabling superior sodium storage performances. *Small,* **14**, pp. 170347–170354.

62. Wu, S., C. Han, J. Iocozzia, M. Lu, R. Ge, R. Xu, and Z. Lin (2015) Germanium-based nanomaterials for rechargeable batteries. *Angew. Chem. Int. Edit.*, **55**, pp. 7898–7922.

63. Tan, H., D. Chen, X. Rui, and Y. Yu (2019) Peering into alloy anodes for sodium-ion batteries: Current trends, challenges, and opportunities. *Adv. Funct. Mater.*, **29**, pp. 1808745–1808776.

64. Mei, J., T. Liao, L. Kou, and Z. Sun (2017) Two-dimensional metal oxide nanomaterials for next-generation rechargeable batteries. *Adv. Mater.*, **29**, pp. 1700176–1700200.

65. Hu, Z., Q. Liu, S. L. Chou,and S. X. Dou (2017) Advances and challenges in metal sulfides/selenides for next-generation rechargeable sodium-ion batteries. *Adv. Mater.*, **29**, pp. 1700606–1700629.

66. Sun, Q., Q.-Q. Ren, H. Li, and Z.-W. Fu (2011) High capacity Sb_2O_4 thin film electrodes for rechargeable sodium battery. *Electrochem. Commun.*, **13**, pp. 1462–1464.

67. Chen, L., M. Fiore, J. E. Wang, R. Ruffo, D.-K. Kim, and G. Longoni (2018) Readiness level of sodium-ion battery technology: A materials review. *Adv. Sustain. Syst.*, **2**, pp.1700153–1700189.

68. Yu, L., L. P. Wang, H. Liao, J. Wang, Z. Feng, O. Lev, J. S. C. Loo, M. T. Sougrati,and Z. J. Xu (2018) Understanding fundamentals and reaction mechanisms of electrode materials for Na-ion batteries. *Small*, **14**, pp. 1703338–1703359.

69. Y., Cao (2020) The opportunities and challenges of sodium ion battery. *Energy Storage. Sci. Tech.*, **26**, pp. 2095–4239.

70. Boebinger, M. G., J. A. Lewis, S. E. Sandoval, and M. T. McDowell (2020) Understanding transformations in battery materials using in situ and operando experiments: Progress and outlook. *ACS Energy Lett.*, **5**, pp. 335–345.

71. Li, W.,G. Chu,J. Peng, H. Zheng, X. Li, J. Zheng, and H. Li (2014) Fundamental scientific aspects of lithium batteries (XII)-Characterization techniques. *Energy Storage Sci. Tech.*, **3**, pp. 2095–4239.

72. Clément, R. J., P. G. Bruce, and C. P. Grey (2015) Review-manganese-based P2-type transition metal oxides as sodium-ion battery cathode materials. *J. Electrochem. Soc.*, **162**, pp. A2589–A2604.

73. Bella, F., A. B. Muñoz-García, G. Meligrana, A. Lamberti, M. Destro, M. Pavone, and C. Gerbaldi (2017) Unveiling the controversial mechanism of reversible Na storage in TiO_2 nanotube arrays: Amorphous versus anatase TiO_2. *Nano Res.*, **10**, pp. 2891–2903.

74. Cao, Y., L. Xiao, M. L. Sushko, W. Wang, B. Schwenzer, J. Xiao, Z. Nie, L. V. Saraf, Z. Yang, and J. Liu (2012) Sodium ion insertion in hollow carbon nanowires for battery applications. *Nano Lett.*, **12**, pp. 3783–3787.

75. Duan, W., Z. Zhu, H. Li, Z. Hu, K. Zhang, F. Cheng, and J. Chen (2014) $Na_3V_2(PO_4)_3$@C core–shell nanocomposites for rechargeable sodium-ion batteries. *J. Mater. Chem. A*, **2**, pp. 8668–8675.

76. Wang, X., L. Fan, D. Gong, J. Zhu, Q. Zhang, and B. Lu (2016) Core-shell Ge@Graphene@TiO_2nanofibers as a high-capacity and cycle-stable anode for lithium and sodium ion battery. *Adv. Func. Mater.*, **26**, pp. 1104–1111.

77. Whiteside, A., C. A. Fisher, S. C. Parker, and M. S. Islam (2014) Particle shapes and surface structures of olivine $NaFePO(_4)$ in comparison to $LiFePO(_4)$. *Phys. Chem. Chem. Phys.*, **16**, pp. 21788–21794.

78. Lu, J., T. Wu, and K. Amine (2017) State-of-the-art characterization techniques for advanced lithium-ion batteries. *Nat. Energy*, **2**, pp. 17011–17023.

79. Lu, Y., L. Li, Q. Zhang, Z. Niu, and J. Chen (2018) Electrolyte and interface engineering for solid-state sodium batteries. *Joule*, **2**, pp. 1747–1770.

80. Peled, E. and S. Menkin (2017) Review-SEI: Past, present and future. *J. Electrochem. Soc.*, **164**, pp. A1703–A1719.

81. Lin, F., Y. Liu, X. Yu, L. Cheng, A. Singer, O. G. Shpyrko, H. L. Xin, N. Tamura, C. Tian, T. C. Weng, X. Q. Yang, Y. S. Meng, D. Nordlund, W. Yang, and M. M. Doeff (2017) Synchrotron X-ray analytical techniques for studying materials electrochemistry in rechargeable batteries. *Chem. Rev.*, **117**, pp. 13123–13186.

82. Gong, Z. and Y. Yang (2018) The application of synchrotron X-ray techniques to the study of rechargeable batteries. *J. Energy Chem.*, **27**, pp. 1566–1583.

83. Bak, S.-M., Z. Shadike, R. Lin, X. Yu, and X.-Q. Yang (2018) In situ/operando synchrotron-based X-ray techniques for lithium-ion battery research. *NPG Asia Mater.*, **10**, pp. 563–580.

84. Fuoss, P. H., K. S. Liang, and P. Eisenberger (1992) *Synchrotron Radiation Research: Advances in Surface and Interface Science*, Ed. Bachrach, R. Z. Vol. I, "Techniques," (Applied Materials Inc. Santa Clara, California and Stanford University Stanford, California), pp. 2–9.

85. Balerna, A. and S. Mobilio (2015) *Synchrotron Radiation: Basics, Methods and Applications*, Ed.Mobilio, S., Boscherini, F., and Meneghini, C., "Introduction to Synchrotron Radiation." Synchrotron Radiation. (Springer, Berlin, Heidelberg), pp. 3–28.

86. Zhao, Z. (2019) *Synchrotron Radiation in Materials Science Light Sources, Techniques, and Applications*, Ed. Fan, C. and Zhao, Z., Vol. I, "Synchrotron Light Sources." (Wiley-VCH), pp. 1–30.

87. Lee, D. H., J. Xu, and Y. S. Meng (2013) An advanced cathode for Na-ion batteries with high rate and excellent structural stability. *Phys. Chem. Chem. Phys.*, **15**, pp. 3304–3312.

88. Nam, K.-W., S.-M. Bak, E. Hu, X. Yu, Y. Zhou, X. Wang, L. Wu, Y. Zhu, K.-Y. Chung, and X.-Q. Yang (2013) Combining in situ synchrotron X-ray diffraction and absorption techniques with transmission electron microscopy to study the origin of thermal instability in overcharged cathode materials for lithium-ion batteries. *Adv. Func. Mater.*, **23**, pp. 1047–1063.

89. Pramudita, J. C., R. Aughterson, W. M. Dose, S. W. Donne, H. E. A. Brand, and N. Sharma (2014) Using in situ synchrotron X-ray diffraction to study lithium- and sodium-ion batteries: A case study with an unconventional battery electrode (Gd_2TiO_5). *J. Mater. Res.*, **30**, pp. 381–389.

90. Yang, Z., L. Trahey, Y. Ren, M. K. Y. Chan, C. Lin, J. Okasinski, and M. M. Thackeray (2015) In situ high-energy synchrotron X-ray diffraction studies and first principles modeling of α-MnO_2 electrodes in Li-O_2 and Li-ion coin cells. *J. Mater. Chem. A*, **3**, pp. 7389–7398.

91. Bleith, P., W. van Beek, H. Kaiser, P. Novák, and C. Villevieille (2015) Simultaneous in situ X-ray absorption spectroscopy and X-ray diffraction studies on battery materials: The case of $Fe_{0.5}TiOPO_4$. *J. Phys. Chem. C*, **119**, pp. 3466–3471.

92. Crumlin, E. J., Z. Liu, H. Bluhm, W. Yang, J. Guo, and Z. Hussain (2015) X-ray spectroscopy of energy materials under in situ/operando conditions. *J. Electron Spectrosc.*, **200**, pp. 264–273.

93. Liu, X., D. Wang, G. Liu, V. Srinivasan, Z. Liu, Z. Hussain, and W. Yang (2013) Distinct charge dynamics in battery electrodes revealed by in situ and operando soft X-ray spectroscopy. *Nat. Commun.*, **4**, pp. 2568–2567.

94. You, Y., X. Yu, Y. Yin, K.-W. Nam, and Y.-G. Guo (2014) Sodium iron hexacyanoferrate with high Na content as a Na-rich cathode material for Na-ion batteries. *Nano Res.*, **8**, pp. 117–128.

95. Zhou, Y.-N., J.-J. Ding, K.-W. Nam, X. Yu, S.-M. Bak, E. Hu, J. Liu, J. Bai, H. Li, Z.-W. Fu, and X.-Q. Yang (2013) Phase transition behavior of $NaCrO_2$ during sodium extraction studied by synchrotron-based X-ray diffraction and absorption spectroscopy. *J. Mater. Chem. A*, **1**, pp. 11130–11134.

96. Liu, Q., J. Gao, C. Cao, G. Yin, Z. Jiang, M. Ge, X. Xiao, W.-K. Lee, and J. Wang (2019) Insights into enhanced sodium ion storage mechanism in Fe_3S_4: The coupling of surface chemistry, microstructural regulation and 3D electronic transport. *Nano Energy,* **62**, pp. 384–392.

97. Wang, J., Y. C. Chen-Wiegart, and J. Wang (2014a) In operando tracking phase transformation evolution of lithium iron phosphate with hard X-ray microscopy. *Nat Commun,* **5**, pp. 4570–4579.

98. Wang, J., Y. C. Chen-Wiegart, and J. Wang (2014) In situ three-dimensional synchrotron X-ray nanotomography of the (de)lithiation processes in tin anodes. *Angew. Chem. Int. Edit.,* **53**, pp. 4460–4464.

99. Gibot, P., M. Casas-Cabanas, L. Laffont, S. Levasseur, P. Carlach, S. Hamelet, J. M. Tarascon, and C. Masquelier (2008) Room-temperature single-phase Li insertion/extraction in nanoscale $Li_{(x)}FePO_4$. *Nat. Mater.,* **7**, pp. 741–747.

100. Diaz, J., O. R. Monteiro, and Z. Hussain (2007) Structure of amorphous carbon from near-edge and extended X-ray absorption spectroscopy. *Phys. Rev. B,* pp. 76–87.

101. Uchaker, E., Y. Z. Zheng, S. Li, S. L. Candelaria, S. Hu, and G. Z. Cao (2014) Better than crystalline: Amorphous vanadium oxide for sodium-ion batteries. *J. Mater. Chem. A,* **2**, pp. 18208–18214.

102. Yu, P., L. Wang, Y. Xie, C. Tian, F. Sun, J. Ma, M. Tong, W. Zhou, J. Li, and H. Fu (2018) High-efficient, stable electrocatalytic hydrogen evolution in acid media by amorphous Fe_xP coating Fe_2N supported on reduced graphene oxide. *Small,* **14**, pp. 1801717–1801725.

103. Liu, H., F. C. Strobridge, O. J. Borkiewicz, K. M. Wiaderek, K. W. Chapman, P. J. Chupas, and C. P. Grey (2014) Batteries. Capturing metastable structures during high-rate cycling of $LiFePO(_4)$ nanoparticle electrodes. *Science,* **344**, pp. 1252817–1252823.

104. Orikasa, Y., T. Maeda, Y. Koyama, H. Murayama, K. Fukuda, H. Tanida, H. Arai, E. Matsubara, Y. Uchimoto, and Z. Ogumi (2013) Transient phase change in two phase reaction between $LiFePO_4$ and $FePO_4$ under battery operation. *Chem. Mater.,* **25**, pp. 1032–1039.

105. Cao, C., M. F. Toney, T.-K. Sham, R. Harder, P. R. Shearing, X. Xiao, and J. Wang (2020) Emerging X-ray imaging technologies for energy materials. *Mater. Today,* **34**, pp. 132–147.

106. Chen, J., D. H. C. Chua, and P. S. Lee (2019) The advances of metal sulfides and in situ characterization methods beyond Li ion batteries: Sodium, potassium, and aluminum ion batteries. *Small Methods,* **4**, pp. 1900648–1900673.

107. Harks, P. P. R. M. L., F. M. Mulder, and P. H. L. Notten (2015) In situ methods for Li-ion battery research: A review of recent developments. *J. Power Sources,* **288**, pp. 92–105.

108. Palomares, V., P. Serras, H. E. A. Brand, T. Rojo, and N. Sharma (2015) Structural evolution of mixed valent (V^{3+}/V^{4+}) and V^{4+} sodium vanadium fluorophosphates as cathodes in sodium-ion batteries: Comparisons, overcharging and mid-term cycling. *J. Mater. Chem. A,* **3**, pp. 23017–23027.

109. Yang, J., S. Muhammad, M. R. Jo, H. Kim, K. Song, D. A. Agyeman, Y. I. Kim, W. S. Yoon, and Y. M. Kang (2016) In situ analyses for ion storage materials. *Chem. Soc. Rev.,* **45**, pp. 5717–5770.

110. Brant, W. R., S. Schmid, G. Du, Q. Gu, and N. Sharma (2013) A simple electrochemical cell for in-situ fundamental structural analysis using synchrotron X-ray powder diffraction. *J. Power Sources,* **244**, pp. 109–114.

111. Liu, D., Z. Shadike, R. Lin, K. Qian, H. Li, K. Li, S. Wang, Q. Yu, M. Liu, S. Ganapathy, X. Qin, Q. H. Yang, M. Wagemaker, F. Kang, X. Q. Yang, and B. Li (2019) Review of recent development of in situ/operando characterization techniques for lithium battery research. *Adv. Mater.,* **31**, pp. 1806620–1806676.

112. Ren, Y. and X. Zuo (2018) Synchrotron X-ray and neutron diffraction, total scattering, and small-angle scattering techniques for rechargeable battery research. *Small Methods,* **2**, pp. 1800064–1800076.

113. Wei, X., X. Wang, Q. An, C. Han, and L. Mai (2017) Operando X-ray diffraction characterization for understanding the intrinsic electrochemical mechanism in rechargeable battery materials. *Small Methods,* **1**, pp. 1700083–1700101.

114. Borkiewicz, O. J., B. Shyam, K. M. Wiaderek, C. Kurtz, P. J. Chupas, and K. W. Chapman (2012) The AMPIX electrochemical cell: A versatile apparatus for in situ X-ray scattering and spectroscopic measurements. *J. Appl. Crystallogr.,* **45**, pp. 1261–1269.

115. Brant, W. R., D. Li, Q. Gu, and S. Schmid (2016) Comparative analysis of ex-situ and operando X-ray diffraction experiments for lithium insertion materials. *J. Power Sources,* **302**, pp. 126–134.

116. Nikolowski, K., C. Baehtz, N. N. Bramnik, and H. Ehrenberg (2005) A Swagelok-typein situcell for battery investigations using synchrotron radiation. *J. Appl. Crystallogr.,* **38**, pp. 851–853.

117. Rosciano, F., M. Holzapfel, H. Kaiser, W. Scheifele, P. Ruch, M. Hahn, R. Kotz, and P. Novak (2007) A multi-sample automatic system for

in situ electrochemical X-ray diffraction synchrotron measurements. *J. Synchrotron Radiat.*, **14**, pp. 487–491.

118. Dose, W. M., N. Sharma, J. C. Pramudita, H. E. A. Brand, E. Gonzalo, and T. Rojo (2017) Structure-electrochemical evolution of a Mn-rich P2 $Na_{2/3}Fe_{0.2}Mn_{0.8}O_2$ Na-ion battery cathode. *Chem. Mater.*, **29**, pp. 7416–7423.

119. Lee., J. Lu, Y. Ren, X. Luo, X. Zhang, J. Wen, D. Miller, A. DeWahl, S. Hackney, B. Key, D. Kim, M. D. Slater, and C. S. Johnson (2014) Layered P2/O3 intergrowth cathode: Toward high power Na-ion batteries. *Adv. Energy Mater.*, **4**, pp. 1400458–1400465.

120. Fang, Y., Q. Liu, L. Xiao, Y. Rong, Y. Liu, Z. Chen, X. Ai, Y. Cao, H. Yang, J. Xie, C. Sun, X. Zhang, B. Aoun, X. Xing, X. Xiao, and Y. Ren (2018) A fully sodiated $NaVOPO_4$ with layered structure for high-voltage and long-lifespan sodium-ion batteries. *Chem.*, **4**, pp. 1167–1180.

121. Wang, D., X. Bie, Q. Fu, D. Dixon, N. Bramnik, Y. S. Hu, F. Fauth, Y. Wei, H. Ehrenberg, G. Chen, and F. Du (2017) Sodium vanadium titanium phosphate electrode for symmetric sodium-ion batteries with high power and long lifespan. *Nat. Commun.*, **8**, pp. 15888–15896.

122. Pramudita, J. C., S. Schmid, T. Godfrey, T. Whittle, M. Alam, T. Hanley, H. E. Brand, and N. Sharma (2014) Sodium uptake in cell construction and subsequent in operando electrode behaviour of Prussian blue analogues, $Fe[Fe(CN)_{6}]_{(1-x)}.yH_2O$ and $FeCo(CN)_6$. *Phys. Chem. Chem. Phys.*, **16**, pp. 24178–24187.

123. Wang, W., Y. Gang, Z. Hu, Z. Yan, W. Li, Y. Li, Q. F. Gu, Z. Wang, S. L. Chou, H. K. Liu, and S. X. Dou (2020) Reversible structural evolution of sodium-rich rhombohedral Prussian blue for sodium-ion batteries. *Nat. Commun.*, **11**, pp. 980–987.

124. Lee, M., J. Hong, J. Lopez, Y. Sun, D. Feng, K. Lim, W. C. Chueh, M. F. Toney, Y. Cui, and Z. Bao (2017) High-performance sodium-organic battery by realizing four-sodium storage in disodium rhodizonate. *Nat. Energy*, **2**, pp. 861–868.

125. Yang, C., S. Xin, L. Mai, and Y. You (2020) Materials design for high-safety sodium-ion battery. *Adv. Energy Mater.*, pp. 2000974–2000990.

126 Yabuuchi, N., I. Ikeuchi, K. Kubota, and S. Komaba (2016) Thermal stability of Na_xCrO_2 for rechargeable sodium batteries; Studies by high-temperature synchrotron X-ray diffraction. *ACS Appl. Mater. Interfaces*, **8**, pp. 32292–32299.

127. Didier, C., M. Guignard M. R. Suchomel, D. Carlier, J. Darriet, and C. Delmas (2016) Thermally and electrochemically driven topotactical

transformations in sodium layered oxides Na_xVO_2. *Chem. Mater.,* **28**, pp. 1462–1471.

128. Tai, R., J. He, Y. Huang, J. Wang, X. Gao, T. Xiao, and J. Ma (2019) *Synchrotron Radiation in Materials Science: Light Sources, Techniques, and Applications*, Ed. Fan, C. and Zhao, Z., Vol. I, "Synchrotron Radiation Experimental Techniques." (Wiley-VCH), pp. 61–116.
129. Guignard, M., D. Carlier, C. Didier, M. R. Suchomel, E. Elkaïm, P. Bordet, R. Decourt, J. Darriet, and C. Delmas (2014) Vanadium clustering/declustering in P2-$Na_{1/2}VO_2$ layered oxide. *Chem. Mater.,* **26**, pp. 1538–1548.
130. Sava Gallis, D. F., H. D. Pratt III, T. M. Anderson, and K. W. Chapman (2016) Electrochemical activity of Fe-MIL-100 as a positive electrode for Na-ion batteries. *J. Mater. Chem. A,* **4**, pp. 13764–13770.
131. Sottmann, J., M. Di Michiel, H. Fjellvag, L. Malavasi, S. Margadonna, P. Vajeeston, G. B. M. Vaughan, and D. S. Wragg (2017) Chemical structures of specific sodium ion battery components determined by operando pair distribution function and X-ray diffraction computed tomography. *Angew. Chem. Int. Edit.,* **56**, pp. 11385–11389.
132. Wang, X., S. Tan, X.-Q. Yang, and E. Hu (2020) Pair distribution function analysis: Fundamentals and application to battery materials. *Chinese Phys. B,* **29**, pp. 028802–028812.
133. Xiang, K., W. Xing, D. B. Ravnsbaek, L. Hong, M. Tang, Z. Li, K. M. Wiaderek, O. J. Borkiewicz, K. W. Chapman, P. J. Chupas, and Y. M. Chiang (2017) Accommodating high transformation strains in battery electrodes via the formation of nanoscale intermediate phases: Operando investigation of olivine $NaFePO_4$. *Nano Lett.,* **17**, pp. 1696–1702.
134. Xu, S., J. Wu, E. Hu, Q. Li, J. Zhang, Y. Wang, E. Stavitski, L. Jiang, X. Rong, X. Yu, W. Yang, X.-Q. Yang, L. Chen, and Y.-S. Hu (2018) Suppressing the voltage decay of low-cost P2-type iron-based cathode materials for sodium-ion batteries. *J. Mater. Chem. A,* **6**, pp. 20795–20803.
135. Brennich, M., M. Cardenas, H. Castillo-Michel, M. Cotte, and M. Sztucki (2016) *Pharmaceutical Nanotechnology: Innovation and Production.* "Nanoparticle Characterization Methods: Applications of Synchrotron and Neutron Radiation: Innovation and Production." (Wiley-VCH Verlag GmbH and Co. KGaA.)
136. Clément, R. J., J. Billaud, A. Robert Armstrong, G. Singh, T. Rojo, P. G. Bruce, and C. P. Grey (2016) Structurally stable Mg-doped P2-$Na_{2/3}Mn_{1-y}Mg_yO_2$ sodium-ion battery cathodes with high rate performance: Insights from electrochemical, NMR and diffraction studies. *Energy and Environ. Sci.,* **9**, pp. 3240–3251.

137. Pang, W. K., S. Kalluri, V. K. Peterson, N. Sharma, J. Kimpton, B. Johannessen, H. K. Liu, S. X. Dou, and Z. Guo (2015) Interplay between electrochemistry and phase evolution of the P2-type $Na_x(Fe_{1/2}Mn_{1/2})O_2$ cathode for use in sodium-ion batteries. *Chem. Mater.*, **27**, pp. 3150–3158.

138. Aquilanti, G., M. Giorgetti, R. Dominko, L. Stievano, I. Arčon, N. Novello, and L. Olivi (2017) Operando characterization of batteries using X-ray absorption spectroscopy: Advances at the beamline XAFS at synchrotron Elettra. *J. Appl. Phys.*, **50**, pp. 074001–074011.

139. Doeff, M. M., G. Chen, J. Cabana, T. J. Richardson, A. Mehta, M. Shirpour, H. Duncan, C. Kim, K. C. Kam, and T. Conry (2013) Characterization of electrode materials for lithium ion and sodium ion batteries using synchrotron radiation techniques. *J. Vis. Exp.*, pp. 50594.

140. Marco, G. and S. Lorenzo (2017). *X-Ray Absorption Spectroscopy Study of Battery Materials*, Ed. Mehdi K. and Luca P. "X-ray characterization of nanostructured energy materials by synchrotron radiation" (Intech), pp. 51–75.

141. Giorgetti, M. (2013) A review on the structural studies of batteries and host materials by X-ray absorption spectroscopy. *ISRN Mater. Sci.*, 2013, pp. 1–22.

142. Liu, X., W. Yang,and Z. Liu (2014) Recent progress on synchrotron-based in-situ soft X-ray spectroscopy for energy materials. *Adv. Mater.*, **26**, pp. 7710–7729.

143. Lachowski, E. (1988) *X-ray Absorption Principles, Applications, Techniques of EXAFS, SEXAFS and XANES*, Ed. Koningsberger, D. C. and Prins, R. (Wiley, New York).

144. Chen, M., S. L Chou,and S. X. Dou (2019) Understanding challenges of cathode materials for sodium-ion batteries using synchrotron-based X-ray absorption spectroscopy. *Batteries & Supercaps*, **2**, pp. 842–851.

145. Gu, Q., J. A. Kimpton, H. E. A. Brand, Z. Wang, and S. Chou (2017) Solving key challenges in battery research using in situ synchrotron and neutron techniques. *Adv. Energy Mater.*, **7**, pp. 1602831–1602837.

146. Crumlin, E. J., Z. Liu, H. Bluhm, W. Yang, J. Guo, and Z. Hussain (2015) X-ray spectroscopy of energy materials under in situ/operando conditions. *J. Electron Spectrosc.*, **200**, pp. 264–273.

147. Wanwisa Limphirat, N. W., S. Tonlublao, S. Chaichoy, P. Pruekthaisong, S. Duandmanee, P. Kamonpha, D. Kaewsuwan, N. Meethong, R. P. Poo-arporn, P. Songsiriritthigul, J. Hormes, and Y. Poo-arporn (2020) The current status of time-resolved XAS beamline at SLRI and application on in situ experiments. *Radiat. Phys. Chem.*, **171**, pp. 108750–108756.

148. Hwu, Y. K., G. M. Chow,W. C. Goh, T. S. Cho, and C. K. Lin (2000) *Nanostructured Films and Coatings*, Ed. Chow, G.M. "Phase differentiation and characterization of nanostructured composites by synchrotron radiation techniques." (Springer Netherlands), pp. 203–214.

149. Cao, M.-H., Y. Wang, Z. Shadike, J. Yue, E. Hu, S. Bak, Y. Zhou, X. Yang, and Z. Fu (2017) Suppressing chromium disproportion reaction in O3-type layered cathode material for high-capacity sodium-ion batteries. *J. Mater. Chem. A,* **5**, pp. 5442–5448.

150. Panda, M. R., A. Raj, A. Ghosh, A. Kumar, D. Muthuraj, S. Sau, W. Yu, Y. Zhang, A. K. Sinha, M. Weyland, Q. Bao, and S. Mitra (2019) Blocks of molybdenum ditelluride: A high rate anode for sodium-ion battery and full cell prototype study. *Nano Energy,* **64**, pp. 103951–103963.

151. Pang, W. K., S. Kalluri, V. K. Peterson, N. Sharma, J. Kimpton, B. Johannessen, H. K. Liu, S. X. Dou, and Z. Guo (2015) Interplay between electrochemistry and phase evolution of the P2-type $Na_x(Fe_{1/2}Mn_{1/2})O_2$ cathode for use in sodium-ion batteries. *Chem. Mater.,* **27**, pp. 3150–3158.

152. Huang, W., B. Li, M. F. Saleem, X. Wu, J. Li, J. Lin, D. Xia, W. Chu, and Z. Wu (2015) Self-assembled alluaudite $Na_{(2)}Fe_{(3-x)}Mn_{(x)}(PO_4)_{(3)}$ micro/nanocompounds for sodium-ion battery electrodes: A new insight into their electronic and geometric structure. *Chem. Eur. J.,* **21**, pp. 851–860.

153. Zhang, K., D. Kim, Z. Hu, M. Park, G. Noh, Y. Yang, J. Zhang, V. W. Lau, S. L. Chou, M. Cho, S. Y. Choi, and Y. M. Kang (2019) Manganese based layered oxides with modulated electronic and thermodynamic properties for sodium ion batteries. *Nat. Commun.,* **10**, pp. 5203–5314.

154. Wei, S., C. Wang, S. Chen, P. Zhang, K. Zhu, C. Wu, P. Song, W. Wen, and L. Song (2020) Dial the mechanism switch of VN from conversion to intercalation toward long cycling sodium-ion battery. *Adv. Energy Mater.,* **10**, pp. 1903712–1903719.

155. Keller, M., T. Eisenmann, D. Meira, G. Aquilanti, D. Buchholz, D. Bresser, and S. Passerini (2019) In situ investigation of layered oxides with mixed structures for sodium-ion batteries. *Small Methods,* **3**, pp. 1900239–1900248.

156. Li, C., Q. Yang, M. Shen, J. Ma, and B. Hu (2018) The electrochemical Na intercalation/extraction mechanism of ultrathin cobalt(II) terephthalate-based MOF nanosheets revealed by synchrotron X-ray absorption spectroscopy. *Energy Storage Mater.,* **14**, pp. 82–89.

157. Yabuuchi, N., M. Kajiyama, J. Iwatate, H. Nishikawa, S. Hitomi, R. Okuyama, R. Usui, Y. Yamada, and S. Komaba (2012) P2-type $Na_{(x)}[Fe_{(1/2)}Mn_{(1/2)}]O_2$ made from earth-abundant elements for rechargeable Na batteries. *Nat. Mater.*, **11**, pp. 512–517.

158. Wang, P.-F., H.-R. Yao, X.-Y. Liu, Y.-X. Yin, J.-N. Zhang, Y. Wen, X. Yu, L. Gu, and Y.-G. Guo (2018) Na^+/vacancy disordering promises high-rate Na-ion batteries. *Sci. Adv.*, **4**, pp. 6018–6027.

159. Rahman, M. M., J. Mao, W. H. Kan, C.-J. Sun, L. Li, Y. Zhang, M. Avdeev, X.-W. Du, and F. Lin (2019) An ordered P2/P3 composite layered oxide cathode with long cycle life in sodium-ion batteries. *ACS Mater. Lett.*, **1**, pp. 573–581.

160. Lin, C.-C., W.-H. Lin, S.-C Huang, C.-W. Hu, T.-Y. Chen, C.-T. Hsu, H. Yang, A. Haider, Z. Lin, U. Kortz, U. Stimming,and H.-Y. Chen (2018) Mechanism of sodium ion storage in $Na_7[H_2PV_{14}O_{42}]$ anode for sodium-ion batteries. *Adv. Mater. Inter.*, **5**, pp. 1800491–1800500.

161. Keller, M., T. Eisenmann, D. Meira, G. Aquilanti, D. Buchholz, D. Bresser, and S. Passerini (2019) In situ investigation of layered oxides with mixed structures for sodium-ion batteries. *Small Methods*, **3**, pp. 1900239–1900238

162. Li, Q., R. Qiao, L. A. Wray, J. Chen, Z. Zhuo, Y. Chen, S. Yan, F. Pan, Z. Hussain, and W. Yang (2016) Quantitative probe of the transition metal redox in battery electrodes through soft X-ray absorption spectroscopy. *J. Appl. Phys.*, **49**, pp. 413003–413014.

163. Massel, F. (2019) *Digital Comprehensive Summaries of Uppsala Dissertations from the Faculty of Science and Technology 1835.* "Anion redox processes in novel battery cathode materials investigated by soft X-ray spectroscopy" (Uppsala: Acta Universitatis Upsaliensis).

164. Yang, W. and R. Qiao (2016) Soft X-ray spectroscopy for probing electronic and chemical states of battery materials. *Chinese Phys. B*, **25**, pp. 017104–017112.

165. Yang, W., X. Liu, R. Qiao, P. Olalde-Velasco, J. D. Spear, L. Roseguo, J. X. Pepper, Y.-D. Chuang, J. D. Denlinger,and Z. Hussain (2013) Key electronic states in lithium battery materials probed by soft X-ray spectroscopy. *J. Electron Spectrosc.*, **190**, pp. 64–74.

166. Li, Q., S. Yan, and W. Yang (2020) Interfacial properties in energy storage systems studied by soft X-ray absorption spectroscopy and resonant inelastic X-ray scattering. *J. Chem. Phys.*, **152**, pp. 140901–140909.

167. Nanba, Y., T. Iwao, B. M. D. Boisse, W. Zhao, E. Hosono, D. Asakura, H. Niwa, H. Kiuchi, J. Miyawaki, Y. Harada, M. Okubo, and A. Yamada (2016) Redox potential paradox in Na_xMO_2 for sodium-ion battery cathodes. *Chem. Mater.*, **28**, pp. 1058–1065.

168. Niwa, H., K. Higashiyama, K. Amaha, W. Kobayashi, and Y. Moritomo (2018) Electronic states in oxidized $NaCoO_2$ as revealed by X-ray absorption spectroscopy coupled with ab initio calculation. *J. Power Sources,* **384**, pp. 156–159.

169. Qiao, R., K. Dai, J. Mao, T.-C. Weng, D. Sokaras, D. Nordlund, X. Song, V. S. Battaglia, Z. Hussain, G. Liu, and W. Yang (2015) Revealing and suppressing surface Mn(II) formation of $Na_{0.44}MnO_2$ electrodes for Na-ion batteries. *Nano Energy,* **16**, pp. 186–195.

170. Rahman, M. M., Y. Xu, H. Cheng, Q. Shi, R. Kou, L. Mu, Q. Liu, S. Xia, X. Xiao, C.-J. Sun, D. Sokaras, D. Nordlund, J.-C. Zheng, Y. Liu, and F. Lin (2018) Empowering multicomponent cathode materials for sodium ion batteries by exploring three-dimensional compositional heterogeneities. *Energy Environ. Sci.,* **11**, pp. 2496–2508.

171. Rahman, M. M., Y. Zhang, S. Xia, W. H. Kan, M. Avdeev, L. Mu, D. Sokaras, T. Kroll, X.-W. Du, D. Nordlund, Y. Liu, and F. Lin (2019) Surface characterization of Li-substituted compositionally heterogeneous Na $Li_{0.045}Cu_{0.185}Fe_{0.265}Mn_{0.505}O_2$ sodium-ion cathode material. *J. Phys. Chem. C,* **123**, pp. 11428–11435.

172. Xiao, W., Q. Sun, M. N. Banis, B. Wang, J. Liang, A. Lushington, R. Li, X. Li, T. K. Sham, and X. Sun (2019) Unveiling the interfacial instability of the phosphorus/carbon anode for sodium-ion batteries. *ACS Appl. Mater. Inter.,* **11**, pp. 30763–30773.

173. Li, J., X. He, S. Ostendorp, L. Zhang, X. Hou, D. Zhou, B. Yan, D. M. Meira, Y. Yang, H. Jia, G. Schumacher, J. Wang, E. Paillard, G. Wilde, M. Winter, and J. Li (2020) Tin modification of sodium manganese hexacyanoferrate as a superior cathode material for sodium ion batteries. *Electrochim. Acta,* **342**, pp. 135928–135935.

174. Li, C., Q. Yang, M. Shen, J. Ma, and B. Hu (2018) The electrochemical Na intercalation/extraction mechanism of ultrathin cobalt(II) terephthalate-based MOF nanosheets revealed by synchrotron X-ray absorption spectroscopy. *Energy Storage Mater.,* **14**, pp. 82–89.

175. Wu, J., J. Song, K. Dai, Z. Zhuo, L. A. Wray, G. Liu, Z. X. Shen, R. Zeng, Y. Lu, and W. Yang (2017) Modification of transition-metal redox by interstitial water in hexacyanometalate electrodes for sodium-ion batteries. *J. Am. Chem. Soc.,* **112**, pp. 18358–18364.

176. Yan, G., S. Mariyappan, G. Rousse, Q. Jacquet, M. Deschamps, R. David, B. Mirvaux, J. W. Freeland, and J. M. Tarascon (2019) Higher energy and safer sodium ion batteries via an electrochemically made disordered $Na_3V_2(PO_4)_2F_3$ material. *Nat. Commun.*, **10**, pp. 585–596.

177. Cheng, C., S. Li, T. Liu, Y. Xia, L. Y. Chang, Y. Yan, M. Ding, Y. Hu, J. Wu, J. Guo, and L. Zhang (2019) Elucidation of anionic and cationic redox reactions in a prototype sodium-layered oxide cathode. *ACS Appl. Mater. Inter.*, **11**, pp. 41304–41312.

178. Kim, E. J., L. A. Ma, L. C. Duda, D. M. Pickup, A. V. Chadwick, R. Younesi, J. T. S. Irvine, and A. R. Armstrong (2019) Oxygen redox activity through a reductive coupling mechanism in the P3-type nickel-doped sodium manganese oxide. *ACS Appl. Energy Mater.*, **3**, pp. 184–191.

179. Li, M., X. Bi, K. Amine, and J. Lu (2020) Oxygen-based anion redox for lithium batteries. *Acc. Chem. Res.*, **53**(8), pp. 1436–1444.

180. Li, N., S. Sallis, J. K. Papp, J. Wei, B. D. McCloskey, W. Yang, and W. Tong (2019) Unraveling the cationic and anionic redox reactions in a conventional layered oxide cathode. *ACS Energy Lett.*, **4**, pp. 2836–2842.

181. Mortemard de Boisse, B., G. Liu, J. Ma, S. I. Nishimura, S. C. Chung, H. Kiuchi, Y. Harada, J. Kikkawa, Y. Kobayashi, M. Okubo, and A. Yamada (2016) Intermediate honeycomb ordering to trigger oxygen redox chemistry in layered battery electrode. *Nat. Commun.*, **7**, pp. 11127–11406.

182. Dai, K., J. Wu, Z. Zhuo, Q. Li, S. Sallis, J. Mao, G. Ai, C. Sun, Z. Li, W. E. Gent, W. C. Chueh, Y.-c. Chuang, R. Zeng, Z.-X. Shen, F. Pan, S. Yan, L. F. J. Piper, Z. Hussain, G. Liu, and W. Yang (2019) High reversibility of lattice oxygen redox quantified by direct bulk probes of both anionic and cationic redox reactions. *Joule*, **3**, pp. 518–541.

183. Hertz, H. (1887) Uber einen einfluss des ultravioletten lichtes auf die electrische entladung. *Ann. Phys.*, **267**, pp. 983–1000.

184. Ratner, B. D. and D. G. Castner (2009) *Electron Spectroscopy for Chemical Analysis. Surface Analysis: The Principal Techniques*, 2nd Edition, Ed. Vickerman, J. C. and Gilmore, I. S. "Electron Spectroscopy for Chemical Analysis" (Wiley), pp. 47–112.

185. Ciosek Högström, K., S. Malmgren, M. Hahlin, M. Gorgoi, L. Nyholm, H. Rensmo, and K. Edström (2014) The buried carbon/solid electrolyte interphase in Li-ion batteries studied by hard X-ray photoelectron spectroscopy. *Electrochim. Acta*, **138**, pp. 430–436.

186. Hori, H., M. Shikano, H. Kobayashi, S. Koike, H. Sakaebe, Y. Saito, K. Tatsumi, H. Yoshikawa, and E. Ikenaga (2013) Analysis of hard carbon for lithium-ion batteries by hard X-ray photoelectron spectroscopy. *J. Power Sources,* **242**, pp. 844–847.

187. Lebens-Higgins, Z. W., H. Chung, M. J. Zuba, J. Rana, Y. Li, N. V. Faenza, N. Pereira, B. D. McCloskey, F. Rodolakis, W. Yang, M. S. Whittingham, G. G. Amatucci, Y. S. Meng, T. L. Lee, and L. F. J. Piper (2020) How bulk sensitive is hard X-ray photoelectron spectroscopy: Accounting for the cathode-electrolyte interface when addressing oxygen redox. *J. Phys. Chem. Lett.,* **11**, pp. 2106–2112.

188. Philippe, B., M. Hahlin, K. Edström, T. Gustafsson, H. Siegbahn, and H. Rensmo (2016) Photoelectron spectroscopy for lithium battery interface studies. *J. Electrochem. Soc.,* **163**, pp. A178–A191.

189. Song, J., B. Xiao, Y. Lin, K. Xu, and X. Li (2018) Interphases in sodium-ion batteries. *Adv. Energy Mater.,* **8**, pp. 1703082–1703105.

190. Panaccione, G. and K. Kobayashi (2012) Hard X-ray photoemission spectroscopy: Variable depth analysis of bulk, surface and interface electronic properties. *Surf. Sci.,* **606**, pp. 125–129.

191. Mogensen, R., D. Brandell, and R. Younesi (2016) Solubility of the solid electrolyte interphase (SEI) in sodium ion batteries. *ACS Energy Lett.,* **1**, pp. 1173–1178.

192. Mogensen, R., J. Maibach, A. J. Naylor, and R. Younesi (2018) Capacity fading mechanism of tin phosphide anodes in sodium-ion batteries. *Dalton Trans.,* **47**, pp. 10752–10758.

193. Mogensen, R., J. Maibach, W. R. Brant, D. Brandell, and R. Younesi (2017) Evolution of the solid electrolyte interphase on tin phosphide anodes in sodium ion batteries probed by hard X-ray photoelectron spectroscopy. *Electrochim. Acta,* **245**, pp. 696–704.

194. Doubaji, S., B. Philippe, I. Saadoune, M. Gorgoi, T. Gustafsson, A. Solhy, M. Valvo, H. Rensmo, and K. Edstrom (2016) Passivation layer and cathodic redox reactions in sodium-ion batteries probed by HAXPES. *ChemSusChem,* **9**, pp. 97–108.

195. Ma, L. A., F. Massel, A. J. Naylor, L.-C. Duda,and R. Younesi (2019) Understanding charge compensation mechanisms in $Na_{0.56}Mg_{0.04}Ni_{0.19}Mn_{0.70}O_2$. *Commun. Chem.,* **2**, pp. 1–9

196. Zhang, S. S. (2006) A review on electrolyte additives for lithium-ion batteries. *J. Power Sources,* **162**, pp. 1379–1124.

197. Park, S.-K., J. Shim, J. H. Yang, C.-S. Jin, B. S. Lee, Y.-S. Lee, K.-H. Shin, and J.-D. Jeon (2014) Effect of inorganic additive sodium pyrophosphate

tetrabasic on positive electrolytes for a vanadium redox flow battery. *Electrochim. Acta,* **121**, pp. 321–327.

198. Wagner, R., B. Streipert, V. Kraft, A. Reyes Jiménez, S. Röser, J. Kasnatscheew, D. R. Gallus, M. Börner, C. Mayer, H. F. Arlinghaus, M. Korth, M. Amereller, I. Cekic-Laskovic, and M. Winter (2016) Counterintuitive role of magnesium salts as effective electrolyte additives for high voltage lithium-ion batteries. *Adv. Mater. Inter.,* **3**, pp. 1600096–1600106.

199. Park, H. and H.-J. Kim (2020) Diphenyl diselenide as SEI-forming additive for a high-voltage $LiCoO_2$/graphite battery. *J. Electrochem. Soc.,* **167**, pp. 070555.

200. Jo, C.-H., J. U. Choi, H. Yashiro, and S.-T. Myung (2019) Controllable charge capacity using a black additive for high-energy-density sodium-ion batteries. *J. Mater. Chem. A,* **7**, pp. 3903–3909.

201. Jo, J. H., J. U. Choi, Y. J. Park, J. K. Ko, H. Yashiro, and S.-T. Myung (2020) A new pre-sodiation additive for sodium-ion batteries. *Energy Storage Mater.,* **32**, pp. 281–289.

202. Dahbi, M., N. Yabuuchi, M. Fukunishi, K. Kubota, K. Chihara, K. Tokiwa, X.-F. Yu, H. Ushiyama, K. Yamashita, J.-Y. Son, Y.-T. Cui, H. Oji, and S. Komaba (2016) Black phosphorus as a high-capacity, high-capability negative electrode for sodium-ion batteries: Investigation of the electrode/electrolyte interface. *Chem. Mater.,* **28**, pp. 1625–1635.

203. Cheng, L., E. J. Crumlin, W. Chen, R. Qiao, H. Hou, S. Franz Lux, V. Zorba, R. Russo, R. Kostecki, Z. Liu, K. Persson, W. Yang, J. Cabana, T. Richardson, G. Chen, and M. Doeff (2014) The origin of high electrolyte-electrode interfacial resistances in lithium cells containing garnet type solid electrolytes. *Phys. Chem. Chem. Phys.,* **16**, pp. 18294–18300.

204. Lin, F., I. M. Markus, D. Nordlund, T. C. Weng, M. D. Asta, H. L. Xin, and M. M. Doeff (2014a) Surface reconstruction and chemical evolution of stoichiometric layered cathode materials for lithium-ion batteries. *Nat. Commun.,* **5**, pp. 3529–3537.

205. Lin, F., D. Nordlund, I. M. Markus, T.-C. Weng, H. L. Xin, and M. M. Doeff (2014b) Profiling the nanoscale gradient in stoichiometric layered cathode particles for lithium-ion batteries. *Energy & Environ. Sci.,* **7**, pp. 3077–3085.

206. Lin, F., D. Nordlund, T. C. Weng, D. Sokaras, K. M. Jones, R. B. Reed, D. T. Gillaspie, D. G. Weir, R. G. Moore, A. C. Dillon, R. M. Richards, and C. Engtrakul (2013) Origin of electrochromism in high-performing nanocomposite nickel oxide. *ACS Appl. Mater. Interfaces,* **5**, pp. 3643–3649.

207. Maibach, J., F. Jeschull, D. Brandell, K. Edstrom, and M. Valvo (2017) Surface layer evolution on graphite during electrochemical sodium-tetraglyme Co-intercalation. *ACS Appl. Mater. Interfaces,* **9**, pp. 12373–12381.

208. Feng, J., L. Wang, D. Li, P. Lu, F. Hou, and J. Liang (2018) Enhanced electrochemical stability of carbon-coated antimony nanoparticles with sodium alginate binder for sodium-ion batteries. *Prog. Nat. Sci-Mater,* **28**, pp. 205–211.

209. Yuca, N., H. Zhao, X. Song, M. F. Dogdu, W. Yuan, Y. Fu, V. S. Battaglia, X. Xiao, and G. Liu (2014) A systematic investigation of polymer binder flexibility on the electrode performance of lithium-ion batteries. *ACS Appl. Mater. Interfaces,* **6**, pp. 17111–17118.

210. Zhao, H., Z. Wang, P. Lu, M. Jiang, F. Shi, X. Song, Z. Zheng, X. Zhou, Y. Fu, G. Abdelbast, X. Xiao, Z. Liu, V. S. Battaglia, K. Zaghib, and G. Liu (2014) Toward practical application of functional conductive polymer binder for a high-energy lithium-ion battery design. *Nano Lett.,* **14**, pp. 6704–6710.

211. Maibach, J., F. Jeschull, D. Brandell, K. Edstrom, and M. Valvo (2017) Surface layer evolution on graphite during electrochemical sodium-tetraglyme Co-intercalation. *ACS Appl. Mater. Interfaces,* **9**, pp. 12373–12381.

212. Dahbi, M., T. Nakano, N. Yabuuchi, T. Ishikawa, K. Kubota, M. Fukunishi, S. Shibahara, J.-Y. Son, Y.-T. Cui, H. Oji, and S. Komaba (2014) Sodium carboxymethyl cellulose as a potential binder for hard-carbon negative electrodes in sodium-ion batteries. *Electrochem. Commun.,* **44**, pp. 66–69.

213. Boebinger, M. G., J. A. Lewis, S. E. Sandoval, and M. T. McDowell (2020) Understanding transformations in battery materials using in situ and operando experiments: Progress and outlook. *ACS Energy Lett.,* **5**, pp. 335–345.

214. Li, W., M. Li, Y. Hu, J. Lu, A. Lushington, R. Li, T. Wu, T.-K. Sham, and X. Sun (2018) Synchrotron-based X-ray absorption fine structures, X-ray diffraction, and X-ray microscopy techniques applied in the study of lithium secondary batteries. *Small Methods,* **2**, pp. 1700341–1700361.

215. Sakdinawat, A. and D. Attwood (2010) Nanoscale X-ray imaging. *Nat. Photonics,* **4**, pp. 840–848.

216. Miao, J., T. Ishikawa, I. K. Robinson, and M. M. Murnane (2015) Beyond crystallography: Diffractive imaging using coherent X-ray light sources. *Science,* **348**, pp. 530–535.

217. Schröder, D., C. L. Bender, T. Arlt, M. Osenberg, A. Hilger, S. Risse, M. Ballauff, I. Manke, and J. Janek (2016) In operando X-ray tomography for next-generation batteries: A systematic approach to monitor reaction product distribution and transport processes. *J. Appl. Phys.,* **49,** pp. 404001–404012.

218. Wang, L., J. Wang, and P. Zuo (2018) Probing battery electrochemistry with in operando synchrotron X-Ray imaging techniques. *Small Methods,* **2,** pp. 1700293–1700312.

219. Bilderback, D. H., P. Elleaume, and E. Weckert (2005) Review of third and next generation synchrotron light sources. *J. Phys. B-At., Mol. Opt. Phys.,* **38,** pp. S773–S797.

220. Raines, K. S., S. Salha, R. L. Sandberg, H. Jiang, J. A. Rodriguez, B. P. Fahimian, H. C. Kapteyn, J. Du, and J. Miao (2010) Three-dimensional structure determination from a single view. *Nature,* **463,** pp. 214–217.

221. Taiwo, O. O., D. S. Eastwood., P. D. Lee, J. M. Paz-García, S. A. Hall, D. J. L. Brett, and P. R. Shearing (2015) In-situ examination of microstructural changes within a lithium-ion battery electrode using synchrotron X-ray microtomography. *ECS Trans.,* **69,** pp. 81–85.

222. Wang, J., Y. C. Chen-Wiegart, and J. Wang (2013) In situ chemical mapping of a lithium-ion battery using full-field hard X-ray spectroscopic imaging. *Chem. Commun.,* **49,** pp. 6480–6482.

223. Wang, J., Y.-C. Chen-Wiegart, and J. Wang (2014) In situ three-dimensional synchrotron X-Ray nanotomography of the (de)lithiation processes in tin anodes. *Angew Chem. Int. Edit.,* **53,** pp. 4460–4464.

224. Wang, R., C. Xu, J. Sun, L. Gao, and H. Yao (2014) Solvothermal-induced 3D macroscopic SnO_2/nitrogen-doped graphene aerogels for high capacity and long-life lithium storage. *ACS Appl. Mater. Inter.,* **6,** pp. 3427–3436.

225. Zhu, H., Y. Huang, H. Zhu, L. Wang, S. Lan, X. Xia, and Q. Liu (2019) In situ probing multiple-scale structures of energy materials for Li-ion batteries. *Small Methods,* **4,** pp. 1900223–1900252.

226. Wang, L., J. Wang, F. Guo, L. Ma, Y. Ren, T. Wu, P. Zuo, G. Yin, and J. Wang (2018) Understanding the initial irreversibility of metal sulfides for sodium-ion batteries via operando techniques. *Nano Energy,* **43,** pp. 184–191.

227. Wang, J., C. Eng, Y. C. Chen-Wiegart, and J. Wang (2015) Probing three-dimensional sodiation-desodiation equilibrium in sodium-ion batteries by in situ hard X-ray nanotomography. *Nat. Commun.,* **6,** pp. 7496–7484.

228. Selvaraj, B., S.-S. Huang, C.-E. Wu, Y.-H. Lin, C.-C. Wang, Y.-F. Song, M.-L. Lu, H.-S. Sheu, and N.-L. Wu (2018) Micrometer-sized nanoporous Sb/C anode with high volumetric capacity and fast charging performance for sodium-ion batteries. *ACS Appl. Energy Mater.,* **1**, pp. 2317–2325.

229. Yu, Z., J. Wang, L. Wang, Y. Xie, S. Lou, Z. Jiang, Y. Ren, S. Lee, P. Zuo, H. Huo, G. Yin, Q. Pan, and J. Wang (2019) Unraveling the origins of the "unreactive core" in conversion electrodes to trigger high sodium-ion electrochemistry. *ACS Energy Lett.,* **4**, pp. 2007–2012.

230. Wang, L., J. Wang, X. Zhang, Y. Ren, P. Zuo, G. Yin, and J. Wang (2017b) Unravelling the origin of irreversible capacity loss in $NaNiO_2$ for high voltage sodium ion batteries. *Nano Energy,* **34**, pp. 215–223.

231. Lee, D. H., J. Xu,and Y. S. Meng (2013) An advanced cathode for Na-ion batteries with high rate and excellent structural stability. *Phys. Chem. Chem. Phys.,* **15**, pp. 3304–3312.

232. Yoshida, H., N. Yabuuchi, K. Kubota, I. Ikeuchi, A. Garsuch, M. Schulz-Dobrick, and S. Komaba (2014) P2-type $Na_{(2/3)}Ni_{(1/3)}Mn_{(2/3-x)}Ti_{(x)}O_2$ as a new positive electrode for higher energy Na-ion batteries. *Chem. Commun.,* **50**, pp. 3677–3680.

233. Xie, Y., H. Wang, G. Xu, J. Wang, H. Sheng, Z. Chen, Y. Ren, C.-J. Sun, J. Wen, J. Wang, D. J. Miller, J. Lu, K. Amine, and Z.-F. Ma (2016) In operando XRD and TXM study on the metastable structure change of $NaNi_{1/3}Fe_{1/3}Mn_{1/3}O_2$ under electrochemical sodium-ion intercalation. *Adv. Energy Mater.,* **6**, pp. 1601306.

234. Wang, J., L. Wang, C. Eng, and J. Wang (2017a) Elucidating the irreversible mechanism and voltage hysteresis in conversion reaction for high-energy sodium-metal sulfide batteries. *Advanced Energy Mater.,* **7**, pp. 1602706.

Index